INSTITUT FRANÇAIS DU PÉTROLE PUBLICATIONS

Bernard BIJU-DUVAL
Professor at ENSPM-Formation Industrie
Institut Français du Pétrole

SEDIMENTARY GEOLOGY

Sedimentary Basins
Depositional Environments
Petroleum Formation

Translated from the French
by J. Edwin Swezey and Traduclair Translation Company

2002

Editions TECHNIP
27 rue Ginoux, 75737 PARIS Cedex 15, FRANCE

ÉCOLE DU PÉTROLE ET DES MOTEURS
IFP SCHOOL

FROM THE SAME PUBLISHER

Dynamics and Methods of Study of Sedimentary Basins
ASSOCIATION OF FRENCH SEDIMENTOLOGISTS PUBLICATION

Geodynamic Evolution of Sedimentary Basins
F. ROURE, N. ELLOUZ, V.S. SHEIN, I.I. SKVORTSOV, Eds.

Kerogen. Insoluble Organic Matter from Sedimentary Rocks
B. DURAND, Ed.

Main Types of Geological Maps. Purpose, Use and Preparation
FRENCH OIL AND GAS INDUSTRY ASSOCIATION PUBLICATION

Applied Petroleum Geochemistry
M.-L. BORDENAVE, Ed.

Basics of Reservoir Engineering R. COSSÉ

Geophysics for Sedimentary Basins G. HENRY

Geophysics of Reservoir and Civil Engineering
J.-L. MARI, G. ARENS, D. CHAPELLIER, P. GAUDIANI

Best Practices in Sequence Stratigraphy
For Explorationists and Reservoir Engineers
P. HOMEWOOD, P. MAURIAUD, F. LAFONT

Signal Processing in Geosciences (CD-ROM)
F. GLANGEAUD, J.-L. MARI

Signal Processing for Geologists and Geophysicists
J.-L. MARI, F. GLANGEAUD, F. COPPENS

Translation (reviewed Edition) of
« Géologie sédimentaire. Bassins, environnements de dépôts, formation du pétrole » B. Biju-Duval

ISBN 2-7108-0802-1
ISSN 1271-9048

FOREWORD

This introductory manual to sedimentary geology was originally intended to accompany and complement J. Guillemot's excellent book of basic geology (entitled *Éléments de géologie*, fourth edition of 1986) used by generations of engineering students at the French *École du pétrole et des moteurs*. It is larger than Guillemot's book, as it was designed to extend the students' foundations in the *geology of sedimentary basins* by introducing and discussing new concepts that are widely used today in petroleum exploration. So it serves as basic documentation and an introduction to more in-depth education in geology, and refers the reader to more specialized books and articles in different fields as each field is developed.

The material for the manual comes from introductory courses in sedimentology and geology available to engineering students at the Petroleum exploration and economics centers, and from a graduate course on geodynamics entitled "Quantitative methods in the geosciences."

The first chapter presents the primary driving forces that act and interact on the planet in a perspective of dynamic geology, along with the ideas of three-dimensional objects and the time evolution of geological processes.

The second lays out the general framework of sedimentary geology with a description of the various basins where petroleum-forming sediments are likely to accumulate.

The third chapter is the core of the book. Here, deposition mechanisms and sedimentary environments are described in greater detail. Continental environments and their main types of deposits are first described, drawing heavily on material updated from *Les grès du Paléozoïque inférieur au Sahara*, by S. Beuf, B. Biju-Duval, O. de Charpal, et al., Éditions Technip, 1971).

The chapter then goes into oceanic environments with the variety of deposit tracks that can be found there. This draws largely on material from *Oceanologie* (B. Biju-Duval, 1994, in the *Geosciences* collection published by Editions Dunod, ed. J. Aubouin).

With the fourth chapter, geology is addressed more in its historical dimension, with a quick review of short- and long-term environmental variations, events, time scales, modern dating methods, sequence analysis and correlations, and paleogeographic reconstructions.

The fifth chapter describes phenomena affecting sediments in the course of their history: diagenesis, in which these sediments are transformed into rock during their burial, and the tectonic deformations that then affect the rocks and bring it back to the surface, usually in mountain chains.

The books ends with a chapter giving the basic elements of petroleum systems: sedimentation of organic matter, kerogens, parent rock, the genesis of hydrocarbons, migrations toward reservoirs, capping and closure of the fields (with extensive references to *Applied Petroleum Geochemistry*, by Bordenave et al., 1993, published by Éditions Technip).

B. BIJU-DUVAL

As this book is intended for engineering students in geology and geophysics, drillers, producers, and economists, it voluntarily leaves out certain aspects of geology such as mineralogy, the geology of crystalline basements, and metamorphism. It essentially deals with sedimentary geology and was designed as a teaching support with an emphasis on the basics and the language used in the profession.

It is based on geological observations on different scales, and is largely illustrated from *field observations and analysis*. The illustrations were prepared from my own experience and draw many examples from studies conducted in the framework of research projects at the *Institut français du pétrole* (IFP) and in the petroleum exploration work mentioned in the references.

The reader will find figures and examples borrowed from applied geophysical techniques routinely used in exploring sedimentary basins, but not discussed here. These are developed in manuals and collections recently published by Éditions Technip:

- *Geophysics for Sedimentary Basins*, G. Henry, 1997
- *Seismic Surveying and Well logging*, S. Boyer and J.-L. Mari, 1997
- *Signal Processing for Geologists and Geophysicists*, J.-L. Mari, F. Glangeaud, and F. Coppens, 1999.

B. Biju-Duval

ACKNOWLEDGMENTS

This book would never have been written without the initial encouragements of Michel Lavergne, who was director of the Exploration center at the *École du pétrole et des moteurs* at the time work began on the book.

I would also like to express my gratitude toward my two colleagues, Paul Trémolières and Alain Mascle, for their very constructive cooperation and the pertinent feedback they gave me on the initial manuscript.

Special thanks go to Claude Laffont for his very active help in choosing the illustrations taken from the Exploration center's documentary fund. Pierre Bot, Étienne Brosse, Bernard Colletta, Patrick Duval, Rémi Eschard and Alain-Yves Huc at the IFP and at the School, and Christian Montenat at the IGAL, also provided precious assistance in the preparation of many illustrations. And Bernard Durand, the Exploration center's director, encouraged me to finish the work.

Final thanks go to the supportive and available technical team who took care of everything from production to publication, from secretarial services, graphics, page layout, to bibliography: Mesdames Bertocchini, Bertrand, Darrigade, Mangion, and Rio, and Messrs. Arenne, Darrigade, and Henry. And for this English version, I should add my compliments and recognition to the translator, J. Edwin Swezey, and to the Traduclair translation company for its steadfast support, supervision, and their work in generating the English index. This book is also theirs.

B. BIJU-DUVAL

TABLE OF CONTENTS

Chapter 2
CONTINENTAL AND OCEANIC BASINS

Chapter 3
SEDIMENTARY DRIVING MECHANISMS AND ENVIRONMENTS

Chapter 4
TIME EVOLUTION: SEDIMENTARY SEQUENCES, STRATIGRAPHY

Chapter 5
FROM SEDIMENTS TO SEDIMENTARY BASIN ROCKS AND MOUNTAIN CHAINS

Chapter 6
PETROLEUM SYSTEMS

Chapter 1

BASICS OF DYNAMIC GEOLOGY

1.1 DEFINITIONS

The word **Geology** comes from the Greek *Geo*, for "earth", and *logos*, for "sciences", and has always encompassed all of the Earth sciences together. Today, however, the very high degree of specialization in certain fields like geophysics has relegated the term Geology to the older, more conventional aspects of this science, replacing it with expressions like Earth Sciences or Geosciences, which are now understood to include the entire discipline. In this book, "geology" is understood in its older, broad, non-restrictive sense, because it necessarily covers the modern approaches in which the naturalist's approach overlaps those of physics, chemistry, and mathematics.

Geology aims to:

- Describe the composition and structure of the globe on different scales
- Understand the biological, physical, and chemical mechanisms of natural phenomena occurring in the course of time as the earth's structure evolves
- Reconstruct the history of this evolution by determining events, cycles, and crises
- Define rules by which useful materials form, concentrate, or accumulate, and provide the indispensable support data for any construction on land or on the seabed.

So geology has complementary aspects. It is a science that is descriptive, analytical, dynamic, historical, and applied. Recent progress in geology has emphatically demonstrated that our globe is a living planet in which many processes operate together, so the perspective we adopt here is one of **dynamic geology**. Geology extends beyond just the geosphere, because mechanisms of the atmosphere, biosphere, and hydrosphere are all at play in geological processes, with many interactions. The geologist thus has to use concepts and data from other fields like biology, chemistry, climatology, and physical oceanography, to name just a few. This is why we often refer to Sciences of the Universe.

As will be seen in every chapter, geology concerns **fluids** every bit as much as it does **solids**.

That is, geology usually conjures the idea of just the rock the earth is made of; but this rock is a locus of major interactions between fluids and solids. Fluids firstly concern the environments of sedimentary deposit, which differ according to location: salt ocean water, continental fresh water, atmospheric air, to name a few. Then the movements of the water tables in the beds running along accidents, evidenced by water springs, for example, are important phenomena. The oil and gas fields that petroleum explorers look for are yet another illustration of how fluids participate in geological processes. Many other aspects where fluid interactions are crucial in geology will also be mentioned and detailed here.

Geology comprises a number of specialized branches that often border upon other sciences. Here are some examples:

- Petrography, mineralogy, and crystallography are the study of rocks, minerals, and crystals, respectively.
- Mineral and organic geochemistry are the science of elementary constituents and reactions with time.
- Radioactive elements are studied in nuclear or isotope geology.
- Paleontology and micropaleontology study animal and vegetal fossils preserved in rock.
- Sedimentology and stratigraphy are concerned with the processes of stratum accumulation and layering.
- Tectonics and rock mechanics study the deformation of rocks in reaction to stresses.
- Pedology is the study of environments specific to land and to soil formation.
- Oceanology studies the ocean and the mechanisms occurring in the seawater column.
- Geodynamics, geomorphology, and geodesy all study the mechanisms that shape the planet.
- Seismology and earth physics, or geophysics, study the potential fields of the earth and experimental investigations.

Sometimes we hear the terms "petroleum geology", "hydrogeology" or, today, "environmental geology". All of these designate particular aspects of geology applied to specialized fields where fluids play a major role, as we have just seen.

Geology is applied in finding, using, and managing natural subsoil resources, and also in forecasting natural hazards (in vulcanology and seismology).

There is always a need for complementary techniques. In the final analysis, it may be said that geology is a science that aims to define both the present state of the planet and its past history. With recent works, and especially with the development of plate tectonics, geology has assumed a global dimension, which implies approaches at different scales.

The methods used in the various disciplines of geology are also varied. What we can say here is that geology uses observation, measurement, experimentation, and modeling.

And lastly, the tools used in geology are complementary. Field work and cartography are the essential basics on land, followed by analyses in the laboratory, geophysical operations, well drilling and measurements (subsurface geology), and regional syntheses.

Seismic data has in turn become essential today, especially in petroleum geology where three-dimensional (3D) tools are used to generate cartographic representations of the subsoil that are beyond the reach of surface studies alone.

Airplanes, and now satellites, serve as platforms for direct and indirect measurements, as is now known to the general public by Landsat and Spot images. And at sea, specialized ships, various other vehicles and underwater observatories are all tools that have considerably broadened our knowledge in a geodynamic perspective.

This voluntarily brief introductory chapter on geology does not refer the reader to the many works on the subject. Aside from the bibliographical references mentioned with certain figures, the reader is referred to the treatise of J. Dercourt and J. Paquet (1990), the manuals of J.M. Caron et al. (1992) and J. Angelier et al. (1992), a special issue of the CNRS newsletter, the book by J. Debelmas and G. Mascle (1993), and the Géosciences *collection of works recently published by Dunod (see end of chapter).*

1.2 OBSERVATION AND MEASUREMENT SCALES. TIME SCALES

In its different approaches of study, from observation to modeling, dynamic geology aims to define, clarify, and quantify a number of essential parameters of the earth's structure and evolution. These parameters cover a wide range of scales, illustrated by **geometric** reconstructions like geological sections and maps; the space and time evolution of **potential fields** like the thermal gradient or stress distribution; or the definition of a **movement** like that of fluid circulation in a porous medium, the dispersion of a turbidity current, or the displacement of a lithospheric plate.

The first step in geology is to observe different objects, to analyze their type, structure, and mutual relations. Crystalline lattices, crystallinity, pore dimension, or the state of organic matter are observed in the laboratory using optical or scanning microscope. In the field, however, the geologist will observe rocks, sedimentary and tectonic structures, or morphologies by the naked eye, magnifying glass, and sometimes with binoculars (Fig. 1.1).

With in situ or remote measurements and with experimentation, techniques range from the laboratory microprobe or microscanner to worldwide seismological networks and satellite data all representing very different orders of magnitude in the data generated. In analysis, this scale variability is reflected in analog and especially in numerical modeling, where it translates into petrophysical reservoir heterogeneities, the hydrodynamics of a sedimentary basin, or thermal cells in the mantle.

1.2.1 3D Space

One of the purposes of fundamental and applied geology is to describe geological objects three-dimensionally. While it may be natural to think of objects in three dimensions, it is nonetheless difficult to assess them this way, and more so to quantify them. A geological

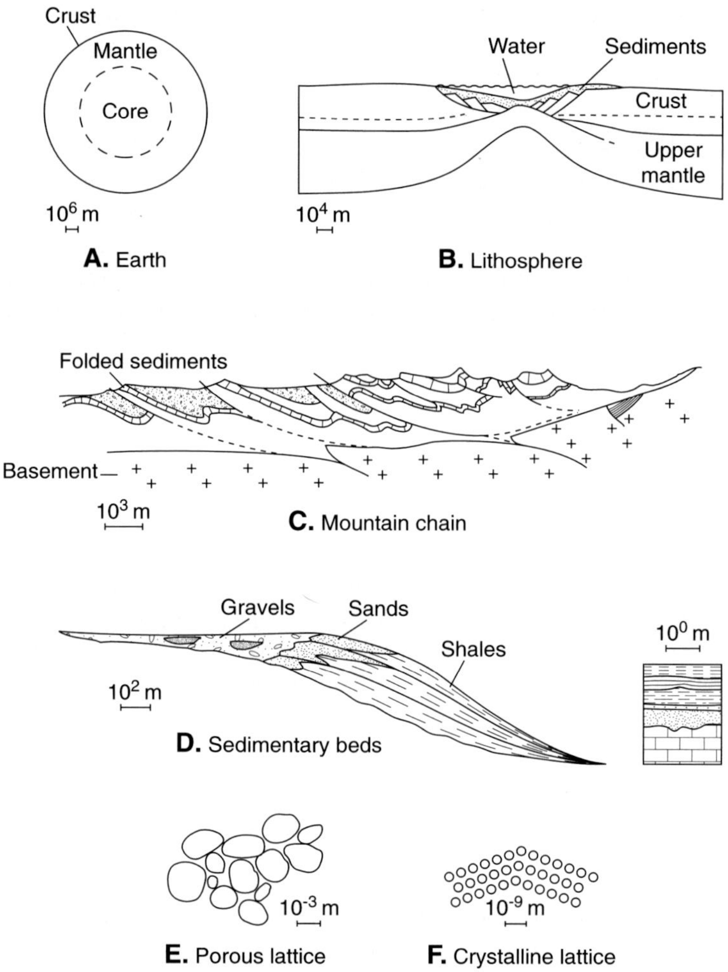

Fig. 1.1 Different scales in geology (modified from J.M. Caron et al., 1992). **A.** In geophysics and seismology, it is the whole Earth that is considered. At this scale, the crust cannot even be represented. **B.** Geodynamics is concerned with the structures of the crust and lithosphere, where sediments correspond to no more than a thin film at the surface. **C** and **D.** In sedimentary geology, the objects represented on the scale of basins and mountain chains range in size from some ten meters to several kilometers. **D** and **E.** In sedimentology and reservoir geology, the structures studied will often be less than a meter in size. This is "high resolution". **F.** In crystallography and detailed geochemical analysis, it is the crystalline lattice that is considered.

landscape is usually viewed in a space with the structures cropping out at the surface as eroded (the present topographical surface with its mountains, hills, river cuts, and other features). This surface pattern is guided directly by the geometry of the structures, with variations depending on the erosion agents (see the section on erosion in Chapter 3) and, in simple cases, it expresses the geological object well in three dimensions (Fig. 1.2).

But a three-dimensional view is not always possible. Objects are often first analyzed in one or two dimensions, such as a sequence of different beds in an outcropping or drilling core (Fig. 1.3). A **geologic column** or lithologic log is developed from this analysis. This will express the bed thicknesses and types, grain size distributions, and mineralogical constituents, for example. The representation scales may vary too, which is an important point that will be illustrated repeatedly here.

Remembering that this one-dimensional data is sometimes the only information available (from drilling, for example), we will go into the various possible two-dimensional (2D) representations below. Going back to the geological landscape of the block diagram we have just seen, this can be represented either in **geological sections or maps**.

A **geological section** expresses the relations in a vertical plane of given orientation. The examples of Fig. 1.1 illustrate various sectional representation scales, from the whole earth structure down to a sample. On the sample scale, different sectional planes may sometimes be used, depending on what the objective is (e.g., to study the porous medium). Geological terrain sections, seismic sections and models generally represent vertical sections (Fig. 1.4). This type of 2D representation also provides an easy way of illustrating concepts like sedimentation dynamics, for example, or fluid circulation.

A **geological map** expresses the relations of geological objects on a horizontal plane. This can be done by overmarking directly on a topographical map, as is commonly done with aerial geology maps (Fig. 1.5A), or by plotting lines of constant bed depth or thickness, or other parameters such as organic matter maturation or chemical element content (Fig. 1.5B).

It is clear that these 2D representations are inadequate. The ultimate aim of geological analysis is to determine the 3D arrangement of the objects, but this is more difficult to represent. Perspective images, or block diagrams, can be reconstructed by combining a series of parallel or secant sections into a block diagram (Fig. 1.6) as in medical **tomography**, to give an image of a total volume. With today's advances in experimental seismic, and especially in the capacity of computers to process large volumes of data, 3D imaging is coming into ever greater use in petroleum exploration (Fig. 1.6B).

Once again, the new approaches possible with today's technical progress concern very different scales, from the porous lattice (scanner tomography) to the oil field.

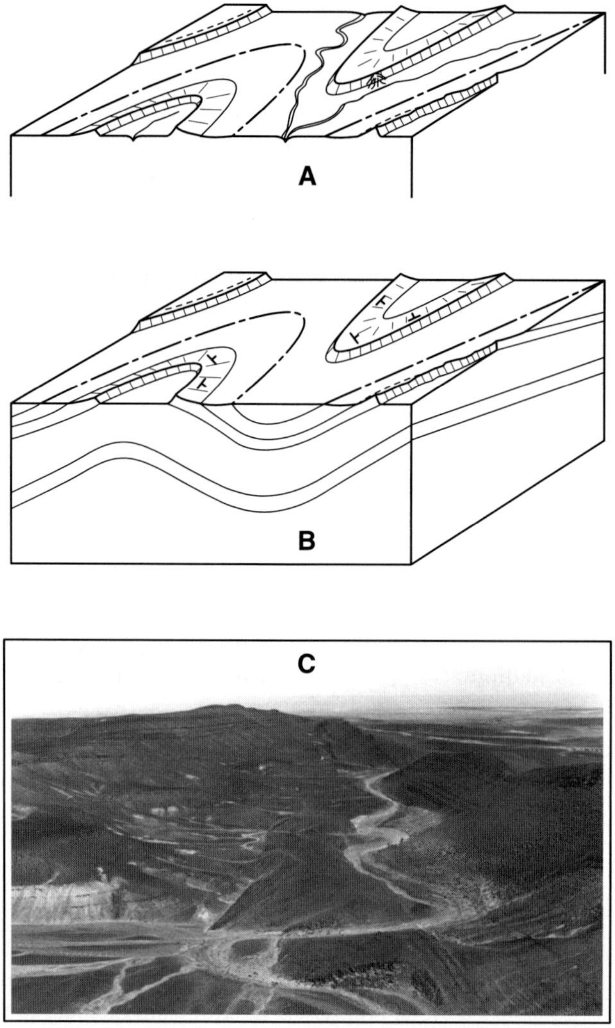

Fig. 1.2 Natural landscape: a three-dimensional perspective.
The geological structure of a region is revealed in contrasted topographical features. Hard beds (calcareous bars here) will form sharper relief as they are more resistant to erosion, while the softer beds of clay or sand are easily grooved by erosive agents. The landscape geometry thus varies in space because of the different kinds of underlying beds. Here, the curved forms of landscape (**A**) can be interpreted as folded geological structures (**B**). The photo (**C**) illustrates a geological landscape where desert erosion brings out the various beds (IFP photo).

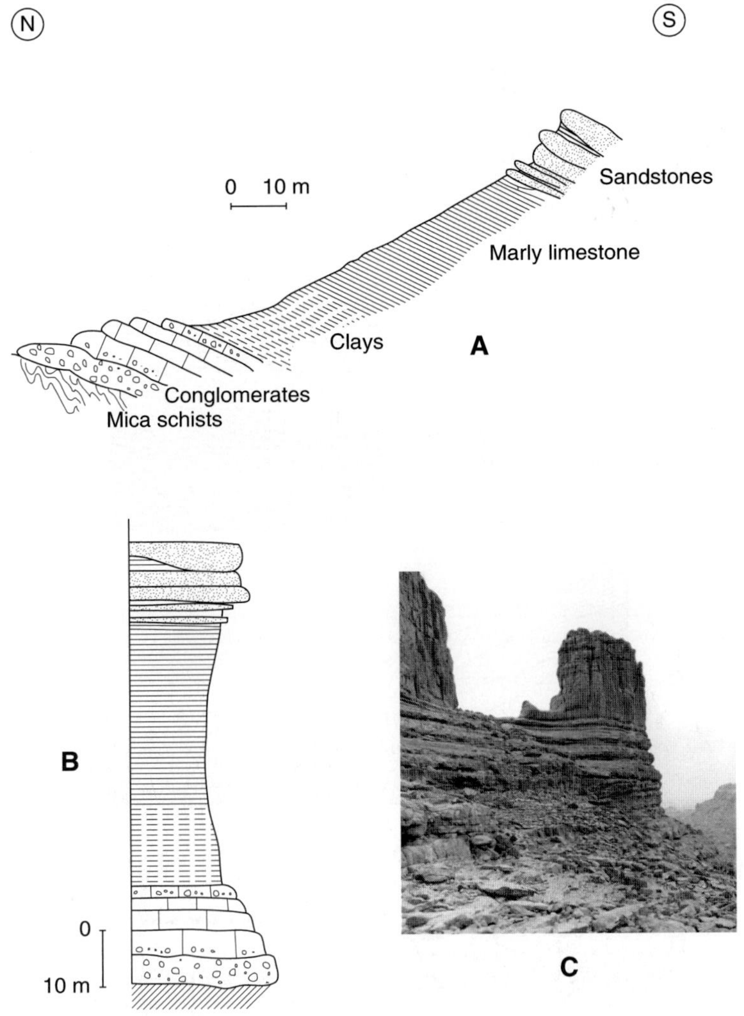

Fig. 1.3 Geological stratum sequence.
A. An outcrop as it would occur naturally in a mountain or at the edge of a marine structure. The various beds are represented in a natural linear (or 1D) geological section. **B.** 1D stratigraphic column representation of section A, comparable to a vertical drilling "log". **C.** Natural outcrop in three dimensions. The example is the Saharan Ordovician with a sequence of sandstone beds cut by recent erosion (IFP photo).

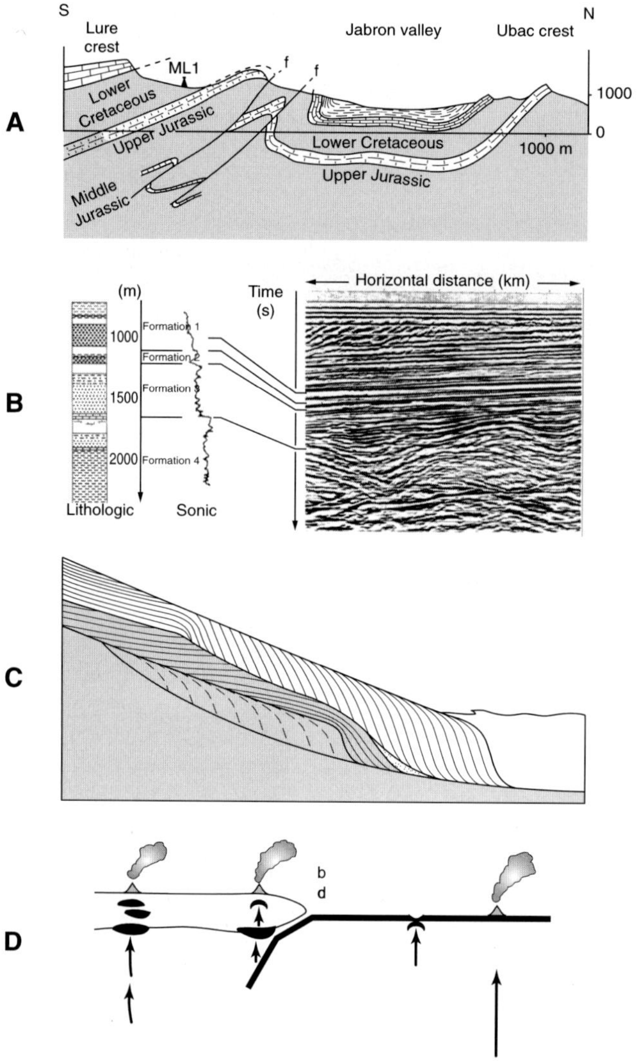

Fig. 1.4 Representation of geological sections.
A. Soil section from the Lure mountain to the Ubac crest, calibrated by drilling data (ML1). **B.** Seismic section (vertical time scale) calibrated by drilling results (lithologic log, sonic log) (from S. Boyer and J.L. Mari, 1994). **C.** Schematic section: a model of the geometric arrangement of successive beds. **D.** Geodynamic section representing a thick continental crust, a thin oceanic crust, and volcanoes.

A

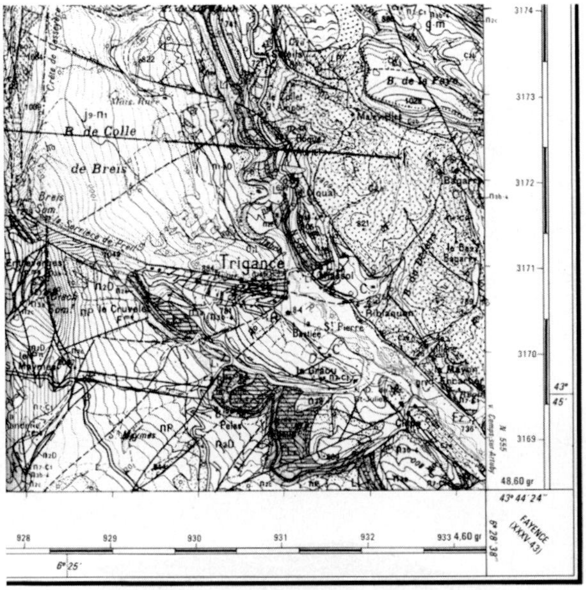

B

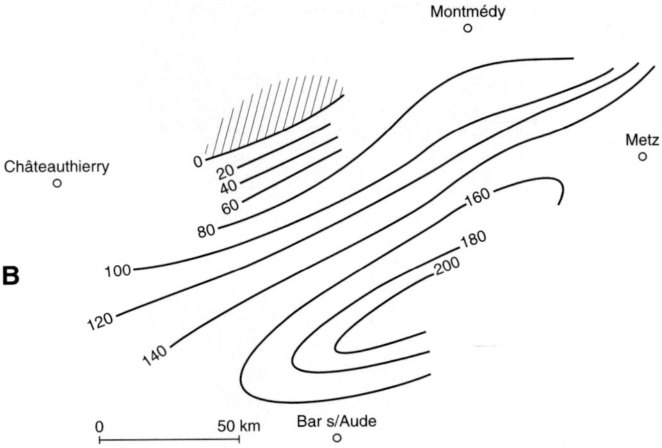

Fig. 1.5 A. Extract of geologic map. **B.** Example of isopach map. Isopachs are lines of equal thickness of a given geological bed. The geological stage of the example here is the lower Keuper in the Paris basin (see Chapter 4), where major thickness variations (from 0 to 210 m) are observed. The map thus represents a volume, with lines of equal thickness drawn on a horizontal plane (from J. Guillemot, 1986).

A

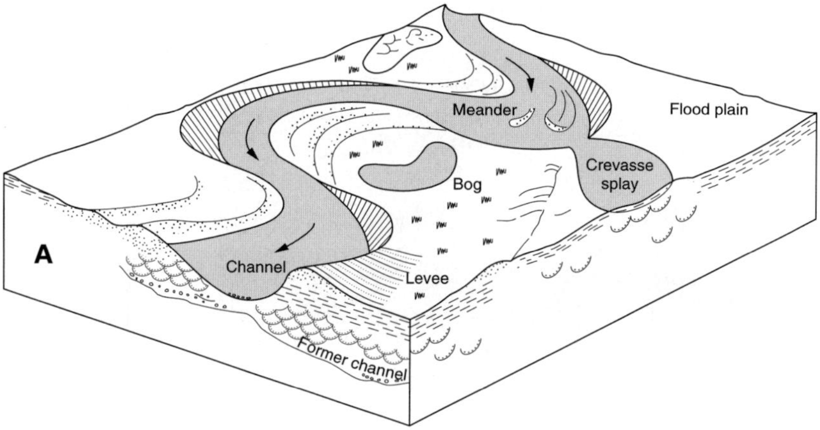

B

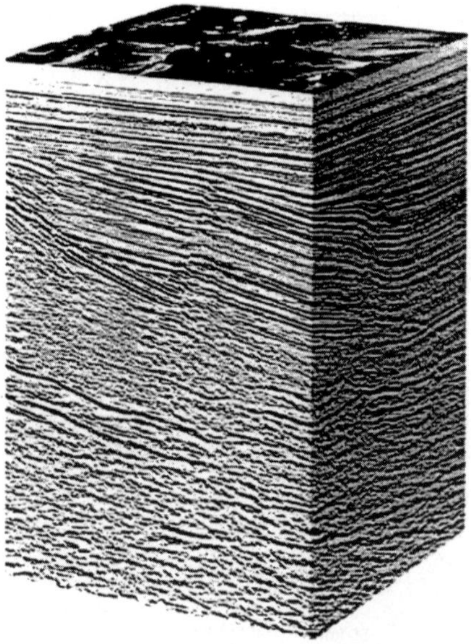

Fig. 1.6 Example of three-dimensional representations.
A. Block diagram illustrating meandering deposits. **B.** Representation of the 3D seismic in two vertical sections (in two orthogonal directions) and one horizontal section.

1.2.2 Time Scales

Dynamic geology does not begin with the Big Bang 15 billion years ago, which is a bit remote when it comes to applications to petroleum geology. Sedimentary geology is concerned, however, with at least the most recent billions of years in which identifiable geological processes have left an interpretable trace on the earth's surface. For most sedimentary basins, and especially those of petroleum geology, the portion of geological time of interest to us is the last 540 million years, though this is simplifying matters because some Precambrian basins are notable exceptions, for specific reasons. Figure 1.7 recapitulates this geological time scale, which will be discussed in greater detail in Chapter 4. This time dimension is basic to geology, of course, as one of its objectives is to reconstruct earth history.

Several remarks are in order here:

- Time is a measurable quantity. We will be speaking of sediments deposited over 10 000 years; or rocks dating from 2 billion or just 20 million years; granites laid in place 140 million years ago; or deformations that occurred between 65 and 60 million years ago, for example. We are always trying to obtain **absolute dating** by various techniques (see Chapter 4). This is the task of chronostratigraphy.
- Often enough, there is no way of defining such a chronology with absolute precision, and we then have to use **relative dating**. For example, sediments can be dated by their deposition before, during, or after some major event or minor landmark, such as rifting or folding.
- **Time scales** will vary considerably depending on the object being studied. In seismology, quakes are recorded on the scale of a second, while it is the daily tidal cycle that will be considered when working on the formation of deposits on the continental shelf. Or if we are trying to interpret glacial-interglacial cycles, the order of magnitude will be different again, ranging from several thousand years to several tens of thousands. If we are reconstructing the evolution of the Gulf of Gascony or the Alps, the time scale is in millions and tens of millions of years (Fig. 1.8).

We will return later to the idea of brief **events** such as the eruption of a volcano or the impact of a meteorite. These events are practically instantaneous on a geological scale, and fall within a history of continuous evolution guided by cycles at different time scales, ranging from tides to the establishment of a lithospheric convection cell.

The idea of time is still capital in seismology and experimental seismic, where wave arrival times, velocities and slowness, are calculated.

1.2.3 Kinematic Reconstruction

Kinematic reconstruction defines the evolving geometry of layered or interlocked geological objects, along with the mechanisms that govern this changing structure with time. The

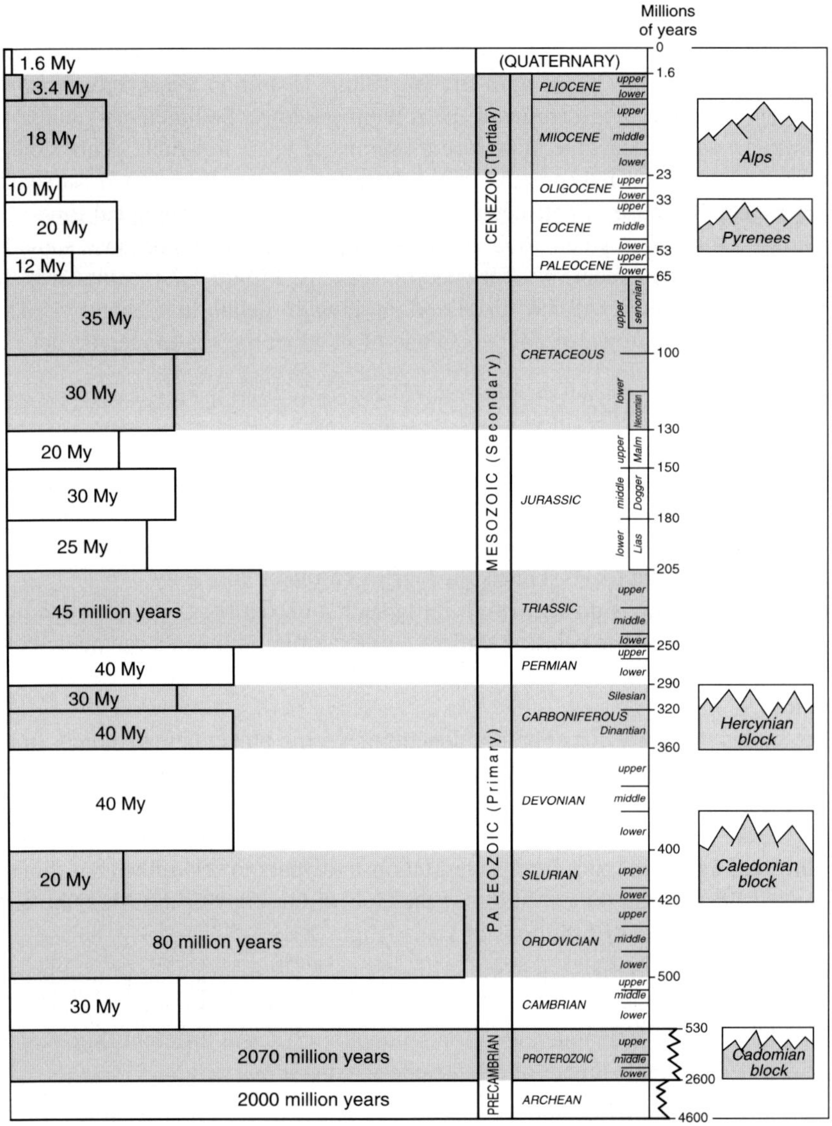

Fig. 1.7 Geological time scale (from the geological map of France, BRGM).
This representation is not proportional to time, expressed in millions of years (My). All Precambrian history, in particular, is highly compressed. Shorter times can be expressed as fractions of millions of years (ky).

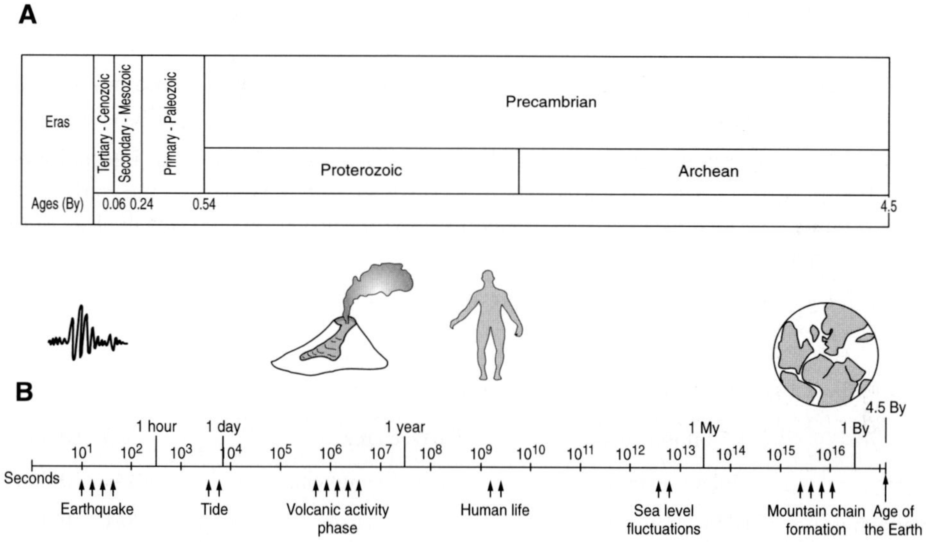

Fig. 1.8 Duration of geological processes.
A. Linear time scale (ages according to G. and C. Odin, 1990). Time is expressed here in billions of years (By). **B.** Variable duration of different geological processes (according to J.M. Caron et al., 1992).

geologist generally has only the last frame of this historical movie to work with – the current geometry – along with a few intermediate landmarks. By direct observations in the field or laboratory, and by simple analogical reasoning, experimentation and simulation, he then proceeds to characterize the different states at different transitory periods. This analysis may cover the stages of formation of a basin, as we will see in Chapter 2, or the development of reefs on a carbonate platform (see Chapter 3), the paleogeography of a sedimentary basin (Chapter 4), the evolution of a stress field or the formation of a fold (Chapter 5), or the migration of hydrocarbons (Chapter 6). As different scenarios may all lead to the same result, the geologist has to find indicators that will allow him to choose the most likely one.

To wind up this presentation of space and time perspectives in geology, the ideas of **uncertainty** and possible **error** come to the fore. While geology is becoming more and more quantitative, it is still a field where there is usually more than one hypothesis to choose from, so some estimation of the margin of error is needed.

1.3 EARTH STRUCTURE: GEODYNAMIC FRAMEWORK

In the following, we will begin with the essentials of internal geophysics and build up to the basins and porous media that will be of more special interest to us in the following chapters.

It is important to know what our planet's general structure is, and what internal mechanisms are at play, if we want to understand how basins evolve. Models of the Earth's structure are coming into better and better focus now, with the various geophysical techniques used on land and sea and from space.

Table 1.1 gives the Earth's basic structural data.

Table 1.1 Basic figures concerning the Earth's structure.

Age	4.5 billion years	Surface area	$510 \cdot 10^6$ km^2
Shape	Oblate spheroid	Area of continents	$149 \cdot 10^6$ km^2
Polar radius	6357 kilometers	Area of oceans	$361 \cdot 10^6$ km^2
Equatorial radius	6378 kilometers	Mean density	5.52
Volume	1000 billion km^3	Average surface temperature	15°C

1.3.1 Shape

Because of its rotation within the solar system, the Earth is shaped something like an ellipsoid of revolution, i.e., flattened at the poles and bulging at the equator. This ellipsoid of revolution differs somewhat (by a few hundred meters) from the **geoid**, which is defined as the equipotential surface of gravity coinciding with the equilibrium surface of the seas extended beneath the surface of the continents.

The latest studies have shown finer variations yet in the gravity field. The equilibrium surface of the seas itself varies in direct relation with the distribution of masses deeper in the Earth, which themselves vary in time with the geodynamic evolution of the globe (Fig. 1.9).

1.3.2 General Makeup

Like the other telluric planets close to the Sun—Mercury, Venus, and Mars—, Earth has a high density. Like Venus and Mars, it has an atmosphere that accounts for an infinitesimal part of its total mass but which, as we will see later, plays a fundamental role in geological processes. The abundance of water in its different states is a special feature on Earth, making it the **blue planet** where biological phenomena play a major role.

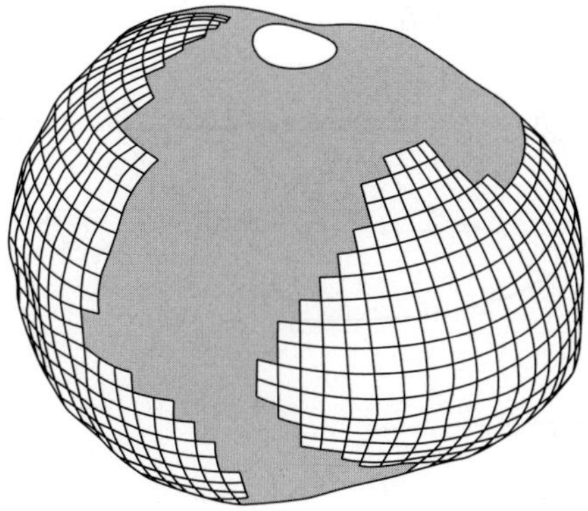

Fig. 1.9 Shape of the Earth.
The geoid is flattened at the poles due to the Earth's daily rotation, and also has two large bulges opposite each other over the Pacific and Atlantic. These result from the internal distribution of matter, which is determined by the internal dynamics of the globe (from A. Cazenave, 1990, in *Courrier du CNRS*). The amplitude of the undulations is exaggerated by a factor of 105 here, with respect to the Earth's radius.

Although our basic assumptions are still changing today, it is generally thought that the radioactive, chemical, and thermal phenomena that existed in the essentially homogeneous primitive Earth gradually induced a differentiation of the deep magmas, thereby releasing the water and atmosphere that was initially locked into the minerals. These were then ejected at the surface by volcanism, in the form of steam and gases. The essential role of the biosphere was initiated several billions of years ago by the internal geodynamic mechanisms of the early atmosphere and hydrosphere. We must therefore remember that the composition of the Earth's outermost shell has changed with time, especially when analyzing the oldest rocks (Fig. 1.10).

The initial atmosphere of 4.5 By ago was either the remnant of some primeval nebula of hydrogen and helium, or was released from the solid parts of the Earth, mainly by degassing in the form of water vapor and carbon dioxide (and other trace gases such N_2, NH_3, *and CO). It then evolved more or less rapidly to its current composition of nitrogen and oxygen. The evolution of this atmosphere and its sibling hydrosphere can be determined by analysis of sediments and the fossils they contain (Fig. 1.11).*

Today's astronomical observations and calculations tell us that the Earth's overall density is 5.5, and it is generally assumed that the materials making it up increase in density with depth. The density of the surface rocks that will be of interest to us in the following chapters is about 2.8 on the average (Table 1.2).

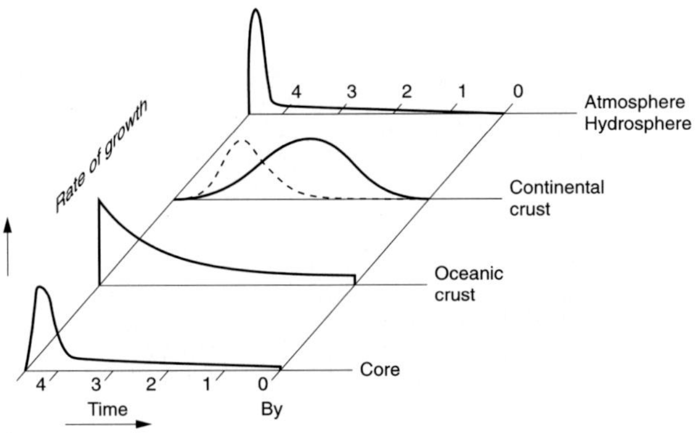

Fig. 1.10 Evolution of the Earth's major geochemical reservoirs with time (from B. Dupré and E. Lewin, *Courrier du CNRS*, 1990).
Here, the assumption is that the mantle degasses rapidly and the metal core segregates, while the oceanic crust continues to renew incessantly. Two different models are shown for the continental crust, and both show permanent erosion balancing accretion today.

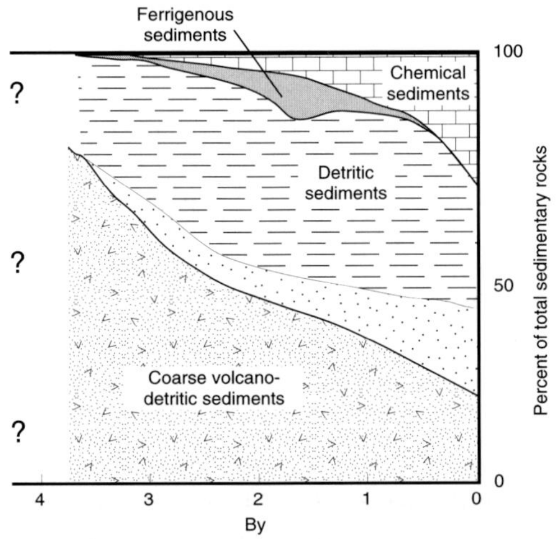

Fig. 1.11 Time variation in the relative proportion of different types of sediments (from P. Thomas in Angelier et al., 1992).
The relative proportion of different broad classes of sediments reflects the composition of the atmosphere and hydrosphere. For example, the abundance of sediments rich in ferric oxides would seem to indicate a change from a reducing to an oxidizing atmosphere about 2 By ago, while the increase in chemical and biochemical deposits in the last 500 My is due to the rapid growth of the biomass in the sea and then on land.

Table 1.2 Earth's internal structure.

	Lithosphere	Crust (continental/oceanic)	2.5 to 2.9	g/cm^3
100 km		Upper mantle	3.3 to 3.4	g/cm^3
300 km	Asthenosphere			
700 km			3.3 to 5.6	g/cm^3
2900 km	Mesosphere	Lower mantle		
	Core	Outer core Inner core	5.6 to 9 up to 12	g/cm^3

1.3.3 Deep Earth

The general cross-section of the Earth given in Fig. 1.12 is based on observations, measurements, and experiments (conducted on the Earth's surface, in the atmosphere, and from space) of the gravity field, heat flow, and seismic wave propagation. The **core** accounts for 33% of the Earth's mass and 16% of its volume. It is about 3400 km in diameter, and its density and temperature are very high (from 10 to 13 g/cm^3 and 4000 to 5000 K), at pressures of 3000 to 3600 kbar. It seems to consist essentially of iron and nickel, with a few other elements such as sulfur, silicon, and potassium. Seismological studies have shown that the solid **inner core**, or **seed**, is anisotropic because of its higher propagation velocities pole-wise, whereas the liquid **outer core** is a very homogeneous medium that does not propagate shear waves, doubtless because of the stirring by internal convection currents. The Earth's dynamo results from the interactions of fluids in motion carrying electric currents in this outer part. What should be remembered here is that the electric currents circulating in the core generate nearly all of the Earth's magnetic field. We will return to this later.

The Earth's most abrupt seismological discontinuity is located at the upper limit of the core. This discontinuity seems to be of irregular topography, with undulations at wavelengths of between 2500 and 5000 km and amplitudes of a few kilometers (might this hint that there is some mechanical coupling between mantle and core?). The thickness of this "D" layer is variable, up to 200 km. Above it, major chemical variations appear that might be the cause for the upward movements of magma and the hot spots (discussed further on) observed at the Earth's surface.

The **mantle** envelops the core in a thickness of nearly 3000 km, rising up to the Mohovicic seismic discontinuity, called the **Moho**. The mantle and crust, which are discussed below, together account for 67% of the Earth's mass and 84% of its volume, with a density varying from 3.3 to 5.6 g/cm^3. The mantle has its lower and upper parts, separated by a transition zone at a depth of 650 to 750 km under the surface. Each of these parts of the mantle may be driven by its own convective motions, in several stages, but this is still a matter of conjecture. A zone characterized by low seismic velocities, which may correspond to a partial melting zone, is found in the upper mantle at depths between 100 and 300 km. This is the

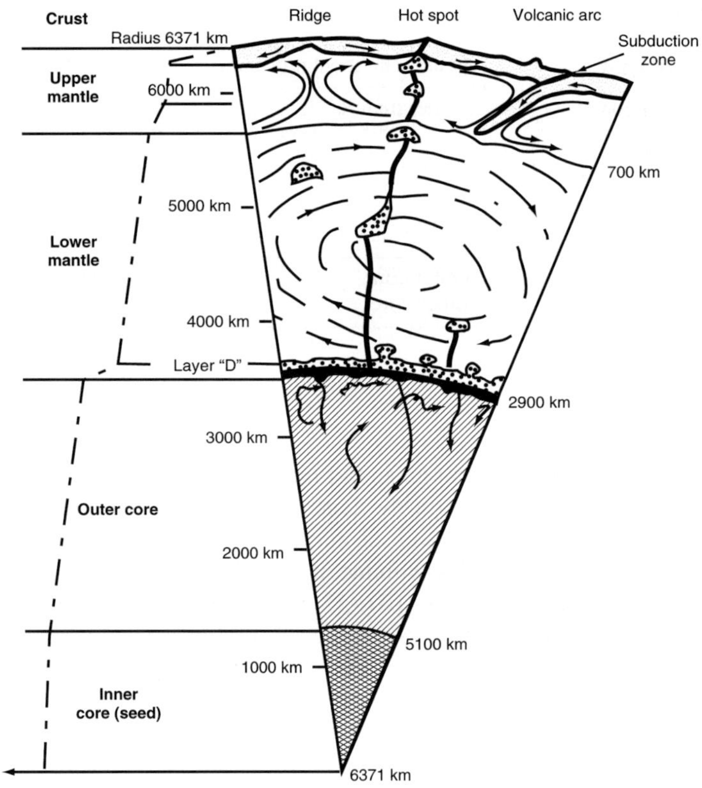

Fig. 1.12 Deep structure of the Earth (from V. Courtillot, *Courrier du CNRS*, 1990).
The globe consists of several concentric spherical layers with a very thick (more than 3400 km) core at the center. Then there is the mantle (2900 km) with the convective movements shown here, and lastly the very thin surface crust, from 6 to 60 km thick.

viscous **asthenosphere**. On top of this, we find the **lithosphere**, which is characterized by high wave propagation velocities and rigid and elastic properties, enveloping the upper part of the upper mantle and crust.

The **crust** is the Earth's outer integument, varying in thickness from 6 to 60 km (Fig. 1.13) between the Moho discontinuity and the surface. It consists of a low-density (2.6 g/cm^3) granitic layer overlying a denser (2.7 to 2.9 g/cm^3) basaltic crust, the mean composition of which is given in Table 1.3. The sedimentary basins that are our essential concern in this book develop at the surface of this crust, but we can see that their origin and evolution are guided by the deep convective mechanisms in the mantle.

Table 1.3 Mean chemical composition of the crust.

Oxygen	46.5%	Iron	5%
Silicon	27.5%	Calcium, sodium, potassium, magnesium	11%
Aluminum	8%	All other elements	2%

1.3.4 The Blue Planet

The main characteristic structure of the Earth's crust is the distinction between continents and oceans. We will learn of the continents that they are not limited to what emerges from the water, and that the crust under them is generally 35 to 40 km thick, while it is an average of 6 km thick under the oceans, not counting the depth of the water itself (Fig. 1.13). We thus have a granitic and basaltic continental crust and a basaltic oceanic crust. It may happen that two continental crust domains overlap in certain mountain chains (Alps, Himalayas), and we then have thickness anomalies of up to 60 km.

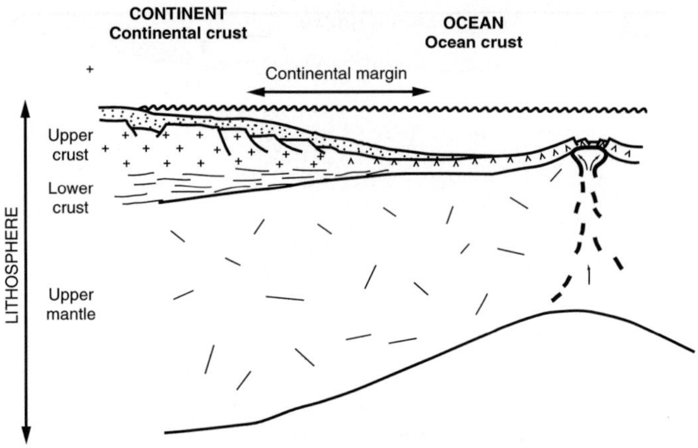

Fig. 1.13 Continental and ocean crust.
Crust thickness varies considerably in the zone called the continental margin, between the thick continental crust and the ocean crust.

The Earth is thus characterized by the extent of the globe's surface covered by saltwater seas and oceans, accounting for 70% of the planets surface area in all, mainly in the southern hemisphere (Figs. 1.14 and 1.15).

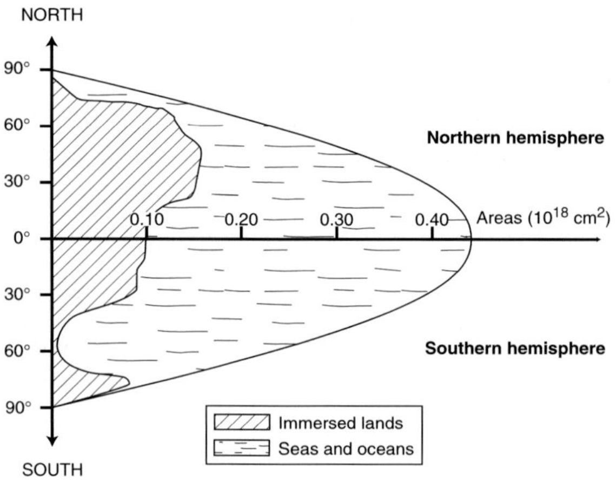

Fig. 1.14 Distribution of immersed land and ocean areas (from Tardy, in J.M. Caron et al., 1992). The areas are represented in a latitudinal distribution.

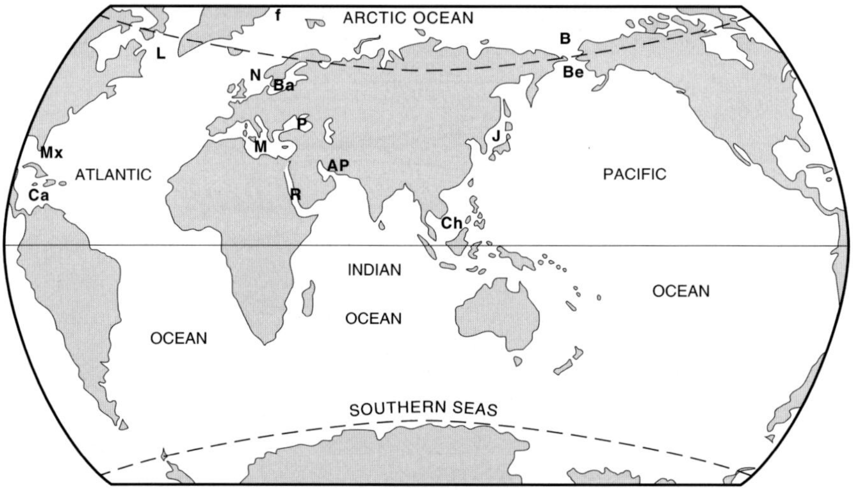

Fig. 1.15 The Earth's broad oceanic domains.
The three largest oceans communicate with each other at the high southern latitudes. The Arctic Ocean is cut off from the Pacific at the Bering Straits (B) and from the Atlantic at the Fram Straits (f). Vast maritime spaces are appended to these principal oceans: the Gulf of Mexico (Mx), the Caribbean (Ca), the Sea of Labrador (L), the Mediterranean (M), the North Sea (N), the Baltic (Ba), the Black Sea (P), Red Sea (R), the Arab-Persian Gulf (A-P), the China Sea (Ch), the Sea of Japan (J), the Bering Sea (Be).

The greater part of the free water in the hydrosphere is contained in the oceans, amounting to about 1372 × 10^6 km^3. But the water contained in rocks (sediments, crust, and mantle) no doubt accounts for about 25%.

While the world ocean is characterized by salinity, an organized hydrodynamic system, and major biological populations to be considered in geological study, what distinguishes it geologically from the continents is the structural criteria stemming from its origin. Certain maritime portions of the thin-crust oceans are submerged portions of continent, such as the Baltic, the North Sea, and the Channel in the European area. The shoreline boundary between the domains above and below water is an important environmental boundary that has fluctuated in the course of time (see Chapter 4). But the boundary between the continental and oceanic domains, called the **continental margin**, is generally located at the foot of the continental slope, with a more or less extensive transition zone (thinned crust, see Fig. 1.13). We will see that several types of continental margins are defined by their history and structure.

The **continental shelf** (Fig. 1.16) is a platform extending outward from the coast to an edge, which generally appears suddenly at a distance (average 70 km) from the coast, and at a depth of between 120 and 200 m. Starting from this edge, there is a steep transition of 4 to 5% toward the oceanic domain, and sometimes much more along certain escarpments and canyons. This is called the **continental slope** or **apron**.

The **deep ocean floor** consists of abyssal plains and hills at depths of between 4000 and 6000 m. In the heart of the oceans, we find more brutal relief with **ridges** and **rifts** and certain underwater plateaus and volcanic ridges, as well as **trenches** as deep as 10 km along many island arcs.

The most common ocean depth is about four kilometers. The average altitude of the continents is about 1 km, and the morphological regions are plains, plateaus, and mountain chains (Fig. 1.16).

The following two points should be remembered here:

- By definition, sedimentary basins are cup-shaped or structurally low, which is why they serve as sedimentary receptacles. The ocean is therefore their preferential location, and continental plains and broad lacustrine hollows are a second preferential category of basins. However, we will see further on that the thickest sedimentary basins, and those of greatest interest in petroleum geology, are, for various reasons, not in the great ocean depths but on the continental margins and continents where most of the sediments are trapped.
- Continents and oceans interacts in many ways, but the coupling between the ocean (97% of the hydrosphere) and the atmosphere is very strong (see Chapter 3).

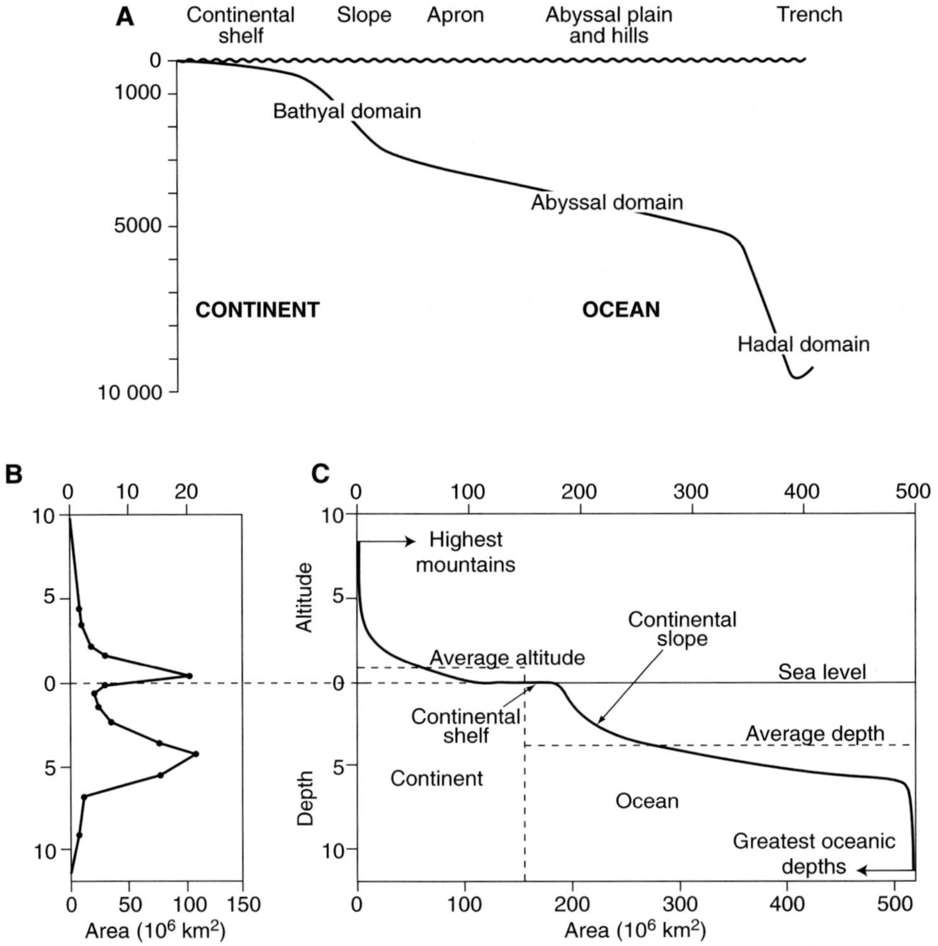

Fig. 1.16 Bathymetric profile of the ocean.
A diagrams the broad bathymetric regions from the coast to the hadal zones. The hypsometric curves in **B** and **C** illustrate the relative distribution of land altitudes and sea depths, in the forms of a frequency histogram and cumulative curve, respectively (from Willie, in Boillot, 1990).

1.3.5 Plates and Hot Spots

The importance of the mantle's convection phenomenon has already been discussed. The result of this phenomenon at the surface is **lithospheric plate motion**. These plates are systems that are mobile in time. The idea of plate tectonics was first conjectured at the end of the sixties, on the basis of more and more observations and measurements in the oceans and along the continental margins, showing rheological stratification (lithosphere-asthenosphere), dispersed mechanical energy at the Earth's surface along clearly localized seismic belts, associated volcanism, and ocean expansion generated at ridgelines.

As plate tectonics gradually developed into a unifying concept, the idea of **global tectonics** came to explain the tectonic and seismic activity detectable at the surface. This was considered to be the result of the interaction among a few **rigid plates** moving along **identified plate boundaries** (Fig. 1.17). Arguments were eventually found to prove this theory, but it excluded the idea of any absorption of the deformation processes within the plates themselves. Today, the foundations of this theory, sometimes refined with the idea of intraplate movements, for example, are universally accepted and widely used, even in the industry.

The following reviews the basics of this theory, which will be of use later.

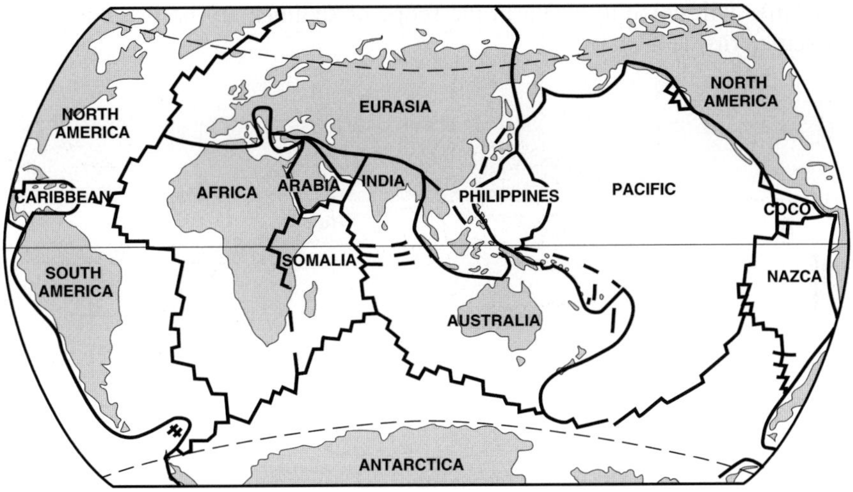

Fig. 1.17 Lithospheric plates.
Some plates are entirely oceanic (such as the Pacific, Philippines, Cocos, Nazca) while others are purely continental (Arabia). But the typical plate includes both oceanic and continental lithosphere, corresponding to the Gondwana super-continent which has been breaking apart for 180 My now into a number of "ocean opening" zones: the American plates; Africa; Eurasia; Antarctica; Indo-Australia. There are more complex cases: remnants of former plates (the Mediterranean); new ones being created (Indonesia); and micro-plates in the Pacific (Galapagos, Easter-Rapanui, Juan Fernandez, and others).

1.3.5.1 Plate Characteristics

The lithosphere is not a continuous integument enveloping the Earth's surface, but is rather broken up into a number of spherical caps called plates. Lithospheric material is created at the oceanic ridges over zones of rising convective cells, and from there moves across the surface like a conveyor belt to the subduction zones, where it plunges deep toward the asthenosphere. Each plate is rimmed by a boundary characterized by separation, collision, or slip parallel to the boundary line.

Each plate has its own characteristic rigidity (low flexural but high torsional rigidity), but is defined mainly by its mobility. Instantaneous or average plate motion is on the order of a few centimeters per year. On the geological scale, the object is to reconstruct plate kinematics, and the devices can change with time, as the boundaries are unstable.

1.3.5.2 Plate Boundaries

Plate boundaries will come up again in the scope of sedimentary basins, but a few basics should be settled first (Fig. 1.18).

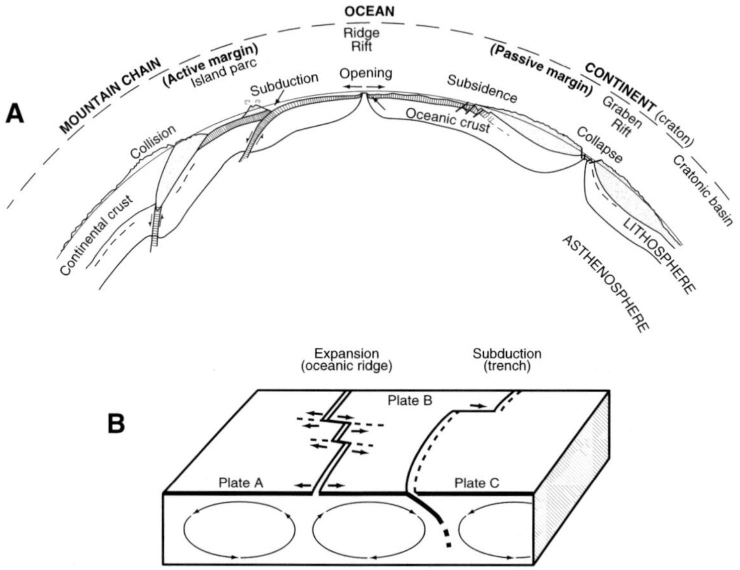

Fig. 1.18 Convection cells and plate boundaries.
In the lithospheric divergence zone of asthenospheric rise, the plates move away from each other, expanding the ocean and rifting to form the oceanic ridges. In the convergence zone, the lithosphere plunges toward the asthenosphere. This is subduction. **A.** General section (from J. Guillemot, 1986). **B.** Perspective.

One type of boundary is the **plate separation**. This is driven by lithospheric **divergence**, in which the continental crust is stretched to the breaking point, creating **rifts**. The rift opens and, as the continental edges separate, the mantle rises and cools in the opening. This is the process of **oceanic expansion** in line with a ridge characterized by earthquakes, the rise of basaltic **magmas** forming the ocean crust, and segmentation along transform fractures and faults. In the course of this divergence, the basaltic crust thus formed gradually spreads away from the ridgeline (Fig. 1.19). A passive, or stable, **continental margin** then develops around the edges of the ocean as it opens. Stressing processes (thinning, breakup, subsidence) and thermal processes vary greatly throughout this evolution. This will be dealt with in the chapter on sedimentary basins stemming from these mechanisms, which we call **rift basins**.

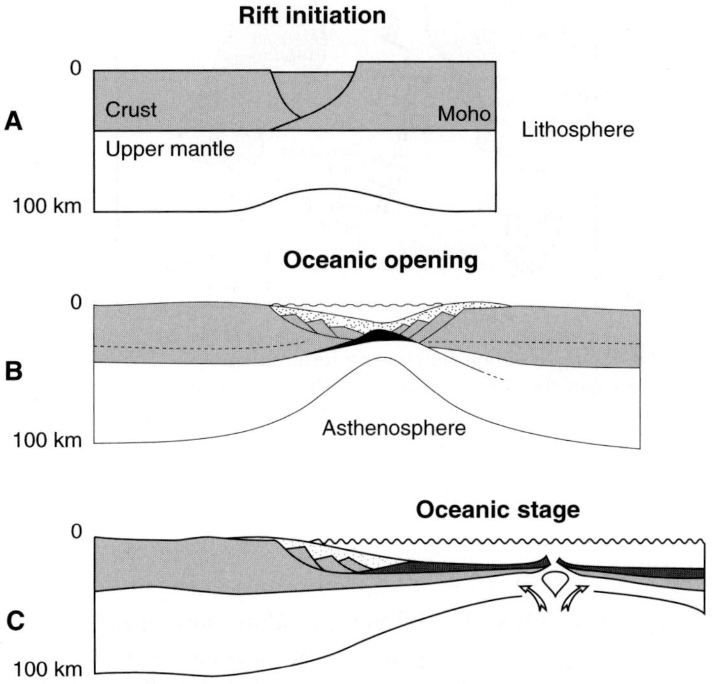

Fig. 1.19 Divergent boundary: from rift to ocean basin.
A. The rift appears as the crust stretches and breaks. **B.** Then, as the ocean opens, it allows the basaltic magmas to rise and create margins to either side of the former rift. **C.** At a later stage of ocean maturity, the continental margin becomes the transition between continent and ocean.

The ocean accretion or **expansion rate** can vary considerably from one plate boundary to another in the course of time and in the same ocean, as can be seen in present oceans: figures range from less than 1 cm/year to 20 or 30 cm/year.

For as yet unexplained reasons, this expansion can come to a halt and leave a fossil ridge in place of the active one. The plate boundary then disappears (Fig. 1.20).

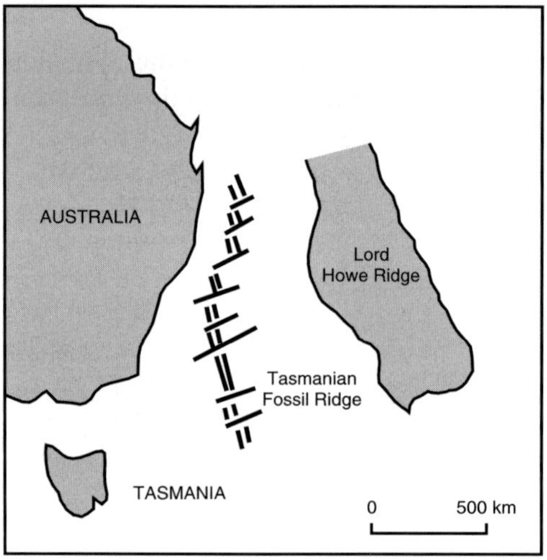

Fig. 1.20 Example of non-functioning former ridge.
The Tasmanian sea, southeast of Australia, was created by ocean expansion in the Cretaceous, and the Howe ridge contains a fragment of Australia. This is a fossil ridge that is no longer active today, and is not the sole example of the type. Others are known in the Indian Ocean, the China Sea, and elsewhere (from Biju-Duval, 1994).

The second major type of boundary occurs with **plate convergence**. This produces **subduction**, in which a slab of lithosphere plunges to great depths with effective seismic activity to a depth of 250 km (Fig. 1.21). The stresses and strains in these plate convergence zones are manifold and spectacular. Convergence can become a **collision**, which is an extreme mode of subduction and is considered to be the origin of most of the Earth's mountain chains. The term **active margin** is generally used for these convergence zones.

It is also considered that the ocean crust absorbed into the subduction zones is equivalent to that produced along the ridgelines, and there is no increase in the volume of the Earth. (It is estimated that 3 to 3.5 km² of ocean surface is created or destroyed each year, or 300 to 350 km³ of lithosphere.) According to certain hypotheses, part or even all of the upper mantle could have been recycled over the past 1000 My.

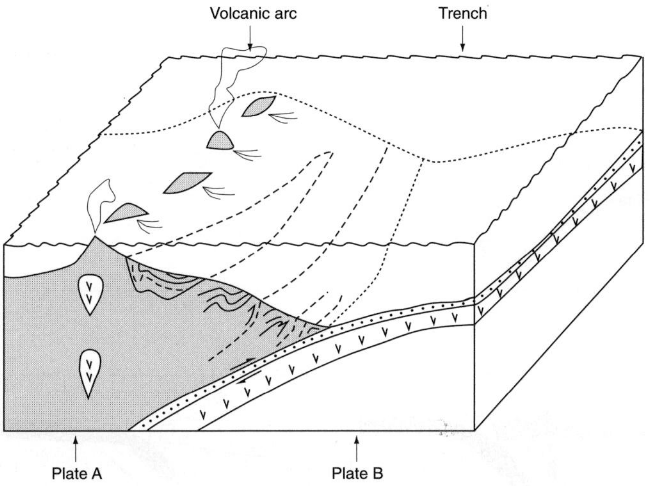

Fig. 1.21 Convergent boundary: lithospheric subduction.
This example shows an ocean plate plunging in subduction under an insular arc formed by a row of volcanoes. The beds ahead of the arc are highly deformed at the front with the overlapping plate along the trench.

Several types of subduction have been identified: abutting ocean plates (Mariannas), overlapping continental plates (Andes), continental plate overlapping with insular arc (Japan), and colliding continental plates (Alpine and Himalayan arcs). Sometimes a distinction is made between B-type subduction (B for "Benioff") and A subduction ("Ampferer"), depending on whether it is the oceanic or continental crust, respectively, that is being subducted (Fig. 1.22).

In subduction, the plate that subducts is folded to plunge at a variable angle, thereby creating a **flexural basin** with subsidence and thermal processes that differ from those of pull-apart basins (see Chapter 2).

If part of the sediments cannot be subducted for rheological or mechanical reasons, the sediments are detached, deformed, and tectonized into an **accretionary prism** or wedge in which **forearc basins** of different types will develop, depending on the evolution (Fig. 1.23). Fluids play a crucial role here in constructing sedimentary wedges, as it is the fluids that allow **décollement** in the first plate. Sometimes the material is accreted at depth, which is called **underplating**. But it is also considered that **tectonic erosion** and delamination of the crust occur at depth. This is how active margins of the accretionary type (Barbados, Nankai, Makran) are distinguished from those of the ablative type (Isu-Bonin, Tonga, Peru).

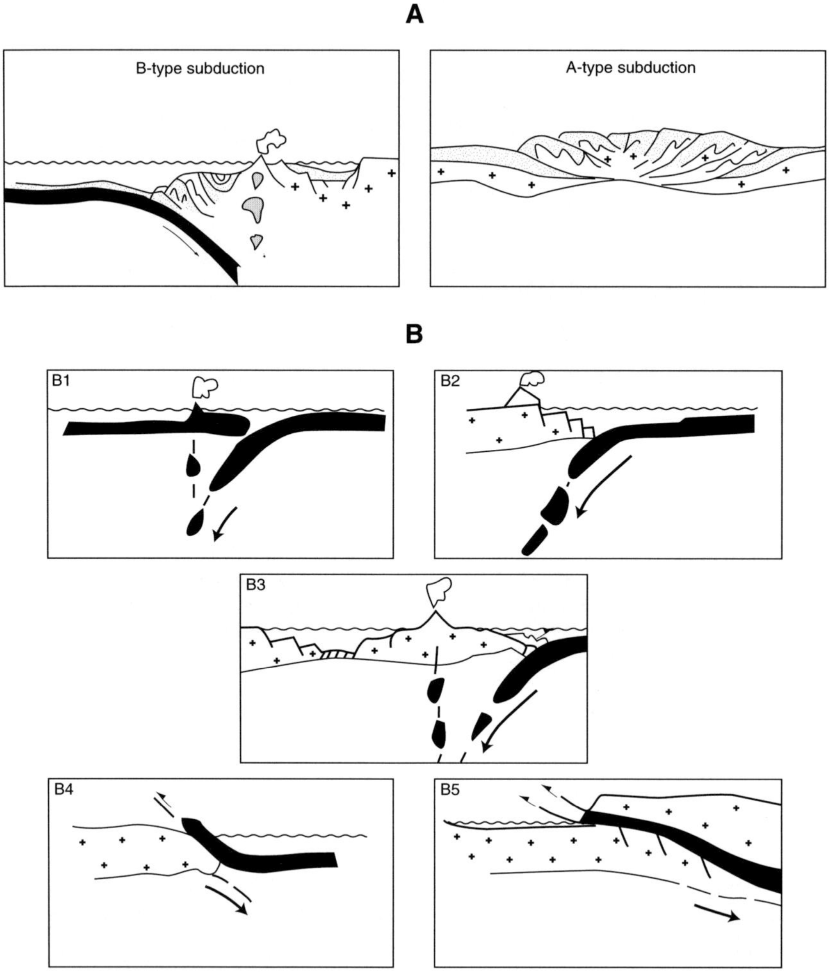

Fig. 1.22 Different subduction processes (from Biju-Duval, 1994).
A. "Ampferer" A-type continental subduction and "Benioff" or B-type oceanic subduction. Two different modes of plate convergence. **B.** Different types of subduction: **B1.** Abutting oceanic plates (e.g., Marianna islands). **B2.** Ocean plunging under a continent (Andes and Cordillera type). **B3.** Ocean plate plunging under an insular arc separated from the continent by a back-arc basin (e.g., Japan). **B4.** Obduction (overlapping) of the ocean crust onto continental crust (e.g., Taiwan). **B5.** Continental collision, with two continental plates abutting or overlapping (e.g., Tibet).

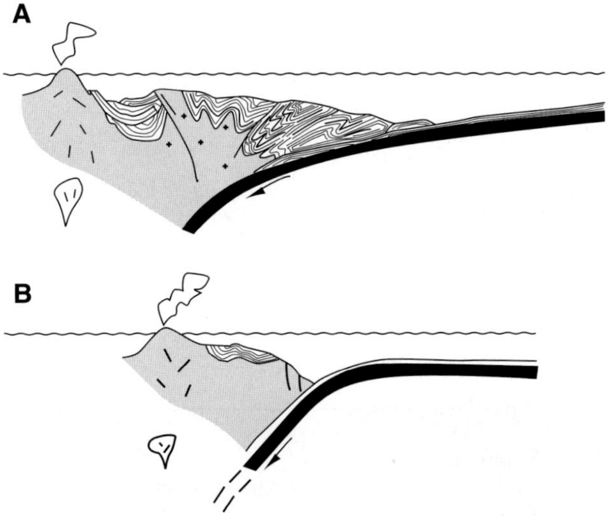

Fig. 1.23 Two types of active margins: accretionary and ablative.
A. The sediments from the trench are not subducted, or are only partly so, but are rather accreted into a sedimentary wedge atop a décollement zone at the front of the overlapping plate. **B.** The sediments are subducted and, with active erosion of the front, a great deal of material can be entrained downward.

Behind the front is the **volcanic arc**, evidenced either as a string of islands or on the continental margin, emitting calc-alkalic products that disrupt the sedimentation, including in the back-arc that stretches out behind it.

Once the oceanic space is completely absorbed, **continental collision** begins. In certain cases, the ocean crust or some uneven parts of it are not subducted. It then participates in frontal tectonic scaling and is transported over the neighboring plate in a process called **obduction**.

Quite often, the motion of the plates is not orthogonal to the subduction zone boundary. Major sliding then occurs.

Between the two boundary types described above, which are often designated as **passive** and **active margins**, there is a third type: the **transform plate boundary**, which is called **shear plate margin** when speaking of a continental domain. Examples can be given of **transform faults** that shift the ridges (divergent boundaries) in the equatorial Atlantic domain and Indian Ocean, and those of the Pacific domain that link subduction zones that operate in opposition to each other (Hebrides-Tonga) or in the same direction (San Andreas). Pull-apart and/or compression processes and volcanism can coexist or alternate alongside these active boundaries, marking a very sharp border between continental and oceanic crust (Spitzberg, Cordilleras). These strike slips can range in length over several hundred kilometers (Fig. 1.24).

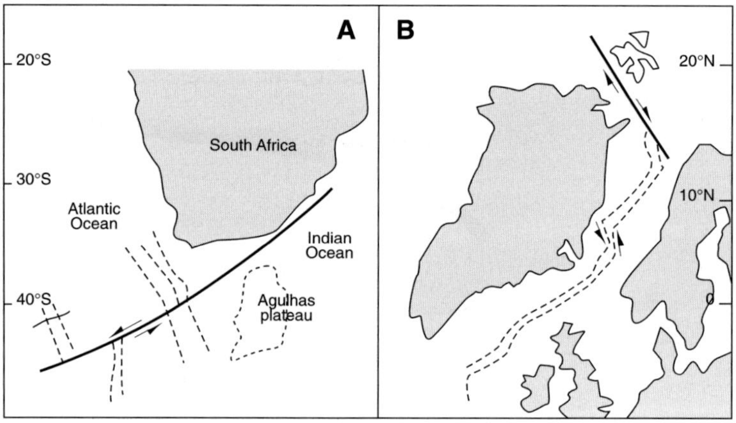

Fig. 1.24 Transform plate boundaries.
A. Great transform fault of the South African margin (present position).
B. Reconstruction of the Spitzberg shear plate margin before the opening of the North Atlantic (from Biju-Duval, 1994).

Lastly, we should mention the triple junctions at the meeting point of three plates. There are several types of these (Fig. 1.25), which are sometimes unstable in time.

Chapter 2, and then Chapter 4, will show how this geodynamic framework determines not only the type of sedimentary basin, but also its evolution in time. That the plate boundaries are unstable has already been mentioned. We know fairly well how to reconstruct their history over the past 200 My (today's oceans are no older than 180 My). Certain continental rifts like the North Sea have not evolved into a continental margin edging an ocean; once-active oceanic ridges have "flamed out"; whole plates like Farallon have almost entirely disappeared; divergent margins have been swallowed up in subduction (Alps); convergent zones have been arrested (Amirantes ridge in the Indian Ocean) while others have produced super-collisions (Himalayas); certain mountain chains are a patchwork of small continental fragments that have drifted with the plate motion.

1.3.5.3 Intraplate Volcanism and Hot Spots

The ocean covers 70% of the Earth's surface, and plate tectonics explains how the ocean is formed by the magmatic accretion process along ocean ridges or wrinkles (asthenospheric rise lines). However, a large part (estimated at 20% today) of the ocean surfaces are the result of other processes that are grouped under the term of **intraplate volcanism**, which also occurs in the continental domain (such as the French Massif Central and Dekkan).

This volcanism generates different types of underwater relief—plateaus, volcanoes, and table knolls—some of which may reach into the air, others just to the surface (atolls), while

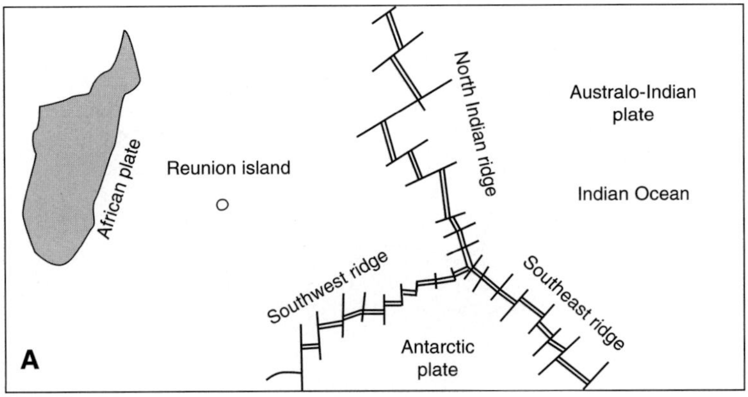

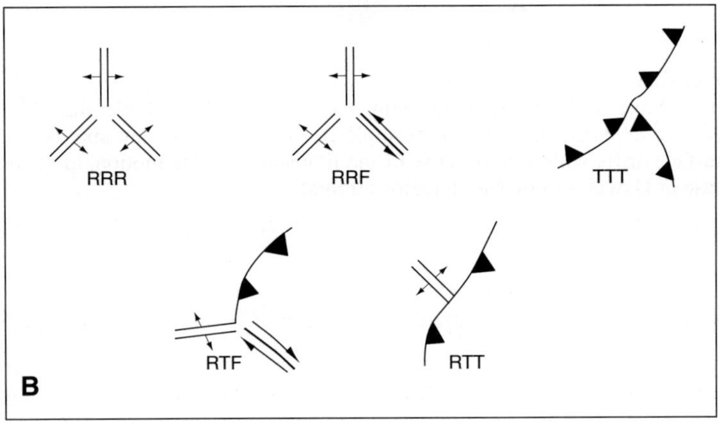

Fig. 1.25 Different types of triple junctions between three plates.
A. The triple junction in the Indian Ocean illustrates the divergence of three plates along three active ridges. **B.** Different processes at work: R: oceanic accretion ridge (divergence) T: subduction along a trench (convergence) F: transform fault (strike slip).

others yet lie deep. There are some one hundred active ocean volcanic systems that are interpreted as surface manifestations of **hot spots**. These hot spots are generally assumed to originate at great depths (see Fig. 1.12). The well-studied example of Hawaii is explained by the lithospheric displacement of the Pacific plate over a deep, passive hot spot that has gradually generated a string of volcanoes of various ages, lying in the same direction as the plate motion (thereby providing a way of evaluating this motion, Fig. 1.26). This is also true of the Fournaise on Reunion island.

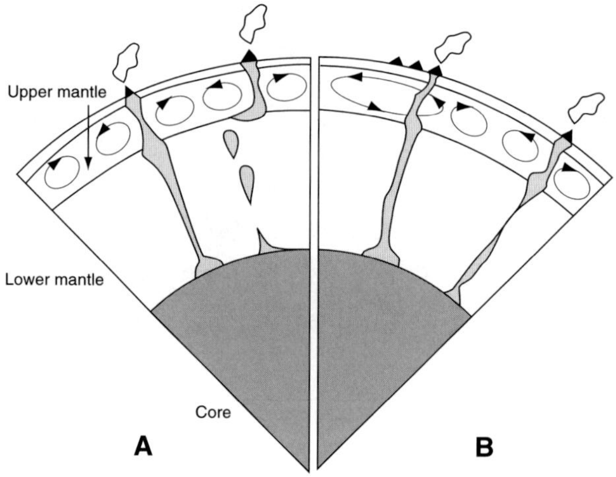

Fig. 1.26 Deep source of hot spots.
In example **A**, the hot spot emanating from the base of the lower mangle is passive in space and time. In example **B**, the trace of the hot spot on the surface drifts with time because of the lithospheric plate motion, as in the case of Hawaii (according to many authors).

It is generally thought that hot spots are associated with vast bulges in the lithosphere, local flexuring, and pronounced gravimetric signs. The stability of hot spots with time is still being studied.

The chemical composition (especially the isotope ratios) of the alkaline basalt that forms in the hot spot is what distinguishes them from the basalt of the ridges (Mid-Oceanic Ridge Basalt, or "MORB") and from the calc-alkalic products of the active margins.

1.3.6 Sedimentary Basins

We have briefly seen that the lithospheric plate motion and the rise of hot plumes create thermal and mechanical stress fields, which, as they evolve in time, originate all of the oceanic and continental sedimentary basins and mountain chains. The sedimentary basins themselves are the outermost shell, giving the Earth its present morphology and geography.

Now, addressing sedimentary basins, we will define them first from a geometric perspective as low-lying **depressions**, **structural lows**, **troughs**, or **cups**, which are all hollows in the Earth's crust where sediments are accumulating or where they once did, filling up the hollow (Fig. 1.27). The **sediments** themselves, as will be developed in the remainder of this chapter and especially in the following ones, are either solid particles eroded from neighboring rock, transported by various means and then deposited, or they are biochemical precipi-

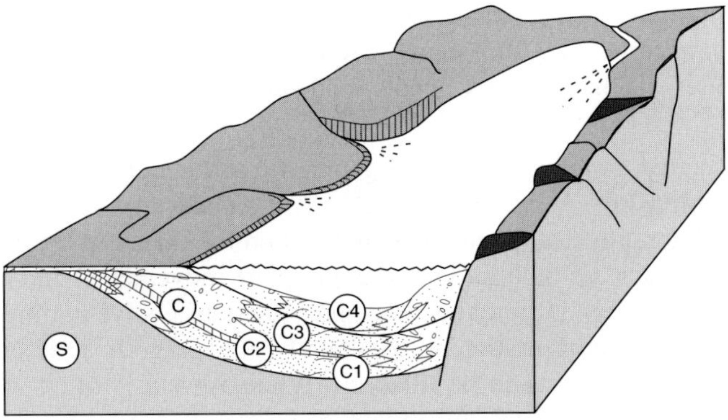

Fig. 1.27 Diagram of a sedimentary basin.
A sedimentary basin is a hollow in the relief. Erosion products accumulate in it and gradually fill it up. Its size varies considerably, from lake to ocean. The bottom of the basin is called the basement or substratum (S). The sedimentary fill, or cover (C) is a sequence of layers of different kinds (C1, C2, C3, C4). The deepest layers are the oldest, lining the bottom of the initial hollow. This is a functional basin.

tates carried in solution and derived either from leaching of terrain or from the aquatic environment itself (lakes, seas, or oceans).

This idea of a low or hollow is essential. It explains how sediments or deposits **accumulate** and partially or completely **fill** a **basin**, in conjunction with the major dynamic factors of transport (generally from a higher to a lower point), driven by the gravity, which is inescapable for biogenic "buildups" too (such as reefs that grow from the bottom upward).

In this erosion-transport-deposit sequence, it is generally only the deposit that is observable and all the rest of the processes have to be deduced and interpreted from this.

The accumulated sediments are generally referred to as **sedimentary cover**, as opposed to the **basement** or **substratum**, which is the hollow receptacle that receives the sediments.

The general principle of layer superposition goes along with this idea of filling a basin. The deposits at the bottom of a basin are those that are deposited first, and are thus the oldest. Each successive overlying layer is younger and younger.

The geological series that have been observed show that considerable volumes of sediments (millions of cubic kilometers) can be accumulated over long periods of time (hundreds of millions of years). It can thus be said that a sedimentary basin is a region where a sedimentary coverage has been deposited within a definite time span.

The present geography of the Earth as expressed on a globe or map (**actualism**) clearly illustrates this basin idea. The variety of sizes, configurations, and environments of present sedimentary basins can be seen in the different structural lows from the great ocean depths to the continental margins, and even in the heart of the continents with the great alluvial plains, closed "endorheic" depressions, and lacustrine cups. A **functioning basin** is one in which a cup is still accumulating sediments. The variety of climatic locations, sedimentation environments, water depth, transport agents, and other basin characteristics, will be discussed further on.

The last perspective, ranging beyond the geometric and geographical aspects of a basin, is its **historical dimension**. A sedimentary basin is the result of a sequence of events controlled by chemical and physical phenomena that act on the planet in the course of geological time. These processes are both internal (mainly plate movements) and external (climate and sea level, among others). Different geotectonic situations evolve in time. This is the way a basin "lives", and its configuration is continually changing. A structural low is never formed once and for all, waiting to be filled up. When a basin is filled with thousands of meters of sediments, it cannot be assumed that the cup was so deep to begin with. The shape of the depression and the water level may have been moderate all along the filling process, with the basin simply sinking gradually over the course of millions of years, leading to its present deep-cup configuration. Such a gradual settling, guided by internal dynamics, is called **subsidence**. It is closely dependent on **tectonics**, which will introduce stresses and strains that will also change the basin's shape. In contrast to functioning basins, then, there are **structural basins** whose present cup shape does not necessarily reflect the initial geometry (Fig. 1.28). This is true of very many basins at the Earth's surface, which can be considered **remnant basins** left over from formerly vaster systems (Fig. 1.29). The extreme case of basin deformation is the **formation of a mountain chain**, where the initial cup shape has, of course, been obliterated.

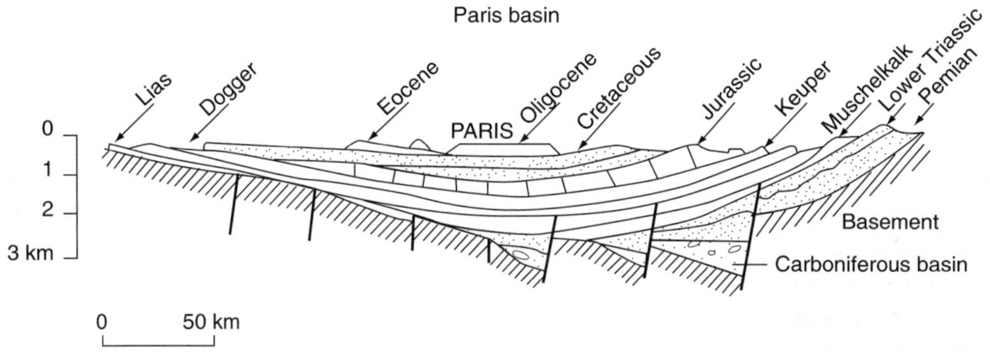

Fig. 1.28 Example of structural basin.
Schematic cross-section of the Paris basin showing a sequence of nested layers eroded at the present edges, which therefore do not correspond to the basin borders at different periods of its operation. This is a remnant basin.

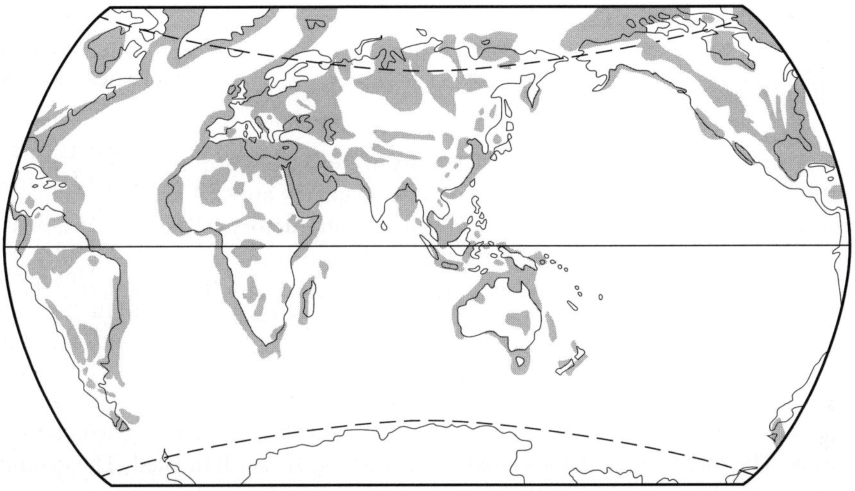

Fig. 1.29 Earth's major sedimentary basins (modified from A. Perrodon, 1980).
The ocean zones and continental margins are functioning basins. The continents usually exhibit residual basins.

This will be developed in the following chapters. But let us first recall how sedimentary basins are filled, and how this filling is conditioned by the combined action of internal factors, which are decisive in the evolution of the cup, and external factors that determine its filling.

1.4 DRIVING MECHANISMS

Geodynamics is subdivided into internal geodynamic fields and mechanisms, and external geodynamics (mainly the effects of the atmosphere, weather, and ocean physics). Since we are focusing on the sedimentary basins that make up the thin film around the planet where the geosphere, hydrosphere, atmosphere, and biosphere all interact, we will recall here the general parameters that condition their content and form. Basin **driving mechanisms** are highly interactive and act in complex combinations. The following will first address the internal drives or factors, and then the external.

1.4.1 Internal Drives

1.4.1.1 Earth Dynamo

The Earth's **magnetic field** is generated mainly by electric currents flowing in the fluid outer portion of the Earth's core. The dynamo mechanisms are still not fully known, as the rapid progress being made in this field is at times contradictory. However, observations of the field's time variations, combined with theory, hint that these variations are not small-scale turbulent movements but are in some way organized. There is no doubt that the field at the surface moves with some semblance of symmetry, and seismology indicates that this movement is probably a manifestation of organized deep motions.

The **global field** is defined by its intensity, *F*, and two angles: its declination, *D*, and inclination, *I*. Intensity is measured in teslas (T) or nanoteslas (nT, 10^{-9} T), which used to be called Gauss. The intensity of the horizontal component, *H*, is often used. The origin of the global field is therefore mainly (for 99%) an internal dipole, while the rest is due to an external source: the solar wind of ionized plasma. The field's sphere of influence is called the magnetosphere. Its order of magnitude at the Earth's surface is about 50 000 nT, but this varies from point to point (Fig. 1.30). These spatial variations of the global field at the Earth's surface are due to disturbances by local or regional **static anomalies**: magnetized rock in the crust, and certain bodies that have acquired magnetization may in themselves constitute a specific, permanent magnetic field source (Fig. 1.31). This is why, in petroleum geology, the field heterogeneities recorded on **anomaly charts** are of use in determining a basin's deep structure (substratum). These charts are generally obtained by aeromagnetic surveys taken by airplane or satellite (Fig. 1.32).

Geomagnetism is a very old branch of geophysics that also tells us something about the field's **time variation**. Variations can be rapid, such as those of externally-induced magnetic storms (some 500 nT in a few hours). But they can also be slower. There are external periodic variations due to the 11-y solar cycle, and seasonal variations within this; but there are also internally induced variations, which are the essential source of slow changes. **Secular variation** is a slow drift of a few tens of nanoteslas per year, with occasional sudden changes prompted by altered **electric currents** in the core. These variations induce telluric electric currents in the subsoil, which in turn induce a magnetic field. The variation observed at the surface therefore has two components. The existence of these telluric currents sparked the development of a basement prospecting method called the **magnetotelluric method**.

Magnetic field reversals are an even more spectacular form of variability. Normally, the magnetic North is located near the geographic South Pole (and vice versa). However, it has now been clearly established by many measurements that the field has, in the past, reversed suddenly (i.e., in a few thousand years), and the chronological scale of these reversals is relatively precise, at least for the last 100 My. This is the field of **paleomagnetism**, which is well described for the Mesozoic and Cenozoic by analyses of oceanic materials (the ridges created over the course of this time provide a perfect record of these successive

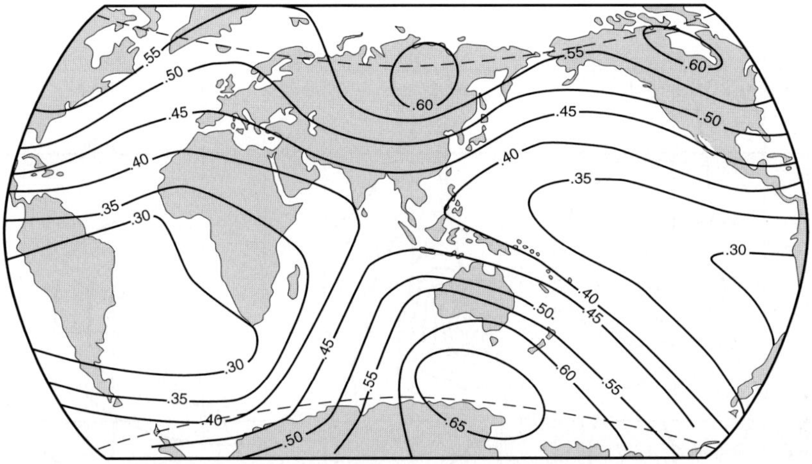

Fig. 1.30 Map of the Earth's total magnetic field intensity (from Leaton, 1971).
The Earth's magnetic field in gauss, based on data from many observatories at different points on the Earth's surface.

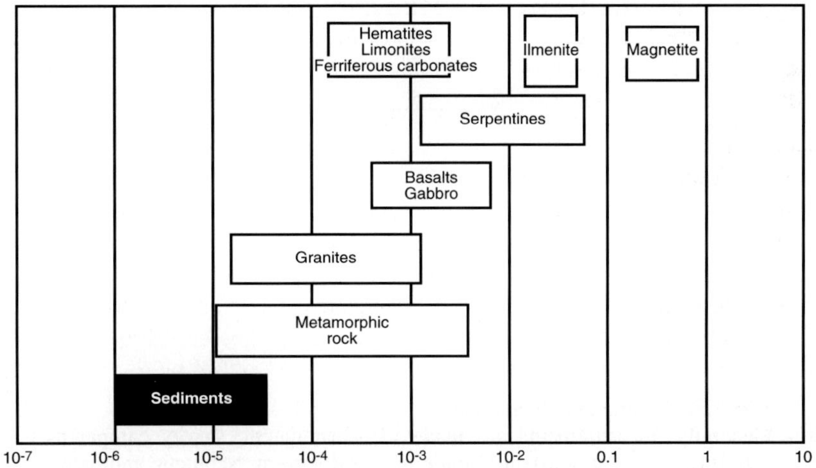

Fig. 1.31 Magnetic susceptibility of rocks.
Thermoremanent magnetization (TRM, which is acquired by a rock containing ferromagnetic minerals as it cools to below the Curie temperature of 585°C), and depositional remanent magnetism (DRM) are due to preferential orientation of magnetic grains. Magnetic susceptibility (K) varies considerably depending on the type of rock. It is expressed here in c.g.s. units (from Sheriff, 1978).

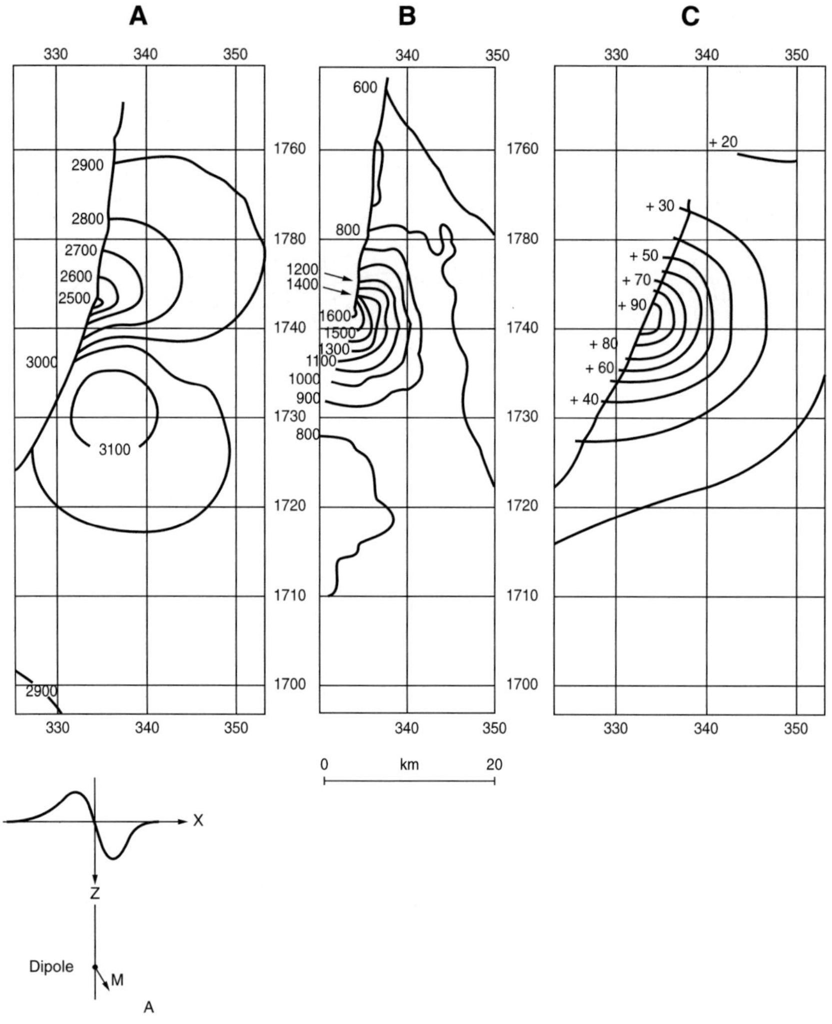

Fig. 1.32 Magnetic field anomalies.
Generally, measurements are made in aeromagnetic survey campaigns covering a zone at constant altitude in a grid pattern. Magnetic anomalies then appear as a dual anomaly: one positive and one negative (**A**). These are due to magnetized bodies of varying shape and depth. Once corrected (by comparison with ground station records) for any time variations, they are sometimes extended upward or downward. The chart is then reduced to the pole (**B**) by mathematical filtering, to eliminate the bipolarity of the anomaly, and compared with the corresponding gravimetric chart (**C**). The map here represents an anomaly on the Atlantic coast of Senegal (from Nettleton, 1976).

reversals), but is still imprecise for the Paleozoic (Fig. 1.33). Correlations today are tuned by **magnetostratigraphy** whenever other markers are lacking (see Chapter 4).

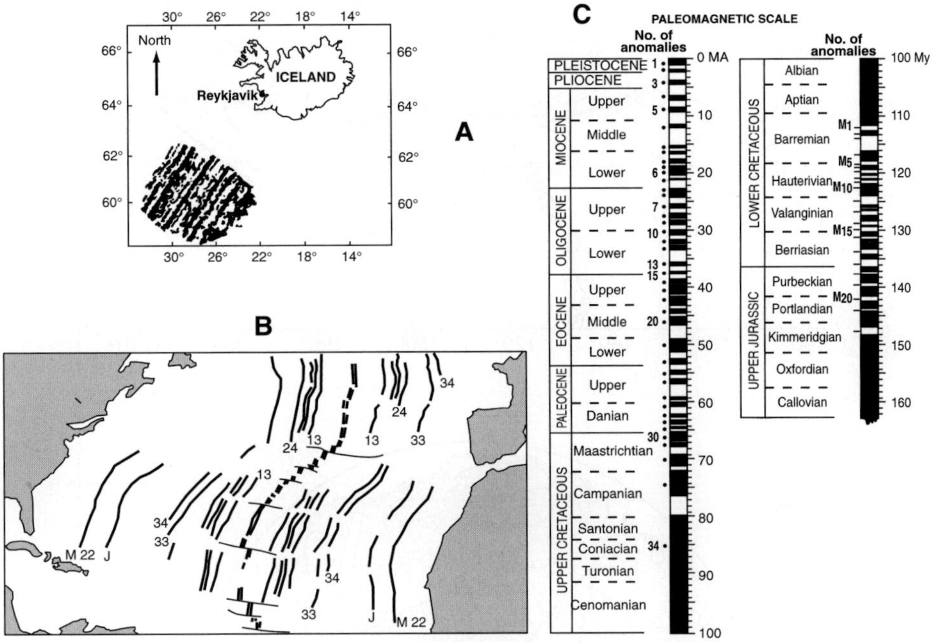

Fig. 1.33 Record of magnetic reversals on the ocean floor.
A is a record of magnetic anomalies to the southwest of Iceland in the North Atlantic. Each black strip corresponds to basalt with "normal" magnetization. For each white strip, the magnetization is "reversed". The pattern is symmetrical to either side of the ridge, where present magnetization (age 0) is the normal one. Each of the strips can then be attributed to a particular period. **B** is a schematic distribution of a few of the anomalies found in the median and central Atlantic. The figures refer to the geological and paleomagnetic scale given in **C** (from various authors).

Paleomagnetism also allows us to reconstruct the latitudinal positions of the continents, to determine the apparent polar drift, and plate motion (Fig. 1.34).

In conclusion, let us remember that spatial (local and regional) and time (reversals, tellurism) anomalies are natural signals used in applied geology. While the signatures may vary depending on the rocks, their uses in the study of sedimentary basins vary too: prospecting methods, basement structure, stratigraphic sequence, paleogeographic reconstructions.

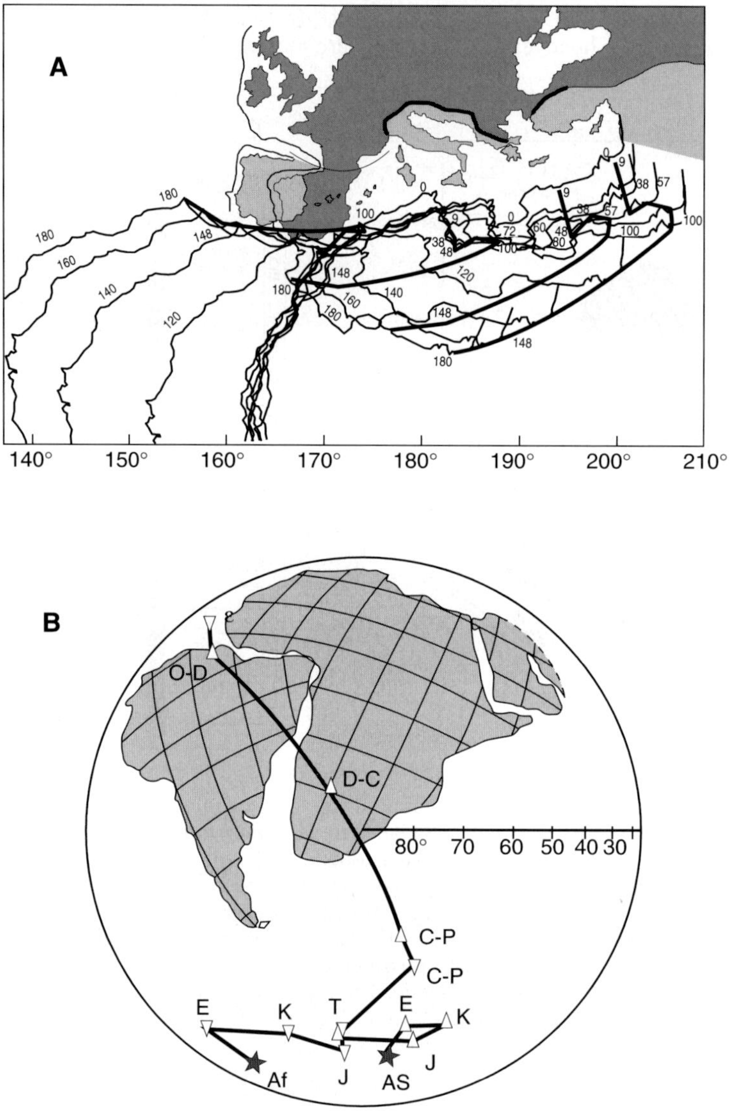

Fig. 1.34 Paleomagnetism, a precious tool used in paleographic reconstructions.
A. Reconstruction of relative positions of Africa in the course of geological time by tracking the magnetic anomalies of the Atlantic. Eurasia is considered to be stable (from Biju-Duval et al., 1978). **B.** Example of reconstruction of continental drift by measurements made on different samples. Here, the apparent drift of the pole is represented (from various authors).

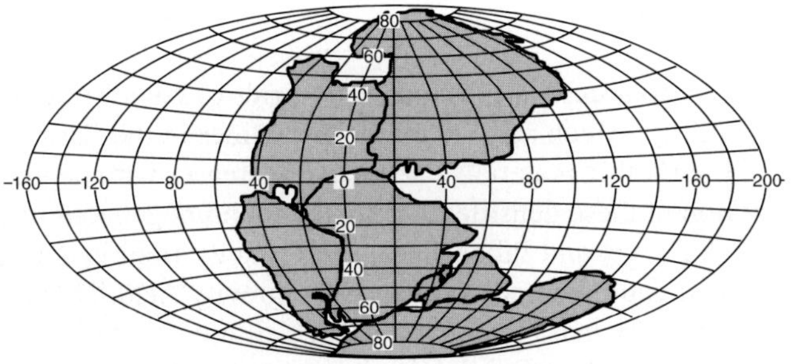

The single Pangea continent in the Jurassic, 200 My ago.

C

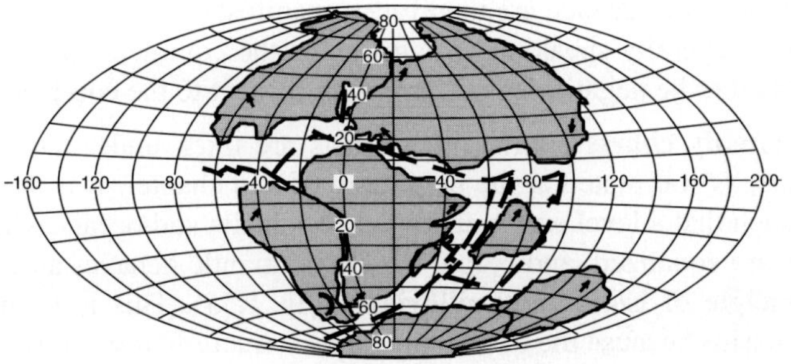

End of the Jurassic, 135 My.

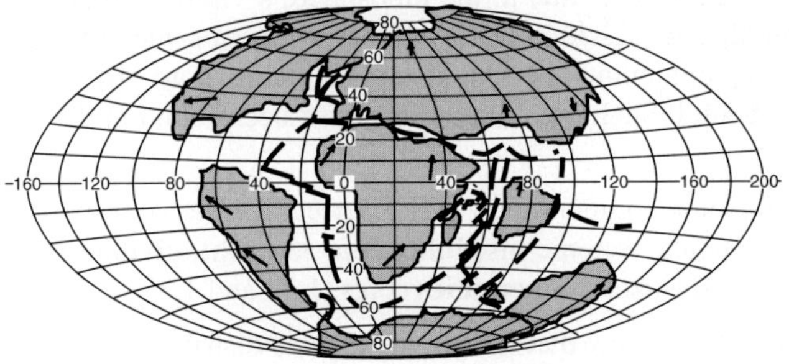

End of the Cretaceous, 65 My.

Fig. 1.34 (cont'd) Paleomagnetism, a precious tool used in paleographic reconstructions.
C. Example of global reconstruction using paleomagnetic data of the continents and oceans (from various authors).

1.4.1.2 Gravity Field

The gravity field, *g*, is an acceleration expressed in gals (1 cm/s^2) or milligals in gravimetric prospecting and geodesy, which are two major disciplines in the Earth Sciences. Remember that the density of the Earth's materials varies, and increases with depth (Table 1.2), but that the spatial distribution of these densities is actually not homogeneous (refer to the discussion on the structure of the crust and lithosphere). The gravity field is not homogeneous either: the geoid, which is an equipotential surface of the gravity field, is not spherical. This non-spherical character reflects mass irregularities, which are detectable at the surface but are due not only to topographical relief, but mainly to subsurface irregularities (see Fig. 1.9). The geoid's undulations can be observed at different wavelengths. The undulations observed at short spatial scales are due essentially to the structure of the lithospheric plates. The **geoid's anomalies** at the surface are therefore a manifestation of deep mechanical and thermal phenomena. Generally speaking, *g* is lower on the continents than it is on the oceans. Anomalies are said to be negative or positive with respect to the equipotential surface.

Bouguer **anomaly charts** confirm that crusts are thick under the continents and thin under the oceans, as was stated at the beginning of this chapter. It is thought that the mass distribution is such that a level exists under the continents and oceans where all of the lithostatic pressures are equalized, and below which the mantle behaves as a hydrostatic liquid. This is the principle of **isostatic equilibrium** (Fig. 1.35). This is an important principle when studying basins because the basin, since it is a structural low, constitutes a density deficit and can thus develop only if there exists an excess density at depth to compensate the lighter basin, thereby adjusting the **subsidence**. This isostasy can be local or regional.

Isostatic compensation is due to the lithosphere's viscous properties. It operates by **isostatic adjustment**. The classic example of re-adjustment is the one caused by the melting of Scandinavian ice, lightening the shield and allowing it to rise (Fig. 1.36). This is also referred to as rebound. Such rapid variations have been found in the past, on the geological time scale.

Generally, isostatic compensation is not perfect for the very reason of the time fluctuations (plate movements, tectonic instability, and so forth), giving rise to a lack or excess of compensation, called **isostatic anomalies** (Figs. 1.37 and 1.38).

To conclude, we may say that the gravity field is distributed irregularly. For purposes of studying basins, it should be remembered that basin shape will evolve in reaction to isostatic re-balancing at depth. For the sediments that fill these basins, it obviously follows from the principle of filling structural lows that the deposit of any particle is subject to the gravity field at the base, so the role of gravity is essential in wearing down relief and filling basins.

1.4.1.3 Heat Machine

Temperature gradients are probably the overriding internal drive mechanism, insofar as they directly influence the stress fields (i.e., the lithospheric convection mechanism) and dynamo

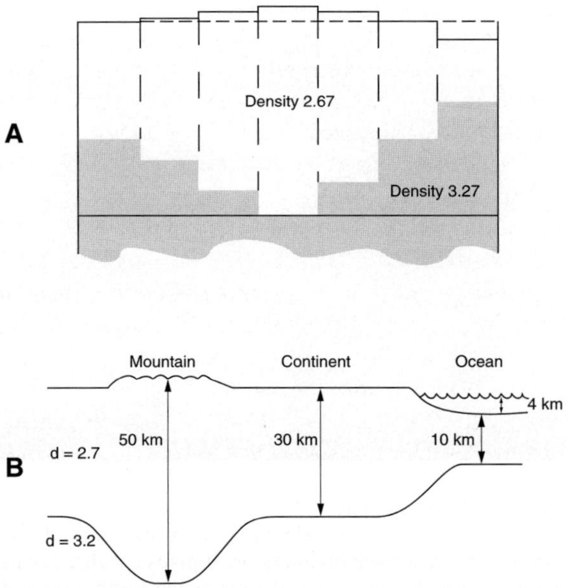

Fig. 1.35 Isostatic equilibrium.
A. Airy's model (from Coulomb, 1972) showing columns with different densities. **B.** Simplified geological application: crust thickness variations between continent, ocean, and mountain chain.

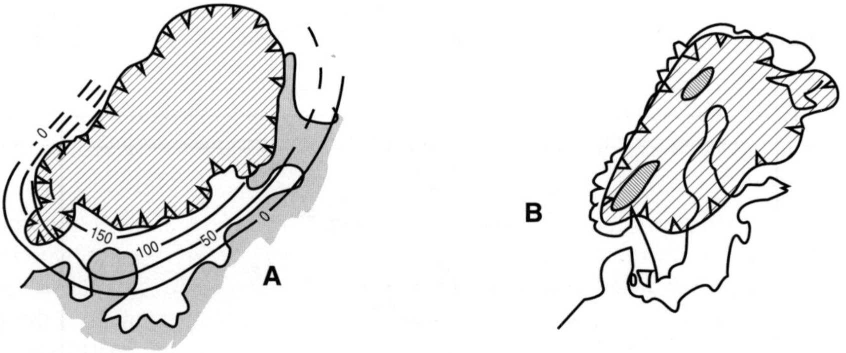

Fig. 1.36 Isostatic rebound of the Scandinavian shield.
A. The curves marked 0, 50, 100, and 150 trace the points of the former Yoldia sea coast (primitive Baltic Sea), which were quickly raised to the indicated altitudes after the melting of the ice cap. **B.** Extent of the inland ice 10 000 years ago (broad hatch marks) and 8000 years ago (tight hatch marks) (from various sources).

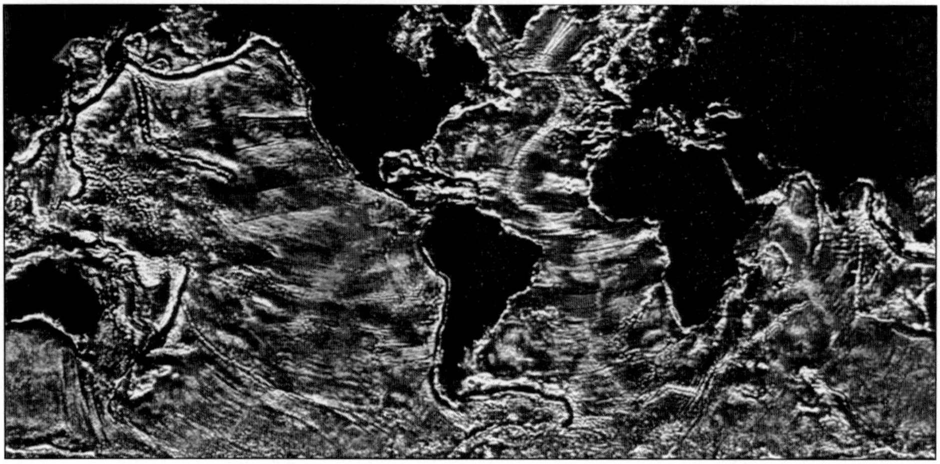

Fig. 1.37 Gravity field anomaly.
This chart is a record of short-wavelength gravity anomalies deduced from satellite records of sea surface undulations. It gives a rather faithful image of the seabed topography, which itself reflects the Earth's deep structures (from W. Haxby).

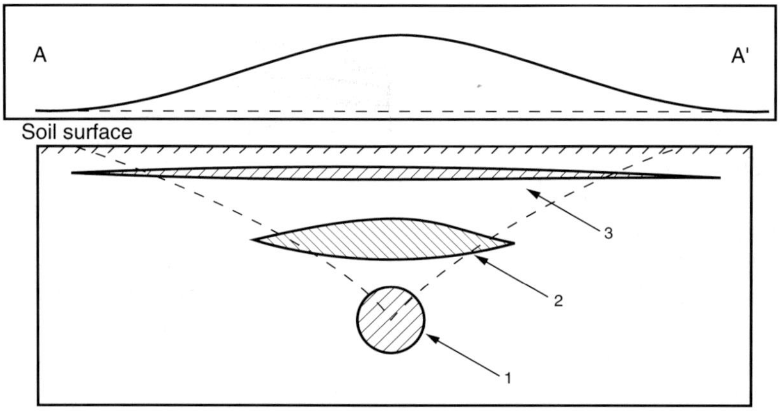

Fig. 1.38 Depth and origin of gravity anomalies.
This theoretical example illustrates the fact that there is no unequivocal interpretation of a gravitational anomaly, represented by profile A-A' here. Solutions 1, 2, and 3 all correspond to dense bodies of different depths and shapes that might be causing the same anomaly.

operation and, at different scales, determine plate movements, hot spot operation, subsidence mechanisms, thermal transfers in the basins, hydrothermal sources, mineral diagenesis, and the maturation of organic matter (Fig. 1.39).

Here are a few temperatures: 3900 to 4000°C in the core, 700°C at the base of the continental crust, 350 to 400°C in the ocean bottom's hydrothermal sources and sedimentary basins. This thermal machine has an internal source, and an external one too: solar energy, which we will speak of later. The internal source consists of the **initial energy** produced by conversion of kinetic energy into thermal, and **energy caused by radioactive disintegration**.

All modes of **heat transfer** are at work:

- **Convection** by mass transfer, which is important in the core, mantle, and lithosphere. This is a major part of **geothermal flow** and basin driving mechanisms.
- **Conduction** by molecular agitation, which is the dominant mode in sedimentary basins, as heat conductivity varies greatly depending on the type of rock.
- **Radiation**, which is negligible compared with the first two modes.

Measurements taken at the surface indicate that the temperature increases with depth, which is called the **geothermal gradient**. This gradient varies from 15 to 80°C/km starting at an average of 30°C/km at the surface, and is smaller at depth.

Heat flow is power, or energy per unit time, transferred through a unit area, and ranges from 30 to 300 mW/m^2 around an average of 70 mW/m^2, or 1.4 $mCal/cm^2/s$.

Different heat flow units are used, including $J/m^2/s$, W/m^2, and $mCal/cm^2/s$, which is the "HFU". Conductive transfer obeys Fourier's law, which is Flow = Conductivity × Gradient.

It should be remembered that the temperature field varies greatly at the Earth's surface. There are "hot" and "cold" basins. Variations in the thermal field and, more generally, its evolution in the course of basin formation and filling, are now characterized by paleothermometry.

1.4.1.4 Stresses, Deformation and Breakup

In speaking of the Earth's structure, and especially plate tectonics, it was said that the Earth's crust is subject to a stress field that evolves in time, and which is capital in the formation and evolution of sedimentary basins (Fig. 1.40). We will wait until Chapter 5 before developing the ideas of the different Earth sciences (rock mechanics and physics, seismology, tectonics) and will recall here only that the **stressing** process generates spectacular **deformations** at the plate boundaries, and frequently in the intraplate domain, which depend on the behavior of the materials in accordance with their own intrinsic properties. This is a matter of **rheology** and external physical conditions.

The stresses in the lithospheric plate system can generate either **extension**, **compression**, or **strike slip** (Fig. 1.41). The present stress field explains why extensive basins are the seat of stretching and **subsidence**, and explain mountain chains as **shortening** with uplift. The

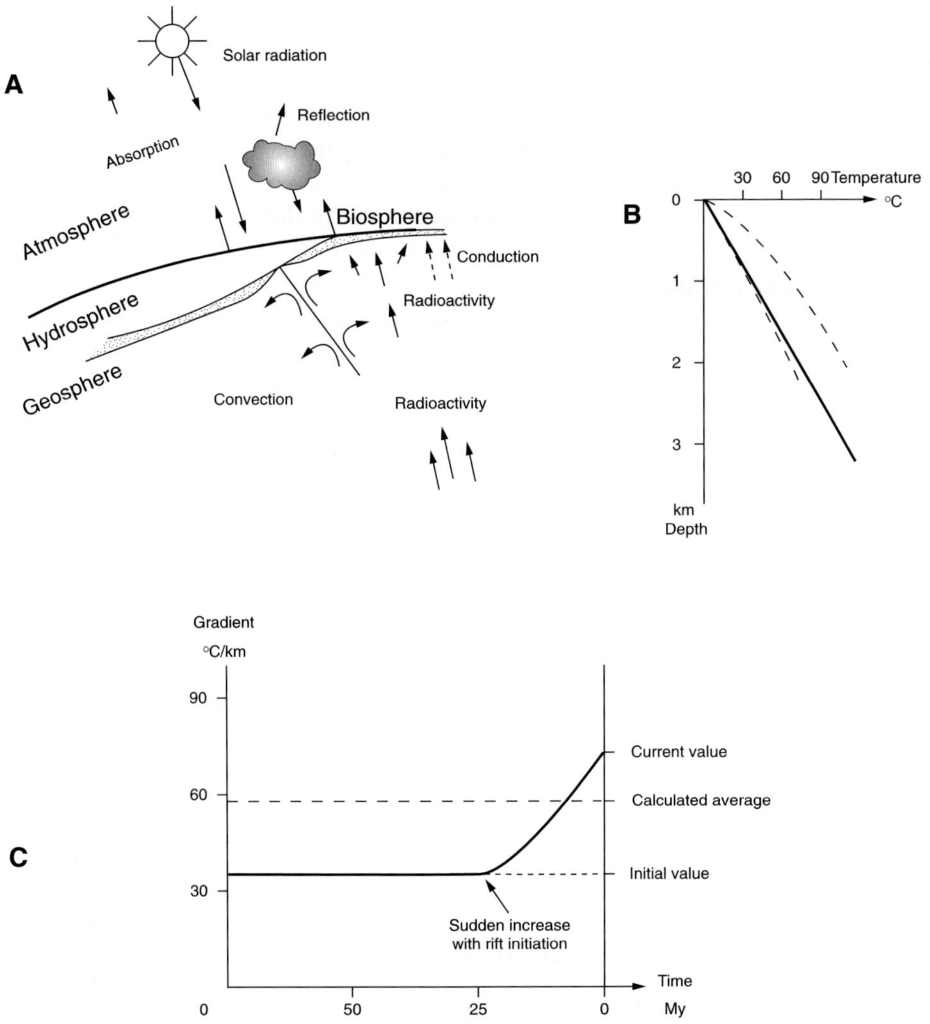

Fig. 1.39 Earth's heat machine.
A. External (solar radiation) and internal heat sources (natural radioactivity, dissipation of initial heat) have major effects on sedimentary basins. The principal heat transfer modes are deep convection (with transfers and cells in the mantle) and conduction in the basins. The ocean is an enormous regulator of these mechanisms. **B.** The geothermal gradient. The temperature rises regularly with depth in sedimentary basins, at an average of 30 to 33°C per kilometer. **C.** An example of time variation of geothermal gradients: the case of a recent rifting episode.

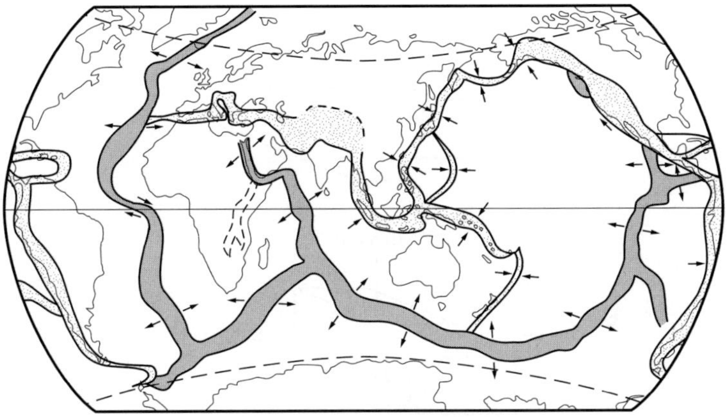

Fig. 1.40 Present stress field: seismicity chart.
The stress field on the global scale is determined by the motion of the plates, and is most apparent on a chart of seismicity distribution and deduced seismic motions. Certain earthquakes are associated with divergence (gray zones corresponding to ocean ridges) while others are associated with convergence (dotted lines corresponding to active margins and recent mountain chains). The arrows indicate the motions more simply.

stress field is clearly reflected in the **seismic** record of earthquakes, which are the result of breakup phenomena at the surface and underneath.

We will return to these aspects in Chapter 5 on basin evolution, but let us first remember a few essential basics. From a mechanical viewpoint, rock deformation proceeds in several stages: elastic deformation; plastic deformation; breakup. The exact sequence depends very much on the type of rock. Some, like salt, may exhibit viscous properties and creep. In addition to the intrinsic nature of the rock, the geostatic pressure also plays a role. Deformation is rarely homogeneous and isotropic. On the large scale, it may result from translations, rotations and distortions, and be manifested concretely by special **tectonic structures** at different scales: bulges, folds, faults, stretch, tension fractures, and other geometries. From all of the configurations at different scales, not only is the geometry of the stress (or rather "paleostress") field characterized finely, but its intensity (qualitatively) as well, and its direction (quantitatively) too.

When beds are buried deep, the combined pressure/temperature field becomes such that the constituent rocks are rearranged, and then **metamorphism** begins, with mineralogical transformations that will not be discussed here. It should be remembered that many sedimentary basins may be affected by this, and their interest for petroleum then declines considerably.

In conclusion, the stress field governed by convection in the mantle is crucial, though relatively discreet, in the development and life of basins. If the mechanical stress generates

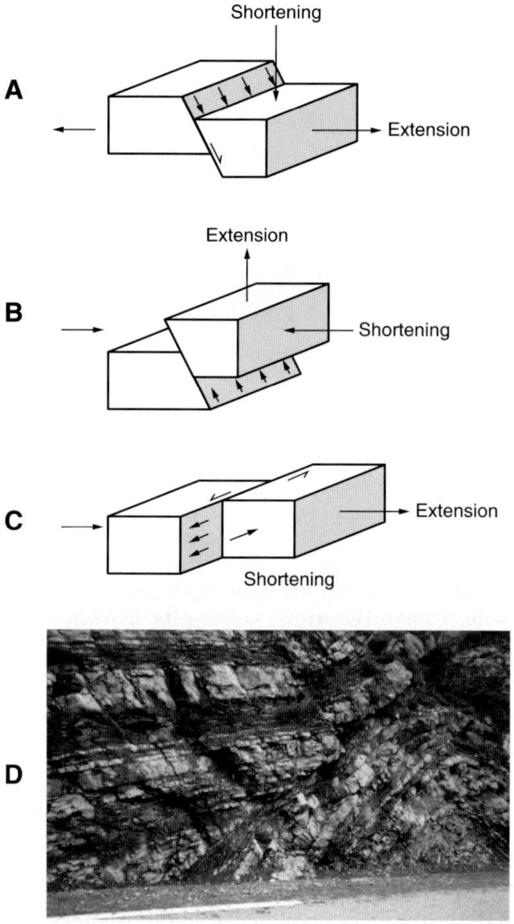

Fig. 1.41 Different types of deformation with breakup.
A. Extension. **B.** Compression. **C.** Sinistral strike slip. **D.** Example: folded beds with a shift, to either side of a fault.

deformations quickly, the resulting basin is said to be **syntectonic**, with a clearly expressed tectonic effect on the sedimentation. Generally, the important effect is basin **subsidence**, since there is always a more or less pronounced **crustal thinning** under the basins, by definition (see the discussion of the gravity field).

As will be seen further on, continuous or more or less permanent processes are to be distinguished from sudden events such as earthquakes and volcanic eruptions, which can be spotted by a clear signal in the basin sedimentation. This gives rise to different subsidence and uplift rates.

1.4.2 External Drives

In addition to the internal geodynamic forces at play in the evolving process of filling sedimentary basins, there are external factors. We have already mentioned the magnetic field component due to solar wind, and the intensity of direct solar radiation from the Sun (insolation). But to these factors we must add universal gravitation, ocean-atmosphere interactions, the climatic situation, the role of surficial fluids, and other external factors that will influence the filling of basins, which is capital for the geology of water, hydrocarbons, and many other substances of use to mankind.

1.4.2.1 Orbital Parameters

The Earth's climate is neither uniform nor stable. Much work is being done today to determine its recent evolution and probable changes in the near future. Here, we will point out a few essential climatic parameters affecting the geology of sedimentary basins.

The varying insolation at different latitudes between the poles and the equator results in more or less contrasted seasons. This insolation depends on the planet's orbital parameters, which are essentially the distance from the Earth to the Sun and especially the inclination of the Earth's axis of rotation. We know that today's climate, with the polar ice caps, is one of the essential factors determining global ocean surface and undercurrent circulation, and these currents have a direct incidence on sedimentation mechanisms. The climate also directly influences continental and marine organic or biomass productivity, which is another important factor of sedimentation. These two aspects will be investigated in greater detail in Chapter 3, along with the continental alteration and erosion guided by climatic factors, which are essential to the origin of the materials fed into the basins. We will also see the extent to which the Earth's present climatic situation is unstable, and that its past fluctuations have been large. **Seasonal variations** are superimposed on the above **astronomical variations**, which cover much longer periods of time. The **eccentricity** of the Earth's orbit, the **obliqueness to the ecliptic** (currently 23° 27'), which is the inclination of the equator with respect to the orbital plane, and the **precession of the equinoxes** (by movement of the rotational axis) are three parameters that fluctuate with periods of 100 000, 40 000, and 20 000 years, respectively (Fig. 1.42). These cyclic parameters are combined in the theory of Milankovitch to explain variations occurring at these frequencies in marine sediments and polar ice over the last 200 million years.

It is reasonable to believe that these variations were also at play before this point, but the Earth's climate has also been subject to major variations over longer periods. Four major glacial periods have been logged over the past 600 My. Of course, astronomic-type cyclic variations can be found for each of these. And between the cold periods, the Earth has known hot periods (in the Cretaceous, for example) that were determined by these same fundamental Earth-Sun system characteristics.

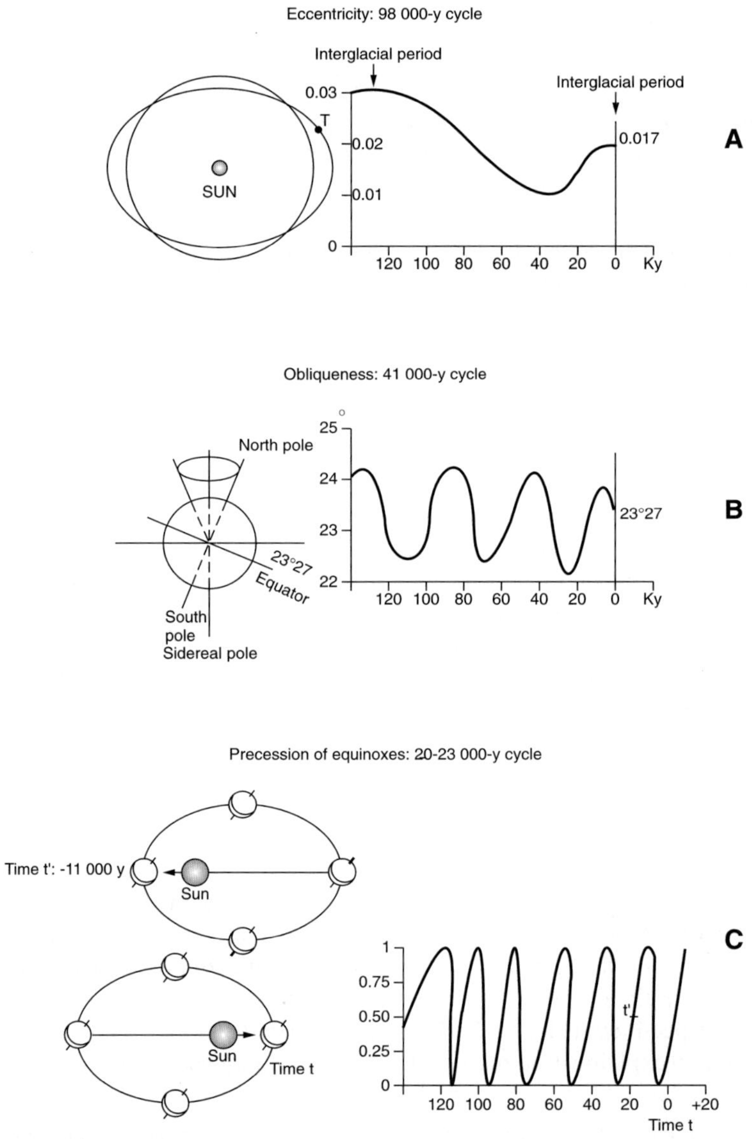

Fig. 1.42 Milankovitch cycles: astronomical variations.
The Earth's rotation about the Sun varies with time. The eccentricity **A**, changes, as does **B**, the obliqueness, with respect to the ecliptic and **C**, the precession of the equinoxes. These all combine to generate cyclic changes which, it is thought, have guided sedimentary variations.

Today, it is thought that the Earth's orbital parameters act as a metronome that times the major climatic cycles, but that these cycles might have been amplified by other factors such as variations in radiation by the Sun itself, and greenhouse effects.

The link between the climate and the hydrosphere is crucial not only for the weathering process, biological productivity, and ocean currents, but also in its effect on sea-level variations, as will be detailed in Chapter 4.

Tides are another direct consequence of the Earth's coupling with astronomical parameters, as will be seen further on when discussing the importance of deposits influenced by them.

1.4.2.2 Ocean-Atmosphere Coupling

Generally, hydrosphere-atmosphere coupling should be remembered. These two spheres have a common origin and their evolution has always been closely tied together on the small and large scales. The phenomena discussed below are at times also highly dependent themselves on the Earth's orbital parameters.

The conditions in a given basin are determined by **climatic zoning** from the ice caps to the equatorial and intertropical regions. The large temperature gradient between high and low latitudes is the principal parameter of climatic zoning. At high latitudes, the ice caps and frozen ocean form an effective barrier to the atmosphere; sedimentation is reduced, phytoplanctonic production is limited to a few months out of the year, glacial erosion can be considerable, while the cold waters descend and induce the deep oceanic circulation. The temperate and tropical zones, on the other hand, are greatly influenced by the general oceanic circulation: water temperature and salinity are major factors that will determine the areas of sedimentary distribution and organic productivity. This climatic zoning can be recognized in geological series, by periglacial, evaporitic, and carbonaceous belts that have evolved with time (Fig.1.43).

The **tidal cycle** is one of the important parameters to be considered when defining geological environments. The tide is a well-known process, resulting from the law of universal gravitation. The gravitational effects of the Moon, which is dominant, combine periodically with those of the Sun, from their conjunction to opposition. There are different types of tides (diurnal, semidiurnal, composite), but what is most important to remember for geology is that tides are very sensitive to the geometry of the coast, and that their amplitude varies with the geographical area. The irregularity of the coast, with its varying orientations, capes, bays, and so forth, has observable effects that cause the formation of tidal currents. Tidal ebb and flow operate on the foreshore, which is defined as the intertidal zone alternately covered and uncovered by the sea; but it also affects broad areas that are always covered by water (the sublittoral domain). These ideas will come up again in Chapter 3 in the discussion of geological responses to these phenomena. At this point, we will simply mention the seasonal variations in the tide (high and low tides, equinoxial tides), and point out that some coasts are preferentially marked by tidal influences, whereas others are dominated by wave and storm effects.

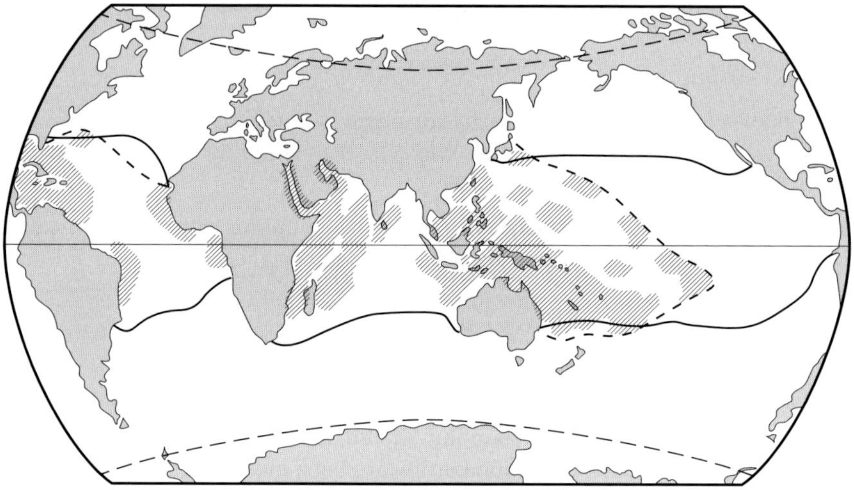

Fig. 1.43 Example of climatic zoning.
This shows a geographic distribution of reef formation (carbonaceous constructs by corals) that develops in warm ocean waters (the solid line is the 20°C isotherm) (from various authors).

Large- and small-scale **ocean circulation** is closely coupled to atmospheric circulation (Fig. 1.44). Chapter 3 will detail the various types of currents at work in the geological processes of erosion, transport, and deposit, from the coast to the high seas. These physical mechanisms are extremely important in explaining accumulations on platforms and basins, and they are not as yet fully known in the present geographical configuration (much work is being done here by physical oceanographers). Their instability in time is beginning to come into focus, and reconstructions are being proposed for various geological periods. In petroleum geology, while close attention is paid to general circulation mechanisms, the emphasis is especially on shallow coastal mechanisms and transfer currents on the margins: cold waters drawn by gravity to great depths, and then upwelling. The current dynamic generates erosion (ablation of deposits) and sedimentary hiatuses, which will be discussed again in Chapter 4.

1.4.2.3 Chemical Environment and Role of the Biological World

The atmosphere, considering its present composition and its water and carbon cycles, is an essential element in geological analysis. It conditions transfers at the Earth's surface, first by feeding surface weathering phenomena on the continents, then in the gaseous transfers at the ocean-atmosphere boundary (Fig. 1.45).

These considerations will be developed in subsequent chapters. Here, the point to be emphasized is that living organisms play a major role, not only in the biological, geological, and chemical cycles, but also in the formation of sediments, the filling of sedimentary basins

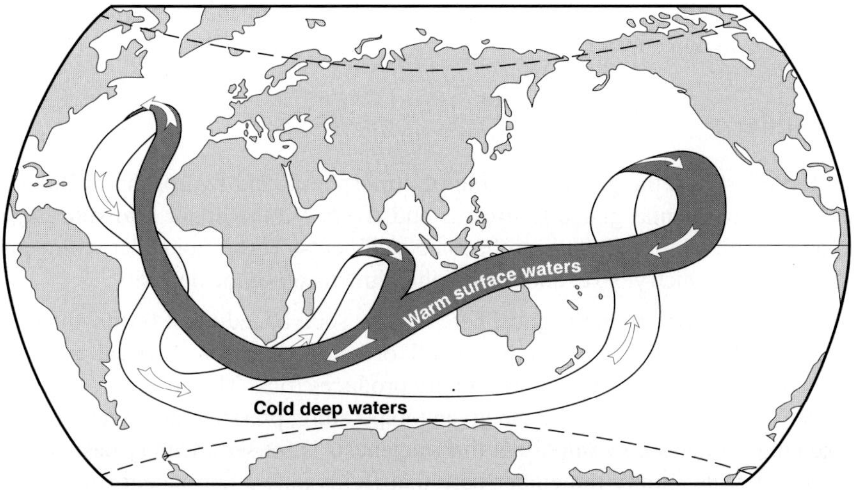

Fig. 1.44 Surface and deep ocean currents.
The figure shows the circulation of warm Pacific waters through the Indian Ocean to the Atlantic. After plunging deep in the north Atlantic, they join the course of cold, deep waters that return to the southern ocean (from M. Fieux, 1994, to be compared with maps 3.94 and 3.95 in Chapter 3).

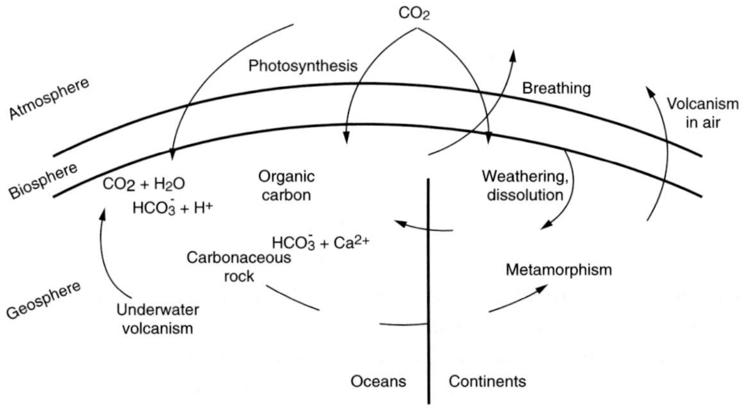

Fig. 1.45 Carbon cycle (from Allegre, in Caron et al., 1992).

(for more than two billion years) and later in the evolution of these sediments (the role of micro-organisms in diagenesis).

1.4.2.4 Fluids

It is easy to understand the geological role of atmospheric and hydrospheric fluids. Wind, ocean currents, continental glaciers, streams and rivers are the major **transport agents** of particles and solutions that are later to become sediments. These fluids determine and control the structure, geometry, size, and geographic dimension of the deposits.

Their role is a major one in the geosphere too. As soon as sediment forms, the interstitial water becomes the environment in which all of the sediment's physical and chemical changes occur in the **diagenesis** that eventually produces rock. The chemical makeup of the fluid gradually does this work under the given temperature, pressure, and other conditions. It will be seen in Chapter 5 how important this diagenesis is for sedimentary basins and petroleum systems. Let us simply remember here that fluids play a permanent role in all of the surface and deep geological phenomena (Fig. 1.46):

- Initial expulsion from sediments
- Circulation in basins and groundwater aquifers
- Continental and marine hydrothermal systems
- Folded strata, hydraulic fracturing, and mountain chains
- Dehydration and magmas.

1.4.3 Rhythms, Cycles, Events

It has been said since the outset of this book that geology is an historical science. The Earth's evolution in time is characterized by a sequence of rhythms, cycles, events, and crises.

It has already been mentioned in the discussion of internal driving mechanisms that the magnetic, gravity, thermal, and stress fields all exhibit notable variations in geological time. These internal geodynamic variations may or may not be periodical. What determines the periodicity of the dynamo, mantle cycles, and other varying phenomena is not well known. One remarkable example of a cycle is the **Wilson cycle** (Fig. 1.47). This is the **ordered sequence of events** in the global scale of plate tectonics, from continental separation or rifting (extension, fracturing, tectonic subsidence), then oceanic opening with expansion of the basalt floor, and finally the ocean closing stage with continental collision and folding of mountain chains. Curves and laws to scale can be determined for a basin over several tens of millions of years.

Sedimentologists recognize sedimentary cycles (Fig. 1.48) on the same scale, or generally for shorter periods. One such cycle, for example, is a basin series beginning with an encroachment, or **transgression**, of the sea on a continent and ending with its withdrawal, or **regression**, some time thereafter.

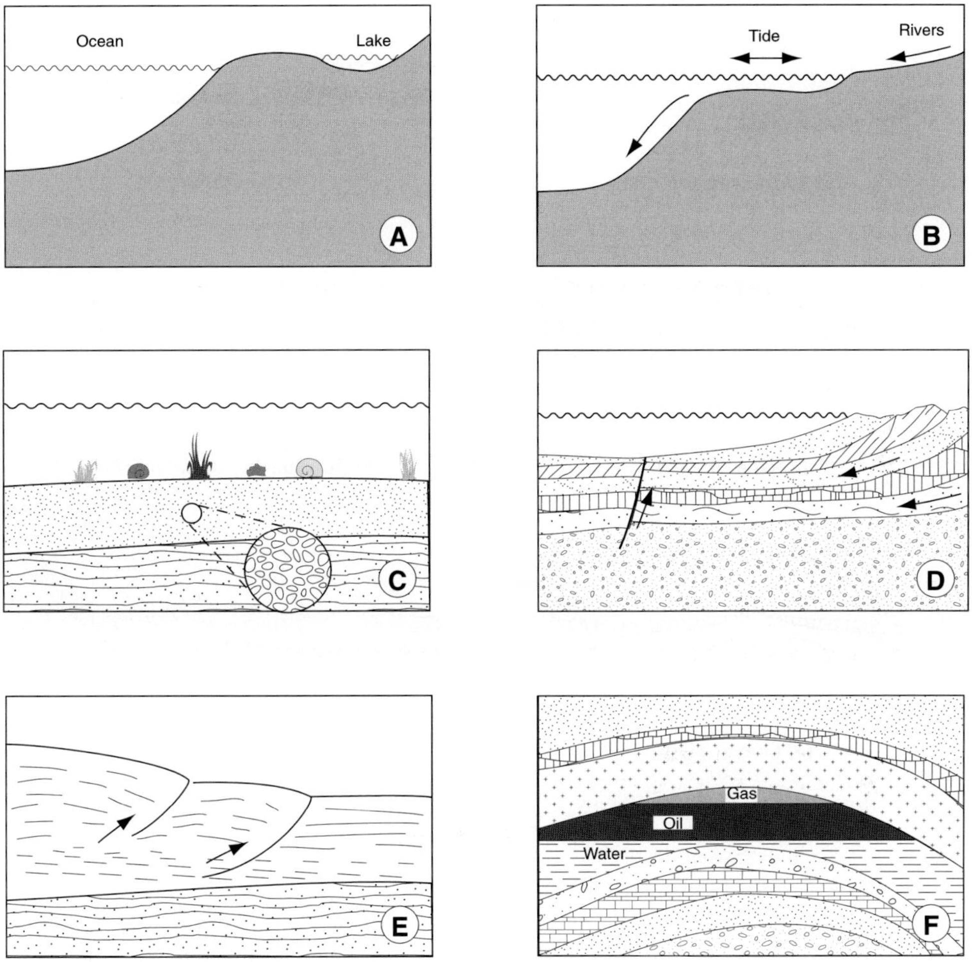

Fig. 1.46 A few examples of the role of fluids in geology.
A. More or less saline sedimentation environment. **B.** Transport agent for particles and solutions. **C.** Interstitial liquid, locus of biochemical transfers. **D.** Circulation in sedimentary beds or along faults. **E.** Lubrication. Here, lubricating a tectonic fold. **F.** Accumulation of hydrocarbons in an anticlinal trap.

Chapter 4 will go into greater detail on time reconstructions based on the sequential organization of deposits.

We have seen that different types of cycles appear with varied periods in geological phenomena, from the Milankovitch (100, 40, 20 Ky or composites) to tidal cycles (diurnal,

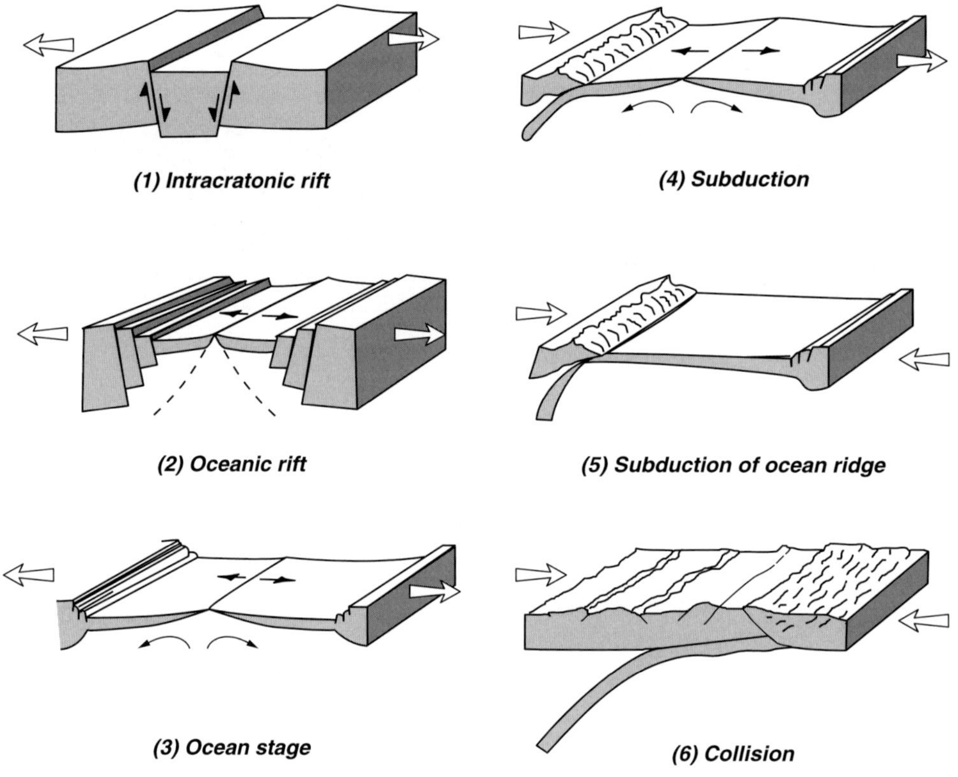

Fig. 1.47 Wilson cycle.

semidiurnal, composite, seasonal). Recent studies have attempted to define tectono-magmatic cycles in the mantle processes of the ocean ridges. What causes this cyclic activity is still not known.

Variations are not only periodical, though. First of all, a sequence of events, which may or may not follow a given rhythm may lead to a gradual **drift** or a more or less regular one-way evolution (such as the initial composition of the atmosphere or subsidence of a basin).

Lastly, we will note the **episodic aperiodic variations** or events. An earthquake, which may exhibit some periodicity when precisely analyzed, is one such particular event for the geologist, as is the tsunami (gravitational sea wave) it generates, a particular storm process (hurricane David, for example), a turbidity current (Nice in 1979 or Newfoundland in 1929), flash flood (Vaison la Romaine in 1992, Egypt or Piedmont in 1994), sudden volcanic eruption (like that of Pinatubo in 1990), unexpected rainfall, or a major meteorite impact. These are all different examples of relatively instantaneous phenomena with major geological consequences leaving a strong record in sedimentary basins (Fig. 1.49).

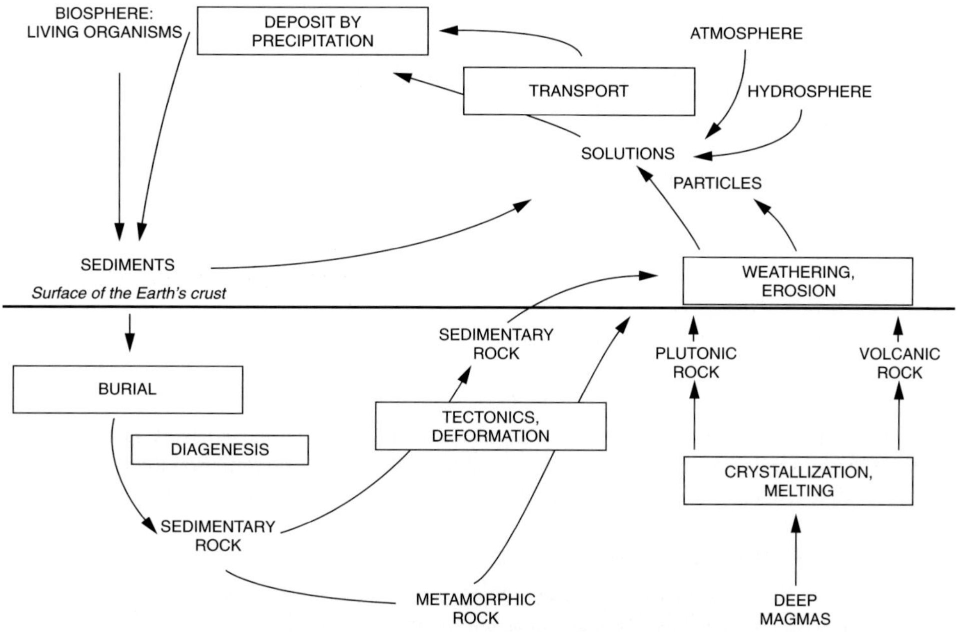

Fig. 1.48 Sedimentary cycle mechanisms.

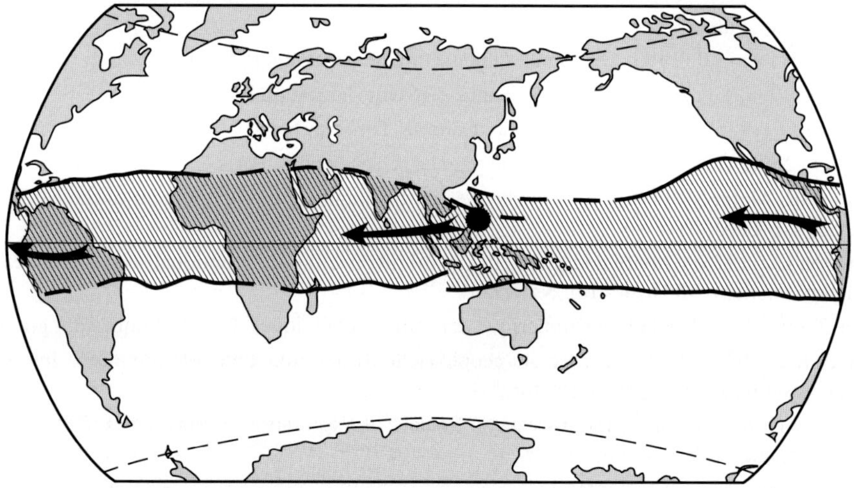

Fig. 1.49 Example of an event in geology.
When Pinatubo suddenly erupted in 1991, it spewed a cloud of volcanic dust that was carried in a broad equatorial belt around the Earth for several weeks.

Other events occur on another much less "instantaneous" scale: the closing of a threshold isolating a basin, paroxysmal volcanic crisis, sudden extinction of fauna and flora, or ice age. All of these ideas will be taken up again in Chapter 4 on sedimentary sequences.

BIBLIOGRAPHY

- ✦ La terre de l'observation à la modélisation (1990) Courrier du CNRS, dossiers scientifiques, 76, juillet 1990.
- ✦ Angelier J, Bartintzeef JM, Chauve P et al. (1992) Enseigner la géologie au collège et au lycée. Nathan, Paris.
- ✦ Biju-Duval B (1994) Océanologie. Dunod, Géosciences, Paris.
- ✧ Biju-Duval B, Montadert L (1977) Structural history of the Mediterranean basins. Histoire structurale des bassins méditerranéens. Congrès, assemblée plénière de la commission internationale pour l'exploitation scientifique de la Méditerranée, symposium international, 25, 25-29 october, 1976, Split, Yougoslavie. Éditions Technip, Paris.
- ✦ Boillot G (1990) Géologie des marges continentales. Masson, Paris.
- ✦ Boyer S, Mari JL (1994) Sismique et diagraphies. Éditions Technip, Paris.
- ✦ Cara, M. (1989) Géophysique. Dunod, Géosciences, Paris.
- ✦ Caron JM, Gauthier A, Schaaf A et al. (1989) Comprendre et enseigner la planète Terre. Ophrys, Paris.
- ✦ Coulomb J, Gobert G (1972) Traité de géophysique interne. Masson, Paris.
- ✦ Debelmas J, Mascle A (1993) Les grandes structures géologiques. Enseignement des sciences de la terre, Masson, Paris.
- ✦ Dercourt J, Paquet J (1990) Géologie : Objets et méthodes. Dunod, Paris.
- ✧ Fieux M (1994) L'océan planétaire. Sciences et avenir, hors série 98.
- ✦ Guillemot J (1986) Éléments de géologie. Éditions Technip, Paris.
- ✦ Jolivet L, Nataf H (1998) Géodynamique. Dunod, Géosciences, Paris.
- ✧ Leaton BR, Malin SR (1967) Recent changes in the magnetic dipole moment of the earth. Nature 213, 5081, p 1110.
- ✧ Nettleton LL (1976) Gravity and magnetics in oil prospecting. International series in the earth and planetary sciences. McGraw-Hill, New York.
- ✦ Odin G, Odin C (1990) Échelle numérique des temps géologiques. Géochroniques 35, pp 12-21.
- ✧ Sheriff RE (1978) A first course in geophysical exploration and interpretation. International humain resources development, Boston.
- ✧ Yilmaz O (1987) Seismic data processing. Society of exploration geophysicists, SEG, investigations in geophysics 2, Tulsa.
- ✦ *Books or articles of general interest.*
- ✧ *Source of one of the figures used, cited in the figure caption.*

Chapter 2

CONTINENTAL AND OCEANIC BASINS

2.1 BASIN DEFINITIONS AND DIVERSITY

The general overview of the Earth's structure and its plate tectonics in the preceding chapter has shown that **sedimentary basins** develop in the upper part of the lithosphere, i.e., in the outermost shell of the crust. Basins were first defined as structural lows or cups in which sediments accumulate and have accumulated in the past. This geometric idea of a hollow in the Earth's crust, as we have said, is essential as it defines (Fig. 2.1):

- A receptacle, or container, which is the basin **substratum**, generally called the **basement**
- The container fill, or content, which is the accumulation of deposits or **sedimentary cover** resting on the basement.

This basin idea can be visualized by examining the Earth's entire surface on a planisphere. Many types of basins can be found that are generally filled with water but that differ in size, geometry, depth, environment and sedimentation medium, climatic position, and volume of sedimentation (Fig. 2.2). The morphological hollows on the Earth's surface thus vary greatly. Lacustrine depressions on the continents, like those of the Aral Sea, Lake Baikal, Lake Chad, and the lakes of east Africa, are vast water tables supplied from a **drainage network** within a **catchment area** that can vary considerably in size (Fig. 2.3). Certain broad "dry land" endorheic plains such as the Sahara and Australia are less obvious, but nonetheless constitute continental cups accumulating sediments distributed by a supply network.

River networks generally lead to the ocean, which is itself a set of basins of different depths, sometimes connected in sequence with each other. It is clear that the Black Sea basin, which is almost completely closed, is not the same as the Baltic or North Sea basins, that of the Red Sea or of the Caribbean, as the water depth, salinity, currents, and other phenomena vary considerably from one to the other. Examples of this are endless. The **diversity of all of these basins** requires an explanation, which will be given in genetic terms further on.

Before going into this classification, the **historic dimension** of the basins should be recalled. The present **functional basins** just mentioned can be viewed only in their geologi-

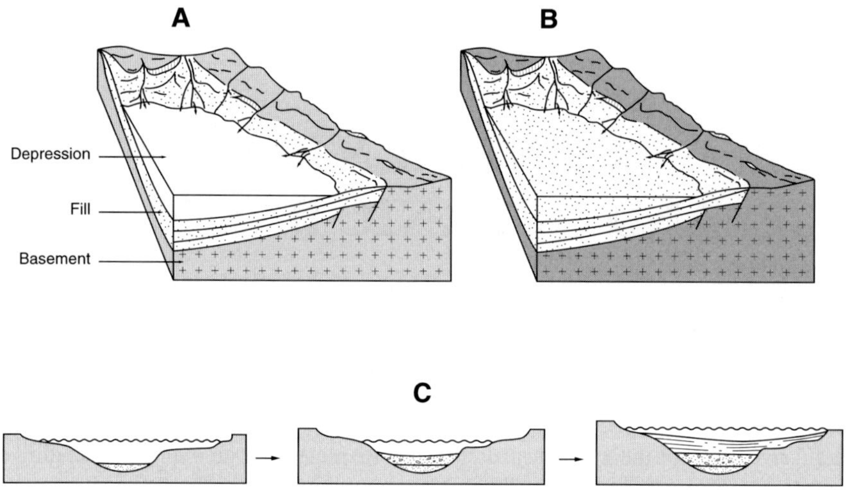

Fig. 2.1 Basin diagram: cup fills with sediments (no scale).
The topographical hollow is a trap for sediments that gradually deposit in the course of time. Once the cup is filled, the basin is no longer functional (**A**) but residual (**B**). **C** represents gradual filling. Note that the basin floor (substratum) is represented here as being stable, which is not generally the case. The water level may vary, which changes the dimensions of the functional basin.

cal perspective. Lake Baikal, the Chad depression, the Red Sea, the Baltic and the Caribbean each have their own history. Their present situation is the result of a sequence of events guided by physical and chemical mechanisms operating in the course of time. The north Atlantic basin has been operating as such for at least 150 My; the south Atlantic for 100 My; the western Mediterranean and Red Sea are much younger, initiated 25 My ago. And while certain basins of various ages are functioning hollows, others are no longer active even if they still have the geometrical cup shape of a receptacle. These are **residual basins**. The cup's present shape may also be inherited from a tectonic history, a large part of which may have occurred after the period of basin operation. These are **structural basins** like those of Paris and the Sahara (Fig. 2.4). We will see that each of them was once part of a larger system in the past, such as the Sahara basins or the American Rockies basins. And to introduce what will be detailed in Chapter 5, a large portion of the ancient sedimentary basins has been incorporated into the deformed mountain chain systems. The ancient basin that existed prior to the formation of the Pyrenees and the Alps, for example, can be reconstructed from the Aquitaine basin and the southeast basin of France (Fig. 2.5).

Different internal and external factors govern basin **filling**, as discussed at the end of the previous chapter. The conditions under which sediments accumulate in the hollow receptacle will be discussed in Chapters 3 and 4.

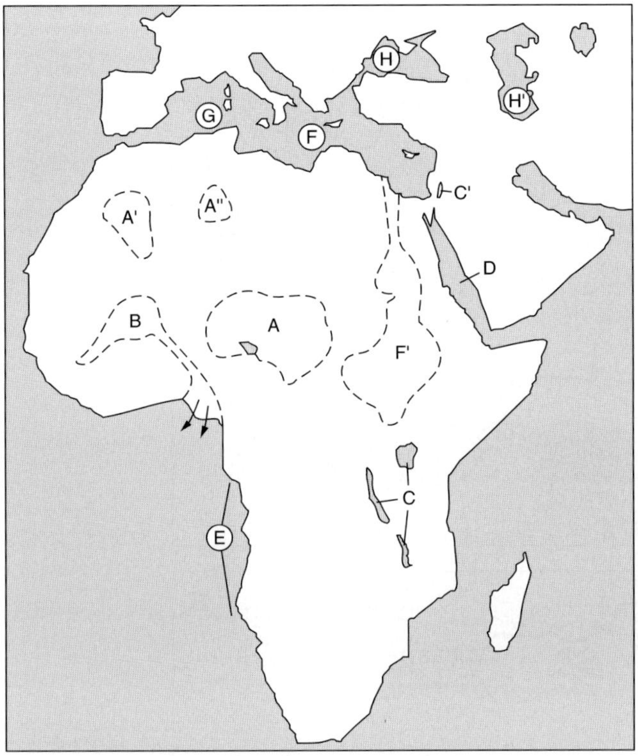

Fig. 2.2 Examples of diversity in present functional basins.
This map of Africa and southern Europe illustrates the variety of localizations, sizes, environments, and ages of formation of different basins. **A.** Chad basin: a recent intracratonic depression with a broad drainage network, a small area covered by a lacustrine water volume with high evaporation, fluctuating in time, probably superimposed on rifts, with no outlet to the ocean (and thus "endorheic"). **A'.** Saoura basin: a recent endorheic intracratonic basin entirely out of water (no permanent water table) superimposed on a former Paleozoic basin. **A''.** Great eastern erg: a depression covered by dune complexes that reverse the initial cup topography into a hump. **B.** Inner cup of the Niger: a slightly subsident continental basin with a delta and outlet allowing the transfer of sediments to the ocean. **C.** Great east African trough lakes: deep depressions with faulted edges along a recent continental crust breakup system. **C'.** Dead Sea, Jordan valley: open depression along a major slip fault accident. **D.** Red Sea: an ocean being formed by separation of the Arabian plate of Africa, after rifting that began 30 My ago (Tertiary). **E.** Africa's Atlantic margin: deep basins from a former margin (140 to 100 My, Mesozoic), with thick sedimentary wedges at the river mouths, normal ocean salinity, and special current processes. **F.** Eastern Mediterranean: deep basin initiated 200 to 140 My ago (beginning of Mesozoic) and partly filled by recent contributions from the Nile and shortened by convergence with Europe (subduction of Cyprus and the Aegean); high salinity, no tide. **G.** Western Mediterranean Basin, Tyrrhenian Basin, Egean Basin: deep basins linked with the formation of Alpine chains, high salinity, no tide. **H.** Black Sea: a closed basin isolated from the saline ocean, with a very broad drainage network, major sedimentary contributions, high confinement, recent subsidence (50 My?), superimposed on a former basin. **H'.** South Caspian basin: an ocean remnant between colliding Alpine chains.

Let us stress here the importance of **subsidence** and **sedimentary input** (Fig. 2.6). Subsidence is the gradual settling of a basin with time. The depression floor is not stationary, and the available volume for accumulating deposits therefore varies accordingly. The sedimentary input and the quantity of products deposited in the bottom of the basin consists either of detritic material inherited from the edges and added, or of precipitates from the water column. If input exceeds subsidence, the basin fills up and dies and is no longer func-

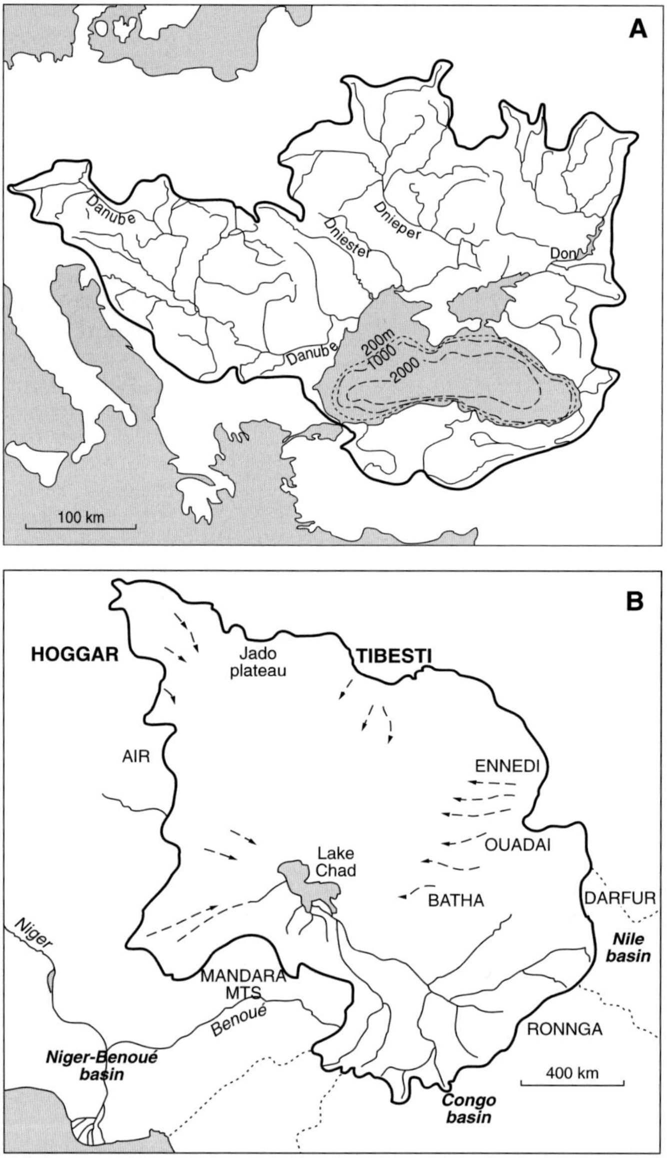

Fig. 2.3 Examples of basin drainage network areas.
A. Black Sea: ocean basin and major contributions from the catchment area.
B. Lake Chad: intracontinental basin with limited contributions.

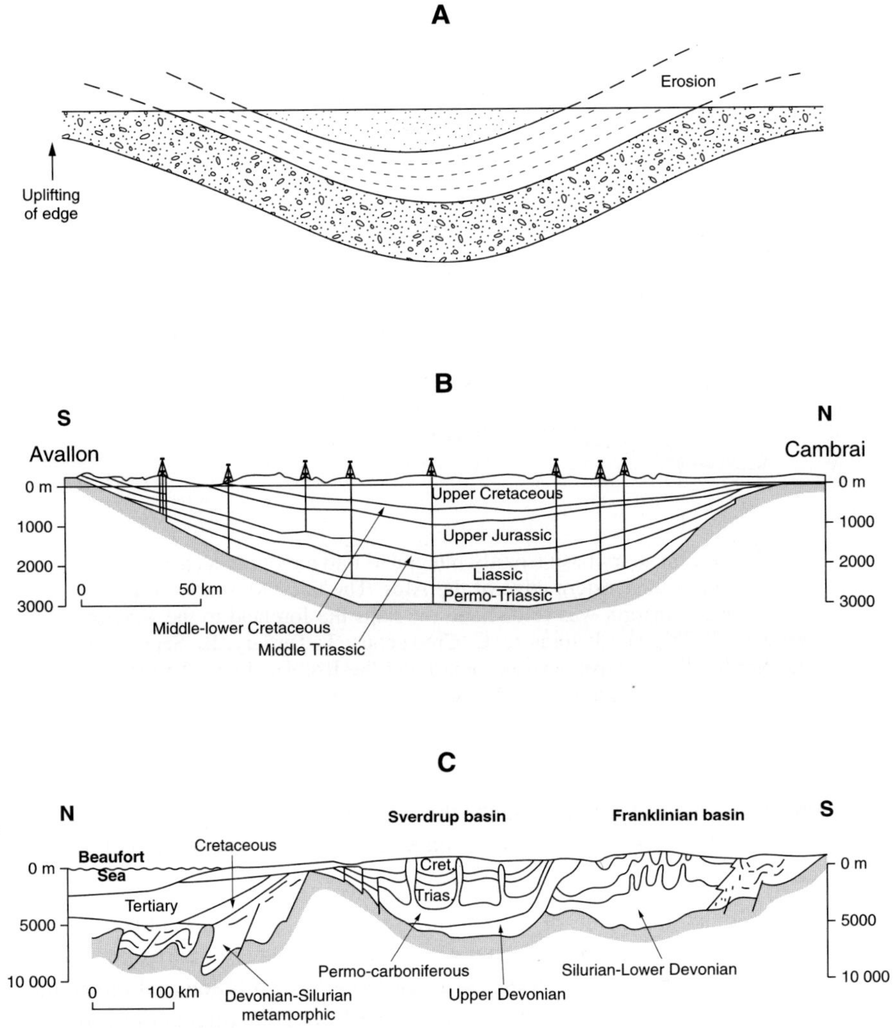

Fig. 2.4 Examples of structural basins.
The present topographical cup shape is due to structural events (deformation and erosion). The present edges or margins do not necessarily correspond to the geological edge of the basin at the time it was filled. **A.** Structural basin. The basin's edge is an erosion limit. **B.** Section of the Paris basin (heights multiplied by 12). Note that the different beds are largely eroded on the current edges of the basin (from A. Perrodon, 1983). **C.** Section of Canada's Sverdrup basin with heights multiplied by 10. Note here the erosion of the beds at the edges, but also their deformations within the basin (from various authors).

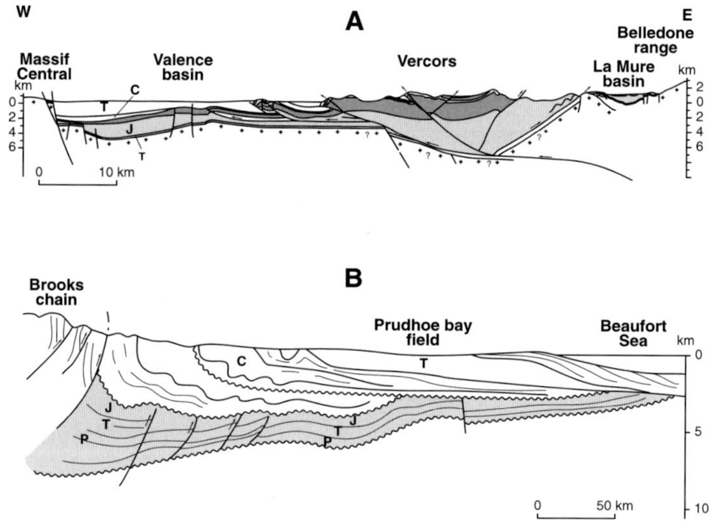

Fig. 2.5 Sedimentary basins partly integrated into a mountain chain. **A.** Section of the western edge of the Alps. The thick Mesozoic series of the European margin was folded and overlaps the foreland to the west of Vercors. T: Triassic; J: Jurassic; C: Cretaceous; T: Tertiary. **B.** Section of the North Alaskan basin (Arctic slope) and the Brooks chain (the vertical scale is highly exaggerated). P: Paleozoic; other letter codes as above.

tional. If there is a deficit of input, though, the available space will increase. We will see later how variations in the basement level or sea level can also change this volume available for sedimentation.

The **depocenter** is the point of maximum sedimentary accumulation in the basin. Depending on the sedimentary flow, this depocenter may or may not correspond to the zone of maximum subsidence. It is rarely centered at the bottom of a functional basin, but rather at one of its edges where the inputs are the greatest. So there is no direct connection between the thickness of a basin's sedimentary coverage and its depth.

In conclusion, a sedimentary basin can be defined as a particular zone of subsidence that has allowed the accumulation of sediments several hundred meters thick.

The reader may refer to more detailed works on this subject, or specialized articles such as those of Ensele (1992), Miall (1984), Perrodon (1988), Burrus (1989) as well as AAPG papers Nos. 29, 34, 51 and ff., and the ASF paper (1989).

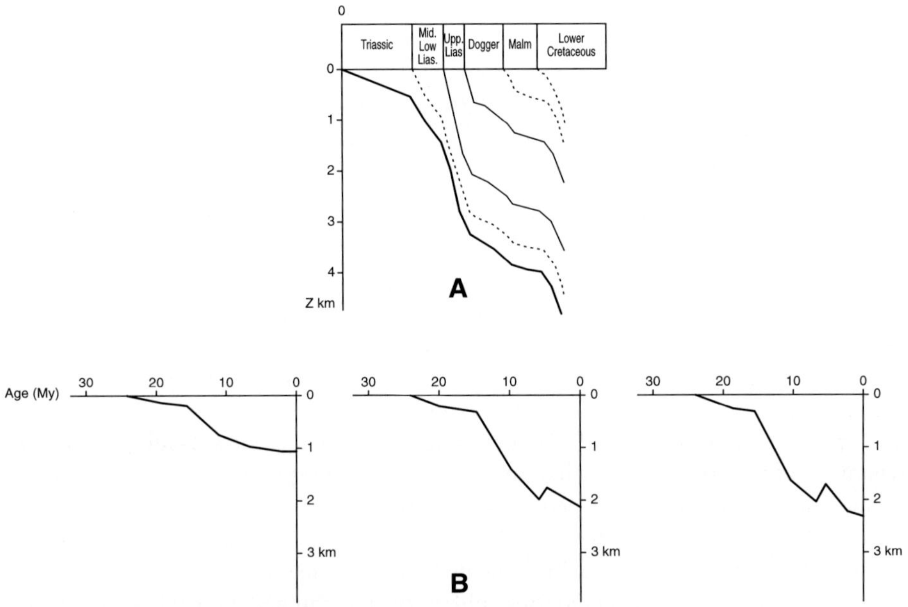

Fig. 2.6 Subsidence: gradual settling of the basin.
A. The curves represent the gradual settling of the substratum (thick line) and successive sediments (thin and broken lines) (from Brunet et al., 1985). The sinking is initiated by some geodynamic or tectonic cause, and the sedimentary load then comes in addition to this. **B.** Depending on the basin position, the subsidence curve may vary considerably.

2.2 BASIN CLASSIFICATIONS

Sedimentary basins are classified differently by different authors, depending on their viewpoints. For example, basins used to be defined by their **depositional environment**, i.e., by type of filling: **lacustrine**, **continental,** shelf or deep **marine basins** (Fig. 2.2). This is an important way of describing the sedimentary environment (the dominant environment, in any case, because this may vary with time) because it conditions the type of basin filling. But it is too general for describing the variety of situations (the Great Salt Lake is quite another thing from the Lake of Annecy) where internal and external driving mechanisms necessarily interfere with each other. Chapter 3 will return to this.

Basins also used to be described by their structural development, rather than by external factors. They were situated in their **stratigraphic** and tectonic **framework** in a cycle between two orogenies (major deformations) and were referred to as Precambrian (such as Africa's Taoudeni), Paleozoic or Primary (north Sahara basins, for example), Mesozoic

(Paris and Aquitaine basins), Cenozoic or Tertiary (such as the western Mediterranean basin). The distinguishing feature was the basin's principal age of operation. This perspective still provides an easy way of comparing basins, their content, and their specific features stemming from the dominant conditions particular to a given period. It can be seen that basins will be described differently depending on whether the viewpoint is sedimentological, structural, or stratigraphic.

Sometimes, a number of basins may lie atop each other in time or, more simply speaking, they may have a **multiphase history** (Fig. 2.5). These are called **superimposed basins**.

While these useful and complementary perspectives still hold, the classification generally used today is genetic, which means that it is based on the concepts of global tectonics. This type of classification, which is the one used in this book, is based on our evolving knowledge of continents and oceans. It was at first strongly influenced by the environmental and crustal differences between the continental and oceanic domains, with a distinction between cratonic and geosynclinal domains. Then, as plate tectonics developed, the general geodynamic framework was specified and this **genetic perspective** provided a way of distinguishing among basins by their geographic distribution:

- Basins located within plates on a relatively rigid, thinned lithosphere, are called intraplate basins and essentially exhibit the effects of extension.
- Basins located at plate boundaries, either over a thinned lithosphere or in a more complex configuration, often associated with mountain chains where extension-compression-slip mechanisms act together or successively.

Following the broad lines defined by various authors[1], we find (Fig. 2.7):

- **Continental** and **epicontinental** basins in broad continental crust areas. These are often referred to as stable platforms or **cratonic basins**. The water cover has never been very deep, but their surface area is generally considerable. Present examples are Amazonia, the Chad basin, and western Siberia. Ancient examples are the Paris basin, the Sahara Paleozoic basins, China's lake basins, and American interior basins.
- **Rift type basins**, which are intracontinental grabens or aulacogen rifts constituting an early stage of continental extension, which may or may not have led to the formation of an ocean domain. The extension involved is linear and limited. Present examples are the east African rifts, Lake Baikal, the Rhine graben, the Red Sea. Ancient examples are the Alsace graben, Russian aulacogens, and the early stages of passive ocean margins.
- **Passive** or divergent **continental margins** due to ocean opening after rifting. These are linear, segmented basins whose width may vary greatly depending on the initial structure and sediment flow. Margins can be broad or narrow, fat or starved. Present examples (but with ancient operation) are the Armorican and American margins, African margins (with or without deltas), and the Gulf of Lions, and ancient examples are the Alpine margin, the Vocontian trough, and the American Cordillera margins (reworked in later orogenies).

1. Bally and Snelson, 1980; Bois et al., 1982; Perrodon, 1988.

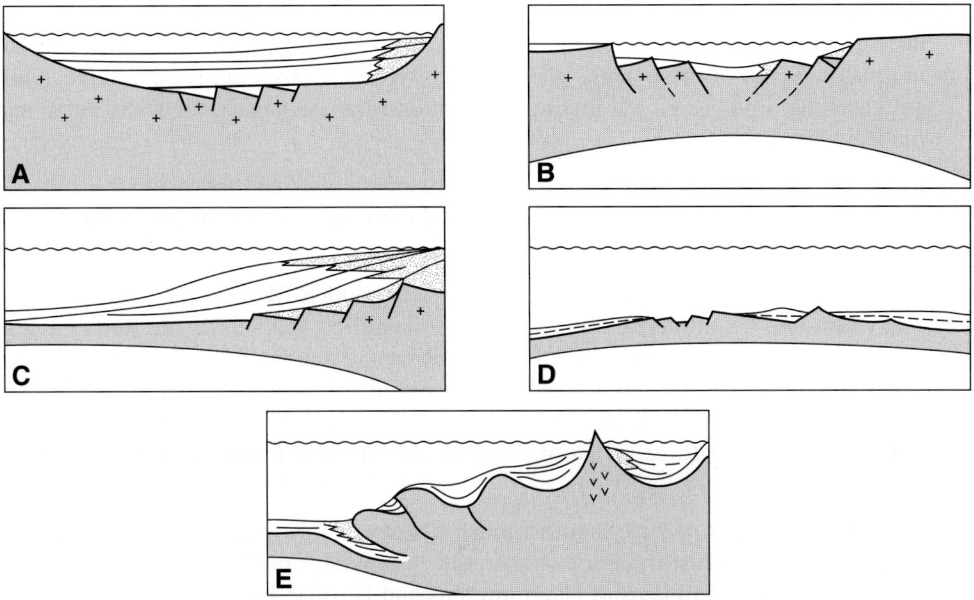

Fig. 2.7 Classification of sedimentary basins.
A. Intracontinental basin. **B.** Rift basin. **C.** Continental margin. **D.** Deep ocean basin. **E.** Different basins of orogenic zones.

- The **great ocean basins**, which often cover considerable area, are generally deep but usually have a thin sedimentary cover of less than 1000 meters even when they are ancient, unless volcanism has contributed to the input, and are of little interest for petroleum (e.g., the Pacific and Indian Oceans or the Madera abyssal plain). Some are characterized by special situations such as volcanic island strings or vast underwater plateaus.

All of these basins form on a thinned portion of the lithosphere (the lithosphere of the first basin pictured above is thinned only slightly, but the lithospheric thinning increases with the latter basins). A second category of basins are those that form on a thickened lithosphere, which may or may not be rigid:

- **Active margin basins** and **foreland basins**, associated with subduction zones where the ocean is gradually receding. These convergence zones, characterized by compression, harbor several types of basins of limited extent: the deep trench, forearc and back-arc basins with highly variable characteristics depending on the processes at work. If the continent is close to the subduction zone, the basin is then referred to as "foredeep". These foredeep and foreland basins are flexural, often broad, and often economically important. More generally, they are termed **perisutural** basins. Present examples are the Caribbean, Japan, the Mediterranean, Indonesia; and ancient examples are the Alps, Turkey, and the Iranian foothills. Today, basins transported piggyback on deformed, overthrust series are being investigated with special attention.

- **Fold belt basins**, sometimes called **episutural** or **intermontane**, often made of former back-arc basins that develop in mountainous areas created by approaching and colliding plates. Some of these act as shear basins where compression is still active. They are generally small (e.g., the interior Turkish and Iranian basins, the Californian and Indochinese basins).

This general classification has the advantage of being simple and following the unifying concept of global tectonics. However, as we have already said, basins **evolve in time** whatever their genetic situation may be.

A basin may thus very well evolve from one type to another. What begins as a continental rift may turn into a continental margin after a few million years (e.g., the Red Sea) and mature more and more (Angola). A passive continental margin will in some cases be reworked by compression due to subduction, becoming an active margin, and will then participate in the construction of a mountain chain such as the Alps; or a cratonic type basin may gradually become a foreland basin, such as the foreland-foredeep of the American interior basin (Fig. 2.8).

A particular region may have a **multiphase history** with a patchwork of elementary basins, each with its own history. For example, the current edge of the Alps in the southeast basin of France exhibits remnants of Carboniferous and Permian basins (fold belt extended over the Hercynian chain), the epicontinental and perisutural edge of broad Mesozoic basins on the Tethyan margin continued by the Alpine chain, and locally, the Oligo-Myocene basins created by the fragmentation-extension of western Europe.

This multiphase history can be shifted in space (Fig. 2.4C). A very broad platform stage has been found among certain cratonic and epicratonic basins (such as the Saharan slab of the Lower Paleozoic and the American Great Plains). The basin is then broken up into several sub-basins during tectonic phases (Fig. 2.9).

A distinction must then be made between **functional** or **depositional basins** existing prior to or at the time of deposit and whose genetic structure thus determines the filling, and **structural basins**, whose present cup structure is due to a deformation occurring after the sedimentation. This latter structure, like the Paris basin, can rather be termed a **syneclise**.

The Sahara sub-basins were separated this way at the time of the Hercynian deformation, in a set of syneclises and anticlises (Fig. 2.9). For these basins, the possibility should not be excluded that the cup structure inherited from the late deformation corresponds to a gradually differentiated subsidence in the course of the filling, over long periods of time.

Lastly, we will see that we can simplify and distinguish two broad categories of basins by thermal and mechanical processes (Fig. 2.10):

- **Pull-apart basins** where the heat flow is variable and the hydrodynamic activity and erosion is low
- **Flexural basins** where the heat flow is stable but the hydrodynamic activity and erosion are always high.

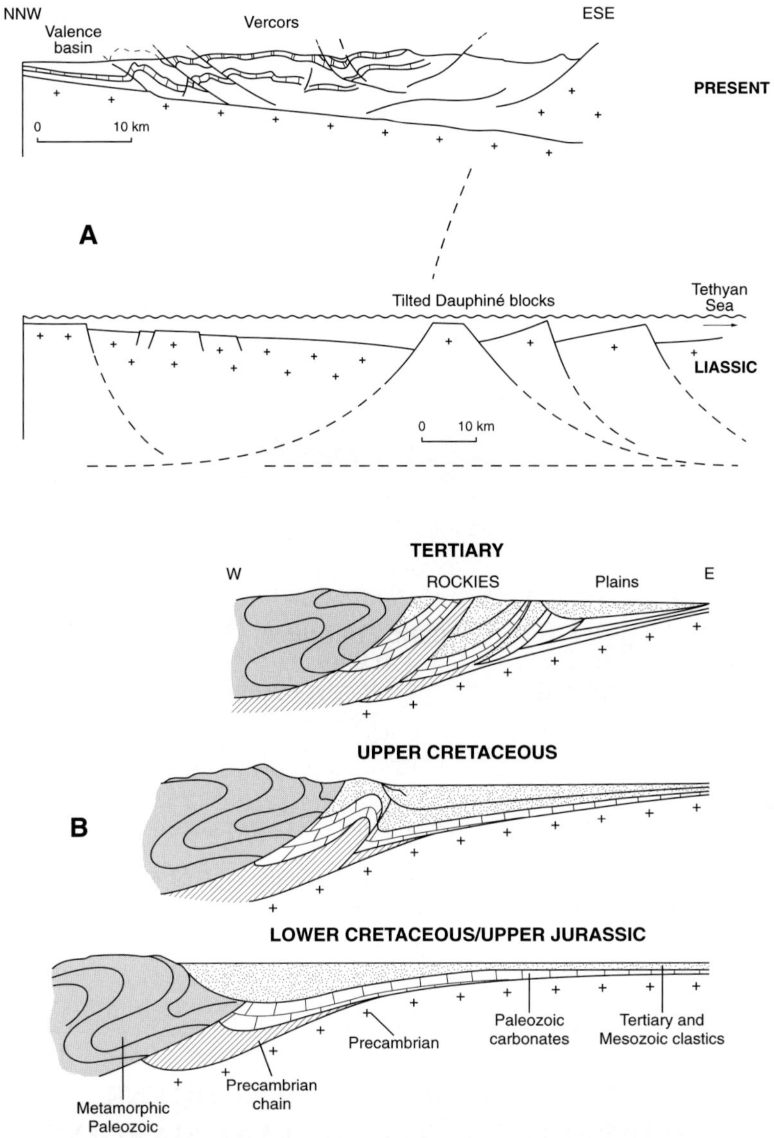

Fig. 2.8 Basin multiphase history, incorporation in a mountain chain.
A. Example of the southeast basin of France tectonized during the formation of the Alps. The farthest part of the Mesozoic rift (here in the Liassic) was itself deformed at the time of the latest Alpine compression.
B. Example of the American Great Plains flexural basin.

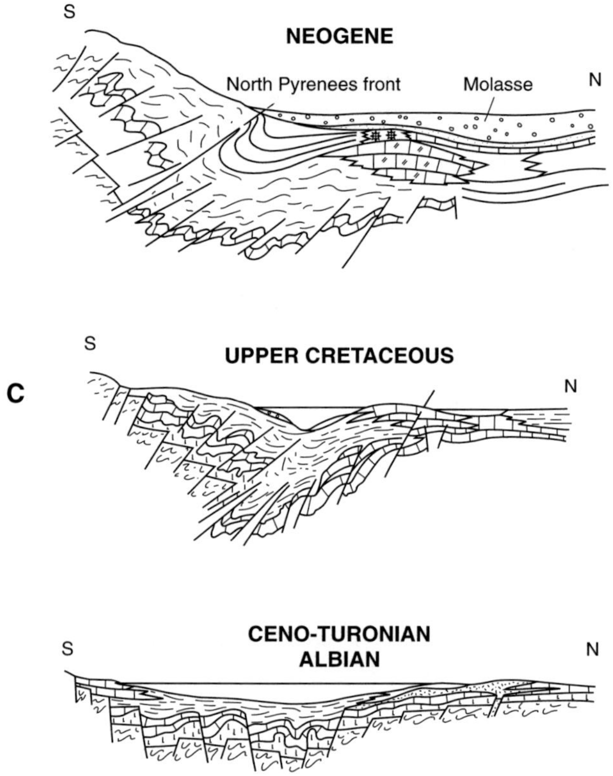

Fig. 2.8 (cont'd) Basin multiphase history, incorporation in a mountain chain.
C. Example of Aquitaine basin, of which only the original northern margin has been conserved with slight deformation, while most of the functional basin was folded during the gradual uplifting of the Pyrenees (from a number of authors).

2.3 TROUGHS, RIFTS, AULACOGENS, AND DIVERGENT CONTINENTAL MARGINS

2.3.1 Definitions

Fault troughs are a very special type of basin. They are long and relatively narrow, and are bordered by roughly parallel normal faults, resulting from a stretching of the upper part of the crust and extension at the surface (see Chapter 5). This extension may be localized and

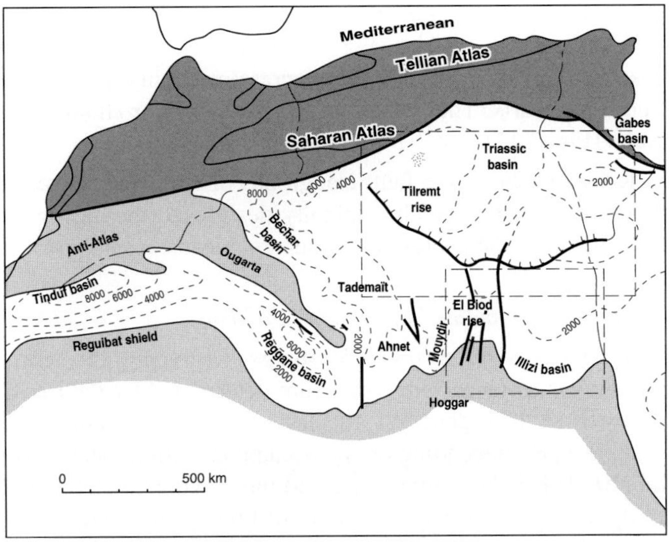

Fig. 2.9 The Sahara's Paleozoic basins: structural syneclises on the African shield.
The Sahara Triassic basin can be seen superimposed on the different contours (to the northeast of the barbed line).

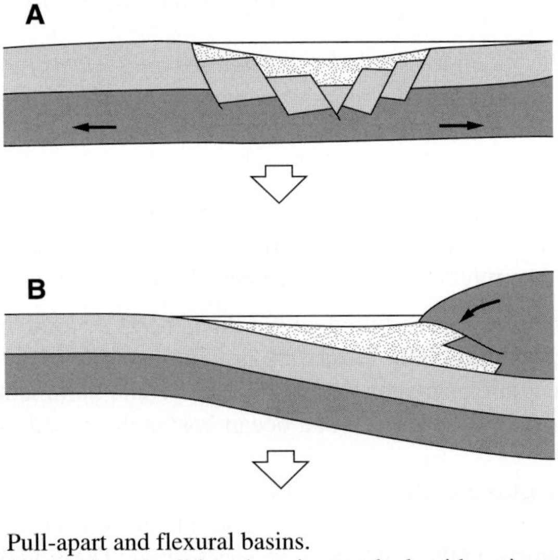

Fig. 2.10 Pull-apart and flexural basins.
A. Typical extension: the lithosphere is stretched, with active subsidence in the resulting rift. **B.** Flexuring of lithospheric material by overloading, with variable elastic thickness.

surficial but, when the troughs extend over several hundreds or even thousands of kilometers on the continental scale, as in east Africa (Fig. 2.11) and western Europe, it is clear that they reflect major lithospheric breakup. A collapsed depression of this type is called a **graben**, as opposed to uplifted **horst** parts. The graben also has benches or shoulders of **tilted blocks** along the main faults (Fig. 2.12).

Following the Russian school, the fault trough has often been referred to by the practically equivalent term **aulacogen**, which originally designated the narrow basins with very deep sedimentary filling over the depression of ancient basins, as is known in Russia and on the North American craton.

The synonym "taphrogeosyncline" has fallen into disuse.

Rifts are also fault troughs, but different meanings are given to this term:

- **Oceanic rift**. This is a typical ridge structure determined by the rate of ocean expansion (see Chapter 1). It is generally the last stage of evolution of continental rifts. There are several types, depending on the oceanic accretion mechanisms and rates.
- **Continental rift**. This is becoming more and more synonymous with the fault trough, but is generally used either when the history of lithospheric extension in the (cratonic) continental domain can be clearly identified, or when this rift has continued its evolution toward a continental margin with a period of **oceanic opening** following the **rifting** period (see further on in this chapter), which is explained by the lithospheric convection system (Fig. 2.13). The Red Sea is a good illustration of rapid rift evolution.

Rifts are sometimes said to be ***impactogenic*** *(west European Oligocene system, Rio Grande rifts, Asiatic rifts in the north of India) when they result from continental collisions.*
They can reasonably be differentiated into Atlantic type systems (considered here), back-arc rifts and rifts due to strike slip (see further on in this chapter), and special post-orogenic and Basin and Range systems.

When continental fragmentation increases to the point where the two lips of the graben edging the rift finally draw apart and separate, the lithosphere is broken and an oceanic crust gradually develops at the center of the rift. A **continental margin** then appears on either side (Fig. 2.14). This term thus refers to the transition zone between continental and oceanic crusts.

Several types of margins can be found within the general tectonic framework:

- **Divergent margins**, originally formed by gradual opening of an initial rift, but exhibiting no major tectonic or magmatic activity today. The dominant phenomenon here is extension. The rims of the deep broad ocean basins discussed later are **passive margins** of this type.
- **Convergent margins** are characterized by shortening and disappearance of the crust by subduction, and also by high seismicity and intense volcanic activity. The compression and deformation on these margins are spectacular. These **active margins** exhibit a number of basin types, including those in extension, as will be seen further on in this chapter.

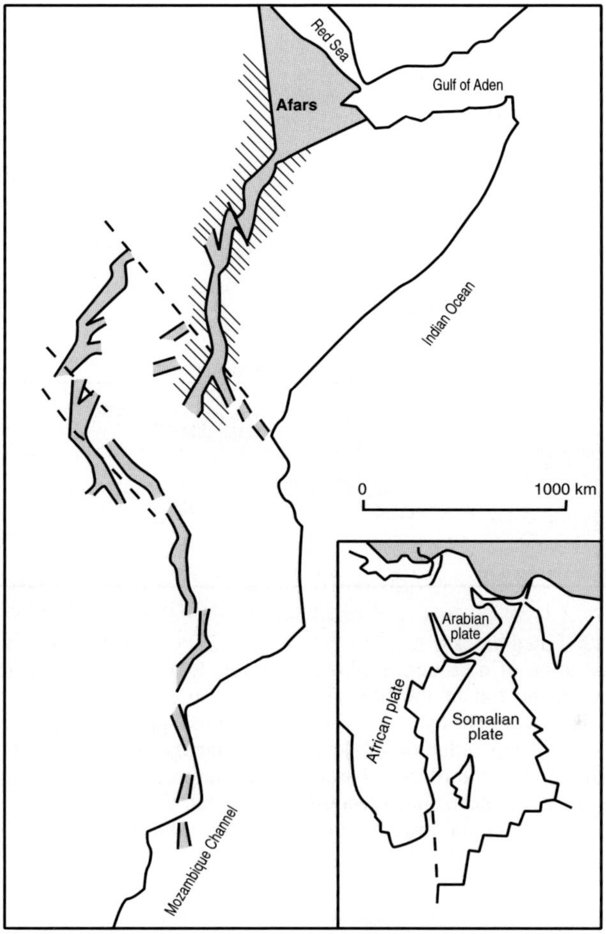

Fig. 2.11 East African continental rift system.
The east African continental lithosphere is torn by a complex system of rifts and fault troughs (making large lakes) over several thousand kilometers ranging from the Afars region in the north to the Mozambique Channel in the south. These rifts and troughs are the initial step toward future detachment of the Somalian plate. They are recent breakups that are comparable to the initial functioning of the Red Sea at the time of the separation between Africa and Arabia, farther to the north. The hatch marks indicate the corresponding volcanism.

- **Shear plate margins** or **transform plate boundaries**, where the boundary between continent and ocean neither extends nor shortens, but is rather marked by shear and slip. These margins are generally abrupt and are thus not the seat of any broad, thick

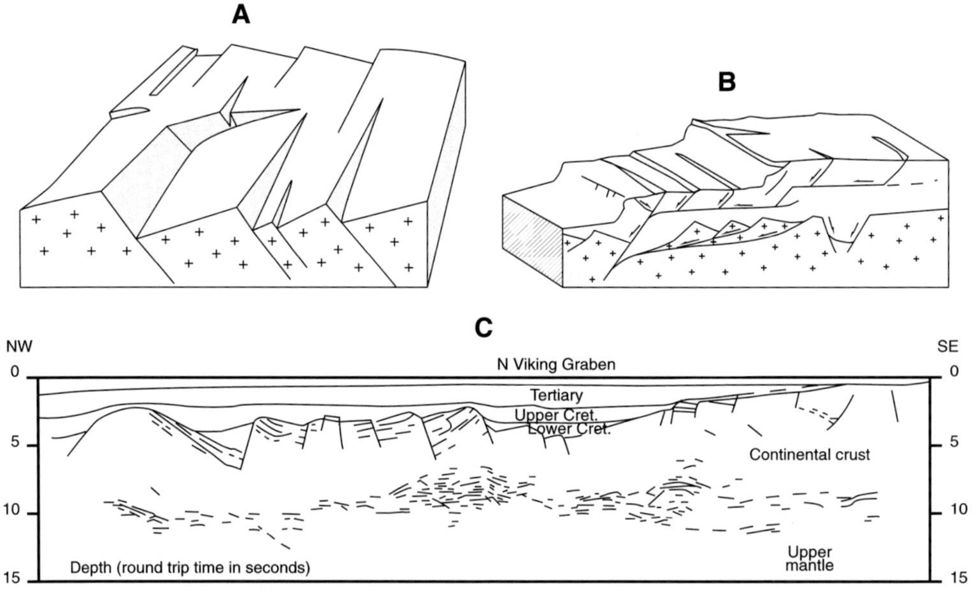

Fig. 2.12 Tilted blocks.
A. Illustration of a sequence of normal faults leaving a series of leaning dominos in the top of the lithosphere. The top of a block is called its "nose". **B.** Model showing several levels of breakup in the upper part of the crust, with separation or slip levels (from Jarrige, 1992). **C.** Section of the Viking graben (North Sea) based on the interpretation of a seismic section showing the mechanical decoupling between the fragile upper part of the crust and its ductile lower part (from Ziegler, 1989).

sedimentary deposits. The **pull-apart rifts** often found along major transform faults are spectacular.

Lastly, there is the special case of the Great Basin or Basin and Range running between the Sierra Nevada and the Colorado plateau in the North American craton, which is a spectacular example of large-scale extension where a broad basin is created on detachment faults accompanied by major magmatism (Fig. 2.15).

2.3.2 Formation Mechanisms

This section discusses the formation of the rifts occurring in the broad continental domains called cratons.

The following is a review of this continental fragmentation in its two successive phases. The first is the **rifting** phase as the continent extends to the breaking point. The second is the

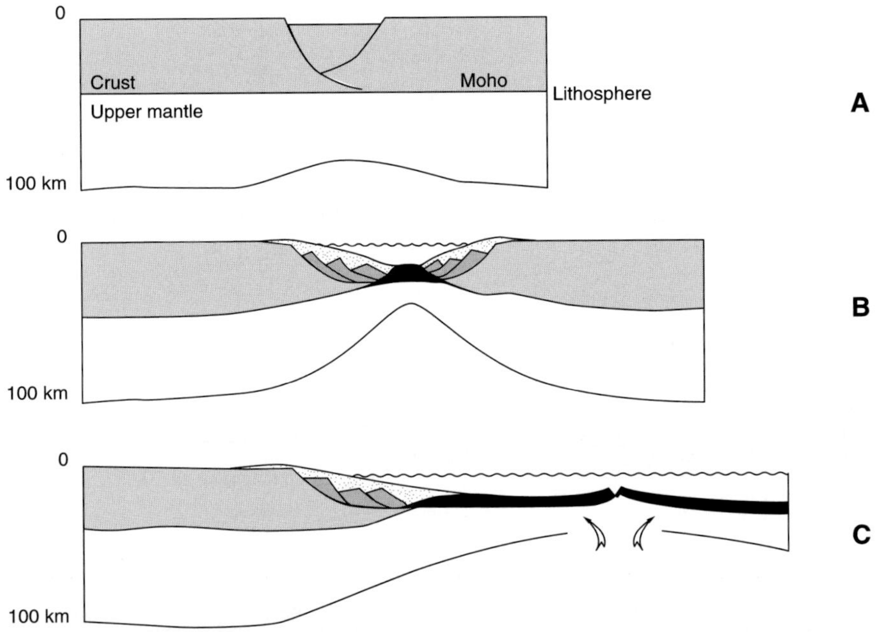

Fig. 2.13 From continental rift to ocean rift.
A. Rift initiation by stretching and breakup of the lithosphere. **B.** Ocean opening, rise of the asthenosphere. **C.** Oceanic stage: accumulation of sediments on the margin and operation of the oceanic rift.

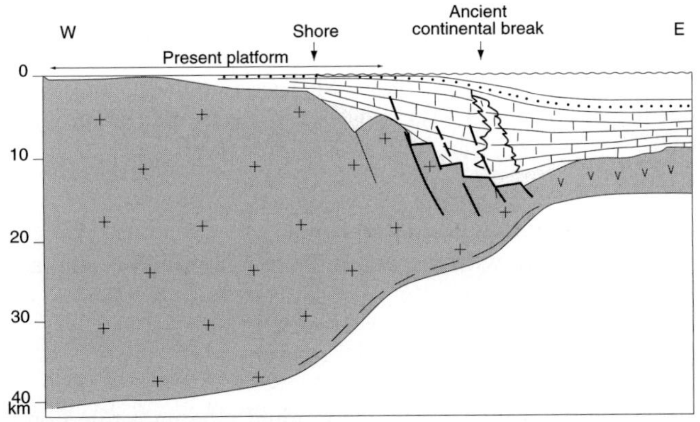

Fig. 2.14 Example of continental margin.
Diagram of American Atlantic margin with rapid thinning of the crust going from continental to oceanic.

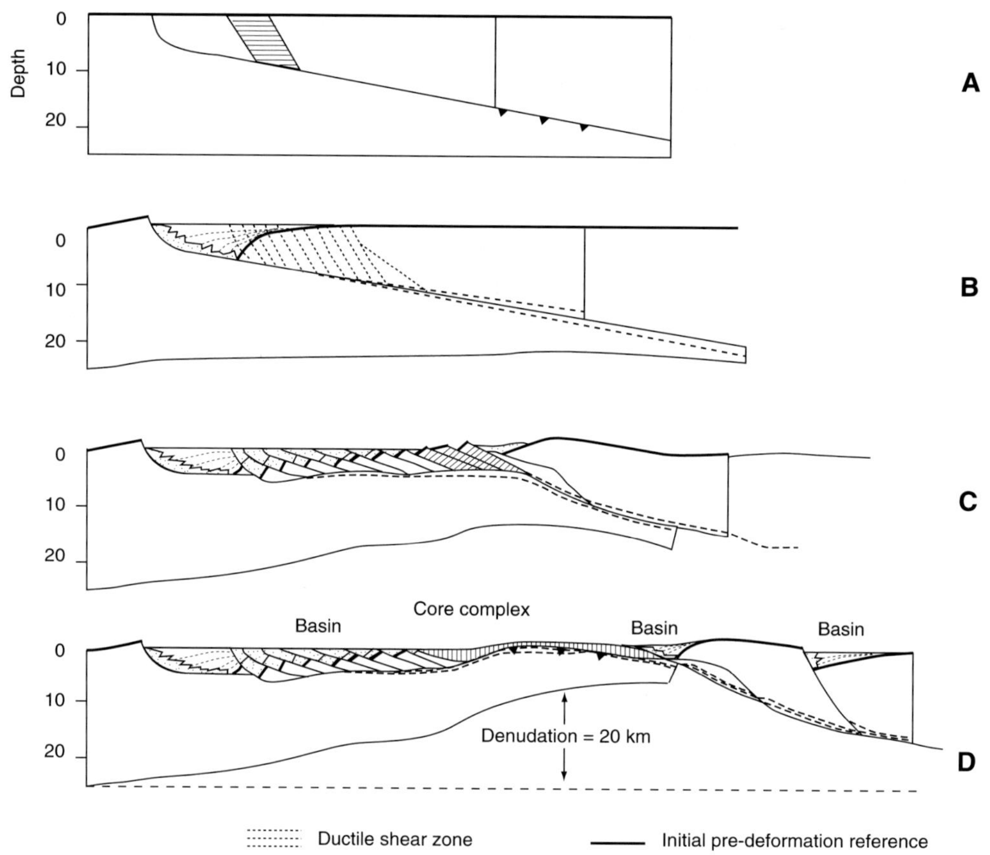

Fig. 2.15 Development of the Basin and Range system (from Wernicke, 1985).
A. Initial state. **B.** Sevier Desert stage 3 My later. **C.** Eldorado stage (+ 8 My). **D.** Raft River stage (+ 14 My): sequence of basins (the Basin) and relief (Range).

gradual evolution of a **passive margin** in the course of a spreading phase with oceanic drift (the margin is said to be passive with respect to the neighboring oceanic accretion).

2.3.2.1 Rifting

Our knowledge of rifting drive mechanisms is gradually being refined on the basis of thermal and mechanical concepts. What has been studied most is the top of the lithosphere, where surface observations and geophysical imaging tell something about the different stages of evolution on a case basis.

Even though little is known of the exact geodynamic processes occurring at depth, the depressed structure along the great normal faults, the crustal extension, negative gravimetric anomaly, high heat flows, and volcanic activity are very important data. The deep structural levels, on the other hand, are rarely accessible. Working hypotheses can only be compared with models. The general principles defined by plate tectonics (see Chapter 1) serve as a guide in examining the mechanical and thermal processes that occur in the rift during stretching and subsidence.

The process may start off with **thermal intumescence**. This idea that there is an initial **swelling** or **doming** phase is based on the observation of major erosion in the basement along the line of the future rift. This phase then seems to be followed by a corresponding detumescence of the asthenosphere. It is felt that millions to tens of millions of years are needed for thermal conduction to be effective on a normal thickness of lithosphere, whereas upswelling would be faster with high lateral temperature gradients.

But it is generally thought today that the edges are raised by heating after the stretching begins. In some cases, a gravity-driven extensive spreading mechanism seems to be at work on the top of a thick crust of some previous tectonic structure (this is one way of explaining the Great Basin).

When the rift enters its active phase, a **graben** or **fault trough** forms. **Extension** causes **fault** planes to appear in the basement while the rift at the surface becomes a valley running along the fault, lined with rather sharp walls. The overall geometry is a sequence of **tilted blocks** leaning often asymmetrically on each other with a number of transverse faults. The Gulf of Suez is a famous example of this (Fig. 2.12). Chapter 5, on tectonics, will show that different geometric motions (translation, rotation, and others) are possible in this breakup process.

On the scale of the lithosphere, this active extension phase, which is seen at the surface as a latticework of faults affecting the brittle upper crust, corresponds to **stretching** of the ductile layer at the base of the crust, generally causing it to **thin** out (Figs. 2.2 and 2.16). Whatever dynamic and geochemical processes are actually at work deeper in the mantle remains an open question.

The **temperature regime** is an essential parameter. The **geothermal gradient** varies from 15° to 60°C/km of depth in the basins, averaging 30°C/km.

The **heat flow** is the energy transferred per unit time and surface from the hot internal zones toward the colder surface, and is expressed in joules per square meter per second ($J/m^2/s$, or W/m^2) or in standard Heat Flow Units (1 HFU = 41.8 mW/m^2).

The flow from radioactive sources accounts for 40 to 70% of the total, and averages an estimated 70 mW/m^2. But heterogeneities in the crust may vary this radioactive flow. In the continental domain, it is of the order of 30 to 80 mW/m^2, while the asthenospheric rise in the rift domain sometimes increases it to as much as 300 mW/m^2.
The specific heat capacity per unit volume ($J/m^3/K$, in which K is the Kelvin temperature) or per unit mass (J/kg/K) is another parameter affecting the temperature regime.

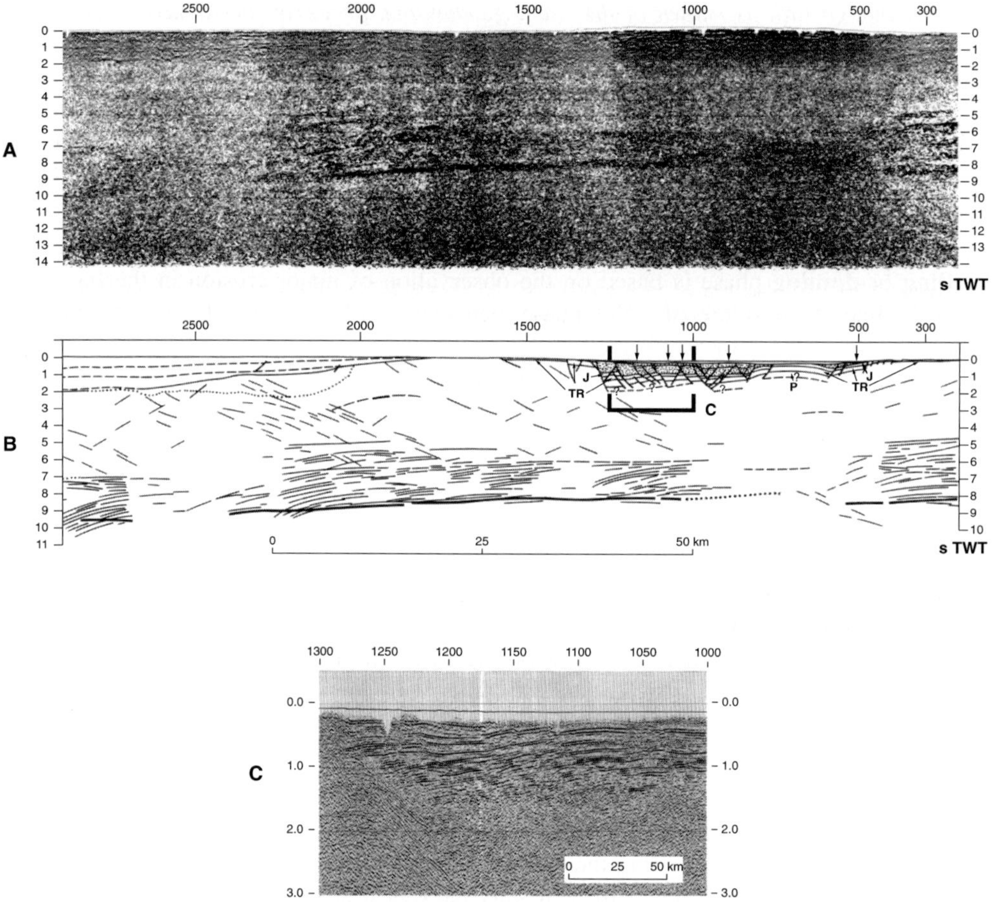

Fig. 2.16 Crustal section of the Alsace trough.
A. Raw seismic section. Note the lengthy sounding time (up to 14 s round trip, or "two-way time", TWT). **B.** In the section interpretation, the rift basin occupies the surface of the crust and the lower part is characterized by layered reflectors. **C.** The western edge of the graben and the sedimentary fill affected by the faults (from B. Colletta).

The time history of the heat flow and subsidence in rift development is **modeled** by combining the crustal thinning with the thermal disturbance it creates. It should be remembered that the flow through the crust may vary spatially, due to geodynamic effects and heterogeneities, but also in time: it is estimated that a thermal event is dissipated within a time constant of some 50 My.

Heat conductivity and the convective process are calculated for sedimentary basins, but the third (radiative) heat transfer mode is neglected. Conductive transfer by molecular agitation varies with the physical characteristics and burial depth of the rock (Fig. 2.17). Remember that the temperature gradient is mainly affected by compaction, that it is difficult to define a single average gradient, and that sedimentation has an effect, in that the input delivered to the basin decreases the flow.

Viewed in the steady state, the flow is considered to be constant throughout the basin thickness, but unbalanced "transient" regimes also exist.

Chapter 6 will show the importance of properly assessing the thermal history of a basin in order to reconstruct the origin of its hydrocarbons.

Convective heat transfer is a mass movement process (by a moving mass of water, for example) and may at times be large. This is the case with the forced convection in the compaction of sediments, free convection by thermal expansion of water (warmed water is lighter and tends to rise, and vice versa), and regional convection due to loading (e.g., in uplift or emersion).

Mechanical properties are also to be considered in modeling basin formation. The crustal heterogeneity of the basement rock will be discussed further on in this chapter and then in Chapter 5. The fracturing and stretching of the crust depends very much on its initial makeup and fluid content.

There are a number of different models today for explaining what is observed at the surface (such as listric faults) and at depth (seismic reflectors at the base of the crust) (Fig. 2.18).

These are the parameters used to delineate the geometric history of deformations and kinematic laws governing basin evolution. Bed compaction and decompaction are also considered, of course, as are the fault kinematics and the subsidence or uplift rates. The paleohydrodynamic components of fluid pressures and circulation over geological time are also a factor, and doubtless more so for the flexural basins analyzed further on.

Crust **stretching** and **thinning** has been explained by different models[1].

2.3.2.2 Uniform Extension Model

The uniform extension model explains initial subsidence[2] as a fragile, ductile extension of the lithosphere (Fig. 2.19). A **pure shear** mechanism[3] then appears, with uniform homogeneous extension. The lithospheric stretching is defined by an extension coefficient β, which is taken to be the same for crust and mantle. When the model was first developed, the extension was considered to be instantaneous; but a later model[4] integrates non-instantaneous

1. A general review of these is given in Burrus, 1989.
2. Seveston, 1976.
3. McKenzie, 1978.
4. Jarvis and McKenzie, 1980.

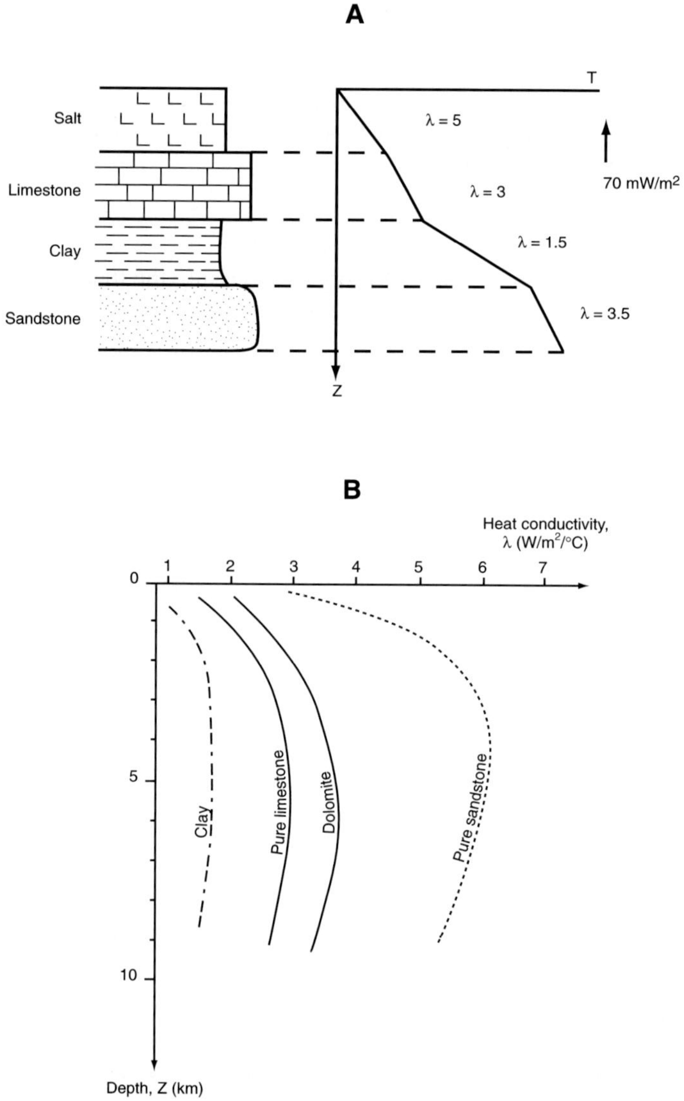

Fig. 2.17 Variation of rock heat conductivity in a basin (from J. Burrus, ENSPM document).
A. Variation of heat conductivity l with lithology. Note that high conductivity corresponds to a low thermal gradient for a given heat flow. **B.** Heat conductivity variation with depth for various rock types. This increases down to 3000 m, levels off, and decreases beyond 6000 m.

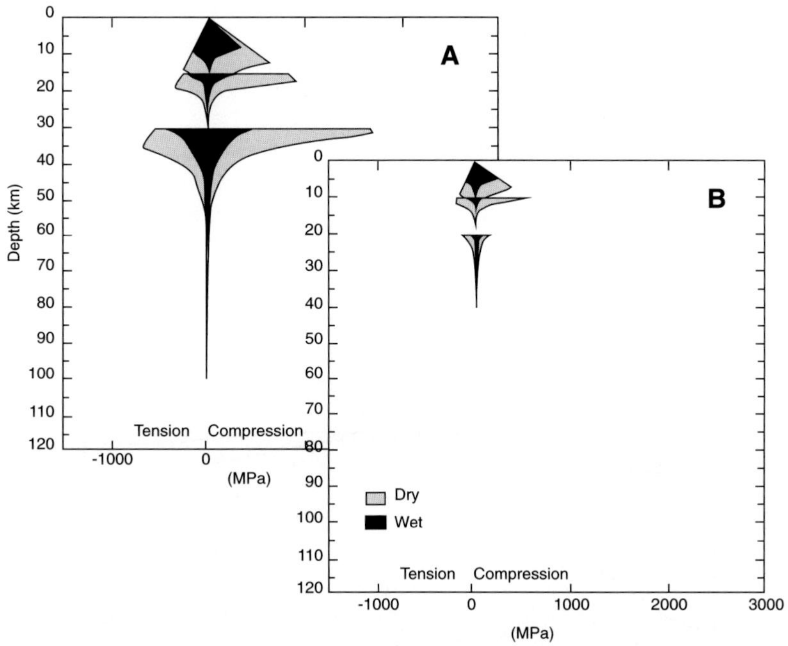

Fig. 2.18 Rheological models as a function of depth (30 km) (from P. Ziegler).
A. Normal continental lithosphere. **B.** Stretched lithosphere (20 km).

uniform extension at a finite extension rate. Part of the heat is diffused before the end of extension, and the resulting thermal anomaly is reduced.

In fact, while extension is considered to be uniform on the scale of the lithosphere, it is heterogeneous in the upper layers of the crust where generally asymmetrical grabens develop. This model is often used to explain large intra-continental rifts like that of the North Sea (Fig. 2.20).

Crustal stretching and heating are generally considered to be the first phase of rifting, the consequence of which is the thermal subsidence as the lithosphere returns to its initial thermal equilibrium after breakup and separation of the two continental edges.

2.3.2.3 Depth-Dependent Extension Model

This model[1] resembles the previous one except that the lithosphere is considered to be a series of ductile and brittle layers, and the mantle is considered to extend more than the crust

1. Royden and Keen, 1980.

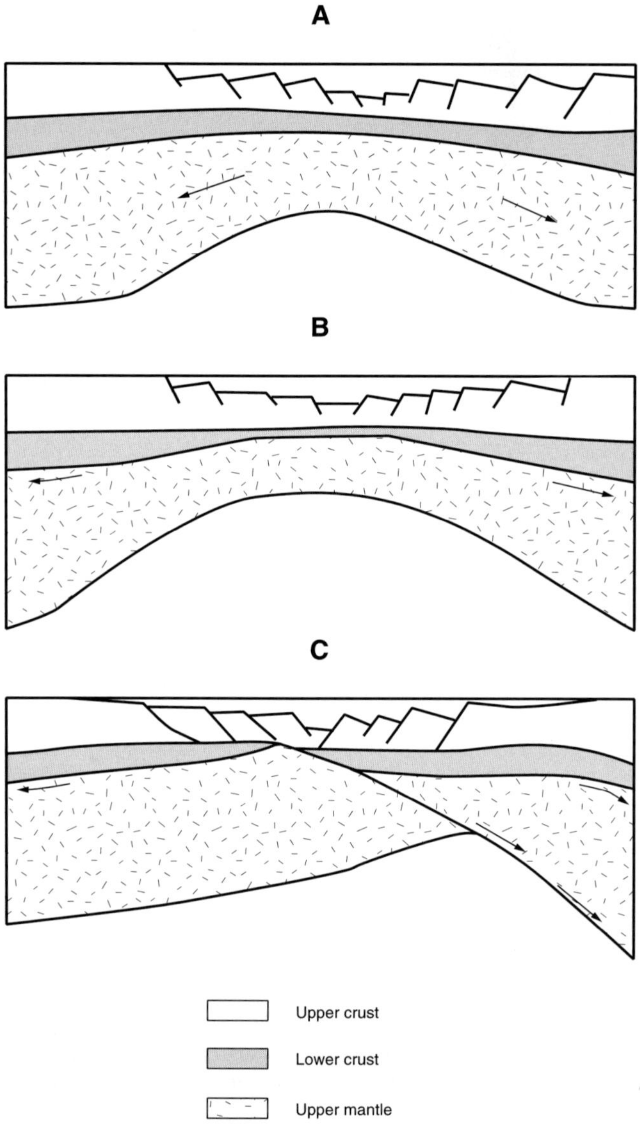

Fig. 2.19 Main lithospheric stretching models.
A. Uniform homogeneous extension model with the same thinning in the crust and lithosphere. **B.** Discontinuous, inhomogeneous extension model in which the crust stretches less than the upper mantle. **C.** Non-uniform extension model with simple shear.

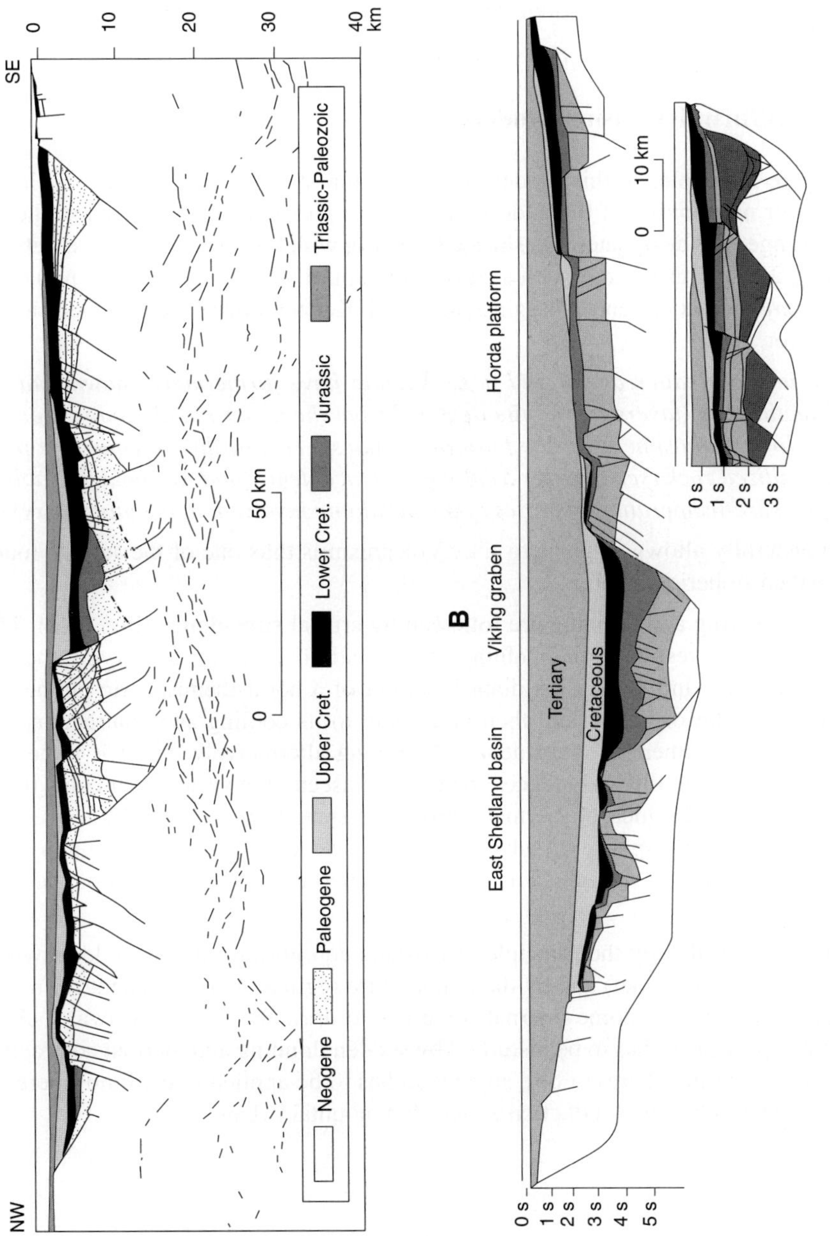

Fig. 2.20 Section of the North Sea crust.
A. Deep geophysical section (from IBS). **B.** Another section parallel to A, showing lateral variations (from Faure, 1990).

(Fig. 2.19). There is less tectonic subsidence during the extension, but most importantly there is a lateral heat flux that would explain the uplifting of the basin edges. This phenomenon has often been observed, and initial doming seems to be a rather unlikely alternative explanation for it.

2.3.2.4 Non-Uniform Extension Model[1]

The shearing here is simple. In this model, the maximum stretching is by shearing in the crust, with little or no thinning of the underlying mantle. This model stems from studies of the Basin and Range, and designates this **simple shearing** with low angle of dip as a **detachment fault** (Fig. 2.19). This has met with a certain amount of success in application to often asymmetrical conjugate continental margins, which McKenzie's pure shear model did not clearly account for.

Work done in the Mediterranean and near Atlantic have furthermore shown that the ascent of ductile deep layers (peridotite denuded from the upper mantle) might be windows of the lower lithosphere (in the form of wrinkles). This further assumes the presence of mylonitized rock (see Chapter 5) along such accidents (and we should be able to track the pressure-temperature histories of the peridotite involved, in the laboratory).

Extension generally allows magmas to rise. Volcanism is thus one of the surface manifestations of asthenospheric ascent.

The crustal stretching and thinning are followed by crustal **subsidence** (Fig. 2.21). This sinking of the crust is a response, maintaining the isostatic balance in the varying thermal field. Assuming that the initial cause is diapiric ascent of a hot asthenospheric plume, the initial thermal expansion of the system then gives way to its cooling and contraction, and therefore the settling or "thermal" subsidence. Assuming alternatively that it is rather the stretching that prevails initially and induces the thermal ascent, the system still cools thereafter and we again find the idea of thermal subsidence spread out in time. Moreover, we have already said that the response of subsidence can vary considerably depending on the model used, with uniform or non-uniform extension, since the cooling conditions are not considered to be the same.

Subsidence is controlled by the principle of isostatic equilibrium (there is a level where all lithostatic pressures equal out) and by the temperature regime. A basin cannot form on a plate in isostatic equilibrium. Some internal force has to draw the basin downward: either the density of the lithosphere has to be disturbed by sudden thinning and thermal contraction (true of the basins examined here) or else some load has to be applied to the lithosphere (as in peri-cratonic flexural basins, subduction zones, and mountain chains).

1. Wernicke, 1985.

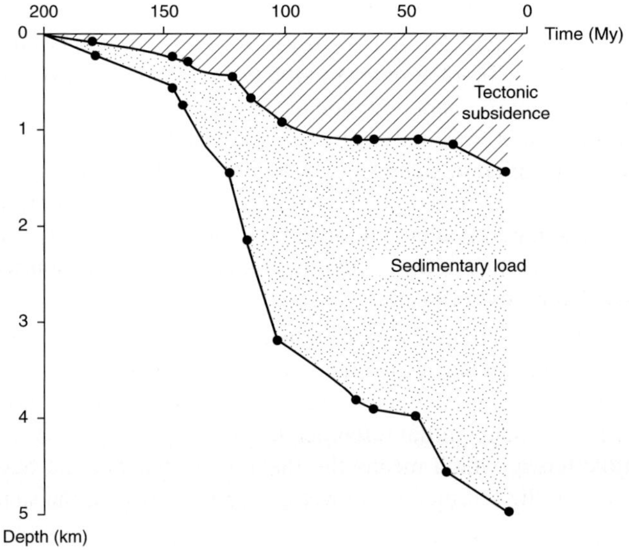

A

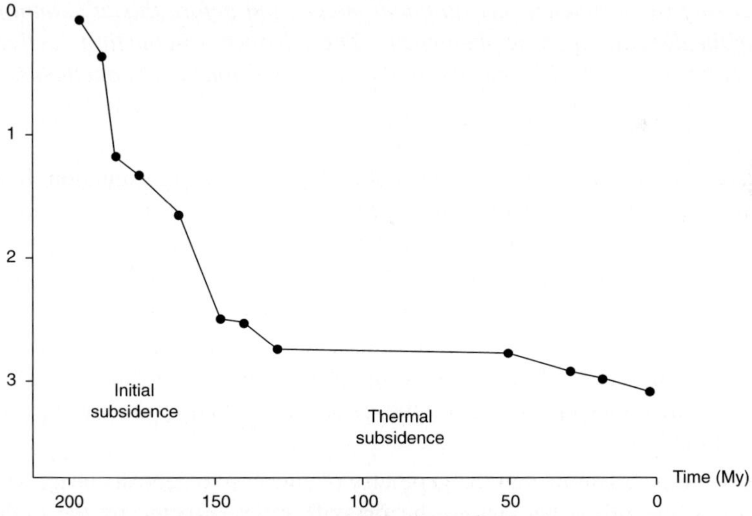

B

Fig. 2.21 Theoretical subsidence curves.
A. The component due to sedimentary load is separated from that due to tectonic subsidence. **B.** This brings out the change in subsidence rate, which is rapid for initial subsidence and slower for thermal cooling.

Simplifying, it can be said that there are two kinds of subsidence:
- **Initial** subsidence is rapid and clearly manifested by the sinking of the rift and only the rift. It is stronger at the block foot than at the nose, and this will very closely determine the sedimentation environment and type of deposit.
- **Thermal** subsidence is slow and is due to the re-establishment of the thermal and isostatic balances. This kind of subsidence is essentially the same whether the continental rift enters an oceanic expansion phase or if it evolves toward a continental margin. The effect on sedimentation, though, will differ in either case. The end of the rifting may be marked by a sudden break in the sedimentation called a **breakup unconformity**, separating **syn-rift** from **post-rift** series.

That is, the formation of a fault trough is immediately followed by fairly extensive filling with sediments, which add a load. A distinction is thus made between:
- **Total subsidence**, which designates the full distance the basin bottom has actually sunk with respect to some initial reference level.
- **Tectonic subsidence**, which means the theoretical distance the basin bottom would have sunk without the sediments and water column, but with the same excess mass at depth.

Chapters 4 and 5 will resume the discussion of sedimentary filling and basin evolution when speaking of subsidence measurement means; but before this, it is important to get an idea of the duration of these phenomena. The subsidence in the first "active" phase of rifting may be called "rapid", but it does last a few million years nonetheless, and sometimes much more. The Triassic rifting prior to the opening of the Atlantic Ocean, for example, is considered to have taken 60 My.

The duration of the rifting varies considerably depending on the situation. It seems to be briefer for rifts that evolve toward oceanic opening (15 to 100 My) than for aborted rifts in the cratonic domains (30 to 200 My). During these periods of more or less rapid sedimentary filling, discontinuous phases of **volcanic activity** may occur that are often important when the rifting initiates, but are sometimes recurrent over long periods of time.

It is generally considered that an equilibrium is reached beyond 100-120 My as the subsidence attenuates in a thermal or "post-rifting" phase. Calculated (theoretical) **subsidence curves** exist for various basins illustrating these concepts, and they are always used in petroleum studies of the basins.

Chapters 4 and 5 will show that the geometry of successive deposits bears witness to the fact that some of the filling belongs to the **pre-rift** series affected by the fault tectonics, whereas another **syn-rift** series corresponds to active filling during the initial subsidence, and **post-rift** series correspond to sediments deposited later (Fig. 2.22).

2.3.2.5 Passive Margins

Once the continental rift is initiated by lithospheric extension, an ocean zone can then open up. Here, an oceanic ridge begins to function and the divergence becomes effective, with

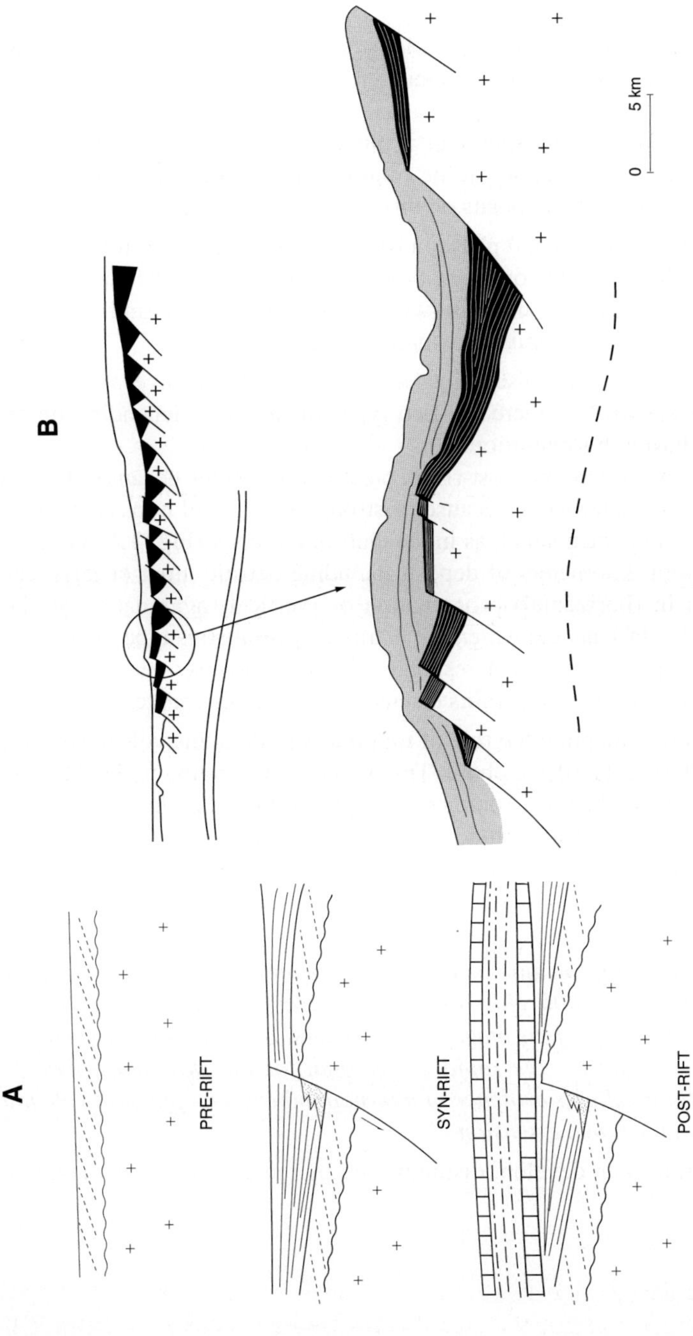

Fig. 2.22 Pre-rift, syn-rift, and post-rift series.
A. General scheme. A syn-rift series fans out into an active fault, causing syndepositional tilting. **B.** Example of the Gulf of Gascony margin with tilted blocks cut out of the pre-rift substratum, the fan configuration of syn-rift series in the semi-grabens (black), and the draping of post-rift series (in gray).

maximum stretching. The two former edges of the old rift decouple and gradually drift apart, and the previous mechanisms give way to:

- **Magmatic ascent** from the mantle, expressed along the rift line by what now becomes an oceanic ridge (with continued thermal intumescence along the line)
- A specific oceanic rift extension process
- Gradual **cooling** of the oceanic lithosphere as it moves away from the ridgeline, and prolongation of the **thermal subsidence** at the continental margin, with the additional effect of the sedimentary load that deposits on this continental margin as it is created.

Different stages occur between the initial phase immediately following the rifting, where the ocean is still narrow and the incipient continental margin is still close to the plate boundary (as in the Red Sea), and the phase of maturity where the ocean is broad and where the margin then represents a former plate boundary, as in the Atlantic.

We will simply refer to our previous discussion of how subsidence evolves, to say that oceanographic studies have shown that there are two types of margins, depending on the volume of sedimentary fill during this evolution:

- **Lean** or **starved margins** where the post-rift deposit is thin, either because of a low sedimentation rate (see Chapter 4) or because of strong erosion and bypass phenomena. The sedimentary load is then small, as in the Gulf of Gascony (Fig. 2.23A).
- **Fat margins** with several kilometers of deposit including detritic input or large carbonaceous production in (horizontal) progradation or (vertical) aggradation as discussed in Chapter 4. The African and American margins illustrate this type of situation (Fig. 2.23B). The lithosphere may then respond by elastic or visco-elastic flexural folding, as in the case of the **flexural basins** discussed later in this chapter.

It is important at this point to mention the role of **magmatism**. Alkaline volcanism from deep in the mantle appears during the rifting phase. The African rifts ranging to the Afar, the Red Sea, and the Gulf of Suez, are the most common examples of this[1], but the volcanism of the Rhine graben in Europe or of the Rio Grande on the North American craton can also be mentioned. The crustal fractures will become even more effective in the ocean opening phase, and the basaltic magmas from the magma chamber will rise to make the new ocean floor in the core of the rift.

It is generally felt today that this volcanic activity is not continuous but cyclic. The periodicity is still being debated—50 000 years or a million years—as are the reasons for the more or less ephemeral nature of the magmatic chambers. But it is widely recognized that magmatic activity is cyclic for the rapid ridges and, for the slow ones, that the roughness of the ocean floor relief is evidence of alternation between abundant volcanic production and episodes of tectonic extension.

Two major types of margins can develop, distinguished by the extent of this magmatic activity:

1. Debelmas and Mascle, 1993.

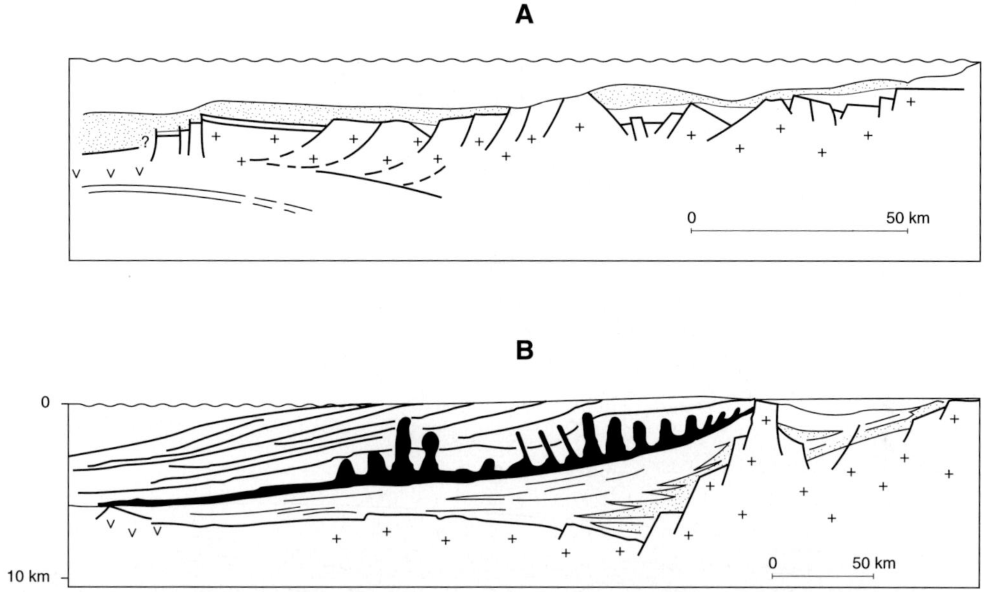

Fig. 2.23 Examples of continental margins covered by variable thickness of sediments.
A. Starved margin. This example of the North-Gascony margin (Channel entrance) shows little sedimentation during the different periods of its history. **B.** Fat margin. Inputs were large during the various stages of Angolan margin development.

- **Volcanic rifted margins** like the one off Norway (Fig. 2.24) where reflection shooting has revealed dipping reflectors that are interpreted as volcano-sedimentary series. *The subsidence may then be greatly affected by a very special thermal regime and by the volume of the volcanic beds that may reach the surface or intrude the base of the crust as underplating.*
- **Non-volcanic margins** (or those with little volcanic) where the bulk of the magmatic activity is diverted to the ridge axis.

This volcanism is expressed at the surface by dikes, outflows, and volcanoes running along different segments of a margin. The rift may furthermore exhibit plumes or hot spots as it opens, and the magmatic activity will then be spectacular, as it is in the African rifts.

The case of the North Sea (Fig. 2.20) is rather special because it consists of a rift system with major crustal thinning, while the whole zone is still part of the Eurasian continent. This is very similar to certain intra-cratonic basins that have developed on aborted rifts.

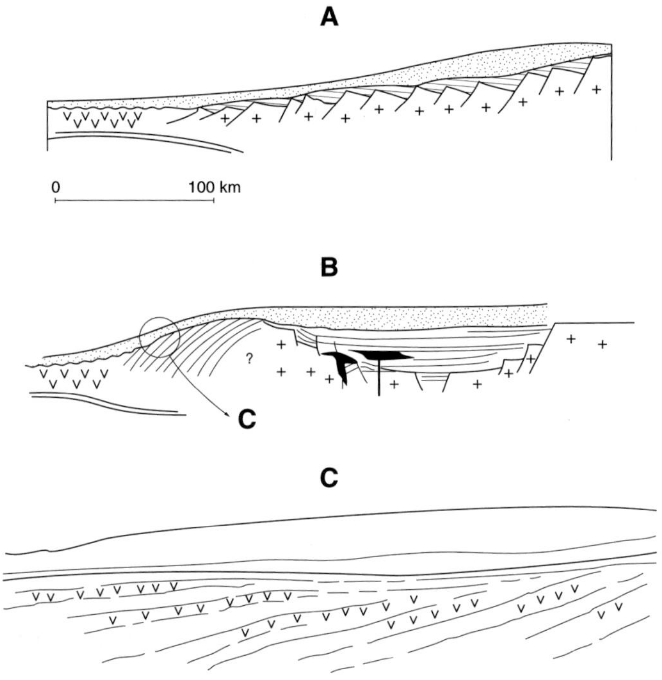

Fig. 2.24 Non-volcanic and volcanic continental margins.
In contrast to the usual non-volcanic margin **A**, the volcanic margin (**B** and **C**) is characterized by a thick volcanic series at the continent-ocean boundary.

2.3.3 General Features

Rift collapse basins at the continental margin can be characterized as follows.

- **Strongly subsident basins**. The subsidence curves reflect the rate of subsidence. The functional basin is generally (though not always) deep, as illustrated by the present example of the Red Sea (Fig. 2.25A) and many ancient examples such as the Tethyan edge and the North Sea. While inputs are considerable, the subsidence may be continually compensated by filling, with the trough remaining in continental or shallow-water conditions (Fig. 2.25B).
- **Major segmentation heritage**. The continental drift develops in a heterogeneous substratum and the tilted-block formation does not extend very far laterally. The system is segmented along lines of weakness inherited from its previous history (Fig. 2.26A), and this same segmented arrangement is found again in the oceanic rifts (Fig. 2.26B).
- **Uneven filling**. As we have seen, some rifts develop on the continent in an environment that is also continental, with fluvial or lacustrine deposits that are often confined

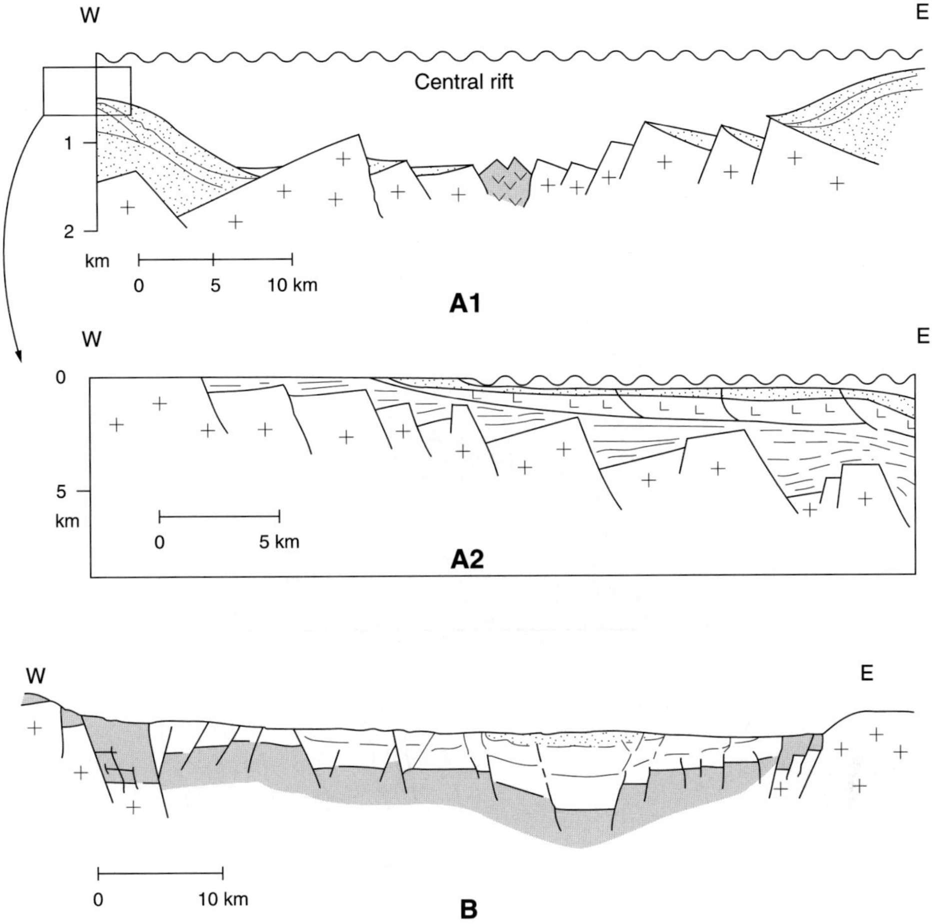

Fig. 2.25 Rift basin subsidence and filling.
A. Red Sea. Rift subsidence, followed by oceanic opening, occurred more rapidly than the sedimentary filling. **A1** shows a vacuity in the central part, while **A2** shows that the margin of tilted blocks is covered by a thick Mio-Plio-Quaternary series of clays, sands, and evaporites. **B.** Alsace. Here, the Rhine trough between the horsts of the Vosges to the west and the Black Forest to the east is completely filled by the (white) Oligocene series and (dotted) Plio-Quaternary series resting on the (gray) faulted Mesozoic (from J. Debelmas and G. Mascle, 1993).

by the narrowness of the initial basin. Other rifts may be invaded immediately by the sea, and the evaporation conditions are then such that large salt deposits develop. As we have said, sedimentation rates may vary greatly depending on the climatic situa-

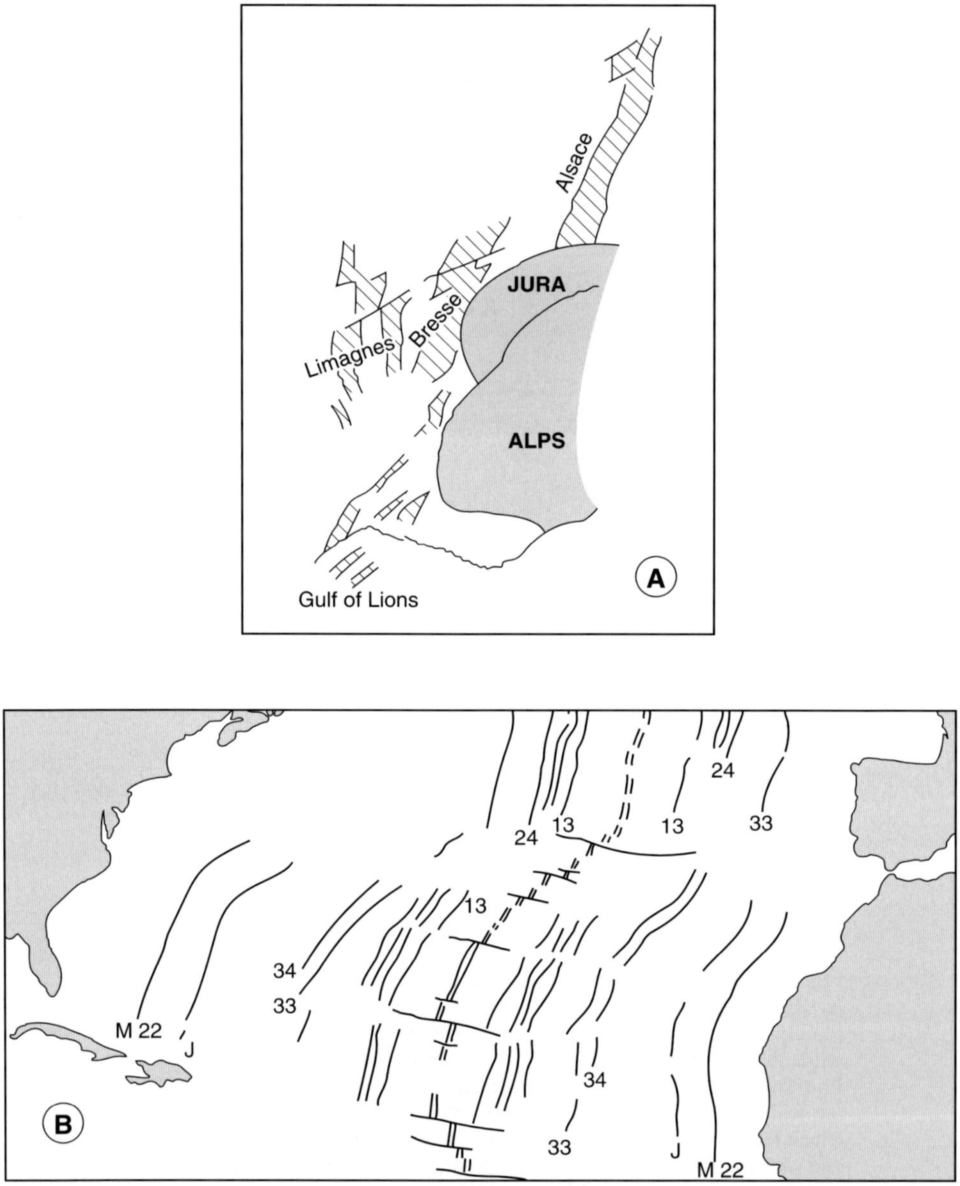

Fig. 2.26 Rift segmentation.
A. The Oligocene troughs of Western Europe, from the Mediterranean to the Rhine. **B.** The Atlantic ridge and transform faults.

tion, inputs, topographical barriers, and other factors. Confinement will often permit the preservation of the organic matter (see Chapters 3 and 6). In the final analysis, rifts and mature continental margins can be filled in many different ways (Fig. 2.27).

- **Variable volcanism**. If we compare the Afar and the East African rifts with the Gulf of Suez, for example, the situation is highly contrasted. The volcanism is abundant in some parts, while elsewhere it is discreet. We have already mentioned that one margin may be very rich in magmatic products (Greenland) while, on the same ocean, another is very lean (Gascony).

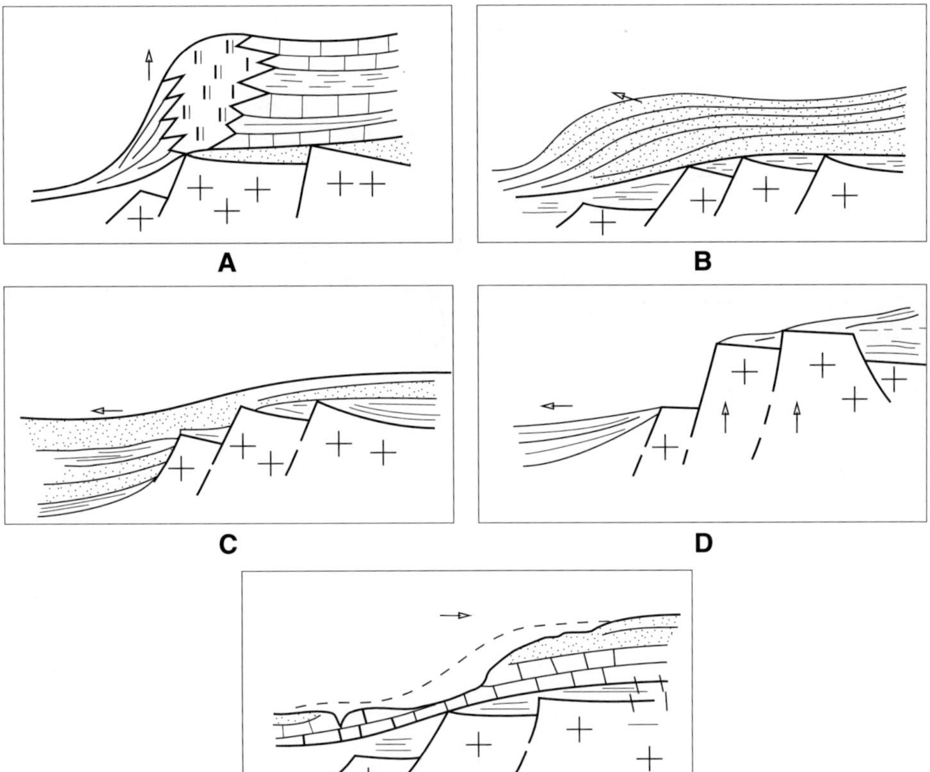

Fig. 2.27 Various sedimentary filling modes on continental margins.
A. Aggradation of a carbonaceous platform edged by a reef. **B.** Seaward progradation of a detritic prism. **C.** The margin is bypassed, leaving no deposit, and sediments accumulate massively at the foot of the margin. **D.** Tectonic uplift and differential trapping. **E.** The margin withdraws by erosion.

Special Cases

- **Pull-apart basins**. Four-sided rhombohedral basins sometimes develop along a major transform fault (Spitzberg and San Andreas faults) or a major slip fracture. These are extension basins prompted by a general longitudinal slip movement. The Jordan trough (Fig. 2.28) is a classic example of this, and ancient examples can be found among the Benoue-type African rifts and certain Hercynian basins of Western Europe.

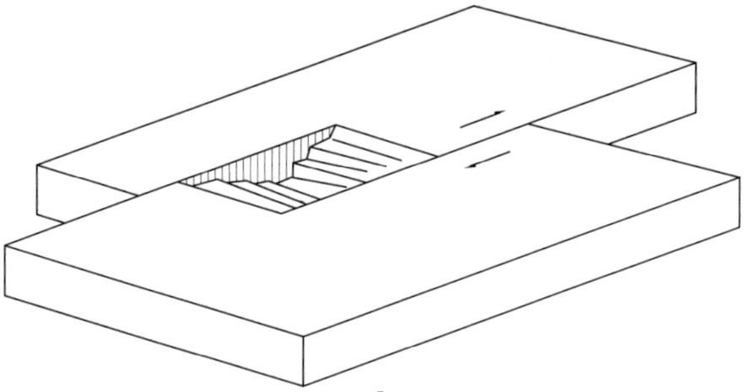

A

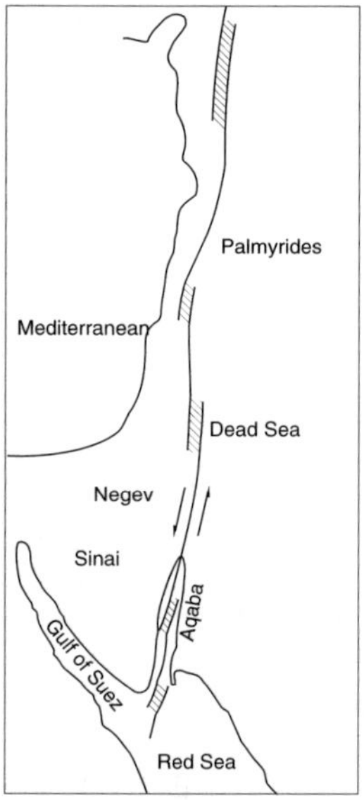

B

Fig. 2.28 Pull-apart basin.
A. Theoretical scheme: The fault edge follows a broken, irregular line.
B. Example of Levant slip system: The Levant fault is discontinuous, and smaller subsident basins of varying depth appear along the line at Aqaba, the Dead Sea, Jordan, the Sea of Galilee, and onwards.

- **Multiphase rifting basins**. As has already been pointed out a number of times, a basin may have a multiphase history, including rifting phases. The Aquitaine basin features one period of rifting with Triassic-Liassic thinning and subsidence, and a second extensive event at the end of the Lower Cretaceous, which thereafter led to the opening of the Bay of Biscayne.

2.4 CRATONIC, CONTINENTAL, AND EPICONTINENTAL BASINS

2.4.1 Craton and Cratonic Basins

Cratonic, continental, and epicontinental basins are vast basins localized at the top of the rigid continental crust, and constitute a large part of the Earth's sedimentary basins. They are also the seat of a large part of our hydrocarbon reserves.

If we consider the oldest (Precambrian, Paleozoic) geological series of the continents, we find **ancient stable platforms** where relatively thin sedimentary series have accumulated without undergoing much deformation.

As Beuf et al. (1971) recalled in their work on cratonic sedimentation, the term ***craton*** *(taken from the Greek kraton for "force") comes from the "kraton" used by Stille (1941) drawing upon Kober (1923), who distinguished "kratogenic" from "orogenic", or folded zones. This term was taken up again by the Russian school and by Kay (1947), and was extended to all interior continental basins. See, for example, the A.A.P.G. memoir No. 51, 1990.*

The Paleozoic basins of the Sahara illustrate the idea of the cratonic basin. The Sahara's sedimentary cover, with outcroppings of the basement in **shields**, is an extensive, fairly thin, platy **platform** with slight deformations in **syneclises** and **anteclises** the contours of which indicate the craton's structural evolution rather than the functional cup shape it had at the time of deposit. Structural basins are smaller than the original sedimentation basin (Figs. 2.9 and 2.29). The present extension of the basins toward the shield reflects a recent erosion limit which should not, of course, be confused with the boundary of the functional basin at the time of deposit. Chapter 3 will show how important sedimentological reconstructions are for defining this idea of a basin's edge. Other examples in the Paris basin and North American craton also illustrate this idea: the fluctuating coastline of the marine basin passed to the east of the Paris meridian in the Early Cretaceous.

Some basins can be found where the basement has settled more at the basin's present uplifted edge than at its center (i.e., the center of the topographical cup that can be seen at the surface today), thereby offsetting the basement cup with that of the free surface. This is even more pronounced in certain foreland basins.

Craton **sizes** and **boundaries** vary with time. In the above example of the Sahara, the Taoudeni basin syneclise (Fig. 2.29) on the West African craton was separated from the Nilotic craton in the Precambrian. After the Pan-African chain was constructed at the end of the Precambrian, the sedimentation area was unified from Senegal to Arabia. Another example is the European post-Hercynian craton where the idea of a craton becomes a relative one. The boundary between cratonic and peri-cratonic basins is blurred both in space (where is the real boundary with the continental margin?) and in time (a basin may evolve from one type to another, as was said above).

Some cratonic basins are closely tied to (and in fact are continuous with) peri-cratonic basins, such as in the Arabian platform illustrated in Fig. 2.30. Others are entirely interior basins. Different terms have been used for the latter: intra-cratonic, interior, continental, epi-continental, or sag basins. We will now take a brief look at the variety of situations with different examples.

2.4.2 Formation Mechanisms

The initial definition of global tectonics assumed that plates are rigid and undeformable, and that the only major deformations to be found in them were at their boundaries. If the continental lithosphere is thus taken to be rigid, how can we explain the formation of cratonic basins?

This question can be answered in a number of ways:

- A cratonic area already inherits the history that led to the construction of its basement and, once stabilized, it is still not inert. The Paleozoic basins of the Sahara again serve to illustrate this Precambrian **inheritance**, especially that of the Pan-African chain, as does the Paris basin developed on a Hercynian substratum which itself was cut up by Permian troughs.
- It is felt today[1] that **intraplate deformations** are not only possible but are far from negligible. They develop in response to major events (e.g., oceanic accretion, subduction, collision, obduction) occurring at each of the plate boundaries, yielding recognizable deformations and motions at great distances from the boundaries.
  For example, the effects of the collision that originated the uplift of the Pyrenees in the course of the Eocene (50 My) had recognizable effects in the form of moderate folds and reverse faults a thousand kilometers to the north, in the North Sea.
- Heat flow, to name just one driving force, can vary in time and space.

Even if we do not always know why a basin has developed in a given region, plate tectonics will not give us much information about the internal mechanisms of continental cratons that generate cups and depressions (basins), nor about the length of time these basins will operate. What should be remembered is that some of the large cratonic basins overlie zones that are inherited from the previous history – especially rifts and active or aborted aulacogens.

The overriding mechanism here is **subsidence**, which was originally defined to describe the collapse of carboniferous basins. If a cup appears, this implies that the initial flat surface

1. See Cloetingh, 1987.

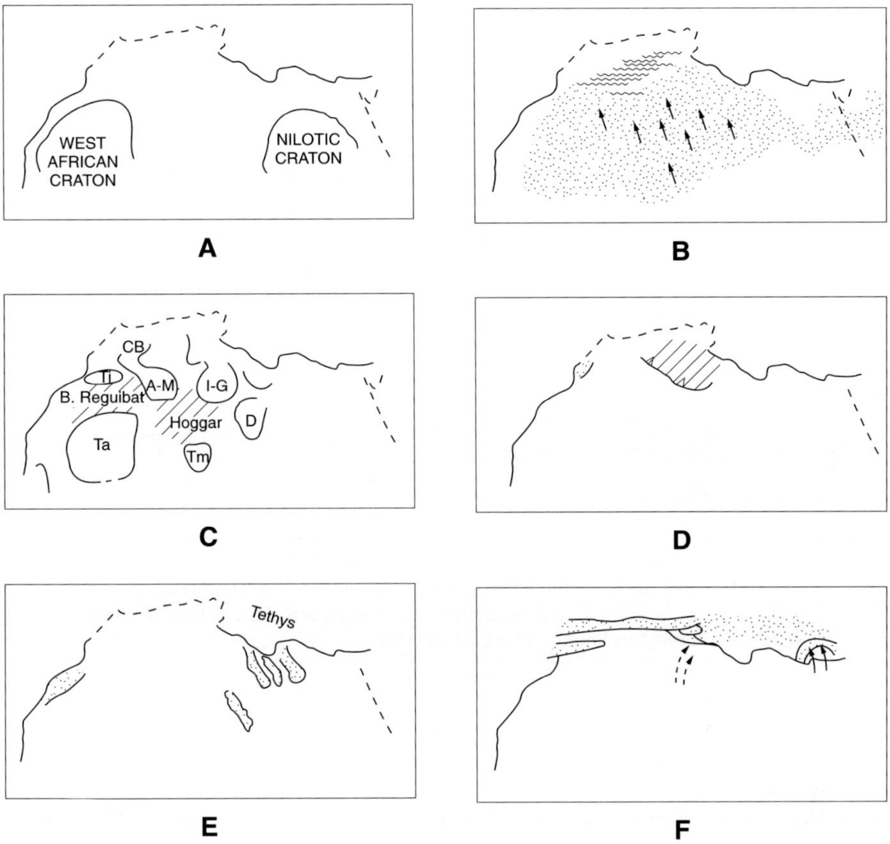

Fig. 2.29 Example of evolution of a cratonic area: North Africa.
A. Situation in the Precambrian: Two cratons separated by the mobile Sahara zone (tectonized, granitized oceanic domain) at the end of the Precambrian. **B.** Ordovician: A vast subsident platform with northward-transiting fluvial deposits. **C.** Structural basins inherited from the Hercynian orogeny. Ta: Taoudeni; Tm: Tamesna; Ti: Tindouf; CB: Colomb-Bechar; A-M: Ahnet-Mouydir; I-G: Illizi-Ghadames; D: Jado. **D.** The Triassic North-Saharan open toward the Tethyan rift system. **E.** Subsident basins of the Late Cretaceous due to African fragmentation. **F.** Peri-Alpine episutural basins of the Cenozoic and the Nile Delta.

has sagged, whence the term "sag basin". As subsidence is defined as a gradual descent of the basin substratum within the craton, this assumes there is an extensive or thermal (cooling) event and a time history from event initiation to senescence.

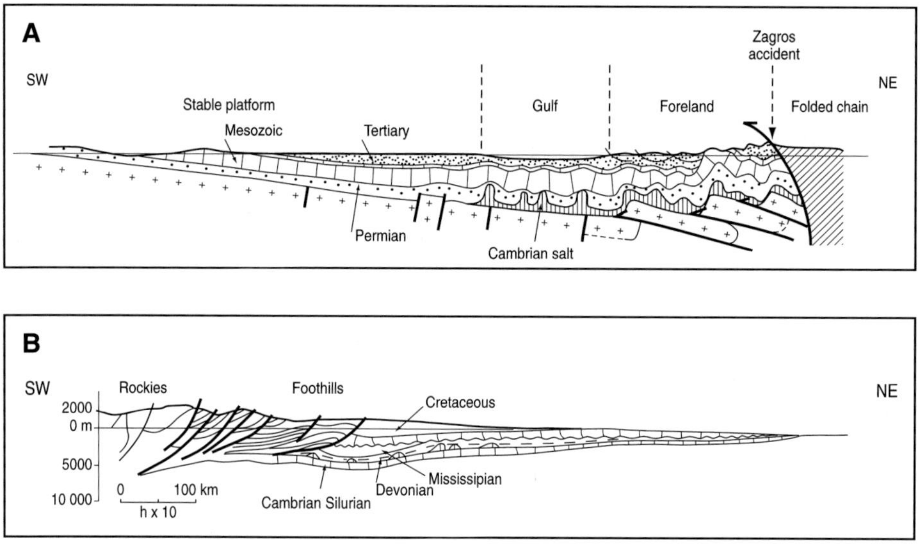

Fig. 2.30 A and **B** Cratonic basins, peri-cratonic foreland, and rifts.
A chain front is often found at the transition between the cratonic domain and the flexured foreland (sections **A** and **B**).

The thinning and the stress deviator in these intracratonic sag basins are probably not great enough to cause any major fracturing, so the basin subsides and the edges lift up with no spectacular trace of surface mechanics. When the heat source is exhausted, thermal contraction causes the subsidence. The active rifting then evolves into a phase where **subsidence** is the mechanical response for maintaining isostatic equilibrium with a **cooling** process.

As we have seen concerning rifts, several **models** have been developed for the initial stages of the basins. Certain cratonic basins did in fact develop on rifts, and these rifting models then apply; but they do not always.

Whatever the case, remember that subsidence is determined by more or less interactive complementary factors, which are the thermal history, the stretching and horizontal extension of the lithosphere, the rise of the upper mantle (or its thickening, inversely), the sedimentary overload, and phase changes in the crust.

While the initial basin creation phase is largely due to lithospheric thinning, which is dominant in rift history, it seems that **lithospheric cooling** is then the major factor for intracratonic basins (note that, if the lithosphere is heated, its volume is increased and may cause an uplift opposed to subsidence).

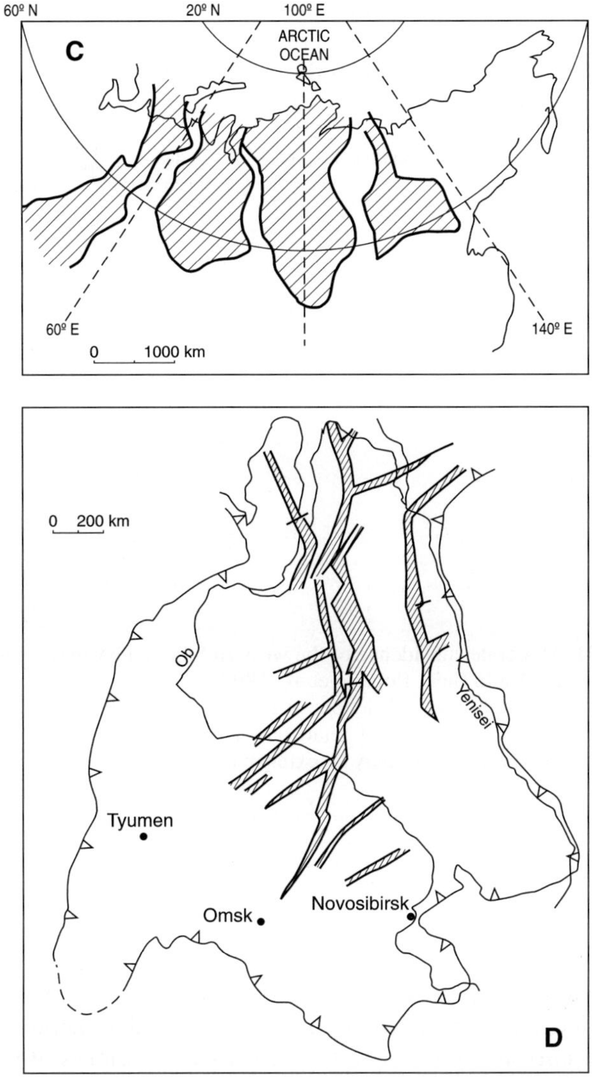

Fig. 2.30 C and **D** Cratonic basins, peri-cratonic foreland, and rifts. But cratonic basins are often superimposed on rift systems (from various authors).

Subsidence curves such as those of Fig. 2.31 can be developed to separate out the isostatic compensation of the overload due to the sedimentary and water column in the basin. To calculate the evolution of the subsidence, then, the paleo-bathymetry has to be determined as precisely as possible for each period, as well as its compaction history and other effects. Generally, it can be said that large cratonic basins are the seat of moderate subsidence compared to the rift basins or those of the continental margin.

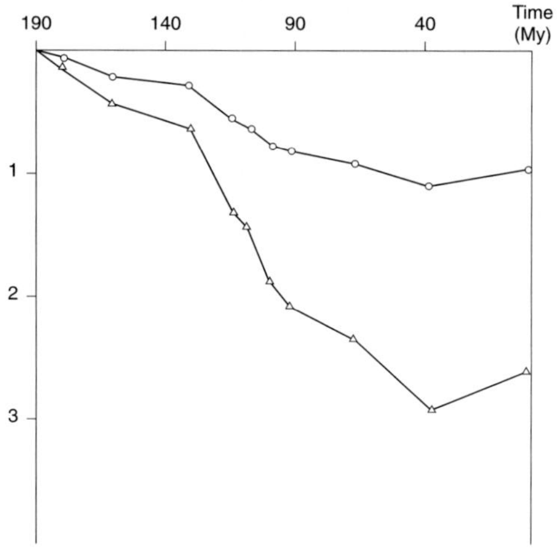

Fig. 2.31 Moderate subsidence of the western Siberian basin (according to data from the Kalin field, Lopatin et al., 1992).
The sedimentation and subsidence rates are low in the course of the Jurassic, and increase in the Cretaceous and Lower Paleogene. Erosion (since the Oligocene, here) may be extensive.

2.4.3 General Features

Examples are given hereafter to illustrate the following general features of intracratonic basins.

Shallow topographical depressions. The functional basins are never deep, so what accumulates in them will be dominated by continental, littoral, or neritic processes:

- **Broad fluvial outwash**, such as in the ancient red sandstones of the Devonian or the Saharan Cambrian-Ordovician (Fig. 2.32A), or
- **Endorheic lacustrine basins** with a peripheral fluvial drainage network, such as that of Chad. The lacustrine cup may vary in time, such as in the Aral Sea or the ancient Chinese basin of Sangliao (Fig. 2.32B), or
- **Broad eolian networks** like the present dunes of the Saharan ergs and Australia, and ancient American examples (Navajo and others) (Fig. 2.32C), or
- **Shallow epeiric seas** as on the China and North Sea plateaus, with ancient examples of interior seas (African Cenomanian, the Cretaceous seas of the American Great Plains) (Fig. 2.32D).

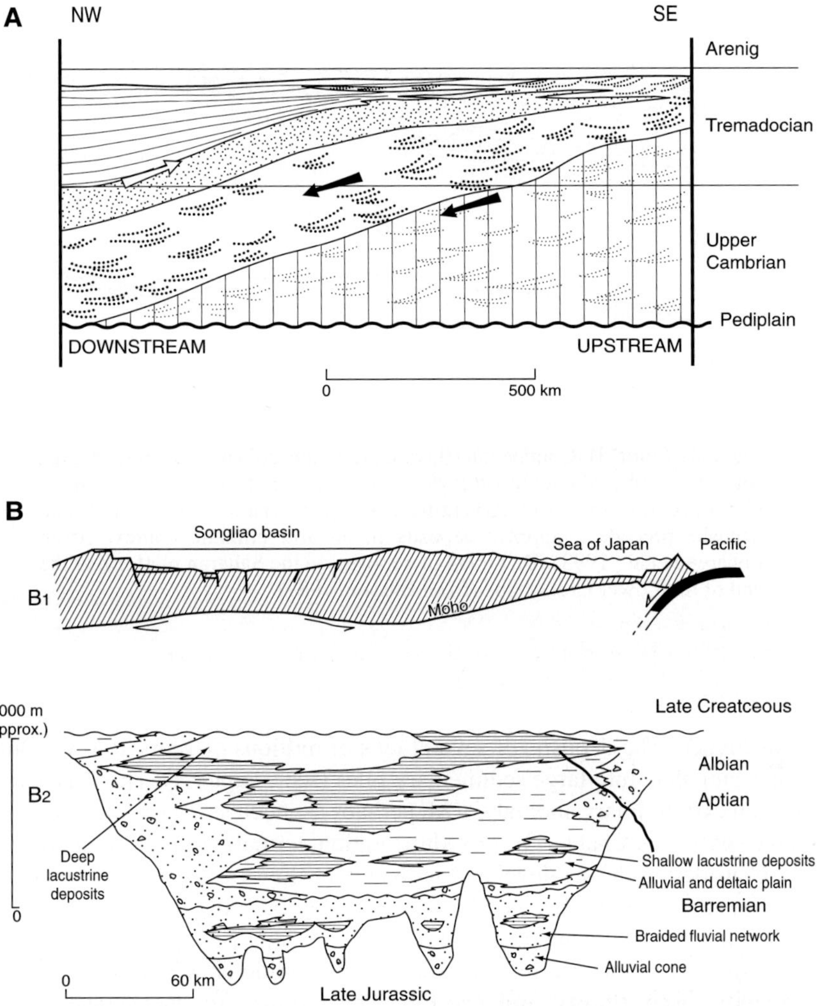

Fig. 2.32 Continental (fluvial, lacustrine, eolian) or shallow marine deposits distributed over broad surfaces: a characteristic of cratonic basins. **A.** Saharan Cambrian pediplain. The fluvial deposits upstream are constantly reworked and redistributed downstream (solid arrows), covering a considerable area. The very gradual marine encroachment (white arrow) is reflected in the onlap of deposits in the direction opposite to that of the fluvial progradation (from Beuf et al., 1971). **B.** China's lacustrine Songliao basin. A vast continental basin formed on an aborted rift behind the Sea of Japan and the subducting Pacific plate (B1). The environments have changed with time: fluvial network, shallow lake, deep lake (from various authors).

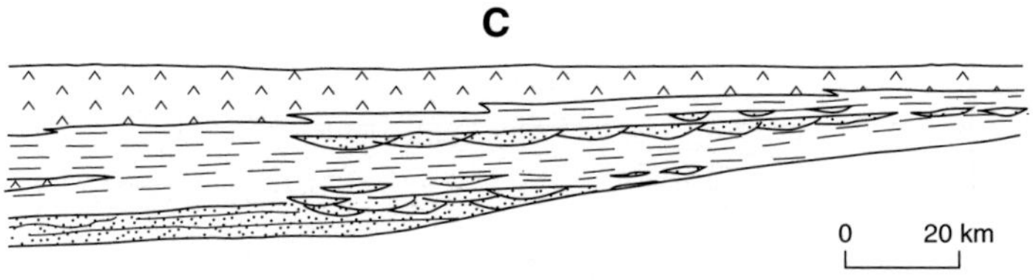

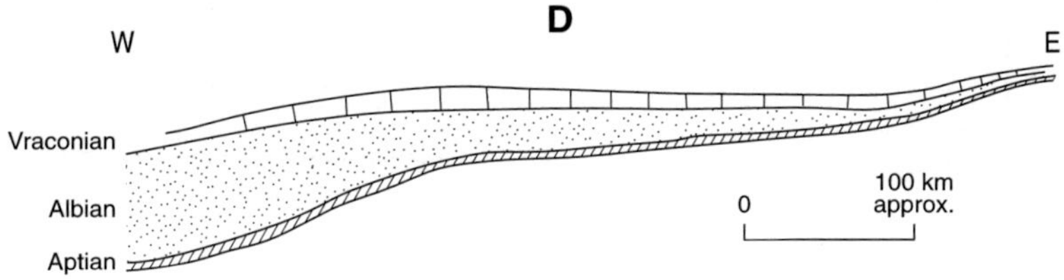

Fig. 2.32 (cont'd) Continental (fluvial, lacustrine, eolian) or shallow marine deposits distributed over broad surfaces: A characteristic of cratonic basins. **C.** Conceptual model of Argentina's Neuquen basin where thick eolian deposits precede evaporitic deposits in an arid climatic context (from Legaretta et al., 1993). **D.** Onlap of deposits on the Saharan platform at the end of the Lower Cretaceous (from Busson, 1989).

Low subsidence. The deposits of several tens of millions of years at low sedimentation rates do not accumulate into large sedimentary piles (see Chapter 4). The very thick Precaspian basin is an extraordinary anomaly, with perhaps as much as 20 kilometers of sediments. The available space (see Chapter 3) for sedimentation normally remains moderate over time (Fig. 2.33). Considering the sea-level and sedimentary flow variations, this entails:

- Many **variations of facies**. The basin will easily bear the trace of coastline fluctuations, with successive encroachments and regressions and a rapid variation of facies lines in space (i.e., the different environmental conditions, with time) (Fig. 2.33).
- **Frequency discontinuity and erosion**. The sedimentary mechanisms that will be analyzed in Chapter 3, such as the physical sedimentary transport factors, will often generate erosion with lateral redistribution of the eroded material. This is the way discontinuities are created, with gaps of different extents observed in the time sequence of deposits (Chapter 4). The sudden change from one environment to another can bring about major changes in the diagenetic processes (Chapter 5).

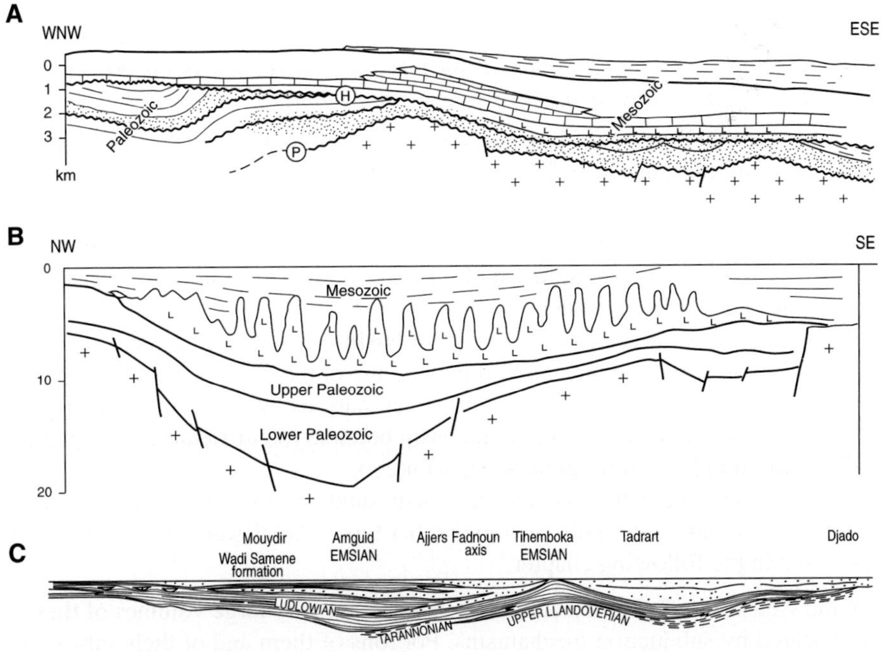

Fig. 2.33 Subsidence, variation of facies, and discontinuities in intracratonic basins.
A. Section of North-Saharan basins: Moderate subsidence for the general Paleozoic and Mesozoic series, separated by the Hercynian discordance (H). The variations of the Paleozoic facies overlying the base discordance on the Precambrian (P) are slow and are not expressed on the section, where many internal discontinuities and discordances are found. **B.** The very strong subsidence of the Precaspian (Paleozoic and Mesozoic) basins, the multiphase history of which can be seen here with several active subsidence zones overlying each other. Note the saliferous deformations. **C.** Slow, gradual variations in facies: Example of the Lower Devonian sandstone on the edge of the Hoggar (central Sahara).

2.5 OCEANIC BASINS

2.5.1 Definitions

In terms of the area they cover on the Earth's surface, ocean basins are the largest basin category: as was said in the previous chapter, salt water seas and oceans cover 70% of the planet's surface, mainly in the southern hemisphere.

In the immersed domain, consisting essentially of the Atlantic, Pacific, and Indian oceans, broad maritime expanses of submerged continental basins have been found, with a lithosphere that is sometimes thinned, such as in the North Sea. Some basins are hypersaline (Mediterranean) while others are hyposaline (Baltic).

Remember here that the **ocean basin** develops atop a thin **ocean crust**. This excludes the many cases of small marginal seas developed behind island arcs. These will be dealt with in the following chapter on active margins.

Recalling the actualist approach to the sedimentary basin concept at the beginning of this chapter (Fig. 2.2), all the **depressions of the major oceans** fall into this classification. A number of points need then be made:

- Oceanic cups are segmented into a set of **sub-basins** separated from each other by ridges.
- There is generally little sedimentary filling because of the distance from the input sources.
- Oceanic basins are downstream of continental margins, with one often running continuously into the other, so the functional basin boundary is difficult to define and in fact varies with time (with migration of depocenters).
- In those zones where the oceanic basement subducts toward an active margin, the basin also evolves continually toward another type—the flexural basin—which will be discussed in the following chapter.

From the viewpoint of global tectonics, it is clear that very large volumes of these basins have disappeared by subduction mechanisms. Portions of them and of their substrata can in some cases be found in mountain chains. This is true of the deep series associated with the ophiolites (Fig. 2.34). These ancient oceanic basins can sometimes be reconstructed on the basis of terrain or paleomagnetic analyses (Fig. 2.35), along with their bathymetric characteristics and current configuration (see Chapter 4).

2.5.2 Formation Mechanisms

The **thermal subsidence** that drives the rifting process and the development of continental margins is also what drives the formation and evolution of oceanic basins.

Without going into the mechanisms themselves again, it can be said that the depth of the broad cups, expressed by their bathymetry, indicates this subsidence: The increase in the depth of the ocean crust roof goes hand in hand with the aging of the lithosphere. The depth varies as a simple function of distance from the active ocean ridgeline, which is a direct relation of the age according to the thermal cooling model mentioned above (Fig. 2.36).

Special anomalies should be pointed out, corresponding to the sometimes considerable effect of intraplate magmatism that was mentioned in the previous chapter. Basins can develop in provinces, including uplifted plateaus, where the filling will be strongly dominated by volcanic input.

And in some cases, the structural cause is dominant. Small faulted basins sometimes develop at a distance from a ridge or parallel to it, or in connection with large transform faults.

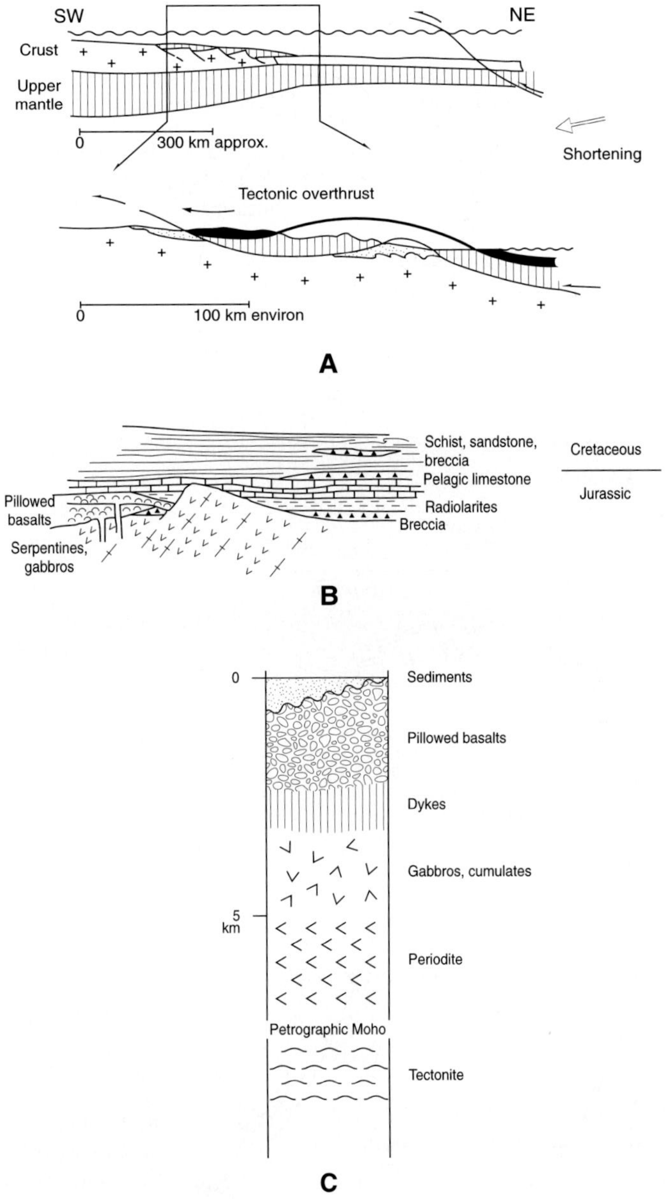

Fig. 2.34 Sedimentary deposits associated with oceanic basalts.
A. Oman example of the tectonic emplacement scheme for deep oceanic series. **B.** Deep oceanic deposits of the Alps before they were tectonically integrated into the chain (from M. Lemoine, 1984). **C.** Theoretical section of an ophiolitic sequence.

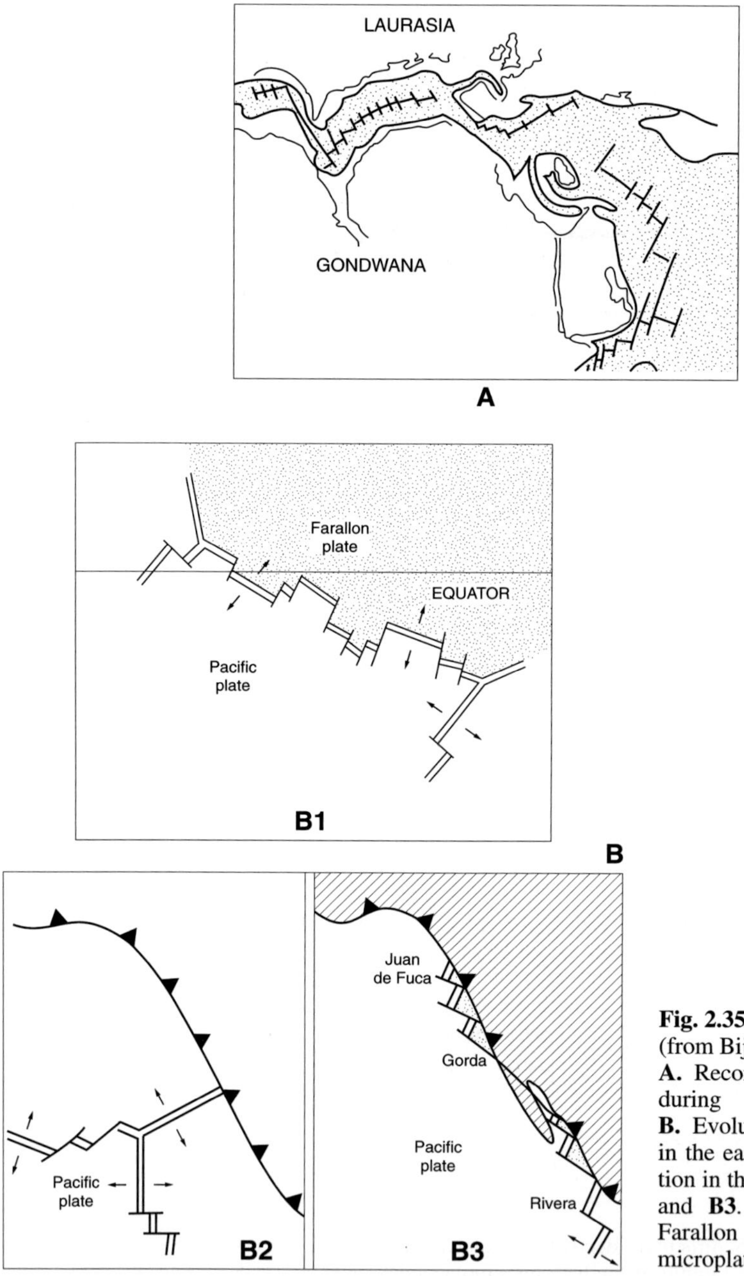

Fig. 2.35 Ancient oceanic basins (from Biju-Duval, 1994).
A. Reconstruction of Tethys during the Upper Jurassic. **B.** Evolution of oceanic domains in the eastern Pacific. **B1**. Situation in the Lower Cretaceous. **B2** and **B3**. Disappearance of the Farallon plate and associated microplates by subduction.

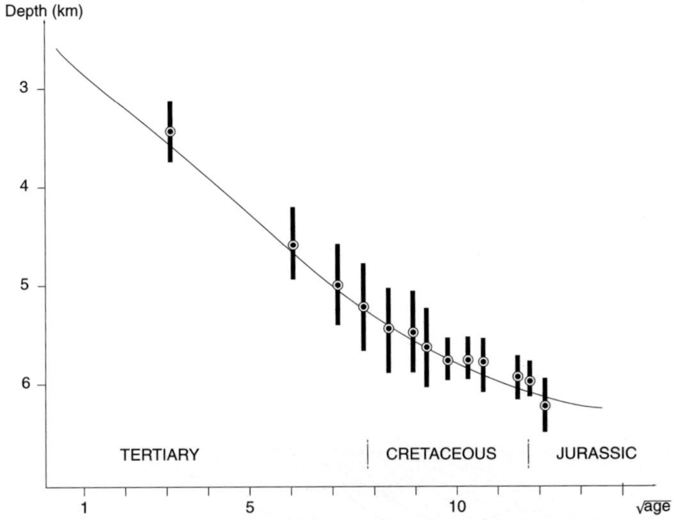

Fig. 2.36 Thermal subsidence of oceanic basins.
The curve represents the gradual settling of the oceanic basalt floor. A basin of Jurassic age is deeper than a Cretaceous or Tertiary basin (from Parsons and Sclater, 1977).

2.5.3 General Features

- **Stability**. The deep oceanic basins on the roof of a lithosphere marked by thermal subsidence are more or less stable and have undergone little deformation.
- **Sedimentation rates** are generally **low** except for the broad turbid outspreads from the margins. Far from the terrigenous flows, basins between the oceanic ridge and continental margins are generally the seat of low sedimentation (pelagic input) and the effect of sedimentary load is then almost negligible. The exception of the broad detritic fans like those of the Amazon, the Niger, and the Ganges, should be noted. This will be discussed in Chapter 3 (Fig. 2.37).
- Oceanic basement of **variable age**. Oceanic basins are characterized by a substratum and fill of differing ages (Fig. 2.38).

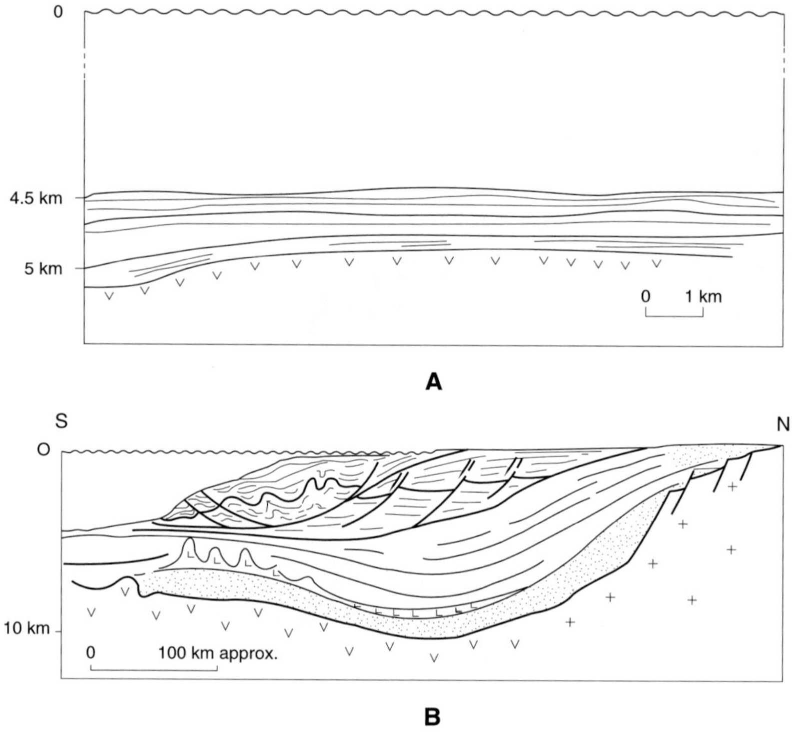

Fig. 2.37 Sedimentation rate in the oceanic domain.
A. The general case, with low sedimentation, condensed series, and sometimes many hiatuses (Atlantic abyssal plain). **B.** Example of the exceptional rich and highly prograding margin (the deep delta of the Niger) with a thick series deposited on an oceanic basement.

2.6 BASINS ASSOCIATED WITH ACTIVE MARGINS AND FOLDED BELTS

2.6.1 Different Types

Basins that occur in regions where plate motion is convergent, with an accompanying subduction and construction of **orogenic** folded mountain chain systems, differ by the subduction process at work, and the type of megasuture[1]:

1. See especially Bally and Snelson, 1980.

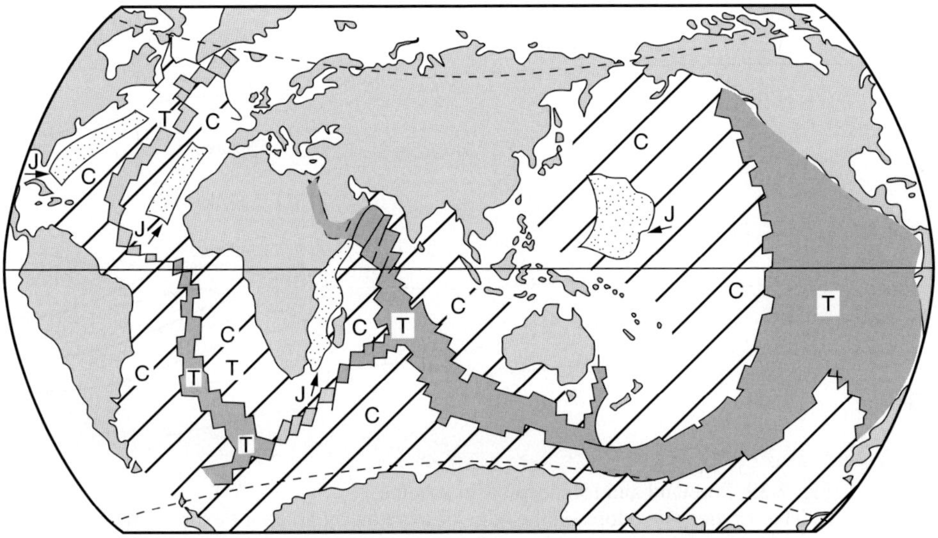

Fig. 2.38 Simplified map of ocean bed ages.
Jurassic (J: fine dots); Cretaceous (C: diagonal hatch marks); Tertiary (T: dark) (from Biju-Duval, 1994).

- **Perisutural basins**. These are the continental foretroughs in which we find the flexural basins lying between the craton and the folded chain. The emphasis here will be on the **foreland** or **foredeep** basins at the continental edges, which are considerable in size and economic importance (Fig. 2.39).
  It should be noted that many of these are the result of a cratonic or continental margin history, as in the case of the American basins on the Rocky Mountain front, or those of Venezuela and Zagros.
- Active margin basins bear direct traces of convergence and are said to be **episutural** in the broad sense, but they do include:
  - Oceanic **trough** basins in the **forearc** zone, **intra-arc** basins, and **backarc basins** in the usual subduction zone arrangement (Fig. 2.40), so that the chances of these basins being conserved in the geological structures are generally slim. Some of these basins in present and ancient configurations are accompanied by major shearing witnessed by the pull-apart type basin as in Anatolia and California.
  - **Episutural basins** in the strict sense, in the folded chains stemming from mountain chain creation mechanisms with more or less developed continental collision, adjusted to the irregular geometry of the plate boundaries (Fig. 2.41). Some **intermontane basins** of the Alpine-Himalayan system, the Cordilleras, and the Rockies (of which we have seen the very special case of continental extension in the Basin and Range). These basins vary greatly in size (Pannonian basin, Chinese basins).

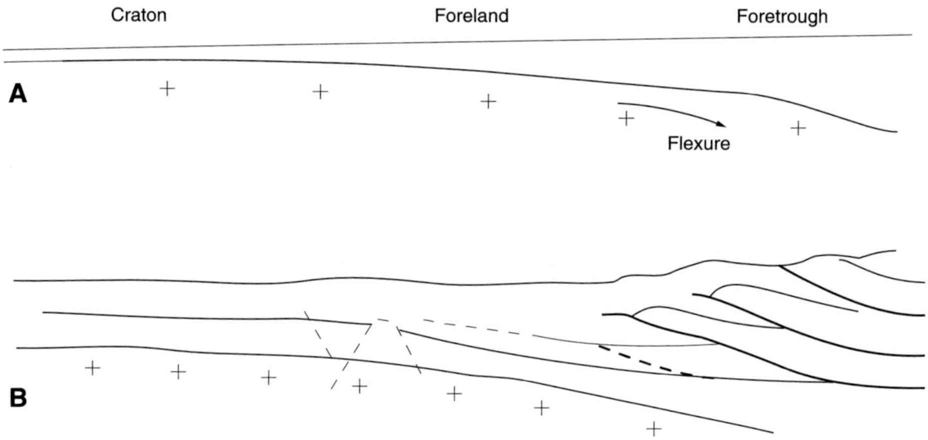

Fig. 2.39 Foreland and foredeep basin scheme.
A. The foredeep develops on the edge of a craton by gradual flexure of the crust. **B.** This folding intensifies with the overload due to deformation. The foreland is the transition zone with intermediate subsidence and accumulation.

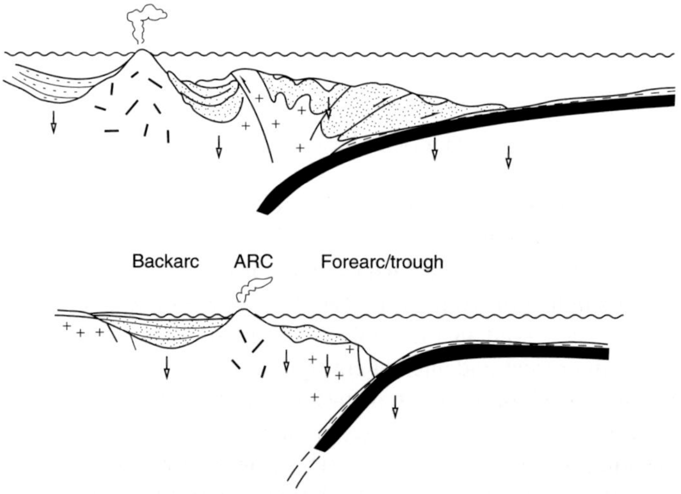

Fig. 2.40 Trough, forearc, intra-arc, and backarc basins stemming from oceanic subduction. The figure shows two extremes of the many possible arrangements.

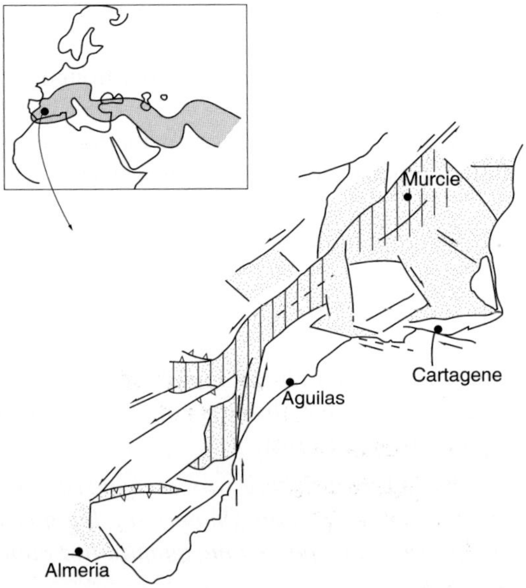

Fig. 2.41 Molassic mountain basins of the Betic Cordilleras (from Montenat et al., 1987) in the Alpine megasuture (inset).
Different basins in extension (dotted) and compression (hatched) can be found along major strike slips.

So the range of sizes, geodynamic situations, and driving mechanisms is very wide and is amplified by the fact that many of these basins possess a multi-phase history with heat flow and stress regimes that vary widely in time.

The case of the sedimentary accretion prisms that sometimes develop in a forearc zone will be dealt with separately as the basin morphology is not the usual one, but rather consists of troughs with highly asymmetrical "filling". The same discussion will bring out the special configuration of piggyback basins of widely varying sizes.

2.6.2 Formation Mechanisms

The prime mechanism forming the great foreland basins is lithospheric bending. The following takes up a few special cases of basins directly related to folded systems, but will make no further mention of the crustal extension mechanisms generating the backarc basins, as these are similar to the basins described under rift and margin formation, even if the underlying causes (especially the heat flow) do differ somewhat.

The formation processes will vary greatly depending on the type of basin, involving a combination of:

- **Flexure** of the broad perisutural basins and oceanic troughs, where the effect of the sedimentary load is large
- **Complex mechanisms** involving magmatic and thermal ascent (arc domain), sedimentary loads (accretion prism), and surficial slip-extension (forearc domain and episutures)
- **Extension** and crustal thinning of the backarc basins.

2.6.2.1 Flexure

Local or regional loading of the crust by additional mass (in the form of water, ice, sediment, or volcanic material) or unloading (as ice melts or sediments erode) rapidly elicits an elastic, or rather visco-elastic, response from the crust.

This idea of adjustment by regional flexure was recognized very early concerning the Scandinavian inland ice in the course of the Quaternary, but it is mainly the work on the great American foreland basins that sparked the quantitative approach to flexural basin evolution during the eighties.[1]

Bending, in foreland basins, is **caused primarily by the load** of folded and transported sediments continually being added to the chain front as it is constructed. The lithosphere, in response to this load, is bent at a certain wavelength that depends on the plate rigidity. This creates a basin bordered by a bulge (Fig. 2.42). This phenomenon may occur on the edge of a continent with a thick, ancient, rigid, and relatively cold lithosphere, such as in the Canadian basins, or on the edge of an ocean with a thin, young, and still-warm lithosphere (such as the peri-Pacific troughs). This rheological response will differ depending on the initial situation. If the crustal response is elastic and the load and plate strength are taken to be invariant, the deformed geometry is acquired immediately after the load is added and the hollow and its bulge will not change in time. But this model does not really square with the geological data of the basins. A visco-elastic model is better, with temperature-dependent visco-elasticity. Even if the load remains the same, the trough deepens and the bulge migrates and rises, prompting more active erosion. The initial elastic response is practically instantaneous, and gradual subsidence then sets in.

It should be noted that flexural basins are not associated with a thermal or internal crustal anomaly, but rather depend on some lateral cause (the crustal origin of which may be more or less remote, as in the case of subduction). The deep thermal regime of these basins is thus generally simpler than that of extension basins, and is not directly related with the subsidence rate. In the elastic model, the subsidence does not change. In the more appropriate visco-elastic model, it does. However, the bending forces and rheological parameters are poorly determined and so, usually, is its evolution.

1. See Beaumont, 1981, and the AAPG memoir of 1992.

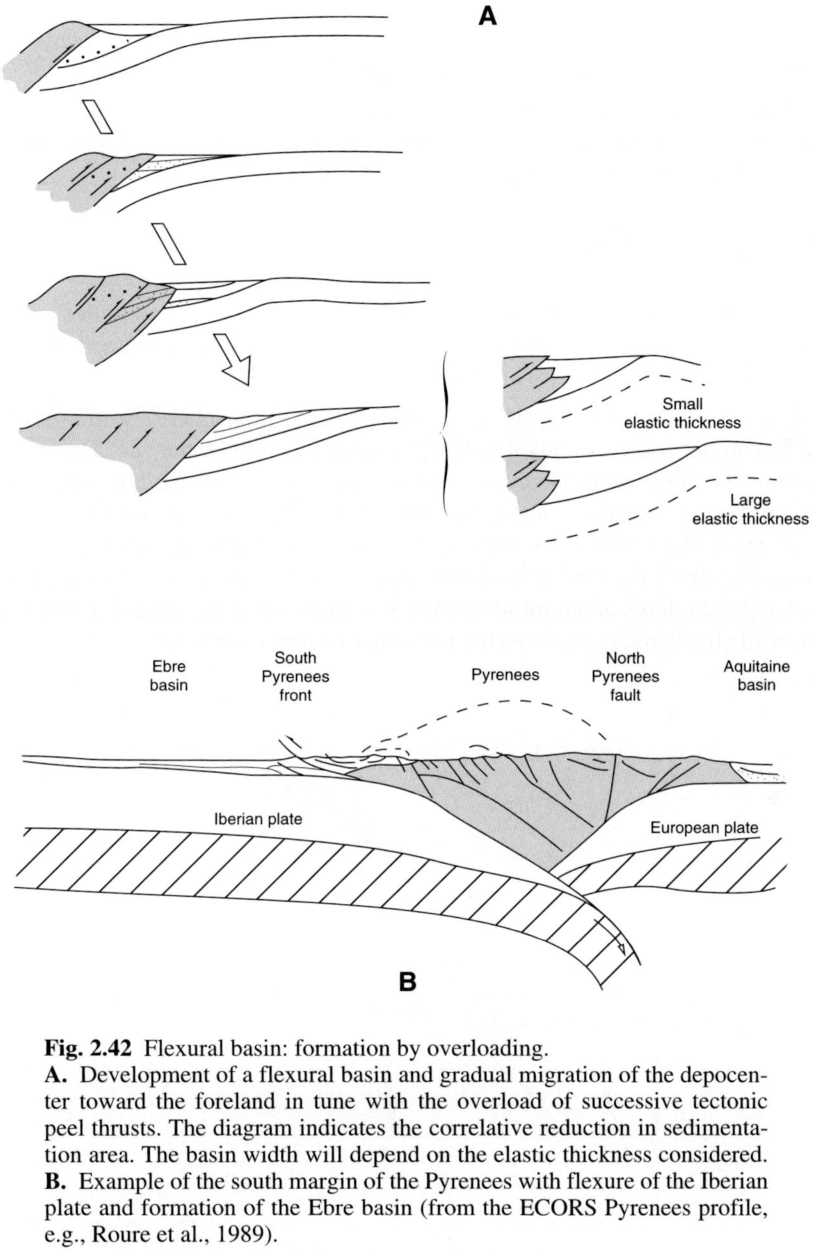

Fig. 2.42 Flexural basin: formation by overloading.
A. Development of a flexural basin and gradual migration of the depocenter toward the foreland in tune with the overload of successive tectonic peel thrusts. The diagram indicates the correlative reduction in sedimentation area. The basin width will depend on the elastic thickness considered.
B. Example of the south margin of the Pyrenees with flexure of the Iberian plate and formation of the Ebre basin (from the ECORS Pyrenees profile, e.g., Roure et al., 1989).

Of course, a model is only a simplification of what can happen naturally. Among other things, the underlying plate is not continuous, the inherited history creates potential mechanical discontinuities, and the thermal evolution depends on the initial situation.

Depending on the type of flexural basin, then, significant differences will appear. The situation in the foreland of the Rockies cannot be confused with the Alpine molasse, the Zagros foreland, or the trough of Japan.

2.6.2.2 Subduction

The major flexural basins are in perisutural position and are associated with a major lithospheric mechanism: subduction. So a closer look should be taken at this geodynamic mechanism which was defined in the previous chapter.

Subduction is by definition a plunging of part of the lithosphere. So the lithosphere is already in flexure regardless of any overloading mechanism there may be. This is confirmed in certain troughs where there is no sedimentary load. Studies of earthquakes and the phenomena occurring at their focus show that the plunge angles of the subduction planes (the "planes" are more like a portion of conic surface, since we are speaking of arcs on a sphere) vary considerably from the surface to depth (Fig. 2.43). The angle is almost always small at the surface, but then dives much more steeply. It is known that this places a part of the crust in tension, while the convergence mechanism causes compression. Gravity may have a large effect too.

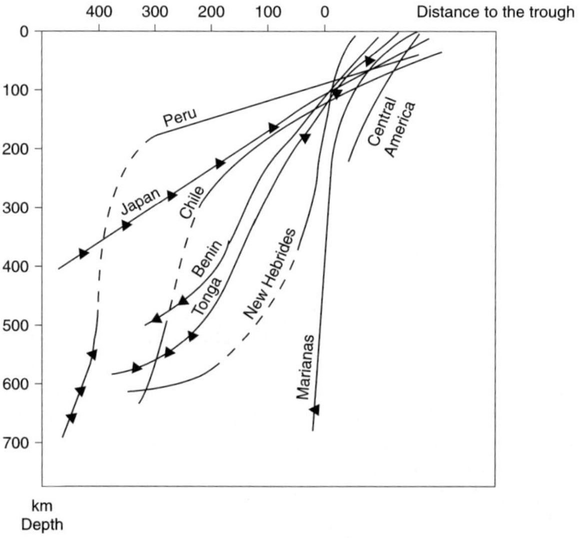

Fig. 2.43 Various subduction plane angles in different regions (from Isaacs and Molnar, 1971, in Biju-Duval, 1994).

There are various subduction configurations. Ampferer (A) subduction is when the continental crust plunges, as opposed to Benioff (B) subduction where it is the ocean crust that subducts.[1] Among the examples of B subduction (Fig. 1.22) we find the following types:

- **Intra-oceanic** subduction, as in the example of the Marianas trench, and subduction with **insular arc** separated from the continent by a string of volcanic islands, as in the case of Japan and Indonesia.
- **Andes-type subduction** directly under the continent, with the examples of the American Cordilleras and the east Pacific.

There are many other special cases that differ by their stage of subduction development and maturity. Taiwan is one of the best-known examples of a continental margin that plunges under the ocean, and the Alps and Himalayas are two colliding continental crusts with all of the initial ocean space gone.

Obduction is when the initial ocean floor is raised, contrary to the normal model of subduction where it is drawn toward the asthenosphere.

So, as the crust plunges and carries its deposit of sediments with it, the subduction process normally causes flexure in a direction concurrent with the plunge. The complex thermal and mechanical transfers that enter into this process boggle the imagination.

The previous chapter pointed out two types of margins: those where major **sedimentary accretion** occurs with a maximum loading effect on the edge of the overlapping plate (the sediments that are not entrained in the subduction then undergo tectonic imbrication at the margin foot), and those where **crustal erosion** occurs and the sediments themselves are actually subducted (Fig. 1.23). This important point will come up again in the discussion of **accretion prisms**.

A number of examples are also known in the general framework of active margins, where reverse polarity occurs behind the chain or in the backarc position, if the flexure polarity is in the direction of the subduction. This can be found in the Alaska's Arctic Slope basin and in the Andes basins.

It is also very important to remember that the flexure may gradually migrate, displacing the **depocenter** (Fig. 2.42).

2.6.2.3 Arc Basin Development

Subduction is the result of plate convergence. It operates at variable rates and angles, but essentially gives rise to a narrow, deep depression, very high seismicity, intense calc-alkalic type volcanism, and major deformations of the sediments subject to complex tectonic processing. Basins are generated by complex mechanisms in different portions of the active margin (Fig. 2.44).

---

1. See Bally and Snelson, 1980.

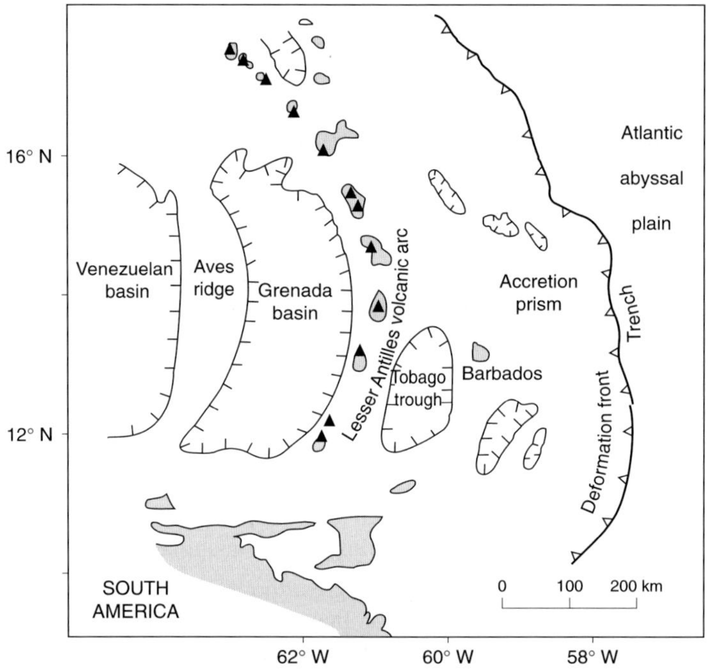

Fig. 2.44 Arc basins in the Lesser Antilles zone.
From east to west: the Atlantic abyssal plain and the trench (partly filled by the Barbados accretion wedge); the forearc basins, including the Tobago trough; the small intra-arc basin at the north of the volcanic string (where the arc splits); the Grenada basin in backarc position; then, beyond the Aves ridge, the Venezuelan basin (see section in Fig. 2.47).

The **deep** or **oceanic trench** lies in line with the oceanic crustal flexure. Contrary to the usual pericratonic flexural basins, it is very deep, ranging from 4.5 to 11 km. This depth clearly reflects the major folding that has occurred in the ocean floor, the average depth of which is about 3.8 km. This type of basin with its pronounced bathymetry is generally narrow and highly asymmetrical. Its inner edge (i.e., towards the arc or continent) is steeper than the outer slope, and follows an irregular topographical pattern (Fig. 2.45). If the input is large, the trench will be the seat of thick sedimentation. The outer bulge sometimes bears traces of extension on its extrados (see Chapter 5), in the form of small horsts and grabens.

Sometimes, in addition to the general flexuring of the trench as it is subducted, the sediments undergo **tectonic** deformation due to the sedimentary overload. Beyond the sharp border of the **deformation front** of the overlapping plate, we come to the other part of the basin where the morphology is highly varied (Fig. 2.46). In the present geodynamic situation, this type of basin is well known. But except for the very special case where the subduction has

A

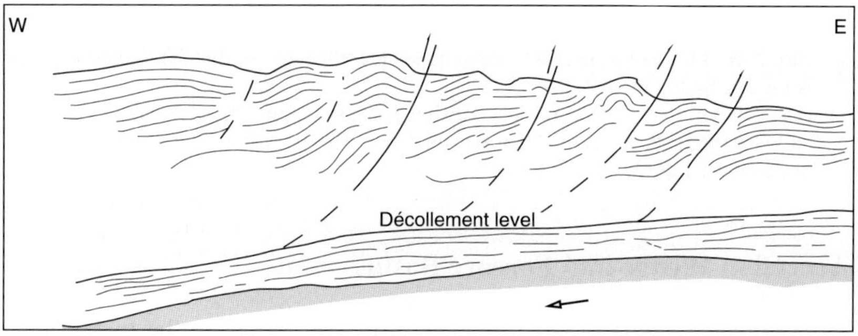

B

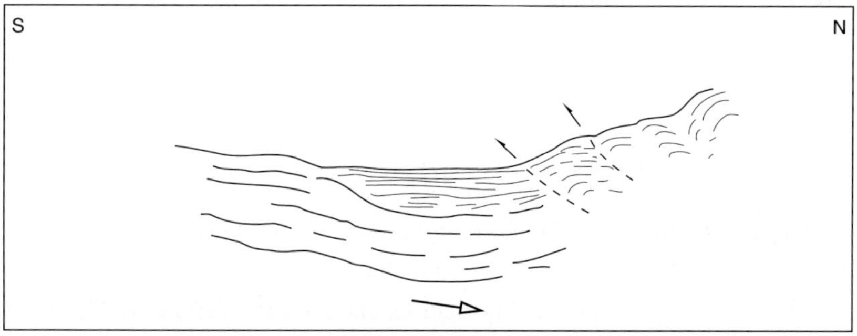

C

Fig. 2.45 Trench basins ahead of the arcs (schematic sections based on seismic profiles.
A. Chile trench. **B.** Nankai trench (Japan). **C.** Los Muertos trench (South Hispaniola).

stopped and the configuration has remained about the same, ancient basins of this type are not easy to recognize. They have been incorporated into mountain chains, raised high, and more or less eroded away.

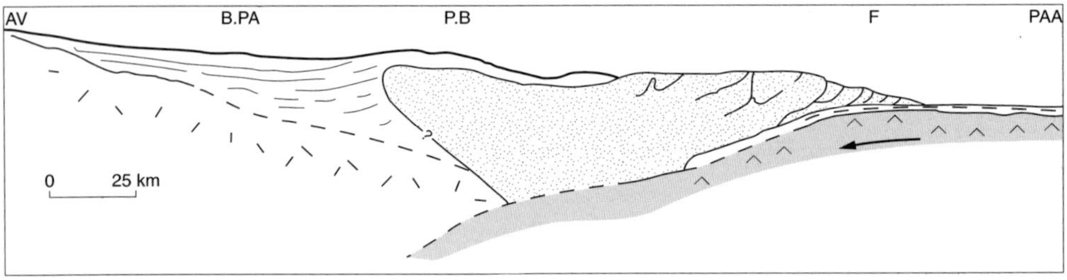

Fig. 2.46 Deformation front ahead of an accretionary wedge. The example is the Barbados wedge.
AV: volcanic arc; BPA: Lesser Antilles forearc basin; PB: tectonized Barbados wedge; F: present deformation front; PAA: Atlantic abyssal plain. See map of Fig. 2.44.

2.6.2.4 Accretion Wedge and Forearc Basins

Taking the well-investigated example of the Antilles arc between the trench and volcanic arc, there is a zone of variable width ranging up to several hundred kilometers, in which the following features can be found:

- An **accretion wedge** or prism has gradually formed on one edge of the trench where part of the sediment has escaped subduction, along a **decollement level** where the subducted and accreted parts are mechanically decoupled. It is thought that this is an initial phase in the formation of a certain number of mountain chains, and such systems are recognized at the front of folded belts rimming large flexural basins, such as the Makran in southern Iran. The size of the prism depends, of course, on the amount of sedimentary input.
- Small **forearc basins** separated by high points can develop on and behind the wedge as it develops, guided by complex tectonic mechanisms.

These basins are created by prism deformation, and are then tectonized. Different types of deformation (Fig. 2.47) can thus be seen in them, and the effects of undercompacted series are at times spectacular (mud lumps) (more details in Chapter 5). Some may be highly subsident while others exhibit an erosional vacuity. As the accretion prism slowly migrates toward the trench by propagation of the detachment, **piggyback basins** appear (Fig. 2.48). This piggyback idea was defined for certain special cases of basins overlying gliding nappes of the Alpine orogeny. The mechanisms forming accretion prisms and the very surficial basins overlying them are always largely dominated by the effect of **fluids**. These basins are always found in the supracrustal part where slip, extension, and compression are in complex interplay.

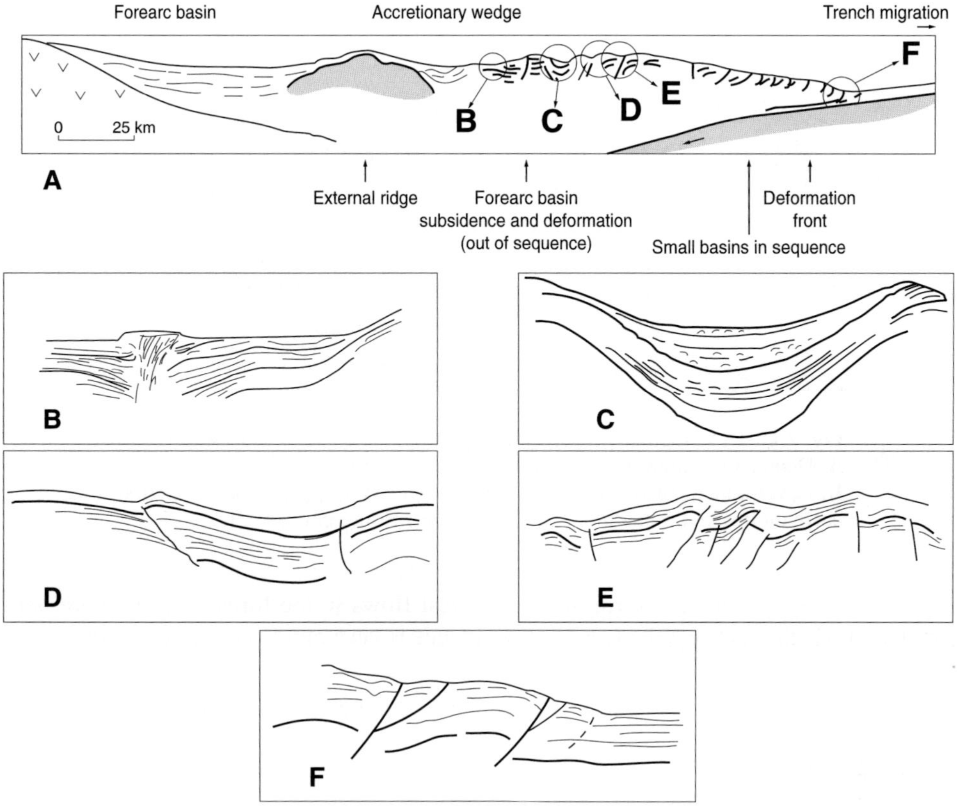

Fig. 2.47 Examples of deformations on an accretionary wedge.
A. General section of the Barbados prism. **B.** Diapirism. **C.** Subsidence, slumping, and thickening. **D.** and **E.** Folds and reverse faults. **F.** Deformation front (from Biju-Duval, 1982).

As the main **forearc basin** is generally "sitting" between the arc and trench, its subsidence is the result of interplay between the plunging subduction basin, magmatic arc construction and cooling, and accretion prism deformation. The example of the Lesser Antilles basin (Fig. 2.46) illustrates this competition of processes well. If the subduction zone is stable in time, the basin can operate for millions of years, contrary to the small forearc or piggy-back basins, which are much more temporary.

If the volcanic arc migrates in time and space, as it does in the case of the Lesser Antilles, then **inter-arc basins** will form. These are rather similar to the **intra-arc basins** that develop between the high portions of volcanic constructs (Fig. 2.44). This type of basin of limited extent can also be found in the Cordilleras. It is clear that a very special thermal regime must be at work, but no simple model is available to explain the subsidence and

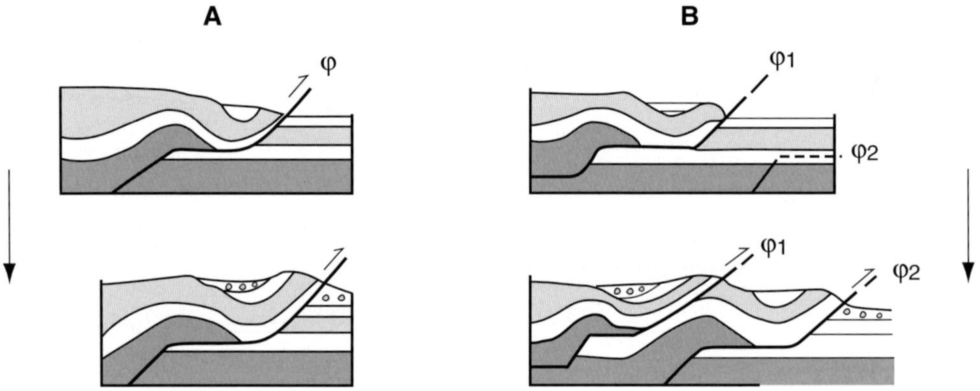

Fig. 2.48 Development of piggyback basins.
A. Depocenter migration and tilting of the onlapping beds.
B. Development of a new basin in sequence with the tilt while the first basin on the first onlap tilts and rotates (from Roure et al., 1991).

deformation mechanisms that form them. The heat flows in the forearc position are below normal, and it is thought that the tectonic subsidence is very rapid, accelerated, and comes in fits and starts.

When the active margin's history is long and folded systems appear, the basins fall into the episutural category.

2.6.2.5 Backarc Extension

The depression behind the arc, often called a **marginal** or **backarc basin**, is one of the characteristic structures of subduction zones (Fig. 2.44).

Examples around the edge of the Pacific (Sea of Japan, Okinawa basin, Indonesian basins) illustrate the variety of situations that can occur behind many island strings, and the Mediterranean example (western or Ligurian-Algerian basin, Tyrrhenian sea, Aegean Sea, Pannonian basin) also show considerable differences marked by their positions in the collision between the African and Eurasian plates (Fig. 2.49). The backarc basins of the Cordilleras exhibit still other differences where extension and flexure often go hand-in-hand.

The elementary driving mechanisms of the backarc basin are the thermal anomaly and extension, with crustal thinning. After a period of rifting, this can lead to the formation of an oceanic expansion line with the development of tectonic and thermal subsidence, as was seen before with rift-type basins.

It is thought that a marginal basin develops 20 to 40 My after subduction begins, depending on the plate convergence rate (the arc's volcanism is initiated when the subducted lithosphere reaches a depth of at least 100 to 150 km).

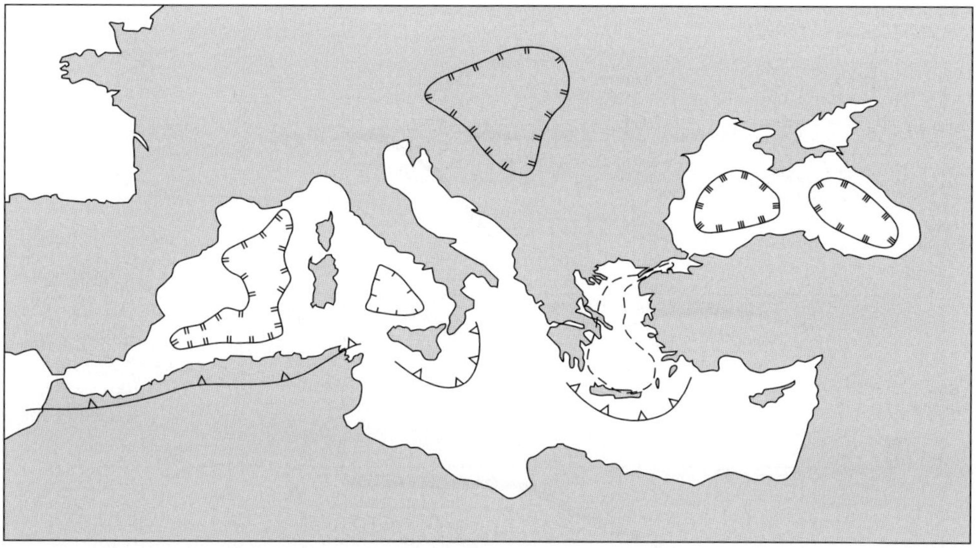

Fig. 2.49 Mediterranean marginal basins.
Small basins have appeared in thin or oceanic crust in backarc position at different periods along the subductions and collisions of the Alpine arc: Black Sea (Cretaceous-Eocene); western Mediterranean (Oligo-Miocene); Pannonian basin, very simplified here (Miocene); Tyrrhenian sea (Late Miocene); and the currently forming Aegean basin.

Backarc basins vary in size. Some (such as the Fiji basin) continue to evolve into oceanic systems proper. Others, at the plate boundary, are very special. The rifting geometry is rarely simple. It is segmented, of course, but also several rifting centers can at times be found, sometimes lying in parallel. Arc jumps may exist, with the development of a marginal system on an ancient arc (such as the Antilles).

A number of processes can be seen at work:

- A marginal basin opened in the Oligo-Miocene in the Gulf of Lions-Ligurian basin, combined with the creation of the continental rift system of Europe and the west (Fig. 2.50).
  Here, we see both the crustal thinning and extension typical of an extension basin, along with surficial extension (thin skin tectonic) superimposed on the ancient structures of the Cevennes margin.[1]
- In the South China Sea, the oceanic opening behind the insular arcs of the Philippines and Borneo is bounded by the large strike-slip faults of the Indochinese peninsula arising from the punching of India against Eurasia (Fig. 2.51).

1. E.g., see Vially and Trémolières, 1995.

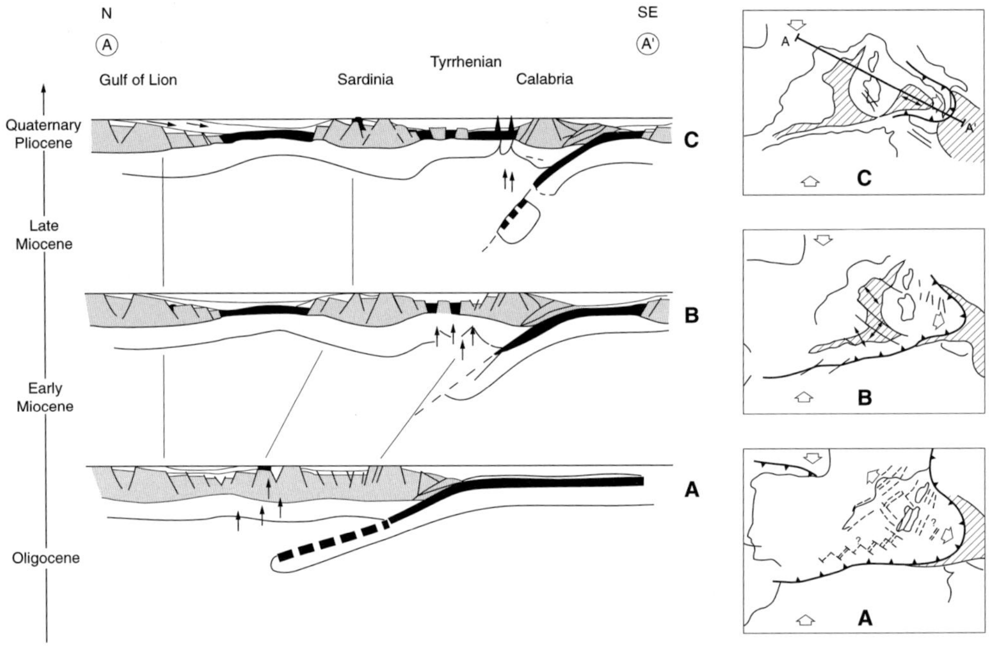

Fig. 2.50 Western Mediterranean basins.
This oceanic crustal basin formed behind the arc of the Appenines and Tell during the Oligocene extension and rifting. Originally (**A**), it was part of the west European rift system stemming from the Alpine collision. Once it finished opening (**B**), the continental collision was blocked and only the part facing the Ionian Sea could still open in extension (**C**). This was the initiation of the Tyrrhenian Sea behind the Calabria arc, which is still active today.

- China's Songliao basin lay on thin crust during the Mesozoic in backarc position to the west of the Pacific subduction, but probably stopped functioning in the Tertiary even though the heat flow was still high.
- The Pannonian basin in the heart of the Carpathians is another example of a large basin associated with a mountain chain. The elementary basins ranging from that of Vienna to that of Transylvania (Fig. 2.49) exhibit a surficial extension that is directed by the deep structure of the Alpine megasuture with strike slips breaking the basins into sub-basins (of the pull-apart type, which can be linked to the intermontane episutural basins).
- Some large backarc basins have, after an extension phase, behaved like flexural basins because of the sedimentary overload caused by backthrusting tectonics.

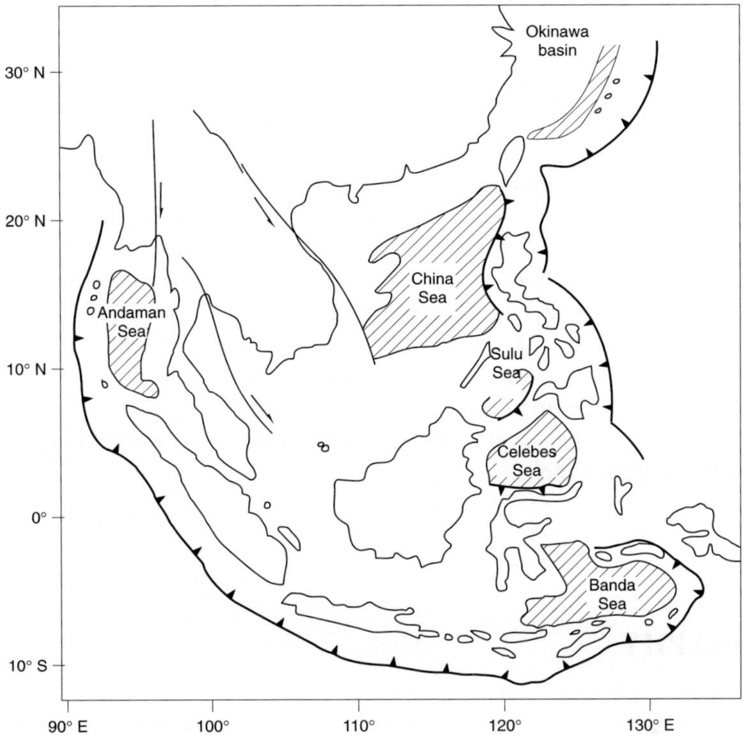

Fig. 2.51 Marginal basins of Southeast Asia: a mosaic of basins in a complex structural framework with a combination of subduction, continental collision, and strike slip stemming from continental punching.

2.6.2.6 Strike Slip and Episutural and Intermontane Basins

There are many examples of episutural basins organized along strike slip faults in mountain chains, in a variety of sizes, depths, and fill. The exact mechanisms controlling their evolution are often very poorly known, and no simple model has been proposed for them as has been done for the other great basins. A number of **pull-apart basins** along major strike slip lineaments have been grouped under this heading, such as the Neogene basins of the Betic Cordilleras (Fig. 2.41), which are narrow intermontane basins. It may also be true of Venezuela's offshore basins along the Oca/El Pilar system.

2.6.3 General Features of Active Margin Basins

- These are often deep **topographical depressions** with sharp relief features and a sedimentary record of tectonics in the course of development. The functional basin is characterized by sharp slopes and major level changes with frequent erosion and surficial gliding. Deposit variations are rapid.
- They are **elongated** and generally **narrow**, while those exceptional marginal basins that occupy broad areas are the ones of greater interest for petroleum.
- **Rapid**, irregular **subsidence**, not always compensated by a major flow of sediments, and often countered by the uplifting of the arc.
- **Vigorous deformation**. The basin lifetime is relatively short and it normally evolves toward a folded chain, except for the episutural basins.
- The features of **perisutural basins** match those of the epicratonic basins they often derive from, with a high level of homogeneity in the facies except in the proximity of the maximum uplift zone.

BIBLIOGRAPHY

✦ Association des sédimentologistes français (1989) Dynamique et méthodes d'étude des bassins sédimentaires. Éditions Technip, Paris.

✦ Bally AW, Snelson S (1980) Realms of subsidence. *In* : Facts and principles of world petroleum occurrence (Miall D, Ed). Canadian society of petroleum geologists, Calgary, pp 9-94.

✦ Beaumont C (1981) Foreland basins and fold belts. The geophysical journal of the Royal astronomical society 65, pp 291-329.

✦ Beuf S, Biju-Duval B, de Charpal O et al. (1971) Les grès du Paléozoique inférieur au Sahara : sédimentation et discontinuités. Évolution structurale d'un craton. Éditions Technip, Paris.

✧ Biju-Duval B (1994) Océanologie. Dunod, Géosciences, Paris.

✧ Biju-Duval B, Mascle A, Le Quellec P et al. (1982) Multibeam bathymetric survey and high resolution seismic investigations on the Barbados ridge complex (eastern Carribean): a key to the knowledge and interpretation of an accretionary wedge. Tectonophysics 86, 1/3, pp 275-304.

✦ Boillot G, Montadert L, Lemoine M et al. (1984) Les marges continentales actuelles et fossiles autour de la France. Masson, Paris.

✦ Bois C, Bouché P, Pelet R (1982) Global geologic history and distribution of hydrocarbon reserves. AAPG bulletin 66, 9, pp 1248-1270.

✧ Brunet MF, Beaudoin B, Fries G et al. (1985) Évolution de la subsidence dans les séries argilo-carbonatées. Journée sur les argiles. École des Mines de Paris, Paris.

✦ Burrus J (1989) Review of geodynamic models for extensional basins; the paradox of stretching in the Gulf of Lion (Northwest Mediterranean). Bulletin de la Société géologique de France 5, 2, pp 377-393.

✧ Busson G, Cornée A (1989) Données sur les paléoclimats déduites de la sédimentation continentale du Mésozoïque saharien. CIFEG, Publication occasionnelle 1989/18, Paris.

✦ Cloetingh S (1987) Intraplate stresses: a new element in basin analysis. Kleinspehn KL, Paola C. New perspectives in basin analysis. Springer, New York, pp 205-230.

✦ Debelmas J, Mascle A (1993) Les grandes structures géologiques. Masson, Enseignement des sciences de la terre, Paris.

✧ Durand B, Masle A (1995) Integrated basin studies. Final report. Document GERTH. Unpublished.

✦ Einsele G (1992) Sedimentary basins: evolution, facies and sediment budget. Springer, Berlin.

✧ Faure JL (1990) Failles normales, coupes et subsidence dans les bassins en extention. Le bassin Viking (mer du Nord) et le domaine briançonnais (Alpes occidentales) au Jurassique. Thèse de doctorat.

✧ Heirtzler JR, Bolli HM, Davies TA et al. (1977) Indian ocean geology and biostratigraphy. Washington, American geophysical union. Studies following deep-sea drilling legs 22-29.

✧ Jarrige JJ (1992) Variation of extensional fault geometry related to detachment surfaces within sedimentary sequences and basement. Tectonophysics 215, pp 161-166.

✦ Jarvis GT, McKenzie D (1980) Sedimentary basin formation with finite extension rates. Earth and planetary science letters 48, pp 42-52.

✦ Kay M (1947) Geosynclinal nomenclature and the craton. AAPG bulletin 31, pp 1289-1293.

✦ Kober L (1923) Bau und entsehung des Alpen. Bornträeger, Berlin.

✧ Legaretta L, Uliana MA, Larotonda CA et al. (1993) Approaches to nonmarine sequence stratigraphy-theoretical models and examples from Argentine basins. Eschard R, Doligez B. Subsurface reservoir characterization from outcrop observations. Éditions Technip, Paris, pp 125-144.

✦ Leighton MW, Kolata DR, Oltz DF et al. (1990) Interior cratonic basins. American association of petroleum geophysicists, AAPG memoir 51, Tulsa.

✧ Lopatin NV, Galushkin YI, Makhous M (1996) Evolution of sedimentary basins and petroleum formation. Roure F, Ellouz N, Shein SS et al. Geodynamic evolution of sedimentary basins: proceedings of the international symposium, May 18-23, 1992. Moscow, Russia. Éditions Technip, Paris, pp 435-453.

✦ MacQueen RW, Leckie DA (1992) Foreland basins and fold belts. American association of petroleum geophysicists. AAPG memoir 55, Tulsa.

✦ McKenzie D (1978) Some remarks on the development of sedimentary basins. Earth and planetary science letters 40, pp 25-32.

✦ Miall AD (1984) Principles of sedimentary basin analysis. Springer, New York.

✧ Montenat C, Ott d'Estevou P, Larouzière (de) FD et al. (1987) Originalité géodynamique des bassis néogènes du domaine Bétique oriental (Espagne). Berthon JL, Burollet PF, Legrand P. Genèse et évolution des bassins sédimentaires. Total, Paris, Notes et mémoires 21, pp 11-49.

✧ Parsons B, Sclater JG (1977) An analysis of the variations of ocean floor bathymetry and heat flow with age. Journal of geophysical research 82, pp 803-827.

✦ Perrodon A (1988) Bassins sédimentaires, provinces pétrolières et tectoniques globales. Bulletin des centres de recherches exploration production Elf Aquitaine 12, 2, pp 493-512.

✧ Roure F, Casero P, Vially R (1991) Growth processes and melange formation in the southern Apenines accretionary wedge. Earth and planetary science letter 102, pp 395-412.

✧ Roure F, Choukroune P, Berastegui X et al. (1989) ECORS deep seismic data and balanced cross sections: geometric constraints on the evolution of the Pyrenees. Tectonics 8, 1, pp 41-50.

✦ Royden L, Keen CE (1980) Rifting process and thermal evolution of the continental margin of eastern Canada determined from subsidence curves. Earth and planetary science letters 51, pp 342-361.

✦ Stille H (1941) Einführung in den Bou Amerikas. Bornträeger, Berlin.

✧ Vially R, Trémolières P (1996) Geodynamics in the Gulf of Lion: implication for the petroleum exploration. Ziegler P, Horvath F. Peri-Tethys memoir 2: Structure and prospects of Alpine basin and foreland. Museum d'histoire naturelle, Paris.

✦ Watkins JS, Drake CL (1983) Studies in continental margin geology. American association of petroleum geologists, AAPG memoir 34, Tulsa.

✦ Watkins JS, Montadert L, Dickerson PW (1979) Geological and geophysical investigations of continental margins. American association of petroleum geophysicists, AAPG memoir 29, Tulsa.

✦ Wernicke B (1985) Uniform-sense normal simple shear of the continental lithosphere. Canadian journal of earth science 22, pp 108-125.

✦ Ziegler PA (1988) Evolution of the Arctic-north atlantic and the western Tethys. American association of petroleum geologists, AAPG memoir 43, Tulsa.

✧ Ziegler PA (1992) Geodynamic processes governing development of rifted basins. Roure F, Ellouz N, Shein SS et al. (1996) Geodynamic evolution of sedimentary basins: proceedings of the international symposium, May 18-23, 1992, Moscow, Russia. Éditions Technip, Paris, pp 19-68.

✦ *Books or articles of general interest.*

✧ *Source of one of the figures used, cited in the figure caption.*

Chapter 3

SEDIMENTARY DRIVING MECHANISMS AND ENVIRONMENTS

This chapter enters the field of **sedimentology**, which is the study of sediments or, more generally, of the **formation of sedimentary rock** from the sediments that accumulate in basins. It covers the various aspects of rock genesis, from the **origin** of the sedimentary components through their phases of **transport** and **deposit** to their modification under the effects of burial (which will be discussed in Chapter 5). The constituents are subject to the **physical mechanisms** of lateral transport from their source (generally a high point of relief) to their point of deposit (generally a low point in the basin cup where they can be trapped), traveling in the form of particles or in solution in a fluid. Then they are subject to vertical sedimentation, where gravity is the main force at work. The constituents are also subject to complex interactions in the **chemical system** of the air, fresh and salt water, and ice, where they travel and deposit. This system conditions the evolution of these sometimes unstable constituents. All of these processes lead in time to the construction of a series of beds overlying each other and gradually **burying** themselves in the basin. The bed sequence will appear in **strata** in the sedimentary basin receptacle (Fig. 3.1).

If the processes vary with time, the kind of bed and its structure will differ from one bed to the next, and we then enter the field of **stratigraphy**, which will be discussed in Chapter 4 with the time evolution of the deposits.

The resulting compaction is not reflected in the diagram. This chapter will address the sedimentary generator mechanisms on continents and in oceans. While most of the phenomena originating the particle components are continental and occur at the interface with the atmosphere where chemical mechanisms predominate, the transport processes leading to the deposit are mostly under water, i.e., in the hydrosphere and especially the ocean, where the biological, chemical, and dynamic mechanisms interplay with each other.

Although it is known that conditions of the past have varied considerably with respect to those of today (as will be discussed in the following chapter), a large part of our geological reasoning here will be based on the principle of **actualism**, according to which ancient series are considered to have been formed by processes that are known and analyzable today. "The present is the key to the past" will be our guide throughout this chapter.

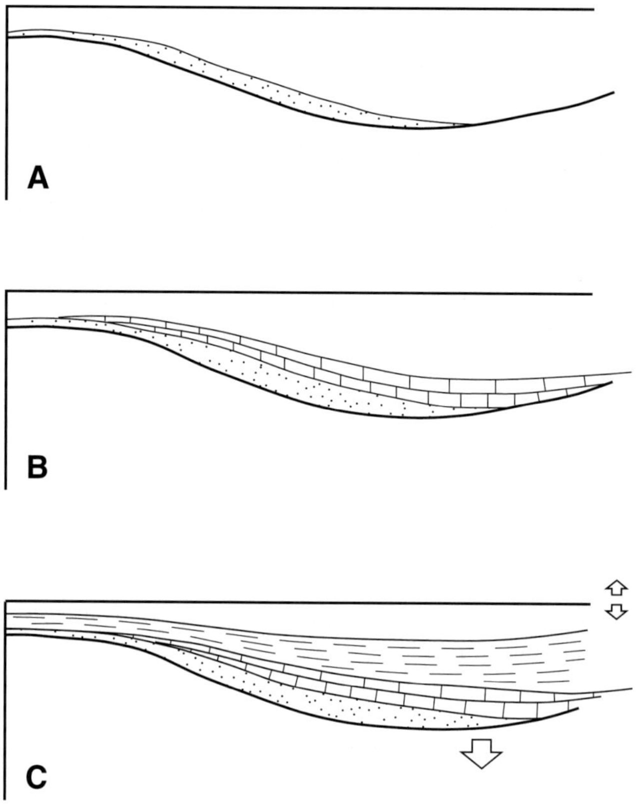

Fig. 3.1 Time sequence of sedimentary beds.
A. Deposit of sands from the edge. **B.** Deposit of limestone in the bottom of the basin. **C.** Clay deposits. The arrows indicate that the subsidence and water level can, among other things, change the basin configuration and bathymetry.

The reader may wish to investigate certain aspects developed in this chapter in greater depth, and is referred to the general works used here, which are those of R.G. Walker and N.P. James (1992), M.E. Tucker and V.P. Wright (1990, 1994), S. Beuf et al. (1971), and H. Chamley (1987).

3.1 SEDIMENT ORIGINS, MODES OF TRANSPORT AND DEPOSIT

3.1.1 Sediment Origins

Sedimentary rocks are made of various constituents. There are **terrigenous clastic** particles deriving from alteration and erosion phenomena on the continent, along with **chemical** and **biochemical precipitates** from solutes in the marine or lacustrine environment and, to a lesser extent, volcanic and hydrothermal products. Lastly, there is the rarer still fallout of extraterrestrial material (Fig. 3.2). Sediment is defined as the set of minerals and particles of organic origin that are deposited in a sedimentation basin.

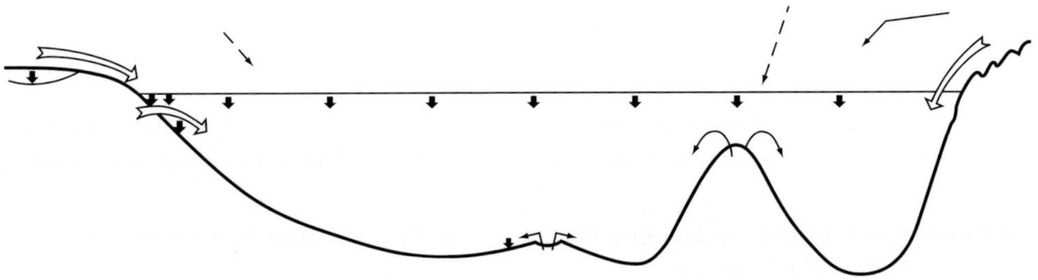

Fig. 3.2 Diversity of sedimentary origins.
The various origins depicted here are the particle input from continental erosion with fluvial (open arrows) transport; volcanic products transported by air or from underwater sources (thin solid arrows); eolian, meteoritic and cosmic dust (dashed arrows); and biochemical precipitation from within the aquatic environment (thick solid arrows).

3.1.1.1 Parent Rock

The major **source** of particles that go into making sediments is all of the existing rock material that can be found on continents and in oceans. These rocks that originate sediments are called **parent rock**, which constitutes the **heritage** of a given sediment. This heritage is useful to know and is in fact fundamental in geology. It will come up again in the following chapters, so the following is a brief look at the various types of rock.

Rocks are assemblages of minerals (elementary particles) studied in **petrography** and **mineralogy**, which will be only briefly summarized here. Rocks are classified as being endogenic or exogenic (Fig. 3.3).

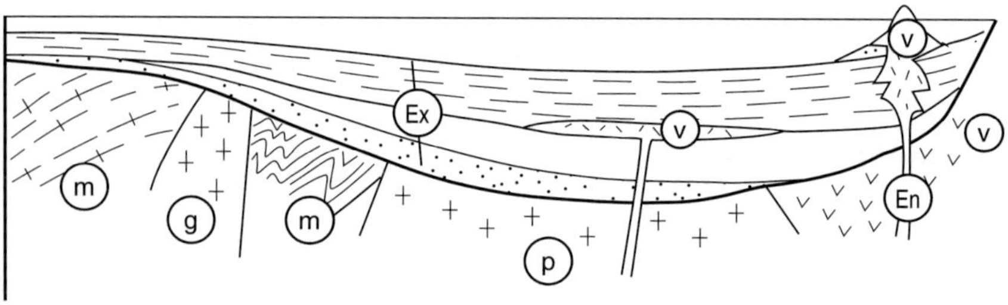

Fig. 3.3 Exogenic and endogenic rocks.
Ex: Sediments deposited in a basin and transformed into rock by diagenesis, such as sandstone, limestone, shales (see Fig. 3.1); En: various types of endogenic rocks making up either the basin substratum or intruding through exogenic rock; g, p: granites, plutonic rock; p: volcanic rock; m: metamorphic rock.

Endogenic rock by definition originates deep within the earth, and generally at high temperature. These rocks are sometimes referred to as **crystalline**, and include the following (Table 3.1):

- **Igneous** or **eruptive rock** comes from the crystallization of molten magma as it cools. These rocks differ by origin:
  - **Plutonic rock** is formed at depths of a few kilometers, and is mostly of granitic composition.
  - **Volcanic** or effusive rock also comes from the depths but is consolidated at the surface, either in open air or on the oceanic bed, forming mainly basalts and adesites.

Igneous rocks are essentially made up of silicate mineral groups of various textures (granulated, microgranulated, microlithic, or vitreous), which is the way they are distinguished (Fig. 3.4). The essential dominant minerals are quartz, feldspaths, feldspathoids, amphiboles and pyroxenes, and olivine. The exact origin of one or another parent rock can often be detected from trace minerals (see Table 3.4 farther on).

Simplifying, we find two broad classes of igneous rock: **acid lineages** typical of continents and **basic** and **ultrabasic lineages** that are much leaner in silica and characterize the large oceans (Table 3.1). The modes in which these families settle will vary considerably depending on the geodynamic mechanisms that have generated them and brought them to the surface.

These will not be detailed here because of their extremely small interest for petroleum, even though there are a few cases of hydrocarbon fields in such rock, in an altered or highly fractured form.

We will simply remember that these igneous rocks are a major **source** of the sediments we will be speaking of, and are also the (crystalline) **basin basement**, where they accumulate. Because of their natural radioactivity, they are a source of heat that will contribute

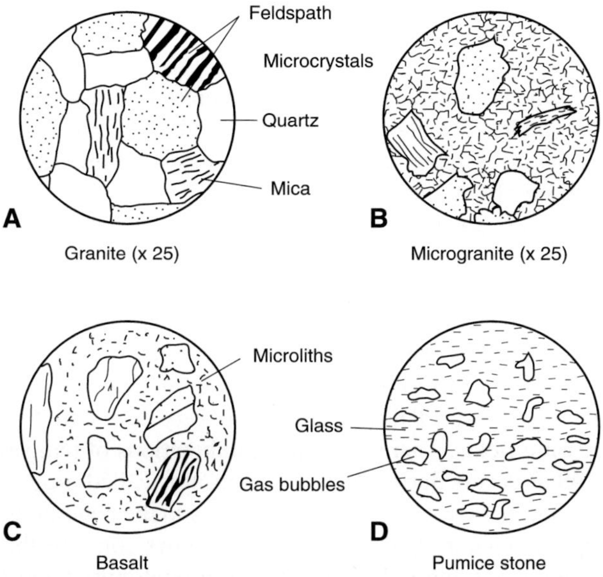

Fig. 3.4 Examples of eruptive rock textures (from Guillemot, 1986). Schematic diagrams of thin rock sections examined under microscope. **A**. Granulated rock. **B.** Microgranulated rock. **C.** Microlithic rock. **D.** Vitreous rock.

Table 3.1 Main igneous rocks (from various authors).

Family		Acids	Intermediates		Basics	Ultrabasics
Silica content		> 66%	66% → 60%	60% → 52%	52% → 44%	< 44%
Essential minerals		Plagioclases Na Feldspath K Quartz	Plagioclases Na Amphiboles	Plagioclases Ca, Na Amphiboles Pyroxenes	Plagioclases Ca Pyroxene Olivine	Pyroxene Olivine
Texture	Granular	Granite	Syenite	Diorite	Gabbro	
	Microgranular	Microgranite	Microsyenite	Microdiorite	Microgabbro Dolerite	Peridotite
	Microlithic	Rhyolite	Trachyte	Andesite	Basalt	Picrite
	Vitreous		Obsidian		Pumice	

significantly to basin evolution and also to the sedimentary content of the basin, especially in organic material.

- **Metamorphic rock** is endogenous, resulting from the transformation of pre-existing igneous or sedimentary rock at depth, under particular pressure and temperature conditions when the burial is large. Chapter 5 will show us how **diagenesis** is the initial transformation term of any sediment.
  The increased pressure and temperature at depth define what is called **metamorphism**, of which there are various types.
  - Dynamometamorphism. The mechanical crushing effects and mineral reorientations (geometric transformations) predominate over mineral transformations.
  - Contact metamorphism. This is localized near the eruptive head with high local temperature rises and iso-chemical transformations (on the scale of a sample).
  - Regional metamorphism. This is produced by very deep burial in growing temperature and pressure fields. There are many combinations of temperature and pressure variations. What occurs during increasing burial is called prograde metamorphism, while the evolution the rock follows as it returns to the conditions of lesser depths is called retrograde metamorphism. Figure 3.5 shows some of the characteristics of these rocks in the varying fields, as they go through burial and then return toward the surface.
    The intensity of the mineral transformations ranges from that of anchimetamorphism (the last term of diagenesis) to ultra-metamorphism. Metamorphism is of the high-pressure or high-temperature type.
    It should be remembered that these metamorphic rocks are formed at depth and are visible at the surface only by virtue of mountain chain formation (see Chapter 5) and erosion, or by rift extension, so they are only very exceptionally of interest for petroleum. They are the basin basements.
- In contrast to eruptive and metamorphic rock, **exogenic rocks** are **sedimentary rocks** that are formed by accumulation of minerals at the surface of the Earth's crust according to various processes that will be analyzed in this chapter. Exogenous rocks that are formed of debris, such as sands and sandstone, are **detritic** or **terrigenous**, while **biogenic rock** is derived from the action of living organisms (coral reefs and chalks, for example) and more purely **chemical rock** is produced by precipitates under certain conditions (evaporites). The term **residual rocks** is used to designate materials resulting from on-site transformation of pre-existing rock, examples of which are coal, paleosols, and bauxite.

Like endogenous rocks, the sedimentary rock formed in a given cycle may become parent rock for new sediments in a later cycle, and a source of new materials if they are subjected to weathering and surface erosion.

As will be seen in Chapters 4 and 5, these sedimentary rocks result from the transformation of materials accumulated in sedimentary basins, the sediments we will be analyzing in the present chapter.

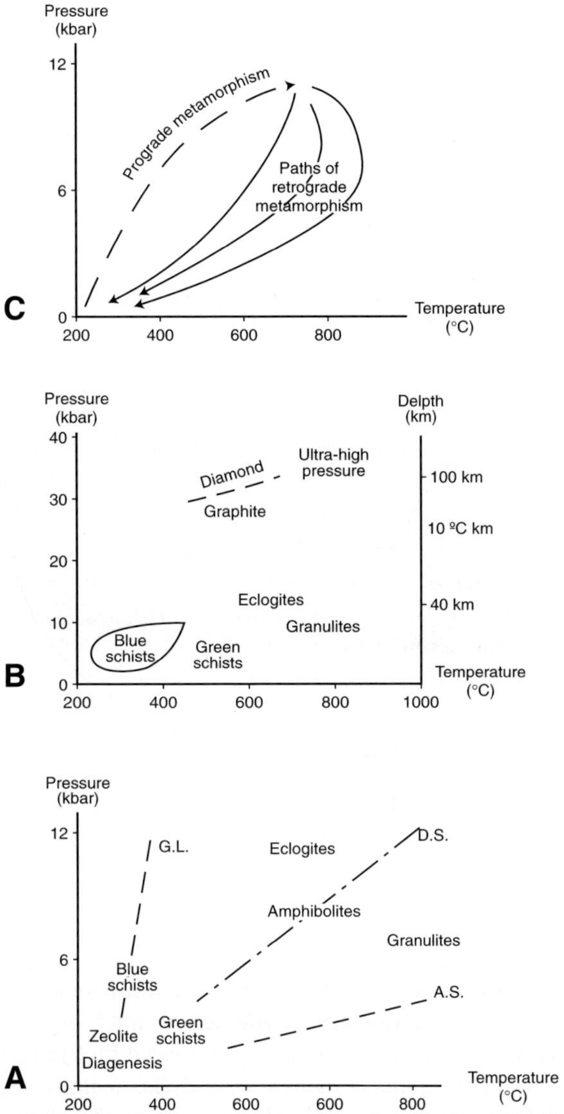

Fig. 3.5 Expressions of metamorphism: mineralogical assemblages corresponding to pressure and temperature conditions (burial, depth, geothermal degree).
A. Types of metamorphism by pressure and depth. GL: low-temperature (glaucophane-lawsonite); DS: average conditions (disthene-sillimanite). **B.** Stability domain of a few mineralogical assemblages. **C.** Paths of evolution (from various authors).

3.1.1.2 Weathering and Erosion

Weathering and **erosion** are mechanisms that affect all kinds of parent rock, whether endogenous or exogenous. There is a difference between the chemical and physical or dynamic processes here, though both are generally closely related. These processes are the first step in the **sedimentary cycle** (Fig. 3.6).

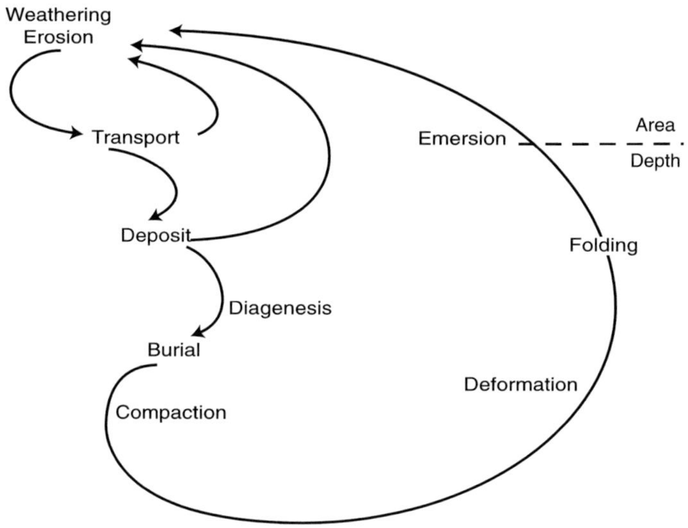

Fig. 3.6 Sedimentary cycle.
Weathering and erosion, which are active mainly in the emerged domain, are the first links in the sedimentary cycle, generating particles and nourishing the solutions of the various sedimentary environments.

Continental weathering is mainly the result of chemical reactions when outcropping rock comes into contact with the atmosphere, hydrosphere and biosphere, and enters equilibrium with surface conditions. The ionic dissociations caused by **hydrolysis** (the commonest and best known of weathering phenomena), acidolysis, salinolysis, and alkalinolysis, increase with the solubility of the minerals. The extent of the chemical weathering will depend on the temperature, the specific surface area of the minerals, the CO_2 content, biological activity (especially microbial), and drainage, which depends on the petrophysical characteristics of the rock. The **weathered zone** becomes a soft material or **alterite**, the upper part of which is the **soil**, and its thickness varies from a few centimeters to several meters or even decameters (Fig. 3.7).

All of the surface transformations (**pedogenesis**) lead to the formation of pedogenic horizons: soils of various types, humus rich in organic material, hardpan, lateritic duricrust and

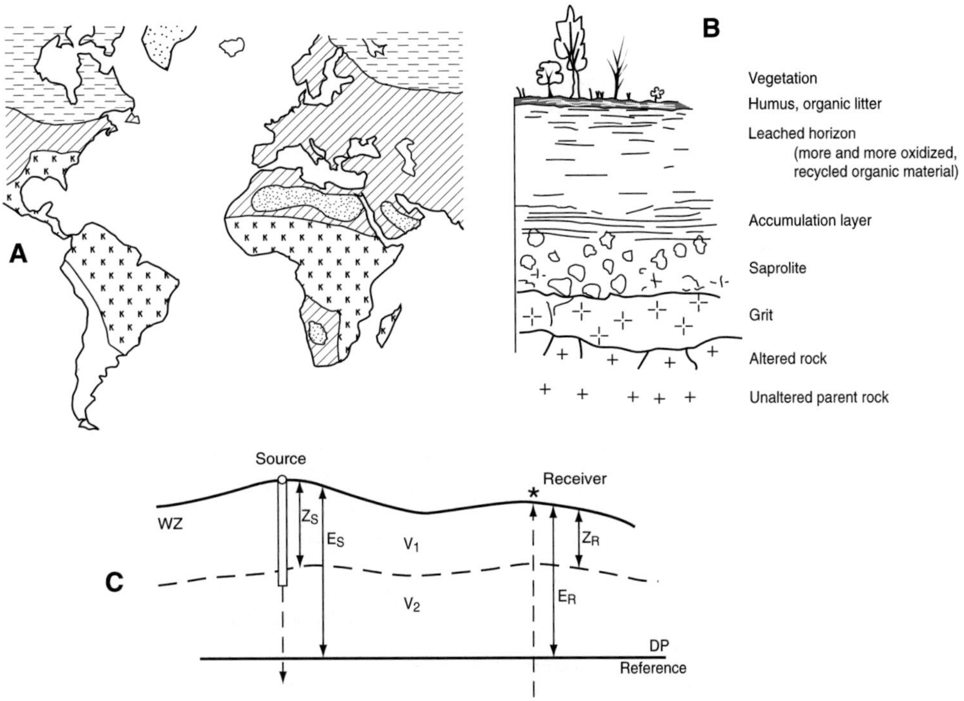

Fig. 3.7 Continental weathering.
A. Different types of soils developed in different climates (from Chamley, 1987): hot and cold deserts (minimum weathering) in fine dots; podzolization in arctic regions in dashed lines; bisiallitization in temperate and Mediterranean regions in dashed lines; kaolinization in tropical zones marked with small k's. **B.** Weathering profile: example of granitic rock. **C.** Example of weather depth picked up in reflection shooting. The weathered zone, denoted WZ, is a surface layer of varying thickness (ZS, ZR) in which the speed of propagation V1 (which may also vary along the profile) differs from the speed V2 in the substrate. "Static" corrections are then made to compensate for the delays introduced by the weathered surface zone.

other crusts, depending on the extent of hydrolysis and the climatic situation, with well-established latitudinal zoning. This is the **pedology** domain where biological action is intense.[1]

These weathered surface zones are of no importance for petroleum geology, but they must nonetheless be taken into account because they seriously disturb the depthward propagation of seismic waves when exploring basins by the reflection shooting method,

1. See Pedro, 1979; Chamley, 1987.

and special processing is needed to erase the disturbances that mar the quality of the information at depth.

Chemical weathering yields varied products: solutes and siliceous particles, the coarse fraction of which is quartz and the predominant fine fraction of which is hydrated silicates. The importance of climate in weathering phenomena is reflected in the latitudinal distribution of sedimentary shales in the oceans (Fig. 3.8).[1]

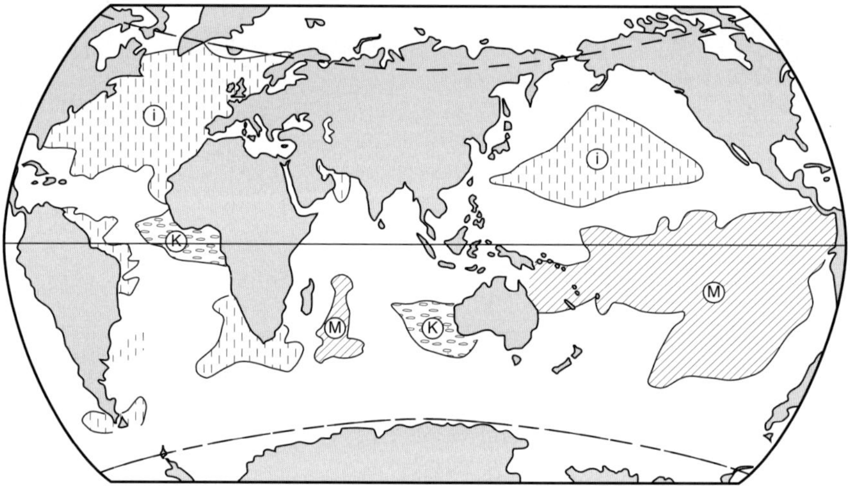

Fig. 3.8 Distribution of different types of clay minerals in the ocean: illite (I), kaolinite (K), montmorillonite (M) and mixtures (blank). (From various authors, including W.B. Berger and H. Chamley.)

One particular and spectacular form of weathering in the carbonaceous medium is **karstification**, which occurs when aggressive waters dissolve the carbonates in the continental and sometimes marine domain. This produces inflow caves, sinkholes, grottoes, underground networks and modifications of the porous lattice characteristics, which is of primordial importance for petroleum.

Weathering is generally complemented by physical and biological processes: fractionation by temperature effect (such as freeze/thaw), swelling by hydration and desiccation, opening of galleries on various scales (burrows, micro-organisms), and can be accentuated today by human activity.

1. See Millot, 1964; Chamley, 1987.

Weathering prompts both vertical (up and down) and horizontal (downstream) migrations, and can lead to mineral concentrations such as supergene metallic deposits of aluminum or nickel.

Of course, continental weathering has occurred in the geological past, and the **paleosols** preserved in a sedimentary sequence, their type, amount of development, and other information that can be gathered from them are extremely important in defining past environments.

Weathering in the marine environment, where the environment is buffered, is much less extensive and much less well known. A few examples can nonetheless be mentioned of mineralogical modifications of sediments that can become a source of new material for erosion or transport. There is the **biological action** of burrowing organisms, especially micro-organisms living in the waterlogged sediment, and also **hydrothermal alteration** that acts rapidly on the crests of oceanic ridges, in certain accretion wedges, and on the flanks of underwater volcanoes and the associated sediments (hydrothermal alterations are due to reactions between the basalt and the seawater circulating under given temperature and pressure conditions). This process is at the borderline of early sediment diagenesis. **Dissolution** is also a very important material recycling factor (before, during and after sedimentation; see the discussion of the compensation depth of carbonates, for example, later in this chapter).

In conclusion, weathering products are **solutions** that are more or less rich in sulfates, carbonates, oxides, chlorides, and minerals that may be transformed and broken up to a certain extent into solid fragments or **particles** under the action of erosion.

Erosion is the result of dynamic actions at the soil-atmosphere interface that set the weathering products into motion. This is where particle transport begins. Erosion agents are wind (general or local atmospheric circulation), rain and runoff, ice (role of large glacial caps), continental and oceanic waters (waves, storms, tides) and living organisms such as those of the lithophagous type. Deflation, subglacial abrasion, the scouring of riverbeds and tidal channels, slope slides (see re-deposition farther on) are varied forms of stresses applied to a more or less lithified sediment or crumbled rock producing permanent load, mechanical friction and shearing (Fig. 3.9).

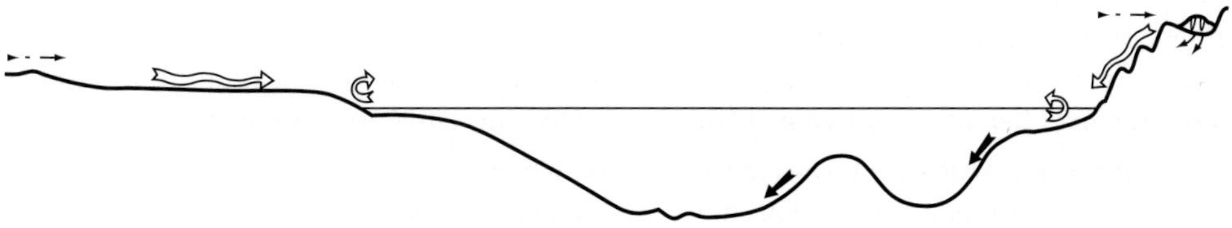

Fig. 3.9 Diversity of erosion agents.
Air deflation is indicated here by broken arrows; subglacial abrasion by small black arrows; runoff and riverbed scouring with the wavy open arrows; the action of waves, storms and tides on the coast by spiral arrows; and underwater erosion on the slopes by thick black arrows.

On different scales, erosion is sometimes tenuous and sometimes spectacular, leaving traces of currents, paleovalleys, subglacial channels, paleocliffs with the associated debris, paleocanyons, gullied areas in an encroachment domain, and sometimes just simple discontinuities like hardened soils (see the sedimentation hiatus farther on). Runoff, solifluction, slippage and landslips all participate in activating the erosion and modeling the catchment areas.

Erosion generally affects rocks, but it also affects sediments in the course of their deposit and burial (Fig. 3.6).

The importance of erosion increases with the lessening cohesion of the material, which is generally the case in the underwater domain. It will be amplified if there is any liquefaction-fluidization.

Biochemical erosion complements physical erosion in the continental and especially the marine environments. This is the role of organisms (poriferans, echinoids, lithophagous mollusks, burrowers) and microorganisms on the continent and in the ocean.

The final result of weathering and erosion is to release **terrigenous material** (solid fragments, siliciclastic or carbonaceous particles, organic debris) and **solutions** (which then allow precipitates to form), and also to determine **landforms** (morphogenesis).

The particle material can be referred to by its grain-size distribution or its composition. The sizes of particles released by erosion vary from blocks measuring several decimeters (from rockslides or volcanic explosions) to the smallest particles found in clays or microcrystals. This coarse-to-fine grain size distribution will partly determine the transport capacity, mode, and distance, and the type of resulting rock, depending on the energy of the medium. The particles may be mono- or polycrystalline mineral debris such as silica and silicates, debris from organisms (bioclasts) or miscellaneous debris and, depending on their composition, will yield sands, calcarenites, or gypsarenites. These detritic products are said to be **clastic**, whence the term **siliciclastic** for the most widespread among them.

For a sedimentary basin, the idea of a **drainage area** defines the zone from which the materials will be supplied and transported through drainage basins to the basin where it will finally settle. Depending on how extensive the eroding oldland is, and on the vigor of the erosion, the particle products may be monogenetic (from a single source) or polygenetic, or more precisely polymictic (a mixture of several sources). This aspect will in fact be largely influenced by the transport characteristics. Generally, when trying to determine parent sources one or more may be found for particulate material, but obviously not for dissolved material, where the initial source is difficult or even impossible to recognize.

Erosion mechanisms lead to **morphogenesis**, which may be continental or under water (see farther on in this chapter). The most spectacular and best-known effects are those of glaciers (leaving striated floors, overdeepened basins, and thresholds), winds (barchans, dune massifs, ergs), runoff and running water (badlands, valleys and interfluve). Littoral morphogenesis, where winds, waves, storms, and tides all come into play, is also marked by the action of organisms. At great depths, the role of gravity-driven streams (canyons) and contour currents is also quite important (see farther on).

Erosion is mainly continental and, if it is active for long periods of time, will go beyond the stage of modeling special morphologies and gradually lead to a **leveling of relief**, which will be mentioned again in terms of various continental environments. This leveling can go as far as wiping out entire mountain chains leaving, for example, Western Europe's Hercynian peneplain.

The final consequences of weathering and erosion, then, are to modify relief and recycle materials that will be available for transport to a sedimentation basin.

3.1.1.3 Chemical and Biochemical Precipitation

Sediments no longer originate directly from a parent rock here, but rather from a more or less saturated, oxygenated, or reductive medium where chemical phenomena are at play. The composition of the medium of course depends very much on the quantity of dissolved matter, coming either from parent rock or from the activity of organisms living in the medium and donating their tests (skeletons) and organic matter at death.

In the (generally oceanic) aquatic medium, there are direct precipitation processes of hydrogenetic origin, but also those dominated by the action of organisms (biogenetic) and combined processes involving several factors, including physical.

Direct chemical precipitation. The natural waters are more or less rich in chlorides, sulfates, and oxides, as Table 3.2 shows in a comparative listing of average sea and river water compositions.

Table 3.2 Comparison of sea and river water composition.

	Sea water (ppm)		River water	
Cl^-	~ 19 000	(1)	7.8	(5)
Na^+	10 500	(2)	6.3	(6)
SO_4^{2+}	2600	(3)	11.2	(4)
Mg^{2+}	1300	(4)	4.1	(7)
Ca^{2+}	400	(5)	15	(2)
K^+	380	(6)	2.3	(8)
HCO_3^-, CO_3^{2-}	140	(7)	58.8	(1)
Br^-	65	(8)	0.02	(n)
H_4SiO_4	1	(n)	13.1	(3)

(From Seibold and Berger, 1982)

Solutions in the oceans do stem from rain, melting ice, and hydrothermal sources, but rivers are by far the main source of additional solutions in the ocean.

Physical factors such as temperature and water agitation will control the precipitation of new sedimentary components, which are precipitates. This precipitation occurs when salts are crystallized by over-saturation, either in the water column itself or in the interstitial water in the top sediments.

The minerals formed among the **carbonates** are **aragonite**, **calcite** and magnesian calcite to **dolomite**.[1] Direct precipitation of carbonates from ions in seawater is a phenomenon that is rarely seen today except in shallow confined tropical environments. Only the unstable aragonite form is likely to precipitate (into milky clouds), while direct formation of calcite and dolomite is much rarer.

In today's ocean, the bulk of carbonate precipitation (aragonite or calcite) is due to the organic and metabolic activity of life forms (see farther on), but conditions for direct chemical precipitation were more widespread in the past. Under certain conditions, calcite can precipitate in a lacustrine domain.

The **saline evaporites** are essentially **halite** (rock salt) or potassium salt, **anhydrite** and **gypsum**, which all form in many present environments (continental and marine, like brine marshes). Normal seawater is under-saturated. In order for the salts to precipitate, evaporation has to be in excess of precipitations and inputs (Fig. 3.10). Depending on conditions, the precipitation will occur at the air-water interface, water-sediment interface, or even within the sediment (see Chapter 5 on diagenesis).

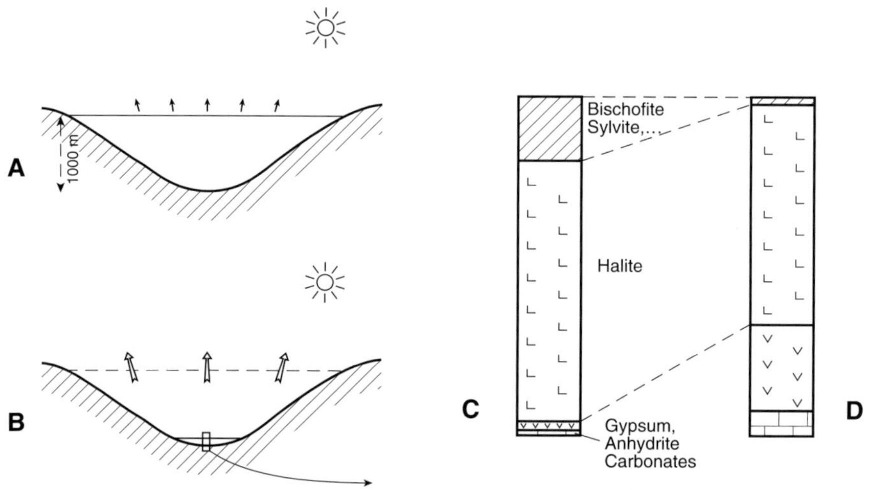

Fig. 3.10 Evaporite formation scheme.
A. A basin 1000 m deep with normal salinity and the beginning of strong evaporation with no renewal of water. **B.** Final situation: basin dried up with 20 m of precipitates, called evaporites. **C/D.** Theoretical possibilities of precipitated salts (**C**). It will be noted that the proportions differ greatly in nature (average of different saliferous formations) (**D**).

1. See Bathurst, 1975; Purser, 1980-1983; and Tucker and Wright, 1994.

Other hydrogenic precipitates may be mentioned, such as **neogenic clays**, zeolites, and **metallic precipitates** (pyrite, sulfides and polymetallic oxides).[1]

Alkalinity, water stagnation, and intense evaporation may cause ***neoformation of clays*** *(although present examples are rare). Restricted or confined environments may allow genesis of ferriferous clays (especially glauconite) by precipitation near the water-sediment interface. Minerals like smectite can be found in metalliferous clays. The metallic deposits of the great depths are part of the complex domain where the role of organisms can become preponderant, as is the case in the formation of* ***phosphates*** *(see farther on). This whole field of "authigenic" neoformed clays and other precipitates that occur very near the sedimentary surface is, in a certain respect, part of early synsedimentary diagenesis.*

Different varieties of ferriferous clays exist, including glauconite, verdite, berthierine, and chamosite ("glaucony" is a term generally used to speak of granules).

Considering the use of these clays in geological reconstructions precisely to determine the deposition environments, let us briefly summarize a few particular points of use in petroleum geology:[2]

- *The importance of the substrates (micro-fossils, fecal pellets and miscellaneous debris) and their size (200 to 500 μm)*
- *Two broad types in present sedimentation: verdite and glaucony*
- *Zonal distribution in latitude and depth. The optimum formation depth is between 20 and 60 m, and up to 200 m at major water stream outlets for verdite; between 150 and 200 m for glauconite*
- *Development in zones of low sedimentation rate and near the water-sediment interface*
- *Possible genesis by transformation of pre-existing silicates (such as in fecal pellets) and/or by precipitation-dissolution-recrystallization*
- *Difficulties in transposing present situations to the genesis of ancient ferruginous oolites and glauconites (see eustatic variations farther on).*

Organic material contributes directly to the creation of **biogenic sediments** in a variety of ways: accumulation of tests (skeletons) of organisms that once lived on the floor (benthic organisms) or in the water column (pelagic) and generating carbonaceous or siliceous sediments; carbonated colonial constructions; decomposition of organic matter; films of microorganisms.

This contribution depends on a number of factors: nutrients (the food chain is based on green algae or phytoplankton); the available light for photosynthesis; temperature; salinity; dissolved oxygen content; and preservation conditions.

The importance of **primary productivity** (i.e., the organic production rate) should be stressed (Fig. 3.11). This is organized along various climatic belts for calcareous or siliceous organisms according to the various factors mentioned above, and is responsible for many

1. See Chamley, 1989.
2. See Chamley, 1989; Odin, 1986.

types of lithogenesis (dominantly carbonaceous). The productivity, which can be interpreted as the biomass, is relatively small in the air environment over continents, except for forests, but is abundant in the marine environment. It is largely guided by atmospheric and oceanic circulation (with the formation of fronts).

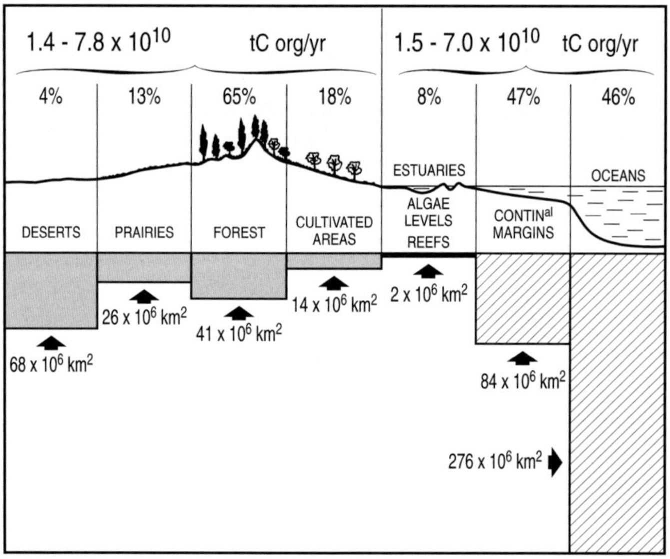

Fig. 3.11 Primary productivity of continents and oceans (from Huc, 1980, in Durand, 1980).
The upper part of the diagram expresses the quantities of organic carbon, and the lower part the surfaces affected, demonstrating the importance of the forest biomass and that of the continental margins.

Practically all groups of organisms participate in rock formation in one way or another. Examples are the reefs of madrepores, stromatolites and algal ridges, pelagic calcareous oozes with foraminifers and coccolithophorids, radiolarian and silicoflagellate oozes, the higher plants of coal, and lacustrine algae of some bituminous schists (Fig. 3.12).

Zooplankton and phytoplankton are a major factor not only for organic matter but also for the mineral material, with the accumulation of carbonaceous or siliceous tests.

The main source of carbonates in present oceans are the green algae that construct a skeletal limestone. Carbonaceous sands are often debris of calcareous algae (Halimeda).

Biochemical precipitation, where the control of the living world is essential, is a quantity of mixed processes affecting very different environments, and which are the source of a large proportion of sediments. Various biochemical and biophysical mechanisms accom-

Fig. 3.12 Coralline limestone.
This example in Morocco's High Atlas Jurassic illustrates the role of colonial organisms (essentially corals, here) in reef building. These edifices form a sequence of small relief structures brought out by erosion, above a regularly layered underlying series.

pany and relay the direct production of carbonaceous and siliceous particles. The role of algae, bacteria, and various microorganisms is a capital one for the carbonate domain, creating algal ridges and stromatolites (which trap detritic input), and micritization. Fecal pellets (dejection of many metazoans), aggregates, pelloids, are also forms of organic activity. Clouds of aragonite in needle form might not come from direct precipitation, but from algal activity. For ooliths, direct precipitation from seawater seems to be negligible. Organic activity is evidenced by organic residues that participate in the concentric layers, and the effect of physical agitation is certain.

Biochemical precipitation is important at the water-sediment interface and within the interstitial domain (to the limit of early diagenesis).

Phosphatic concretions are related to zones of high organic productivity, and the chemical processes are in fact greatly influenced by the biological mechanisms and physical actions (currents, agitation).[1] This mainly concerns calcium phosphate, which is fixed and concentrated by biological processes.

1. See Lucas and Prévost, 1987.

3.1.1.4 Other Sources (Precipitation, Volcanic, Hydrothermal, Cosmic Dust)

There are a number of other special sources of sediments (Fig. 3.13):

- **Atmospheric precipitation** (rain, spouts, fog, among others) is a source of particles and dissolved materials produced by atmospheric washout. These inputs might be much greater than is generally thought, but there is still no precise quantitative evaluation of them.
- The large mass of gaseous and solid material released into the air or under water by **volcanism**, once cooled, can be sedimented directly into a basin by ashfall or lava flows, or indirectly by clastics. Volcanism acts directly, in the form of material flow, but also indirectly by way of the haze and cloud cover affecting solar radiation input, CO_2 level, and hydrothermal processes.
- **Hydrothermal reactions**, the effect of which was thought to be limited until oceanic sources were recently discovered, are actually significant. This is a particular type of alteration by reaction with the seawater circulating at depth. The water is heated by the ascending magma, reacts with it, and then carries metallic solutions to the surface, where they precipitate near the vents or modify the composition and pH of the ocean water.
- **Cosmic dust** from meteorites and micrometeorites and impact tectites should be mentioned even though they account only for an infinitesimal part of the sediments. The meteoritic impact at Rochechouart in France, and the one in the Yucatan, are examples of events, to be discussed in Chapter 4.

3.1.2 Sedimentary Transport and Deposit, Lateral Progradation, Vertical Aggradation

We have seen that sedimentary sources produce two broad classes of sediments: clastic or terrigenous particles, and dissolved particles. The main transport agents are **water**, **wind**, and **ice**, in very unequal proportions. The environments are both continental and marine (Fig. 3.13).

3.1.2.1 Transport in Solution

The main mechanisms that disperse solutions during their transport, before any chemical or biochemical precipitation and subsequent sedimentation occurs in a basin, can be briefly summarized as follows (from upstream down):

- Active water streams with concentration in closed (endorheic) basins of the lacustrine sort or other, and in alluvial flood plains
- Active watercourses with marine outlets, deltas, and estuaries, with special importance for all transfer currents in the coastal fringe
- Tidal ebbs and flows and coastal drift currents, wave currents

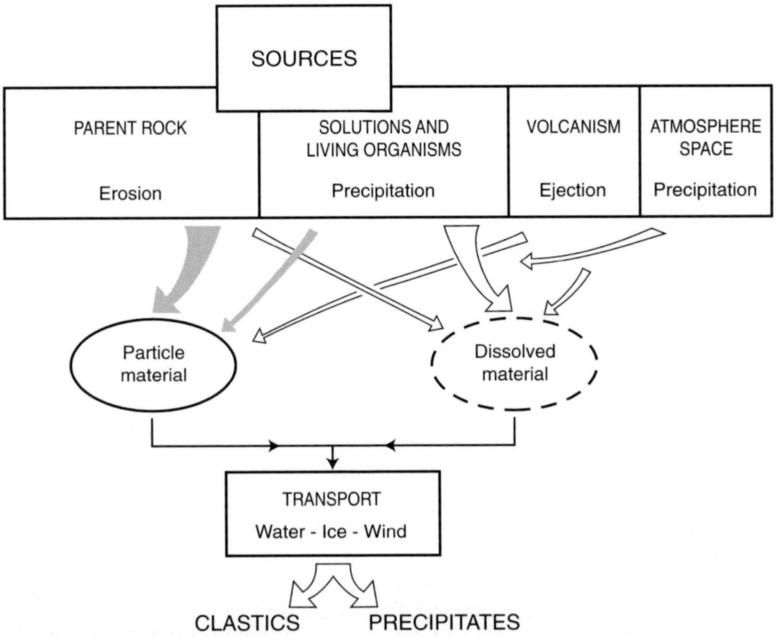

Fig. 3.13 The various sources of sediments.
Most of the particulate and dissolved material available for transport into a basin are products derived from parent rock and solutions (mainly in the ocean) where organisms live. Secondary sources are volcanism, which is very important locally, and atmospheric input, to a lesser extent. This diagram complements figures 3.2 and 3.6.

- Major ocean surface ocean currents, which are tied in with the general atmospheric circulation, and upwelling currents of cold waters rich in nutrients
- Gravity-driven currents, floor and contour currents, Antarctic stream current
- Hot (and cold) hydrothermal plumes.

3.1.2.2 Particle Transport, Lateral Progradation, Vertical Aggradation

All of the dynamic actions mentioned above will play a major role in transporting terrigenous particles of a given grain size class if there is enough energy. Continental and marine water serve as the main vehicle, but particles are also transported by ice and wind.

The relation between the transport velocity and the particle grain size is a simple one (Fig. 3.14) when the vehicle is water or wind. Cohesion also plays a role, as do friction forces and the medium's viscosity.

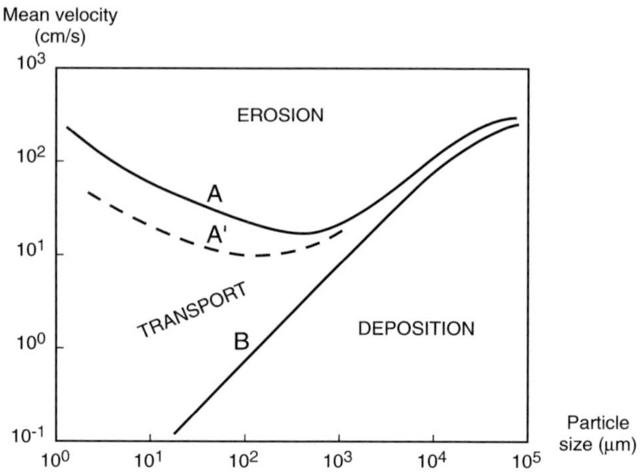

Fig. 3.14 Domains of erosion, transport, and deposition, as a function of particle size and stream velocity.
This Hjulstrøm diagram illustrates stream action. When the velocity is high, the stream is erosive. With decreasing velocity, the transport becomes active and then stops, leading to the deposition of the particles. The limit A is variable (A') and depends on the consolidation and water content of the material subjected to the current.

The particles are transported either in suspension in the medium or are rolled along the bottom, and the proportions between these two modes will vary considerably depending on various parameters. One of the most important is the level of turbulence, as opposed to laminar flow. It should be noted too that certain very energetic but brief (on the geological time scale) modes like a flash flood or turbid dust can carry a considerable volume of sediments.

Transport is also classified as bed load (rolling along the floor), suspended load, saltation (intermittently in suspension) or in extended suspension.[1]

Tractive transport, in which particles roll or skip (saltation) over the bed with a low level of suspension, can be produced by the wind (eolian mode), continental water streams (fluvial and estuarian), ground swells and storm waves (shallow offshore), ebb and flow currents (tidal domain), deep contour and bed stream currents.

The particles are carried onto the bed or to its immediate vicinity in rolling carpets or turbid clouds at a height that generally depends on the water's depth but especially on its velocity. As the velocity decreases, there is no longer enough energy and the particles settle in oblique layers or elementary beds, i.e., with a **lateral progradation** corresponding to a sequence of sandy avalanches, and not a vertical aggradation mode which is the result of another type of deposition or time variation, as will be seen later on (Fig. 3.15).

1. See Reineck and Singh, 1980.

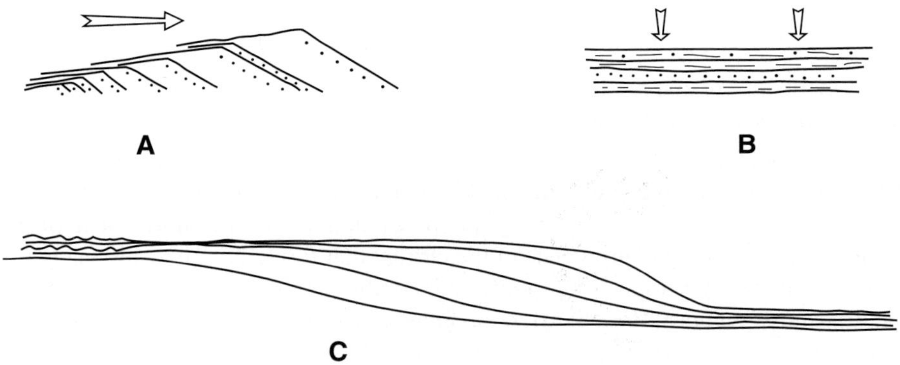

Fig. 3.15 Progradation and aggradation: two complementary modes of accumulation.
A. The main mechanism is transport over the bed with lateral progradation of the deposits. The vertical aggradation is slight. **B.** Settling out of the suspended material is predominant here, with vertical aggradation in a sequence of beds. **C.** Depending on the dynamic of the environment and position within the basin, one or the other mode will predominate. If the prograding material is voluminous and the velocity is quickly reduced, a large prograding sedimentary prism will develop in S-shaped envelopes. The aggradation will be small at the base of the basin, and there may even be sedimentation gaps.

This progradation may be largely mono-directional, as in fluvial deposits, or it may be multi-directional as in the cases of wave-formed or tidal deposition. Sediments transported and deposited by rivers will be truly mono-directional (alluvial cones and braided drainage patterns) or relatively spread out and meandering, depending on the value and regularity of the slope. The amount transported and the extent to which it is spread over the surface will depend directly on the high and low water periods, i.e., on the climate.

In the marine domain, regularly or irregularly oscillating flows will dominate in shallow domains of less than 50 m, while regular contour and bed streams can also generate transport and preferential accumulation in a dominant direction.

The longer the time and distance of transport, the more the particles will be sorted, blunted and rounded by wear and friction. The sediment can thus be typed as proximal or distal, mature or immature, well sorted or unsorted, and its reservoir qualities will depend considerably on these characteristics.

The more energetic the tractive stream is, the more erosion marks it will generate, and these will give clues to the type and strength of the stream (channeling, tides, wave, storm, for example). Subglacial flow in particular, where the pressure plays a considerable role, will leave characteristic figures at the base of the deposits.

Note in passing that glacial deposits, aside from the ground till formed by the transport, are material picked up by the ice and then released and settled after melting.

These marks of erosion are major indicators that will help in reconstructing the deposition environments in ancient series.

When the tractive flow loses its capacity to transport, its particle load will settle out on the downstream slope in **oblique banks** making a **laterally prograding deposit**. The slope and the internal and external geometry of the progradations form a **figure of deposit** that will indicate the type, depth, turbulence of the current that formed it. There are many figures of deposit of all sizes corresponding to different conditions (lower flow regime, upper flow regime, transitory regime). **Reactivation surfaces** can sometimes be found within a given sequence, witnessing to special events. But it should above all be remembered that the deposition mechanism leads to the idea of a **genetic unit**, which will be discussed in Chapter 4.

Most deposits generated by tractive currents occur in surficial conditions, i.e., in the continental or shallow marine domain (tidal and wave currents, storms) (Figs 3.16 and 3.17),

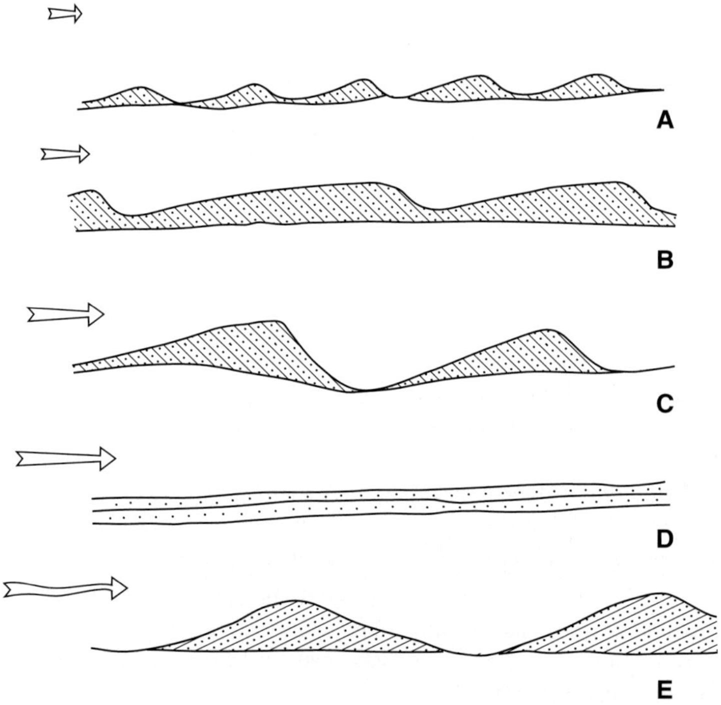

Fig. 3.16 Different types of progradational deposits created by tractive currents, in order of increasing strength.
A. Ripples. **B.** Sand waves. **C.** Dunes or megaripples. **D.** Flat deposits. **E.** Antidunes.

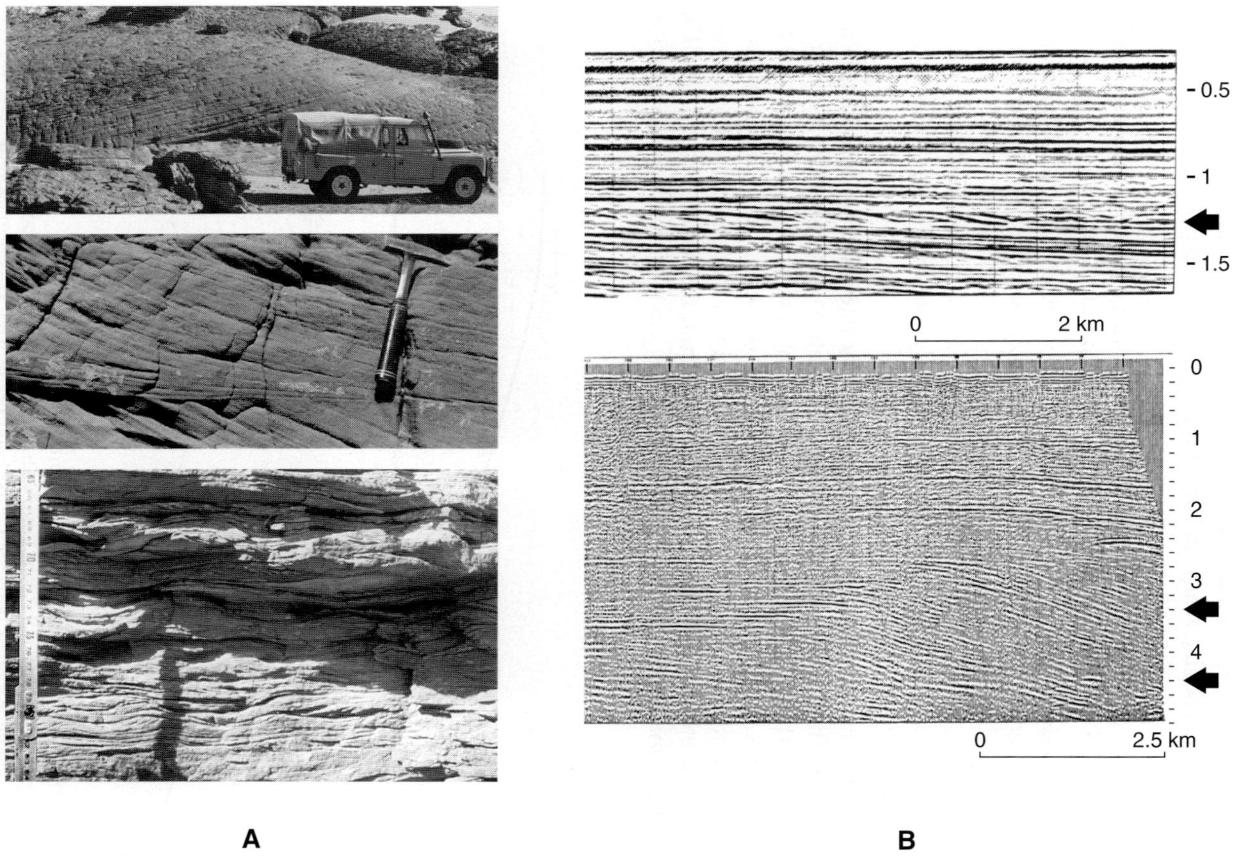

Fig. 3.17 Oblique progradation banks.
A. Examples of terrain on different scales, from megaripple to ripple (IFP photos). **B.** Seismic examples (from ENSPM document, 1986).

but deep sedimentary figures are now coming to light beyond water depths of 4000 m. These figures imitate surface deposits. For example, a stream current would need only to acquire a certain velocity in order to create a subaqueous dune such as the carbonaceous bathyal dunes at the foot of the New Caledonian scarp, and abyssal hemipelagic dunes.

The amount of material available for sedimentation is a very important factor determining the sequence of deposition formed, which will vary considerably depending on a lack or excess of input. The idea of **available space** will be taken up later (Chapter 4).

The importance of this can be seen for the formation of continuous or discontinuous reservoirs. There are certain recognizable cases of ebb and flow or neap-tide alternations, and those of high and low sea levels.
While this tractive mode of transport/deposit on the bottom has been studied largely in terms of eolian, fluvial, and marine sand drifts, it actually concerns all types of particles, both siliceous and calcareous.

Transport in suspension may be partial, continuous, or intermittent, depending on the amount of turbulence. The tractive current may itself generate a counter-current in the proximity of and in front of the sediment progradation, producing a separation bubble where the material in suspension is entrained, forming a sandy fallout or avalanche. In the normal suspension regime, the particles are held suspended by ascensional motions caused by the turbulence, and move in the body of water, sometimes far from the bottom (Fig. 3.18). The particles in the water may be elementary grains (siliceous sand, calcareous fragments) or

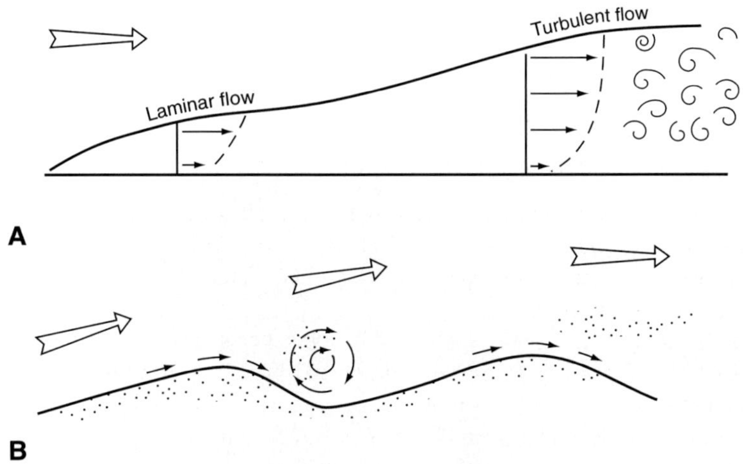

Fig. 3.18 Flow and suspension pick-up process.
A. Diagram of the switch from laminar to turbulent flow with the formation of a boundary layer between the bed and main flow. **B.** Formation of vortices and the dispersion into suspension of material picked up along the subaqueous dune crests.

aggregated flakes (clays). In air, where the lift is weak, the amplitude of the ascending motions and the high velocities can still keep a large volume in suspension, traveling in dust clouds over thousands of kilometers before settling out or returning in the form of red rain. While transport is essentially lateral, the deposit is **vertical fallout** or **decantation**.

The current may act upon the waterlogged sediments in such a way as to remobilize the particles in a turbid layer at the bottom, producing "silty plugs". When fine particles are returned to suspension on the ocean floor, a **nepheloid layer** is formed, ranging in thickness from a few meters to several hundred, and exceptionally to more than a kilometer.

The time these fine particles remain in suspension is called the **residence time**, in the same way as for chemical elements.

Decantation occurs when the vehicle dynamic is no longer great enough to keep the particles in suspension. They will fall out faster or slower depending on their size, which therefore affects their residence time. This sedimentation mode concerns fine terrigenous particles (silts and clays) and the remains of planktonic organisms. This slow, regular fallout will produce **pelagic ooze** on the floor with a dominance of planktonic organism tests, accumulated and still waterlogged, or **hemipelagic ooze**, with a dominance of terrigenous material. Pelagic clays are intermediate between these. If the conditions do not change, these decantation deposits will be characterized by regularity, homogeneous composition, no erosion patterns, and vertical **aggradation**. The whole will be disturbed only by **bioturbation** if the waterbed is oxygenated enough to allow the proliferation of organisms, and sometimes precipitation of orthogenic minerals.

Actually, though, it will be seen later on that conditions can change with time, in terms of primary productivity and terrigenic material inflow, as they do for rolling and jumping deposits, and that different decantation deposits with and without hiatus may alternate. Such deposits may occur in the **lacustrine domain**, in alluvial plains, in a sheltered coastal domain or on the coastal platform, *but the great decantation sedimentation domain is still the* **deep ocean**, from continental margin to abyssal plain.

This decantation process will be partly inhibited by **dissolution** phenomena in the water column. These phenomena differ for silica and carbonates (Fig. 3.19). Silica dissolution is maximum in surface waters, causing a short repetitive fixation-dissolution cycle. For carbonates, it increases with depth (the pH decreases because of the dissolved CO_2 content). The **lysocline** is the critical depth of about 4000 m below which the dissolution suddenly increases. The carbonate compensation depth ("ACD" for aragonite and "CCD" for calcite) corresponds to the zone where dissolution is maximum. This depth varies in the different oceans, depending on the topography and distribution of the water masses (Figs 3.20 and 3.21), and it has varied in the past. CCD (calcite) is the most commonly used acronym.

Gravity-driven currents and deposits. Although in the final analysis all deposits result from the force of gravity, what is meant here is slippage transporting a mass of particles down a slope. This is mainly caused by an overload (Fig. 3.22). There are several processes of this type, which will be developed farther on in this chapter.

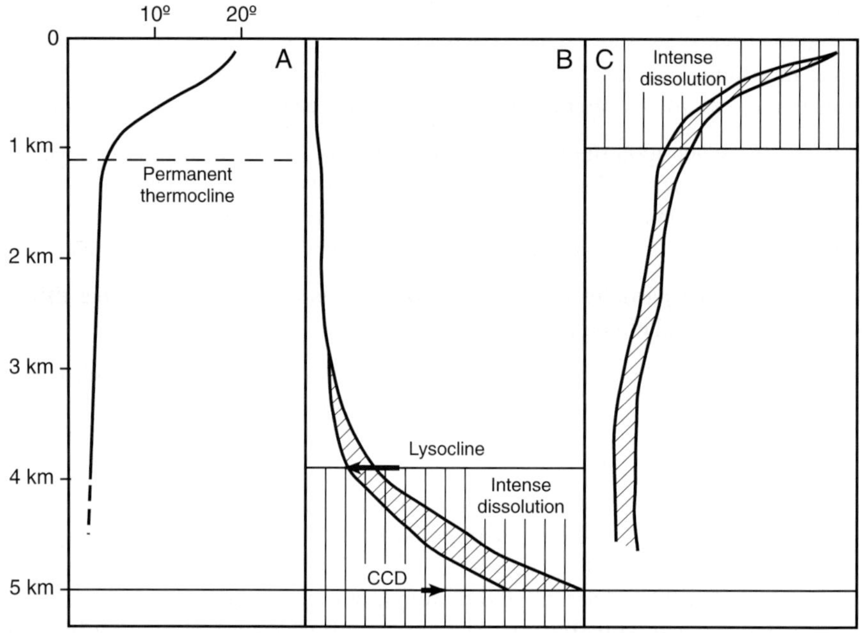

Fig. 3.19 Water solubility and temperature profiles versus depth. **A.** Rapid falloff of temperature under the warm surface layer. **B.** Calcite dissolution curve (see Fig. 3.20) with intense dissolution under the lysocline. **C.** Silica dissolution curve: intense at the surface like the temperature gradient.

- Massive **resedimentation** of a sedimentary layer that is neither well consolidated nor well dissociated, with a more or less intense deformation of the initial structure. This is **slumping**, with the formation of new structures called **slumps**, which may vary in size and extent from a decimeter to several decameters. When the sediment slides apart, it rotates and leaves a lens-shaped landslide scar upstream.
- **Mass flow** by gravity current, entraining dissociated and more or less fluidized sediments. This is of various types: **grain flows**, **debris flows** and **liquefied sediment flows**. They develop consequent to sudden events such as quakes or sedimentary overloads. **Olistostromes** can be included here. These develop in tectonic zones along rifts and mountain chains, and contain **olistoliths** and (very large) **sedimentary klippes** set into the sedimentary **matrix** formed by gravity.
- **Turbidites** deposited by a high-density **turbidity current**. Density currents can be set up by salinity or temperature gradients, but turbidity currents are formed by a layer of water with high particle density in which the velocity is determined by the contrast in density between the layers and the slope water bottom. The velocity is high, the flow is turbulent, and particles of various grain sizes are in suspension, making a dense liquid

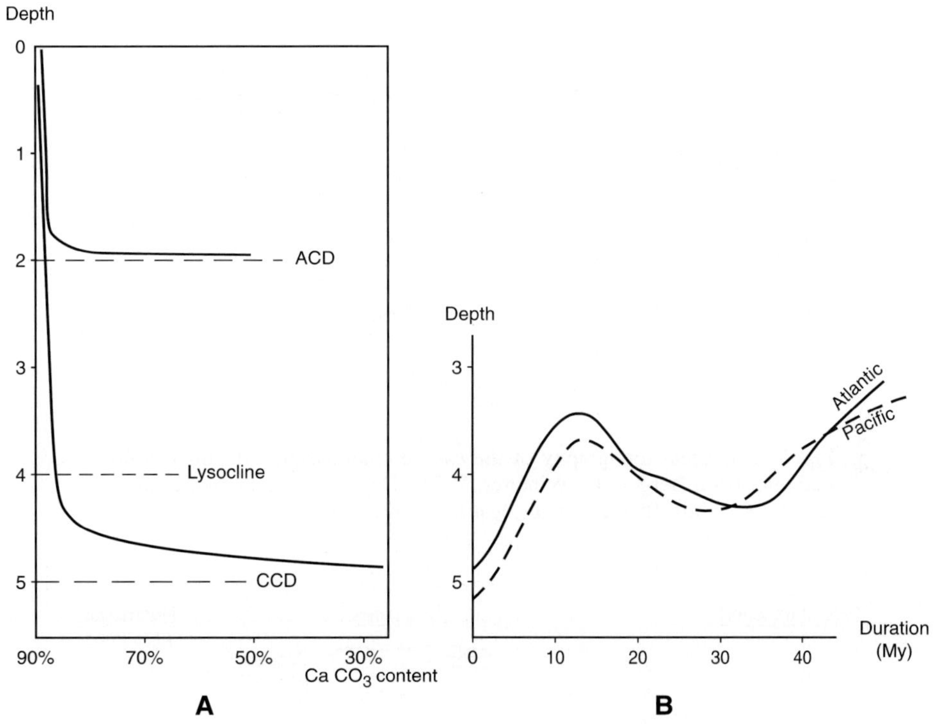

Fig. 3.20 Carbonate dissolution depth.
A. The aragonite compensation depth (ACD) is about 1000 meters. The more resistant calcite begins to dissolve rapidly under the lysocline until the CCD is reached. **B.** The CCD varies from one ocean to another, but has also varied over the course of geological time (curve from Caron et al., 1992).

mass where the particles will be graded by size. When the current slows down, the particles fall out and the deposit will reflect the generator mechanism by an arrangement of facies in graded sequence: a "Bouma sequence" with proximal and distal turbidites, and so forth. Erosion forms are frequent.

3.1.2.3 Forms of Deposit, Structures, and Sedimentary Bodies

If the various modes of transport and deposit are constant enough, they will result in the construction of strata or deposit figures characterized by some internal structure, morphology, and extent in space. Generally, specific environments can be defined from these sedimentary structures. The internal organization of elementary particles in deposit patterns is as varied as the mechanisms that put them there, and these will change with time. This is why the

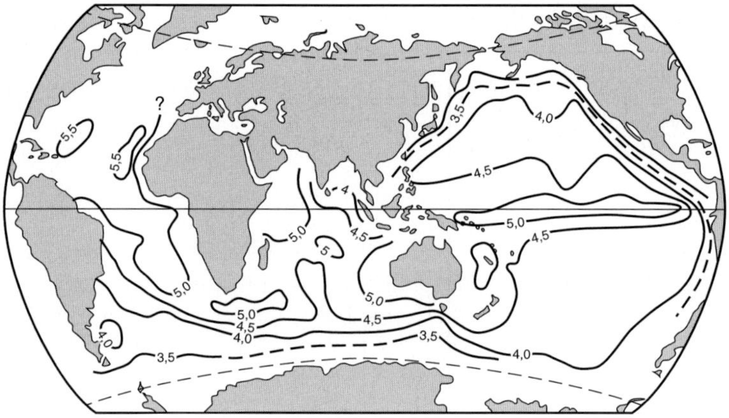

Fig. 3.21 Current topography of the calcite compensation depth (simplified map from Berger and Winterer, 1974). Depths are expressed in kilometers. Note how this level rises near the continents.

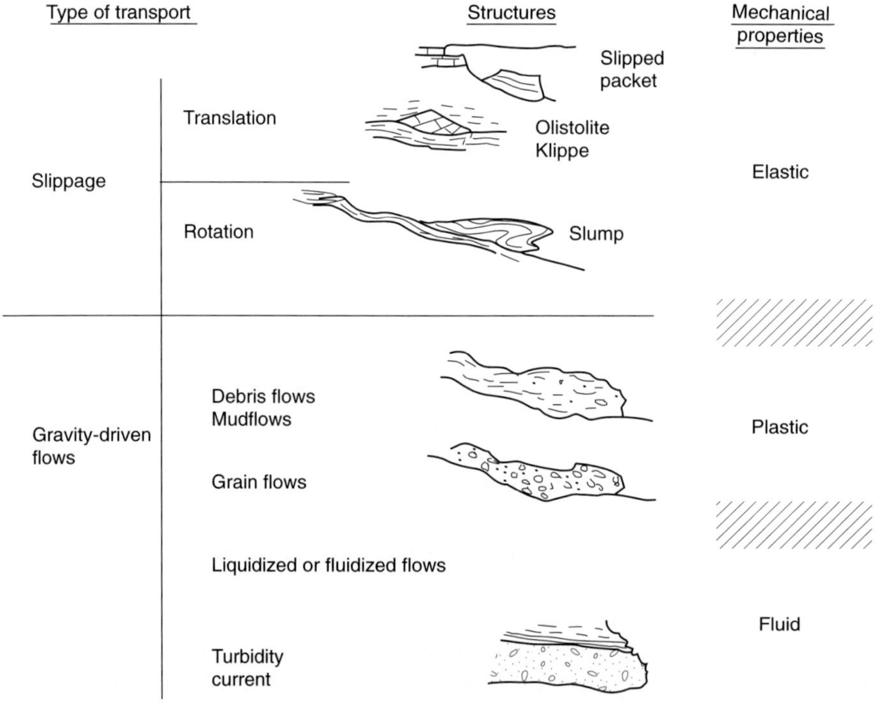

Fig. 3.22 Different types of resedimentation.

most varied **sedimentary structures** can be found in environments where the hydrodynamism changes most. Some materials like quartz sands have an aptitude for conserving deposit patterns, but of course this is not true for the less stable chemical deposits where post-sedimentation processes (diagenesis) will have their effect and the original deposit forms will be partially or totally obliterated. This is true of evaporites and dolomitic rock, for example.

The most characteristic forms of deposit are those produced by rolling transport, the most familiar of which are current **ripples**. The patterns are described as running crosswise to the direction of particle transport, or longitudinally in the direction of flow, with intermediate forms between the two. Depending on the current velocity and flow regime, the pattern will go from small ripples to dunes to flatbeds and finally to antidunes (see Figs 3.16 and 3.23). **Ripples** are distinguished from dunes by their wavelength. Anything smaller than 0.6 m is a ripple[1] while anything larger (sand waves, large ripples, megaripples) is a **dune**.

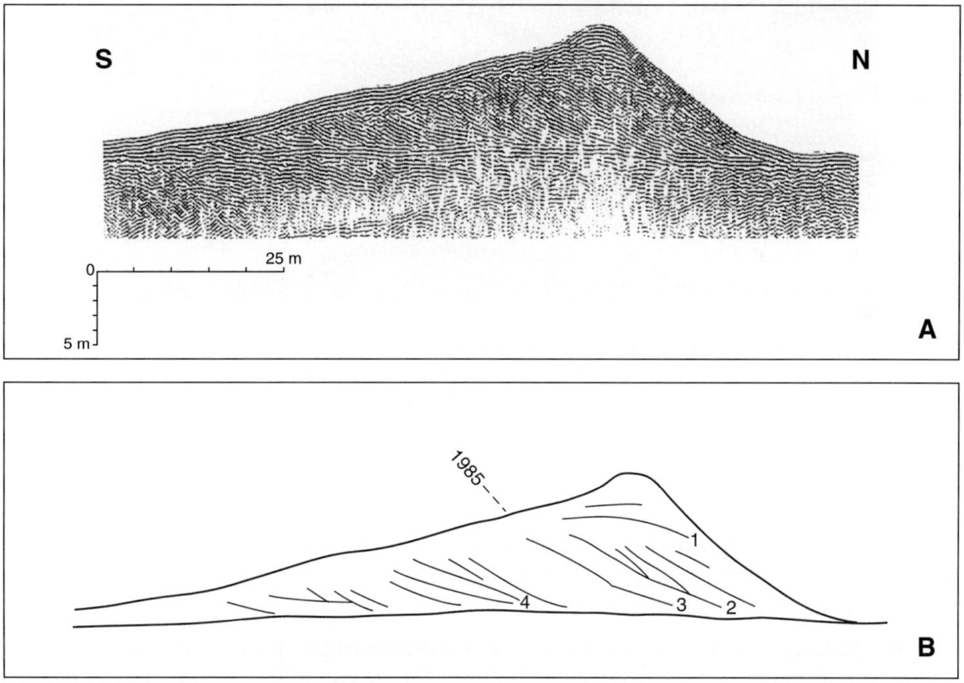

Fig. 3.23 Example of large subaqueous dune: section of a tidal barchan off Surtainville (from S. Berné, 1992).
A. Seismic image (high resolution technique). **B.** Interpretation of the internal structure.

1. According to the classification proposed by S. Berné, 1991.

Subaqueous ripples and dunes can be observed in any water depth from tidal to deep, and in various forms: straight, sinuous, S-shaped, barchan, depending on current speed and depth. These positive structures are in relief and differ from the hollow filling structures of channels, troughs, or cradles (Fig. 3.24).

Fig. 3.24 Channeled sedimentary structure.
Here, a channel was dug out and then refilled by successive sets of sandy avalanches until it was completely refilled (Saharan Ordovician) (IFP photo).

If the current is **mono-directional**, the layering will be flat or concave but regular, and will be disturbed only by reactivation surfaces if any temporary erosion has occurred. The elementary result is **progradation**.

An **oscillating current** will result either in oscillation ripple marks or oblique layering in hummocks with more or less pronounced hollows, or cross-bedding of interleaved alternating stratifications corresponding to the various currents. These forms are specific to tidal domains (tide reversals) and to coastal areas in general, where waves, storms, and currents are important.

Because conditions vary in time and space (with daily and seasonal variations, sedimentary flow variations), the deposition forms left will be the sum of the various elementary forms stemming from these variations. They will leave flaser, wavy, or lenticular bedding

(Fig. 3.25). If the flow is turbulent enough to erode previously formed deposits, different hollow figures of erosion will appear such as flutes, crescents, or grooves, or traces of dragged objects, also leaving flutes, chevrons, or rebound marks. So the original structure can be deformed or completely obliterated by biological activity on the floor (Fig. 3.26), or the escape of the original interstitial water, overloading, or the current. Storms and earthquakes may also leave a record, and many sedimentation gaps (hiatuses) may exist if erosion is resumed.

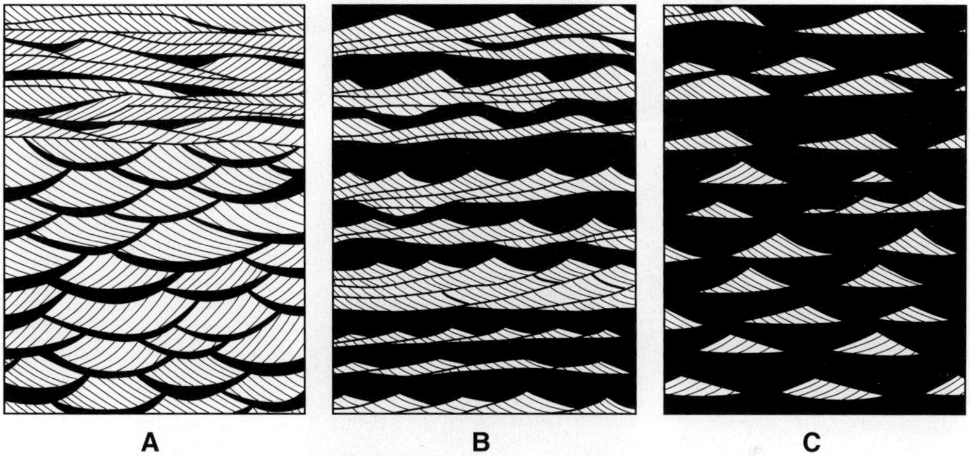

Fig. 3.25 Ripple bedding.
Schematics of wavy and tidal oscillatory current marks. **A.** Flaser bedding. **B.** Wavy bedding. **C.** Lenticular bedding (from J.R.L. Allen, 1982).

Special forms of deposits are created by gravity-driven flows and floor currents at the foot of slopes and at great depths. One such formation is the **deep sea fan**[1] or drifts made up of turbidites organized in "complexes" with slope lobes, deep lobes, channels, and lateral levees (see the ocean environments later in this chapter). Contour currents also construct **sediment drifts**.[2]

As we have said before, the great depths are mainly characterized by regular drapings of pelagic and hemipelagic sediments that may include distal turbidites in no particular structure (Fig. 3.27). This is also the domain of sedimentary **hiatuses** over broad areas where **hardgrounds** will form if the diagenesis is precocious.

1. See Middleton and Hampton, 1986; Bouma and Coleman, 1985; Mutti, 1985.
2. See Hollister, 1983; Stow and Piper, 1984.

A

B

Fig. 3.26 Examples of bioturbation modifying the original texture of a sediment (from Beuf et al., 1991).
A. Traces of a Cruziana type trilobite crawling across a bank surface.
B. Practically joined worm burrows (Tigilites or Scolithus).

The deposition forms therefore depend on the volume of available material, the hydrodynamics, seasonal variations, water depth (accommodation power), which will vary with the tectonics and eustasy (see farther on).

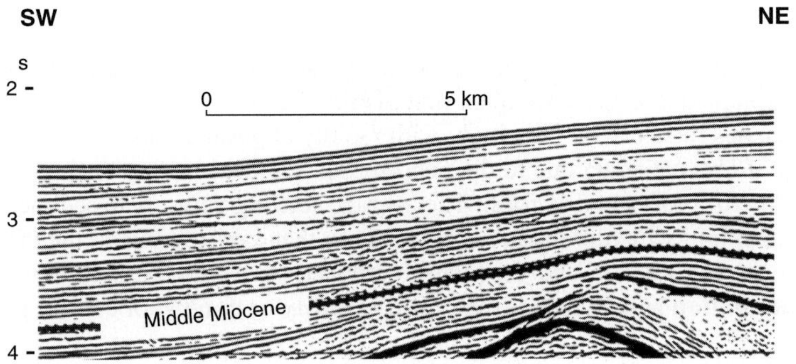

Fig. 3.27 Seismic example of pelagic and hemipelagic draping (offshore Venezuela) (IFP photo).

In the tropical domain, they are highly influenced by the primary production, especially for all of the reef constructions we spoke of before, yielding aggradational systems (built upward), but also progradational and retrogradational, depending on the relative level of the sea.

3.1.2.4 Sedimentation Rate

The **sedimentation rate** for different sedimentary mechanisms will depend on geographic position (climatic aspects), the type of material source (detritic, chemical, constructional), biomass activity, sea-level variations in time, and other factors. In certain cases, a thick **accretionary prism** or **pile** will accumulate, while in others there will be a more or less pronounced **void** or sedimentation hiatus.

A. Flow of Matter

The quantity of dissolved particles the ocean receives each year is difficult to estimate with any precision. **Rivers** are one fairly well known source (with a highly inhomogeneous geographic distribution and a buffer effect near the river mouth). **Eolian transport**, by wind fields, is still poorly quantified. **Precipitation**, by washing out the atmosphere, is a third source of dissolved particle material, which has been fairly much overlooked until just recently. Lastly, there is **primary production**, which is the origin of the ocean's calcareous and organic input. While primary production is well known in its general forms (Fig. 3.11), it does fluctuate, for example with the development of hydrological fronts (dynamic ocean surface phenomena described later in this chapter). So it should be remembered that little is as yet known of the phenomena occurring in the water column and at the interface with the floor.

B. Sedimentation Rate

Keeping the extreme variability of sedimentation rates in mind, according to climatic, tectonic, and other environmental aspects, we must first distinguish between instantaneous and average rates. Instantaneous rates can be very high, with rapidly deposited mass flow or turbidity, creating an array of oblique strata due to the neap/spring tide cycle, or producing carbonate reef (as much as 6 m/10^3 years), and so forth. The average **settling rates**, with sedimentation slowing down or stopping, will vary greatly (Fig. 3.28) so that, considering all the effects in geology, with intervening erosion and especially compaction, it may be preferable to speak in terms of **accumulation rate**, which is expressed in m/My or mm/ky, or even in g/cm^2.

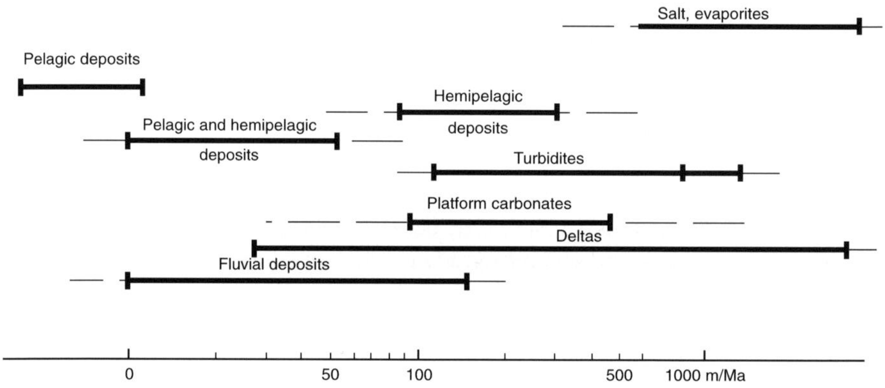

Fig. 3.28 Variety of sedimentation rates.

Sedimentation rates vary according to geographic position and the variation of the sea level in time (with catch-up and keep-up in the carbonates).

Depending on the flow of sediments and the energy of the stream carrying them, the process will change from horizontal progradation to vertical aggradation, from a continuous deposit to a sedimentation **hiatus**, or gap. Depending on sea-level fluctuations, or **eustasy**, the sedimentary response will vary greatly with the **accommodation space** (see Chapter 4).

3.1.2.5 Autocyclic and Allocyclic Phenomena

The physical, chemical, and biological processes of deposition lead to the formation of organized sediments structured in **sedimentary bodies**. A sedimentation body is a set of facies linked to the deposition environment in more or less repetitive units on the decimetric, metric, or decametric scale. This pattern has been termed a **cyclothem** and it has long been considered that the pattern is due to **autocyclic** phenomena, i.e., phenomena specific to the sedimentary system considered: channels and levees, river avulsion, wandering deltas, dune migration. These are said to be **genetic deposition units**. Their arrangement in a stratigraphic sequence in a basin will be discussed in Chapter 4, where it will be seen that the stratigraphic rhythms observed are often due to oscillations in some strong external control, though changes in the deposition conditions themselves must not be overlooked. These strong outside controls are called **allocyclic** driving factors, such as tectonics, eustasy, and climatic variations (Fig. 3.29).

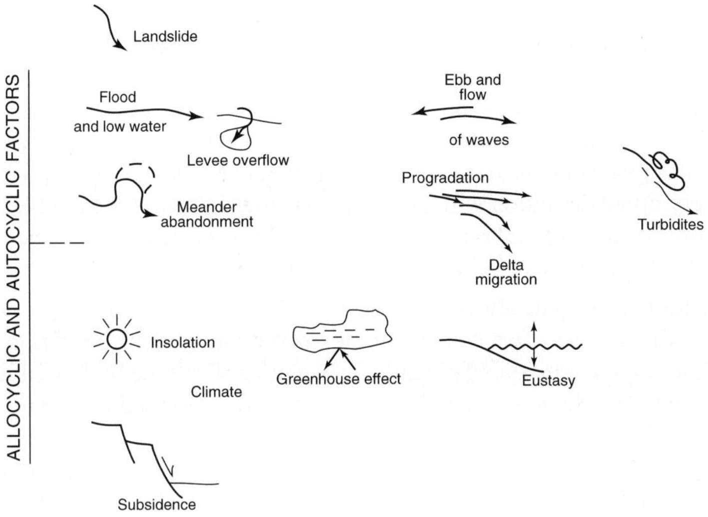

Fig. 3.29 Autocyclic and allocyclic factors.
The sedimentary pattern is created and governed by the deposition process itself (autocyclic), but is also affected by external (allocyclic) factors.

Not everyone believes in the allocyclic origin of deposition units. One alternative discussed today is deterministic chaos theory, with apparently ordered random phenomena and unforced oscillations.

3.2 CONTINENTAL ENVIRONMENTS

On the global scale, sediments that are trapped on the continents are only a minor proportion of all sediments found in sedimentary basins, but they are important in cratonic basins and continental rifts where rapid subsidence conditions allow this trapping to take place. The trapping is often transitory, as much of this sediment is later transported to the ocean by wind and especially by rivers. Although the various processes involved are often associated geographically, the following will deal with the main deposition environments separately and successively, after a few general considerations.

3.2.1 General Characteristics

3.2.1.1 Continental Facies

As will be seen below, continental deposits are extremely varied. They are described by a **facies**, which is the sum of a sedimentary deposit's lithological and biological characters. A facies is defined in terms of sedimentation criteria, i.e., all of the criteria by which the physical, chemical, and biological conditions of the environment are described. The term **continental facies** designates deposits made in the continental domain, as opposed to **marine facies**. This very broad definition will have to be narrowed down, for example by defining a lacustrine or fluvial facies based on more precise criteria. Facies will also be distinguished by climatic indicators, such as glacial facies or sebkha evaporation facies.

A **genetic indication** will always be given. This is based on simple recognizable features, which is why a facies often calls for some interpretation in the definition. The characters are often too imprecise to qualify clearly as one or another type of deposit. For example, a sand bank (sandstone) may not include any elements for determining whether the deposition process was fluvial or marine. The facies term will then be used for a set of genetically related deposits in a sequence (see Chapter 4) including various lithologies in association. As it is generally used in this sense for a bank or group of banks, a smaller scale is sometimes defined. For example, **microfacies** may be defined on the basis of microscopic observations. This will come up again in the discussion of marine deposits.

3.2.1.2 Deposit Zoning

The importance of climate has already been mentioned as concerns the weathering phenomena that occur in the continent. In the Earth's present geographical configuration, glacial and periglacial deposits predominate in higher latitudes while deserts develop in arid and semiarid intertropical zones, where eolian facies and intermittently operating flood plains predominate. Surface weathering phenomena are therefore organized in ordered belts of different types of soils (with podzolization, allitization, and related processes), such as the saline deposits that require maximum evaporation zones (Fig. 3.11). The continents thus exhibit latitudinal bands of **facies belts**. Of course, this distribution is largely modulated by the geographical configuration of the continents and oceans, the extent of the continental areas, and global atmospheric transfers.

This idea of **climatic zoning** (Fig. 3.30) as applied to the past is very useful because it opens the door to the idea of **paleogeography** where other considerations come into play, such as the position of the poles (determined by paleomagnetic data).

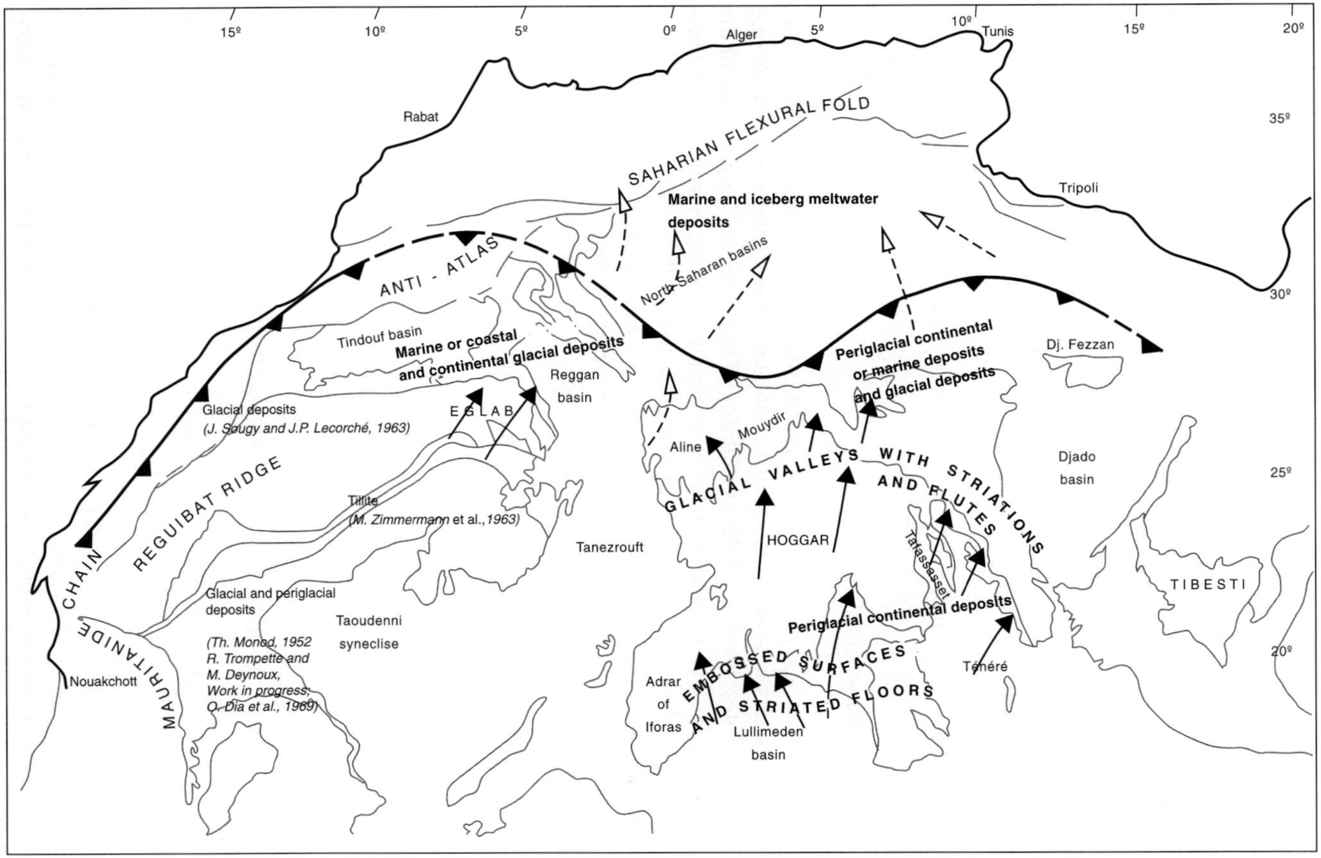

Fig. 3.30 Example of climatic zoning in the past: Saharan glacial Ordovician.
The south pole of the time was located in Africa at that time, and was covered by broad inland ice with periglacial deposits developing on the margins.

Anticipating the following chapter, let us note that today's situation with ice-capped poles and contrasted climates is not necessarily applicable to all geological periods. Climatic balances are in fact extremely fragile.

Altimetric zoning is superimposed on climatic zoning. This is due to the effects of relief, and is thus generated by tectonic processes. Vegetal cover, for example, is **tiered** by altitude. In certain cases (take the example of the island of Hispaniola in the Caribbean, with its contrasted relief), the vegetation changes from an alpine type to tropical, and even to a hyper-arid environment within a few kilometers. It is clear that weathering and erosion conditions are extremely variable over very short distances. Special thermal conditions will affect the deposit facies directly, such as in the high plateaus of the Andes and Tibet, and it is sometimes possible to recognize such situations in the geological past. Sometimes too, temperature and humidity may upset the analysis: with the lack of hydrolysis in desert regions, the weathering processes may seem more typical of high latitudes.

3.2.1.3 Oxidizing Environment

Except for special lacustrine and palustrine environments that will be addressed later on, one of the pronounced characters of sedimentation in a continental environment is the oxidizing medium in which it operates, because of the availability of atmospheric oxygen. One of the major results of this situation is the low probability of organisms being conserved. If organisms could be fossilized, they would provide precise indicators of the prevailing ecological conditions. But even though the continental biomass is considerable, only a small part contributes to continental sedimentation and the rest is recycled through the general carbon cycle. One of the characteristics of continental facies, then, is that it is devoid of life (azoic) in the same way as for the geographically more extensive eolian and fluvial deposits, to the point where past geological series are sometimes referred to as **azoic facies**, a negative character that is used to qualify it as a continental deposit.

This is a dangerous approximation that can be avoided by finer analysis, especially by palynology (spores and pollens), which has shown that certain life forms can in fact be preserved.

The question has also been raised as to what atmospheric composition prevailed during the oldest Precambrian, as this might have differed greatly from the well-established composition used freely for all series since the Cambrian.

These ideas will be taken up again later, especially in the discussion of the preservation of organic matter.

Oxidizing conditions are also reflected in the development of **rubifaction** (reddening of the soil) as the ferriferous minerals undergo heavy oxidation (Fig. 3.31). This enters upon the effects of diagenesis, which will be covered in Chapter 5. Here, it is simply pointed out that the distinction between those primary characters linked to an initial continental environment and those acquired during some later geological history (with emersion and return to open air) is often difficult to make. Many "red" deposits do, however, provide a classic illustration of continental series, such as in the European Permian.

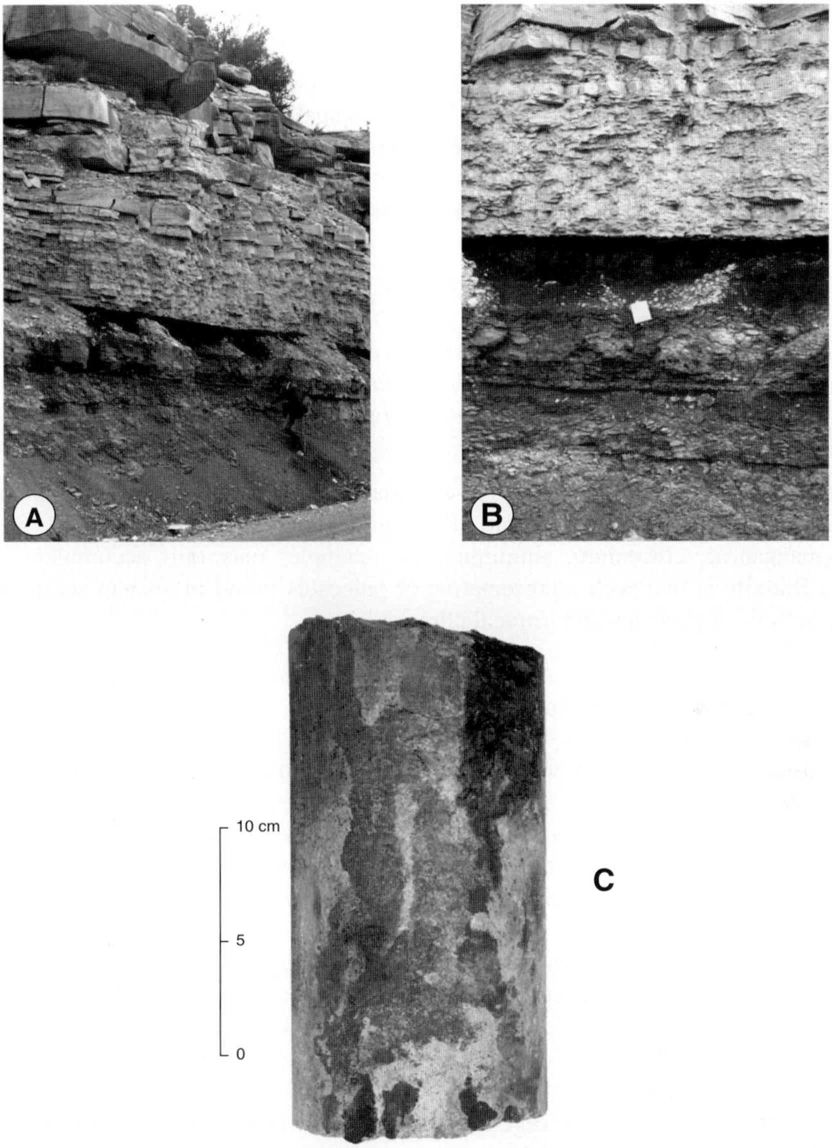

Fig. 3.31 Ancient continental deposits affected by rubifaction.
A and **B.** Fluvial deposits of the coastal plain (sands, clays, coals) show up in dark in the photo, with overlying transgressive marine sands in light. **C.** Rubifaction of karstic fill (dark ferriferous clays) in a light carbonaceous rock (IFP photos).

3.2.1.4 Soils of Different Types, Paleosols

The continental weathering already discussed above (see Fig. 3.7) will lead to different types of soil, depending on the climate. In the tropical zone, the weathering mantle can develop into a thickness of a hundred meters, constituting a reddish **lateritic mantle** or **laterite**. This is the result of intense leaching by meteoritic atmospheric water, with a residual formation of kaolinite-type clay beds of oxides and hydroxides of iron, aluminum and manganese.

The transition from the bottom of the laterite to the parent rock is gradual, with unaltered relics appearing simultaneously within the weathering products (coarse saprolite levels). Toward the top, the solid rock structure is lost little by little, with different generations of neoformation appearing if the duration of weathering is long. Lateritic soils can evolve toward tropical paleosols or lead to accumulations of iron oxide in what is called "ferricrete".

Ferricrete sometimes develops over considerable areas: large lateral variations may often be prompted by a lateral dynamic of leaching, and concentrations of fairly immobile metals (manganese, chromium, aluminum, for example) may thus accumulate into ore deposits. **Bauxite** is one such representative of paleosols found in ancient series, and is a valuable archive of these ancient tropical climates.

3.2.1.5 Vegetation and Photosynthesis

Another characteristic of continents is the vegetation that exists in many forms. This prompts a number of remarks:

- The situation is evolutionary. It is known that there was no such vegetation 500 million years ago, and the massive colonization of the continents with a plethoric diversification of the vegetal kingdom dates only from the early Devonian, about 400 million years ago.
- Vegetation plays an extremely important role in regulating weathering and erosion phenomena (Fig. 3.32), by stabilizing river banks, for example. The type of circulation and organization of atmospheric and fluvial water will also depend largely on the type of vegetal population. The comparative situations of the Ordovician and Cretaceous provide an example of this.
- The continental biomass of vegetal origin is largely dominant and is about equivalent to the marine biomass (Fig. 3.11).
- Photosynthesis plays a major role. Its coupling with the atmosphere is expressed in the carbon cycle (Fig. 3.33). All combustible materials of the coal family and a large share of the hydrocarbons formed derive from this chlorophyll function of land plants (the rest derive from marine algae). We will return to this point in greater detail in Chapter 6, in recognizing the different origins for kerogens stemming from the continental and marine environments.

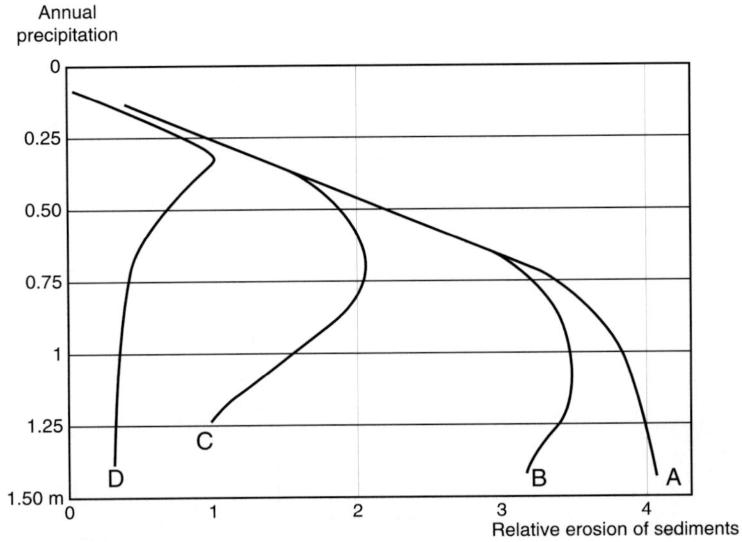

Fig. 3.32 Role of vegetation in erosion processes. Curves showing the theoretical relation between erosion and precipitation (from S.A. Schumm, 1968).
A. Before land vegetation appears. **B.** After primitive vegetation appears. **C.** After flowering and coniferous plants appear. **D.** After grasses appear (Gramineae).

3.2.1.6 Biomarker Fossils and Crises

Although there is little chance of organic remains being preserved in the continental domain, a number of noteworthy cases and circumstances are exceptions to this rule, notably in relation to singular events:

- Lacustrine stromatolites and other algae
- Special groups including reptiles and dinosaurs showing spectacular (or gradual) extinctions
- Amber containing insects and other organisms
- Relatively abundant mammal fossils in more recent settlements (Pliocene-Quaternery).

The continental bridges by which faunal groups effected transfers and migration are important for delineating palaeogeographic configurations.

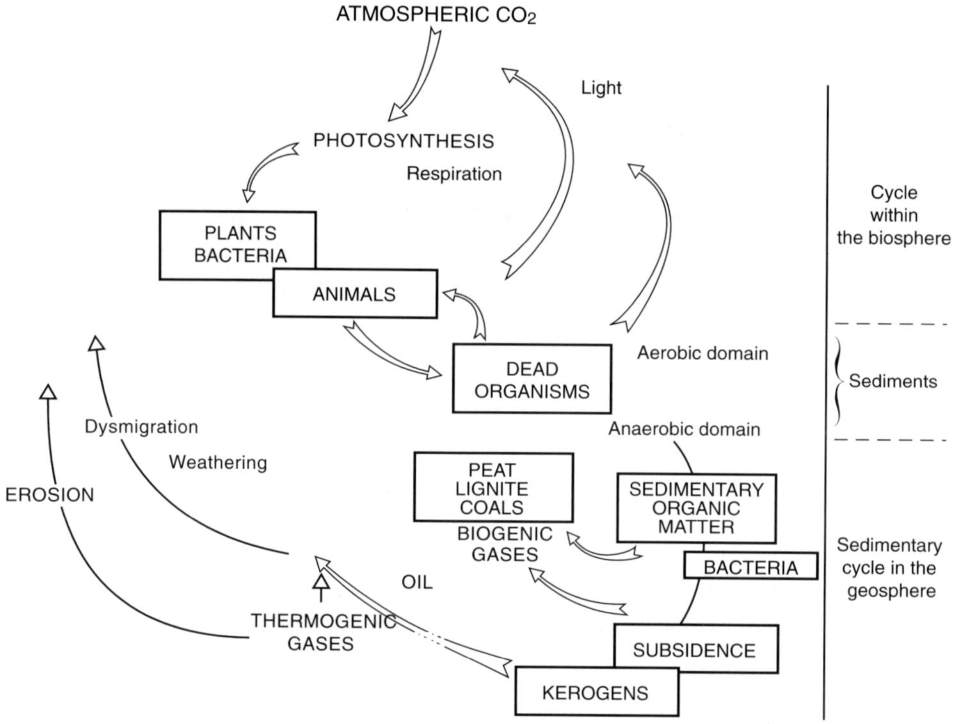

Fig. 3.33 Geochemical cycle of organic carbon.

3.2.1.7 Continental Morphologies

Characteristic morphologies are fashioned on the continental surfaces by air, running water, and ice, acting as erosion and transport agents and countered by the fixation due to vegetation. This **morphogenesis** is part of the field of **geomorphology**, which designates the shaping processes that are often precise indicators of driving mechanisms.

As relief evolves, it normally tends to flatten out gradually, and the role of run-off and running water is essential. A river's **equilibrium profile** (equilibrium between erosion and deposition) is a surface that gradually flattens out toward sea level, which is generally considered to be the **erosion baselevel** (discussed in following chapter). The exact topographical **landform** largely depends on climatic control.

Several models have been proposed for this. Remember the **peniplain** formation (Fig. 3.34), which is the way a long fluvial erosion phase will complete its work on an originally uneven topography in temperate regimes, as opposed to the **pediplain**[1], which is valid

1. See Davis, 1989; Penck, 1924.

A

B

C

Fig. 3.34 Example of ancient pediplain.
The surface is remarkably flat, viewed both from the air (**A**, **B**) and from the ground (**C**). The Precambrian basement of the Hoggar is eroded and the Paleozoic sits in discordance on this erosion surface (IFP photos).

for desert, semi-arid and dry intertropical origins of broad inclined plains called **pediments**[1] (Fig. 3.34). Geology books generally mention the Infracambrian and Hercynian peniplains as such broad surfaces separating major sedimentary cycles. But the type of surface has to be determined too, in geomorphological analysis. In petroleum geology, paleo-relief can be found that leave **geomorphological traps** (Fig. 3.35) as discussed in Chapter 6.

The intermediate shaping stages yield transitory landscapes that are unlikely to be conserved by fossilization in a geological series. The famous Badlands, for example, are an instantaneous expression of an erosion process. But on this scale of observation, the **valley** system or even **fluvial channels** are morphologies that can be identified in the field in alluvial plains or subsurface series (Fig. 3.36). These morphologies have been described on various scales in very many geological series.

There is a kind of competition between erosion, which flattens relief, and the tectonic uplift that recreates them. Uplifting increases the volume available for erosion, as does the lowering of the sea level.

Wind is also important. In addition to forming dune massifs and dunes as discussed later on in this chapter, wind action results in characteristic erosion patterns. The **yardang** is one typical form of such erosion in which regularly parallel crests and ridges of hectometric dimensions develop on rocky blocks. Another is the **hydro-eolian depression** of desert cups, depending on the water table position. The forms observed today in desert regions do not seem to have been described in past series.

Lastly, there are the well known glacial and periglacial morphologies known from studies of the Quaternary. These morphologies are characterized at different scales ranging from the broad abrasion surfaces forming glacial valleys at the base of the inland ice to the forms observed on the centimetric to metric scale in frost cracks and polygonal soils. Past morphologies have been studied intensively[2] and are clearly illustrated in the terrain of the late Saharan Ordovician. Deep **U-shaped valleys** are sometimes formed, with **till** and **mackerel surfaces** or **roches moutonnées** (embossed rock) where pressure-scouring by ice and subglacial torrents will dig out overdeepened **umbilicus cavities** and leave crosswise **rock bars** (or "riegels") and, on a large scale, **subglacial channels** with steep walls exhibiting forms of cavitation. The embossed surfaces themselves are characterized by **alignments**, mounds or **drumlins**, **flutes** and **scratches**, grooves or furrows (Fig. 3.37) from which the direction of ice flow can be reconstructed. The pull-out figures of crescentic gouges and lunate fractures on the centimetric scale are **glacial chattermarks**, but on the kilometric scale we can also sometimes find **glacio-tectonic ridges** piled up by the ice as it presses forward (Fig. 3.38). Glacial and periglacial accumulation forms are themselves characteristic and will be described later on.

1. See discussion in Beuf et al., 1971.
2. See, for example, Beuf et al., 1971.

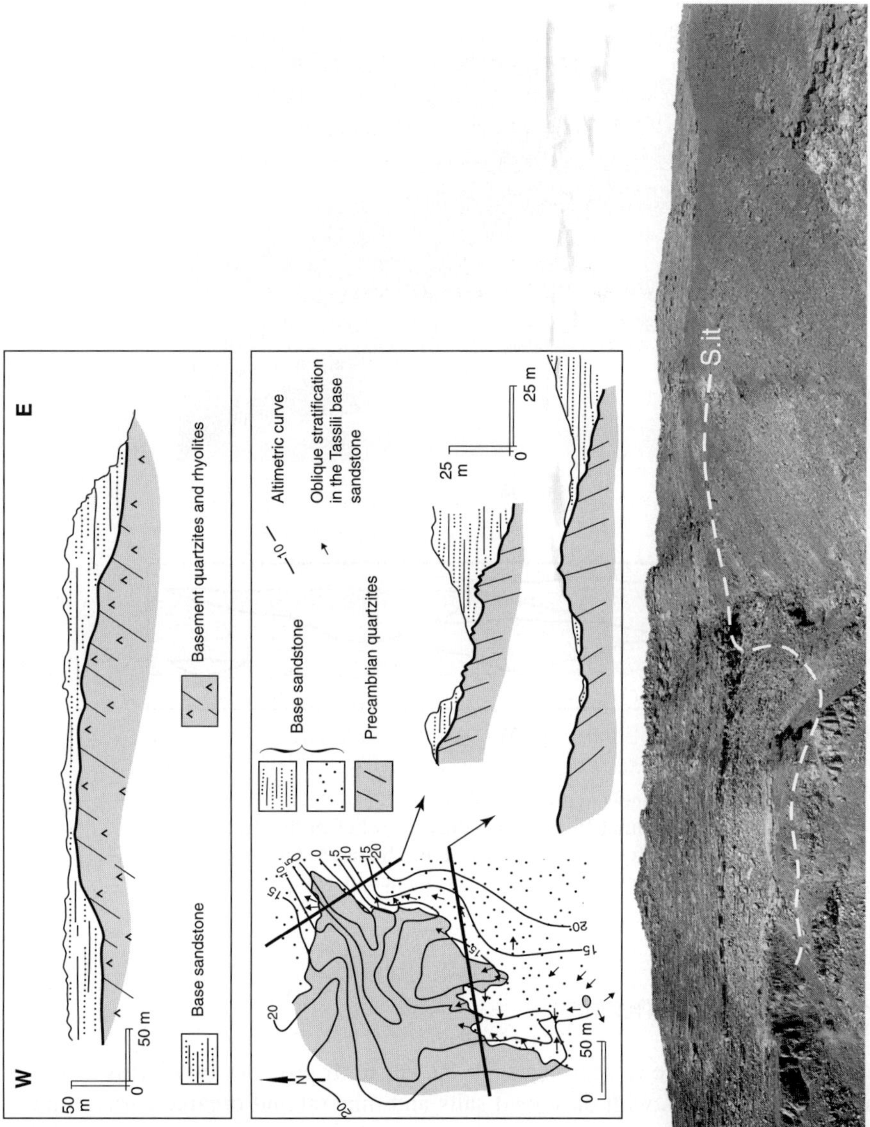

Fig. 3.35 Example of paleorelief buried in the sediments: the Cambro-Ordovician base sandstone onlaps an irregular surface (IFP photo).

A

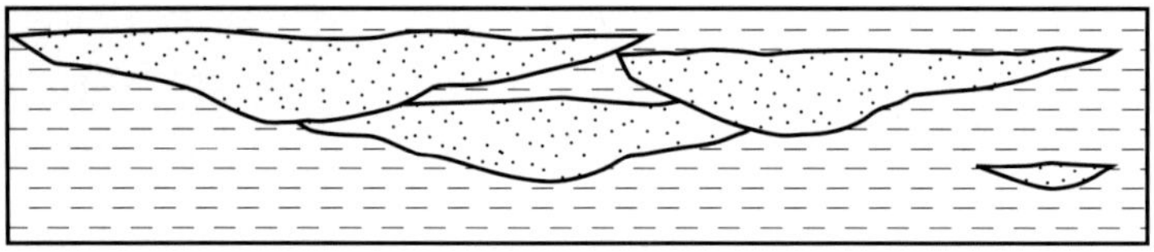

B

Fig. 3.36 Fluvial channels in alluvial plains.
A. Ancient valley channel filled with sands (in relief in the present landscape because of differential erosion). Late Saharan Ordovician (IFP photo). **B.** Nesting of several channels in sequence in a coastal plain (cross section).

3.2.1.8 Transfers with the Ocean

The continent is a major source of sediments for the ocean. It provides two forms of input: solutions charged more or less with dissolved salts and mineral and organic particle material, which are the ultimate products of the climate-regulated continental weathering and erosion processes.

Great rivers are the major source of continental input (Fig. 3.39). The large organic input induces large and varied traffic networks in the transfer zone. The abundance of nutritive inputs determines a very high rate of primary productivity, and a high sedimentation rate by organic matter. The estuarian and deltaic domains are zones of particular interest that will be taken up again later in this chapter.

Fig. 3.37 Examples of glacial pavement (Late Saharan Ordovician). **A.** Striae or furrows. **B.** Grooves and flutes. **C.** Drumlins (IFP photos).

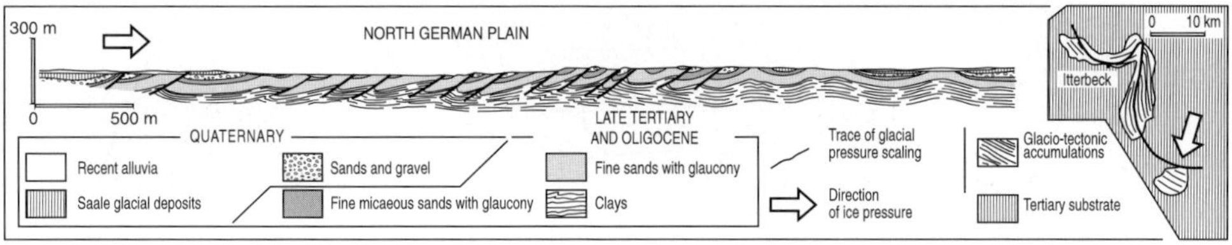

Fig. 3.38 Example of glacio-tectonic feature in the Quaternary of northern Germany (from W. Richter et al., 1950, in Beuf et al., 1971).

The eolian transfers from the continent to the ocean are far from negligible. The major sources of fine particles are the desert belts, and notably the Sahara (Fig. 3.40). Winds can carry clay material over considerable distances along with the finest grain sizes of a number of minerals (quartz, feldspath of granites, and even calcite). Both past and present distribution areas can be very large.

Volcanic emissions in the atmosphere are a source of material from the continent, in the form of ashfall. Tropospheric transport can cover very large distances (Fig. 3.41) and deposit global markers used in relative dating (**tephrochronology**), which will be spoken of in Chapter 4.

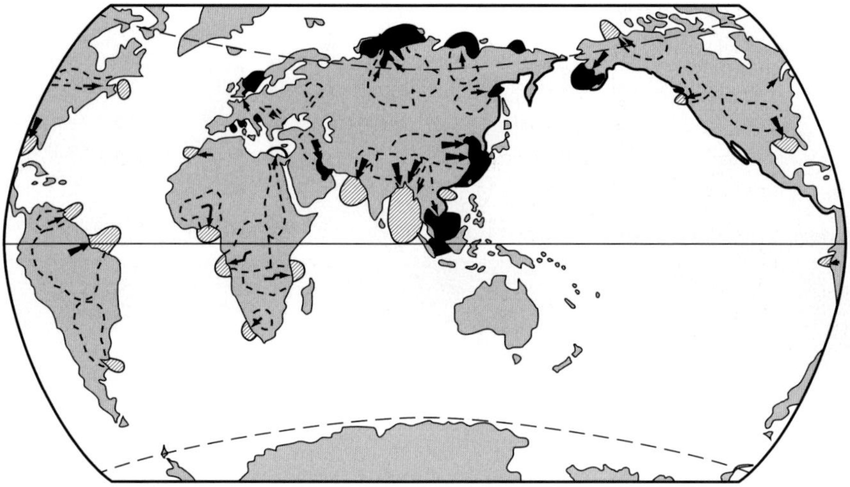

Fig. 3.39 The Earth's major fluvial systems and their platform (black) and deep (hatched) underwater sands.

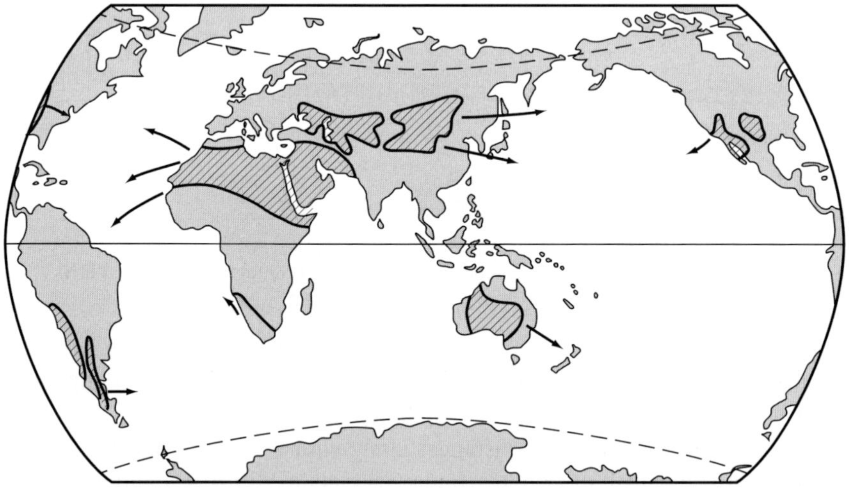

Fig. 3.40 The Earth's broad desert zones, sources of deflation by air and input to the ocean.

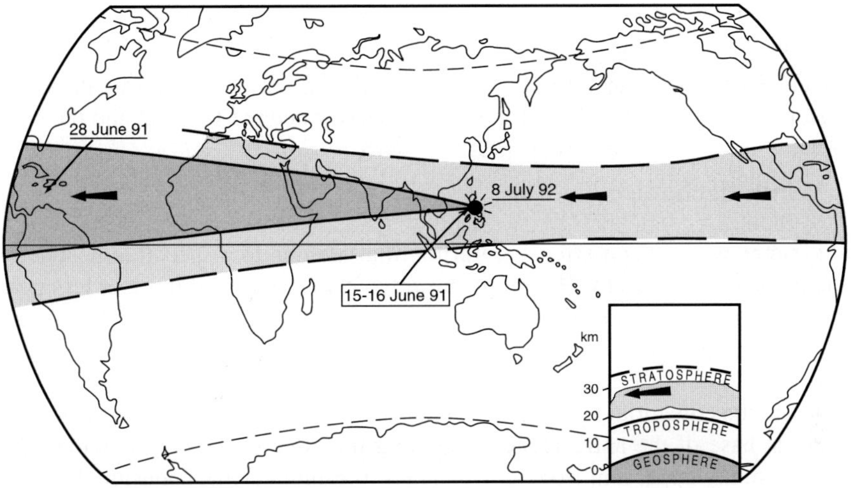

Fig. 3.41 Eolian progradation of volcanic ash with the eruption of Pinatubo. The stratospheric cloud took a little more than a year to circle the Earth.

3.2.2 Eolian Systems and Deposits

In the present geographic situation, we find hot deserts (Sahara, Kalahari) and cold deserts (Gobi), deserts of small extent (Chile, the United States) and those of near-continental size (Australia, Sahara). A desert is a region "where water in the liquid state is rare or practically absent"[1], i.e., a zone of aridity of even hyper-aridity. It is in these zones where rainfall is very low that wind will be dominant as the fluid transport/deposition mechanism, making **eolian deposits**. Although deserts today cover nearly a third of the Earth's un-iced emersed land, which is enormous (Fig. 3.40), they are not the only domain where sizable wind forces act as geological deposit generators. Atmospheric and climatic conditions are such that the edges of the great ice caps (inland ice sheet) and also the barrier island and coastal fringe zones under different climates are also the locus of wind action leading to considerable accumulations that can be observed at present and have been recorded in past series.

Before entering into the subject here, remember that the climatic mechanisms that lead to aridity and hyper-aridity vary quickly in time. The balances are fragile, and the boundaries fluctuate and are fuzzy. It is known, for example, that short pluvial periods have alternated with aridity. On the human scale, exceptional episodic rains can prompt remarkable flux in the great wadis, such as that of the Saoura in the western Sahara. In

1. Rognon, 1989.

other words, eolian deposits may "cohabit" or alternate with fluvial type accumulations (see further on).

In ancient European series, the Permian is a period when great accumulations of eolian sand developed into what are now the gas field reservoir beds of Holland and the North Sea.

3.2.2.1 Wind Mechanisms

The arid systems where wind effects are primordial are due to a special atmospheric situation in the broad desert regions. Here, it is the anticyclone regime of more or less permanent high pressures that is accentuated in the vicinity of the tropics. This positive anomaly of pressures reaches up 10 to 15 km to the upper limits of the troposphere. If we simplify the example of the Sahara, we have a dynamic anticyclone (Fig. 3.42) rimmed by interfering antitrades from the equatorial vortex, trade winds and lobes of the "polar" air front that boost the pressure at base of the anticyclones, including the effects of the trade winds. The very fast jet-stream of the upper atmosphere strengthens the anticyclonic cohesion. High-pressure cells also exist at high latitudes, generating cold winds at the edge of the inland ice, carrying blizzards.

There are different modes of particle transport near the ground:

- The finest mineral particles (clays, silts, very fine sands) are stripped from the ground and placed in **suspension**. This is the eolian **deflation** process that causes sandstorms, dry haze, aerosol-charged twisters, lithometeors of the middle atmosphere. Micrometric particles can be transported over considerable distances (several thousand kilometers) before settling out to sediment when the wind carrying them comes up against another front, causing red rains (or even colored snow) that can be seen as far as the temperate and tropical regions. One particular form of accumulation of this type is **loess** (covered by a weathered part called lehm), which is an accumulation of calcareous dust carried by wind from the periglacial domain or cold steppe. This can be found, for example, in the Quaternary of northern Europe and central Asia, from the Caspian to the north of China. This "steppelike" facies often includes continental mammal fauna and traces of roots, along with a few mollusks from wetter regions.
  Northern Europe's loess is actually a series of cold deposits (loess in the strict sense) and soils (lehm) formed in interglacial periods. They are subject to intense erosion and their chances of fossilization are poor.
- Drainage and saltation **transport mechanisms** are similar to the subaqueous processes seen in the first part of this chapter, and the forms of deposit are also similar, depending on the wind velocity and particle size. High eolian **dunes** can accumulate with the material prograding in front of successive and often very steep avalanches (Figs 3.43 and 3.44).[1]
- The **erosion mechanisms** mentioned above are at play, with deflation by the hydroeolian depressions, formation of yardangs, with maintenance of broad pebbly surfaces

1. Bagnold, 1960; McKee, 1965; Bigarella, 1972; Wilson, 1972.

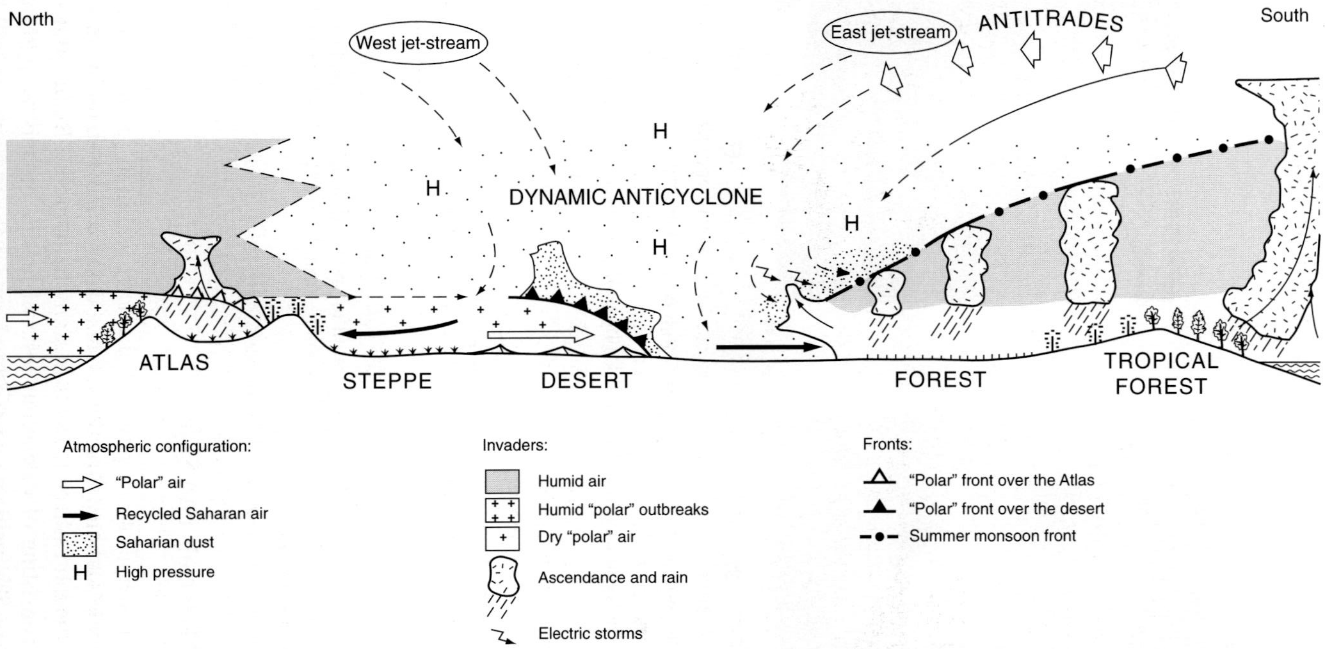

Fig. 3.42 Anticyclonic atmospheric configuration of the Sahara (from P. Rognon, 1989).

and lag deposits (Saharan ergs) or intermediate sandy **fedj** and special shapes found on quite another scale, such as the famous dreikanter and faceted pebbles.

A B

Fig. 3.43 Oblique stratifications of eolian deposits.
A. General view: Jurassic, Chelly canyon, New Mexico. **B.** Close-up of Entrada formation, New Mexico (IFP photos).

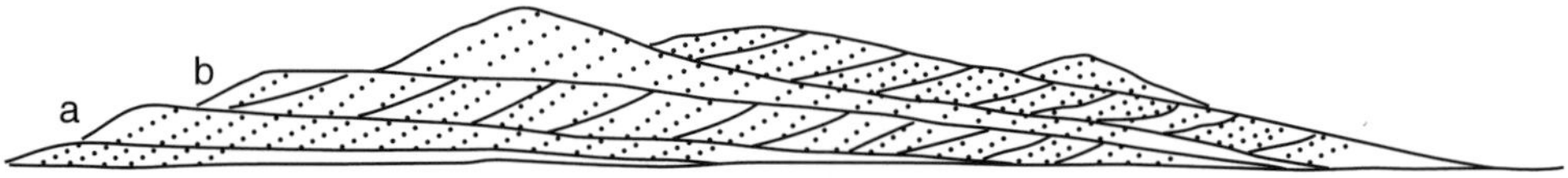

Fig. 3.44 Superposition of various arenaceous accumulations (eolian dunes). The vertical scale is exaggerated to show (a) simple avalanches of dune sands with steep slopes and (b) more complex systems with second-order erosion levels.

3.2.2.2 Deposit Materials and Shapes

The wind can carry material of any physical character within its grain size capacity: clays, silts and siliceous sands, carbonate fractions. Great desert wind systems are characterized by well-graded fine-to-medium siliceous sands in shiny blunt grains. But coastal eolian deposits are known too, consisting of calcareous tests of marine organisms, gypsum, and other material depending on the available source (and sea-level variations, as will be seen later on). The faunal and floral content is null in hot deserts, but may be very high in insular or coastal fringes.

It should be remembered that, contrary to fluvial subaquatic dunes, the statistical slopes of the oblique stratification sets have no paleo-slope significance. They only reflect the wind direction and not the regional topography.

On the small scale, for the large dune massifs (Saharan **ergs**), the large sand accumulations are found in the large topographical depressions corresponding to the idea of a sedimentation basin. The neighboring relief is of great importance on all scales.[1]

Different accumulation forms are customarily distinguished by size: ripples, dunes, and megadune draa (Fig. 3.45), each corresponding to a particular relation between the grain size and the length of the undulation.[2] These various structures can overlay each other. The hectometric scale of the **draa** is explained by the fact that the atmospheric layer that forms it can reach three kilometers in thickness. In the ergs, the dunes organize along wind lines (Fig. 3.46). Between dunes, when the underlying water table intersects with the topography, a sort of depression forms that is called **playa** or **sabkha** (at times very linear, becoming a fedj) where evaporation can lead to the formation of crusts. Dune forms are highly varied: barchans, barchanoid ribbons, linear or starred ripples.

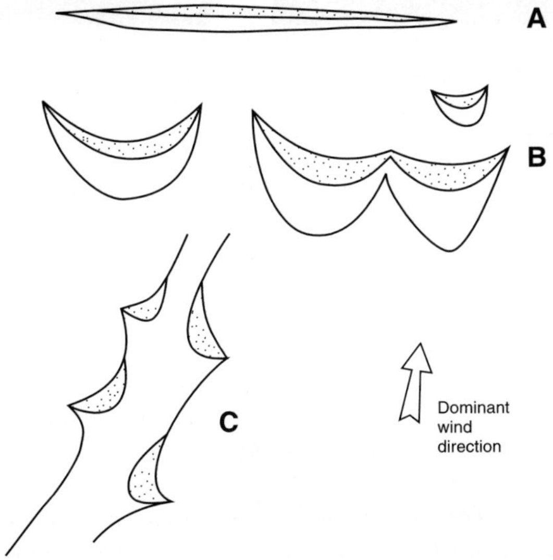

Fig. 3.45 Schematics of eolian dunes (plan view).
A. Transverse dune. **B.** Barchans. **C.** Elongated ribbon dune.

1. Mainguet and Collot, 1974.
2. Wilson, 1972.

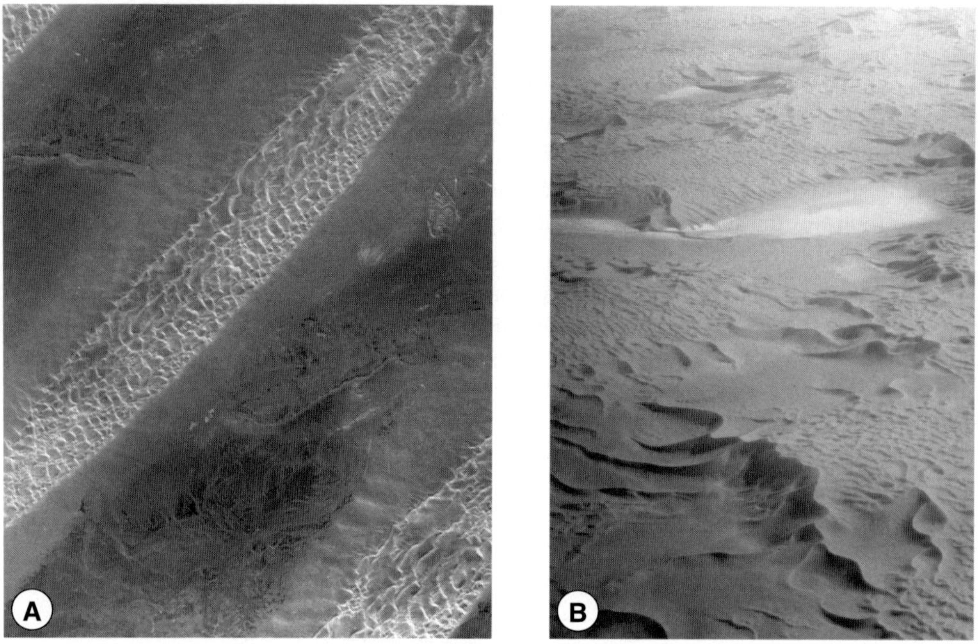

Fig. 3.46 Eolian sand ribbons.
A. Chech erg in the west Sahara (IGN photo). **B.** Issaouan erg at the Libyan border.

Less well organized sand sheets can also be found at the edge of the dune massif (a border that wanders with time fluctuations), or draa. These sand sheets are characterized by oblique sets laid at low angles, distinguishing them from the steeply inclined foresets (up to 34°) of the dunes.
As mentioned before, an outer domain exists where the eolian actions are more modest and where the intermittent draining of wadi-floods will dominate.
Fixed or anchored dunes are generally distinguished from shifting dunes that can migrate, though some of the latter are in fact very stable, such as the great dune ridges of the Saharan Chech erg. The chances of dune system survival by fossilization is an important question because many forms are ephemeral. The conditions of precocious diagenesis are therefore essential, and particularly the role of water table fluctuation.

The desert landscape where eolian action dominates therefore includes a series of different elements where wind deposits will be different and will be found in association with other facies.

Coastal dunes are of different types (barchan and long ribbons) and are made of different materials (littoral drift, ice, estuarian input), notably with frequent vegetation, which anchors the dune. When the reworked littoral material is calcareous, the rock is often termed "eolianite", which is a misleading abbreviation for "eolian calcarenite". These calcareous dunes of bioclasts, pellets, ooids, and foraminifers, are doubtless formed preferentially during a period of low sea level when some carbonaceous stock is available for reworking by the wind. Early diagenesis develops quickly here.

In the periglacial domain, dunes are very flat and broad and often have a veneer veined by ice cracks. They are called **niveolian dunes**. Eolian deflation can generate enough material to fill these deep ice cracks when they thaw, producing sand wedges such as can be found in the Antarctic.

3.2.2.3 Great Eolian Deposition of the Past

Eolian deflation can entrain material into the marine domain and thereby participate in ocean sedimentation, mixing its fine particles with the pelagic deposits. The northeast Antarctic and north Pacific are two broad zones where eolian influence is important today. Continental eolian events are recorded this way, thereby providing information on the Earth's ancient climates.

This lateral (and vertical) superposition of facies can of course be found in past sedimentary series like the north European Zechstein and the Rotliegend Permian.

Oil fields are known in the Precambrian, Permian and Jurassic (Fig. 3.47). Deposits make excellent reservoirs because of the excellent grain size grading that induces high porosity and permeability (see Chapters 5 and 6). But it is, of course, a highly unfavorable environment for producing and conserving organic matter.

3.2.3 Lacustrine Environment

In terms of volume, lacustrine environments do not constitute large deposits in geological series. They can be found in the large intracratonic basins (lake Chad and the North American Great Lakes) and in the small intramontane basins of collision zones (Swiss lakes and French Alps) as well as along the major depressions of ancient and active rifts (Baikal, African lakes). Depending on the geotectonic situation, lacustrine basins will exhibit moderate or very active subsidence. In the variety of environments, it will be noted that certain lakes are deep while others are quite shallow.

The general character of the lacustrine environment is confinement, with generally precise boundaries and often high sedimentation rates.

3.2.3.1 Varied Processes, Varied Facies

There are two broad classes of sedimentation mechanisms that often work together or in alternation.

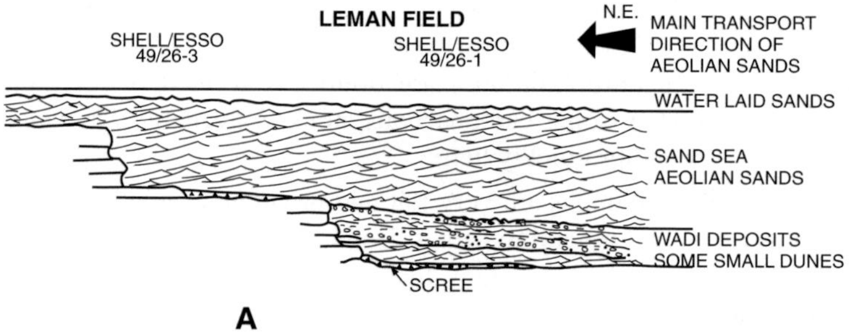

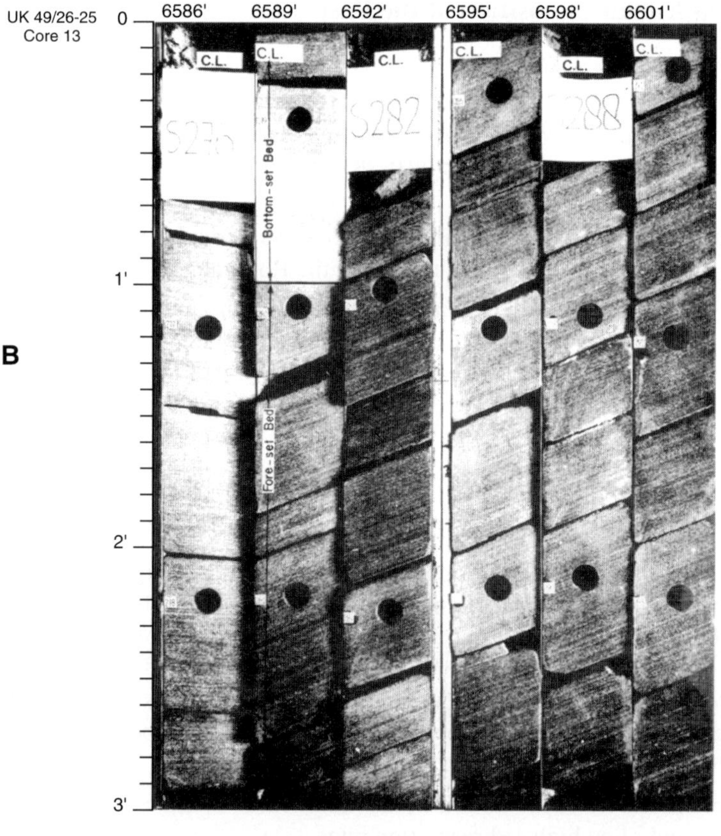

Fig. 3.47 Example of eolian facies (northern Europe's Rotliegend). **A.** Cross-section of Leman oil field. **B.** Example of oblique laminations. (From Petroleum and the continental shelf of northwest Europe, Applied Publishers Ltd, 1975, in CSRPPG, 1986).

If the detritic, terrigenous, allochthonous flow is abundant in the aquatic domain, the following processes will dominate:

- The **fluvial input dynamic** will form small sublacustrine **deltas**, due to the slope of the input stream mouths. If the edge of the lake is steeply sloped, a prograding prism of large dimensions will form of the famous "Gilbert delta" type (Fig. 3.48).
- **Gravity-driven processes** will occur if there are any major unbalances due, for example, to exceptional floods. Turbidity currents will then form, with a deposit of **turbidites**. It will be recalled that this type of deposit was first defined for the Swiss lakes, and was then generalized to the continental margins.
- The wind may induce major sea-level oscillations called "seiches", causing waves and even some littoral drift comparable to what is observed along the edge of the marine domain. **Beach (foreshore)** and **wave deposits** can be formed by reworking the initial input material on the borders.

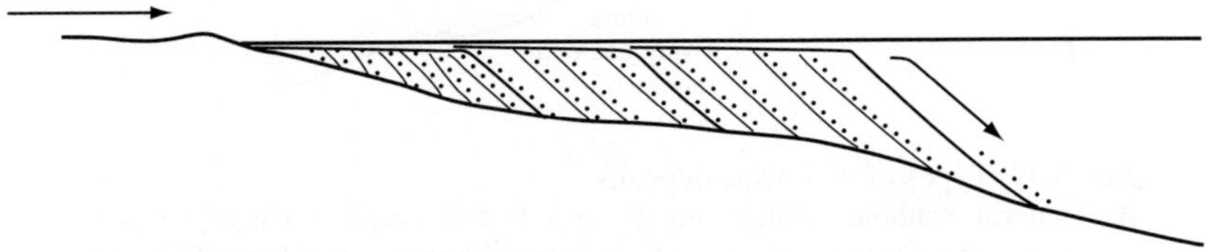

Fig. 3.48 "Gilbert delta" progradation scheme.
The slope is steep enough for the particle load to accumulate in a prograde series of arenaceous avalanches that eventually fill up the edge of the lake.

Let us remember that, because of their small size, lakes are not sensitive to tides, which are insignificant even for the largest of them.

The other great sedimentation mechanism is autochthonous **biochemical sedimentation**, which is characteristic of well confined lacustrine environments if the terrigenous input is low. This is very much influenced by climatic and organic productivity conditions. Under certain conditions, algae and lacustrine zooplankton and phytoplankton will explode and the biomass production can vary rapidly in time. Lacustrine calcareous oozes and clay-muds may form by decantation (Fig. 3.49). Depending on the depth of the lakes, the water body may be stirred enough to become homogeneous or, if there is a circulation deficit, thermal stratification of the water may occur with a special thermocline level separating the warm upper surface waters from the cold deep water. When the stratification is stable, very special anaerobic conditions may occur at the bottom of the lake, making it possible **to conserve the sedimentary organic matter**. This organic matter may be autochthonous (originating from algae) or allochthonous vegetal debris carried in from elsewhere in the catchment basin, and would lead to the formation of limestones and bituminous schist or lignites in different cases.

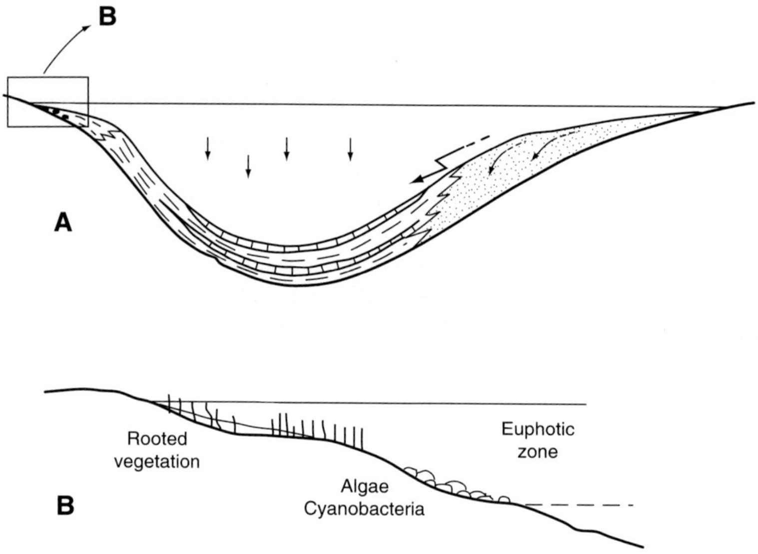

Fig. 3.49 Types of lacustrine deposits.
A. General scheme: deltaic inputs and turbid currents (heavy broken arrows), pelagic sedimentation (small vertical arrows) and bioturbed ooze (vegetation, algal mat). **B.** Euphotic zone with the abundant development of a biomass.

Depending on the climate (arid or humid), the confinement, the evaporation, and the type of input in solution, various types of deposit may occur: algal limestone, stromatoliths, dolomite, organic (sapropelic) clay oozes, trona (an exceptional form of NA_2CO_3 in the Wyoming Eocene), rock salt (Dead Sea), evaporitic crests (sebkha, beaches).

These lacustrine deposits are generally said to be **limnal**. When the salinity exceeds 5000 ppm/liter, we have a **salt lake** (Tuz Gulu in Anatolia, American Great Salt Lake). Salt lakes are characterized by a concentric arrangement of the various precipitates (Fig. 3.50). The similar coastal lagoon environment, when very arid, is referred to by the nearly equivalent terms of **playa** and **sebkha**, where brine concentrations produce the same sedimentary mechanisms.

All lakes are extremely **sensitive to climatic variations** (or, more generally, to environmental variations, as we have seen for the Aral Sea affected by overdrawing from its catchment basins). If the lake is deep, it remains permanent. Otherwise, it may be ephemeral. In all cases, the least variations will modify the deposits in the fragile equilibrium of this environment, and the result is an alternating sequence of **laminites** and seasonal varve deposits, which are good indicators of global or local climatic fluctuations (see Chapter 4).

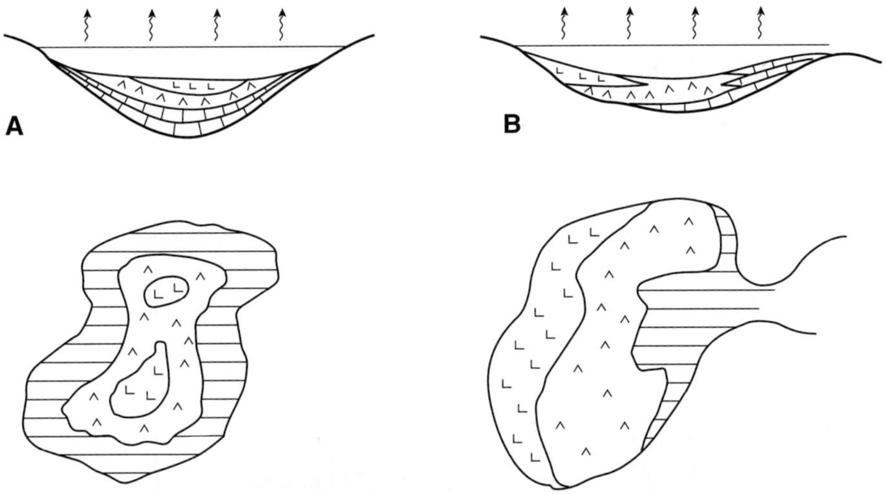

Fig. 3.50 Distribution diagram of evaporitic deposits in salt lakes with high evaporation.
A. Closed depressions (section and plan view). **B.** Open depressions.

3.2.3.2 Ancient Deposits

To conclude, remember that a number of large lacustrine basins of the geological past are known in the United States, China, and Australia (Fig. 3.51). These are large oil provinces by their wealth of parent rock (see Chapters 5 and 6), and rift-type lacustrine basins can also constitute major elements in oil systems as they evolve into continental margins (Chapter 2). Much smaller basins are known in France (Alès, Manosque, among others), offering good conditions for observing the facies mentioned.

3.2.4 Fluvial Domain and Alluvial Deposits

3.2.4.1 Processes and Driving Factors: Definitions

The term **fluvial** or **alluvial deposit** or **alluvium** refers to sediment deposited in a continental running water system anywhere from the source zone to the outlet into a basin, which is generally marine but sometimes lacustrine. **Eluvium** designates weathering products that have not been moved very far by runoff or slippage along the relief. This is an initial stage that is rarely conserved by alluvial sediments. Deposits found at the bottom of a slope after very limited transport are also termed **colluvium**, while the process by which they accumulate is called colluviation.

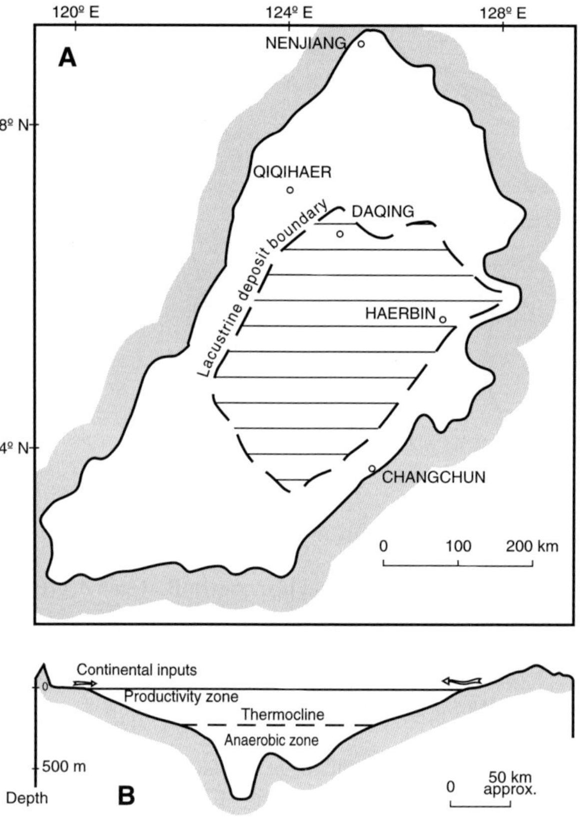

Fig. 3.51 Songliao lacustrine basin (China) (from various authors).
A. General basin map with the permanent lacustrine basin boundary at the Aptian-Albian (see Fig. 2.31). The Daqing oil field is located to the north of the basin. **B.** Schematic cross-section (altitudes highly exaggerated) indicating the anoxic conditions at the bottom of the lake.

The major parameters controlling fluvial systems are the stream slope, the scale of the drainage network, the climate, tectonics, sedimentary load, and vegetation.

Generally, sedimentary processes are first driven by **gravity**, so that **slope** is one of the dominant control factors and the transport is from upstream to **downstream**. If the point of deposit is near the source, the deposit is **proximal**. If the deposit is made far from the source, it is said to be **distal**. This applies to the marine domain (turbidites, for example). This transport by watercourses from a high point to a low leads to sedimentary **progradation**, which is sometimes called lateral accretion as defined at beginning of this chapter. This formation of deposits over long distances is sometimes called **fluvial spreading**. The main characteristic resulting from this mechanism is that the depositional forms assume an orientation that is determined in statistical terms. The stream is said to be **monodirectional** in the direction of steepest slope, as opposed to periodic oscillating currents. The fluvial domain is the one that best illustrates the elementary erosion-transport-deposit relation controlled by stream velocity. Alternating forms of erosion and deposit will thus occur in any fluvial domain, leaving sedimentary gaps on all scales.

The idea of erosion is important. The essential effect of run-off and streams is that they transport particles and thereby help flatten the relief. Here we must define the concept of **equilibrium profile** and **baselevel** (Fig. 3.52).

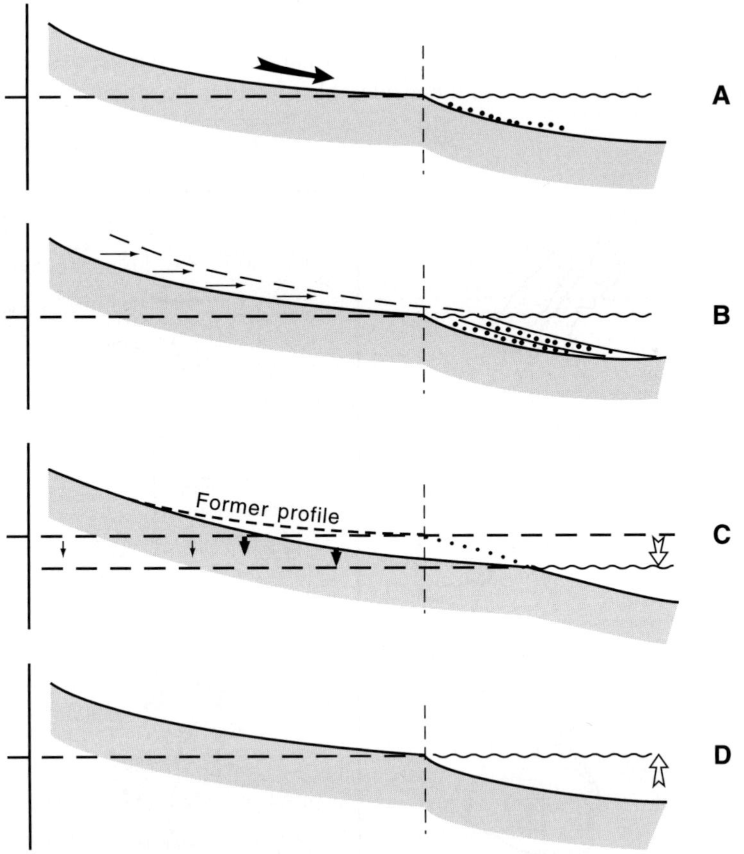

Fig. 3.52 Theoretical variations of the equilibrium profile.
A. Initial situation with river erosion and accumulation in the coastal zone. **B.** Progradation of a wedge as the sea retreats and the equilibrium profile migrates (narrow arrows). **C.** A river profile (short heavy arrows) scoured out to compensate for a general drop in sea level (dashed). **D.** River slope is reduced as the sea level rises.

The streambed generally describes a curve of gradually decreasing slope, intersecting asymptotically with the level of its actual outflow into a lake or sea. This is its equilibrium profile. While it is said to be in equilibrium, the system is actually unstable since it can be modified by tectonics or basin subsidence. Even if the imaginary surface separating the domain where erosion predominates from the domain where sedimentation is possible is defined as the **baselevel**, the baselevel is a reference altitude that is generally taken to be sea level.

One major parameter in the fluvial environment is the size of the domain considered: the drainage area, or **catchment basin** (Fig. 3.53A). Some fluvial domains are continental in extent (Amazon, Ienissei, Missouri, Mississippi, Yellow River) while others are much more limited (Var, Jordan), in accordance with the geotectonic situation.

Another essential parameter controlling the formation of alluvial deposits by fluvial systems is the **climate**, as this will have a considerable effect on the amount of water available

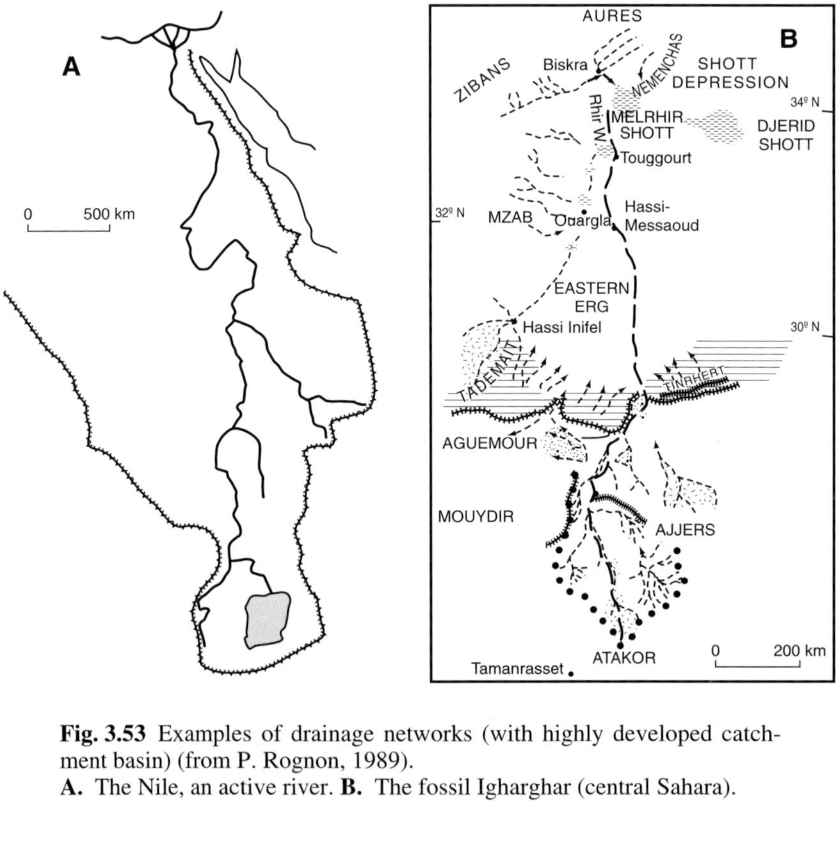

Fig. 3.53 Examples of drainage networks (with highly developed catchment basin) (from P. Rognon, 1989).
A. The Nile, an active river. **B.** The fossil Igharghar (central Sahara).

as water yield.[1] In temperate and humid regimes, the water yield is high and **inflows** can be large (see Table 3.3). The regime may also be regular or intermittent. There may be **flood** periods with an input that depends on rainfall, and **low water** periods where the level is minimum. These fluctuations are themselves more or less contrasted. In semi-arid and arid systems, watercourses may be temporary, intermittent, or operate only exceptionally (Fig. 3.53B). On the geological scale, though, they are still important.

Independently of the rate at which it is flowing, river water will permeate the alluvium along its course and create an **alluvial water table** (Fig. 3.54), which then alternately feeds and drains the river. This underflow has its own flow rate, sometimes running very close to the soil surface and even above that of the river, making a marshy or **palustrine** domain.

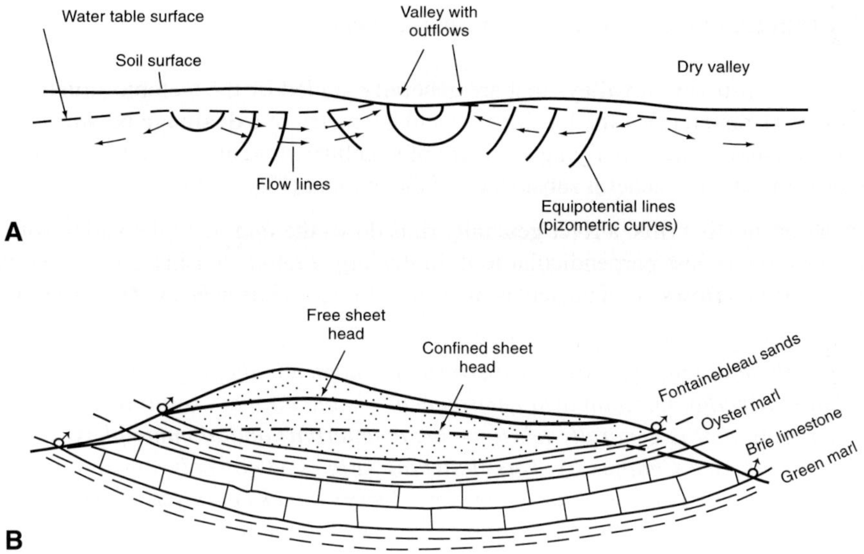

Fig. 3.54 Free ground water (from de Marsily, 1981).
A. Valley ground water. The infiltrated water saturates the environing rock up to a certain level called the "free surface". The saturated zone is called the blanket or sheet. The deepest valleys dominate the sheet. The outflow is not at a localized point, and is more like a slippage face. **B.** Perched free water table (sands of Fontainebleau) and confined groundwater aquifer (Brie limestones). In the latter case, the waterhead is greater at the roof of the sheet (dashed line), and the water contained in the sheet is compressed. If the pressure in the confined sheet is high enough, the sheet is artesian (whence the term "artesian well").

1. Schumm, 1973, 1983.

These alluvial sheets should not be confused with the water tables where water infiltrates the rock to saturation between the free surface at the top and some impermeable zone underneath. In the alluvial blanket or sheet, the waters circulate toward outlets constituting sources and streamheads. In arid regions, the infiltration is much lower than the evapo-transpiration process. In cold areas, when the soil remains frozen underneath, the underflow is limited to the upper portion and a slow "solifluction" flow of surface material occurs. Also note the very special case of calc-tufa calcium deposits and sinter appearing at the point of emergence of some sources.

The sedimentary load and the role of vegetation will be discussed later, under alluvial deposition forms.

3.2.4.2 Structural Control and Geomorphology

Rivers flow downstream in **valleys** that are generally guided by the tectonic structure. Slope is the primary parameter, but the localization and geometry of a **drainage network** is generally closely related to the underlying structure of synclines, fold and fracture networks favorable to erosion, and the general subsidence of the domain (Fig. 3.55).

In linear basins like rifts, a river generally runs down the middle of the valley with tributary streams more or less perpendicular to it. In the large cratonic basins, convergent valleys separated by **underflows** are themselves made up of major **channels** and **floodplains**.

Whatever the geomorphology, the structural and climatic guidance of the watercourses is what will determine the associated erosion phenomena. Shaping of erosion surfaces—peneplain or pediplain in an arid environment—is the ultimate result of successive erosive ***stripping*** *that is generally felt to be the counterpart and ultimate phase of the orogenic process (see Chapter 6). This flattening (Fig. 3.34) of polygenetic or polycyclic surfaces over long periods of time sometimes leaves a lucky* ***correlative deposit*** *of colluvia, alluvia, or sediments that give an indication of the shaping mode.*

When the slope leads out to a coastal plain, the alluvial domain becomes part of the coastal domain, and then floodplain, palustrine environment, and lagoon may all coexist. Sometimes the system has no connection with the marine domain, constituting an **endorheic system** as in today's Sahara and central Asia (Fig. 3.56). These are relatively common, but are doubtless of little permanence in the intermontane domain.

It will be seen later on that river equilibrium profiles are not guided by structure alone, but tectonic control and baselevel variations (of the sea, as discussed later) do explain:

- Phased or nested arrangements of alluvial terraces related to alternating periods of alluviation and erosion (Fig. 3.57)
- The entrenchment of river meanders in rocky blocks as the blocks are lifted up (Fig. 3.58).

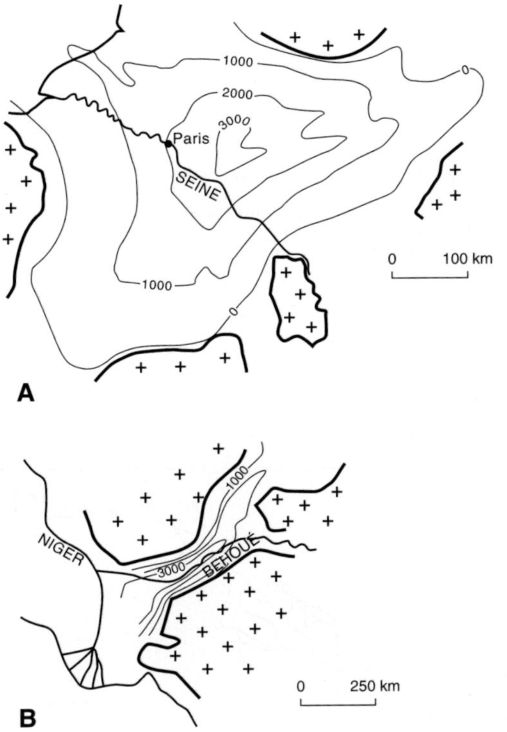

Fig. 3.55 Drainage network and deep structure.
A. The Seine occupies the central part of the Paris basin with no close relation to the tectonic structures. **B.** The trough where the Benoue River flows is a narrow tectonic graben. The fine curves indicate the basement depth.

3.2.4.3 Main Types of Fluvial Deposits at Different Scales[1]

Generally and approximately, it can be said that stream velocity decreases with stream slope. Erosion becomes less vigorous while the stream gradually sheds its bottom load and sediments in suspension. The coarsest material is left in a proximal position and the finest in distal, with a gradual gradation between them. This is **grain size partitioning**.

Fluvial forms fall into broad categories:

- **Channels**, where concentrated flows circulate and deposit the coarsest particles of rounded pebbles, gravel and sand
- **Levees**, which are waterside deposits forming **banks** where vegetation plays an important stabilizing role

1. See Schumm, 1993; Beuf et al., 1971; Miall, 1992.

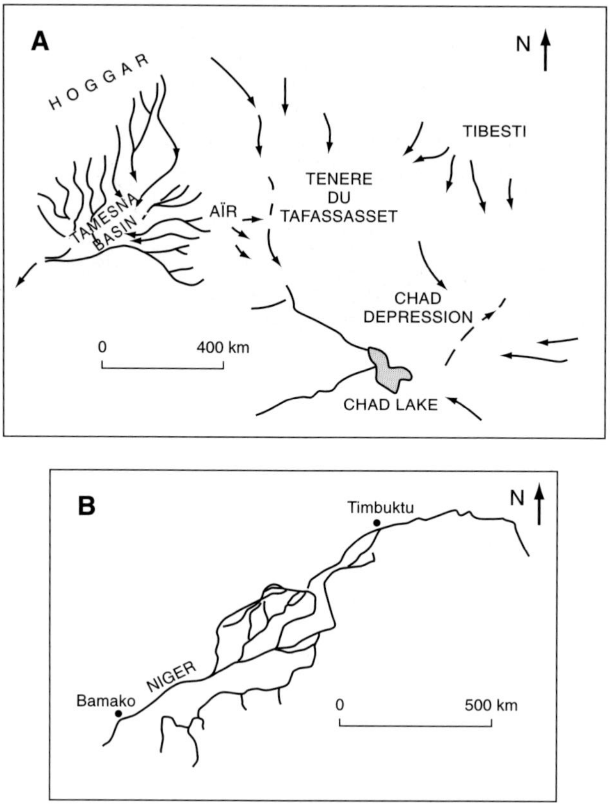

Fig. 3.56 Endorheic networks.
A. The Tamesna and Chad basins between the Sahara and Sahel are more or less closed depressions. **B.** The mean course of the Niger is characterized by a kind of braided inland delta formed in a subsident basin.

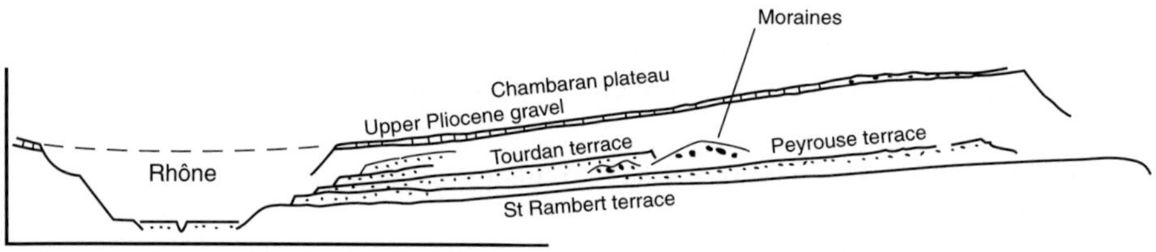

Fig. 3.57 Nesting of successive terraces in the example of the Rhône valley (from M. Gignoux, 1950).

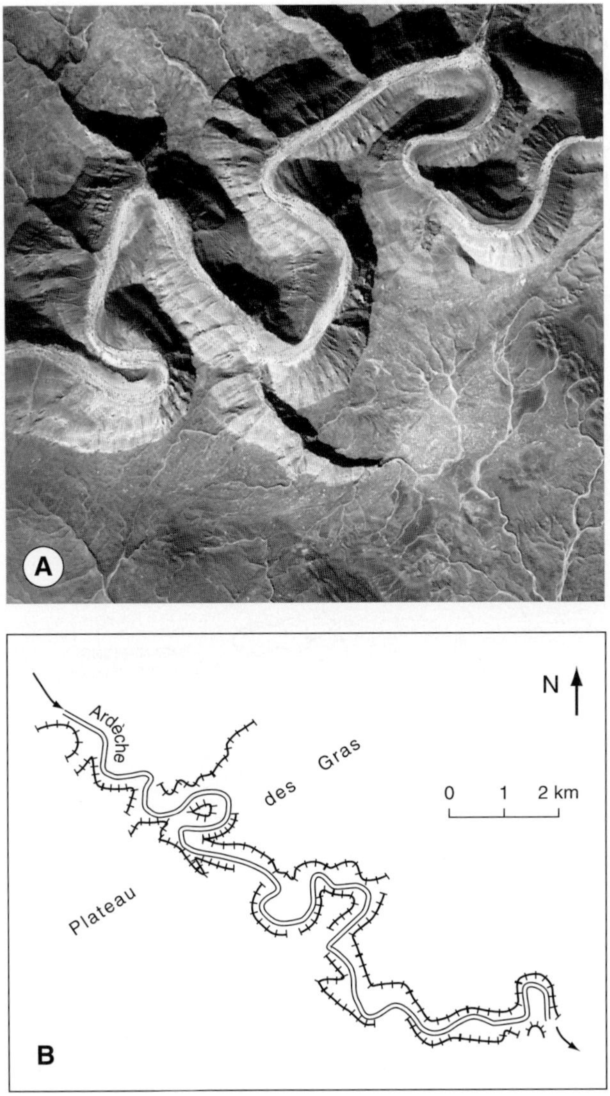

Fig. 3.58 Entrenchment of meanders.
Originally meandering rivers may become deeply entrenched in a plateau following tectonic uplift, with large level differences. **A.** Saharan wadis are deeply entrenched in a sandstone plateau and are inactive today except during exceptional flooding (Devonian Tassili) (IFP photo). **B.** The Ardeche River continues to dig its bed out of the Gras plateau. The gorges cut a level difference of 200 to 300 m.

- **Floodplains** where a great variety of sand, silt, and clay soil is deposited and where facies are fractionated between sandy channel deposits and clay-sand overflow deposits. The **palustrine environment** of marshes and swamps overlies the typically fluvial environment.

Fig. 3.59 gives a schematic review of the major types of fluvial deposits, running downstream, including the following.

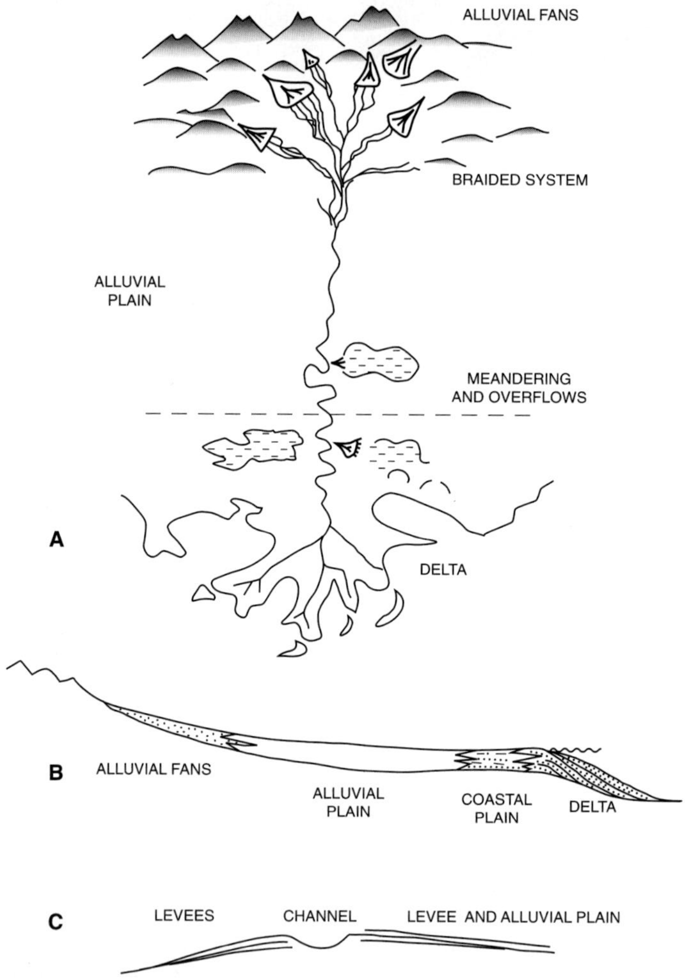

Fig. 3.59 Fluvial network scheme.
A. Evolution from upstream down. **B.** Longitudinal profile.
C. Transverse profile.

- **Alluvial fans** develop at the foot of the initial relief at slopes of 1° to 25°, where gravitational slippage interacts with transport under the stream current. These are found in different climatic conditions, from the glacial (outwash plains called "sandurs") to the arid. Alluvial fans are often aligned in rows along the relief (Fig. 3.60). The material is generally transported over extremely short distances, and the deposit is mainly a coarse, graded **conglomerate** dubbed "fanglomerates".
  The slippage transport mechanisms can be preponderant, with debris flow and grain flow between the avalanche breccia or upstream sediments and sediments rolled downstream along the bottom by a tractive current. The two types of current may occur simultaneously.

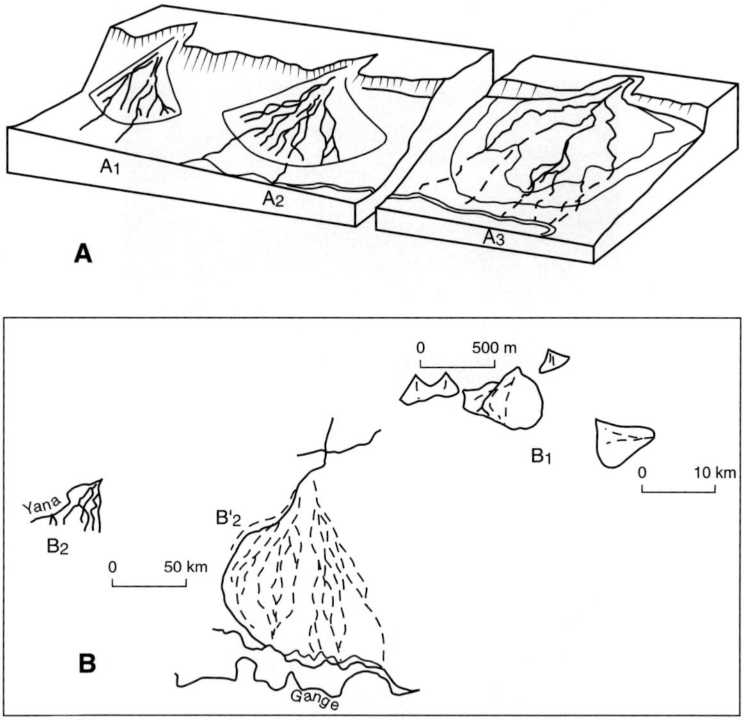

Fig. 3.60 Different types and sizes of alluvial fans (from Miall, 1992, and other authors).
A. Illustration of an alignment of alluvial fans of different shapes and sizes. A_1 sloped fans with slippage of debris flow. A_2 fans with braided channels and sand and gravel banks. A_3 gently sloped fans with more or less meandering braided systems, levees, and palustrine areas. **B.** Current examples. B_1 small braided fans of high-relief desert regions. B_2 example of large fans in Alaska and India.

The fans vary considerably in size, from a few hundred meters to several kilometers. The Kosi fan to the North of the Ganges has developed over 140 kilometers. (The formation found at the foot of seascarps is also often referred to as an alluvial fan because the sedimentary processes at work are nearly identical.)

The larger the fan, the more distal and better graded are the facies. They are sometimes sinuously channeled, so that the oblique stratifications have to be analyzed to determine the stream direction.

It can be said that the slope streamlines are "wild", as opposed to a "straight" or concentrated flow in a well-oriented bed.

- **Braided systems** develop in the upstream area especially when the sedimentary load on the bottom and slope is large and when it fluctuates greatly, giving rise to a system of unstable currents with wandering banks, ephemeral sandbars, and irregular, sinuous courses (Fig. 3.61). Deposits range from gravel to coarse and fine sand, at times with

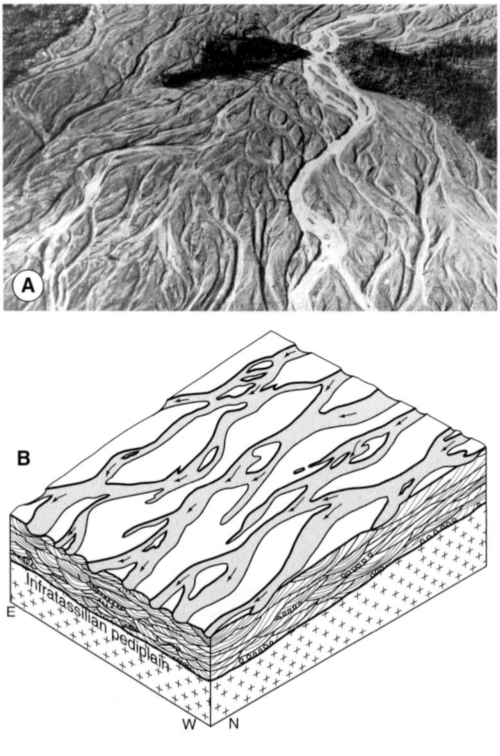

Fig. 3.61 Braided fluvial network.
A. Current example in a periarctic region of Alaska (taken from C. Wahrhaftig, 1965, in Beuf et al., 1971. **B.** Ancient example: reconstruction of the Saharan Cambro-Ordovician sandstone deposition system (Beuf et al., 1971).

several grain size stocks and often with internal erosion figures (reactivation surfaces). There is no flood plain per se separated from the main channel because the continual wandering of the currents homogenizes all of the sandy deposits, which divide up into a set of interleaved channels with high lateral migration.
This type of braided system can be found just as well in periglacial as in arid regions. It is, of coarse, promoted by the lack of vegetation, i.e., the lack of consolidation and fixation of the arenaceous banks, which are thereby incessantly reworked and eroded.

- There is the somewhat different case of **anastomosing systems** in which the channel arrangement is less sinuous and wandering. The braiding effect is still there (Fig. 3.62), but the banks are fixed by vegetation and are stable. Opinions differ on the

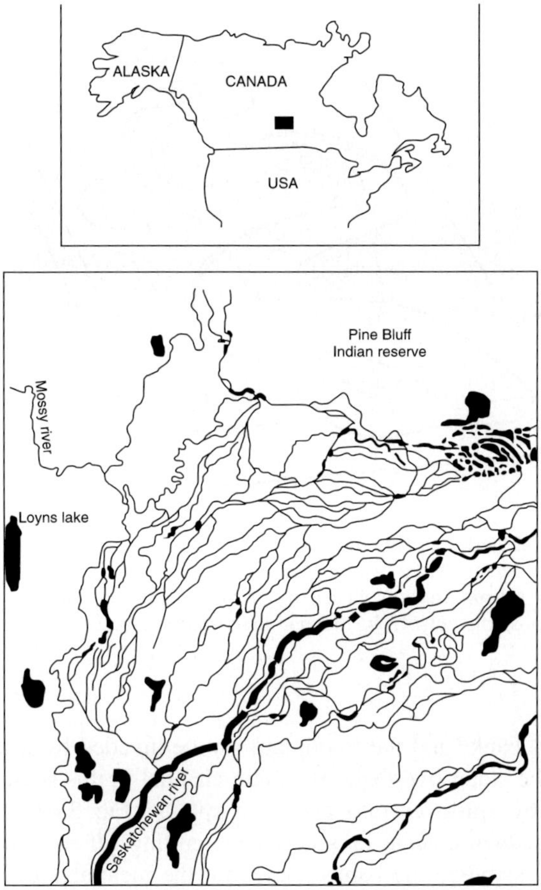

Fig. 3.62 Example of anastomosing system (from EAP Bulletin).
Note the large number of active watercourses and small lakes, with nested fluvial, palustrine and lacustrine domains.

subject, but the slope is generally considered to be less than in the truly braided networks where the courses and banks are unstable.
Forms of deposition vary depending on the scales of observation, and there are many gradations between these forms and those of the other systems. The most pronounced character is the development of channels and sandbars.

- **Meandering streams** are another type of fluvial architecture where the channels exhibit very strong sinuosity (Fig. 3.63), where the sedimentary load on the bottom and the material in suspension coexist in variable proportions, and where the slope is generally lower. The flow is highly concentrated in one major channel and, even though its velocity is lower, the erosion capacity is still high.

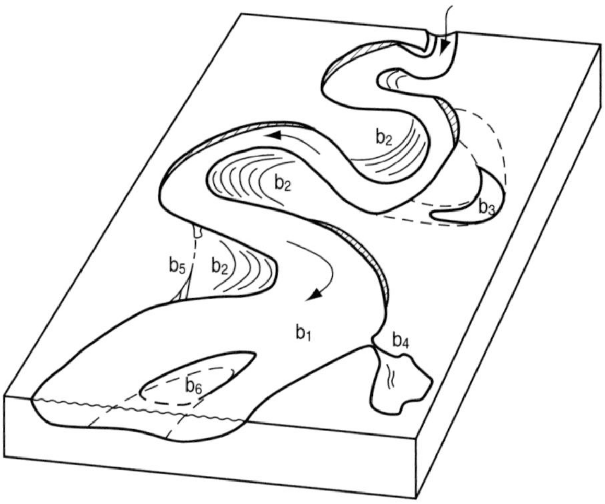

Fig. 3.63 Meandering fluvial system.
Scheme of meandering system elements. b_1 main channel; b_2 point bar; b_3 derelict stream (oxbow lake); b_4 crevasse splay created by flood; b_5 drop and possible cut-off; b_6 mean course sandbar.

The river has its banks and the floodplain can be invaded by water overflow in flood season. When the water overflows the main channel in this season, lateral distributaries may form by spillover into **crevasse splays** (Fig. 3.64) of various sizes and geometries. A sudden change in watercourse cutting off the land is called **avulsion**. Vegetation may stabilize the banks some, but the meanders evolve in time, creating bars and leaving isolated oxbow lakes (Fig. 3.63). The **meander** system characterizes temperate and wet climates rather well.

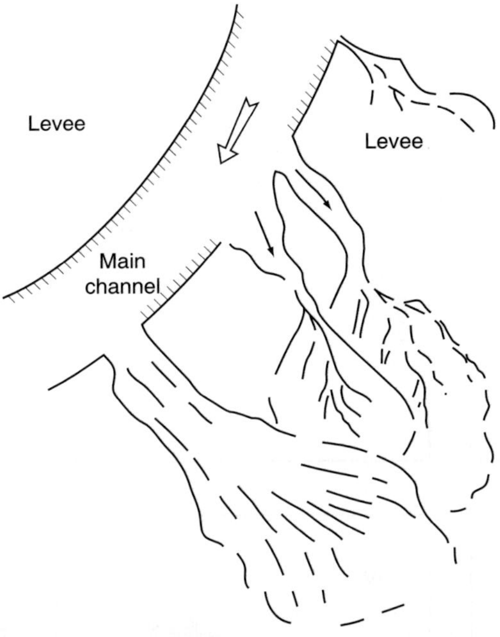

Fig. 3.64 Spillover deposition scheme (crevasse splays) on the edge of fluvial channels. Several splays are shown breaching a levee and overlapping each other.

The sinuosity is such that the paleostreams exhibit patterns of tidal (bimodal) type oscillating currents (Fig. 3.65). The deposition forms are widely varied in sandbars and sandflats, but one of the most characteristic forms is the **point bar** (Fig. 3.66).

In the zones farthest downstream to the coastal plain, when the slope and current velocity are low, we find **channels** and channel borders in the **alluvial plain** filled with the coarser sediments and **argillaceous** plain spillover deposits where **peat beds** (initial stage of coal formation) in the flooded zones appear side by side with small arenaceous crevasse splays and clay plugs from abandoned oxbow lakes.

If sedimentation is active, the river bed may rise up above the level of the plain (Fig. 3.67) resulting in an unstable perched river with a high probability of creating crevasse splay lobes and avulsion.

- **Straight rivers** are actually few and far between. They have stable, regular banks and a single channel where the flow is straight and concentrated and where the load can range from dominant tractive rolling on a bed to maximum suspension (Fig. 3.68). A certain amount of meandering may be recognized, depending on high and low water levels.

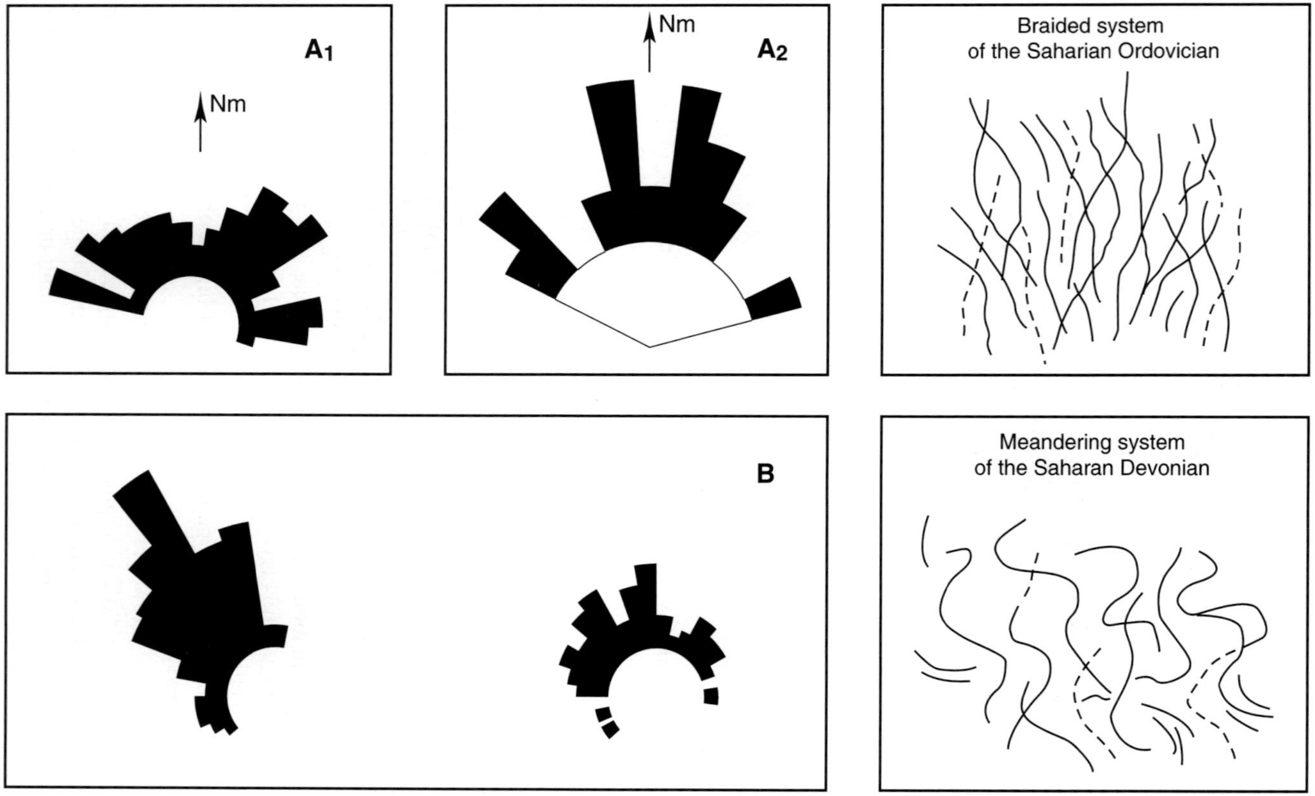

Fig. 3.65 Directions of the sedimentary dip in the obliquely stratified sandstones of the Saharan Paleozoic, corresponding to two types of fluvial systems.
A. Braided Ordovician system. A_1 dip azimuth in classes of ten degrees, terrain outcropping; A_2 same but at depth, after core sample measurements in the Hassi-Messaoud field. **B.** Meandering network of the Devonian: measurements on outcrops.

A

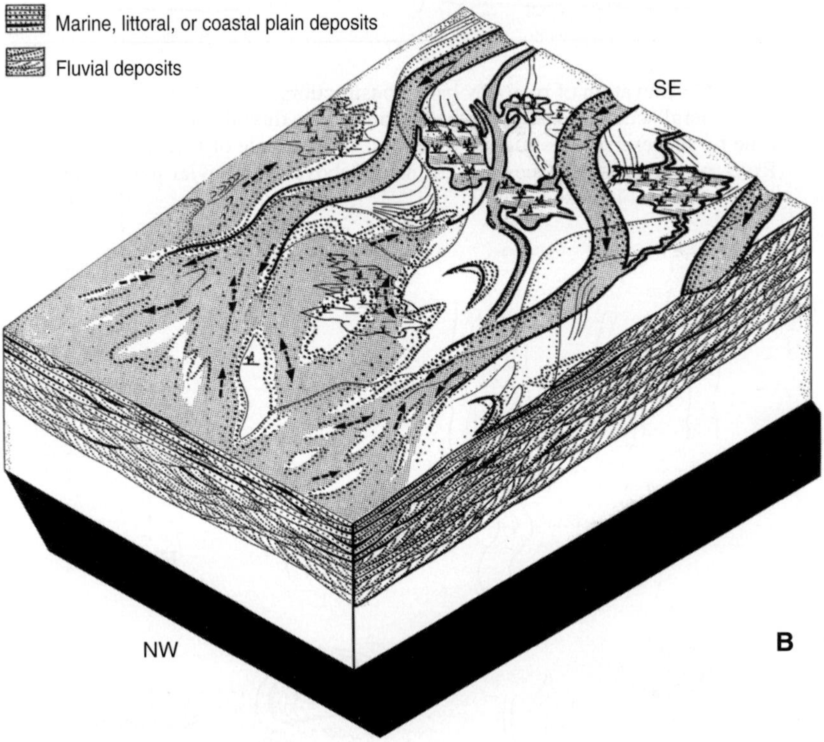

B

Fig. 3.66 Meandering fluvial system.
A. Present example in the periarctic region (Alaska, taken from Wahrhaftig, 1965 in Beuf et al., 1971). **B.** Ancient example: reconstruction of the sandstone deposition system of the Saharan Lower Devonian (Beuf et al., 1971).

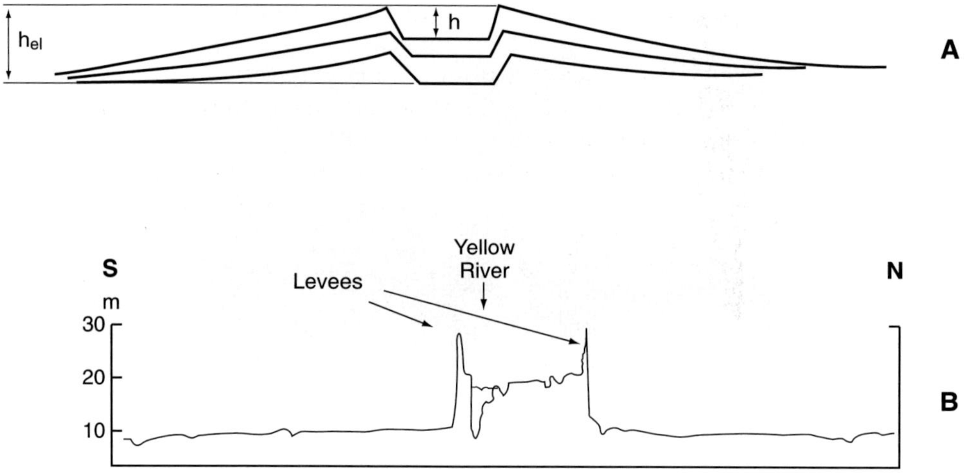

Fig. 3.67 Elevation of rivers by bank construction.
A. General scheme explaining the elevation of a fluvial watercourse over time by successive construction of levees. **B.** Example of China's Yellow River where present levees are 20 meters above the alluvial plain (from L. Shu and B. Finlaysou, 1993).

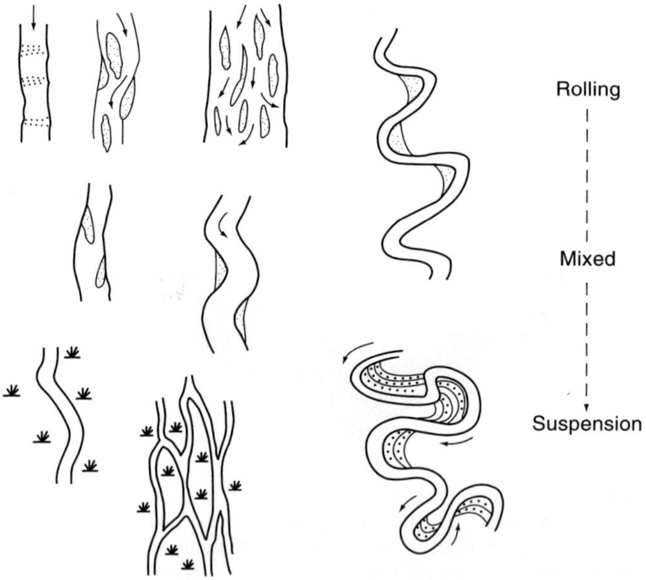

Fig. 3.68 Different fluvial patterns depending on load and slope.

In distal position, the streams flow out into a permanent water column (lake or sea) in **estuaries** or **deltas**.

- **Estuaries** are special zones where the fluvial and marine processes interact. The effects of micro- and macrotidal waves are particularly strong, as is the mixing of fresh and salt water. Estuaries are known in all types of climate. Mainly, they were initiated by the sea-level rise in the Holocene, and sometimes for tectonic reasons. In the well-studied example of the Gironde River, the deposits are arranged in three terms: fluvial, mixed and marine (Fig. 3.69) in tidal channels, baymouth bars, and sheltered bays, making varieties of sedimentary forms that will come up again in the discussion of new marine shore deposits. In wet conditions, the fresh water invades the oceanic domain with the formation of a saltier deep stream. This is estuarine circulation. When there is deficit of fresh water input or excess evaporation, the dense waters tend to flow toward the ocean depths with an anti-estuarine circulation of a cooler surface stream current coming from the ocean (Fig. 3.70).

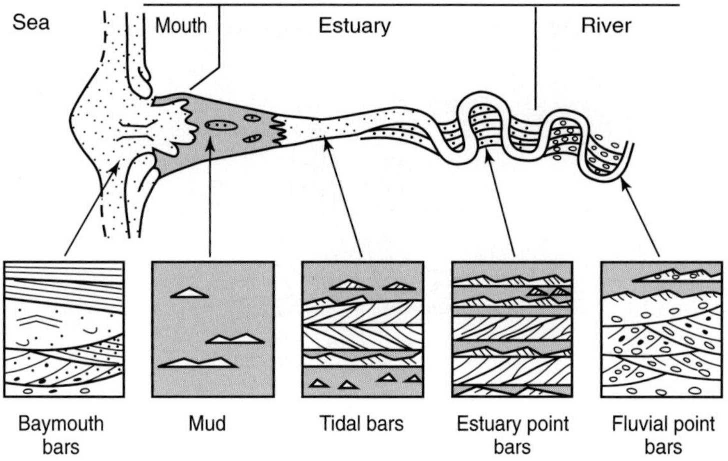

Fig. 3.69 Distribution of different deposits in an estuary, from upstream down. Schematic developed from a study of the Gironde River (from G. Allen, 1991).

- **Deltas** are another form of continental river outflow into lacustrine, lagoonal, or marine water volumes. Contrary to the estuary form, the delta occurs because the sedimentary load from the river is such that the littoral processes cannot immediately take it up and rework it. However, even the largest deltas are influenced by waves and tides, giving rise to different types of deltas with special morphologies (Fig. 3.71).
  When the delta is predominantly fluvial, the morphology is lobed or elongated, with many more or less straight distributaries, levees and crevasse fans or splays. We then

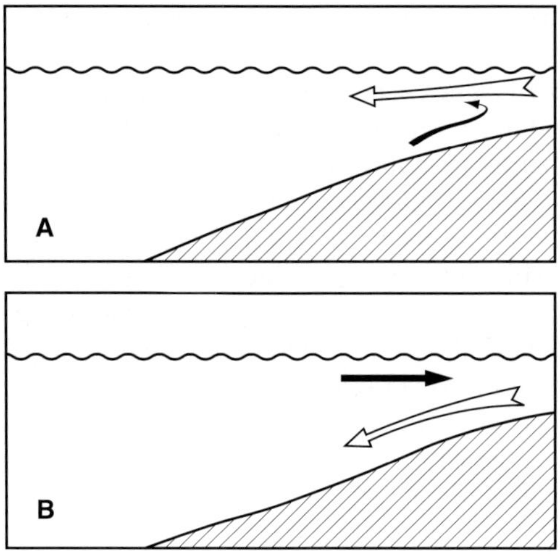

Fig. 3.70 Estuarine (**A**) and anti-estuarine (**B**) circulation.

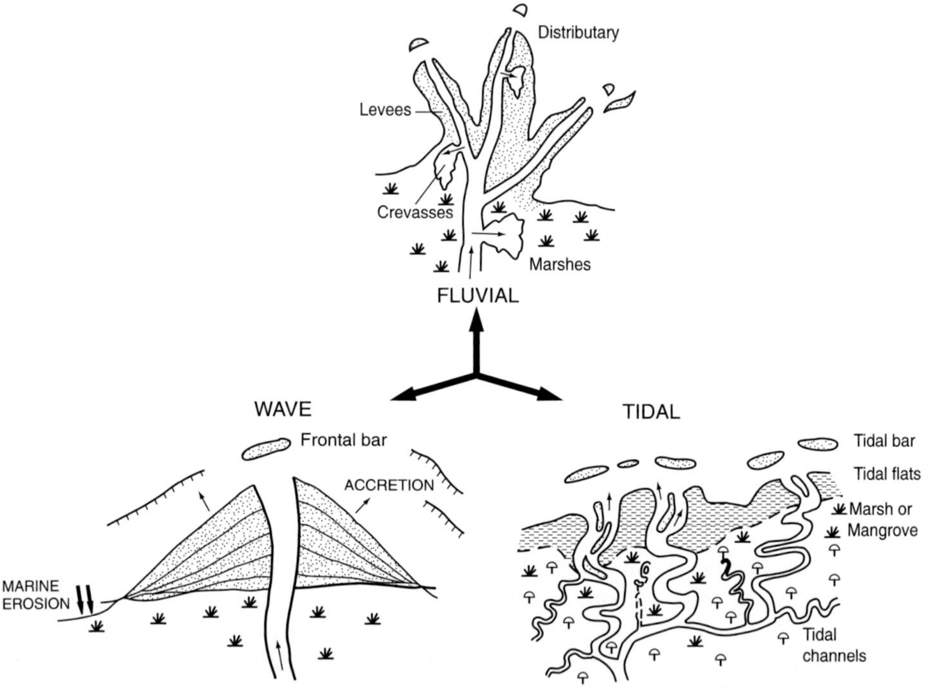

Fig. 3.71 The three major types of deltas depending on the dominant factor (from an ENSPM document).

find an emerged **delta plain** part with channels and marshes, the **delta front** where lobes can develop, and the submerged prodelta, the extent of which will depend on the slope, current velocity, and sedimentary load. Progradation downstream is the major recognizable mechanism in time in the sedimentary structures (Fig. 3.72).

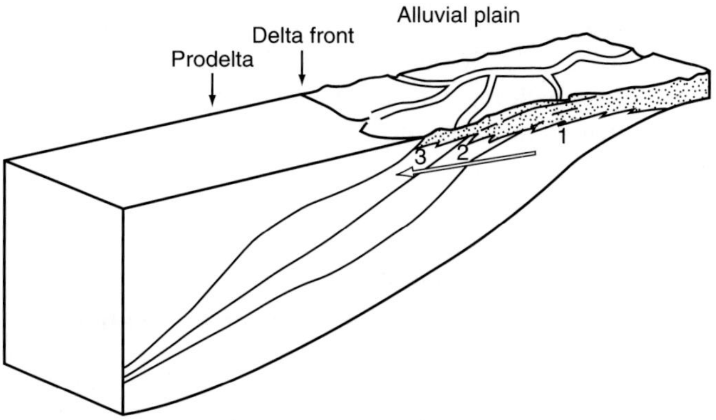

Fig. 3.72 Delta progradation scheme.

As will be seen later on in this chapter, when waves predominate as they do in the Rhone, then different types of baymouth sandbars develop. These are dispersed by the littoral drift, which elongates them.
If the tidal effects are energetic, then tidal channels, sandbars, and mud flats will develop. These sometimes cover large distances (the Ganges, for example), and at times resemble estuarine forms.
A large part of the sediments input by the rivers are then redistributed in a mixed fringe. Autocyclic processes may lead to lateral migration and generate facies variations that are even more pronounced if allocyclic processes are involved.

The importance of deltas for petroleum geology should be emphasized, because it is in the coastal fringe that the **facies fractionation** occurs, allowing the construction of reservoir systems.

It will be noted (see further on in this chapter) that the flow of most great rivers extends out into the sea and supplies gigantic **deep sea fans**.

In a few special cases such as in semi-arid regions, a river may construct a delta in a closed endorheic depression, or even construct an intermediate delta in a floodplain or ephemeral lake (ephemeral because the evaporation is high) before resuming its normal course (example of the Niger, Fig. 3.56).

On another scale, the **sedimentary structures** generated by fluvial transport are highly diversified by the variance systems analyzed above. Briefly, **subaquatic dunes** are an elementary pattern of variable size depending on the water depth, ranging from the centimetric ripple mark to giant ripples of several meters. As the autocyclic patterns repeat and overlay each other (due to fluctuations), **bars** of various shapes are created (transverse bar, tongue-shaped, longitudinal, among others). A succession of bars and depressions in a ripple pool is an ephemeral figure that can, however, be recognized in certain ancient series. The internal architecture is marked by **erosion** or **reactivation surfaces**. The oblique stratifications may be planar, in troughs (criss-crossed) or tabular (Fig. 3.73). Depending on the stream shape, natural accretion may predominate. Gravity-slippage figures are frequent, with **slumps** but also with **overturned cross-bedding**, which are doubtless more a result of sudden flow variations in the main bed than of microseisms.[1]

Arenaceous bodies will coalesce (mainly in braided systems) or be channeled into a predominately clay alluvial plain characterized by spillovers (crevasse splays, avulsions, exceptionally in broad sandy spreads from flooding) and marshes will appear where organic production may proliferate (resulting in peat and coal).

The following points should also be mentioned for the **centimetric to microscopic scale**:

- The coarser materials (rounded pebbles) are not stocked and organized randomly (Fig. 3.74) and can provide a criterion for recognizing the deposition conditions.
- The sandy (siliceous) material includes all grain size classes. These may be well graded or only certain grain sizes may be conserved, and the angularity of the elements will depend on the transport time. The material **maturity** depends on the extent of blunting and grading (the transported material often has a polycyclic history involving several phases of erosion).
- Minor elements such as **heavy minerals** (Table 3.3) are very good indicators of the material source.
- The "fines", or clays, indicate the source heritage. If the streambed cohesion is poor, the bed will itself constitute a secondary source.
- Organic products are either retained in the marsh or bog zone, or are transported floating (debris, colloids) to the marine or the lacustrine domain.

In the final analysis, **watercourses** are more or less heavily charged **solutions** of dissolved silica, salts, and colloidal organic matter, providing considerable input for the mineral matter and the nutrients needed for the development of organisms at sea (see further on).

3.2.4.4 Petroleum Aspect

The fluvial environment is thus an extremely favorable medium for constructing arenaceous reservoirs, with the added interest for the petroleum that they are coalescent and not sharply separated by the permeability barriers that fine sediments would set up. The often poorly graded alluvial fans do not always make good reservoirs. Braided deposits generally make

1. Beuf et al., 1971; Wells et al., 1993.

A

B

C

Fig. 3.73 Examples of oblique stratifications: sectional views.
A. Superposition of several sedimentary bodies in succession. **B.** Large channel fill bank element. **C.** Overturned cross-bedding (probably due to a precocious deformation of the sediments) (IFP photos).

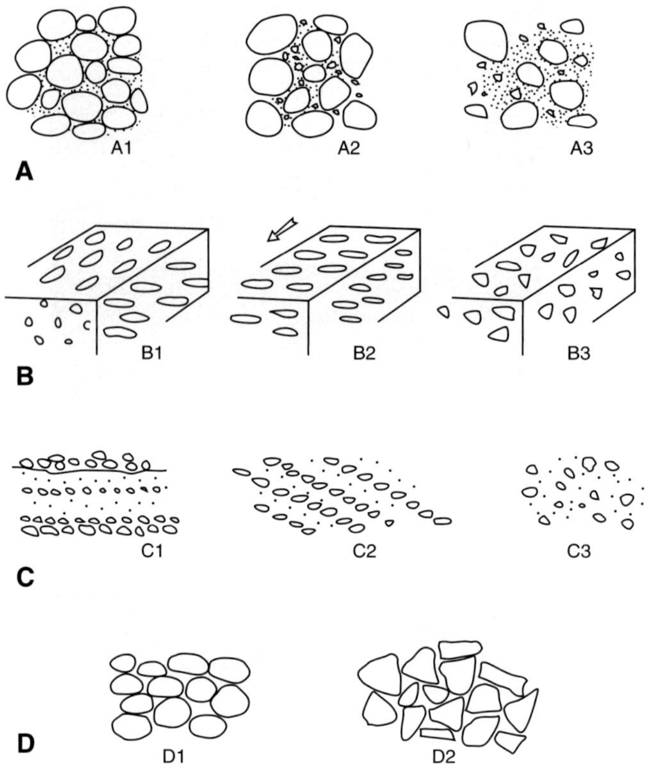

Fig. 3.74 Different types of conglomerate.
A. Grain size: A_1 bimodal (well-graded matrix); A_2 polymodal; A_3 matrix-supported polymodal. **B.** Fabrics: B_1 longitudinal; B_2 transverse; B_3 random. **C.** Stratifications: C_1 horizontal; C_2 inclined; C_3 heterogeneous, non-stratified. **D.** Rounded: D_1 puddingstone; D_2 breccia (or angular elements).

excellent reservoirs if the sand has not been cemented during diagenesis. The petrophysical qualities and dimensions of meandering systems will vary greatly in space as in the examples of the Chaunoy field (Paris basin) and that of Wytch Farm (North Sea) (Fig. 3.75). The fluvial medium too, in an alluvial plain, is a generator of in situ organic matter like coal. Lastly, the river transports terrestrial vegetal debris to the marine environment, constituting one of the large classes of kerogens (see Chapter 6).

Table 3.3 Main heavy minerals.

Opaque	Magnetite, limonite, hematite		(dark)	
	Ilmenite		(gray to black)	
	Pyrite, marcasite		(yellow, bronze)	
	Leucoxene		(white)	
	Glaucony		(green)	
Isotropic	Fluorine	(colorless)	cubic	
	Spinel	(various shades)	cubic	
	Garnet	(various shades)	cubic	
	Blende	(brown)	cubic	
	Diamond	(colorless)	cubic	
Anisotropic	Straight extinction	Colorless or Slightly colored	Sillimanite	
			Zircon	
			Andalusite	
			Apatite	
			Topaz	
			Olivine	
			Anhydrite	
		Colored	Anatase	(blue)
			Rutile	(yellowish brown)
			Staurolite	(reddish brown)
			Siderite	(brownish)
			Hypersthene	(pleochroic)
			Tourmaline	(pleochroic)
	Oblique extinction	Colorless	Disthene	
			Diopside	
		Colored	Glaucophane	(blue)
			Sphene	(beige)
			Monazite	(beige)
			Chloritoid	(green)
			Epidote	(pale green)
			Actinolite	(pale green)
			Hornblende	(yellow-green)
			Augite	(dark green)

3.2.4.5 Time Evolution and Ancient Fluvial Systems

So far, our analysis of fluvial system mechanisms and responses has revolved around the autocyclic driving phenomena, which themselves generate considerable variations. Above and beyond the system's specific dynamic, though, fluvial **outwash** is the result of a time evolution in which allocyclic processes—climatic variation (hydrological), tectonic evolution, changes in sea level—are decisive. If there is enough subsidence over long periods of time, large hydrological systems will leave only the trace of their geomorphological shaping (peniplain, pediplain) and, in favorable cases, only the last fluvial system before eustatic

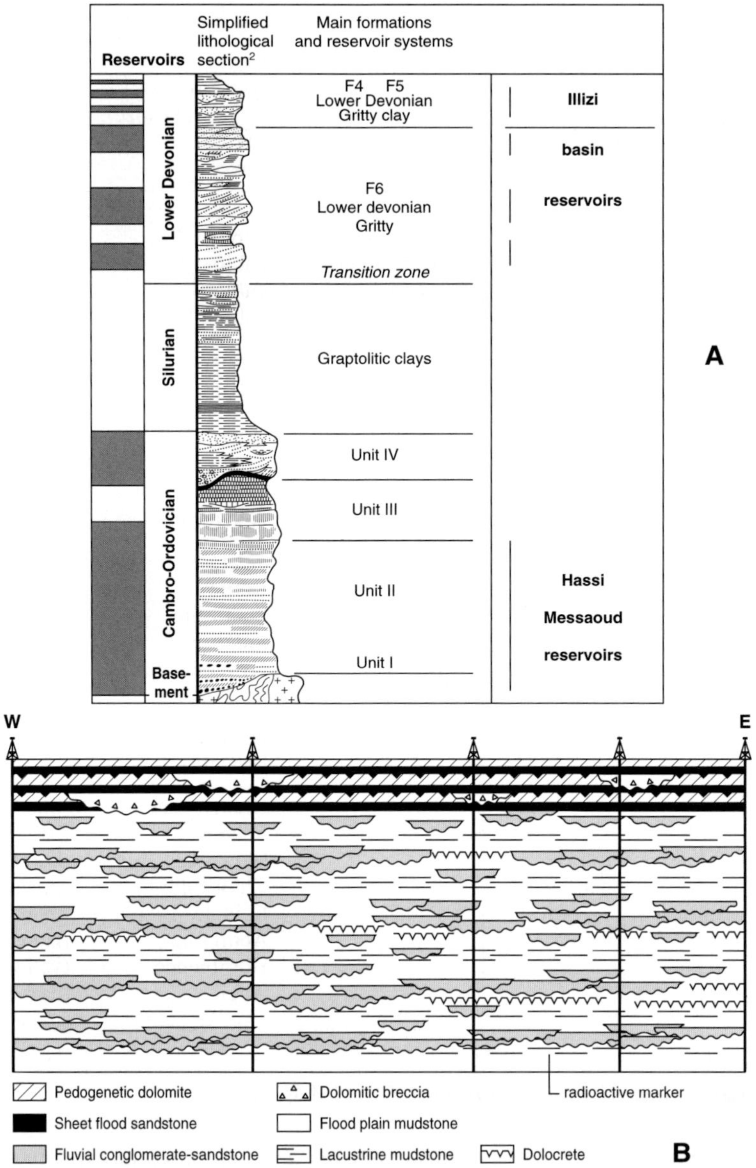

Fig. 3.75 Examples of fluvial and river delta reservoirs.
A. Reservoirs of the Saharan Paleozoic. **B.** Chaunoy reservoir (Paris basin Triassic), from Eschard, 1998.

transgression will be conserved. If the drainage basin subsides enough, leaving a large "available space" (see next chapter), vertical aggradation will be found in the fluvial deposits (Fig. 3.76). If the sea level suddenly drops, the equilibrium profile will be modified with the concomitant erosion and incised valleys will be created.

This is how the great sedimentary piles of the past were constructed. Many of them make good oil reservoirs, like the Saharan Ordovician and Devonian, American Pennsylvanian, North Sea Jurassic.[1] Each system has its own characteristics. Paleobotanic changes control the evolution of the environments. In the Ordovician, no stable bank could exist for lack of vegetation and the braided networks dominate in the permanent river bed. Today's climatic conditions will favor one or another type of flow more.

Statistical measurements of currents can be used to **reconstruct** ancient systems and illustrate these great fluvial outwashes characteristic of certain periods (Fig. 3.77).

We know that the great rivers can be dynamic transport agents over very long periods of time. The Niger delta has been operating on the margins of the African continent since the Cretaceous, and is one of the most voluminous systems in the world (Fig. 2.37). The Nile, like many other great rivers of today, probably dates from the end of the Eocene.

3.2.5 Glacial and Periglacial Environments

Our planet's current geography gives us some valuable indications about the processes, environments, types of erosion, and glacial and periglacial deposits, that also occurred in the construction of the geological series of the Infracambrian, Ordovician, Karroo and Pleistocene.[2]

The close continental glacial environment should be distinguished from the periglacial domain, but also, considering the great instability of the ice ages, what is pertinent to the glacial stages should be distinguished from that of the interglacial.

3.2.5.1 Processes

The processes at work in the broad glaciated continental regions under the polar ice cap, called **inland ice** (as in Greenland and the Antarctic today) are the pressure of the ice column itself and especially the mechanical erosion that the ice effects as it moves. The erosion processes vary greatly with the geometry of the **ice sheet**. Ice has high abrasive power. The surfaces are probably shaped rapidly.

The various forms of **erosion** have been mentioned before (Fig. 3.37). This erosion leaves a glacial **pavement**, which provides a means of reading the scheme of the ice motion

1. See special issue of *Sedimentary Geol.*, 1993.
2. This section will draw largely from the discussions of the book by Beuf et al., 1971 (already cited in this chapter).

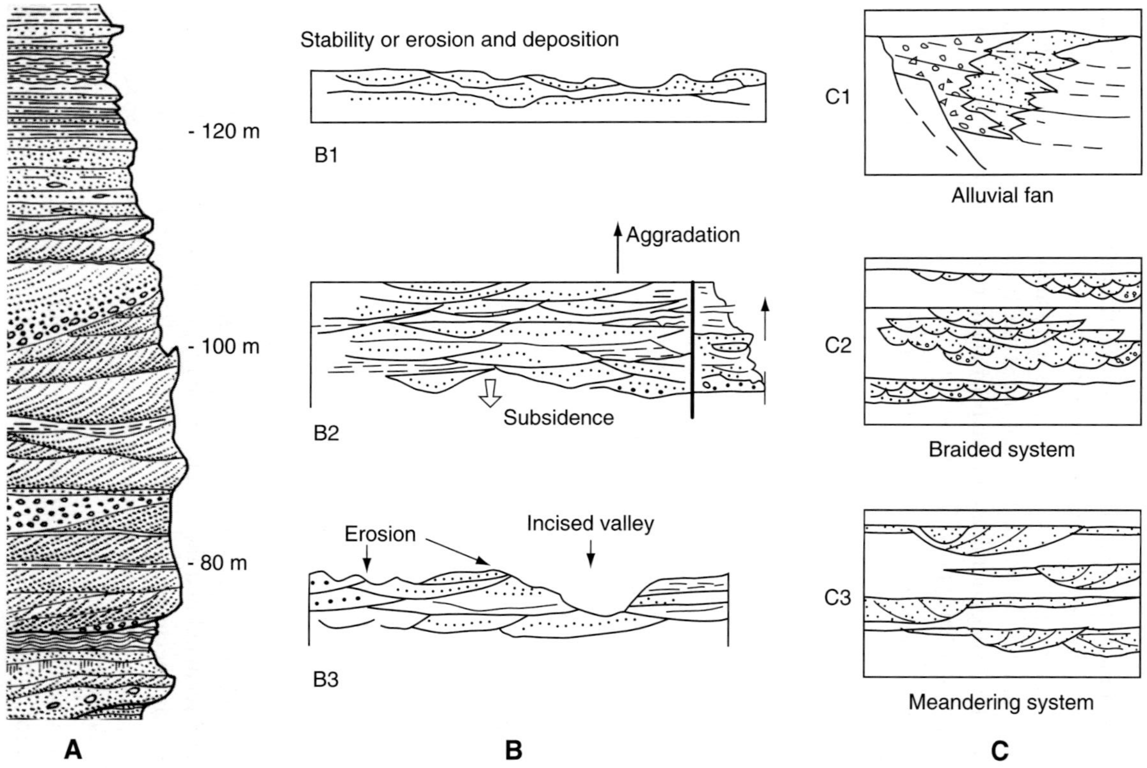

Fig. 3.76 Aggradation of fluvial deposits.
A. Lithological sequence of obliquely stratified sands consisting of nested braided channels of the Saharan Devonian. **B.** Theoretical schemes. B1: Platform with little subsidence, with minimum chances of conservation. B2: Active subsidence, where the deposits are trapped and overlaid in vertical aggradation. B3: The equilibrium profile is modified and deep valleys are cut out by erosion. **C.** Different types of aggradation.

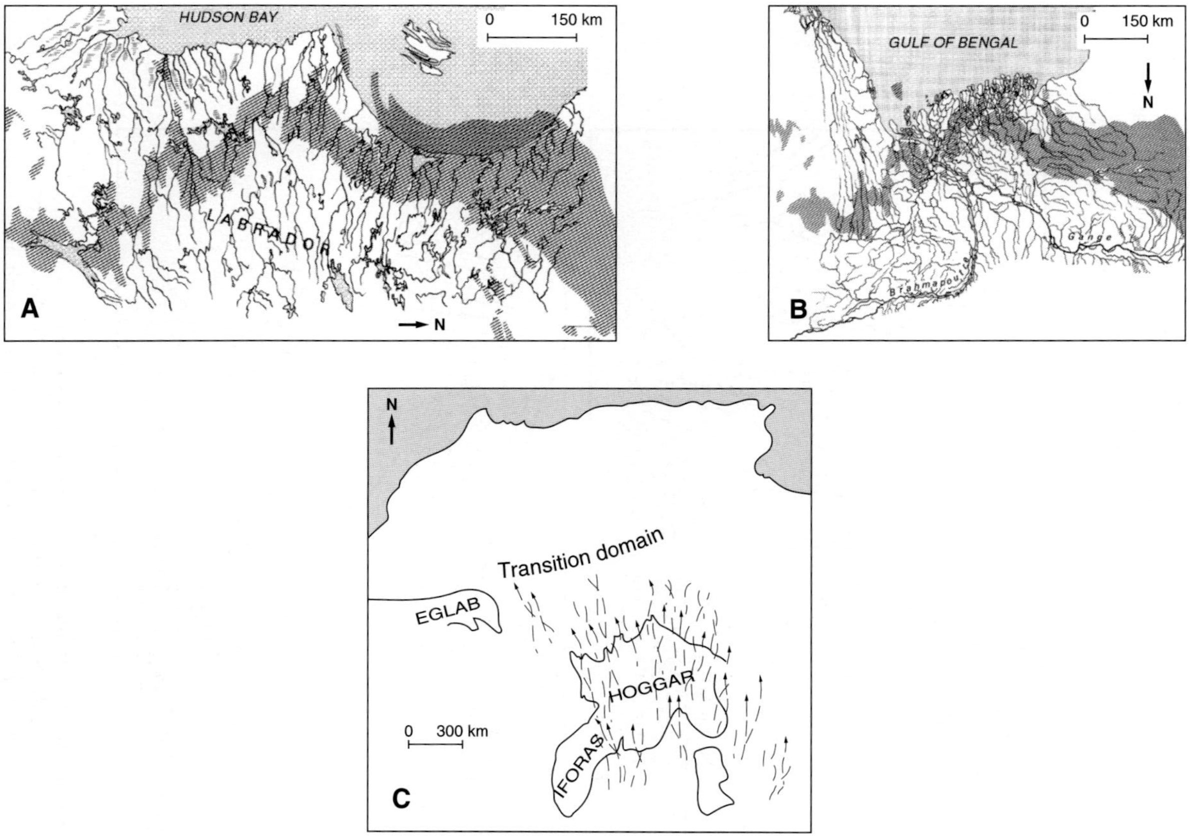

Fig. 3.77 Dimensions of fluvial systems.
A and **B.** Present hydrographic systems. **C.** Ancient system. The braided system of the Saharan Cambro-Ordovician that covered all of the Central Sahara, as attested by observed outcroppings. (The hatched areas in **A** and **B** are the Tassili outcroppings in the north of the Hoggar, superimposed on the same scale.)

on the continental scale (Fig. 3.78), and it also leaves valuable indicators of the environment on all scales. When the solid ice is in contact with the substratum, it leaves highly characteristic figures such as crescent-shaped plucking and markings and bench terrace fractures indicating cold glaciers. The high hydrostatic pressure of the subglacial waters perform intense polishing and cavitation at the same time. With its heavy glacial meal charge of sandy parti-

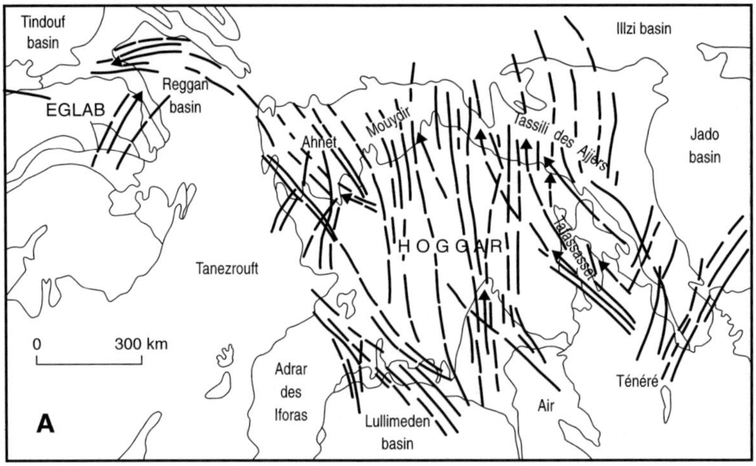

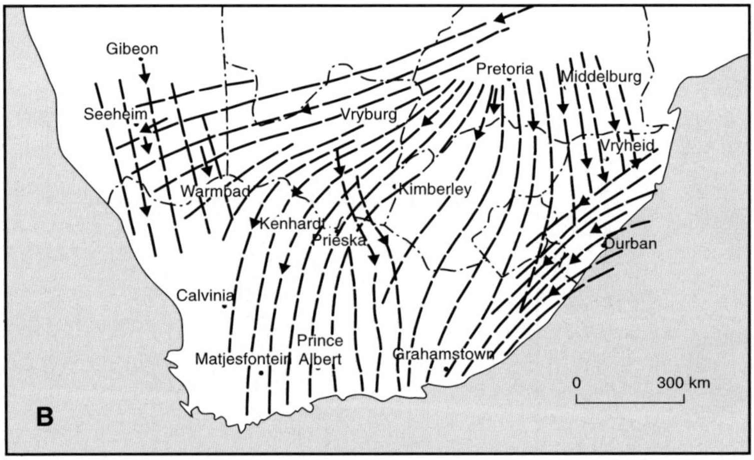

Fig. 3.78 Continental dimensions of inland glacial flows.
A. Example of the Saharan Ordovician (Beuf et al., 1971). It can be seen that the extent of the domain subjected to ice action is much more extensive, as other indicators have been found in Mali and even as far as Saudi Arabia. **B.** Classic example of South Africa's reconstructed Carboniferous ice sheet.

cles, the pressured water digs out **subglacial channels** underneath the glacial pavement (Fig. 3.79). This is the work of "temperate glaciers", though the phenomenon may also occur during melting episodes when the inland ice is momentarily warmed. These channels are then filled with detritus, forming **eskers**. Remember that the tangential pressure of the advancing glacier can raise up very large glacial pressure ridges (see Fig. 3.38). What is **transported** by continental ice will of course depend on the slope and channeling of the

A

B

Fig. 3.79 Examples of Late Ordovician subglacial channels. (IFP photos). **A.** Side view of the gritty fill in a channel dug out of Early Ordovician sandstone. **B.** Perspective of an esker with a visible sandy fill by differential erosion of clays.

substratum. If the substratum is hard, blocks will be plucked out; but if it advances over frozen soil or a sandy or clay sedimentary cover, it will pick up material and form **glacial flour**. Its speed of advance fluctuates greatly.

Melting is an indispensable part of achieving a deposit. Deglaciation phases are dominant here, and different processes are at work:

- With restricted melting under the ice load, the pressured water circulates at high velocity and creates **lodgment till** in the topographical lows.
- When melting occurs at the inland ice front and the **fluvial glacial** or **proglacial** regime is established, ablation till and sand will wash out on the ice cap fringe or further, if the melting becomes generalized. Three-quarters of the ice is transported out to sea in icebergs.

The same processes can be found in ***mountain glaciers****, but this will not be developed here because of the feeble chances of preservation of the associated deposits in the geological series (due to the gradual flattening of the relief).*

In high-latitude **periglacial regions**, the main processes at work are:

- **Freeze** and **thaw**, resulting in the formation of very thick layers of frozen soil called permafrost and special forms like hydrolaccolith (or pingo, Fig. 3.80), kettle (Fig. 3.81), ice wedges and polygonal lattices, solifluction on the slopes, and mud volcanoes.
- The material in the melted ice is carried disordered by the **running water** and forms braided systems and proglacial fans near the glacial front. The systems are better organized throughout the periglacial zone (Fig. 3.82), with channels in the alluvial plains (see above).
- Intense **eolian action** sets up dunes and loess, as has been said before.

3.2.5.2 Glacial and Periglacial Sediments and Depositional Forms

Different types of deposit will be generated depending on the process. One of the major characteristics of glacial deposits is their **poor grading** and their angularity: ice is not a good grain size sorting agent. It is also true that chemical reactions are relatively restricted in a glacial environment and that the transported material is therefore altered little, with little mineral transformation (which is not the case in a wet or a warm oxidizing continental environment). So it is easier to find fragile heavy minerals such as epidote, monazite and apatite, associated with the resistant minerals (tourmaline, rutile, zircon) of fluvial assemblages.

Tills and **moraine** are very poorly graded angular sediments consisting of blocks in a mountain glacier, but essentially sand for inland ice (Fig. 3.83). The "microconglomerate" appearance is a constant character with blocks disseminated in a large arenaceous matrix. A number of moraine types can be found: **lodgment till**, **ablation till**, and the **push moraine** that is more characteristic of the glaciation phases.

Ancient moraine is called **tillite**, and sometimes **diamicton**, though this also designates glacial marine sediments ice-rafted to sea and released by the melting ice.

A

B

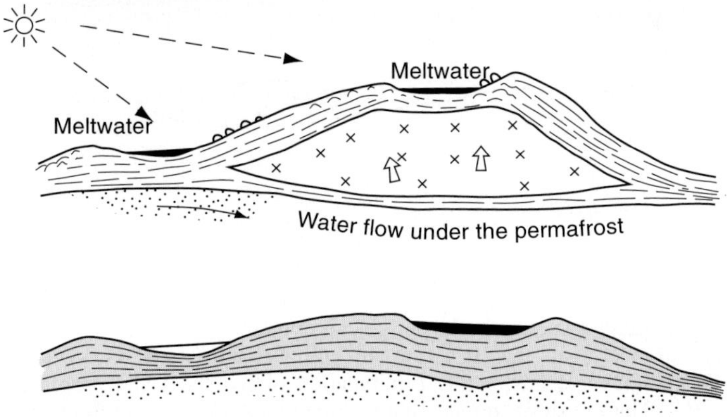

B1

B2

Fig. 3.80 Hydrolaccolith (or pingo) in the periglacial domain.
A. Structure observed in the Saharan Late Ordovician (oblique aerial view). The dome is about 300 m across, and is very similar to the Arctic pingos (IFP photo). **B.** Development mechanism scheme. B1: A lens of ice grows above the permafrost with some melting on the dome slopes exposed to solar radiation. B2: In the postglacial process, the ice and permafrost have disappeared, leaving depressions on the top of the former lens while a crescent-shaped pocket may appear on the exposed slope where peat can develop.

A

B

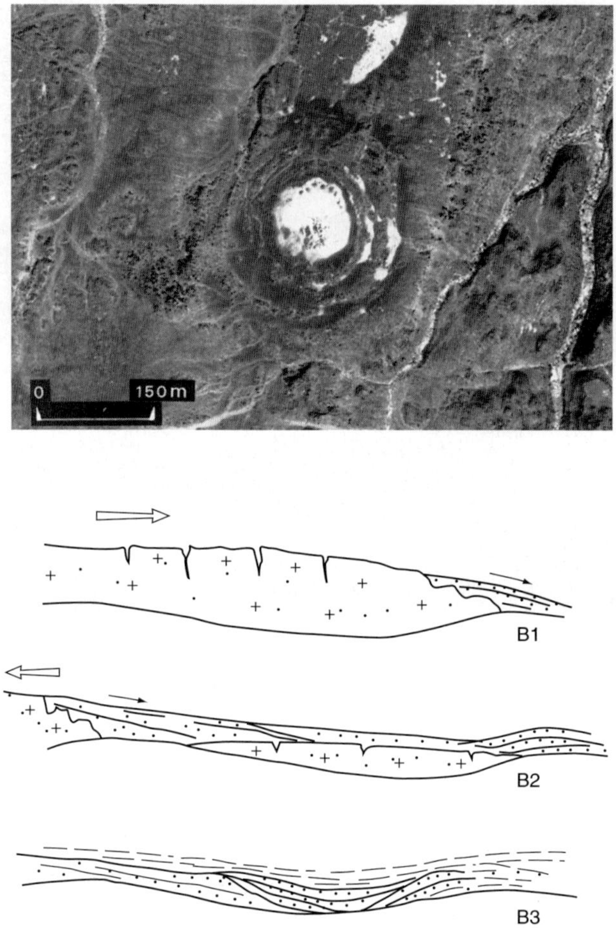

Fig. 3.81 Kettle structure in the periglacial domain.
A. Aerial view of the structure in the Saharan Late Ordovician. **B.** Formation mechanism. B1. Progradation of ice sheet and proglacial fans. B2. The ice sheet retreats, maintaining an ice lens while the lateral deposits collapse. B3. Everything melts and the proglacial sands collapse.

Moraine **outwash sand**, or sandur, is the first sediment reworked at the glacial front. This is arranged in irregular arenaceous fans, in a wandering, disordered network, in variable grain sizes, ranging from the rounded pebble to sand (Fig. 3.82).

With the highly mobile meltwaters and strong erosion, the stratification system is oblique, complex, and prograding. The wild streamlines of the flow become less variable downstream as the fluvial network gradually organizes.

Fig. 3.82 Proglacial outwash sand.
Obliquely stratified deposits with much erosion and small channels dug out (IFP photos).

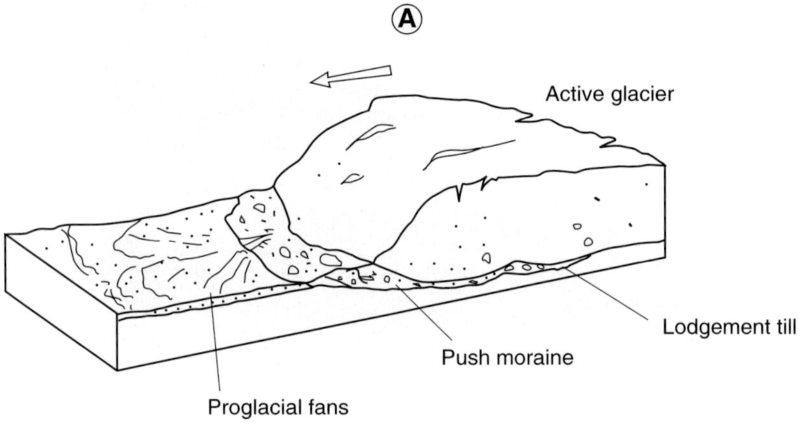

Fig. 3.83 Continental till.
A. General scheme. **B.** The typical moraine with large blocks from mountain glaciers changes here into a largely sandy till from the ice sheet, with disseminated blocks. The example is the Saharan Late Ordovician (Tassili of Tafassasset) with rare blocks of granite in a very poorly sorted sandy matrix (IFP photo).

At the front, gravity-driven flow till can be found with **subglacial channels** and **eskers** (Fig. 3.81) running in elongated ribbons that are filled with poorly sorted sand in the deglaciation phase.

Different fluvial and lacustrine environments affected by the formation of **permafrost** exist together in the **periglacial domain**. Flows concentrated in meandering channels may coexist with more diluted flows in anastomosing systems passing through lacustrine zones. Glacial lake facies are characterized by **varves**, which are seasonal sediments bearing the mark of alternating cold and thaw periods. Varved sedimentation consists of often dark fine clay beds of winter alternating with lighter, sandier summer beds. **Loess** deposits are the expression of the finest eolian deposits, which are transported over long distances and cover large areas.

3.2.5.3 Glacial Epochs and Geological Impact

Glacial and periglacial deposits indicate that the poles were glaciated. As the following chapter will show, this has not been true for all geological periods (Fig. 3.84). In order for a

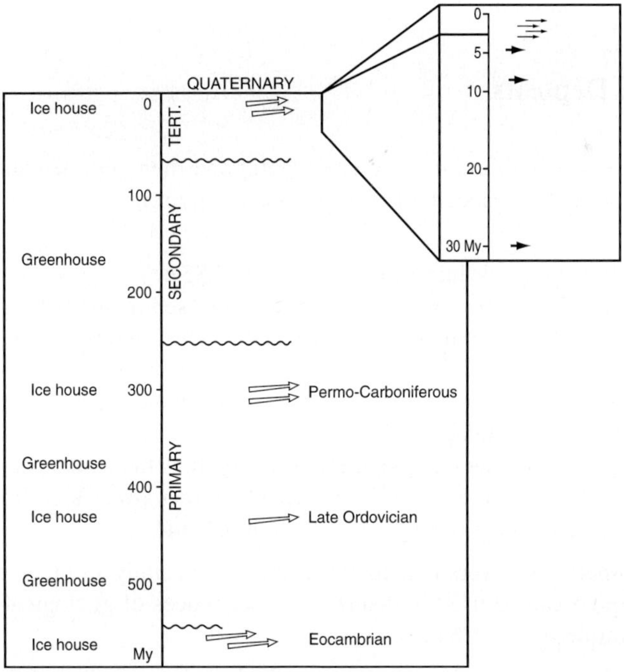

Fig. 3.84 Glacial periods in geological history.
Icehouse and greenhouse periods have succeeded each other through geological time, influencing the sea level by storing ice inland (glacio-eustasy). Ocean drilling has shown traces of cooling and glaciation pools since the Tertiary, beyond 2.5 My, and probably since 30 My in the Antarctic.

large continental glacial environment to exist, the high latitudes must have emerged zones. (Note that the Antarctic cannot serve as an example because, since nearly the entire continent is ice-covered, the only sedimentological studies being done today are in the marine domain.)

But detailed studies of the Quaternary have revealed a succession of glacial and interglacial stages appearing cyclically over short periods. The balances are very fragile. What should be remembered most of all is that the continental ice caps constitute a considerable store of fresh water which, if melted, would have an immediate repercussion of the world's oceans. This explains the major marine transgressions of the Holocene and Silurian, for instance. The direct consequence of deglaciation is therefore a radical change in the areas of marine and continental sedimentation by glacio-eustasy. Inversely, a sudden cooling causes a major stocking-up of free water, which lowers the sea level and adjusts the equilibrium profile of rivers as they cut out deeper incised valleys.

Another direct consequence of the large continental glacial environments is the strong impact on oceanic conditions, and especially on deep circulation, as will be seen at the end of this chapter.

3.2.6 Volcanic Deposits

Although volcanoes are both marine and continental and their sedimentation input is small in terms of volume, a few aspects of volcanoclastic environments are worth mentioning here.

Continental and insular volcanic zones show (Fig. 3.85):

- **Volcanic flows** themselves, which are not actually sediments but can be found interlaid with sediments, sometimes over considerable areas and over long periods of time.
- **Epiclastic debris**, which are the products of weathering and erosion of any volcanic head. The **hydroclastics** that form in breccia under aquatic conditions are one particular form of volcanic product.
- **Pyroclastic** products from explosive eruptions such as the avalanche of **ash** and gases called **nuée ardente**. There are different types of direct ash fallout. Rocks formed from volcanic projections often form graded beds called **tuff**.

Ash fallout sometimes covers considerable areas (especially in the ocean) with tephra. These layers of tephra can be used in determining sequences of geological events, which is called **tephrochronology** (see Chapter 4).

Continental volcanoclastic deposits are usually largely reworked and can be found in slumps, concentrated grain and debris flows, fluvial washout, in slope-dependent stratifications.

Fig. 3.85 Examples of continental volcanic deposits. Massive beds of cinerite, cut out by recent erosion (Turkey, IFP photo).

3.3 MARINE ENVIRONMENTS

3.3.1 Ocean Composition and Dynamic

The first chapter in this book spoke of the blue planet, with the ocean covering 70% of the Earth's surface and containing 97% of the Earth's total water. Today's ocean is divided into three major oceans (the Atlantic, Pacific, and Indian), which communicate largely with each other in the southern seas and much less so in the north, by the Arctic ocean. Each ocean has its own geographic dependencies (such as the Mediterranean, Sulu, and Red Seas). Some, such as the Baltic and the North Sea, overlie emerged continental zones while others are deep (Gulf of Mexico). Some seas are hypersaline (Mediterranean) while others are hyposaline (the Baltic) with respect to the world ocean's normal salinity. The Arctic is frozen part of the year while the Red Sea is overheated. This gives an idea of the variety of oceanic environments.

A number of important geotectonic parameters of the ocean regulate the sedimentary processes of sedimentary basins. Major advances in oceanography have shown the combined role of physical, chemical, and biological processes.[1] Temperature, salinity, pressure due to depth, continental inputs, exchanges with the atmosphere, are all variables to be considered. Of course, today's processes have to be transposed to geological periods, which can only be done with care and adjustment, as will be discussed further on in this chapter and the following one, concerning paleogeographic reconstructions.

Between the atmosphere and the seabed, the liquid mass of the ocean can be defined as a chemical system and a physical system where reactions and transfers take place. This dynamic system is also an environment where the processes of the biosphere interplay, and where sediments from the continent are transported.

3.3.1.1 The Ocean: a Special Chemical Environment

Composition, salinity

The ocean's water is saline. It is a chemical solution of water and a certain number of dissolved salts, major and minor elements, organic nutrients, and trace elements.

The **major elements** are chlorine (Cl), sodium (Na), magnesium (Mg), calcium (Ca), potassium (K), bromine (Br), sulfur (S), strontium (Sr), boron (B), fluorine (F), and silicon (Si). Their abundance ($\geq$ 1 ppm) varies little from one ocean to another or within the same ocean.

Salinity is expressed as the number of grams of dissolved salts per kilogram of seawater. It averages 34.7 ‰ and generally varies only from 33 to 38 ‰. The highest salinity is observed at tropical latitudes, with a minimum at the equator because of the rainfall (Fig. 3.86). In the large oceans, there is a slight vertical variation with layers of relatively homogeneous salinity because of the currents affecting the water column (Fig. 3.87). These concentrations or dilutions are due either to excessive evaporation or to confinement (the Red Sea, for example, registers 40-41 ‰ and the Mediterranean 37-40 ‰), or by a high input of fresh water from continental freshwater precipitation or ice meltwater, producing cases like the Baltic extreme of 3.5 to 10 ‰, the Black Sea at 16-18 ‰, or at river mouths). Strait passages are zones of high salinity gradients (Fig. 3.88).

These examples show that these extreme cases of salinity are due to special geographic configurations, which in turn stem from geological history (opening of an ocean, closing of a strait, climatic evolution).
In the most ancient geological past (Precambrian), the ocean's composition was very different. The chemical constituents from the biological cycle reflect the evolution of living groups. The composition of the atmosphere was not the same either.

Among the ocean's **minor elements** with concentrations of less of 1 ppm, there are helium (He), lithium (Li), boron (B), copper (Cu), zinc (Zn), cobalt (Co), iron (Fe), molybdenum (Mo), and the rare earths. These have little effect on salinity, but often play a major

1. See Minster, 1994; Biju-Duval, 1994.

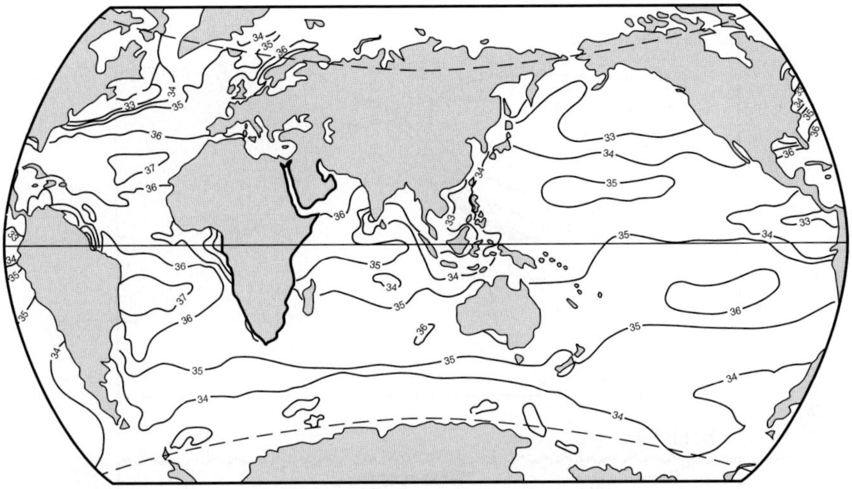

Fig. 3.86 Salinity variations at the ocean's surface (from Sverdrup, 1945, and many other authors).

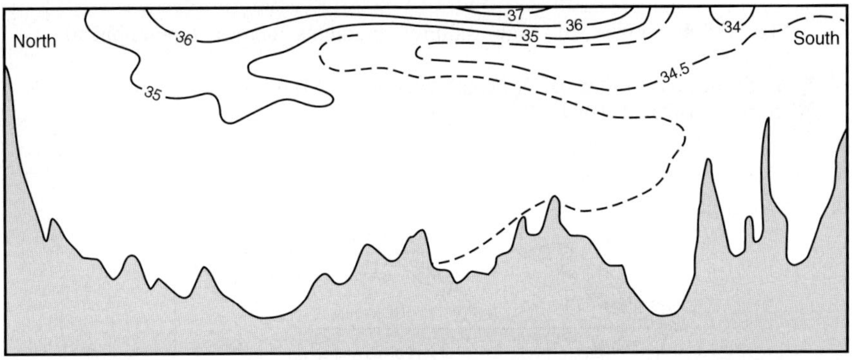

Fig. 3.87 Vertical profile of salinity in the Atlantic ocean (simplified, from many authors). Excess evaporation is visible in the tropical regions.

role in the whole oceanic environment, particularly in its biochemical and thermodynamical reactions and transfers.

The cycle of chemical species depends on their geochemical properties, but also on the depth of the ocean dynamic (stratification by currents or mixing), and biological actions. Biological activity depends very much on the available light in the euphotic layer where the first photosynthesis effects establish the link in the organic chain (Fig. 3.89).

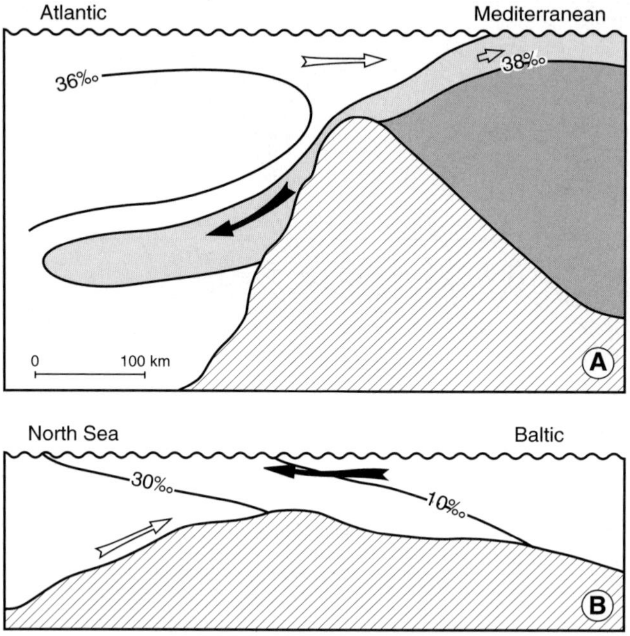

Fig. 3.88 Examples of high salinity gradients across topographical thresholds.
A. Strait of Gibraltar. **B.** Kattegat, between the Baltic and the North Sea (from Biju-Duval, 1994).

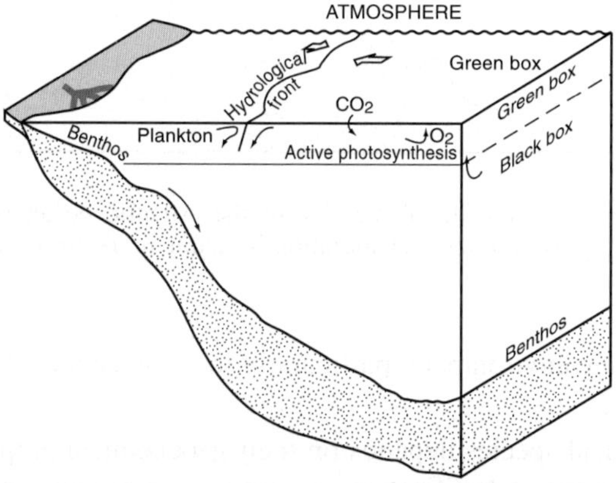

Fig. 3.89 The euphotic layer is a preferential zone for transfers and productivity (from Biju-Duval, 1996).

Special concentrations occur due to biological activity, since organisms will easily concentrate one or another element group.
Fluid circulation along ridgelines also generate variations, especially in the proximity of hydrothermal vents where anomalies of Mn, He, or CH_4 can be found that serve as geochemical guide and are associated with exuberant developments of various benthic forms.

Attention should be drawn to the nitrogen compounds originating from the organic products of biological activity, and to **dissolved gases**, CO_2 and O_2, which are of particular importance in an overall view of the interactions with the atmosphere and the biosphere (Fig. 3.90).

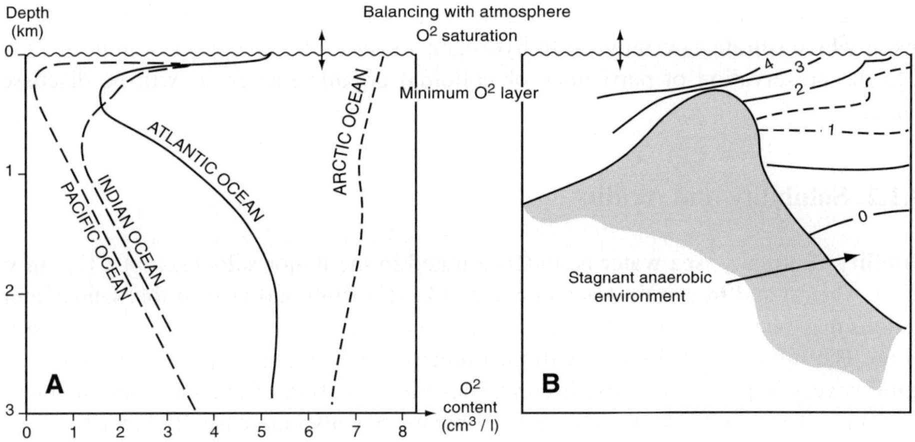

Fig. 3.90 Vertical distribution of dissolved oxygen in the water column (from various authors).
A. Variation with depth. Under the surface layer, which is in equilibrium with the atmosphere and very rich in oxygen, a sudden decrease in oxygen content is observed in all oceans except the Arctic, in the minimum oxygen layer. **B.** Example of a restricted, confined basin with a bathymetric threshold and no deep circulation to allow mixing. The dissolved oxygen content (cm^3/l) curves show the existence of an "anaerobic" environment with an oxygen deficit at depths, in the closed domain.

Concentrations vary with temperature (the solubility of oxygen varies inversely with temperature) and transfers with the atmosphere are continuous. The carbon cycle is more complicated, with a balance between carbon dioxide, carbonate and bicarbonate (discussed further on). In the geological past, CO_2 generation by microbial, phytoplanktonic, and vascular plants has varied greatly.

Nutrients, which are carbon (C), oxygen (O), nitrogen (N) and phosphorus (P), come in widely varied forms (especially bicarbonates, nitrates, and phosphates). They are extracted from the seawater in metabolic reactions, participate in photosynthesis, and then reenter the food chain. Organisms will use certain elements like calcium and silicon in constructing their shells, tests, or skeletons. The nutrients of bacteria, which are primordial in the food chain, are found to vary in a vertical profile, with the surface waters generally deficient and deep waters richer. If there is no mixing between the two, the surface waters can become so impoverished that primary production may decrease rapidly. But advection, diffusion, and convection normally provide effective mixing. As will be seen later on, the upwelling of deepwaters rich in nutrients can generate special situations.

In special cases of confinement for lack of active stirring currents, **chemical stratification** becomes very important. A total lack of oxygen can be found at the bottom of the Black Sea, for example, and in the Gulf of Cariaco (Venezuela). These are **anoxic** conditions in which sulfates are reduced to hydrogen sulfide, which supports conditions favorable to the preservation of particulate or colloidal organic matter, as will be discussed in Chapter 6.

3.3.1.2 Solubility and Acidity

Solubility. Normally, sea water is undersaturated in the major salts (except CO_3– in warm surface waters) and in certain trace elements like Thorium and rare earths. Undersaturated solutions may become saturated under certain conditions, such as in the case of excess evaporation. The sulfates and chlorites will then precipitate, forming evaporites. This idea of saturation is very important because it conditions the formation of chemical sediments on the ocean floor. The solubility of some trace elements may also increase, by formation of complexes. Solubility also depends on other factors like agitation, and also varies vertically in the water column. Carbonates will dissolve at great depths in the calcite compensation zone below the CCD (see Figs 3.20 and 3.21). Transfers with the atmosphere are also possible under certain conditions with the upwelling of deep water.

Acidity is defined by **pH**. The ocean is slightly basic, with a pH of 8 (pure water pH is 7). Variations of pH in the ocean depend on CO_2 content in the atmosphere, and that of H_2CO_3 and CO_3Ca. The ocean's pH can fall to 3 in certain rare cases, like that of hydrothermal sources where the waters are very aggressive. As concerns oxidation-reduction, or **redox potential** (eH), it has already been said that the conditions are generally highly oxidizing. But an oxygen deficit can exist for reasons of biological action or for lack of hydrodynamism in cases of confinement, which lowers the pH to anoxic conditions (Fig. 3.90). These conditions may also occur locally with the stratification of the basin waters (Black Sea). Special periods are known in geological history where confinement conditions have prevailed (producing the "black shales" of the Atlantic in the Cretaceous).

As will be seen later on in the section on transfers, some elements (Mg, Co, and others) can be enriched by **continental input** from streams.

3.3.1.3 Ocean Temperature

The ocean behaves like a gigantic **thermal machine**. Part of the solar radiation that gets through the atmosphere is reflected and another component penetrates the ocean, which absorbs and diffuses it. This heats an average 30-meter thickness of ocean water at the surface (the thickness varies considerably). The heated layer is in turn subject to thermal losses by evaporation.

The ocean's surface temperature of course varies from the high latitudes (30°C) to the low (–2°C), which is very important for the general chemical processes and especially for the repercussions it has on biochemical activity, the total calcium budget, all of the transfers between the atmosphere, biosphere and hydrosphere, sedimentation on the ocean floor, and evaporation (Figs. 3.91 and 3.92). Schematically, it can be said that heat is absorbed under the tropics and returned to the atmosphere in the higher latitudes. The ocean acts like a heat conveyor from the Equator to the poles, and thus as a thermal regulator, constituting a heat reservoir at the Earth's surface.

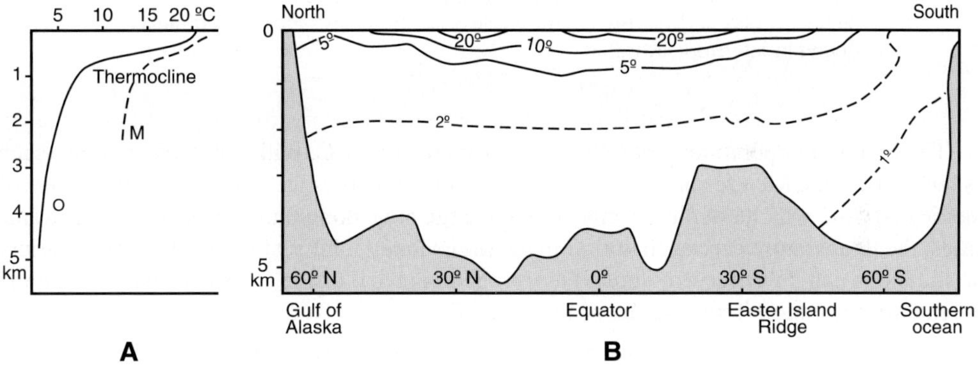

Fig. 3.91 Temperature profiles in the oceans (from various authors). **A.** Vertical profile with a sudden drop down to the thermocline, followed by a slower decrease (O: Ocean; M: Mediterranean). **B.** Longitudinal profile of the Pacific along the Americas.

The temperature also varies in the water column, but this characteristic is largely affected by the general ocean circulation. Figure 3.91 gives examples of vertical temperature distributions in which the permanent **thermocline** can be seen under the warm surface layer that is very sensitive to seasonal variations. This thermocline is a transitional layer where the temperature gradient is very high. Its base level varies but hardly ever goes beyond 1100 m, where the temperature is 6°C. A **summer thermocline** may often occur with another high temperature gradient at 30-50 m.

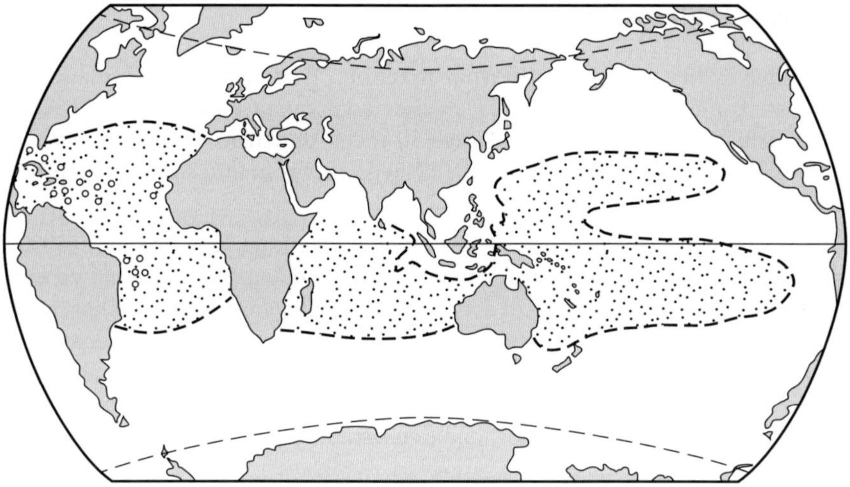

Fig. 3.92 Distribution of today's carbonated sediments in warm waters. Pellets (dots) and ooliths (small circles) are frequent types of sediments found in intertropical waters.

The bottom temperature generally varies from 0° to 2.5°C, with a few exceptions (such as the 13°C in the Mediterranean). Special localized anomalies are due to the **hot springs** of the ridges, from which hot plumes may cover a large area along the ocean ridges. In extreme cases, the temperature can reach 400°C in the immediate vicinity of the vent. A great variety of situations can be observed, depending on the geometry of the hydrothermal system and the rate of circulation in the basalts of the ocean floor.

At high latitudes, the **frozen ocean** at the surface creates a quite special environment that corresponds to the present global climatic situation. This will be discussed later in this chapter as concerns deposition environments. Remember that this situation has fluctuated in the course of geological time (glacial and non-glacial periods, with and without an ocean existing at the pole).

The density balance of the surface water changes when it is cooled by the atmosphere, and it then sinks to greater depths. This phenomenon is amplified with salinity, which opens the subject of ocean dynamics.

3.3.1.4 The Ocean: a Dynamic System

The fluid medium of the ocean is far from being a stationary mass of water. It is a dynamic system driven by the Earth's rotation (Coriolis force); the attraction of celestial bodies (tides); heat transfers from solar radiation, leading to pressure fields, winds, and evaporation; the Earth's internal dynamics (tsunamis, hydrothermal plumes), all with seasonal variations. Schematically, the ocean's motion is controlled by two forces: wind action, which determines the surface circulation, and heat flows at the atmosphere-ocean interface, which generates "thermohaline" circulation.

Different periods of motion can also be defined: very long (general circulation), a few days (drift), a few hours (tidal), a few seconds (waves), or even shorter (small-scale turbulence), and sometimes special currents corresponding to specific events (turbidity currents, for example). The different types of currents are of capital importance for determining sedimentary transport, sorting, and deposition mode, and also condition the forms of erosion recognized in geology.

3.3.1.5 Ocean Circulation

So the ocean is not an immobile volume of water but has its own dynamic at the surface and at depth. This broad advective circulation of ocean water is driven by interaction of the Earth's rotation (producing geostrophic currents at the surface and at depth) with solar energy input, which causes winds, sets up temperature and pressure gradients and density gradients too, by evaporation, which further supplies a thermohaline (temperature-salinity) drive mechanism (Fig. 3.93).

These advective motions are effective, as they transfer water at velocities up to several knots, but diffusion also comes into play on all scales where turbulence is important.

3.3.1.6 Surface Circulation

The ocean and atmosphere form a coupled system. The general atmospheric circulation governs the ocean's **surface circulation**. The winds to either side of the equator move obliquely from east to west in the tradewind pattern and, beyond the high-pressure zone ending at about 40° of latitude, they organize into "westerlies" (Fig. 3.94). The whole system is subject to the Earth's rotational motion and the distribution of the continental masses, resulting in the creation of large cells or "gyres" over the large oceans, turning clockwise in the northern hemisphere and counter clockwise in the southern—a mirror image to either side of the equator. These make the large warm and cold ocean currents with convergence and the divergence zones, particularly the antarctic convergence zone (Fig. 3.95).

Generally, the currents are 20 to 200 m deep at low latitudes, and the Gulf Stream is exceptional with its depth of 1000 m. The currents are broad and slow at the equator and faster and more concentrated at the higher latitudes. The volumes of water transported are large, flowing at 20 to 100 $\times$ 10^6 m^3/s, and the thermal and radiative transport as well as the mechanical friction energy are considerable. Recent studies describe fluctuations in the form of meanders, and eddies on the scale of a hundred kilometers. This highly unstable internal turbulence is reflected in very rapid instantaneous currents in different directions, and large heat transfers.

A

B

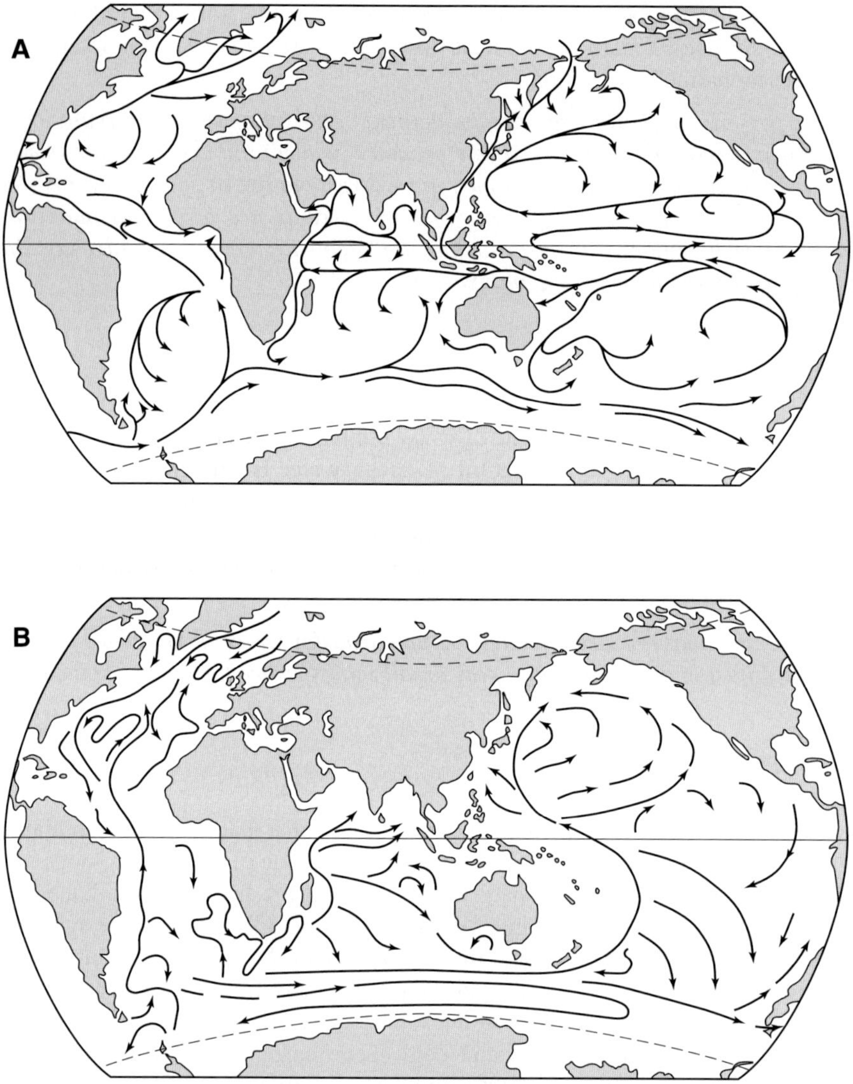

Fig. 3.93 General oceanic circulation (from various sources).
A. Surface circulation (compare with Fig. 3.94). **B.** Deep circulation: cold Arctic and Antarctic waters form bottom streams: North Atlantic Deep Water (NADW) and Antarctic Bottom Water (AABW).

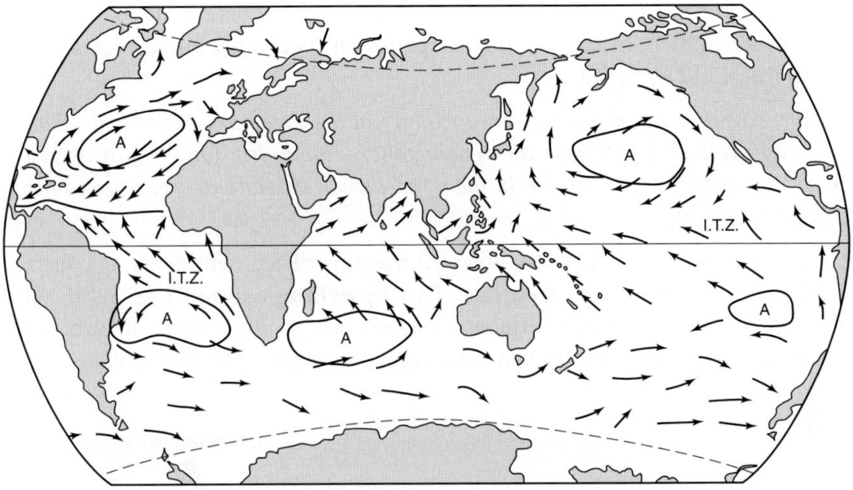

Fig. 3.94 Map of general atmospheric circulation (from various sources). Anticyclonic cells (A) determine the tradewinds and the monsoon to either side of the Intertropical Convergence Zone (I.T.Z).

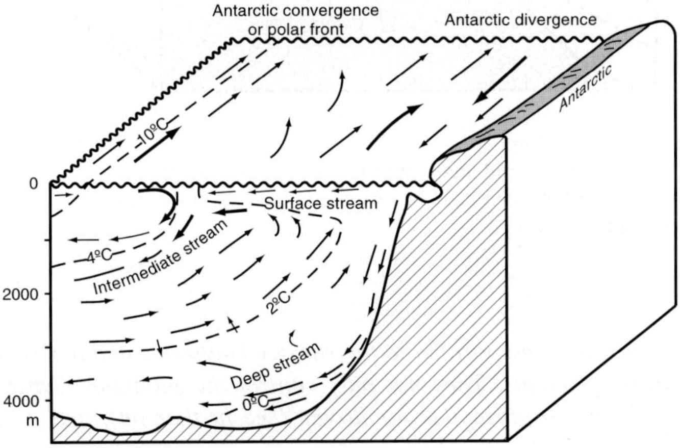

Fig. 3.95 Antarctic convergence and divergence.
The plunging of surface waters and upwelling of deepwater around the edge of the Arctic generates a surface circulation with divergence in the Antarctic zone and convergence in the subantarctic.

It is known that certain currents like the Gulf Stream have been functional since the Oligocene. The establishment of such currents depends on special geological conditions such as when a strait threshold is closed.

Moreover, surface waters do not move exactly in the direction of the wind, but drift 20° to 40° to the right of it in the northern hemisphere due to frictional shearing stress in the "Ekman" boundary layer. At depth, the wind-driven velocity decreases and the angular difference decreases.

Under certain conditions along the western margins of the continents, the surface water is entrained and the deep waters then rise. This **upwelling** current (Fig. 3.96) will be discussed later in the chapter. The continental masses can produce large disturbances in this general arrangement, such as the Indian subcontinent, creating the monsoon.

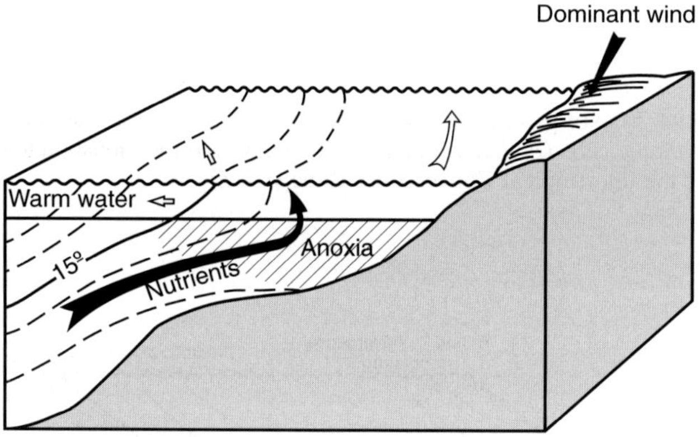

Fig. 3.96 Upwelling.
An upwelling of cold deep waters and meridian circulation along the coasts, with littoral drift, can be observed along certain continental margins (from Biju-Duval, 1994).

Generally, the currents run parallel to the surface isotherms and are stronger along the high-temperature gradients. They not only control the plankton distribution, but also benthic assemblages, in particular for larvae (see further on), and play an important role in reef systems too.

The currents are subject to **seasonal variations** in velocity and volume of water transported, sometimes resulting in a geographical drift.

Irregular *event-type* ***variations*** *may also occur. For example, the El Niño phenomenon suddenly appears when the tradewinds abnormally decrease and upwelling is reduced. The stream transporting warm waters of low salinity southward along the South Ameri-*

can Pacific coast is then extended considerably southward, with disastrous effects on the environment and marine life.

In light of short-term fluctuations (seasonal and decadal turbulence) and paleo-oceanographic studies, it can be seen that the general circulation has varied considerably over the course of geological history due to variations in the geometry and latitudinal position of the continents and global climatic variations. Tests are being conducted today with models to try to simulate what this circulation was and how it influenced the depositional environments in the Mesozoic and even before.

3.3.1.7 Thermohaline Deep Circulation

This is caused by differences of temperature, density, and salinity. Taking the Atlantic model, several stratified layers can be found with their own distinct motions. There is the surface layer, some 200 m deep with large seasonal variations; an intermediate layer (stemming from the Antarctic convergence at latitude 50°C); and the deep layer of cold North Atlantic Deep Water (NADW) and Antarctic Bottom water (ABW) (Fig. 3.97). These density currents correspond to maximum and minimum oxygen concentrations.

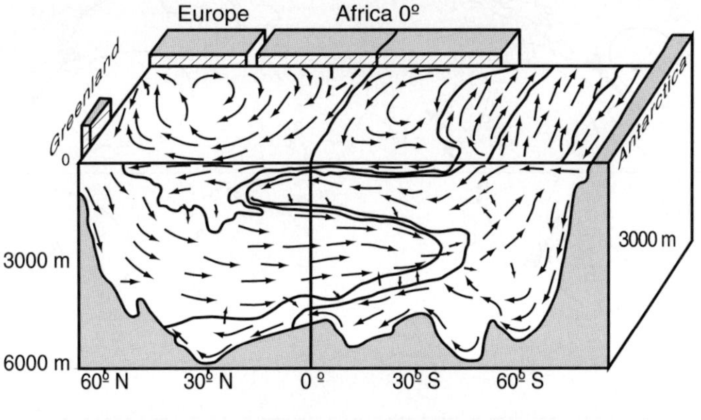

Fig. 3.97 A schematic of water layer motions between the surface and bottom of the Atlantic ocean (from Defant, in Fairbridge, 1966). Note the anticyclonic gyres to the north and south of the equator (see Figs 3.93A and 3.94) and the Antarctic convergence (see Fig. 3.95).

The deep waters form in geographic "windows" such as the Norwegian Sea and the Weddell Sea in the South, where certain conditions are met (such as strong permanent winds and cyclonic oceanic eddies at the surface). The deep cold water streams extend from there to all of the oceans and deeply affect the sedimentation with effects on benthic life, large and small-scale sedimentary accumulations (from ripples to giant deep dunes, deep sedimentary lobes), dissolution of components, and erosion.

Surface currents and thermohaline circulation are not stable. It is certain that these currents have varied according to changing climatic conditions in geological time, such as in the glacial and interglacial episodes.

3.3.1.8 Upwelling Currents

The deflection of the currents on the coast is caused by the Coriolis effect, turning them to the right in the northern hemisphere and to the left in the southern. This is reinforced when the winds come from offshore, entraining the surface waters that are replaced by the deep water rich in nutrients from the base of the thermocline (Fig. 3.98). We then observe a proliferation of the phytoplankton entraining the entire food chain, and a major accumulation of organic matter in the sediments. Repeated anaerobic deposits (laminites) can occur in extreme cases.

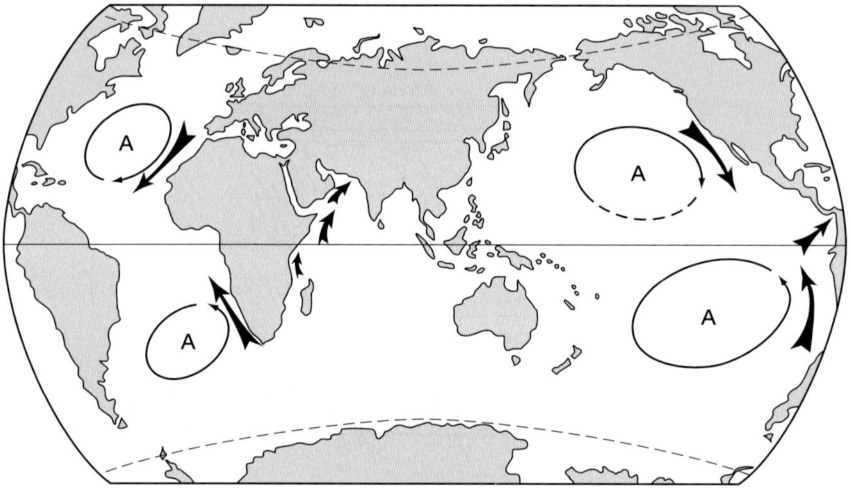

Fig. 3.98 Main upwellings running alone anticyclone cells (A).

The upwelling is not limited to the coastal zones. Equatorial divergence, for example, causes high primary organic productivity and a high sedimentation rate in the Atlantic and Pacific.

3.3.1.9 Tides and Tidal Currents

Tides are certainly the expression of ocean dynamic best known to the general public. They result from the universal law of attraction, or gravity. The predominant effects of the Moon combine periodically with those of the Sun when they are in conjunction and opposition, and variously in between, causing **diurnal**, **semi-diurnal** and **composite** tides (Fig. 3.99), and over longer periods too (semi-monthly, semi-annually). Tides are sensitive to the geometry of the ocean boundaries in capes, bays, and other complex patterns that vary the amplitudes of the tides.

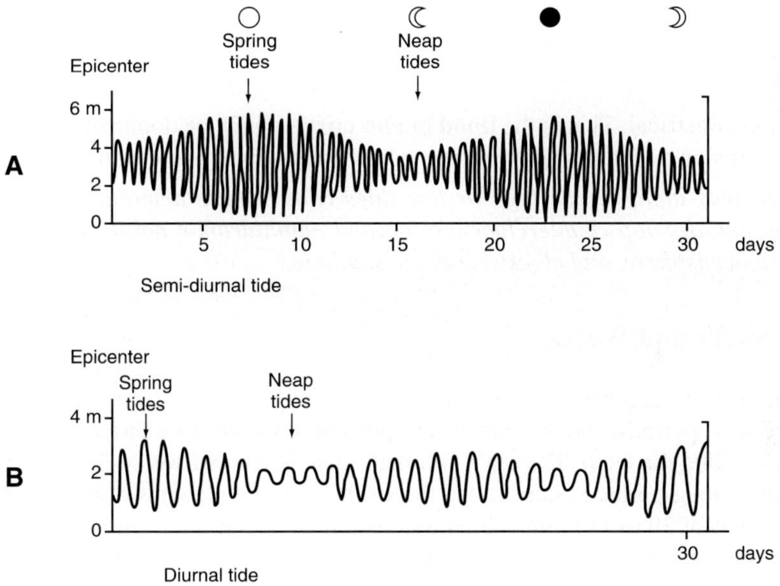

Fig. 3.99 Sea-level variations by type of tide and period in the tidal cycle.

There are spring tides, neap tides and seasonal tides. The **tidal range** is the vertical amplitude, or rise and fall, while the **intertidal zone** is the part of the shore that is alternately covered and uncovered by the sea (Fig. 3.100). A number of sedimentary domains are defined on this basis, namely **tidal** and infra-tidal, and tidal deposits are sometimes called tidalite.

The irregularity of the ocean border geometry, friction, and other parameters, causes **tidal currents**, which are particularly important on continental platforms extending on the scale of a kilometer, but with rare exceptions out to 20 or even 50 km. These currents are not

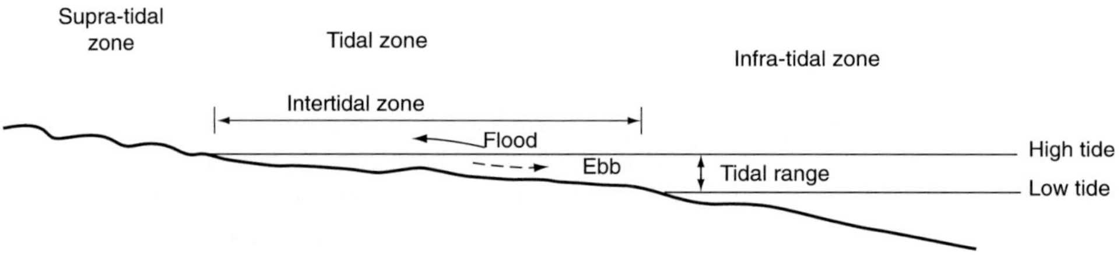

Fig. 3.100 Tidal domain: foreshore and tidal range.
The flood and ebb act on the infra-tidal zone with each tide. This zone varies in size depending on the cycle and seasons and the varying tidal rise and fall.

necessarily symmetrical. That is the **flood** or **ebb current** may be dominant. The tide creates a variety of erosion and deposition forms, such as megaripples and channels.

Co-tidal lines *along which high or low tides coincide are determined by the Earth's rotation and develop counterclockwise around* ***amphidromic points*** *of null tide in the northern hemisphere, and clockwise in the southern.*

3.3.1.10 Swells and Waves

Ordinarily, the wind deforms the sea's surface, setting up **swells** and waves in wave trains of different heights, periods, and directions. The portion of the ocean's surface affected by constant wind is called the fetch. The sea's "normal" state is with **waves** of short period that can organize into a long-period system called the **swell**, with wavelengths of 60 to 400 m travelling at velocities of 30 to 75 km/h, which will remain in action for a while after the wind that created the waves has lapsed. Swells and waves have an effect on the seabed in shallow waters (in fact, to a depth of half the wavelength, which is up to some 50 m and sometimes even more). Seabed sediments are sensitive to these effects. On the coast, the waves are **refracted** and part of their energy is dissipated. If the waves reach the coast obliquely, they induce a current parallel to the coast called **littoral drift** (Fig. 3.101). This is capable of transporting large quantities of material that are sometimes carried out to sea by rip currents.

Wave action on the shoreline is considerable and often destructive, with erosion of the coasts, retreat of littoral cliffs, littoral drift of particles, grading of sediments, formation of current ripples, mineral concentrations (placers) and organic remains. Waves are thus of considerable importance in geology for the entire littoral domain. Their oscillations are at times the cause of **microseisms**.

Cyclones and storms amplify these effects, and the consequent erosion and reorganization of deposits on the seabed can be large, to depths of several tens of meters, leaving sandy **tempestite** deposits. In exceptional cases, the normally emerged domain will be affected and **washover deposits** will appear in the coastal zone.

A

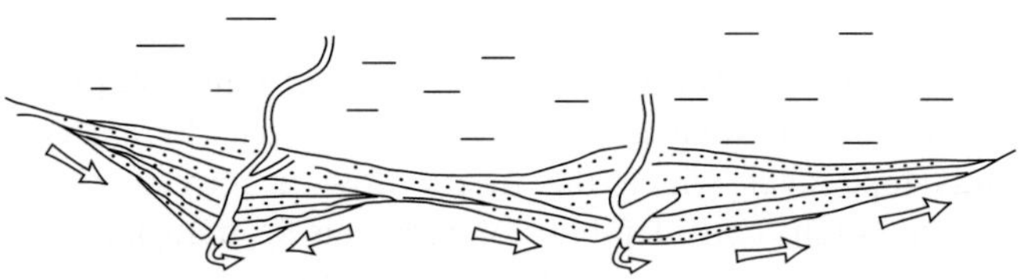

B

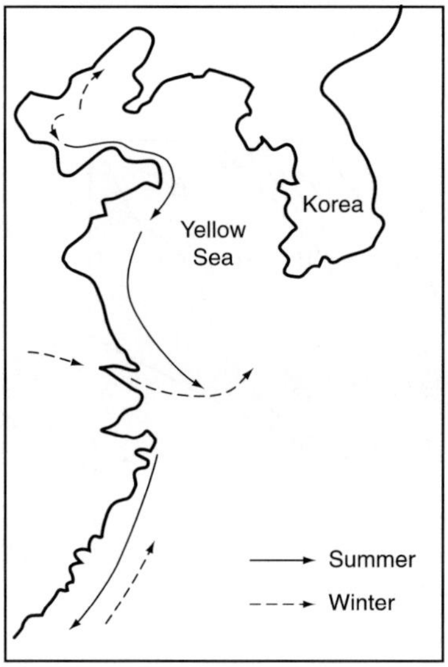

Fig. 3.101 Effects of littoral drift.
A. General scheme: particles transported to sea by the coastal rivers are then reworked and redistributed along the shore, with vigorous wave action. **B.** Example of the China Sea.

Another event is the shock caused by an earthquake, volcanic explosion, or submarine landslide, resulting in a tidal wave or **tsunami**, which is a shock wave travelling at speeds of up to 800 km/h! **Seismite** is the name used to designate this type of deposit.

There also exist **internal waves** between deepwater layers of different densities. The height of these waves ranges from 5 to 20 meters and sometimes much more.

3.3.1.11 Turbidity Currents

Density currents form when two fluids of different densities (because of temperature, salinity, sediment in suspension) lie one on top of the other. When a layer of water with a very high concentration of sediments is introduced, it produces a special density current called a turbidity current. This is an important current in terms of sedimentary transport and deposition in oceanic basins. It can travel at high velocities of up to 20 m/s, depending on the density contrast and slope over which the turbid flow is moving (Fig. 3.102). When the concentration is very high, considerable volumes can be transported over long distances (figures like 3×10^8 m^3 over 500 km). One recently mentioned cause of these currents is earthquakes (Newfoundland, 1929), but violent storms are often recognized as a simpler cause. The consequences are always large, with a rapid, instantaneous (on the geological scale) deposit of **turbidites**. They are of considerable importance in constructing thick detritic piles, as shall be seen later on.

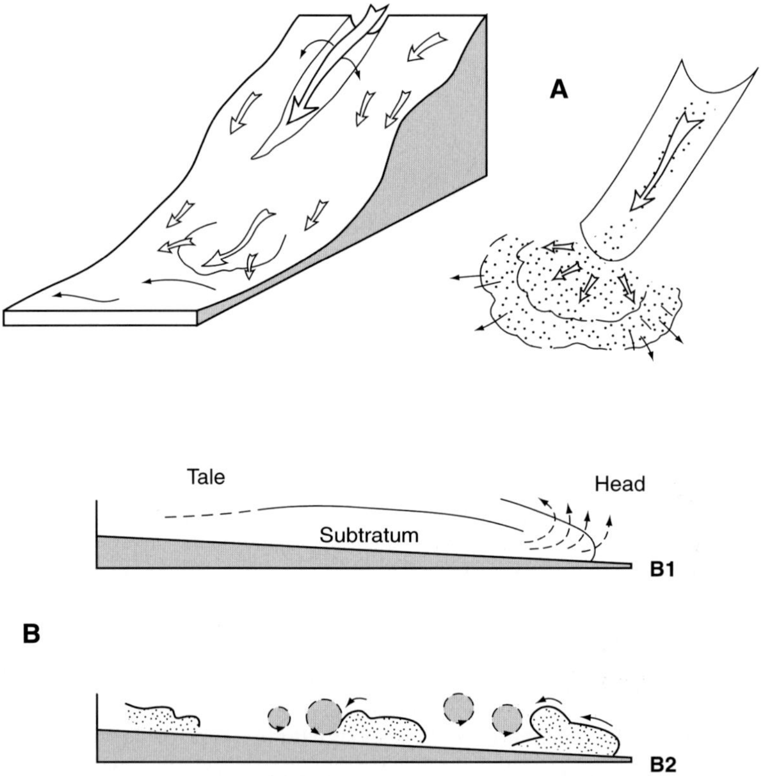

Fig. 3.102 Turbidity current mechanisms.
A. The mechanism can be triggered in various forms on an irregular slope (traction bed, dense flows, channeling) and continues to the end of the slope where the bed currents can rework the material. **B.** Form of the current (B1) and formation of puffs and eddies (B2).

3.3.1.12 Contour Currents

Slow, regular thermohaline bottom currents are initiated by a density contrast and follow the seabed topography (Fig. 3.103), hence the name **contour current**. These currents may have

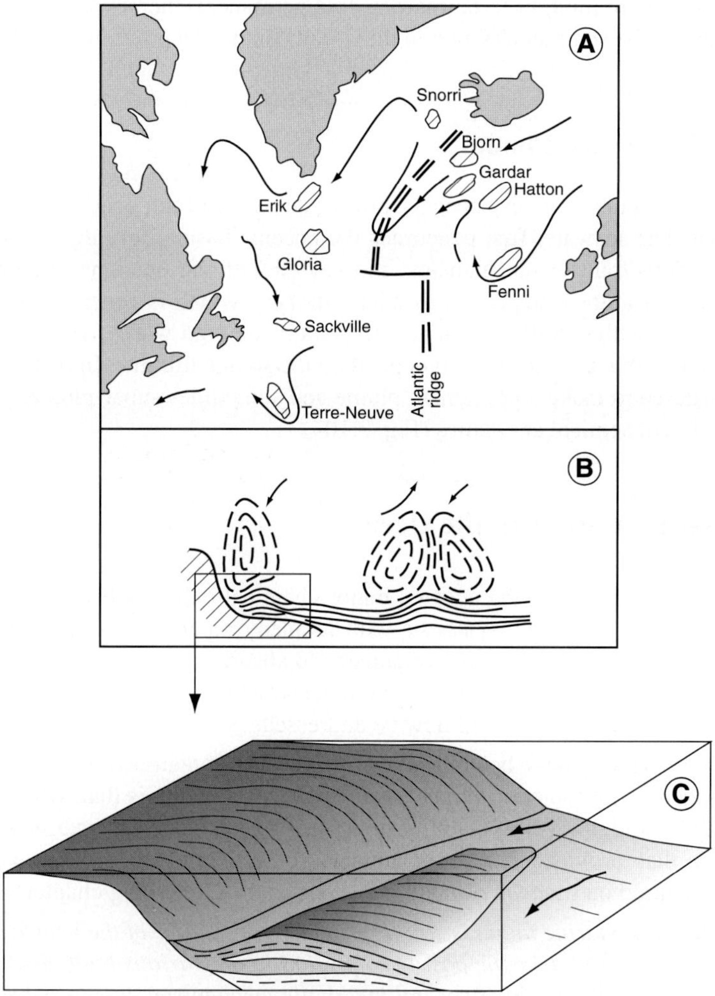

Fig. 3.103 Contour currents and contourite.
Deep ocean currents (**A**) can be active and create sedimentary accumulations called contourites or drifts (examples found in the north Atlantic are Fenni and Gardar drifts and others shown in the figure). **B** is a cross-section schematic of isovelocity lines. Deposit occurs when the velocity falls, creating a giant sedimentary ridge (diagrammed in **C**).

an erosive effect on any material that is not very cohesive, but they mainly provide the drive for transporting and depositing large quantities of fine detritus that has already been deposited, by picking it up and rolling it over the floor or carrying it in suspension, later to form **contourite**.[1]

Contrary to the short, violent turbidity currents, contour currents range in velocity from 0.05 to 0.4 m/s and are thought to be more or less permanent, thereby constituting a means of accumulation of the sedimentary pile in the ocean depths, deep sediment drifts, and hemipelagic draping.

3.3.1.13 Hydrothermal Plumes

The hydrothermal plume is a very special form of density current, originating in hydrothermal circulation. The seawater first penetrates the oceanic basalts (in ridges and volcanoes), is heated by various transfer mechanisms, and loaded with various ions. It comes back out through high temperature vents in a turbid jet (smoker), which is generally highly concentrated and charged with metallic elements, and enters the cold water from the bottom. The extent of the plume that then forms will depend on the source lifetime. In fact, small individual plumes coalesce to make a permanent plume and sometimes super-plume, characterized by a number of geochemical anomalies (Fig. 3.104).

3.3.2 Biological Activity in the Ocean

The role of flora and fauna in the ocean's major chemical balances has already been mentioned. The importance of photosynthesis with its many repercussions in the CO_2 cycle is one aspect, as is the capacity of living organisms to absorb mineral substances to construct their skeleton or use them in metabolism, and to construct edifices that become barriers. The role of organisms in lithogenesis will now be addressed.

Some 160 000 species have been listed for the oceanic domain, which is in contrast with the near-homogeneity of the environment itself. Another paradox is that, while the zoological groups exhibit a high level of stability in time, it was the study of marine fauna in geological history that provided the main support for Darwin's theory of evolution, and **stratigraphy** is based on time variations of the species (see following chapter).

Our knowledge of marine fauna is still limited. Twenty to 70% of the known species have been newly discovered over the past twenty years in the various biological groups, and for certain species the study of special metabolic mechanisms in situ is beginning only now. Much work is still being done on symbiotic associations and ecological niches, and the direct and indirect participation of living organisms in sedimentation is receiving more and more attention.

1. Concept reviewed in Hollister, 1993.

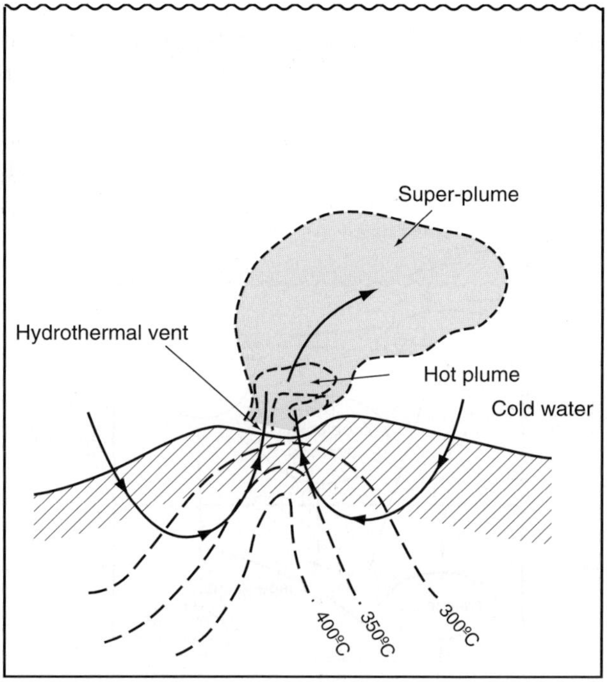

Fig. 3.104 Diagram of hydrothermal circulation and transfers.
The ocean's cold bottom water enters the basalt of the ocean floor and quickly escapes (the form of the rising isotherms figured here recall that of the ridges themselves) and many transfers occur. This water is evacuated along the vent alignments in smokers that are black or white (depending on the mineral content) and generally hot, hence the term hyrdothermal vents (from Biju-Duval, 1994).

Life in the ocean depends not only on seawater but is also affected by transfers with the atmosphere, the substratum, and sediments, with many interactions in different environments. Here, we will mention only those aspects of direct concern for the different environments, and the implications for sediments.

Recent discoveries of special life forms—anaerobic bacteria living at high temperatures in oases around deep hydrothermal vents—are raising questions today about the mechanisms and origin of life on our planet.

Life in the ocean concerns not only the surface layers of water where organic productivity may vary considerably, but the entire water body of nearly 1.5 million km^3 occupied by living organisms. This is considerable compared with the continental population, which lives within a few meters above the ground. However, most organisms live at the surface or on or near the bottom (only a small percentage of the species live in between).

Food chains are longer at sea than on land, with different levels and low energy transfer (Fig. 3.105). Fertility is essentially determined by the richness in nutritive salts that phytoplankton needs in order to grow. This is maximum on the continental shelves and upwelling zones.

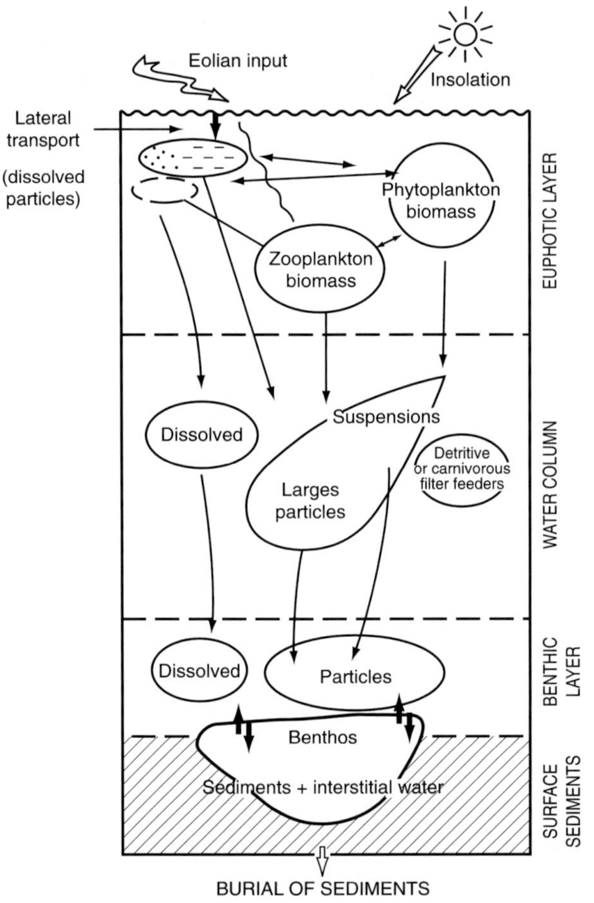

Fig. 3.105 Biological activity in the ocean.
Complex chains and processes occur between the surface and the bottom. These depend on the productivity rate, luminosity, turbidity, temperature, and oxygen content (from an Ifremer document).

3.3.2.1 Major Biological Groups

The living world of the ocean is extremely diversified with vertebrates, invertebrates, unicellular vegetal and animal organisms of the benthos and plankton, and bacteria. These become **marine fossils** when preserved in geological series.[1]

Vertebrates. Most vertebrates belong to different groups of fish such as lampreys, selachians and teleosts, but also to the mammal, reptile, and bird families. While they may be the most spectacular part of underwater life forms, they are not majority contributors to the major biochemical balances or to the ocean's biomass production, and contribute few elements that will incorporate sediment and later provide evidence of the environment. Due to the action of physical and chemical agents and bacteria (see later), the role of the predators is such that their remains after death have little chance of being fossilized. So fossil remains are relatively rare except under special conditions of oxygen deficit, confinement, and quick burial such as in the Manosque Oligocene, the Messinian of Italy, or the Early Jurassic of Bourgogne and Germany.

The soft parts of organic matter are assimilated or dispersed, while the hard parts (skeletons) are generally fragmented, dispersed, and turned into minerals. Shark and crocodile teeth might be preserved, but bones are rare.

A few of today's fauna are veritable living fossils, corresponding to species of the Paleozoic (coelacanth and lung fish, for example). Some, like the jawless agnatha, have existed in the ocean since the Late Cambrian. The tooth-shaped conodonts known from the Cambrian to the Triassic are doubtless the remains of primitive chordates, which are very useful in the stratigraphic analysis of the Paleozoic.

Multicellular **metazoan invertebrates** appear in a variety of marine forms in many zoological groups, and many are still very close to the fossil forms of the Paleozoic and Mesozoic. These are poriferans, cnidarians (medusas, polyps), mollusks (bivalves, gastropods, cephalopods), arthropods (essentially crustaceans), brachiopods, and bryozoans. These are the basis of ancient fossils studied in **paleontology** to develop stratigraphic scales (Fig. 3.106).

Their life mode, environment, and metabolism provide the geologist with useful information for reconstructing the past. While the associations on the continental shelf are generally well known, though, this is not true for the deep ocean. Studies are being conducted today to understand how communities and ecosystems of certain ecological niches can constitute zones of asylum or centers of dispersion on the continental margins.

A distinction is made between **fixed organisms** (such as crinoids and sponges), which live off fallout in suspension or particles of sediment (polychaetes), and **mobile organisms**, some of which are pelagic (floating or swimming). Some are free and others are parasites (like certain ostracods). Some live in symbiosis (worms in hydrothermal colonies). Many

1. Enay, 1990; Bignot, 1988.

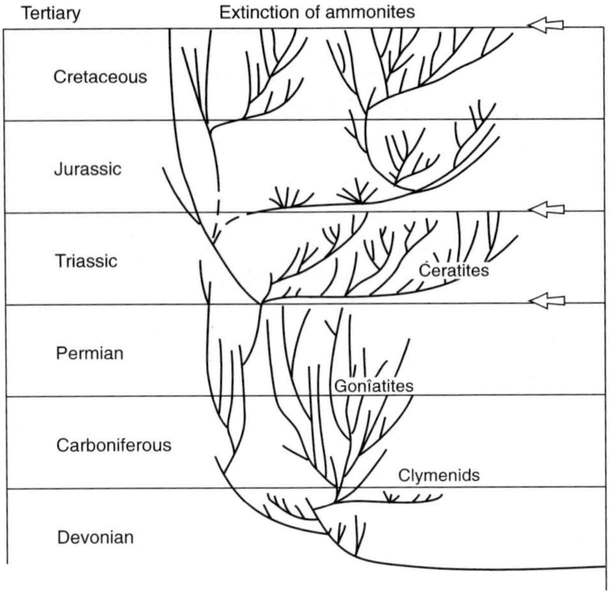

Fig. 3.106 Example of phylogenetic evolution: successive outbranching from the large family of ammonites in the course of geological periods (from Devillers and Chaline, 1989).

are very sensitive to the salinity and temperature of the environment and are thus of use in geological reconstructions for defining deposition conditions. Others are more ubiquist. After death, their bodies may concentrate into **thanatocoenoses**, making coquinite, or pteropod ooze, if they remain dispersed.

Some—trilobites, ammonites, and brachiopods, for example—have evolved rapidly enough to be used now as major stratigraphic markers (see Chapter 4).

Their coprolites in fecal pellets are sometimes abundant, recognizable products. Their habitat (burrows) and mode of locomotion also leave fossilizable traces that can later become precious indicators called **ichnofacies**.

3.3.2.2 Protozoans

These micro- and nano-organisms belong to different phytological and zoological groups. We will recall here only a few characteristic elements, and note that the broad zooplankton and phytoplankton groups that are the majority participants in today's primary production have not always played this role in the geological past. The great **phytoplankton** diversifications occurred in the Jurassic, while the **zooplankton** has mainly developed since the Cretaceous.

Foraminifers (ranging from 50 µm to a few millimeters and sometimes more) are protozoan rhizipods with endoplasm and ectoplasm. They have an intra-ectoplasmic test (or skeleton) that may be agglutinated, microgranular, or porcelaneous, and is usually calcareous. They live in widely varying conditions. Most are stenohaline, but some accommodate to large variations in salinity (sometimes hyperhaline, like the miliolids). They are either benthic or planktonic, and the latter are widely dispersed geographically. Planktonic foraminifers generally live at shallow depths in the euphotic level, though this depends on their life cycle (their tests exhibit a thin layer formed during their life at the surface, and another thick layer constructed commonly at depths of 200 m, and sometimes as much as 1000 m. Benthic foraminifers with calcareous or agglutinated tests are found to range from the platform to the abyssal plains. This group is a major one in today's oceanic biomass, and it has also played a major role in the past. Average figures of 50 to 200 benthic foraminifers are mentioned for 10 cm^2 of sea floor, with a few planktonic foraminifers per liter of water. Precise biostratigraphic zonation has been developed on the basis of the major lines of descent (fusulinids, orbitolinids, nummulites, globigerinids). They have populated the oceans from the Cambrian to the Present, and are one of the major groups of micropaleontology (Fig. 3.107).

A

B

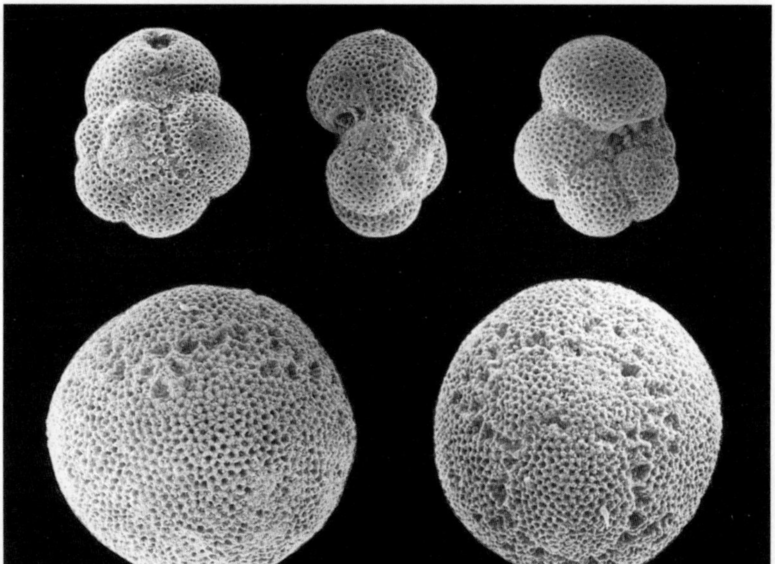

Fig. 3.107 Examples of foraminifers used in stratigraphic dating (Total MEB photos, 120 × enlargement).
A. Globorotaloide Suteri Bolli, 1957, from the Upper Eocene.
B. Orbulina saturalis Bronnimam, 1951, from the Middle Miocene.

Coccolithophores are minute (1 to 50 µm) unicellular algae with a calcareous test that form the major part of the nanoplankton in nutrient-poor, well-lit waters at depths of up to 50 m. They can stand up to cold waters and various salinities (some have been described in the Black Sea waters) and sometimes live at depths of up to 4000 m. A liter of seawater is known to contain an average of 10^3 coccoliths.

While the tests of planktonic foraminifers can reach the bottom in a few days, it probably takes several months or years for coccolithophore tests, which are less sensitive to dissolution effects as they are protected by organic matter. They are often found in abundance in fecal pellets at all depths.

The petrogenetic role of nano-fossils (discoasters and other coccoliths) is very important (Fig. 3.108). The example of chalk has already been mentioned, but its precise role as stratigraphic marker is also noteworthy. The absence of these forms in the Paleozoic is doubtless more apparent than it is real (it is difficult to recognize them in lithified rock).

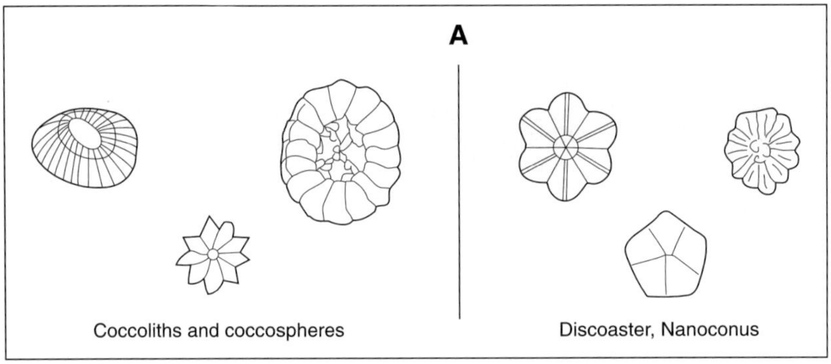

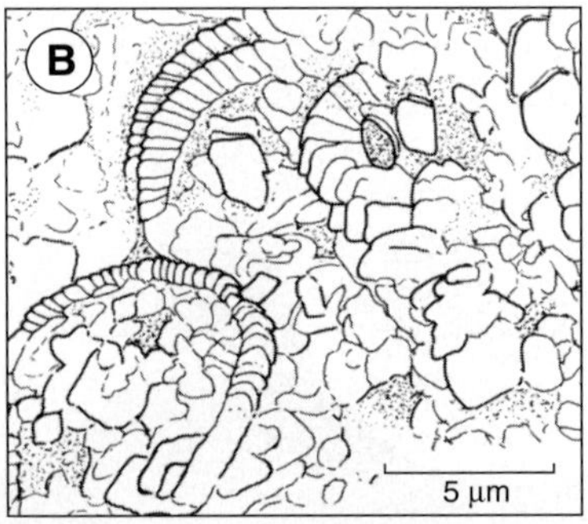

Fig. 3.108 A few nano-planktonic forms.
A. Examples of different forms. The morphological variety is extensive (diagrams from Bignot, 1998). **B.** Association of coccoliths in the chalk of the Late Cretaceous (North Sea, porosity 40%) (from MEB photo, in Guillemot, 1984).

The protozoan **radiolarians** with their siliceous tests measuring from 100 to 400 μm and sometimes up to 2 mm, are planktonic organisms with a particularly short lifetime in water depths of up to 100 m and sometimes more (beyond 500 m). These are particularly abundant (several dozen per liter of water) at high latitudes, at the equator, and on the western edges of the continents or in the water column. If their opal (silica) tests are not dissolved at the surface or in the water column, they are incorporated in the sediments after their death in radiolarian oozes (Fig. 3.109). These have been known since the Paleozoic. They are plethoric in the Mesozoic and, when the compaction and diagenesis were effective, they constitute **radiolarite**. But it should be remembered that only a small proportion (<1%) of the biogenic silica can be found trapped in the geological archives.

Diatoms are unicellular algae with a siliceous bivalve shell (frustule) measuring 50 μm to 2 mm. They can be benthic (fixed or mobile) or planktonic, and are found especially in cold seas at high latitudes. They are doubtless one of the most abundant forms of plankton ($>10^3$ per liter of seawater). They sometimes accumulate into diatomaceous oozes and then form rocks called **diatomites** (Fig. 3.109). These have been known in abundance since the Lower Cretaceous.

Silicoflagellates are planktonic algae of complex test (20 to 100 μm) that can accommodate wide variations of salinity. Few species are living today.

Dinoflagellates are especially abundant in plankton. These are unicellular algae that are bare or protected by a theca (a plate skeleton found on peridinians). They can be autotrophic, parasitic, or even symbiotic like the zooxantella associated with the radiolarians. They bloom intensely under extreme conditions, creating toxic red waters. Acritarchs (50 to 100 μm) are another form subsumed under palynomorphs that have been known since the Precambrian.

Microplankton includes still many other kinds, belonging to different branches such as tintinnids, which are ciliate protozoans with a chitinous shell, and micro-crustaceans: copepods and ostracods. Copepods constitute a major part of the plankton and can live at great depths. They are coccolith predators. Ostracods live in various environments from the open sea to estuaries. Most are benthic, but some are pelagic. They are fossil forms of use in geology for defining depositional environments.

3.3.2.3 Bacteria

Pelagic production, like life on the bottom, is characterized by a large volume of bacteria and cyanobacteria, the life of which depends on cellular dissolution and excretion products. They are major factors in the biomass. It is thought that up to 20 000 bacteria can be found per cubic centimeter or 150 ng C/dm^3, even in waters that are very poor in nutrients (oligotrophic).

The role of bacteria in the ocean is therefore paramount, though much research is still to be done on the exact transfers that occur in the metabolism, in lithogenesis and metallogene-

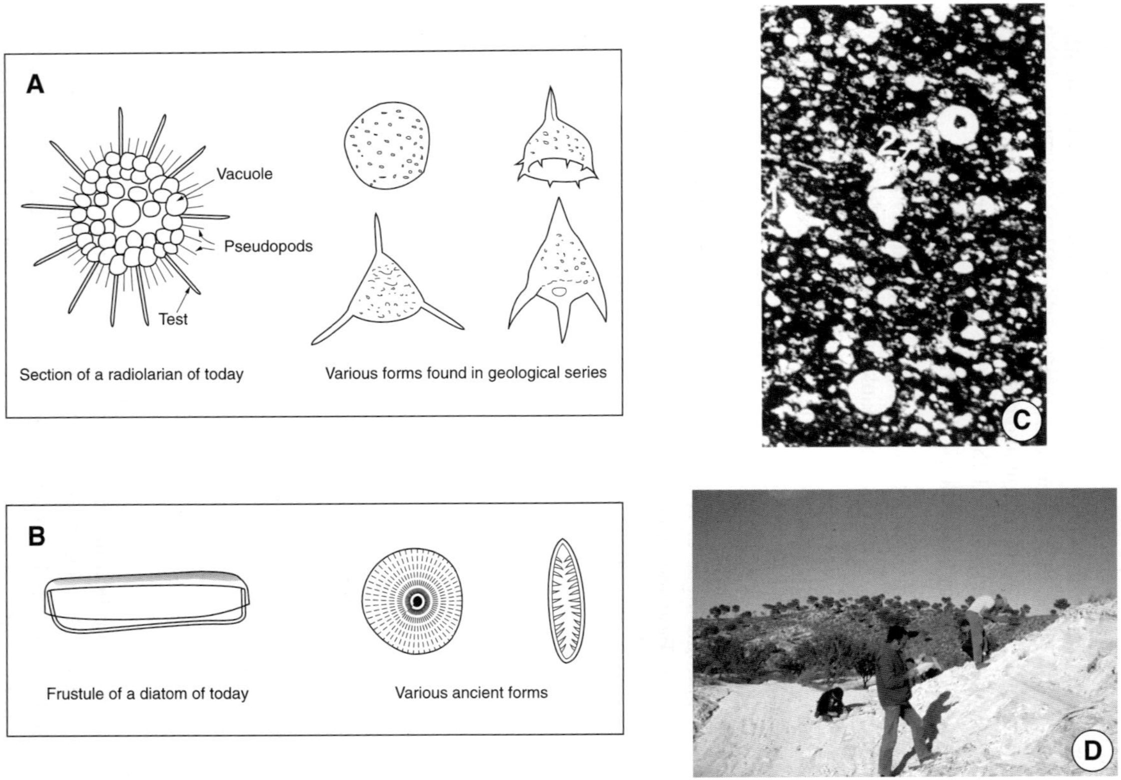

Fig. 3.109 Organisms with siliceous tests constituting non-detritic siliceous rock.
A. Radiolarians. **B.** Diatoms (from Bignot, 1988). **C.** Thin section of radiolarite, 30 × magnification. **D.** Beds of diatomite formed by major accumulation of diatom tests. Here, the diatomite is from the Columbars Messinian (IGAL photo).

sis, for example. It has been very important since the Precambrian and throughout all geological eras.

To conclude this rapid overview, the many **spore** and **pollen** forms should also be mentioned. These can be found in many vegetal groups, and contribute to oceanic sedimentation. The role of algae will be developed a little later in this chapter.

3.3.2.4 Biological Domains

A. Benthic Life

Benthic animal (fauna) and vegetal (flora) communities live on the water's bottom from the **intertidal** zone to the great oceanic depths. These are sometimes divided into **epibiontic** life, which lived in free or fixed contact with the sea bottom, and **endobiontic** life within the waterlogged sediment itself. Organisms living within a few dozen meters of the bottom (such as certain fish with pelvic fins who hover near the turbid bottom layer) are said to be bentho-pelagic. The mechanical actions of the benthos in the sediment are called **bioturbation**.

The benthos is phased in steps from the littoral zone to the great ocean troughs, with each species occupying its own (generally limited) vertical bathymetric stage (Fig. 3.110). These stages are particularly well marked in shallow waters, especially in the intertidal zone with its wealth of living communities. Many nutritive inputs are available there, and the phytoplanktonic productivity is especially abundant. The benthos lives in close relation with the planktonic communities of the euphotic zone.

Fixed sponges and mobile holothuroids that eat detritus and filter the sediments dominate the continental shelf down to the abyssal plains. The geographic ranges of the various

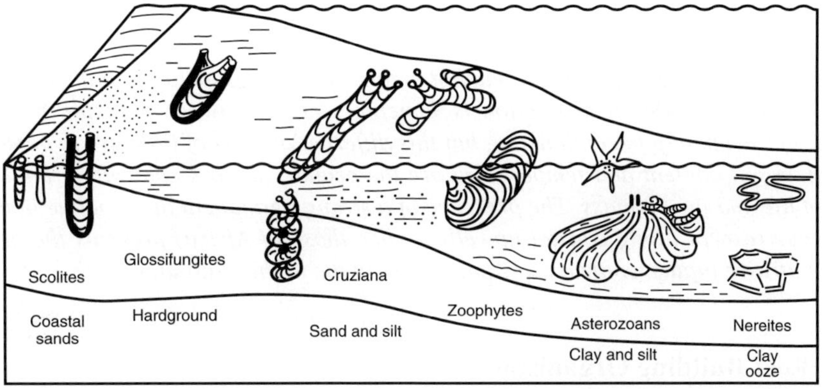

Fig. 3.110 Diagram or bathymetric distribution of various bioturbations from the coast to the deep domain.

species increases (some can be found everywhere in the ocean), bioturbation can be large, and the effect of surface fertility is lessened. In the troughs beyond 6000 m, species are notably endemic (of limited geographic distribution).

For reference, note the large number of animal species, and large population of each, surrounding the deep hydrothermal vents that have been studied since their discovery in 1976. The dense populations of large invertebrates found here constitute veritable oases of life around the vents as they spew hot fluids containing more or less concentrated sulfides. These ecosystems differ greatly depending on whether the vent is hot (with polychaete worms dominating) or cooler vents (with bivalves and vestimentiferans). An original ecosystem of bacteria also exists that can use the hydrogen sulfide to produce organic molecules from carbon dioxide, independently of light energy. This hydrothermal fauna is doubtless ancient (from 150 My to 200 My for certain groups), but more and more new species are being found (invertebrates, bacteria, and archaebacteria). The existence of ultrathermophilic bacteria is still being discussed, as are the metabolic mechanisms that will not be discussed here. What is of greatest geological interest in these populations is their dispersion and speciation with time. Much more research is still needed on this.

B. Pelagic Life

The bulk of primary productivity is found in the **euphotic** or **epipelagic** zone, ranging from the surface to a depth of some 100 meters, where there is enough solar energy to allow photosynthesis. This is where the pelagos or **plankton** develop:

- Phytoplankton: diatoms, dinoflagellates, various microscopic algae and nano-plankton (coccoliths).
- Zooplankton: foraminifers, radiolarians, copepods (with a biomass greater than that of all the other animals), krill (small crustaceans of the upper latitudes), amphipods and others, which form the link between primary vegetal productivity and the species used by man.

Zooplankton extends beyond the euphotic zone to the mesopelagic, bathypelagic, and abyssopelagic depths. A distinction is sometimes made between plankton in the strict sense and swimming nectoplankton, but this differentiation is of little use in geology.
Nutritive salt content and temperature are the primary factors controlling the distribution of life and the biomass. The presence of a strong permanent thermocline in the tropical zone (except along coastal upwellings like those of Africa) prevents the arrival of deep nutritive inputs, contrary to what occurs in the higher latitudes.

3.3.2.5 Reef-Building Organisms

Of the fixed organisms at shallow depth, special mention should be made of reef-builders, for the consequences they have on lithogenesis.

A. Algal Mats and Stromatolite

Algal mats and laminites abound in the intertidal zone, sometimes forming a crust (Fig. 3.111). These mats are flat crusts, often only a few centimeters thick, consisting of hardened film of filamentous cyanobacteria. They are sometimes impermeable enough to trap air and form domes by plastic deformation under pressure and, when desiccated, can

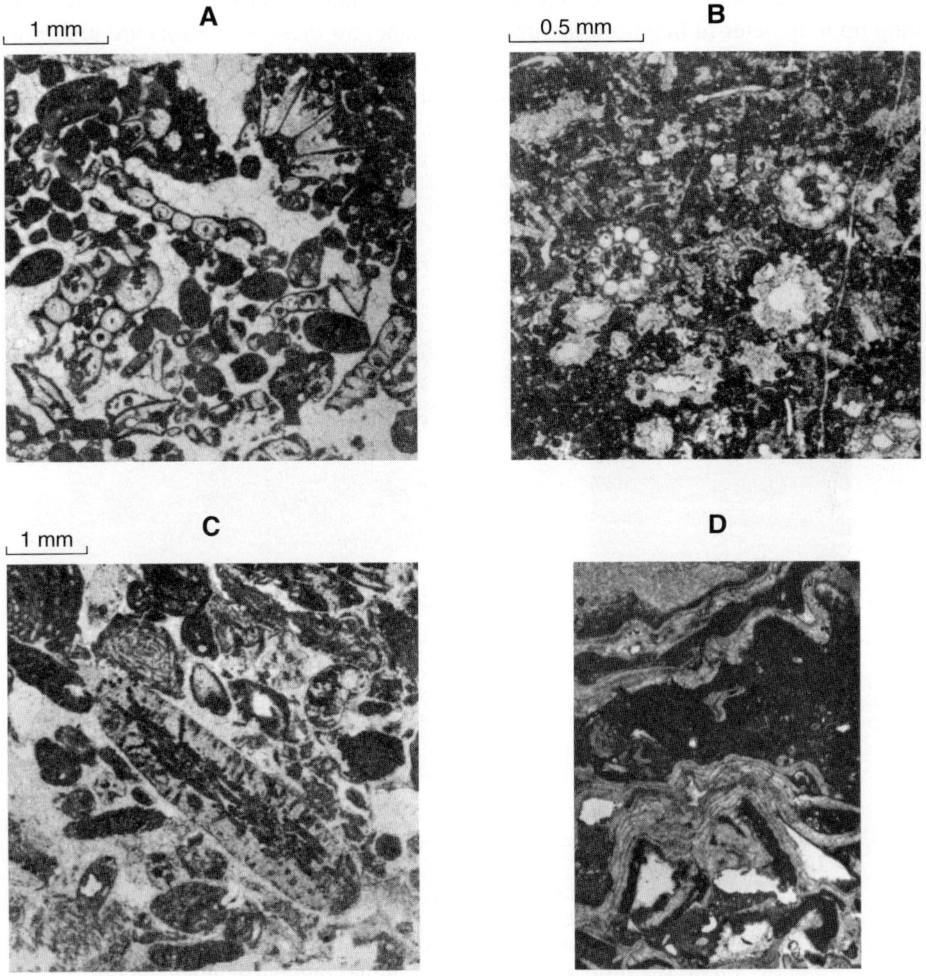

Fig. 3.111 Different types of algae contributing to sedimentation. Cellular structures observed in thin sections: **A.** Dasycladaceans. **B.** Characeans (40H). **C.** Codiaceans. **D.** Example of limestone made of algae (from J.P. Bertrand, 1969).

break up into fragments.[1] Certain "bird's eye" structures can develop on the top of algal mats. These are cavities due to the biogenic gas formed by decomposition of the organic matter. Many fossil laminites may be algal in origin.

Some cyanobacteria in the infra- to supra-tidal zones form **stromatolite** that develop most in confined environments of high salinity. These are actually complex associations of several mucilaginous microvegetals. They can grow rapidly into assemblages of various forms—columns, planes, or domes depending on the type of environment (the controlling parameters are the tidal amplitude, temperature, and water dynamics). These structures can develop up to a meter in height in the infra-tidal zone. So these organisms are highly lithogenic, and there is no lack of fossil examples (domes of collenia, and Conophytons of the Paleozoic and Precambrian). Diurnal and seasonal variations, and those due to storm effects, govern the development of these beds where only the upper parts remain living (Fig. 3.112).

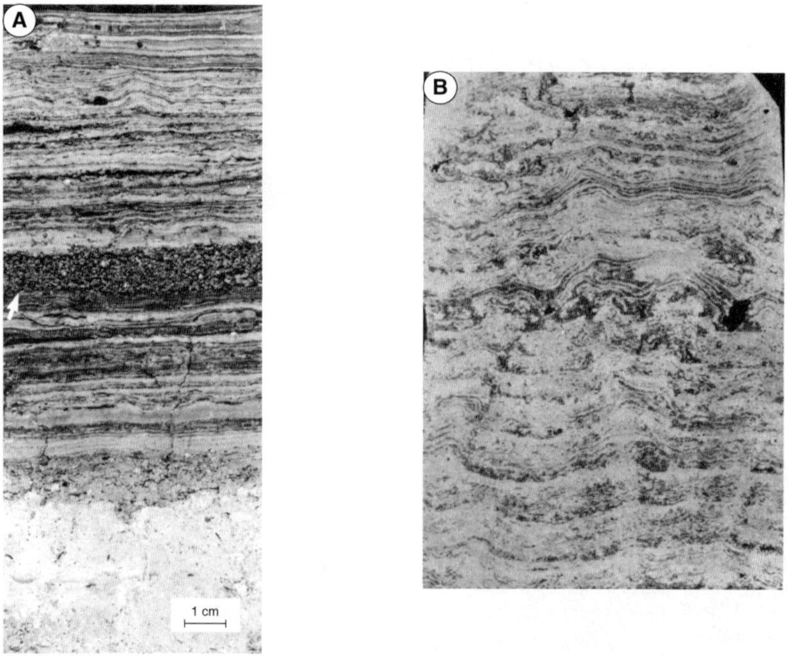

Fig. 3.112 Algal mats and stromatolithic structures.
A. Alternating algal (dark) and detritic (light) laminations broken by a sandy bed laid during a storm. Intertidal zone of Australia's Shark Bay (Purser, 1980). **B.** Stromatolithic structure in the Infraliassic dolomite of Campanie, Italy (J.P. Bertrand, 1969).

1. Purser, 1980.

Algal ridges of dasycladaceans and rhodophyta are another type of algal build-up in the tidal domain with, once again, a precocious diagenesis allowing rapid lithification. The chances of conservation in the wave-beating zone are poor, though.

Various cyanobacterial algae and calcareous red reef-forming algae (rhodophyta in the skeletal or nodal form of rhodoliths) also appear in algal balls making various oolithic sedimentary structures called **oncolites**.

B. Corals and Coral Reefs

Colonies of compound corals (in calcareous skeletons called corallum) in symbiosis with red algae and gastropods (such as Acropora, Littorinacea, Strombus) are characteristic of shallow waters in tropical zones. They construct assemblages of horizontal and vertical sizes, making **reefs**.

Coral reefs develop on shallow platforms and form barriers such as the Australian Great Barrier Reef, or on isolated pinnacles or on underwater mountain tops, forming atolls, of which there are many examples in the Pacific Ocean (Fig. 3.113). The species and the forms (tabular, encrusting, branched, columnar) will differ with the amount of agitation in the environment.

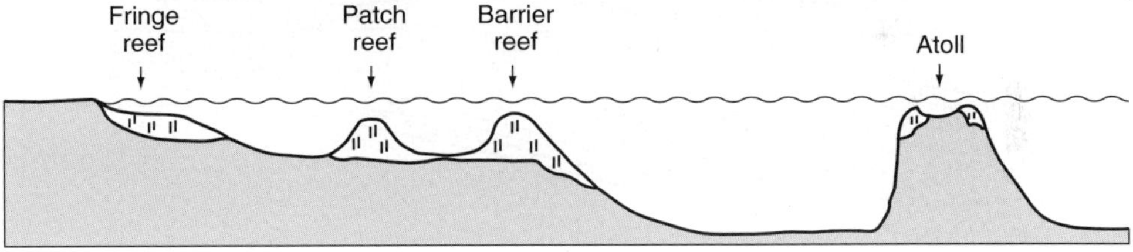

Fig. 3.113 Different types of reefs.

The oceanward side of the reef is the **fore-reef** and the lagoon side is the **backreef** (Fig. 3.114). Various organisms can participate in associations: algae, sponges, bryozoans, bivalves, crinoids. The fallout on the steeper fore-reef mainly consists of bioclastic debris from the reef itself.

Non-reefbuilding (ahermatypic) corals can also live at depths down to 1000 m, and at low temperatures (down to 10°C), though these species are much less abundant.

Reefs are sometimes quick to grow (as fast as 1200 µm/year in the Bahamas) but not always fast enough to follow the sealevel variation. They are extremely fragile and sensitive to environmental variations such as turbidity, for example, or change of sealevel, and thereby constitute a good record of these changes (see following chapter).

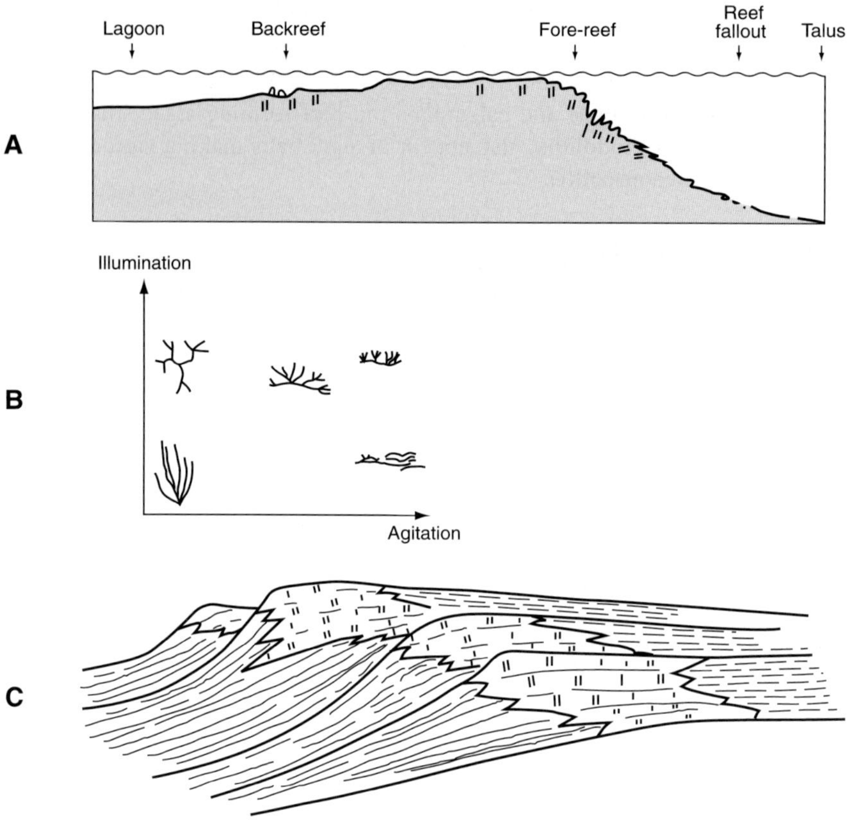

Fig. 3.114 Reef zoning.
A. General section. **B.** Variation of organism morphologies with agitation and illumination. **C.** Progradation of a reef system. This diagram illustrates the spatial migration of a reef body with time (2D-section).

A distinction is made today between ridged skeletal reefs and soft biogenic constructions where organisms trap sediments, in reef and mud mounds. It will be seen later on (Chapter 6) how reefs are petroleum targets because of their reservoir qualities.

Past geological series reveal dome-shaped **bioherms** of standing reefs with their stripping products, and flatter, more stratified **biostroms** witnessing to the reworking of organisms dispersed by currents.

Today's great coral structures differ from what was possible in other geological periods where large reefs were built by various genuses: corals (rugose tetracorals followed by hexacorals), bivalve rudists, bryozoans, algae. But today's environments are of interest for comparison with the constructions conserved on the edges of ancient sedimentary basins.

Such environments for carbonaceous building can be found in temperate and even cold waters, both today and in ancient series.

3.3.2.6 Destructive Organisms

A great many benthic organisms—bacteria, sponges, worms, for example—have a major effect on sediments in the process of lithification, and on rocks. As distinguished from **bioturbation** of soft sediments, rocks undergo **biochemical perforation and erosion**.

Burrowers, holothuroids (sea cucumbers), crustaceans, worms, and other organisms live on the organic matter available at the surface of the sediments or just below it. They leave burrows where they live, and also non-permanent active tunneling and wandering. These **bioturbations** sometimes totally destroy the original texture of the sediment. Under certain conditions, burrows, tunnels, and tracks can be conserved, and this has been used for reconstructing the depositional environments in past geological series (Fig. 3.110), from the coast to the great depths.

Biochemical erosion is especially effective with bacteria and benthic microflora and microfauna, which constantly tend to destroy the calcareous substrata and micritize the grains. It is a very efficient system involving other organisms too, such as microscopic sponges.

Large **perforations** of several centimeters can also be found, caused by lithophagous mollusks (serpulids, pholads, mytilids).

3.3.2.7 Role of Organisms in Sedimentation

Beyond the major role it plays in the food chain, as mentioned above, the world of vegetal and animal nano- and microplankton also contributes actively to sedimentation with siliceous (radiolarians, diatoms) and calcareous tests (coccoliths, foraminifers). And their primary productivity is high, producing radiolarian and foraminifer oozes. Benthic organisms also contribute to sedimentation, not only with built-up facies, but also by accumulation of the debris under the physical action of waves, storms, and currents giving rise to shelly and bioclastic debris (Fig. 3.115).

Generally, the organic matter produced is abundant. Most is destroyed by oxidation in the water and on the bottom, but if the conditions are reductive and the sedimentation rate high, part of it can be conserved and turned into kerogen in early diagenesis, and then converted into hydrocarbonated products (see Chapter 6).

Various types of environment may exist. These will be analyzed later in this chapter, but the following is worth mentioning here:

- The climatic factor is important, with temperature and salinity zoning from the high to low latitudes.
- There are seasonal effects.

A

B

Fig. 3.115 Lumachelle.
This type of deposit consists essentially of more or less fragmented shells of organisms, often with one species dominating. **A.** Beds in the Red Sea Pliocene. **B.** Saharan Paleozoic beds (IFP photo).

- The distribution of fixed and mobile species, their life cycles and geographic dispersions, and distribution areas, are extremely sensitive to physical and geodynamic factors.
- The food chains are complex.

Biological systems have changed considerably in the geological history of the oceans. The rudists that characterized the carbonate platform in the Cretaceous are extinct today, for example, and 96% of the species and 50% of the families disappeared toward the end of the Permian.

3.3.3 Transfer Mechanisms in the Ocean

The ocean is a chemical and physical system where intense biological activity develops, and it is also the seat of material transfers leading to sedimentation. Mineral and organic particles, and elements dissolved in seawater, are subject to the ocean's dynamic regime and, in accordance with the physical and chemical conditions, constitute sedimentary components that will accumulate at the bottom of the water column by various mechanisms.

The following is a discussion of a few of the aspects guiding sedimentation: water depth, origin of sedimentary components, transport and deposition mechanisms. This will give an idea of the diversity of oceanic environments that will be analyzed later in this chapter.

3.3.3.1 From the Coast to the Great Oceanic Depths

Though the ocean's general dynamic is a response to global causes, its characteristics differ according to geotectonic position, from the continent to the bottom of the ocean (Fig. 3.116). The dynamic begins with the floodable **coastal plain** and **littoral zone** starting at the unstable **shore** and including the **tidal zone** where continental inputs (deltas and estuaries) are considerable and tidal effects maximum. Further out is the **continental shelf** or platform with its sometimes simultaneous erosion and trapping of sediments, and especially where carbonates are "manufactured". Then there are the slope and apron of the **continental margin** where great sedimentary wedges often develop, and lastly the **great depths** with their pelagic fallout on abyssal plains and hills, and local accumulations of contourites and turbidites. The deepest troughs and trenches are called hadal zones.

The platform's littoral zones are characterized by:

- Bathymetric zoning, which directly influences the sedimentary mechanisms and is extremely sensitive to any sealevel variation.

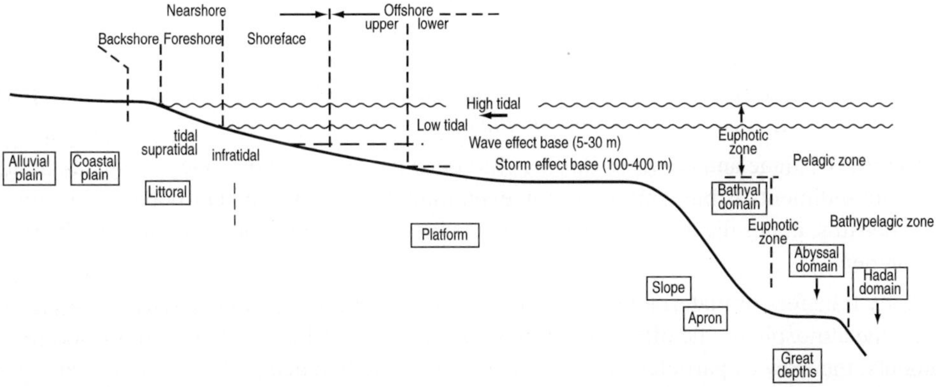

Fig. 3.116 Bathymetric zoning or staging. Water depth terminology.

- The geographic influence (orientation and configuration of the coast) and the latitudinal position.
- The volume of production of calcareous material and concentrated or diluted continental inputs
- Structural control.

The continental margin is characterized by the type of sedimentation in operation. There are thin margins with a few hundred meters of sediment, and fat margins with more than 10 km of sediment on the same vertical. Some margins are prograding and others not. There are escarpments with by-pass and build-up of sedimentary wedges. Some are the seat of upwelling currents and others are quite stable.

In the deeps, the settling of pelagic material gradually drapes the abyssal hills to either side of the ridges and mountains, while deep drift currents construct broad drifts of sediment.

3.3.3.2 Origin of Sediments and Transfers in the Ocean

There are two major sources of particles and dissolved matter received by the ocean: **continental discharge** from rivers, run-off and wind, and vegetal and animal **primary production**.

To a much lesser extent, input also comes from meteoritic input, atmospheric rain and the oceanic transfers of erosion, volcanic production, and hydrothermal processes.

Continental weathering and erosion products are mainly transported by rivers, while the eolian input of various kinds of **aerosols** from deserts and volcanoes is disseminated in different sedimentary domains.

The ocean itself is a source of sediments by **direct precipitation** due to oversaturation (producing evaporites and more rarely carbonates in eugenic clays) but also by **biochemical action**. All **biogenic** sediments are produced in the ocean, either from phyto- and zooplankton, pelagic fallout, or from the benthos.

When organisms die, their test or skeleton is sedimented by gravity and the accompanying organic matter is gradually oxidized. The pelagic fallout is partly filtered by mesopelagic filter-feeding organisms, and partly dissolved in accordance with the solubility curves for silica (at the surface) and for carbonates (at depth). Production is dominantly siliceous or calcareous, depending on the climatic belt.

The role of algae and corals as builders has been developed above. Generally, the bulk of carbonate sediments is biogenic in origin, even though the material later behaves like terrigenous detritus, under the effect of erosion and subject to the same physical laws in the ocean environment.

Many transfers of material and energy occur in the ocean dynamic, closely coupled as it is with the atmosphere, the differentiation of faunal and floral assemblages in different environments, the flow of particles of continental origin, the balancing of temperature and pressure differences, and latitudinal and seasonal variations. The ocean can be looked at as an enormous planetary regulator with a considerable diversity of transfers (Fig. 3.117) that

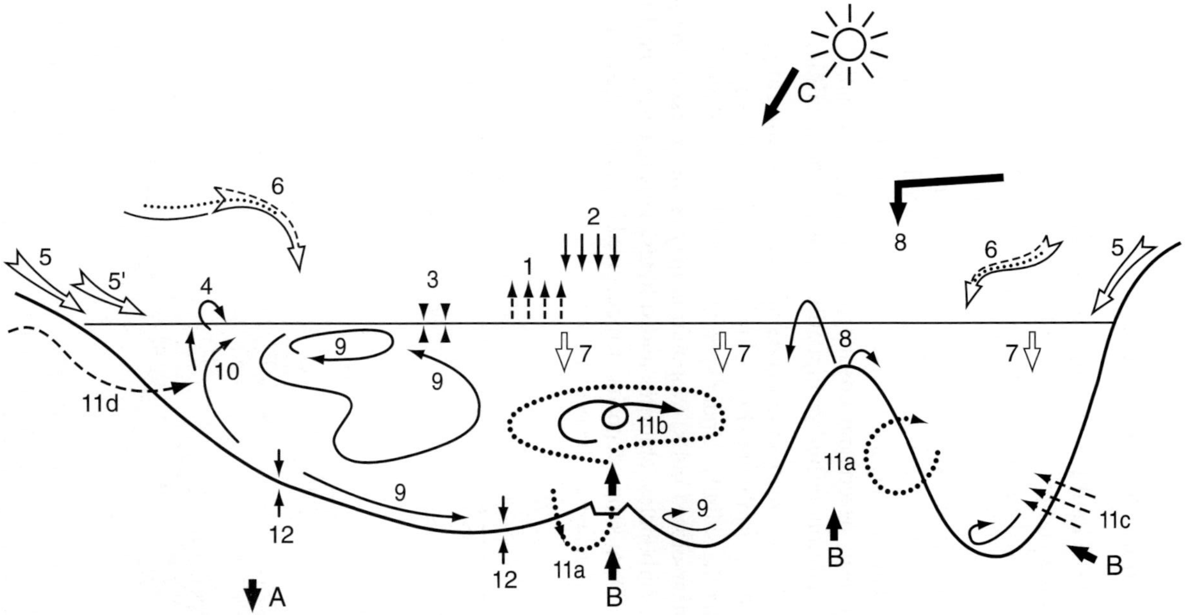

Fig. 3.117 Variety of ocean transfers.
A. Lithospheric cooling. **B.** Thermal gradient (conduction, convection). **C.** Solar and cosmic radiation. 1: evaporation; 2: precipitation; 3: ocean-atmosphere transfers of gases and particles; 4: hydrological front; 5: dissolved and particle inputs from rivers, and 5': ice input; 6: eolian input; 7: pelagic production; 8: volcanic input (atmospheric and subaquatic); 9: general oceanic circulation; 10: recycling of CO_2 in upwelling; 11: fluid circulation (a: ridges, b: plume, c: active margins, d: divergent margins); 12: water-sedimentation transfers, benthic life, early diagenesis (from Biju-Duval, 1994).

always lead to an unstable state of equilibrium subject to variations on the geological scale. All the factors are highly interactive but, to simplify the presentation, it will be broken down hereafter by theme with an emphasis on the consequences for geology.

A. A Multitude of Ocean–Atmosphere Transfers

The ocean's general circulation is **coupled** with that of the atmosphere by the very close relation between the temperature and pressure fields of the two fluid spheres, which determine the broad circulation gyres at the surface and the great masses of deepwater circulation starting at the polar fronts.

The exact **precipitation rates** in the ocean are still poorly known, and even more so their variability in space and time. Precipitation is a source not only of fresh water, but also of dissolved material and particles that will play a role in sedimentation. The dry fallout of various particles—aerosols, volcanic ash, desert and cosmic dust and, exceptionally, meteors—should thus be considered in addition to fluid precipitation of atmospheric origin.

The reduction of salinity due to precipitation is compensated by **evaporation**, which is especially active in the world's warm belts (Fig. 3.118). The effects of evaporation in the ocean are limited to balancing the salinity. Under special conditions of isolation or confinement, such as in lagoon or near-closed gulfs, evaporation may be in excess to the point of allowing the precipitation of carbonates, chlorites and sulfates. This is where most modern evaporatic series come from. Deep saline deposits also occurred in the geological past when oversaturated conditions were reached on the ocean floor.

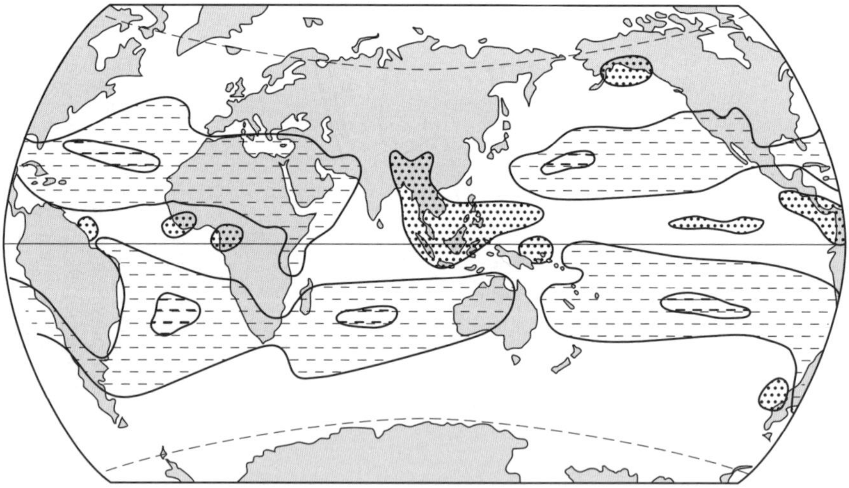

Fig. 3.118 World map of excess evaporation (dashed) and precipitation (dotted) (from various sources).

That part of the solar **light flow** that is not reflected at the water's surface enters the ocean and diffuses through the surface layers, sometimes as far as 150 m into the water column. This is called the euphotic layer, which is the domain of photosynthesis that is crucial to ocean fertility.

Gaseous transfers are abundant at the interface with the atmosphere. All of the change in the ocean's CO_2 content by photosynthesis is balanced by atmospheric CO_2 transfers and flows. Carbon dioxide transfers are controlled by physical (temperature, water agitation) and biological factors in as yet undetermined proportions. The size of the fetch determines the amount of transfers.

Since the solubility of CO_2 is enhanced in cold waters, it is mainly absorbed by polar waters. It is then entrained in the deep circulation and partly returned to the atmosphere at lower altitudes. The greater solubility of carbon dioxide at lower temperatures also means that the ocean absorbs more CO_2 in winter than in summer. The flows depend closely on the phytoplankton, which acts as a pump by fixation in photosynthesis. Certain anomalies may be due to overabundance of zooplankton at the equatorial upwelling. The fixation of CO_2 in the ocean has been estimated at 40 billion tons per year, which is of comparable order of magnitude to the fixation by the higher vegetals on the continent. But planktonic respiration returns part of the gas to the atmosphere.

The oxygen in the ocean's surface layer can be viewed as a product of photosynthesis by the phytoplankton. It is close to saturation and in equilibrium with the atmosphere, and its content decreases quickly through the first hundred meters of depth.

The **heat flow** received from the atmosphere is absorbed at the ocean's surface, and part is dissipated in the evaporation and turbulent transfers occurring there. The transfer is unequally balanced, with variations due to latitude and season.

One important question today is to know how the ocean acts as a thermal and thus a climatic regulator. Discussion of this question enters upon the field of **climatology** in terms of short-period variations such as the "Little Ice Age" and **paleoclimatolgy** with the considerable temperature variations recorded in paleo-oceans.

Non-negligible proportions of aerosols and fine particles can be released in micro-transfers to the atmosphere depending on the sea state (agitation) and evaporation.

A **hydrological front** is a singular zone where masses of surface water meet in an environment of complex dynamic phenomena (turbulence, convergence, divergence) due either to the proximity of the continent (tidal currents, littoral drift, delta flows) or to oceanic transfers with a significant concomitant increase in the biomass (Fig. 3.89). The most important such front is that of the Antarctic convergence, and examples of fronts of lesser importance are those of the Mediterranean and of the Armorican continental plateau off the west coast of France. These zones, where water masses of different origins meet, act as "fertilizers" (and, for that matter, have traditionally been worked by fishermen).

The Antarctic convergence (i.e., the polar oceanic front) corresponds to a narrower spacing between the isotherms at latitudes around 50-60°C, with a rise of deep warm waters and an extension of the intermediate Antarctic waters (Fig. 3.95). It constitutes a sharp biological frontier.

The character of the ocean-atmospheric coupling is marked by general **climatic zoning** by latitude. The importance of the frozen Arctic Ocean and Antarctic margins in the general oceanic circulation has been emphasized because the present zoning with its enormous asymmetrical (between north and south) consequences on the types of sediment can serve in trial reconstructions of ancient oceans. It will be seen that the situation has differed considerably in the course of geological time in reaction to cyclic or event-type changes in the Earth's external and internal driving forces (and especially the CO_2 variation in the atmosphere) and the continent/ocean distribution area (today's Earth-circling oceans allow easy circulation of the Antarctic bottom water).

B. Considerable Continent–Ocean Transfers

The continent is the source of two types of ocean input: more or less heavily charged **solutions** of dissolved salts, and mineral or organic **particles**, carried by rivers, glaciers, and winds. Briefly, the abundance of nutritive input in shallow waters near the coast causes a very high primary productivity there, and a high rate of sedimentation in organic matter. This is especially important for the genesis of hydrocarbons, and will be taken up again in Chapter 6.

Polar ice input is considerable. Three-quarters of all ice derives from the higher latitudes, over an area estimated at 18% of the world's ocean up to 45° latitude, and sometimes more (60°), resulting in a considerable input of fresh water. In geological history, the fluctuation and melting of the ice caps has played a major role in sealevel variations (**glacio-eustasy**).

The sea-floor topography may feature **bathymetric thresholds** forming barriers to these transfers, thereby breaking them or allowing only intermittent passage. The waters can then no longer be renewed (Fig. 3.88). Seasonal variations may occur, as in the example of the Indonesian thresholds between the Pacific and Indian oceans, with their eastward flows in winter and westward in summer.

C. Transfers at the Bottom

Transfers at the bottom of the water vary considerably with the current dynamics, sedimentary flow, benthic life, and confinement. A turbid **nepheloid** layer can sometimes be found near the bottom, carrying non-cohesive waterlogged sediments. If the currents are continuous, however, they can create **hardground** with a **sedimentary hiatus** exceptionally showing figures of erosion. Depending on the oxygenation at the floor, the benthic life will be abundant with burrows, tunneling, and traces of crawling. These are characteristic of one or another group (worms, echinoids, crustaceans) or organic matter, which can be conserved if the environment is **anoxic** (anaerobic life). The way these sediments evolve in **diagenesis** will depend on these conditions, as discussed in Chapter 5.

3.3.3.3 Deposit Zoning

A. *Bathymetric Zoning*

The **littoral** zone is characterized by fluvial input dominated either by wave and storm effects or by tidal effects. Deposit and erosion alternate in space and time. Various forms can be recognized here: beaches and spits, maritime marshes and lagoons, bays, barrier beaches and bars, banks, estuaries and deltas, and tidal flats.

Wave-dominated shores exhibit a characteristic series of dunes, backshore and foreshore, on the infratidal shoreface, and offshore (Fig. 3.116). Storm effects are reflected in washover on the shore and exhibit a certain zoning toward the offshore.

In **tidal**-dominated environments, the effects can be felt far from the coastal zone. Daily and seasonal cycles characterize this **tidal** environment with **flood** and **ebb** effects overlying each other, sometimes with one of them dominating. Transfers occur in the tidal channels, and especially in the tidal deltas and estuaries. Tidal flats are spaces that are regularly uncovered between tides. The zoning here is supratidal, intratidal, and sublittoral. **Subaquatic dunes** from large sand waves develop on the continental platforms with amalgamated banks up to twenty meters high, as in the Channel and North sea, over broad areas that sometimes make regular fields. The different patterns depend on the amount of sedimentary flow.[1] Dunes and tidal flats can also be found in the estuarian domain, as we have seen before.

The coastal and platform environment, whatever the distance from the continent, is also largely characterized by the volume of biomass in the euphotic zone, with a variety of biotopes and vertical benthic zoning from supralittoral (supratidal) to mesolittoral (intratidal), infralittoral (subtidal) and circalittoral (lower part of the phytal system), in what is sometimes called the **neritic zone** over the continental shelf. This is the preferential locus of the carbonate factory (Figs. 3.119 and 3.120).

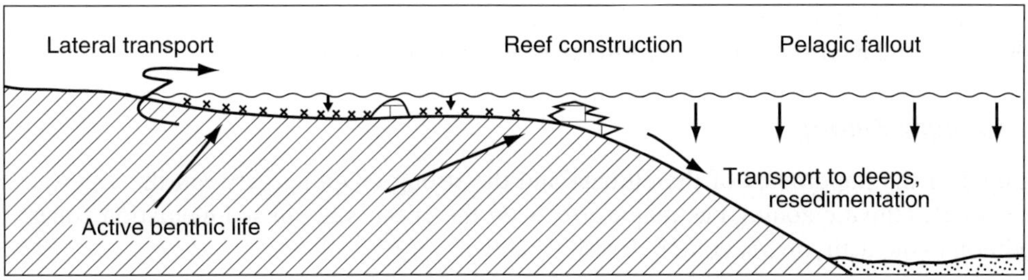

Fig. 3.119 Continental platform: preferential locus of carbonate production.

1. See Berné, 1991.

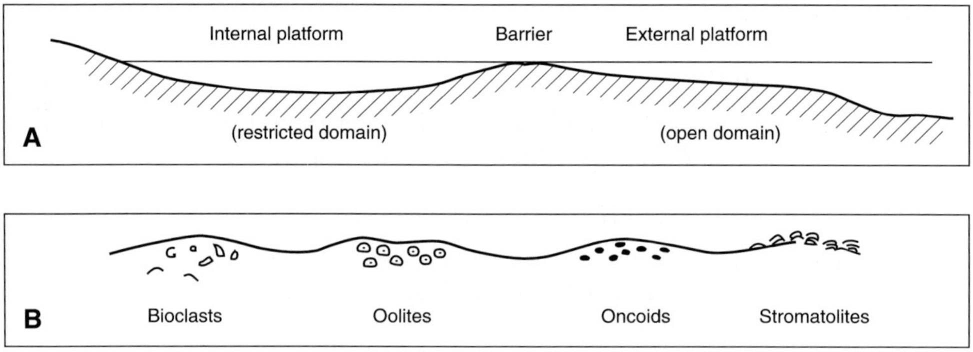

Fig. 3.120 Various domains of carbonate sedimentation on the continental platform.
A. Different platform domains characterized by different agitation, water chemistry, and biological production depending on the profile position.
B. Different types of carbonate production.

The **bathyal** zone from 100/300 m to 1000/4000 m is a euphotic one (lightless) where a limited but varied stenohaline (narrow salinity range toleration) fauna lives on the platform's edge, downgrade, and continental apron. This zone is marked by a number of features: steep **escarpments** unable to retain sediments, sharp slopes allowing mass **slippage** or the formation of breccia, canyons of different depths, sheltered zones drowned by sediments, fans at the slope foot.

These detritic systems can extend and invade the deep ocean's **abyssal domain** with its uniform fauna and low temperature (0-2°C). This zone is the locus of regular **pelagic sedimentation** draping the underlying relief of the ocean crust. The very special biotopes forming **subaquatic oases** around warm and hot hydrothermal sources are anomalies.

Lastly, the **hadal domain** of the great ocean troughs is generally the seat of turbidite sedimentation and intense tectonic deformations at the foot of insular arcs.

B. Climatic Zoning

In light of the importance of the climate and the strong coupling between the atmosphere and ocean, climatic zoning has major effects on the ocean's broad sedimentation zones, in addition to the bathymetric zoning described above.

The high-latitude zone is that of the frozen ocean (Wedell and Ross Seas and Pridz Bay in the Antarctic, and the average 3- to 5-m thickness of pack ice in the Arctic ocean) with very large seasonal variations and an effective barrier with the atmosphere, corresponding to low sedimentation. These **southern** and **northern** oceans are characterized by rigorous conditions, phytoplanktonic production reduced to a few months each year, sedimentation dominated by siliceous oozes, and iceberg-transported detritic residue from glaciers.

The **temperate zones**, like the **tropical**, are largely influenced by the general circulation of the ocean. The water's temperature and salinity are decisive factors determining the distribution areas of the major types of oceanic sediments. The temperature gradient adjusts the organic productivity, salinity variations, degrees and types of continental weathering (and therefore the type of particles and dissolved material delivered to the ocean), and the ocean dynamic.

This results in **belts** of preferential facies (evaporite, reefs, calcareous and siliceous oozes) characterizing a given latitude (Fig. 3.121). Starting from these belts as they are arranged today, the Earth's paleolatitudes can be reconstructed on the basis of sedimentary arguments.

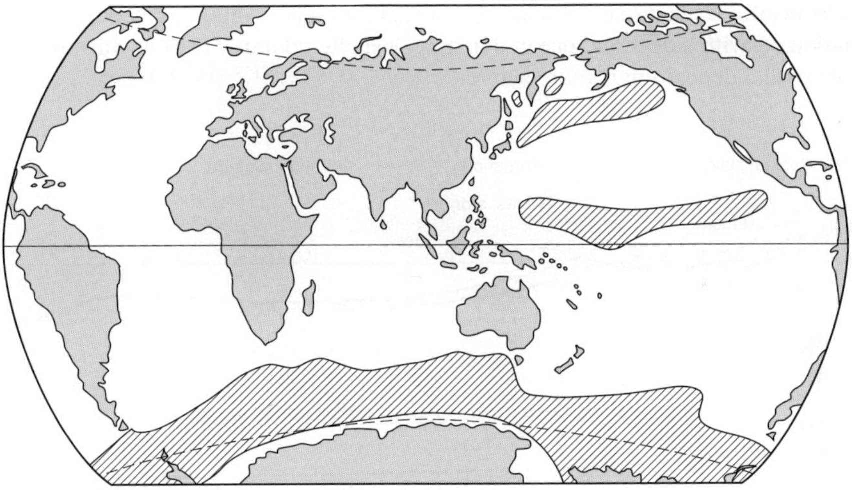

Fig. 3.121 Example of climatic zoning with the distribution of biogenic siliceous sediments in the world ocean. The abundance at the high latitudes and the equator is laid out in belt form.

3.3.4 Littoral and Continental Platform Deposits

The continental shelf is a precise physical environment of shallow water (less than 200 m), but the deposits vary greatly in space with the subaquatic morphology and current distribution, and in time too, considering the sea-level fluctuations.

The different facies that can be found in today's shelves, ranging from the fragile, unstable littoral zone to the shelf edge, reflect the special conditions of estuaries and deltas, fluvial system outlets, the coastal fringe where the detritus is redistributed under the combined action of tides and waves, the plateau domain of the carbonate factory in the euphotic zone, and barrier reefs. These continental shelves with their continental input, fertile waters, and carbonate production, are a locus of remarkable sedimentation, where, with adequate subsidence, a large volume of sediment can be trapped and fossilized to constitute thick geological series. The epicratonic basins and a large share of the flexural basins described in the previous chapter consist essentially of this type of deposit, and a great many oil fields (in the Middle East and the North Sea), for example, are localized in such environments.

3.3.4.1 Detritic Deposits

As the continent progrades toward the ocean, the detritus organizes into a prism or wedge on the littoral fringe (Fig. 3.122) consisting of:

- The specific **backshore** domain with decantation and evaporatic deposits in a bay or lagoon or, with a different ocean dynamic, a sheltered area of storm washover deposits where tidal deposition is more active.

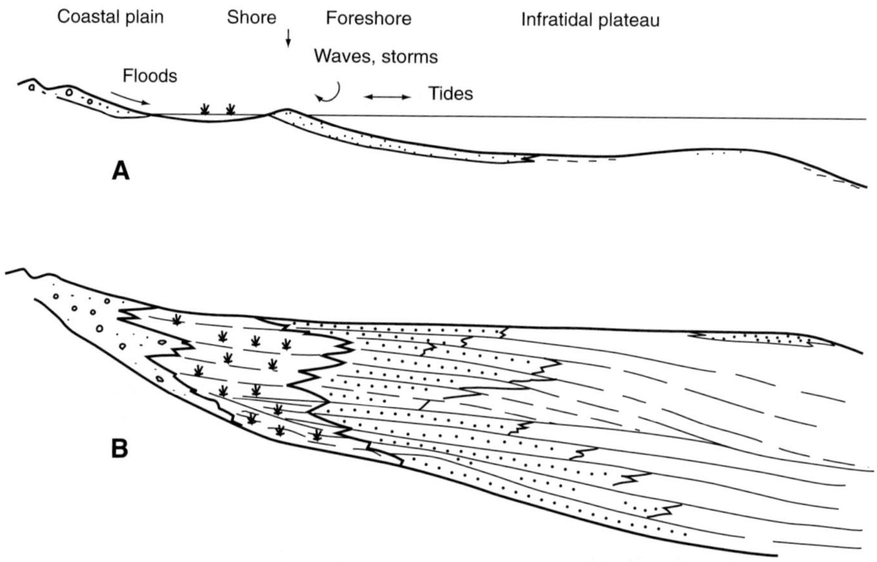

Fig. 3.122 Littoral sedimentary prism.
A. Going downstream, there are the coastal plain deposits that are more or less disturbed by floods and storms, followed by those of the coast (beach) and tidal domain where wave action can predominate and the deposits organize in a prism going from predominantly arenaceous pole to the predominantly argillaceous. **B.** The position of the coastal sands and fine-grain marine sediments may vary in time. Their fluctuations reflect time variations of the deposition systems.

- The **beach** and insular barriers, which is the surface zone in which waves and tides dominate, and the foreshore where wind action also interferes. This is the supratidal and intertidal (or near shore) domain proper.
- The **infratidal zone** or **shoreface** where waves dominate and build up symmetrical, asymmetrical, or crescent-shaped **bars** or sand waves offshore. Depending on the direction of the dominant winds, a longshore drift will organize along the upper shoreface.

Further down in the profile, beyond the area subject to wave effects, begins the prelittoral or **offshore** domain where the main sedimentation is clay ("fines") and sands from storms.

The facies thus show a downstream pattern in the sedimentary structures, from sandy pole to clay pole.

Since the balances are fragile, this prism can shift with time as will be detailed in the following chapter.

Ancient series are also commonly interpreted and reconstructed using a three-pole concept of river-, wave-, and tidal-dominated input effects.

A. *Tidal Detritic Deposits*

The tide is a periodic phenomenon that can be observed in the coastal domain with effects that vary as widely as the amplitudes of the tide and the littoral configuration. Tides result from the law of universal gravitation, with the effects of the Moon predominating over those of the Sun, but with the two combining variously between their **conjunction** and **opposition**. Tides are also affected by the general volume and the depth of the ocean, the exact geometry of the coastline, and resonance effects. This yields different types of tides, namely **diurnal**, **semi-diurnal**, and **composite**, and those with longer semi-monthly and semi-annual periods.

The importance of equinoctial **spring** and **neap** tides is well known. The vertical amplitude is the **tidal range** measured in meters, and the **foreshore** is the area alternately covered and uncovered by the tide. This breaks the area down into a supratidal zone, an intertidal, and an infratidal zone (Fig. 3.100). The system is **macrotidal** if the amplitude is large, **microtidal** if small, and **mesotidal** if moderate.

Tides have very strong effects and can influence estuaries and deltas.

The oscillation between the rising **flood** and descending **ebb** flows is a rigorous environment that can be characterized as follows.

- Particle transport is vigorous and rapid, making the tide an effective sedimentation agent (Fig. 3.123). **Tidal currents** are especially important over many continental shelves. They are linear currents that reverse direction with each tide. Their intensity varies according to the lunar cycles and seasons.
- The current dynamic causes repeated **erosion** and the creation of **tidal channels** that meander and migrate laterally (Fig. 3.124).

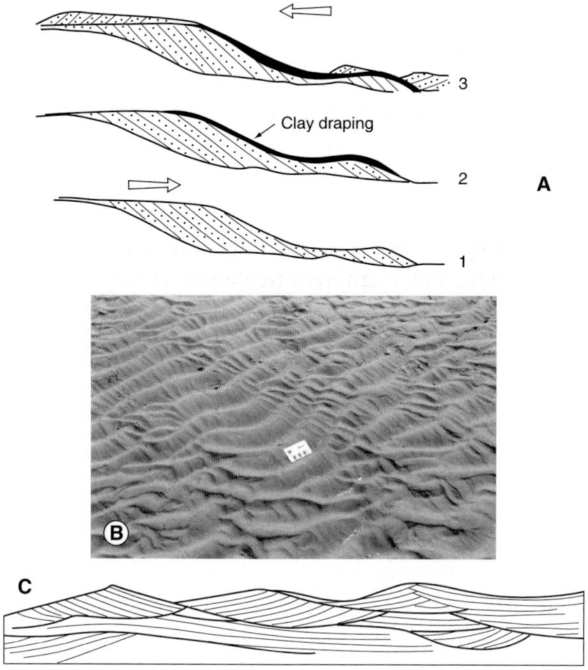

Fig. 3.123 Tidal deposits.
A. Theoretical scheme of tidal current reversal and inverse progradation of elementary dunes. Note that one current (flood or ebb) may dominate the other. **B.** The tidal draining may be variously oriented in the intertidal domain (perspective view). **C.** Example of oblique cross-bedding produced by an oscillatory process.

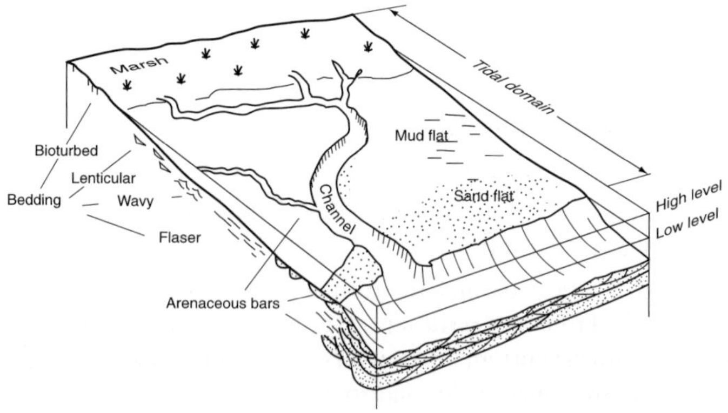

Fig. 3.124 Domains and types of sediments in the tidal zone.

- Conditions here are difficult for living organisms. With the turbidity, erosions, alternation between a saline and air environment (with possible desiccation), there is generally little diversity in the species, though this is at times compensated by a large population, and bioturbation is at times intense.

The tide's periodic mechanisms are recognized by the following criteria:

- Oblique stratifications in opposing directions on different scales, reflecting the current oscillations, and including sets of daily or semi-annual tidal bundles (Fig. 3.125), leaving a **herringbone** stratification that reflects opposing or bipolar directions, sometimes clearly marked by a clay draping. The space-time configuration may be a series of alternating sand-clay layers in a **flaser** ripple or lenticular pattern (Fig. 3.25).
- The ebb and flow oscillations and seasonal variations (with superimposed storm effects) dig channels and **reactivation surfaces** within the arenaceous edifice, as has been seen for fluvial deposits.

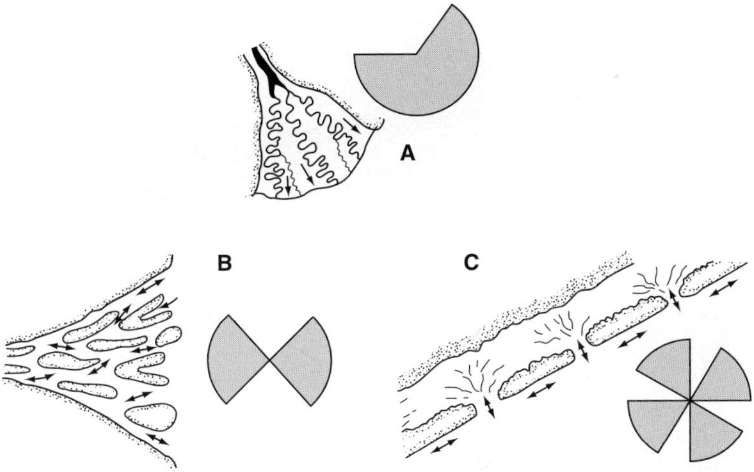

Fig. 3.125 Various types of current orientations in neritic sedimentary environments (from J.R.L. Allen, 1968).
A. Delta with combined fluvial and tidal influence. **B.** Estuary outlet. **C.** Littoral fringe

The **intertidal** and **supratidal** domains become **tidal flats** in the broad foreshore zones when the tidal range is large, as it is on the southern edge of the North Sea. These tidal flats are generally not areas where sediments can accumulate extensively unless the subsidence is large. Clay-arenaceous alternations (flaser) may then occur reflecting lateral variations between arenaceous and clay zones, carved up by deeper sandy channels. These sometimes develop into maritime marshes or bays and lagoons.

The **subtidal** domain below the average low tide level is marked mainly by the development of large sandy **tidal bars** with a crest generally lying perpendicular to the major axes of the tidal current ellipse. These bars are rarely isolated and often stretch out into ribbons or **sand sheets** and **sand bank facies**.

Some present banks on outer platforms are no more in balance than are the present tidal conditions. These are ancient fields, reworked. Some of the structures may originate from the action of internal waves underneath the action of the tide.

Such subaqueous dune fields are known in geological series containing both siliciclastic and calcereous material. Such sedimentary bodies make excellent oil reservoirs if the diagenesis is limited (Fig. 3.126).

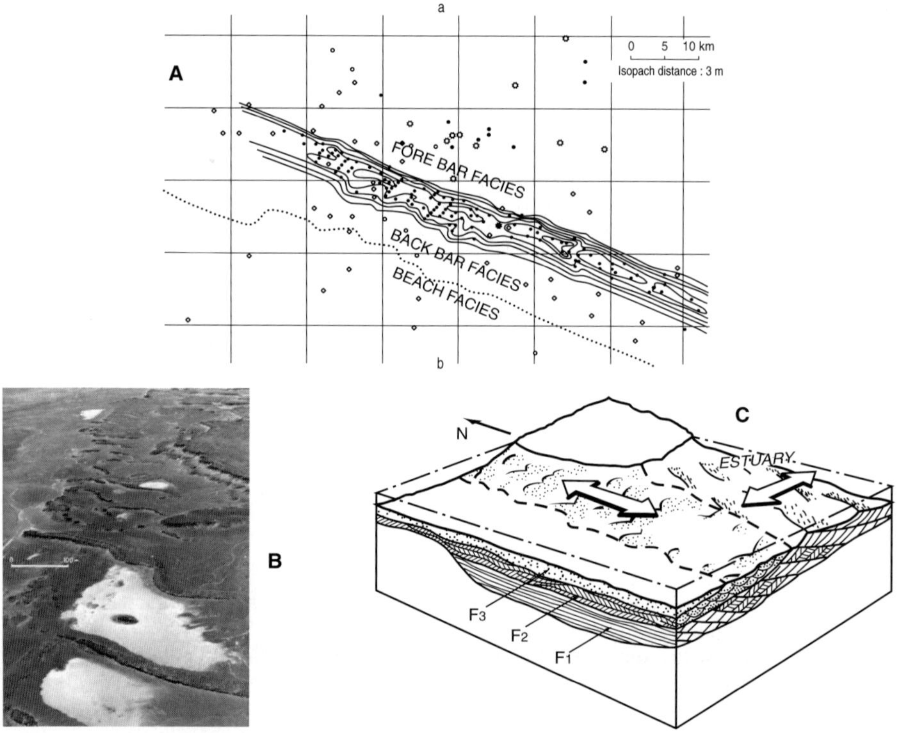

Fig. 3.126 Example of sandbars.
A. Sandbars of the Bisti oil field. Upper Cretaceous of the San Juan basin (USA). Isopach map of sands offshore of the beach species (from Sabins, 1963). **B.** Shoe-strings of the Lower Devonian on the Tihemboka high. This is an aerial view where the shoe-strings run in parallel ribbons through clay, exposed by erosion (IFP photo). **C.** Three-dimensional reconstruction of estuarian (F1) then tidal (F2 and F3) facies of the Permian sands in Rancho Roho, Arizona (from Dalrymple, in Walker and James, 1992).

B. Detritic Deposits from Waves and Storms

The regular effects of wave oscillations and unfurling under regular wind action are visible in microtidal and sometimes in mesotidal environments, with the effects exacerbated under storm conditions. This is a wave-dominated coast, as opposed to a tidal system. Depending on the wave obliquity and dynamics, different types of current will transport the material onto the shoreface and into the surf zone. In the same way as for tidal deposits, though, the material may be siliceous or silicated, especially in the vicinity of estuaries and deltas, or it may be calcareous if there is enough carbonate production on the platform.

The type of deposit will be conditioned by the bathymetric zoning described above, from the coast to the deeper zones beyond those of the storm effect (Fig. 3.116).

In the upstream surf and swash zone, the erosion is often large, especially in the winter season when storms are active (and combined with strong tides). Under extreme conditions, the backshore shows signs of washover with a formation of storm washover fans prograding toward the coastal plain or lagoons.

But generally, waves and storms have the effect of entraining material toward deeper zones (Fig. 3.127). If the waves strike the coast very obliquely, longshore drift currents

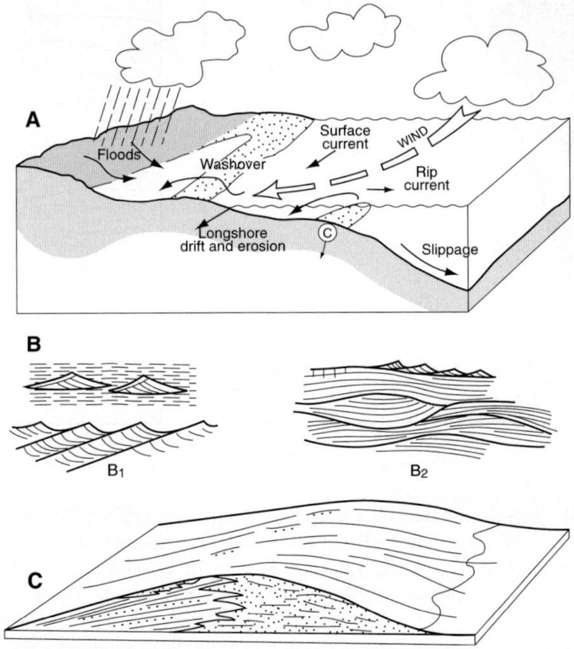

Fig. 3.127 Different wave and storm deposits.
A. General scheme from the coast to the shelf under wave action. **B.** Deposit pattern. B_1: migrating, and climbing wave ripples; B_2: hummocky storm deposits. **C.** Model of an offshore bar formed by storms.

develop and transport the detritus longitudinally along the shore, with very strong erosion due to the instabilities. The amount of material transported can be considerable, and rip currents will return the material to deeper zones.

Sands accumulate in subaquatic dunes (ripple to megaripple) in the infratidal zone, in various symmetrical, asymmetrical, elongated, irregular, or crescent shapes, depending on the depth Fig. 3.128). While wave ripples are frequent and have been found at depths of as much as 160 m, in oscillating and non-oscillating forms, flat parallel lamina have also been observed to form, depending on the wave periodicity, because the hydrodynamic mechanisms are complex and varied. The stratification will vary if the velocity increases.

Gravity mechanisms are also at work, at times (see Fig. 3.127).

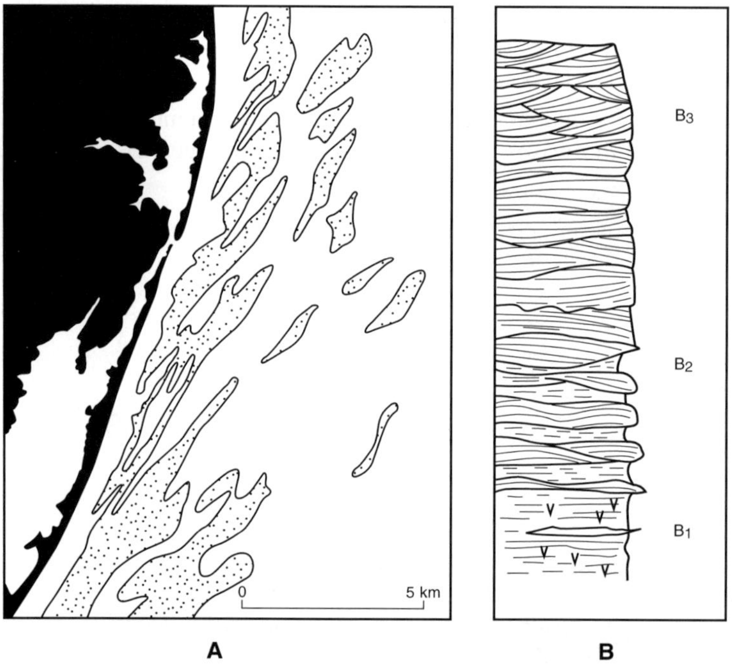

Fig. 3.128 Sets dominated by waves and storms (from ENSPM document). **A.** Example of geographic distribution offshore Maryland. Close to the shore, the ripples run several kilometers long in sandbars along the coast, and are discontinuous and asymmetrical farther offshore. **B.** B_1: more or less bioturbed clay oozes; B_2: arenaceous hummocky cross-stratification (HCS); B_3: arenaceous gullies in swaley cross-stratification (SCS).

Generally, the internal structure of sedimentary bodies can reflect the depositional type and environment. The broad patterns that can be seen are the following (see Fig. 3.129).

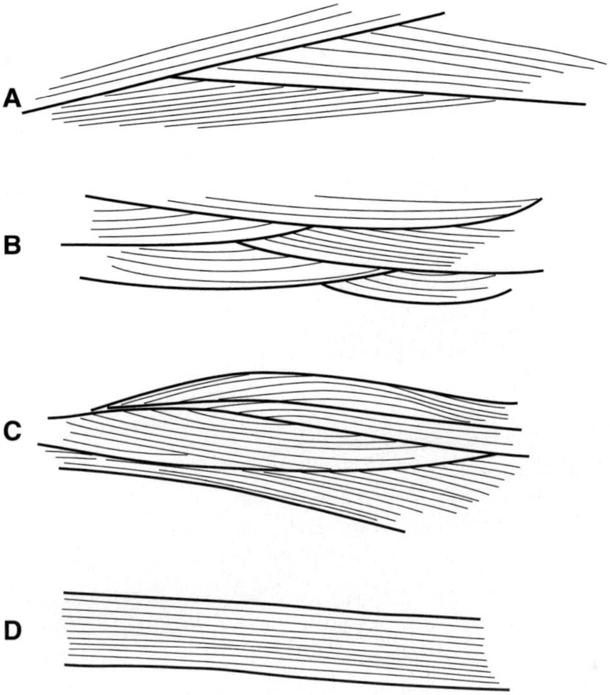

Fig. 3.129 Different internal structures of neritic sedimentary bodies. **A.** Oblique tabular beach stratifications (low foreset angles). **B.** Swaley cross-stratification (SCS) of field depressions under grooves. **C.** Hummocky cross-stratification (HCS) formed under ordinary wave action. **D.** Flat lamina bedding.

- **Swaley cross stratification** (SCS) in the upper zone under the flat beach stratifications. Swales are shorewise depressions that are filled with sand and interleave with each other, generally at low foreset angles of intersection. These facies are often amalgamated and can build up into shoreface or longshore bars several meters thick.
- **Hummocky cross-stratification** (HCS) corresponding to deeper levels under the average wave action base – that is, they reflect storm action.
- **Parallel lamina** of sand/silt/clay, occasionally with base erosion and a development of bioturbation. This is the zone where the lower shoreface becomes offshore.

There are two other noteworthy cases, though the environmental conditions that create them are still not well known:

- Bars dominated by ocean currents, which are a special case on continental shelves, beyond wave action;
- Reworking by tsunamis, which rearrange an initial deposit of arenaceous material.

The relatively narrow zone of the platform where these deposits organize is subject to time variations in the sea level. On the geological time scale, prograding sedimentary wedges can develop seaward, or retrograde landward toward the continent (Fig. 3.130). The deposits are zoned spatially in such a way that paleogeographic variations can be retraced.

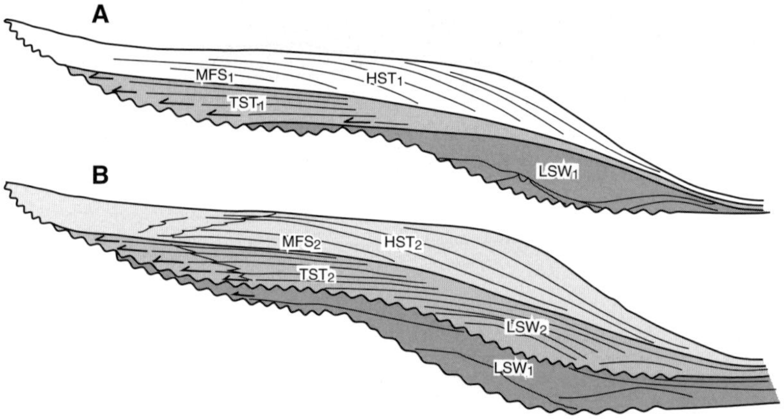

Fig. 3.130 Prograding and retrograding sedimentary wedge reflecting fluctuations in the depositional conditions.
A. Architecture of a depositional sequence according to the Exxon model (from Vail et al., and Haq et al., 1987). A lowstand wedge develops above an unconformity or sequence boundary marking a fall in sea level. When the sea level rises, a flood surface or transgression forms a transgressive system onlapping the margin successively up to a Maximum Flooding Surface (MFS) where the prograding highstand wedge develops. **B.** Time variations are such that several depositional sequences may lie atop one another. The sequence will vary according to the position of the profile.

Considering the sand content and grain size grading effects, these facies make very good reservoirs. Many of today's oil fields are operated in such levels, created at different periods in geological history (Fig. 3.131).

C. Fluvial Detritic Deposits

It happens that the sedimentary flow from rivers is such that deposits of fluvial origin dominate over those of wave and tidal effects. Estuaries, and especially deltas, are very important environments of complex morphology (distributary channels, lobes, and other river characters), and special facies, as was discussed earlier in this chapter.

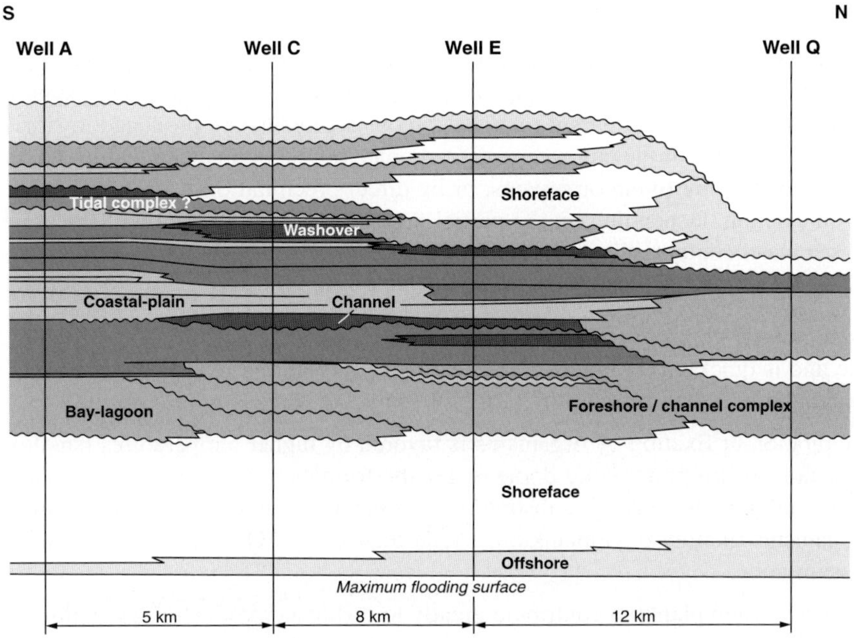

Fig. 3.131 Example of reservoir distribution modeling corresponding to near offshore deposits: the Mesaverde Cretaceous, from an analysis of outcroppings with a representation of fictitious wells (IFP document).

D. Types of Detritic Deposits

The various types of deposit are described by grain size distribution and grading, regardless of the deposition mechanism.

Predominant sands and silicates are referred to as siliciclastic facies of the gravel, sand (and sandstone), silt, and clay types. If the material is calcareous, the grain size distribution will depend on the initial source and dynamic fragmentation agents, and grading.

3.3.4.2 Carbonate Buildup

The following is a discussion of the specific carbonate factory of calcareous shelves (carbonate deposits in the deep ocean are covered later). The carbonate production briefly mentioned at the beginning of this chapter is essentially due to organisms and microorganisms living in this water depth (Fig. 3.130).

A. Review of Elementary Mechanisms

The conditions under which carbonates are produced by biochemical and physical processes on the shelf today are relatively well known. Contrary to siliciclastic sediments that are derived from an often distal parent rock, calcareous sediments are produced in situ, and are thus a direct indicator of the environment. Carbonate production is the result of a fixation of calcium in solution by living organisms, or by direct precipitation if the ocean is oversaturated with calcium. Depending on the partial pressure of carbon dioxide in the atmosphere and the sea water content of it, the reaction triggered is:

$$Ca^{2+} + 2CO_3H^- \leftrightarrow CaCO_3 + H_2O + CO_2$$

Under normal conditions, the carbonate precipitates in the form of aragonite. This is unstable and is quickly converted to calcite. Depending on the initial solutions, this calcite will contain more or less magnesium (1 to 20% $MgCO_3$, and as far as dolomite).

Precipitation or fixation by organisms is favored by higher temperatures (shallow tropical environments are preferential domains for the formation of calcareous deposits), water agitation, sunlight, the type of substratum, and especially the activity of organisms and microorganisms (mainly phytoplankton, which reduce the CO_2 in solution by their chlorophyll assimilation.

The benthos and plankton contribute greatly to carbonate production by building tests or **skeletons**, which are more or less conserved after the death of the organisms, in the form of particle bioclasts, deposited in shelly or crinoidal limestone, for example. The grain sizes vary greatly because the nanoplankton is also a major source with elements so small that they form a **calcareous ooze**, or matrix if mixed with grains. This is sometimes referred to as lithographic limestone (see classification below).

Skeleton grain is of various origins: calcified algae, sponges, encrusting organisms with segmented skeletons, foraminifers (Fig. 3.132). Depending on the fragmentation and size, the calcareous particles produced may constitute a stock transported by tide, wave or wind, depending on their strength, producing the various environments and structures discussed above, from the coast to the infratidal zone. When **bioclasts** accumulate in large numbers, they make **bioclastic limestone**.

Biochemical, physical, and chemical processes also lead to the formation of formless grains having no identifiable organic or skeleton structure, even though the micro organic contribution is assumed to be capital. These form remarkable calcareous deposits in **ooids**, which include **oolith**, **spherulite**, and **bahamite**, and **oncoids** with **oncolite** of algal origin (Fig. 3.133). **Fecal pellets** are another ovoid form, and granular types are intraclasts (debris from eroded calcareous rocks), aggregates (agglutinated grain types), mud balls and micrite (peloids or skeleton grains strongly marked by early diagenesis; see Chapter 5).

The fine carbonaceous fraction is formed by micro- and nanocrystals by direct precipitation, with a major contribution by calcareous **nanoplankton**. Magnesian calcite and dolomite will precipitate under special oversaturation conditions in a sheltered coastal environment or confined lagoon, and yield calcarenite-dolomite and **dolomitic** deposits (Fig. 3.134).

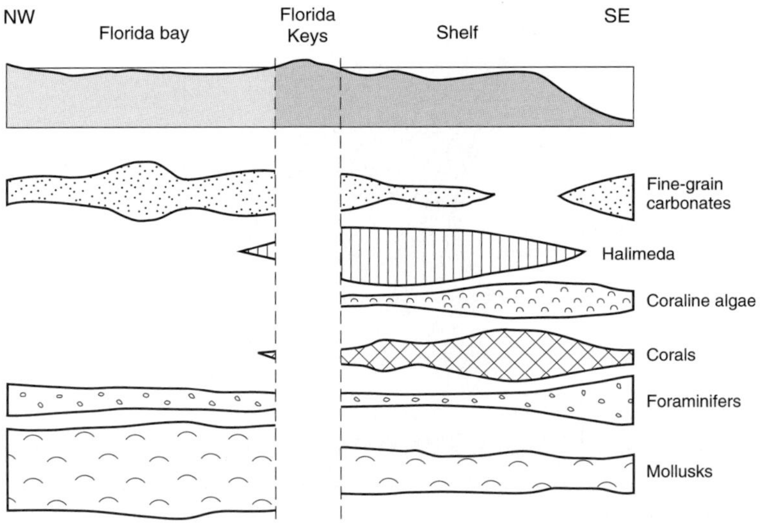

Fig. 3.132 Example of the variety of bioclasts on a carbonate platform. This example from Florida (modified from Ginsburg, 1956) shows the variation in the distribution of different bioclasts along a profile.

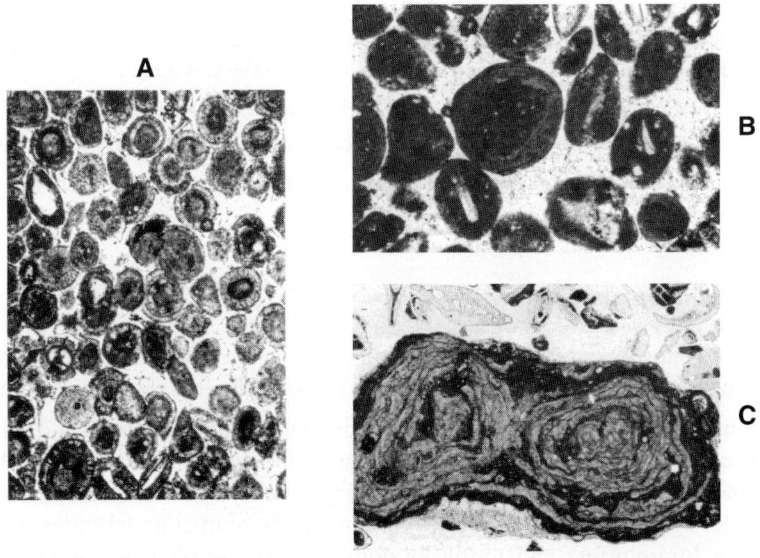

Fig. 3.133 Oolith, ooids and oncolite (photos from Purser, 1980).
A. Oolithic limestone (35 H). **B.** Bahamite type ooids (40 H).
C. Rhodolite type oncolith (20 H).

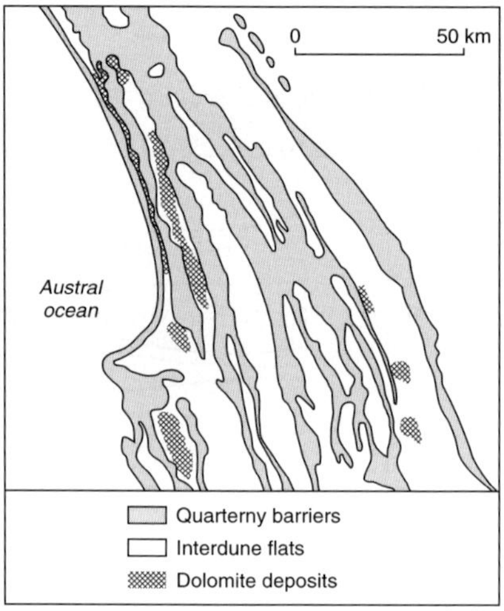

Fig. 3.134 Dolomitic deposits in the littoral environment.
This example (from Van der Borch, 1976, in Tucker, 1990) on Australia's south coast shows zones of dolomite formation (hatched) in intradune flats running between Quaternary barriers (gray).

When the calcareous mode is intensely mixed with a fine detritic fraction, the clays will be a mixed sediment that will later become **marl** (35 to 65% $CaCO_3$).

Secondary forms of carbonates can be found in some cases in the **cement** filling the voids in diagenesis (see Chapter 5), often in the form of crystalline **sparry calcite**.

This distinction between **grains**, **matrix**, and **cement** will be of extensive use to the petroleum geologist in defining reservoir qualities.

B. Climatic Control

Temperature and salinity control the production of carbonates directly, developing associations at the lower latitudes that are different from those of the less fertile high latitudes.

Chlorozoan and **chloroalgal** assemblages dominate in the intratropical domain. These are characterized by corals and green or calcareous algae, which are sharply inhibited if the salinity declines (with fresh water input from large rivers, for example), and the oolithic facies and pellet aggregates extend somewhat toward the temperate latitudes. In temperate waters, benthic foraminifers and mollusks dominate in assemblages called **foramol**. At temperatures less than 15° to 18°C, no non-skeleton calcareous grains will form. This climatic forcing is illustrated in the present distribution of great coralline build-ups (Fig. 1.43).

Most carbonate production occurs in surface waters less than 15 m deep, in the high-temperature euphotic zone. Many of the organisms are very intolerant of temperature variations. One inhibiting factor, of course, is water turbidity.

There is little or no carbonate dissolution at shallow depths, so if we neglect the diagenetic effects (see Chapter 5), the situation can be said to be stable.

The destructive effect of storms should be mentioned. With the resulting mechanical agitation, the fragile carbonated products are broken up and feed the detritic sedimentation either on the continental shelf or at great depths, or resediment in wedges of turbidite on the continental slope.

C. Structural and Bathymetric Control

Carbonate production is largely associated with the continental shelves, banks, and, far from the continental margins, with subaquatic atoll relief. Carbonate deposits range beyond the coastal zones described above, where the available calcareous material can constitute all of the recognizable structures from beach to offshore, under the effect of varied hydrodynamic conditions, to the entire shelf.

The following are the various types of deposit that can be found (Figs. 3.135 and 3.136).

- **Carbonate ramps** where the slope is regular and low (0.5 to 5° seaward) and the energy continually decreases from the coastline. This is the open shelf with various granular deposits (calcarenites) or finer carbonaceous oozes, depending on the relief. Bioconstructions are rare and isolated.
- The **rimmed shelf** with a pronounced break in the slope at an offshore bank, beyond the limit of wave action where the development of carbonaceous **reefs** and oolithic **shoals** is favored.

These colonial or sedimentary build-ups form in the sheltered zones of **lagoons** or, on another scale, **intra-shelf basins**, sometimes with a confined environment.

- **Epeiric shelves**, which are vast plateaus influenced by an attenuated tidal regime with little topographic variation, very sensitive to sealevel variations (causing desiccation or karstification) and storms, and where reefs can develop.
- **Isolated shelves** in shallow waters, but open to the great depths, as in the Bahamas. The reefs are particularly well developed. The platform is very sensitive to tradewinds and storms, so that the shelf has leeward and windward sides.
  The barrier can protect a calm lagoon, which is sometimes deep if the subsidence is strong.
- Drowned shelves, which witness to a time evolution such that the real shelf conditions are no longer stable for reasons of sea-level variation or a reduction of productivity, and the entire zone is then subjected to pelagic carbonate deposits.

These various types recognized in the present global situation are also represented in geological series since the Paleozoic. It should be emphasized that the facies thus built up will vary greatly depending on whether they are sheltered or exposed.

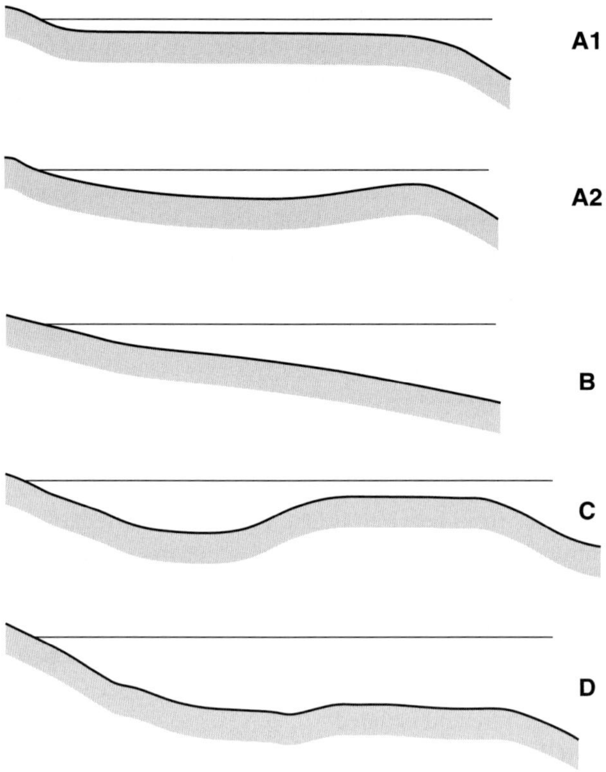

Fig. 3.135 Different morphological types of calcareous platforms.
A. Open (A1) and edged (A2) shelf, which can vary considerably in width. **B.** Homoclinal ramp, all at a greater or lesser slope. **C.** Isolated shelf. **D.** Drowned shelf.

D. Reefs

As mentioned above, builder organisms will construct algal mats, stromatolite and, more especially reefs, and mud mounds. The following are a few special points worth noting here:

- Reef formations are generally found on the edges of continental shelves. They have a pronounced polarity and architecture between the shelf and the continental slope (see Fig. 3.114). They can rise up in pinnacles on the upper part of the slope or on the ramps.

  Reefs do generally develop in shallow water and warm intertropical waters, but there are known exceptions such as the small reefs built at great depths (up to several thousand meters) in the cold waters of the North Atlantic margins.

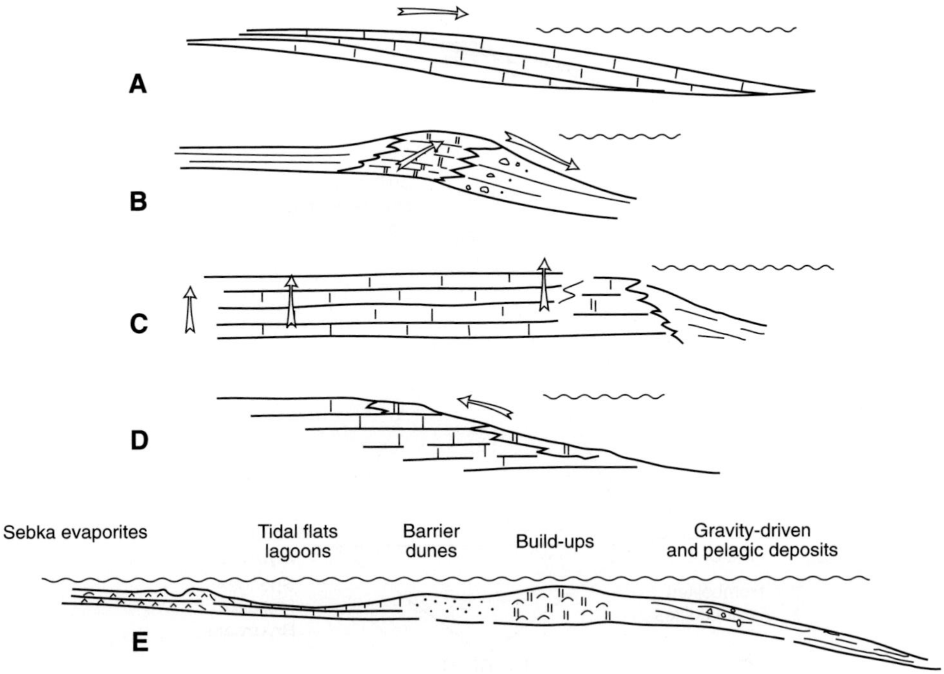

Fig. 3.136 Different carbonaceous shelf build-up and destruction processes.
A. Slow progradation toward the basin (with sigmoid deposits) along the ramp. **B.** Prograding reef edge. **C.** Aggradation: gradual vertical accretion. **D.** Shrinkage by erosion. **E.** General facies distribution scheme.
The processes depend on the initial morphological configuration, variations in sea level, productivity, and the volume of the dynamic.

- Today's reefs are mainly built up by colonies of corals and algae, or in association with other often symbiotic organisms. But in the geological past, other forms have built constructions of different types.
- **Reefs** and **mounds** are different in that a reef is an upward construction consisting mainly of skeletons of large and often colonial organisms and living in an energetic, agitated environment (Fig. 3.137), whereas a reef mound is a low micritic monticule of biological or microbiological origin where small communities of algae, cyanobacteria, and stromatolites have trapped sediments in a calmer environment. A skeletal mound is one in which there are observable skeletons, while the microbial and mud mound is less rigid. Like reefs, however, mounds can sometimes exceed several hundred meters in height.
- The type of internal factory (Fig. 3.138) will differ according to the type of construction, the development and form of the organisms, and the quantity of allochthonous

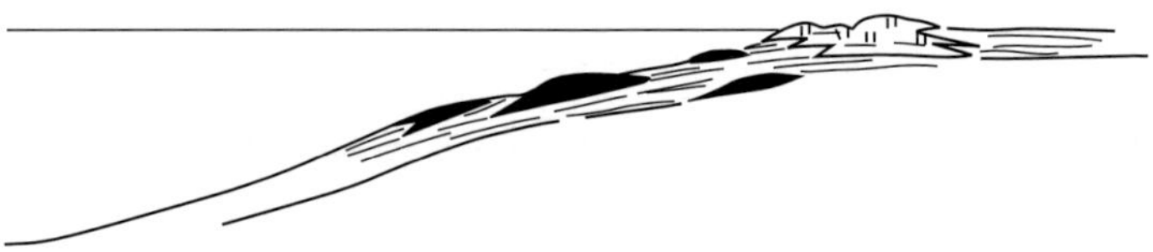

A

B

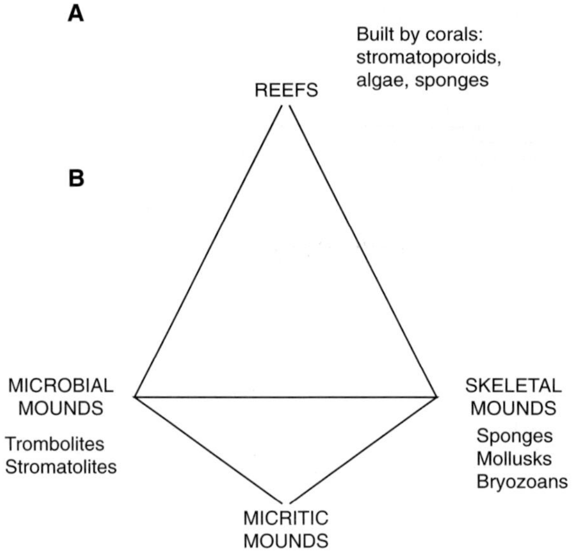

Fig. 3.137 Reefs and mounds.
A. General profile with the reef's position upstream of the mounds, which are localized on the slope. **B.** Classification of different buildups (from R.G. Walker and N.P. James, 1992).

material trapped in the free spaces. Reefs are of interest as oil reservoirs because of the primary and secondary porosities that are preserved there.
- Reef formations are extremely sensitive to relative sea-level variations. If productivity is high enough, though, the reef will be able to grow and maintain itself durably, and large complexes can then develop.

There are many ancient examples from the Paleozoic and Mesozoic (Fig. 3.139).

E. Pelagic Production

A non-negligible part of carbonate production comes from the micro- and nanoplankton living in the euphotic zone. This is the pelagic contribution to sedimentation on the shelves. If the nutrients, water turbidity, temperature, oxygenation, and water motion generate high surface productivity, this contribution will be large. Some shelf deposits do consist mainly of accumulated remains of pelagic organisms, of which chalk is one of the most famous

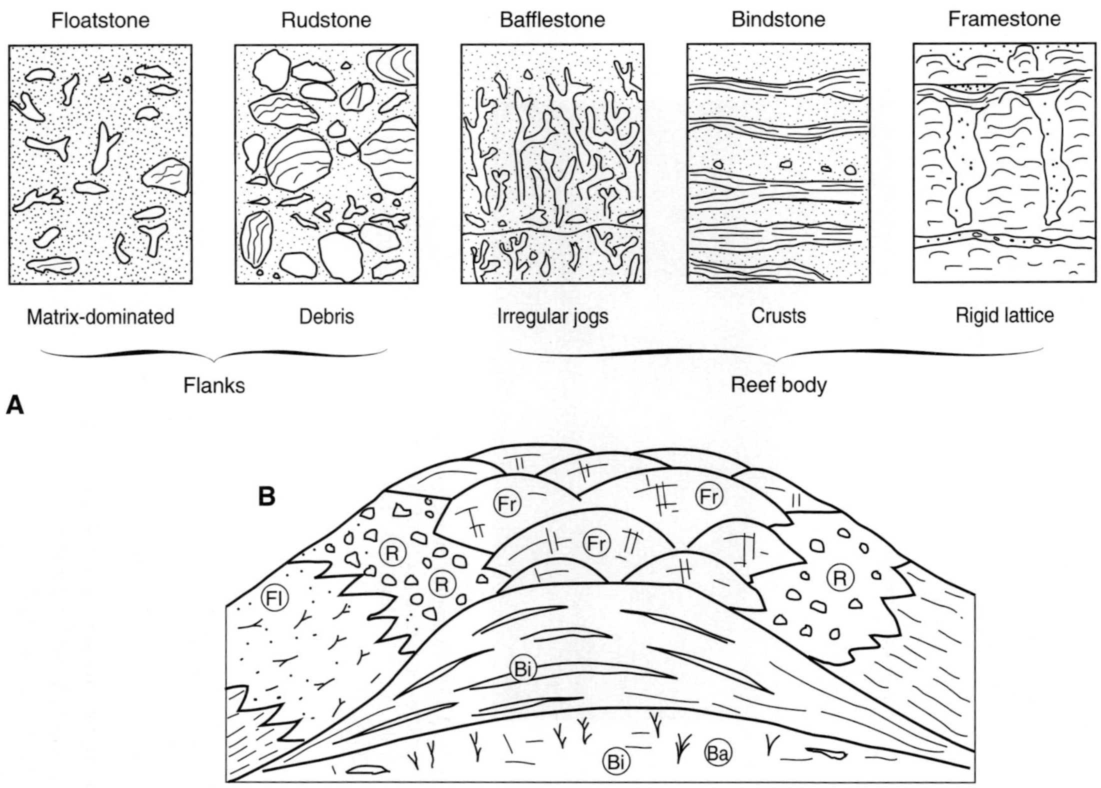

Fig. 3.138 Different textures of reef limestone.
A. Classification (from Embry and Klovan, 1971) of different fabrics. **B.** Fabric distribution in a buildup.

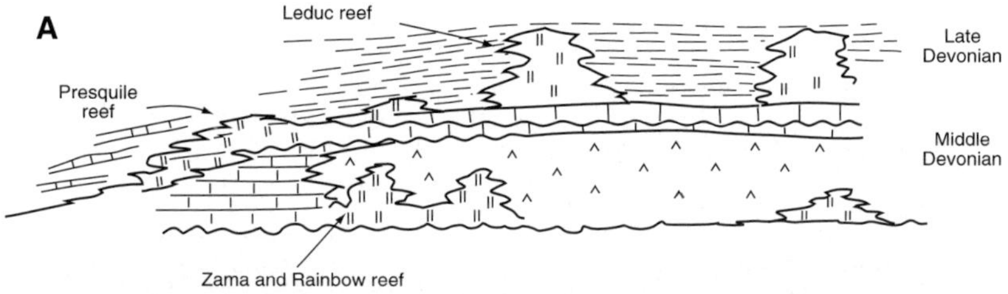

Fig. 3.139 Examples of ancient reefs.
A. The Devonian reefs of Alberta, make excellent petroleum reservoirs (from many authors). **B.** Reefs of the Mediterranean Messinian: Sicily (IGAL photo).

examples (Fig. 3.140). Usually, pelagic production is mixed with fine hemi-pelagic clay particles and more or less argillaceous carbonaceous muds are then formed, making marly limestone and marl, for example.

F. Classification of Carbonates[1]

Various classifications are used in the literature:

- Neritic and pelagic limestone
- Calcerinite and calcilutite
- Crystalline and microcrystalline limestone.

1. See Folk, 1962; Dunham, 1962.

Fig. 3.140 Chalk cliff.
Here, the chalk deposits dominated by coccolith (see Fig. 3.108) are underscored by flint, which outlines the stratification when seen from a distance. The dark blotches correspond to wet spots where vegetation can develop (IFP photo).

Descriptive terms are also used, such as coquinoid, oolithic, gritty or crinoidal limestone (Fig. 3.141), or others referring to the matrix debris: biosparite, oomicrite, oosparite. In petroleum geology, we use Dunham's idea of the proportion of grains to matrix (Fig. 3.142):

- Mudstone: grains <10%
- Wackestone: unjoined grains
- Packstone: joined grains, matrix <10%
- Grainstone: joined grains, no matrix
- Boundstone: no grains, but organic construction
- Crystalline: limestone with original texture unrecognizable.

In practice, the microfacies are described in terms of the sum of characters concerning grains, matrices, types of constituents (Fig. 3.143).

Dolomite is yet another type of carbonate (a dual one, of $CaCO_3$ and $MgCO_3$) that is sometimes of primary origin in a particular environment (see further on in this chapter), but is usually of diagenetic origin, as will be discussed in Chapter 5.

3.3.5 Saline Deposits, Evaporites

Seawater, and lake water too, is generally undersaturated in the salts that go into making evaporitic deposits.

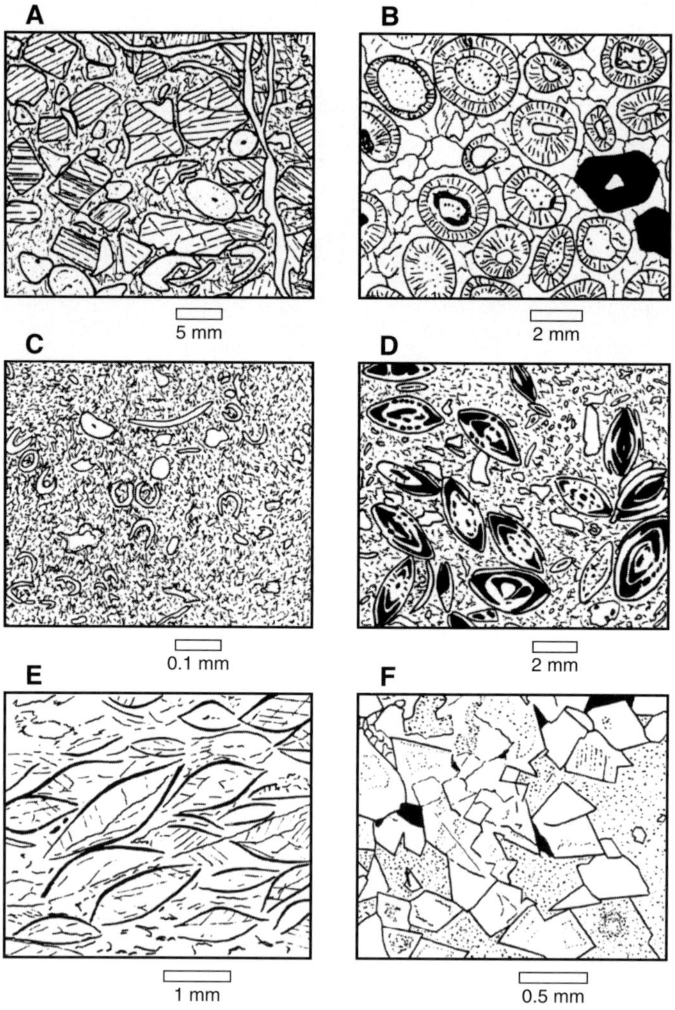

Fig. 3.141 Different types of limestone viewed in thin section.
A. Crinoidal limestone (packstone). **B.** Oolithic limestone (grainstone). **C.** Cryptocrystalline limestone (wackestone). **D.** Nummulitic limestone (wackestone). **E.** Coquinoid limestone. **F.** Dolomitic limestone (from Guillemot, 1986).

Recognizable texture					
Grains < 10%	Grains > 10%	Joined grains	Grains alone	No discernable grains Built-up limestone	Unrecognizable deposit texture
Mudstone	Wackestone	Packstone	Grainstone	Boundstone	Crystalline

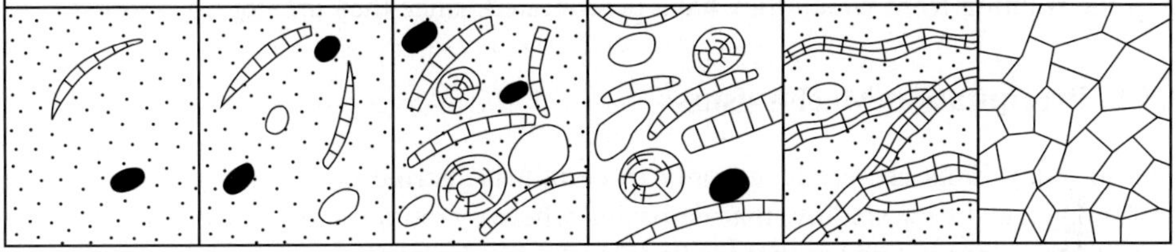

Fig. 3.142 Carbonate classification (Dunham, 1962).

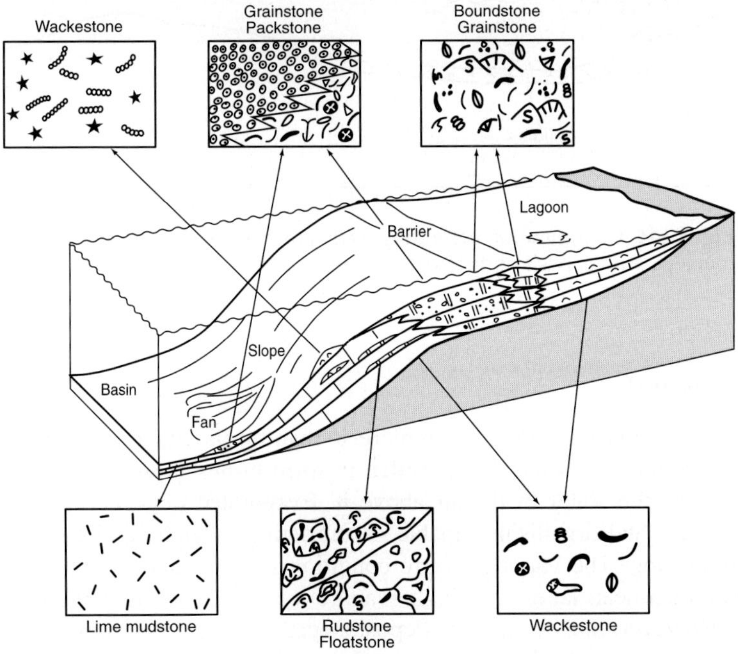

Fig. 3.143 Spatial organization of carbonate facies from the shelf to the great depths (from Sarg, 1988).

So, in order for sulfatic and chloride salts to precipitate, very special conditions need to be met, generally by evaporation in excess of the fresh water input from precipitation or from the continent.

The type of chemical sediments created by this direct precipitation at the bottom of the oversaturated "brine" water column are thus called **evaporites**. When found inside siliciclastic and many calcareous sediments, evaporitic sediments reflect rather well the composition of the solution from which they have precipitated, since they are not transported.

3.3.5.1 Precipitation Mechanisms[1]

In nature today, salts are produced only in certain lacustrine or beach environments, sebkhas, lagoons, and very shallow marine basins where evaporation is so strong that the water reaches oversaturation level in these salts (Fig. 3.144). Crystallization can occur either at the atmosphere/brine interface or in the brine itself, or in the interstitial water of the sediments.

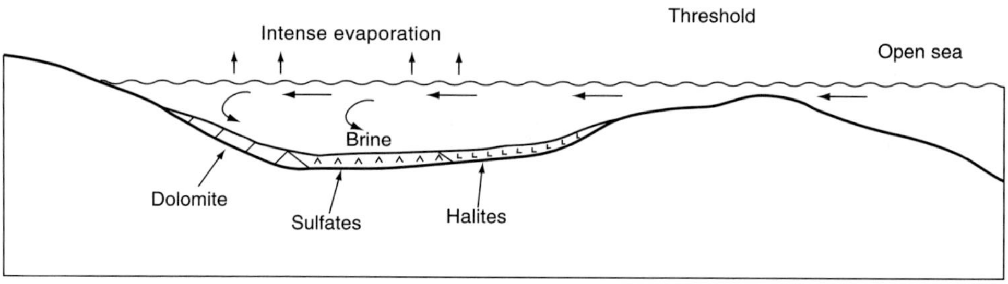

Fig. 3.144 Configuration permitting oversaturation in a tight, closed environment, with deposit of different evaporites.

A. Climatic Control

Climatic control is primordial here, much more so than for the other sediments. The climate will determine the excess evaporation needed to form brines in which the vapor pressure will exceed that of the water in the air above it. Evaporated deposits thus always reflect extreme conditions, and organisms capable of subsisting in such an environment are very rare or totally lacking. The vapor pressure condition implies that aridity is necessary and more important than heat: lakes with high evaporation are known to exist at high latitudes and altitudes. However, most evaporitic deposits occur in warm climates when the internal evaporation is not compensated by fresh water input. This can lead to desiccation.

1. See Kendall, 1994.

B. Tectonic Control

Evaporation conditions are further enhanced the more the zone is confined. Since evaporitic basins are either very constricted or entirely cut off from the open sea, the structural history of the basin will play a major role:

- Narrow basins can be created in rifting zones. Here, the relief will create the necessary aridity, and the hydraulic basins with little renewed water are conducive to flow concentrations, such as in the Dead Sea and in the Oligocene troughs of Alsace, Bresse, and Camargue. Many salt deposits on present continental margins stem from initial histories of this kind.
- Episodic salt deposits can develop under shallow waters in intracratonic basins or very sensitive zones where the sea transgresses the continental plateau, or in concentrations in extensive lacustrine zones such as Lake Chad.
- Some basins correspond to remnants of former ocean basins gradually subjected to desiccation in a mountainous collision zone such as Anatolia (Tuz Galu), Brasil's Altiplano (Uyumi), the Rockies (Great Salt Lake) or Tibet (Qaidom), where the conditions are hyper-arid.
- Deep basins, if closed by thresholds, also exhibit very restricted hydrological conditions in which the brine stagnates.

Structural control is often modeled in one of three types of scenarios: shallow-water barred basin; shallow-water deep basin; deep basin (Fig. 3.145).

3.3.5.2 Different Types of Deposits (Table 3.4)

Natural brine concentration processes are such that the order in which the various mineral species precipitate is not random. The following sequence is generally observed:

- Calcium-magnesium carbonates (mainly dolomite), but also sodium (natron, trona) when the concentration reaches 1.8 times the initial concentration
- Sulfates (mainly gypsum, which transforms into anhydrites) or halides (at concentration 3.8)
- Chlorides (halite, sylvite) (at concentration 10.6).

Marine evaporites are generally deficient in the most soluble phases.

This time spacing depends on the initial composition of the solutions and on the basin's hydrological regime. If there is any hydrothermal circulation, for example, the concentrations would be abnormal.

Very generally, the carbonates-sulfates-chlorides sequence can also be found in the geographic distribution, with the most concentrated deposits in the center of the basin and the others around the edge (Fig. 3.50). This is true for small continental basins and for different types of marine basins.

The forms of precipitation that occur in a very particular chemical environment are relatively unstable, since the forms of dissolution, the metasomatic replacement of one mineral

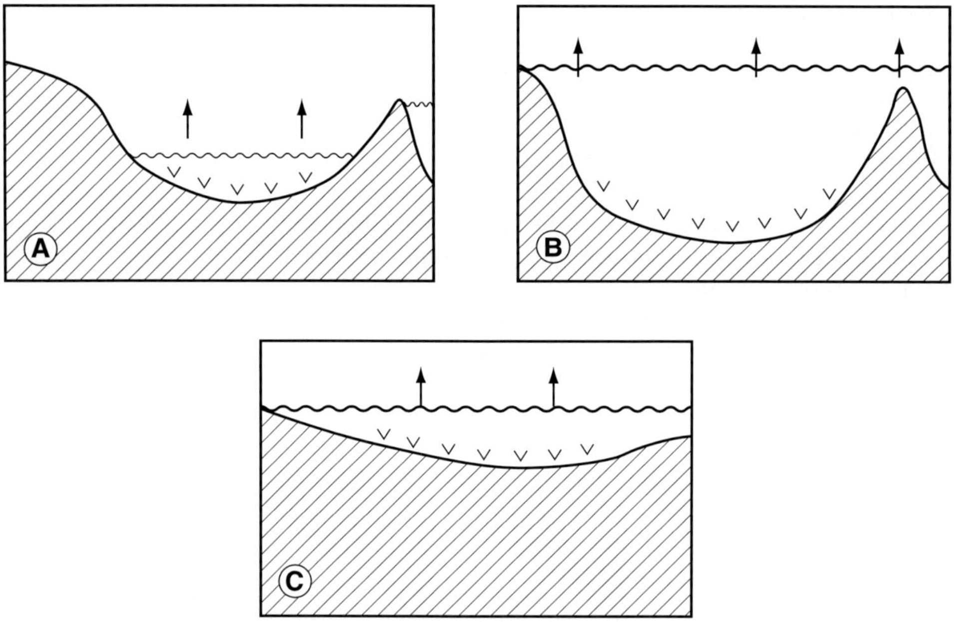

Fig. 3.145 Different evaporite formation scenarios.
A. Shallow-water barred basin (Mediterranean Messinian). **B.** Deep basin with deep brine concentration (Red Sea, Mediterranean, Gulf of Mexico). **C.** Usual shallow coastal lagoon basin.

Table 3.4 Main evaporitic rocks

Carbonates	Aragonite	$CaCO_3$
	Calcite	$CaCO_3$
	Dolomite	$CaMg(CO_3)_2$
Sulfates	Anhydrite	$CaSO_4$
	Gypsum	$CaSO_4$, $2H_2O$
Chlorides	Halite (rock salt)	NaCl
	Sylvite	KCl
	Polyhalite	$2CaSO_4$, $MgSO_4$, K_2SO_4, $2H_2O$
	Bischofite	$MgCl_2$, $6H_2O$
	Carnallite	$MgCl_2$, KCl, $6H_2O$

by another, and bacterial reduction are frequent. As evaporitic deposits are at times subjected to erosion and transport agents, they might be displaced and resedimented as clastic—or detritic—accumulations of evaporitic minerals.

A. *Difficult Recognition (Diagenesis)*

Because of the deposition conditions in fragile systems, and because of their chemical character, evaporitics are subject to early diagenetic transformations that are often more characteristic of them than their geographic situation. Diagenesis is discussed in Chapter 5, but a few remarks are needed here concerning very rapid syndepositional phenomena.

Firstly, a large part of evaporites are actually formed in the very shallow zones of playas, sebkhas, and lagoons, i.e., in the **interstitial waters** circulating in some other type of deposit, such as and especially in supratidal carbonates, as in the Persian Gulf. Certain magnesium limestones can be found containing these **interstitial evaporites**, but the matrix is usually dolomite or sulfate (such as the pink gypsum of sands). When sulfates combine with carbonates, they take various forms such as algal mat plus gypsum, or limestone with nodular anhydrite.

The sulfates may, under certain conditions, undergo bacterial reduction in which the anhydrite is converted to calcite, with a gain in porosity and the production of hydrogen sulfide. Sulfuric acid can also be produced if the sulfides are oxidized, and the carbonates can then be transformed into secondary anhydrite with saliferous clays.

Diagenesis can greatly alter the mineral content and texture of the original deposit to the point of near total disappearance, by dissolution. Most evaporites are, in fact, replacements in the diagenetic process.

Moreover, because of their petrophysical properties, salt and layers of sulfates react in a special way to tectonic stresses. Their plasticity makes it easy for them to creep. An extreme "mobility" is thus observed, especially in the salts, which explains the thickening and décollement as will be detailed in Chapter 5.

B. *Past Examples*

Present evaporitic depositional conditions (still poorly known in deep basins) provide no explanation for the considerable volumes of sulfates and salts in the great geological basins of the past: if today's oceans evaporated entirely, they would leave only a 60-meter deposit of salt. The only explanation for the very thick saliferous concentrations is that very special **subsidence** conditions existed, with periodic renewal of the brine. Evaporating a 100-meter column of water would leave less than 2 m of evaporated material. Assuming the water depths of the deep basin in the Mediterranean Messinian (Fig. 3.146) to have been 1000 m, it would have required a succession of at least 50 desiccation events in order to develop the 1000-meter thickness (and often more) of the great deep basins.

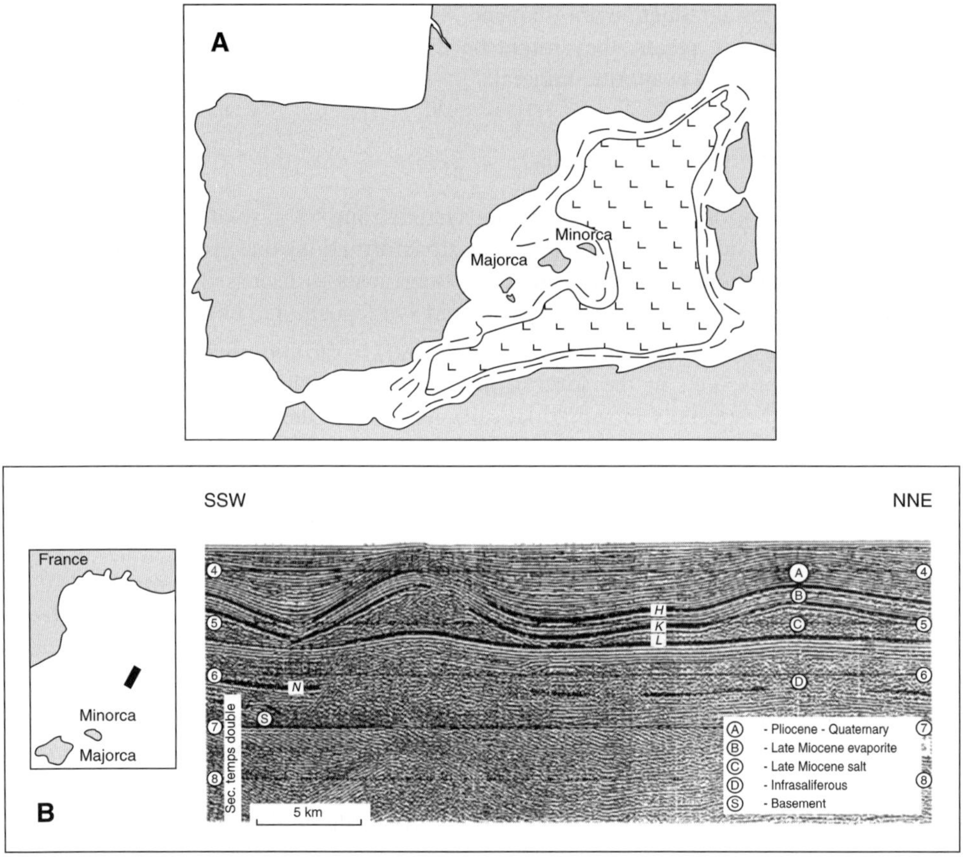

Fig. 3.146 Mediterranean example of Messinian evaporites.
A. Distribution in the west Mediterranean. Dashed line: evaporite deposition boundary; solid line: salt boundary. **B.** Seismic profile in which the thickness of the saliferous set can be observed.

Examples of the oldest periods are:
- The Devonian of Canada
- The Zechsten of northern Europe
- The west European Triassic
- The Jurassic of the Gulf of Mexico
- Brazil's Cretaceous
- The Oligocene of the Paris and Alsace basins, and that of the southwest of France.

3.3.5.3 Interest for Petroleum Geology (see Chapter 6)

Evaporite deposition conditions are so restrictive that an association is often seen with the confined anaerobic (euxinic) deposits rich in organic matter. Evaporitic systems are thus often tied in with **parent rock** of potential economic importance.

Saline deposits, and especially the saliferous beds, are highly impermeable beds of obvious interest for petroleum, because they often serve as excellent **cap rock** that effectively trap the hydrocarbons. This property also makes them useful for storing any kind of substance in the cavities dug out by leaching and salt dissolution.

The saliferous levels, due to their **density** deficit (with respect to the compacted enclosing country rock at depth) and their plasticity, are very sensitive to sedimentary overloads and tectonic stresses, which explains the formation of the **diapirs** of interest in oil exploration. As will be seen in Chapter 6, these are associated with various **traps**.

It should also be said that, beyond their interest for oil, saline beds are substances of direct use to man, providing rock salt, gypsum for plaster stone, and potassic fertilizer, and are also the locus of certain mineral concentrations such as the sulfides from the brine troughs stemming from hydrothermal sources.

3.3.6 Deep Ocean Deposits

The continental margins beyond the continental shelf, at depths of more than 200 m, are zones of special sedimentation ranging from the continental basin to the abyssal plains (Table 3.5). It was said in the introduction to the chapter on basins that some margins are thick, if the sedimentation rate is high, while others are starved if it is low (Fig. 2.23). Depending on the sources and oceanological conditions, various types of deposit can be observed here, and a number of those of importance for petroleum are discussed hereafter.

3.3.6.1 Pelagic and Hemipelagic Sediments

There are three types of fine sediment sources: primary pelagic production of biological origin, terrigenous products (continental and "external" flows), the finest of which are transported very far by ocean streams, and the authigenic products formed in place under special conditions. Different products will sediment by decantation from the material in suspension or in solution.

A. Pelagic Production

The whole ocean is characterized by intense phyto- and zooplankton life in the euphotic zone. Biomass productivity results in the production of carbonated or siliceous tests and depends on climatic conditions (with latitudinal zoning) and local conditions (nutrients, cold water upwelling). When the sediment comprises more than 75% of test debris from micro-

Table 3.5 Recapitulation of depositional environments (from ENSPM document).

SEDIMENTARY DOMAIN / DEPOSITIONAL ENVIRONMENT	CONTINENTAL	COASTAL: Very shallow	Shallow	MARINE: Deeper – Infraphotic	Oceanic
		← Euphotic zone, depth depending on (1) water turbidity (2) temperature (climate) →			
DESCRIPTION AND LOCAL FACTORS	Widespread emerged area	Zone with tidal oscillations, with emersion indicators	Zone covered by the sea with no emersion indicators	Deep part of shelf	Oceanic domain
ENERGY	Low to high	Low to high	Low to high	Low to average	Low to average
WATER CIRCULATION	— No contact with ocean	Confined to free	Very limited to free	Free, sometimes confined	
DEPTH	Often above sea level	> 0- > ± 10 m	< ± 50 m	> ± 50 m < ± 200 m	> ± 200 m
COASTAL INFLUENCE	—	Very pronounced	Nearly or completely absent	Absent	
FAUNA AND FLORA	Non-marine	Mainly benthic fossils of the euphotic zone		Planktonic fauna and deep benthic	
LITHOLOGY	Jointed and un-jointed grain sediments			Autochtonous : un-jointed grains Allochtonous : jointed grains	
	Carbonated domain, barrier position:	Barrier protection:	Built-up barrier or oolithic sandbar	Protection by depths	
Environment-related porosity	Intergranular and vessicular (leaching)		Intergranular		
MAIN SEDIMENTARY BODIES	Deposits: lacustrine, marsh, non-calcareous dunes. Fluvial channel in continental series, natural levees, clay bodies.	**Carbonates.** Deposits: evaporitics, marshes, calcareous and beach dunes, high supracotidals and interco-tidals Intercotidal deposits, built-up limestone, oolithic sandbars WITH EMERSION **Siliciclastics.** Fluvial-marine deposit, channel sandstone, clay	Cotidal limestones with emersion indicators Subcotidal limestones Intercotidal deposits, built-up limestone, oolithic sandbars WITHOUT EMERSION Clay and silt: holomarines and turbidite sandstone	Well-bedded or nodular limestone, tubidite-type reworking Subcotidal limestones Clay and silt: holomarines and turbidite sandstone	**Very deep part** Radiolarites Ophiolites

or nanoplankton, it is called **pelagic ooze** or pelagite. This debris accumulates by decantation at low water velocities, i.e., it falls from the surface. If any cyclic variations occur, as will be discussed in the following chapter, sedimentary beds of different compositions will be deposited in sequences that are sometimes designated laminite or periodite.

The predominance of siliceous organisms (diatoms and silico-flagellates, for example) at high latitudes is such that siliceous oozes form, yielding different types of siliceous rocks including diatomites (originally in the form amorphous opal) (Fig. 3.147) and radiolarites, which designate ancient sediments consisting essentially of radiolarian tests. This type of deposit has been known to develop in different geological periods, and is not limited to the oceanic domain alone (diatom veils are found in the middle of sheltered bays, for example).

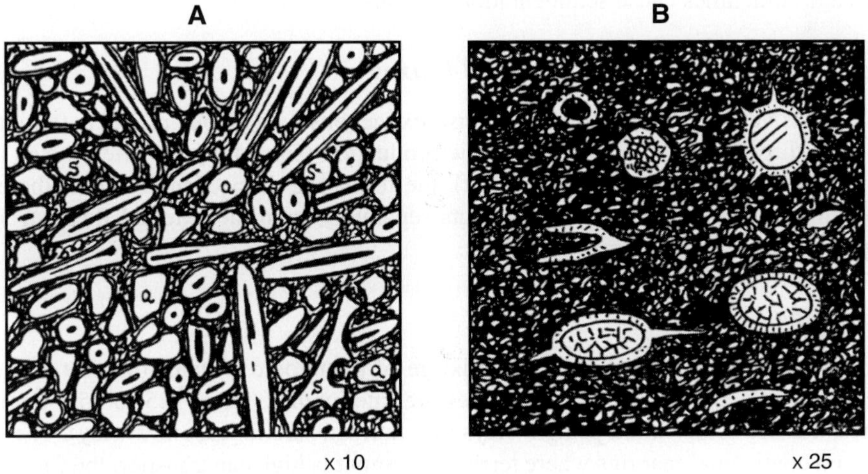

Fig. 3.147 Pelagic siliceous sediments (from Lapparent, 1950).
Under certain conditions, organisms with siliceous tests or skeletons accumulate in the form of siliceous oozes. **A.** Gaize of sponge spicules. **B.** Radiolarian phthanite (viewed under microscope).

Silica can also precipitate in arid, warm climates with high phytoplankton production and intense evaporation.

At temperate latitudes, primary production is dominated by organisms with calcareous tests, especially in the intertropical zone. When these organisms die, their remains become sediments of **calcareous oozes** that generally become fine-grained **pelagic mudstone**. These muds will differ considerably depending on the type of organism (foraminifer, pteropod, coccoliths). If cyclic variations occur and hemipelagic products participate (clays), the resulting **limestone** will alternate with **marl**. It is known that the stability of calcite decreases with depth in the water column (Fig. 3.20), so dissolution may occur: the carbonate, which is not very soluble at the surface, gradually becomes so and dissolves with the increase in relative content of dissolved CO_2 and the correlative decrease in pH. Starting at the **lysocline** level, the dissolution suddenly increases and becomes total (leaving only a few insolubles) starting at a certain depth called the **carbonate compensation depth**, commonly abbreviated **CCD**.

When dissolution characters are found in certain ancient series, they do reflect the **paleobathymetry** in a way. However, remember that the ocean and its topography, and certainly time, will affect the depth of the CCD (Fig. 3.20).

Because of the type of sedimentation (decantation from the surface), these pelagic sediments are arranged in regular **drapes** covering both rises and depressions in the relief (Fig. 3.148), though this is only theory because **floor currents** rework the material (see further on) and sometimes cause sedimentation hiatuses.

B. Hemipelagic Deposits and External Fluxes

Although it is actually difficult to distinguish exogenous terrigenous products fed into the ocean from those that are formed by pelagic production (since they are often mixed as the local hydrodynamic conditions will have it), the following discussion considers the essentially continental fine hemipelagic sediments separately from the external fluxes from the atmosphere: aerosols and volcanic products.

a. Hemipelagic Oozes

A hemipelagic sediment is defined as a mixture of planktonic remains (at least 5%), silts, and clays. Argillaceous (clay) pelagic oozes are intermediate terms (25 to 15% test) with pelagites, and are often called pelagic clays (the horizons are at least 60% clay). The ocean currents on continental margins where terrigenous input is high can transport the finest particles in suspension, which then decant very slowly in regular drapes that are more or less rich in pelagic material and more or less carbonated depending on the surface production (marl-limestone alternations) (Fig. 3.149).

The interest of these sediments with deep pelagic deposits is generally small for petroleum purposes, but it should be remembered that the organic matter from planktonic organisms may not be oxidized and conserved under certain anoxic or anaerobic conditions, which endows this type of deposit with remarkable characteristics as source rock. An example is the famous black shale of the Atlantic and Vocontian basin (Fig. 3.150). Generalizing, it can be said that much source rock in the large petroleum systems stem from these special conditions (Venezuela's La Luna from the Cretaceous, the Kimmeridgian clays of the North Sea, Saharan Paleozoic clays). These regularly draped continuous deposits can also form effective caps for the underlying or interbedded reservoirs (see later in this chapter and Chapter 6).

Fig. 3.148 Pelagic deposits.
A. Seismic section showing pelagic sediments onlapped by turbidites.
B. Outcrop: Priabonian blue marls of the Alpine domain overlaid by onlaps of Annot turbiditic sandstones.

Fig. 3.149 Regular alternation of carbonaceous, calcareous, and marl deposits.
A. Upper Jurassic of the Vocontian basin. **B.** Lower Cretaceous of the Vocontian basin.

Fig. 3.150 Rocks with high organic content: black shales.
Special organic matter enrichment conditions can sometimes be found where black levels of black shale alternate with leaner levels. **A.** The Liassic of Dorset (England) with alternating levels at its base and more massive on top, with organic carbon content of up to 10% (average 3%) (IFP photo). **B.** The Eocene of Barbados (Caribbean) with alternations on the centimetric scale.

b. Aerosols and volcanic products

Various products carried by the wind and sometimes following tropospheric paths of migration can sediment into all of the ocean's depths by decantation and carriage by the ocean's surface and deep currents.

Desert dust is carried over large distances and is mixed with deep sediments over broad surfaces, at times making it possible to determine climatic conditions for ancient series, especially in terms of aridity and the existence of desert areas.

Volcanic products from explosive edifices spend some time in transit and then sediment out in layers of ash called tephra (Fig. 3.151), which are evidence of the events and can therefore be used for relative dating in stratigraphy. The products are at times coarser (in pyroclastics or hydroclastics).

Fig. 3.151 Layer of volcanic ash in the oceanic domain.
Ash lying in thin beds at regular intervals, and strongly bioturbed here, in the chalks of the pelagic Miocene of Barbados.

As was mentioned in the first chapter, the ocean domain is also the preferred locus of volcanism, with submarine volcanoes at the ridges. Although these are negligible for petroleum purposes, submarine volcanic products should be recalled here for their eruption breccia on the slope, pillow lavas, and pyroclastic flows, with the formation of sulfurized products in the warm water upwelling from hydrothermal sources.

Micrometeorites and meteorites, though of great academic interest (for recognizing ancient variations in atmosphere or extraterrestrial formation of kerogens) contribute an infinitesimal fraction to sedimentation.

C. Authigenic Sediments

To fill out the panorama some, it can be said that **neogenic clays** can be generated in the ocean under certain conditions:

- Fibrous clays stemming from the evaporitic deposits discussed above
- Iron clay and glaucony from the continental shelves
- Metallic clays (especially smectite) and fibrous clay (sepiolite and polygoralcite) with a high percentage of iron, which can form on the bottom in an oxidizing regime, producing the **red clays** of the great depths.

Lastly, the sulfurized products from the hydrothermal sources of the ridges and volcanic zones, the **polymetallic** micronodules and **nodules** of the great depths, and the **cobaltiferous** crusts of submarine hills, are other examples of direct precipitation on the seabed (Fig. 3.152).

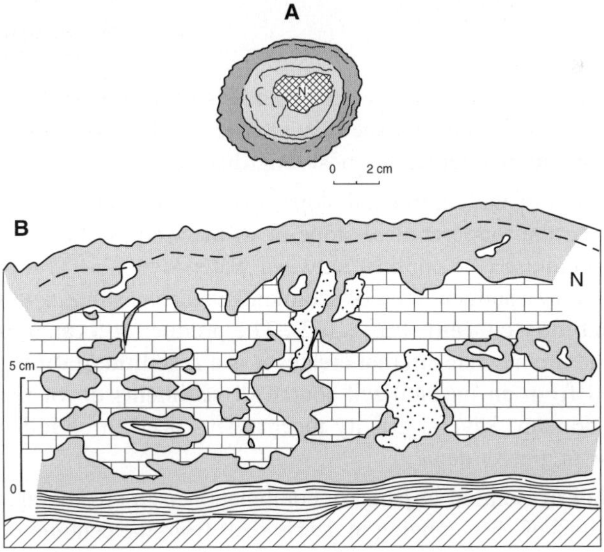

Fig. 3.152 Metalliferous concentrations of nodules and deep crusts in the Pacific.
A. Schematic cross-section of a polymetallic nodule: concentric layers of irregular growth around a nucleus (from Janin, 1987, in Biju-Duval, 1994).
B. Schematic cross-section of a polymetallic crust: a cobalt-rich crust with recent filling (dotted) in the cavities has formed irregularly in a disjunct bank of phosphatized jointed Eocene limestone.

Smectite can form authigenically on the ocean bottom under special conditions of hydrodynamic calm. Metalliferous clays form in the vicinity of the water-sediment interface (preferentially in the vicinity of a hydrothermal environment). Zeoliths (phillipsite) can also form in such an environment, where the major components are oxy-hydroxides of iron and manganese. It should be emphasized that the formation of authigenic minerals is important for the role it plays in the geochemical cycles of certain elements, both in seawater and in marine sediments.

The genesis of deep smectite is still poorly known (low temperature, role of biogenic silica), as is the boundary between sedimentation and early diagenesis, and the role of microorganisms.

Ferriferous clays of different varieties (glauconite, verdite, chamosite) exist in conditions of lesser depth. Their formation depends on the type and texture of the substratum, the water depth (20 to 200 m for verdine, 150 to 300 m for glauconite), the sedimentation rate, and the latitudinal position.

Proximal environments are generally not the locus of major clay formation today, although they were in the past (from the Devonian to the Tertiary).

3.3.6.2 Gravity-Driven Deposits

A. Mechanisms

Sediments on the continental shelf are dispersed and distributed by surface streams generated by wind and tide. Gravity-driven slippage may also occur, depending on the slope, as is true in the continental domain; but generally, most long-distance transport is due to drift currents.

In the deep domain of the continental slope, margin foot, and abyssal domain, **density currents** move along the topographical slope or in the water column. The excess density may be due to temperature (the contrast between water layers, as in the Antarctic Bottom Water), to salinity, or to a sudden input of sediments in suspension, in which case the turbid flow is called a **turbidity current** (Fig. 3.102). The product of these turbidity currents is a **gravity deposit**, and the mechanism that creates it is a **resedimentation** process. The detritus is retransported from an unstabilized source of sediments down the slope by gravity effect to a deeper domain. Several gravity mechanisms need to be looked at separately to explain this process in greater detail.

Slippage and fall of isolated objects. In combination with erosion agents, gravity causes the slippage of portions of cliffs ranging in debris sizes from the centimetric to the kilometric. Such blocks, or pre-existing rock fragments embedded in matrix (the original sediment in which they were enclosed) are generally termed **olistolith. Olistostrome** is a chaotic accumulation of a large number of such olistoliths in a clay matrix acting as mattress (thus the "stroma"). There is often some hint of tectonics, because these slipped and resedimented deposits are observed along fault breccia, in oblique banks, or on mountain chain fronts (Fig. 3.153). Hectometric or kilometric rafts found in allochthonous position are sometimes designated as **sedimentary klippe**. One exacerbated form of olistostromes is the gliding nappe.

A B

Fig. 3.153 Chaotic flows: Olistostrome and olistoliths (A), and debris flows (B).
A. The example here is of carbonaceous blocks in an argillaceous-sandstone matrix, itself lying in a turbidite series (Eocene accretion wedge south of Hispaniola) (IFP photo). **B.** Several debris flows in succession (Oligocene, south Hispaniola) (IFP photo).

Mass slippage. This is a rotational slip of semiconsolidated rock in lumps, with no notable dissociation of particles, on a slope in a subaquatic medium, forming **slumps**. Sometimes there is long translational gliding, or more or less liquefied mud, grain, or debris flows.

Slumps (Fig. 3.154) are structures in which sedimentary beds are deformed and the upstream part is torn away in a detached lump (the form is not always recognizable) and extended. The escarpment produced on the upstream part may then be subject to erosion. The downstream part is characterized by intense deformation of the beds in synsedimentary folds and faults, as the slippage generates compression. Microseisms may prompt the formation of slumps on the slope, or the slope's lack of stiffness to resist an excess sedimentary load. Certain geological series of the past include impressive sequences that witness to active synsedimentary tectonics. Any type of sediment can slump. In certain cases, the slump fold heads are dissociated and the less cohesive material then participates in other types of gravity-driven deposits.

A

B

C

Fig. 3.154 Slumping.
It is relatively common for slumps to form in sedimentary beds if ever they are in disequilibrium. **A.** Carbonaceous banks of the Hauterivian (Vocontian basin). **B.** Carbonaceous slumps of the Alpine Liassic. **C.** Argillaceous-gritty slumps in the turbidite Eocene series of Barbados (IFP photo).

The later chapter on tectonics will discuss the development of the great synsedimentary **normal listric faults** generated by rotational slippage on a level of plastic décollement, where gravity and excess load play an important role. It will be seen that great masses can be transferred in **gliding nappes**.

Cohesive or **liquefied flows** (Fig. 3.155) will differ depending on the cohesion or degree of fluidization, and in fact all of the transition terms can be found with fluidized and turbid flows. The cohesion in a cohesive flow is such that the grains and matrix flow together with friction on the bottom, while liquefied flows organize in a more or less turbulent stream heavily charged in suspended material. These flows are of different sorts:

Debris flows in the strict sense are a plastic process, as opposed to the elastic regime of slumps. The matrix supporting the debris is thick (mud-supported). A **mudflow** is one with no large particulate debris. This flow can move over long distances and low slopes.

Grain flows have no fine matrix and the cohesion is thereby diminished. This is a transition between the plastic and fluid domain, with practically visco-elastic flows in which the grains form a dense mass. The slope has to be rather high for the flow to occur.

Liquefied flows are gravity slips in which there is practically no cohesion. The grains move in the fluid in high concentration (grain-supported) down steep slopes.

Most of these mechanisms lead to deposits in a highly unstable and usually transitory state within a gravity flow, for which the common expression is turbid flow.

Turbid flow or **turbidity currents** are the most fluid fractions of a viscous gravity process. Here, the particles are in suspension in a **turbulent flow** of liquid. The mechanism can be effective over very long distances and very low slopes. Depending on the flow energy and the grain size concentration in the sediments being transported in the liquid mass, vertical and horizontal grading will occur and lead to various deposition forms called turbidites. Flows may be hyper-concentrated or of low concentration, and of high or low density (Fig. 3.156).

Turbidites may correspond to an "instantaneous" surge-flow event or the input may be prolonged (steady flow).

The finest particles are dispersed and transported to long distances by the deep ocean currents (the contour currents discussed later), where they become part of the nepheloid boundary layer and enter into the hemipelagic domain.

B. Types of Deposits

The usual form of turbidite deposits is the deep sea fan, though these deposits are constructed not only by gravity mechanisms but also by decantation and traction over the bottom. These fans develop from one source of sediments: a major fluvial input leading out onto a slope (remember that deep fans may also develop in lakes). The modern examples of large fans on the continental margins from the Rhone, Ebre, Indus, Ganges, Amazon, Mississippi, and others, have all been very closely studied. They vary greatly in size, but gener-

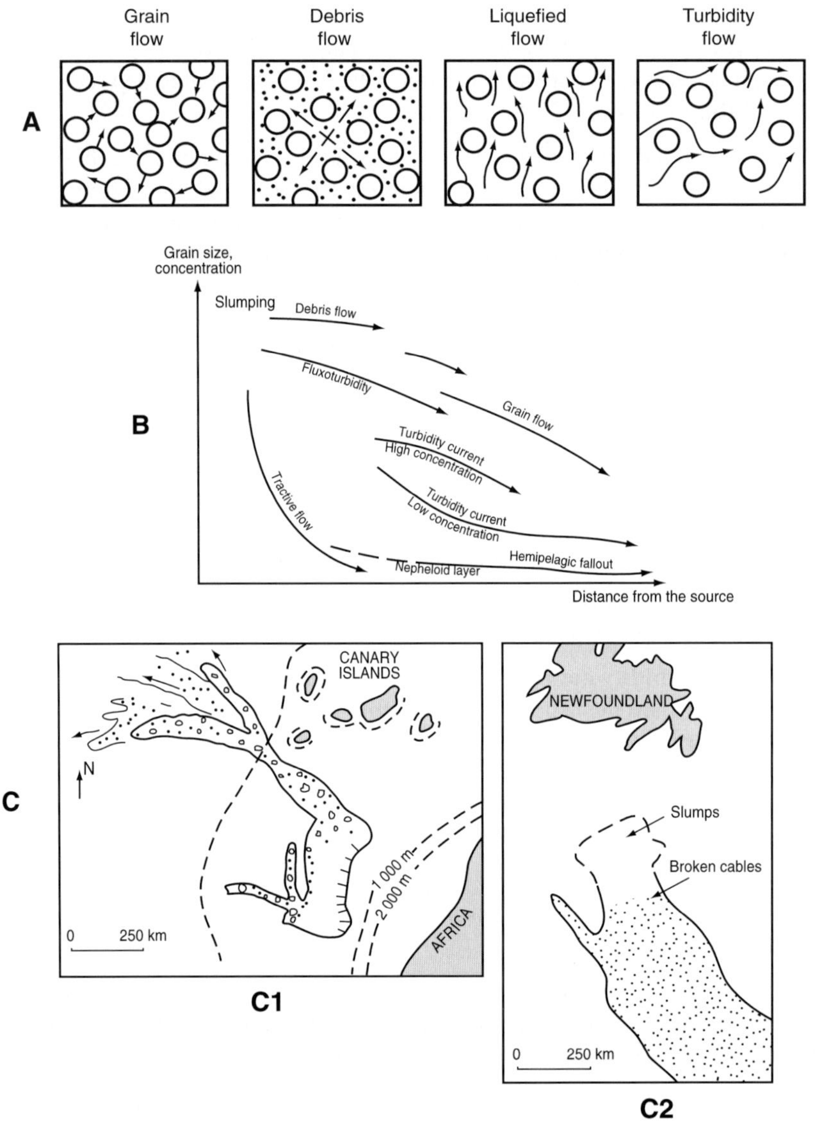

Fig. 3.155 Gravity-driven flow types.
A. General scheme. **B.** Dynamic relations between flow types (from Walker). **C.** Spectacular geographic spread of debris and turbidity flows in the north Atlantic. C1 Debris flow of the African coast; C2 Turbidity flow to the south of Newfoundland in 1929 (from Heezen and Hollister, 1971).

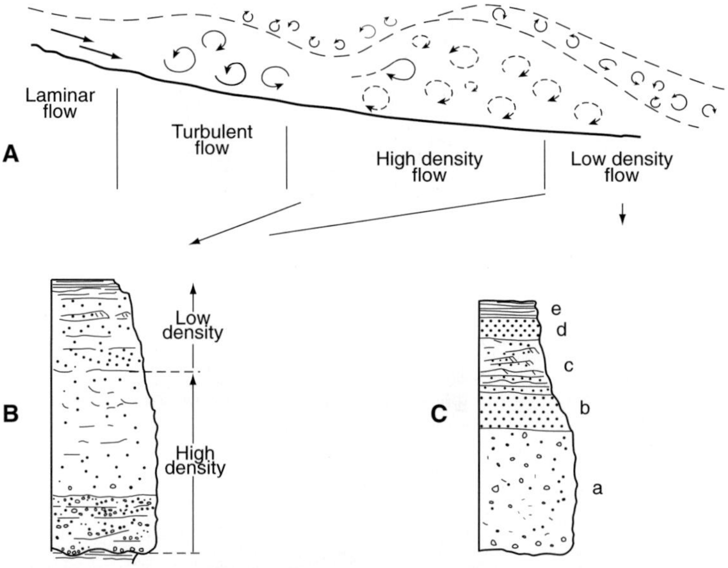

Fig. 3.156 High- and low-density flows and turbidite deposits.
A. General scheme (modified from Mutti, 1992). **B.** Typical high-density sequence with a passage from a tractive carpet to suspension and then to low density (from Lowe, 1982). **C.** Typical low-density sequence: Bouma sequence.

ally exhibit a characteristic morphological sequence with pattern variations that have been modeled (Fig. 3.157).[1]

The **feed channel** through which the fluvial material transits from the continent is very often a canyon dug out at the edge of the continental shelf. The fluvial dynamic can be active quite far out at sea, as has been said before, and the main canyon may have quite a number of tributaries.

Less concentrated input can come from the edge of the continental shelf to the slope laterally to these main distribution sources.

The inner or **upper fan** consists essentially of overwash levees lining the straight or sinuous main channel on a steep slope. This fan is characterized by a variety of often very coarse deposits on the channel's edge. The channel will vary in length from a few kilometers to several hundred.

The **midfan** is a series of **lobes** forming a network of secondary channels that fan out more or less sinuously with their levees, including washover and spreads as in fluvial systems. Various forms of gravity deposits are found here.

1. For example, see Normark, 1979; Walker, 1978; Mutti, 1985.

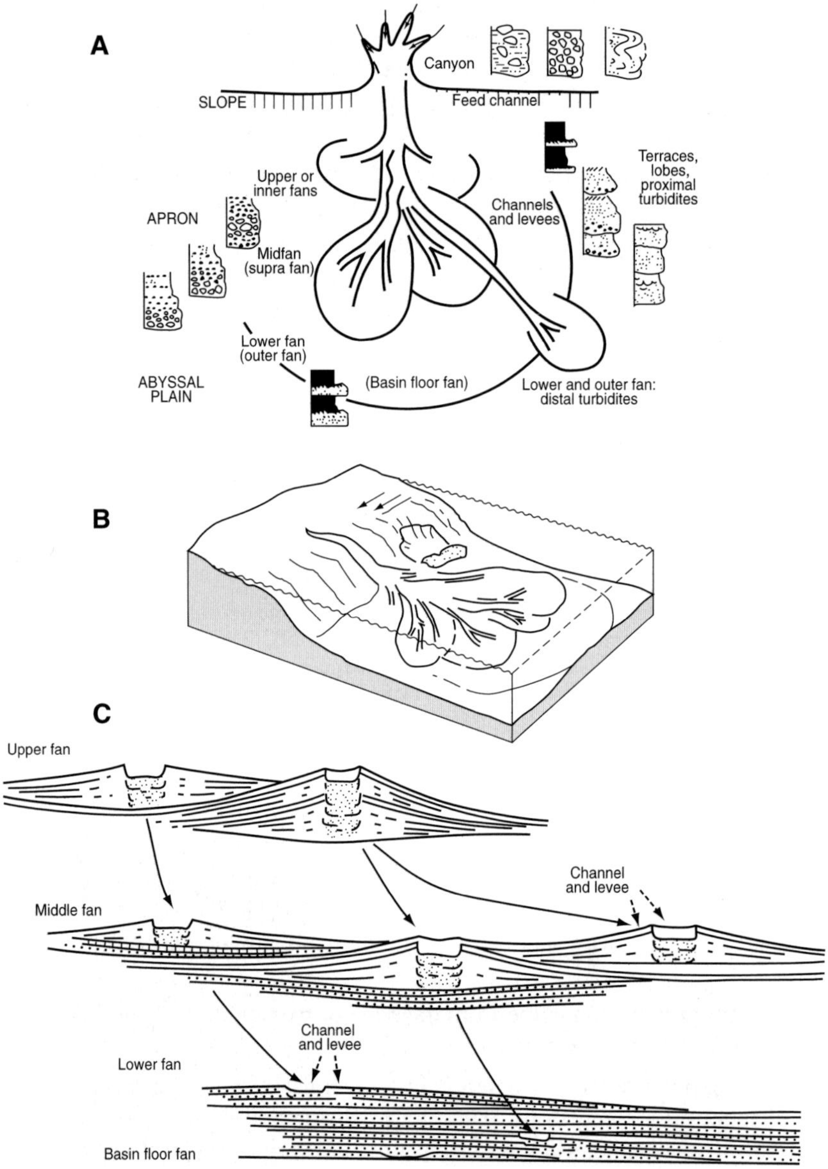

Fig. 3.157 General organization of turbidite deposits.
A. Elements found in a submarine fan (from Walker, 1978). **B.** Three-dimensional view. **C.** Schematic sections of nested channels and levees.

The lower **basin floor fan** develops downstream of the fan at the foot of the continental slope, sometimes very far toward the abyssal plain on very low slopes where the proportion of hemipelagite is large. Channels are rarely observed here.

Generally, the great fans of the present are characterized by high lateral migration of the spread in which the **channel** deposits can be recognized with the **levee** deposits alongside. The gradual construction of the fan generates an **aggradation** followed by avulsion with **migration** and nesting of the lobes. The systems also gradually **prograde** seaward.

The mass gravity transport on the slopes also prompts the buildup of sedimentary edifices. This is often in the form of lobes, but at times it is chaotic. The forms range from the slump and erosion scars left in the upstream zone where slippage begins, to the **irregular draping** and concentrated supply channels in the lower zones (Fig. 3.157).

Large turbidite systems are also known to develop by filling **active margin basins**, as in the troughs of Japan or Iran's Makran, without becoming so organized as the large systems of the passive margins. That is, depressions of all sizes develop in connection with the deformation and concomitant erosion, especially in the great accretionary wedges, namely the trough and the forearc basins where thick turbidite deposits accumulate (Fig. 3.158). This

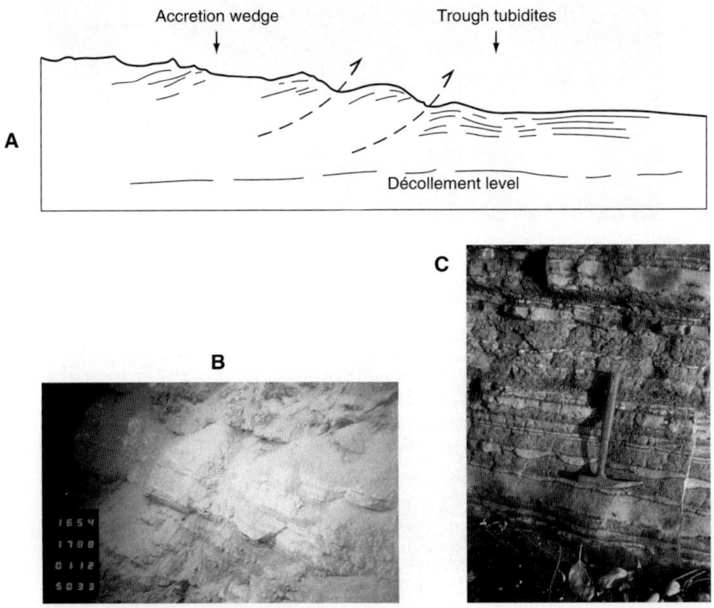

Fig. 3.158 Deposits of turbidite in the deep environments of oceanic troughs.
A. General scheme (as in Japan, Nankai, or south Barbados). **B.** Deep deposits of the trough of Japan subjected to erosion and observed in deep water photography (Ifremer photo). **C.** Fine turbidites of the Eocene observed on Granada (IFP photo).

situation may also prevail in the back-arc basins if inputs from the continent and from the volcanic arc are large. These basins will then be partly filled by alternations or lateral nesting of turbidites, hemipelagites, and pelagites. The example of the Barbados prism is illustrative of the mixed margin where the Orinoco on the passive margin of the Brazilian craton spills its input near the Antilles arc, where it is incorporated (Fig. 3.159).

C. Facies and Organization

Turbidites are major accumulations on the continental margins in a variety of generally siliciclastic facies corresponding to continental terrigenous input. It is nonetheless important to emphasize that turbidite deposits can also be calcareous.

A

B

Fig. 3.159 Folded, tectonized turbidite deposits.
These are relatively deep deposits from the Eocene incorporated in the sedimentary wedge of the Antilles arc (IFP photos). **A.** Barbados. **B.** Granada.

The facies are conventionally termed descriptively – by proportion of sand/clay geometry, internal structure – or from a genetic viewpoint. The typical facies sequence generally called the Bouma sequence is generally used for defining a turbidite (Fig. 3.160). This typical sequence exhibits a pronounced genetic evolution by gradually decreasing energy in a current of average density:

- An **erosive base**, which is witness to the more or less pronounced erosive power of the density current. The various types of figures described, such as flutes and grooves, will show the direction of gravity transport (Fig. 3.161).
- An arenaceous set with generally normal graded bedding (A), very heterogeneous grain size distribution, corresponding to an upper flow hydrodynamic deposition regime (in antidunes).
- An arenaceous set with regular parallel laminations (B) marking the transition to a less turbulent regime.
- An often finer arenaceous set with current ripple stratifications (C), ripples and microlaminations and convolutions.
- A series of parallel silt-clay lamina (D), typically arranged in a weak hydraulic system.
- An argillaceous and hemipelagic set (E) with no apparent deposition structure corresponding to a laminar regime, then final decantation of the finest material. This set is at times bioturbed, which indicates a return to normal bottom conditions.

This theoretical sequence occurring in a few centimeters or a few meters is rarely observed in its totality. If erosion is extensive, the last terms will be removed, leaving amalgamated gritty beds. This situation is more probable in proximal position: the farther from the source, the more the first terms thin out. The general arrangement follows an upstream-to-downstream (or proximal-distal) pattern, and also laterally. In proximal position, the A to C terms will dominate. In distal, the finest turbidites dominate with the C to E terms.

The facies distribution will of course depend on the general mechanism of the fan (Fig. 3.157). Depending on the set of mechanisms at play, more cohesive gravity deposits (olistostromes, slumps, debris flows) can be substituted at any time for the conventional turbidite deposits, as can pelagic and hemipelagic drapings.

Megaturbidites or seismoturbidites correspond to a very special environment, as do **homogenites** and **tsunamites**, which are all customary terms corresponding to particular sedimentation conditions. The term **fluxoturbidite** has also been used to describe very proximal coarse turbidites in a very turbulent regime.

D. Ancient Deposits and Time Fluctuations

Turbidite formations have long been described in ancient geological series. Among the best examples are the alpine **flyshes**, which are very often thick allochthonous detritic formations of various ages transported in nappes (Fig. 3.162). After much research, it was generally thought that these great turbidite series had to do with special structural conditions occurring in the construction of mountain chains. But more recent studies (especially in the ocean) have shown that the accretion wedges that form in the initial stages of folded chains could

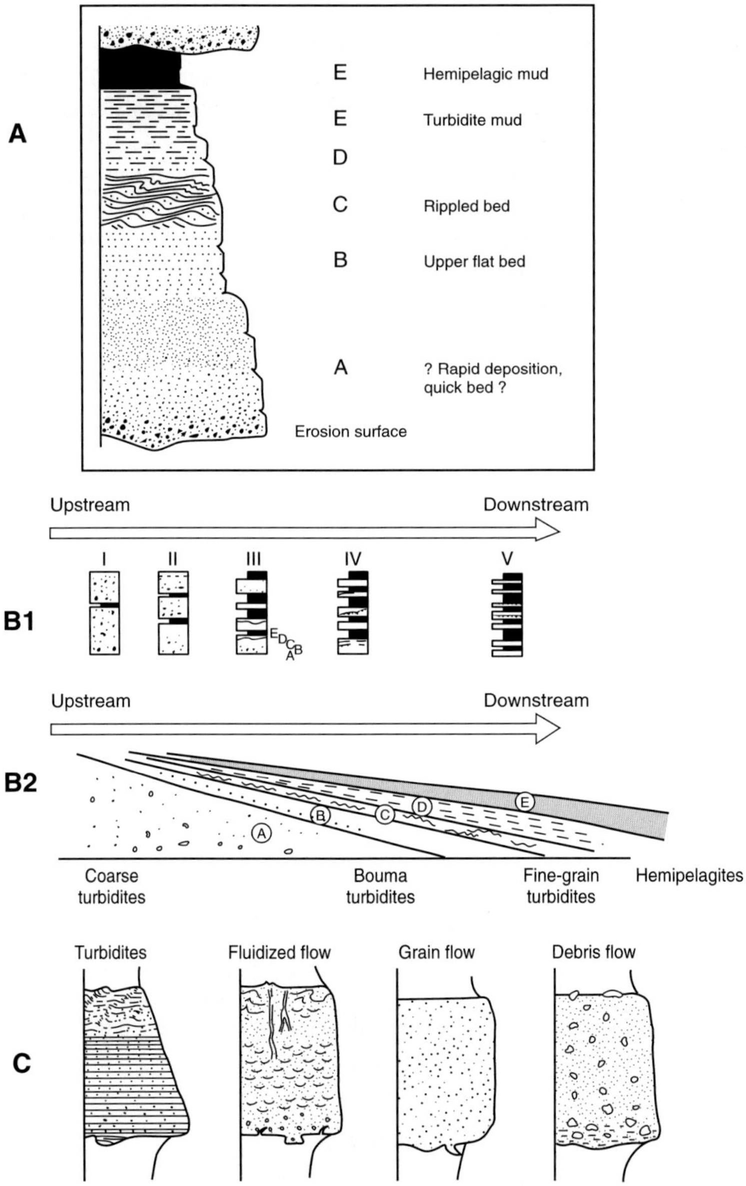

Fig. 3.160 Typical Bouma sequence.
A. Typical sequence. **B.** Lateral evolution diagram (from various authors). **C.** Comparison with other gravity sequences.

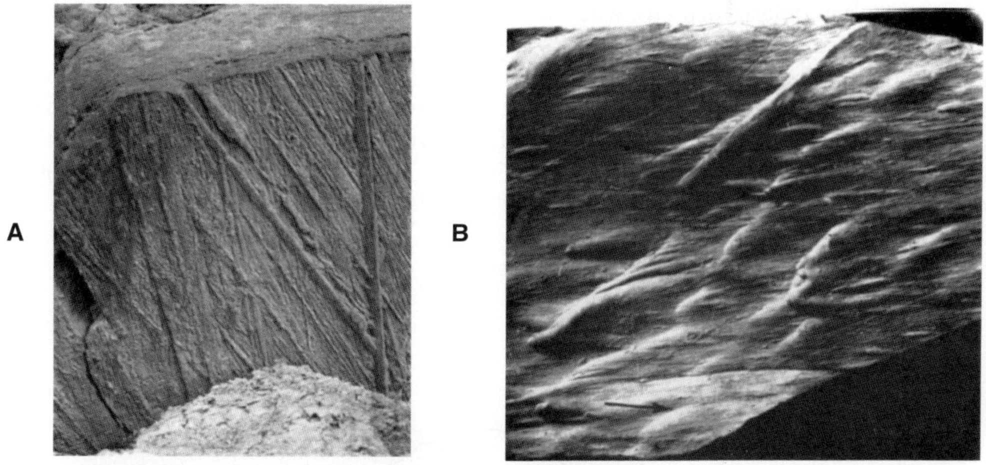

Fig. 3.161 Groove and flute marks in alpine turbidites.
A. Grooves in sandstones of Annot (Late Eocene). **B.** Flutes in Oligo-Miocene flysch of the Carpathians (from Chambre Syndicale des Techniciens du Pétrole).

Fig. 3.162 The sandstones of Annot, an example of turbidite deposits in the alpine domain (IFP photo).

reflect these particular situations when the trough and the transported deformed forearc basins were filled with turbidites. Also, the great complexes were built up on stable margins, and sometimes very far out in the ocean, so they were very sensitive to time variations in the sea level.[1] It is easily imaginable that the detritic sediments would no longer be stored on the continental shelf in a period of **low sea level**, but arrive directly on the slope where they could be entrained by gravity mechanisms (Fig. 3.163). This point will be taken up again and discussed in the following chapter.

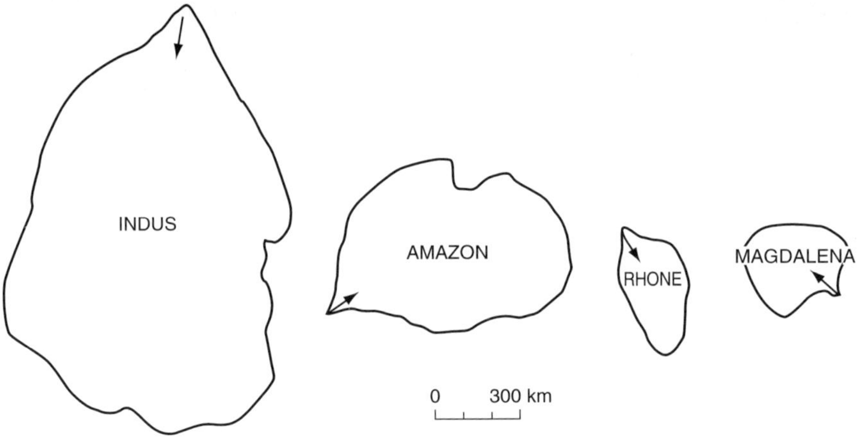

Fig. 3.163 Comparative sizes of a few of the present deep submarine fans. Note that the largest are located on passive continental margins.

E. Interest for Petroleum

Like all of the previous detritic formations, turbidite accumulations can make excellent oil reservoirs. This is true of several petroleum provinces: the North Sea, at its present shallow water depth, the edge of the Gulf of Mexico, and the Brazilian offshore, in deep water going beyond 2000 m (Fig. 3.164).

3.3.6.3 Deep Sedimentary Piles, Contourites

Described for the first time in 1966[2], the geological action of bottom streams is quite effective for moving a large amount of sedimentary material over long distances and depositing large sedimentary bodies of special types. These **contour currents** are essentially large cold deepwater currents (in the Atlantic and Pacific) traveling at high velocities of up to 20 cm/s,

1. See Vail et al., 1977.
2. See the review in Hollister, 1993.

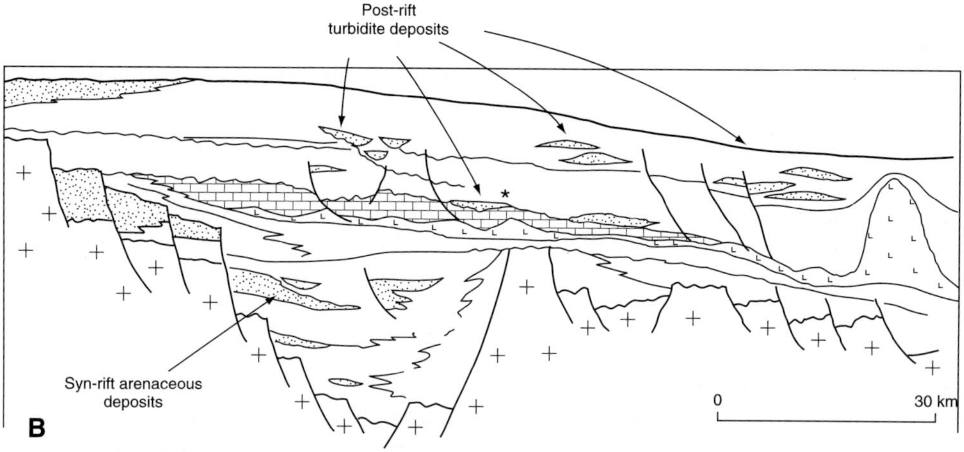

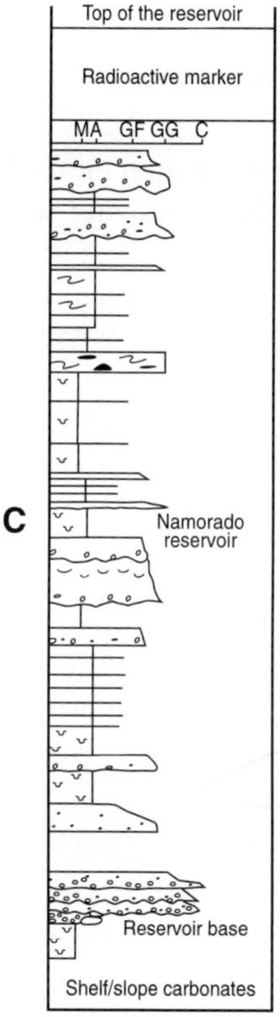

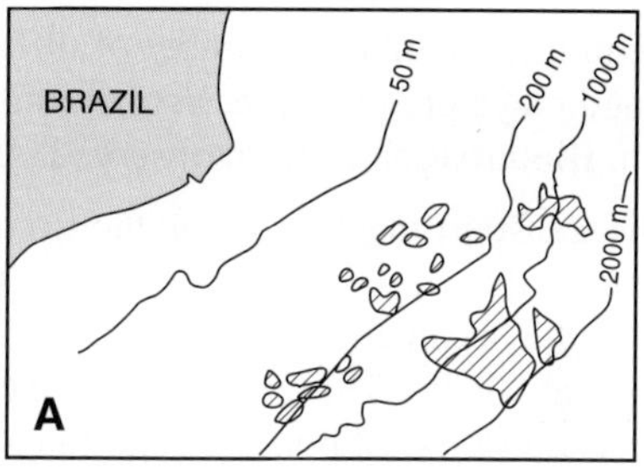

Fig. 3.164 Deep turbidite deposits: the Campos basin offshore of Brazil.
A. Deep basin localization (water depths up to 2000 m). **B.** General section of the margin (from O. Gomez de Souza, 1997). **C.** Stratigraphic sequence of the Namorado field: schematic section from sample core data and log record (from O. Gomez de Souza, 1997).

and their permanence with time allows a considerable amount of accumulation. There are also lesser bottom streams and occasional "benthic storms" that suddenly upset the deep current. The relation with the altimetric variability of the ocean surface has been noted, as the excess energy at the surface can apparently be transmitted to the depths.

Topography is very important, as it guides the current streamlines. Contour currents have a dual effect:

- They transport sediments, but erode them too. Sometimes they simply rework the sediments on site and produce hiatuses, which are always very localized.
- They build up sedimentary piles of tractive deposits called contourites.

The main signs of this activity are generally observed in the relief at the foot of continental margins in frequent relations with the turbidite systems, such that it can often be said that there is a continuum from one type of deposit to the other. The link with the great cold currents, like the Antarctic Bottom Water or the one from the Arctic, implies a direct climatic influence and variations on a long time scale: Holocene build-ups are greater than those of glacial periods.

Contourites contain a number of different types of sedimentary constituents. Fine sands, silts, and clays predominate, but contourite of limestone, radiolarite, and dolomite are also known. Bioturbation is often observed.

The transport mechanism on the floor implies rolling and saltation on the floor and suspension in the current stream, leaving **deep subaquatic dunes** ranging in size from the ripple to the decametric at all depths from 500 to 6000 m, and resembling dune fields on the continental shelf! This elementary pattern occurs in widely varying forms including barchans and sand ribbons. On the longest time scale, there are three broad types of build-ups:

- Giant **sediment drifts** in relief several hundred meters high (and up to a kilometer) and several tens of kilometers (up to 200) long (Fig. 3.165). Examples of this are the Atlantic Ocean drifts (Ferri, Gordon, Erik in the north Atlantic, and Blakes and Bahama in the mid-Atlantic) (Fig. 3.165).
- **Regular draping** with little relief but over considerable areas of as much as 10^6 km^2, with drift amplitudes of 10 to 80 m and wavelengths of 1 to 10 km. Depending on the current velocity and geography, this draping is sometimes difficult to detect.
- **Fans** and **filled channels** that build up in threshold zones such as in the equatorial Atlantic (Kone Gap) where the accumulation will be guided sharply by the submarine relief.

Although good examples are still lacking today in ancient geological series, such environments are known in the Paleozoic and Mesozoic in both terrigenous and calcareous mediums. Considering the general fineness of this type of sediment, its interest for petroleum remains to be proven.

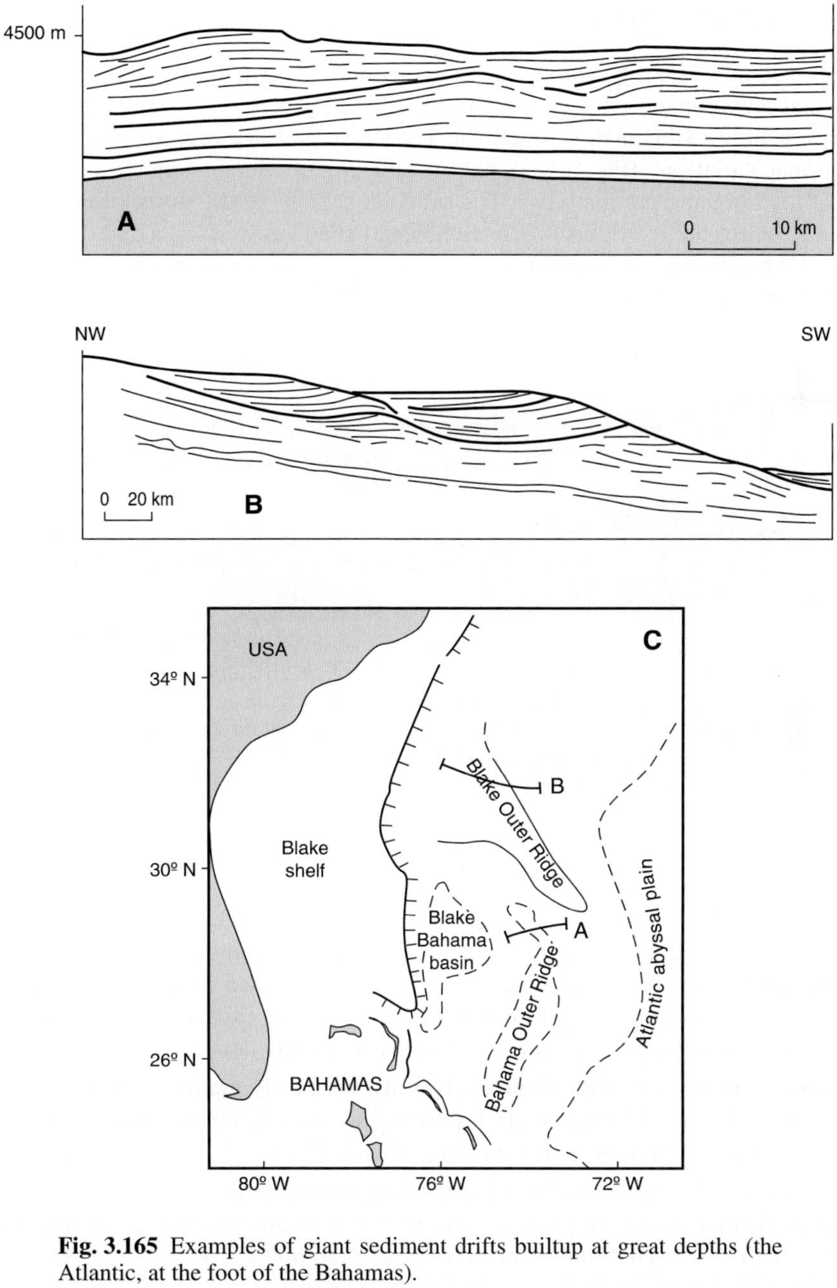

Fig. 3.165 Examples of giant sediment drifts builtup at great depths (the Atlantic, at the foot of the Bahamas).
A. Section of the Bahama outer ridge. **B.** Section of the Blake drift. **C.** Plan view.

3.3.6.4 Glacial Deposits

Much less is known about glacial marine deposits than of continental glacial and periglacial deposits and environments, which have been well described in the geological present and past. There is little sedimentation under ice. The active processes are mainly at the ice front, with seasonal variations (Fig. 3.166) due to temperature fluctuations (summer thaw and periods of high organic productivity). The subpolar regions of the Boreal domain are worth mentioning, where the ice flows on the continental shelf (as in Alaska and the Bering Sea) and the polar region where the ice front rises directly over the continental margin and ocean depths, as in the Antarctic (Fig. 3.167).

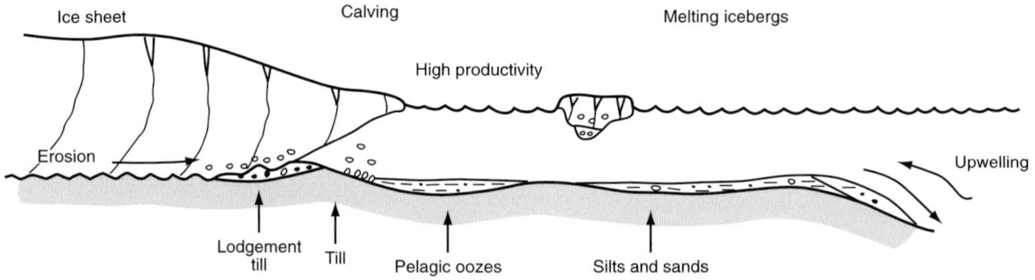

Fig. 3.166 Periglacial sedimentary processes in the ocean.
The detritic material released from calved ice can be effectively transported as far as the temperate zones. This depends on cyclic variations (seasonal or Milankovitch cycle, for example). Organic productivity varies greatly. The effect of deep currents due to thermal variations is capital.

Proximal deposits can also be distinguished from more distal. In proximal deposits, bottom traction is dominant and varies with the delta or channel currents. It carries blocks, gravel and washover sand. More distal deposits are dominated by transport in suspension with a mixture of the detritic material either in another type of sediment (siliceous organic oozes) or resedimentation with slippage, slumps, and gravity deposits in the deep zones.

The detritic deposits are **diamicton**. This is always poorly graded and often coarse sediments corresponding to different ice transport modes: ancient bottom till or thaw products, and iceberg calving (erratic blocks of all sizes) (Fig. 3.168).

Direct glacial influence is evidenced in the long term by:

- Sea-level fluctuations, in glacial-interglacial cycles and oscillations at different periods, which is of capital importance for the past (complex sequences) for shallow water depths.
- Cold ocean deepwater circulation, with sediment transport, especially with the buildup of giant sedimentary drifts and contourite lobes.

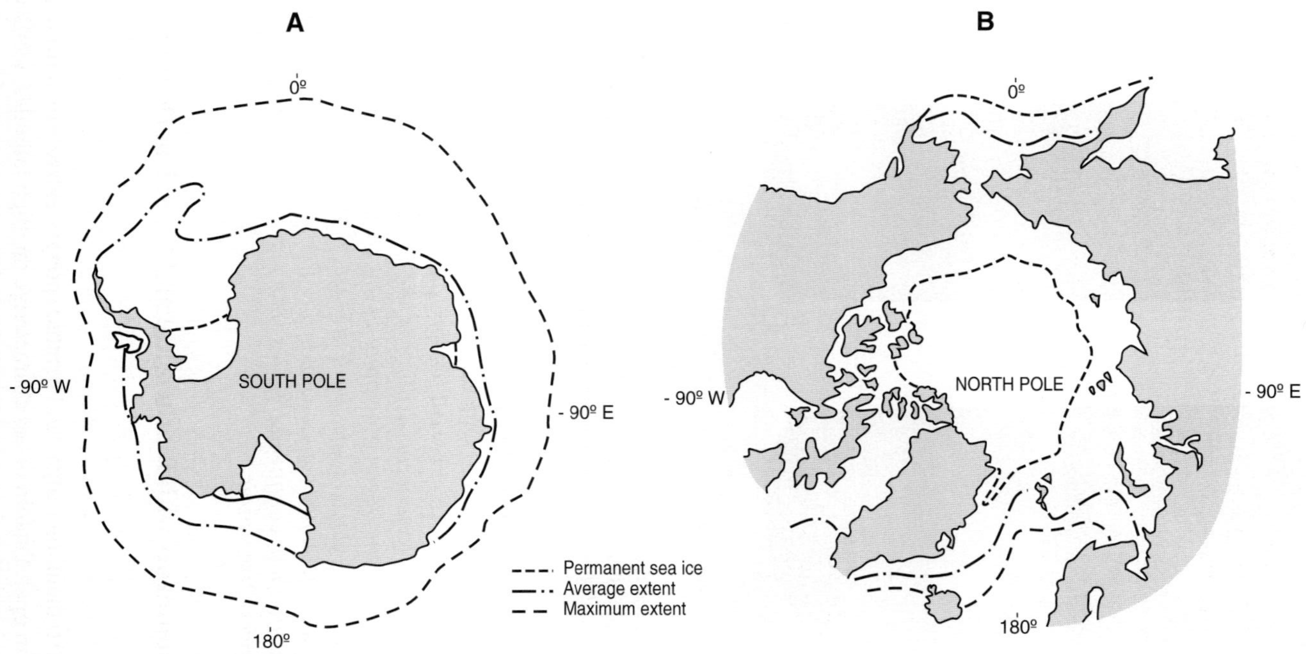

Fig. 3.167 Oceanic periglacial conditions of the South Pole (**A**) as opposed to the situation in the Arctic (**B**) where the extent and influence of marine ice are much smaller but where it acts effectively on the continental margins.

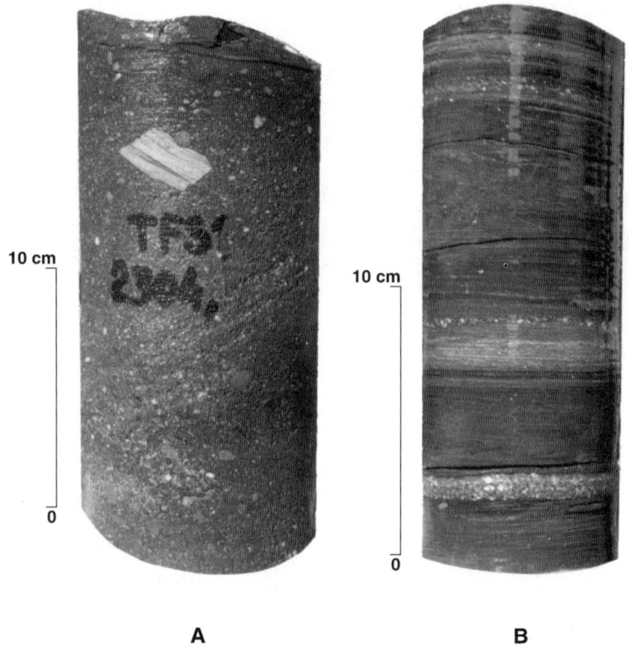

Fig. 3.168 Marine diamicton.
The two examples here show very poorly graded breccia (**A**) and clays of micro-breccia and quartz levels (**B**). Saharan late Ordovician (IFP photo).

Such environments have fluctuated in the Arctic for 2.5 My, but the north Atlantic probably began to cool 4.3 My ago with an increase in terrigenous sedimentation in the Atlantic and the appearance of boreal pollens, and perhaps even 8 My ago (erratic blocks in the Bafin sea). Deep ocean drilling in the Ocean Drilling Program (ODP) has shown that the Antarctic began glaciating in the east, before this extended to the west. The first evidence of glaciers goes back at least 30 My.
In ancient series, the glacial beds of the Gondwana Permian and Saharan Ordovician are better known for their continental facies, but marine deposits are also known there, as in the Upper Proterozoic.

3.3.6.5 Other Environments with Biochemical and Chemical Domination

A. Siliceous Oozes

Diatoms, silicoflagellates, planktonic algae and benthic sponges extract silica from the water to construct a skeleton in opal. Diatom oozes characterize the high latitudes, while radiolarian oozes are localized in equatorial regions (Fig. 3.121). The siliceous deposits correspond roughly to the high fertility zones such as the upwellings and coastal regions, the silica ring around the oceans (with high phosphate values) but, considering their size and light weight, siliceous particles are easily dispersed by the currents and are concentrated in regional or local depressions. A large share of opal is dissolved before reaching the bottom (especially in the upper water level), and it has been estimated that only 1 to 2% of the tests are conserved after burial. The sediments will yield radiolarites, diatomites, porcellanites, tripoli, cherts, jaspers, phthanites, and lydite after diagenetic evolution, with the transformation of opal A to opal CT and then to quartz.

Also, the zooplankton graze on the phytoplankton and a large share of the frustules are thus found in the fecal pellets.

B. Phosphate Environments

Although phosphorus does not have a high level of concentration in seawater where it is not very soluble and is thus largely absorbed, fixed, and concentrated by the plankton, phosphate accumulations do characterize certain environments, particularly on the edges of continental shelves in warm waters, at low latitudes, and in upwelling zones. Phosphatic oozes may reflect events of increased mortality with bioturbation favored by a low sedimentation rate and intense bacterial activity. Marine phosphorites are located at the boundary between hydrogenous and biogenous sediments, and vary in composition. Generally, for phosphatic deposits to form, the phosphorus first has to be captured and concentrated by biological activity. Then the phosphor-rich organic matter has to be trapped and accumulated in a geochemical trap (sediment in a reductive environment) where bacterial diagenesis will allow the formation of apatite. Finally, hydrodynamic reworking is needed, to grade and enrich the sediment in apatite.

Phosphorites found in ancient series are often associated with glauconite and are interpreted as characterizing basins that experience a rising sea level and high organic productivity.

BIBLIOGRAPHY

- ✧ Allen G (1991) Sedimentary processes and facies in the Gironde estuary: a model for microtidal estuarine systems. Smith DG, Reinson GE, Zaitlin BA et al. Clastic tidal sedimentology. Canadian society of petroleum geologists, Calgary, Memoir 16, pp 219-226.
- ✧ Allen JRL (1968) Current ripples: their relation to patterns of water and sediment motion. North-Holland, Amsterdam.
- ✦ Allen JRL (1982) Sedimentary structures: their character and physical basis, 1. Elsevier science, Developments in sedimentology 30A, Amsterdam.
- ✦ Allen JRL (1982) Sedimentary structures: their character and physical basis, 2. Elsevier science, Developments in sedimentology 30B, Amsterdam.

✦ Bagnold RA (1960) The physics of blown sand and desert dunes. Methuen, London.

✦ Bathurst RGC (1975) Carbonate sediments and their diagenesis. Elsevier science, Developments in sedimentology 12, Amsterdam.

✦ Bauling H (1956) Vocabulaire franco-anglo-allemand de géomorphologie. Paris, Les belles lettres, Publications de la Faculté des lettres de l'université de Strasbourg, fascicule 130.

✦ Berger WH (1974) Deep sea sedimentation. Burk CA, Drake CL The geology of continental margins. Springer, New York, pp 213-242.

✧ Berger WH, Winterer EL (1974) Plate stratigraphy and the fluctuating carbonate line. Hsu KJ, Jenkins HC Pelagic sediments: on land and under the sea. Symposium. September 25-26, 1973. Zurich, Suisse. Blackwell scientific, 1974. International association of sedimentologists, Oxford, special publication 1, pp 11-48.

✧ Berné S (1991) Architecture et dynamique des dunes tidales. Thèse de doctorat.

✦ Bertrand JP (1969) Cours de pétrographie appliquée à l'étude des problèmes pétroliers. Éditions Technip, Paris.

✦ Beuf S, Biju-Duval B, de Charpal O et al. (1971) Les grès du Paléozoique inférieur au Sahara : sédimentation et discontinuités évolution structurale d'un craton. Éditions Technip, Paris.

✧ Bigarella JJ (1972) Eolian environments: their characteristics, recognition and importance Rigby JK, Hamblin WK. Recognition of ancient sedimentary environments. Society of economic paleontologists and mineralogists, Tulsa, SEPM special publication 16, pp 12-62.

✦ Bignot G (1988) Micropaléontologie. Dunod, Géosciences, Paris.

✦ Biju-Duval B (1994) Océanologie. Dunod, Géosciences, Paris.

✧ Bordet P, Lapparent (de) A, Lucas G (1950) Les roches : étude pratique. Deyrolle, Paris.

✧ Bouma AH, Coleman JM (1985) Mississipi fan: leg 96 program and principal results. Bouma AH, Normark WR, Barnes NE. Submarine fans and related turbidite systems. Springer, New York, pp 247-252.

✦ Bouma AH, Normark WR, Barnes NE (1985) Submarine fans and related turbidite systems. Springer, New York.

✦ Caron JM, Gauthier A, Schaaf A et al. (1989) Comprendre et enseigner la planète Terre. Ophrys, Paris.

✦ Chamley H (1989) Clay sedimentology. Springer verlag, London.

✦ Chamley H (1987) Sédimentologie. Dunod, Paris.

✦ CSRPPGN – Comité des techniciens (1986) Corps sédimentaires : exemples sismiques et diagraphiques. Éditions Technip, Paris.

✦ Cotillon P (1988) Stratigraphie. Dunod, Géosciences, Paris.

✦ Davis WM (1899) The peneplain. American geologist, 23, pp 207-239.

✧ Devillers C, Chaline J (1989) L'histoire de la vie. Dunod, Paris.

✦ Dunham RJ (1962) Classification of carbonate rocks according to depositional texture. Ham WE Classification of carbonate rocks. American association of petroleum geologists, Tulsa, AAPG memoir 1, pp 108-112.

✧ Embry AF, Klovan JE (1971) A late Permian reef tract on northeastern Banks Island, Northwest territories. Bulletin of Canadian petroleum geology 19, pp 730-781.

✦ Enay R (1990) Paléontologie des invertébrés. Dunod, Géosciences, Paris.

✧ Eschard R, Lemouzy P, Desaubliaux G et al. (1998) Combining sequences stratigraphy, geostatistical simulations, and production data for modeling a fluvial reservoir in the Chaunoy field (Triassic, France). AAPG bulletin 82, 4, pp 545-568.

✦ Fairbridge RW (1966) The encyclopedia of oceanography. Reinhold, Encyclopedia of earth sciences series 1, New York.

✦ Folk RL (1962) Spectral subdivision of limestone types. Ham WE Classification of carbonate rocks. American association of petroleum geologists, Tulsa, AAPG memoir 1, pp 62-84.

✦ Gignoux M (1950) Géologie stratigraphique. Masson, Paris.

✧ Ginsburg RN (1957) Early diagenesis and lithification of shallow water carbonate sediments in south Florida. Leblanc RJ, Breeding JC. Regional aspects of carbonate deposition. Society of economic paleontologists and mineralogists, Tulsa, SEPM, special publication 5, pp 80-99.

✧ Gomez de Souza O (1997) Stratigraphie séquentielle et modélisation probabiliste des réservoirs d'un cône sous-marin profond (champ de Namorado, Brésil). Thèse de doctorat, Université Pierre et Marie Curie, Paris.

✦ Guillemot J (1986) Éléments de géologie. Éditions Technip, Paris.

✧ Heezen BC, Hollister CD (1971) The face of the deep. Oxford university press, New York.

✦ Hollister CD (1993) The concept of deep sea contourites. Sedimentary geology 82, pp 5-11.

✧ Huc A (1995) Géochimie organique. Cours ENSPM. Unpublished.

✦ Jervey MT (1988) Quantitative geological modeling of siliciclastic rock sequences and their seismic expression. Lidz BH. Sea level changes: an integrated apporach. Society of economic paleontologists and mineraligists, Tulsa, SEPM, special publication 42, pp 47-70.

✦ Kendall AC (1992) Evaporites. Walker RG, James NP Facies models: response to sea level change. Geological association of Canada, Toronto, pp 375-409.

✦ Leeder MR (1982) Sedimentology: Process and product. Georges Allen and Unwin, London.

✦ Lees A (1975) Possible influences of salinity and temperature on modern shelf carbonate sedimentation. Marine geology 19, pp 159-198.

✧ Lowe DR (1982) Sediment gravity flows: II. Depositional models with special reference to the deposits of high density turbiditiy currents. Journal of sedimentary petrology 52, pp 279-297.

✦ Lucas J, Prévost L (1975) Les marges continentales, pièges géochimiques : l'exemple de la marge continentale de l'Afrique à la limite Crétacé-Tertiaire. Bulletin de la société géologique de France 7, 17, pp 496-501.

✧ Mainguet M, Collot Y (1974) Air photo study of typology and interrelations between the texture and structure of dune patterns in the Fadin Bilma erg, Sahara. Zeitschrift fur geomorphologie 20, pp 62-69.

✦ Marsily (de) G (1981) Hydrogéologie quantitative. Masson, Sciences de la terre, Paris.

✧ McKee ED (1965) Experiments on ripple lamination. Middleton GV Primary sedimentary structures and their hydrodynamic interpretation. Society of economic paleontologists and mineralogists, Tulsa, SEPM special publication 12, pp 66-83.

✧ Miall AD (1992) Alluvial deposits. Walker RG, James NP. Facies models: response to sea level change. Geological association of Canada, Toronto, pp 119-142.

✦ Miall AD (1996) The geology of fluvial deposits: sedimentary facies, basin analysis and petroleum geology. Springer, Berlin.

✦ Middleton MA, Hampton GV (1976) Subaqueous sediment transport and deposition by sediment gravity flows. Stanley DJ, Swift DJ. Marine sediment transport and environmental management. Wiley, New York, pp 197-218.

✦ Millot G (1964) Géologie des argiles : altérations, sédimentologies, géochimie. Masson, Paris.

✦ Minster JF (1994) Les océans. Flammarion, Paris.

✦ Mutti E (1985) Turbidites systems and their relations to depositional sequences. Zuffa GG Provenance of arenites. Riedel, Dordrecht, pp 65-93.

✦ Normark WR (1970) Growth pattern of deep-sea fans. AAPG bulletin 54, pp 2170-2195.

✧ Odin G (1986) Les formations permiennes, Autunien supérieur à Thuringien, du bassin de Lodève (Hérault, France) : stratigraphie, minéralogie, paléoenvironnements, corrélations. Thèse de doctorat.

✧ Pedro G (1979) Caractérisation générale des processus de l'altération hydrolytique. Bulletin de l'Association française pour l'étude du sol 2-3, pp 93-105.

✧ Penck W (1924) Die morpholgische analyse: ein kepitel der physikatischen geologie. Geographische abhandlungen 2, p 2.

✦ Purser BH (1980) Sédimentation et diagenèse des carbonates nétritiques récents. Tome 1 : Les éléments de la sédimentation et de la diagenèse. Éditions Technip, Paris.

✦ Purser BH (1983) Sédimentation et diagenèse des carbonates néritiques récents. Tome 2 : Les domaines de sédimentation carbonatée néritiques récents : application à l'interprétation des calcaires anciens. Éditions Technip, Paris.

✧ Reineck HE, Singh IB (1980) Depositional sedimentary environments. Springer verlag, Berlin.

✧ Richter W, Schneider H, Wager R (1950) Die saaleizeitliche strand zone von itterbeck velsen (Emsland). Zeitschrift der deutschen geologischen gesellschaft, 11, pp 60-75.

✧ Rognon P (1989) Biographie d'un désert. Plon, Paris.

✦ Sarg JF (1988) Carbonate sequence stratigraphy. Lidz BH. Sea-level changes: an integrated approach. Society of economic paleontologists and mineralogists, Tulsa, SEPM, special publication 42, pp 155-182.

✧ Schumm SA (1972) River morphology. Dowden Hutchinson and Ross, Benchmark paper in geology, Stroudsburg.

✦ Schumm SA (1993) River response to baselevel change, implications for sequence stratigraphy. Journal of geology 101, 2, pp 279-294.

✧ Schumm SA (1968) Speculations concerning paleohydrologic controls on terrestrial sedimentation. Bulletin of the geological society of America 79, pp 1573-1588.

✦ Seibold E, Berger WH (1982) The sea floor: an introduction to marine geology. Springer verlag, Berlin.

✦ Selley RC (1970) Ancient sedimentary environments and their sub-surface diagenesis. Chapman and Hall, London.

✧ Shu L, Finlayson B (1993) Flood management on the lower yellow river: hydrogeological and geomorphological perspectives. Journal of sedimentary geology 85, pp 285-296.

✦ Stow DAV, Piper DJW (1984) Fine-grained sediments: deep-water processes and facies. Blackwell scientific, Geological society, special publication 15, Oxford.

✧ Sverdrup HU, Johnson MW, Fleming RH (1942) The ocean, their physics, chemistry and general biology. Prentice Hall, New York.

- ✦ Tucker ME, Wright VP (1990) Carbonate sedimentology. Blackwell scientific, Oxford.
- ✦ Vail PR, Mitchum RM, Todd RG et al. (1977) Seismic stratigraphy and global changes of sea level. Payton CE. Seismic stratigraphy: applications to hydrocarbon exploration. American association of petroleum geologists, Tulsa, AAPG, memoir 26, pp 49-212.
- ✧ Wahraftig C (1965) Physiographic divisions of Alaska. USA Geological survey, professional paper 482.
- ✧ Walker RG (1978) Deep water sandstone facies and ancient submarine fans: models for exploration of stratigraphic traps. AAPG bulletin 62, pp 932-966.
- ✦ Walker RG (1984) Facies models. Geological association of Canada, Geoscience Canada, reprint series 1, Toronto.
- ✦ Walker RG, James NP (1992) Facies models: response to sea level change. Geological association of Canada, Toronto.
- ✧ Wells NA, Richards SS, Peng S et al. (1993) Fluvial processes and recumbently folded crossbeds in the Pensylvanian Sharon conglomerate in Summit county, Ohio, USA. Sedimentary geology 85, 1-4, pp 63-83.

✦ *Books or articles of general interest.*

✧ *Source of one of the figures used, cited in the figure caption.*

Chapter 4

TIME EVOLUTION: SEDIMENTARY SEQUENCES, STRATIGRAPHY

This book began by defining geology as a science in which one of the objectives is to reconstruct Earth history. The importance of time variations was stressed, as was the need to date events. Evolution on long time scales of up to several tens of millions of years also underlay all the discussion of sedimentary basin classification and formation modes in Chapter 2. The third chapter made constant reference to the inherent fluctuations in sedimentation processes and environments, with some of these fluctuations being cyclic, others nonsteady, punctuated with special events, eventually changing the original conditions.

Although the mechanisms of the Present have constantly served as reference, it has repeatedly been pointed out that environments are unstable in time. Climatic, biological, topographical, structural, physical and chemical environments have changed greatly and over widely varied time scales. Geologists long used to believe in the somewhat stationary idea of Actualism, and it is surely true that the **Present** can be used as a useful **key to the past**. But we also recognize today that the **Past was different from the Present** and **that** the internal and external driving causes examined in Chapter 1 vary in time.

With this in mind, the discussion now turns to **stratigraphy**, which is basically a description of successive strata. In line with the new concepts of mobilism and dynamics, stratigraphy can be viewed as "the study of the space and time arrangement of geological formations and the events they signify, for the purpose of reconstructing Earth history and its various states in the course of time".[1]

We will be dealing with the main time variations and their causes in succession, with the major stratigraphic elements, dating methods, and geological time scales, recent advances in sequential and genetic analysis used in petroleum geology, and lastly some paleogeographic reconstructions on all scales, which is the end purpose of the combined work of stratigraphers and sedimentologists.

1. See the survey document of J. Rey et al., 1997.

4.1 TIME INSTABILITY OF ENVIRONMENTS

Concretely, a sequence of geological beds (Fig. 4.1) is generally analyzed as a logical sequence of different deposits, each being evidence of a particular environment. These sequences are observed in the field, or by drill-core samples, or even by deep seismic sections. The various beds and facies each exhibit their own lithology, and their petrography and geometric relations will be analyzed at different scales. These structures and intrinsic properties constitute the **archives of the past** and are indicators of time variations.

On the basis of this analysis, stratigraphy follows the simple **principle of superposition**, in which the lowest sedimentary beds are older that the ones overlying them (Fig. 4.1). Rare exceptions exist, as in nested fluvial terraces, volcanic intrusions and eruptive veins, which appear after the sedimentary pile in which they are observed (Fig. 4.2).

4.1.1 Major Variations

Shoreline. For the geologist, this is the boundary between the continental and marine domains. At times, it is well marked with a sharp cliff, and sometimes it is fuzzy, with a delta or coastal flood plain. There is always a fixed position at a given moment, but varia-

Fig. 4.1A The geological section: expression of a sequence of beds of different ages. Field example of a sequence of turbidite beds expressed in section B on the next page (IFP photo).

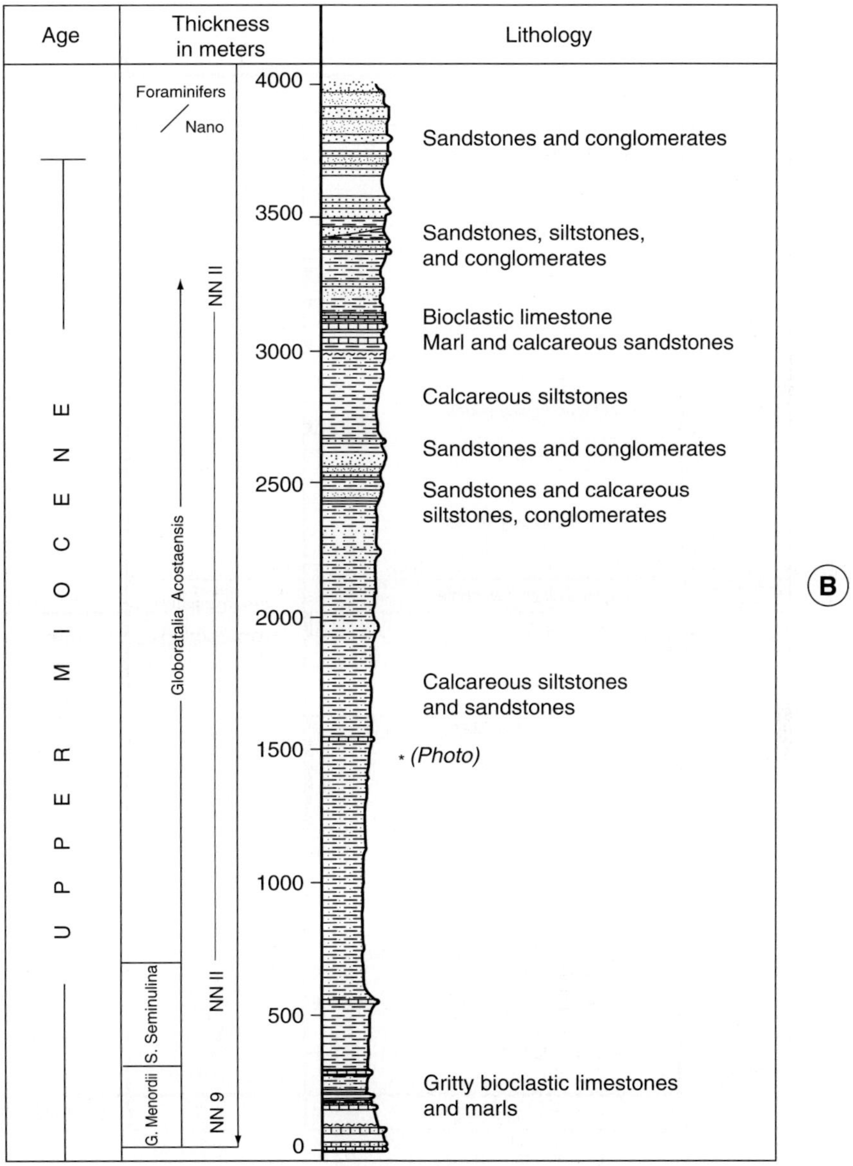

Fig. 4.1B The geological section: expression of a sequence of beds of different ages. Detailed stratigraphic log of a section of the Late Miocene (Hispaniola, Greater Antilles) with the sequence of zones defined by micro- and nanofossils.

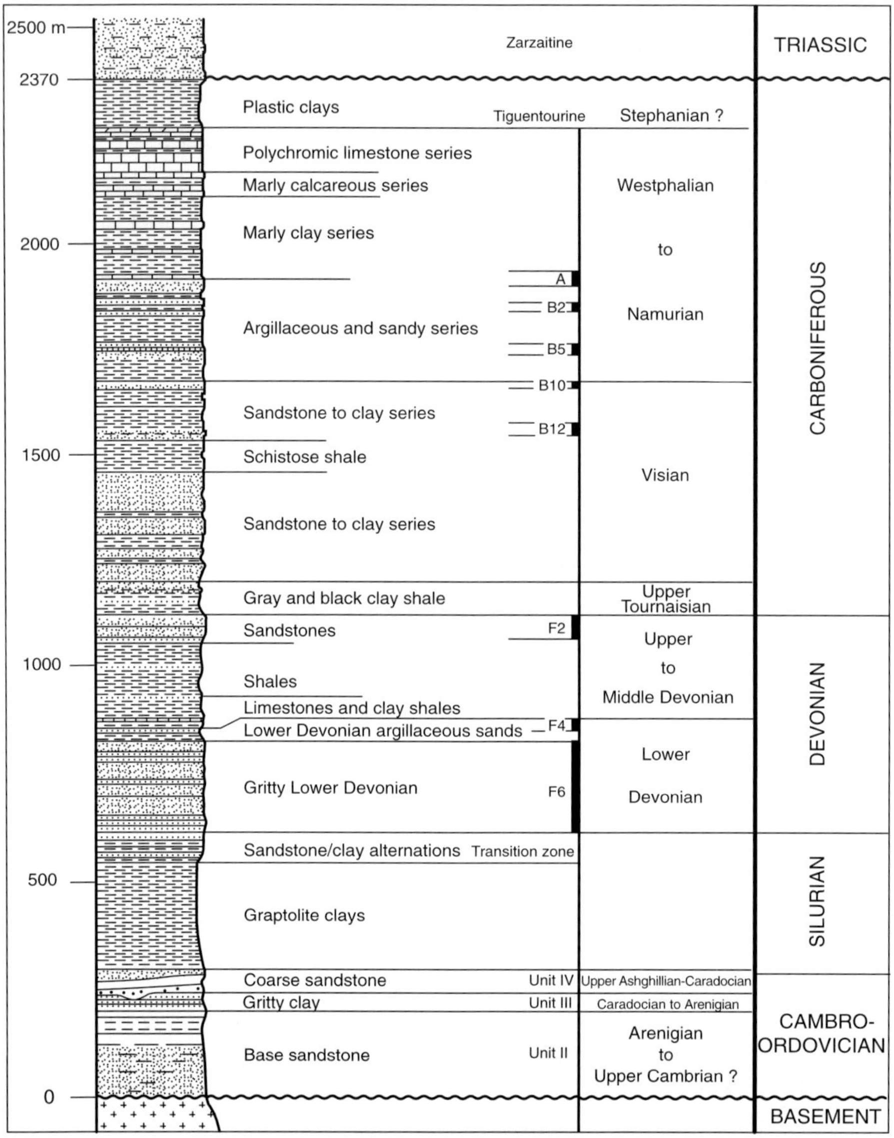

C

Fig. 4.1C The geological section: expression of a sequence of beds of different ages. A Simplified stratigraphic column of the Saharan Paleozoic.

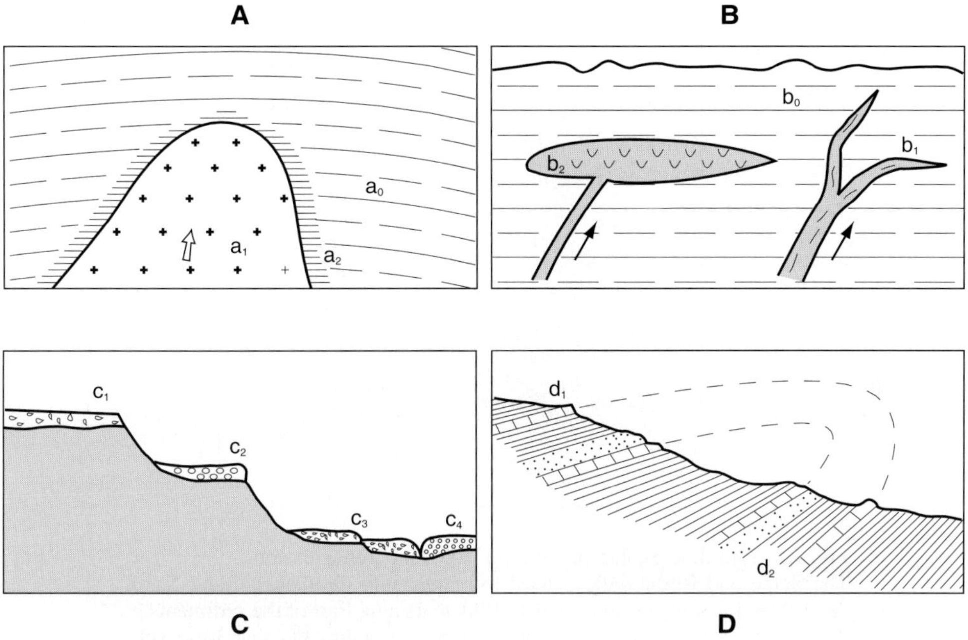

Fig. 4.2 Exceptions to the principle of stratigraphic superposition.
A. Granitic batholite (a_1) with its halo due to contact metamorphism (a_2). This occurs at depth under much older beds (a_0). **B.** Intrusions of dikes (b_1) and sills (b_2) of volcanic rock in a sedimentary series (b_0) older than the volcanic material. **C.** Nested fluvial terraces, in which the most recent (c_4) lie lowest, where the river flowed last, and the oldest (c_1) are the highest (laid before the erosion became active). The terraces between the two mark the sequence of stages. **D.** Normal sequence (d_1) and an inverted sequence (d_2) due to a post-depositional deformation.

tions are at times observable on the scale of a human life or human history: erosion cutting back a range of cliffs, a tsunami generating a sudden marine flood over a flood plain, isostatic readjustment causing beaches to rise, a sea-level variation causing the emergence or submersion of vast territories (Fig. 4.3).

Sealevel variation, the general eustasy of glacial or tectonic origin is doubtless the major cause of variations in the geological past, with various orders of magnitude in amplitude, duration, and rate of variation. Relative changes in sea level (relative in that several driving forces act in combination) cause a change in the **accommodation power** (Fig. 4.4), which will be discussed later in this chapter.

The **Earth's temperature** and climates (of both oceans and continents) have also fluctuated in time. Possible **greenhouse effects** due to variations in the atmosphere's CO_2 and

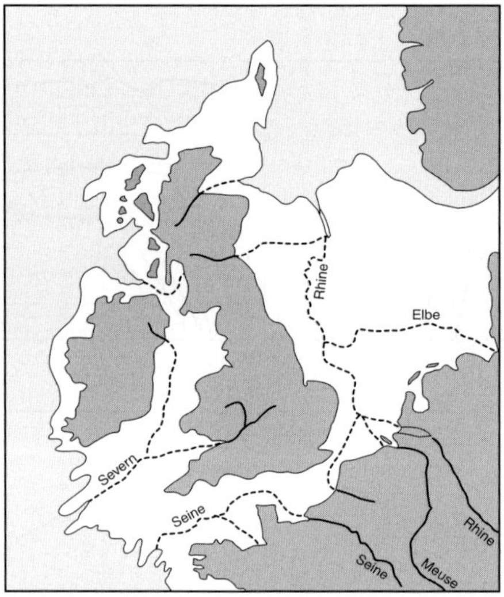

Fig. 4.3 Shoreline evolution with the Holocene transgression.
The Seine and Rhine valleys used to extend into the Channel and North Sea before the last glacial retreat 10 000 years ago. Part of the sediments in these regions corresponds to relic continental sediments that were subjected to marine action during the transgression and are still subjected to it today (from various authors).

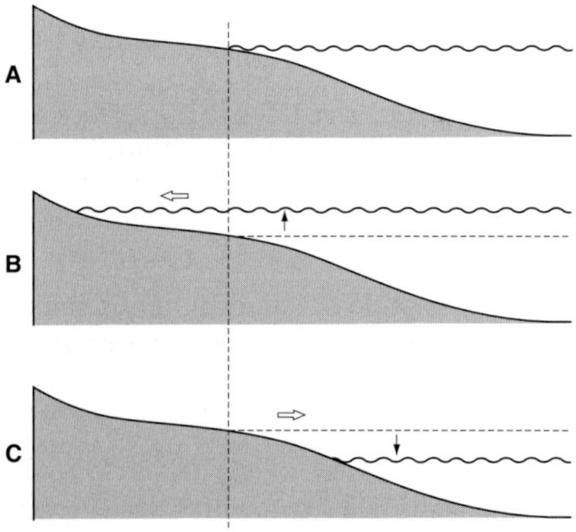

Fig. 4.4 Eustasy: global variation in average sea level.
A. Average situation. **B.** High sea level, where the sea transgresses and floods a large portion of emerged lands. **C.** Low sea level, where the sea regresses and the area of emerged land increases, changing river equilibrium profiles in the process.

methane are being discussed today. Variations are known to have existed in human history, such as the Little Ice Age, and the geological past contains such famous examples as the Permian and Ordovician glaciations. Minor fluctuations also occur within the major cycles, and the effects on the sea level and shoreline are obvious. Changes also occur in weathering and continental leaching conditions, though these are perhaps less spectacular. Various tools such as fauna types or δO^{18} are used to reconstruct these variations.

Warm periods with greenhouse effect are commonly opposed to "ice house" periods when the poles are covered with ice.

Ocean chemistry. It is recognized today that large chemical variations in the global ocean have occurred in Earth history, especially in the early phases (Fig. 4.5). The present

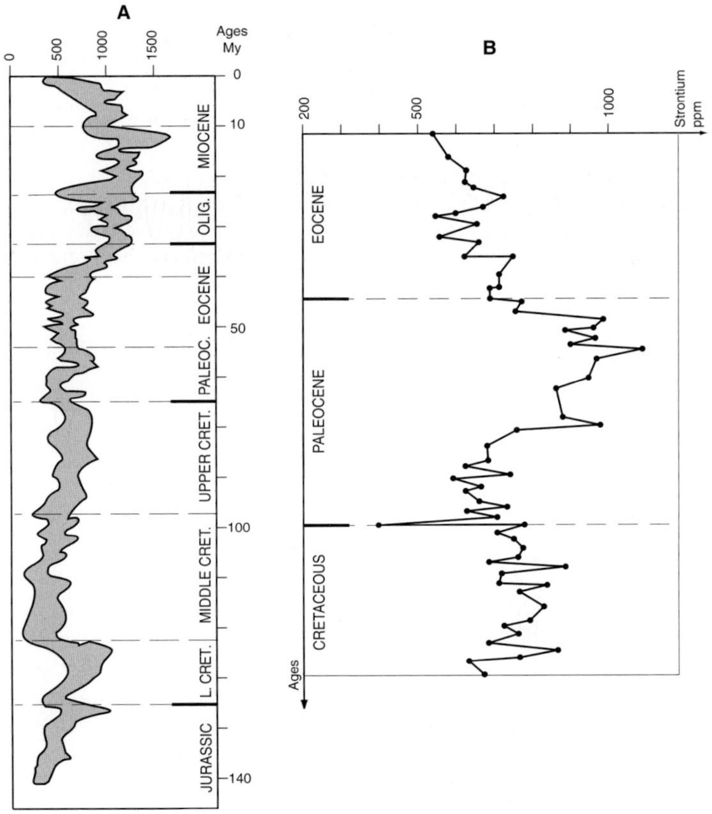

Fig. 4.5 Variation of ocean chemistry over geological time.
A. Variation of strontium content in pelagic carbonates over the past 140 My (from M. Renard, 1985). General curve. **B.** Detailed variation in strontium content in pelagic carbonates from the Gubbio section in Italy (from M. Renard, 1985).

example of hydrothermal plumes on ocean ridges indicates how manganese, helium, or CH_4 anomalies can "contaminate" the world ocean. Tests are being conducted today to develop the use of chemical methods in stratigraphy. This is called **chemostratigraphy**. Variations occur in the major elements, minor oligoelements, rare earth contents, and oxidation reduction (redux) potential.

General atmospheric and oceanic circulation. The Indian monsoon or Gulf Stream system, for example, are not perennial systems even if they were initiated tens of millions of years ago, nor is the general deepwater circulation scheme, which depends on the climate, tectonics, threshold closure, and other factors. These variations will further entail anoxic crises, for example.

Evolution of floral and faunal groups. This is the basis of paleontology and stratigraphy (Fig. 4.6). We recall the example of the end of the Permian where all Fusiline genera

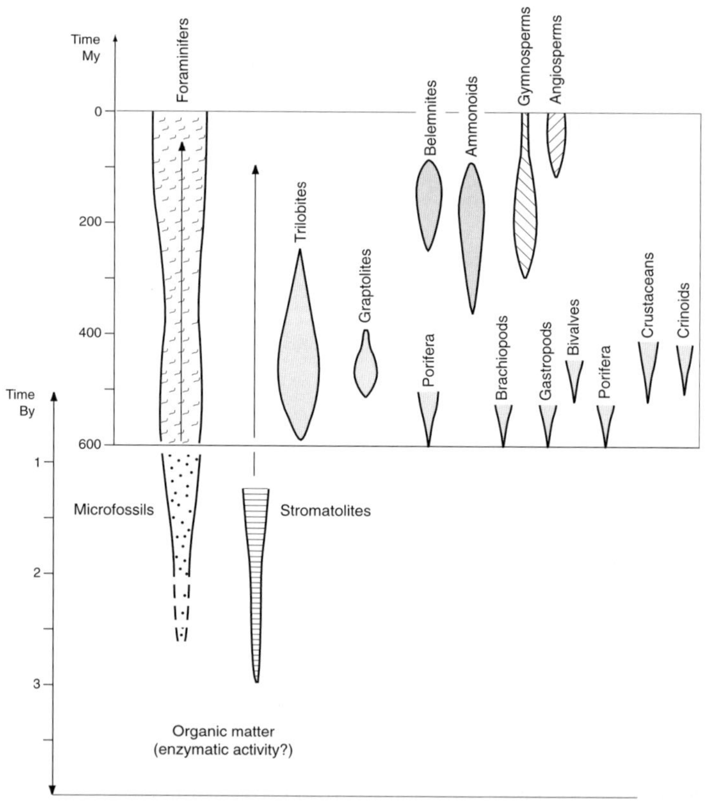

Fig. 4.6 Evolution of faunal and floral groups. A few examples of large groups, with the noteworthy appearance of most of the large branches 600 My ago.

disappeared along with very many brachiopods and whole families of crinoids and corals. The extinction of dinosaurs at the end of the Cretaceous, 65 million years ago, is better known; but the planet's history is populated with more or less sudden crises of this kind.

The **ridge expansion rate** has varied over the course of geological time. It is felt that the ocean floor creation rate can have an effect on the general eustatic variation. The **absolute displacement of the plates** is thus largely variable, and it has already been said how this could have played importantly in the geological past. It can cause a gradual displacement of the sedimentation areas, which are largely controlled by the climate (Fig. 4.7).

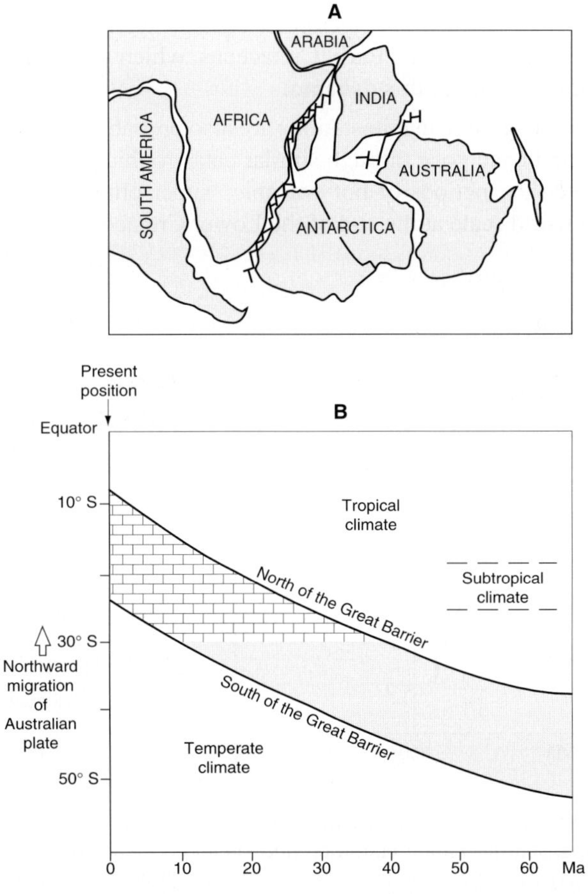

Fig. 4.7 Displacement of lithospheric plates in the course of time.
A. Pangea assemblage scheme prior to breakup into independent plates.
B. The northward migration of Australia in the course of the Tertiary allowed carbonaceous facies to develop starting in the Oligocene (from Davies et al., 1989).

Tectonics due to plate movements. Several broad cycles have been determined. In these cycles, special conditions are associated with each event (rapid subsidence, sudden uplifting, basin isolation) and are recorded in the sedimentation. Thresholds and closures may appear or disappear between basins. The active tectonics of the initial rifts or that of the subduction zones generally change the sedimentation process. Variations in rift facies (alluvial cones, lacustrine deposits, transgressive sands, tectonic breccia, confined environments) are spectacular examples of this, and others can be found along accretion wedges, with small turbiditic basins incorporated in successive tectonic peel thrusts overlain with pelagic draping, resedimentation, and a sudden switch above the CCD.

Earth magnetism. Intensity variations and pole reversals are used in **magnetostratigraphy** to determine ocean ages and global tectonics. Past fluctuations have been large enough to introduce anomalies like that of the Early Cretaceous, which was doubtless due to sudden changes in the operation of the Earth's dynamo.

Magmatism. Deep magmatic manifestations are also probably related to the operation of the Earth's dynamo, and have gone through similar catastrophic periods. Recently, a tie has been proposed between the periods of hot volcanic "super plumes" and the generation of hydrocarbons on the world scale at the end of the Lower Cretaceous (Fig. 4.8).

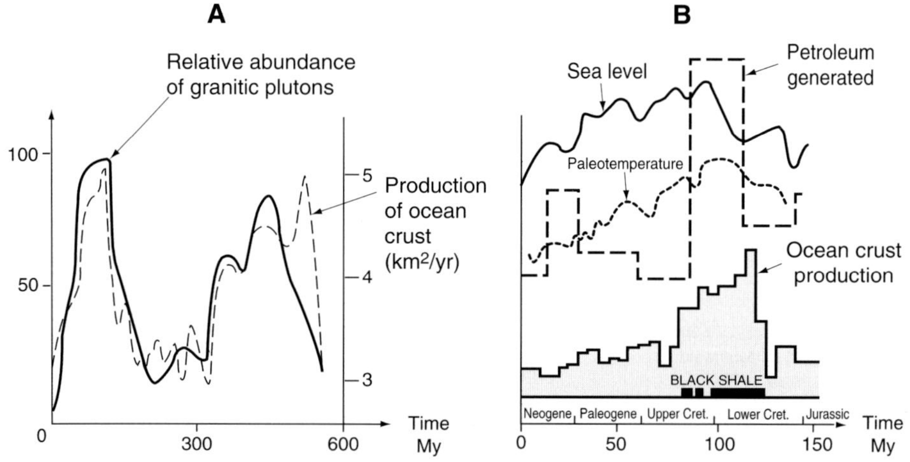

Fig. 4.8 Fluctuations in magmatism in the course of geological time.
A. Comparison of granitic pluton production and oceanic crust during the last 600 My (from L. Hardie, 1996). Note the good agreement of the curves, except for the oldest periods. **B.** Oceanic crust production during the last 150 My has probably controlled the variation in ocean chemistry, temperatures, and eustasy, as well as variations in the production of organic matter and sulfates. The chart represents the great "black shale" periods and the quantities of petroleum generated (according to Larson, 1991).

The **geoid**. The Earth's surface deforms perpetually as the deep masses vary, creating relative sea-level variations that differ by region.

Changes in the relative level can also be generated by climatic fluctuations, as in the case of the El Niño phenomenon.

The Earth's **orbital parameters** are not stable either. The Milankovitch cycles are widely used today to explain lithological variations. These astronomical variations occur at periodicities of 20-21 ky, 40 ky, 100 ky, and 400 ky (Fig. 4.9) due to changes in the Earth's orbital eccentricity, obliquity, and precession. These cycles combine to create other cycles at complex, composite frequencies. Much work is being done today to show the importance of these orbital cycles in the history of the Mesozoic with its limestone-marl alternations, for example, and the Paleozoic.

It can thus be seen that particular internal or external **events** cause variations that are superimposed on the Earth's **gradual evolution** starting with a primitive atmosphere and hydrosphere that were different from what we observe today.

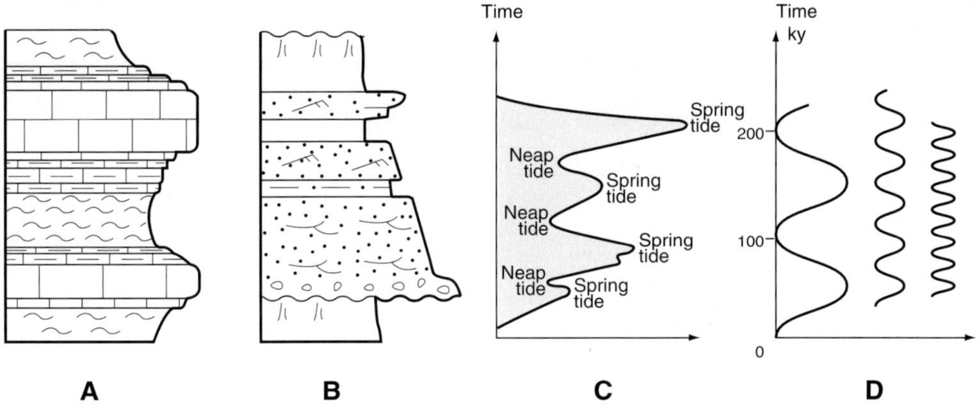

Fig. 4.9 Facies sequences ordered in rhythms and cycles, Milankovitch cycles.
A. Successive cycles of limestone, marly limestone, marl. **B.** Argillaceous sandstone sequence (genetic unit) between two erosion surfaces. **C.** Tidal cycles between spring and neap tides. **D.** Milankovitch cycles: 100 ky, 40 ky, 21 ky. Response to variations in orbital parameters (eccentricity, obliquity, equinoctial precession).

4.1.2 Cyclic Processes and Events

Insolation and the tidal cycle introduce variations on the daily scale that affect such things as photosynthesis, the production of calcite in red algae, or the draining of tidal channels. On the semi-monthly scale, it is the spring-neap tide cycle that will be evidenced, and on the seasonal scale there are other fluctuations such as the monsoon reversal or retreating ice. On a far grander scale still, the forcing by orbital parameters according to the Milankovitch cycle has also been mentioned (Fig. 4.9).

Here, two other types of variations require emphasis: those that result from natural cycles where the climate plays an important role, and those that could be called events, such as a volcanic eruption or earthquake causing a tsunami, where the internal geodynamics is predominant, or meteor impacts which are of external origin. The consequences of these events are sudden, resulting in erosion, material returned to suspension, arrested reef growth, or a sudden drop in primary production.

Geodynamic variations can be aperiodic or periodic (what drives the periodicity is not well known—the geodynamo? mantle cycles?). A variation is simply said to be cyclic when it causes a repeated sequence, and acyclic when it is more of a singular event.

The tectonic cycle is well situated in the former of these categories, such as in the Wilson cycle where a basin rifts and opens by an extension and fracturation mechanism that is followed by subsidence and thermal cooling (Fig. 4.10). Curves and laws can be determined for this on the scale of a basin.

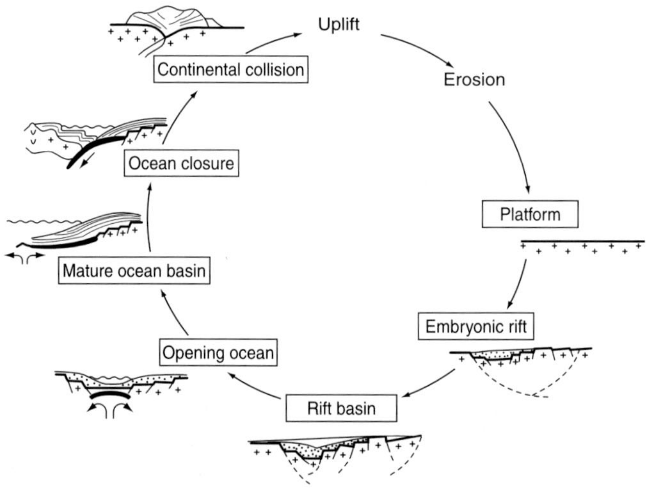

Fig. 4.10 The Wilson cycle: a logical sequence of events in the life of a sedimentary basin.

An example of an acyclic event would be the sudden closure of a threshold isolating a basin (such as the saliferous Mediterranean basin in the Messinian), a paroxysmal volcanic crisis (such as the event in the Middle Cretaceous), or the sudden extinction of fauna (end of the Cretaceous).

4.2 STRATIGRAPHIC ELEMENTS, DATING, AND TIME SCALES

4.2.1 Definitions

A stratum is a sedimentary set of a certain lithology that is different from what can be found at its base, wall, or roof (Fig. 4.11). The sedimentary stratum is generally termed a bed, but also a level, assise, or horizon, all of which mean about the same thing. The outcrop is often called a bank, especially when the rock is hard and forms a surface relief (Fig. 4.11).

A bank or set of banks can outcrop in a massive set that geologists call a bar.

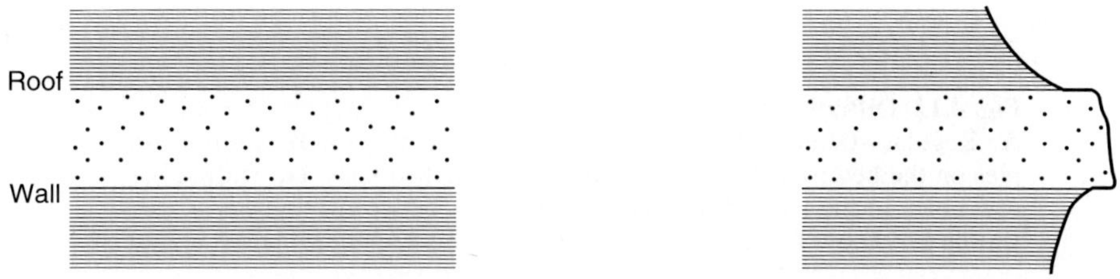

Fig. 4.11 Stratum, bed, bank. These different terms have equivalent meanings. The sand bed here (dotted) lies between two clay beds (straight lines).

The wall or roof of a bed are its **stratal surfaces** where particular **sedimentary figures** are sometimes observed: current ripples, desiccation cracks, traces of current or objects rolled over the bottom. Depending on the type of deposit, some of these figures characterize the bank bases, while others mark the transition to a new bed (Fig. 4.12), a transition that is sometimes sudden, and sometimes gradual.

Sometimes, one bed overlies another very similar one with only a **bedding joint**, in this case called a **dry joint**, to separate them. The two banks are then amalgamated into one. If the bedding joint is a millimetric bed of clay, for example, it is an **argillaceous joint**. A **diastem** is an observed interruption in the active sedimentation process.

Sedimentary figures and structures are also observed within a given stratum. A **folium** or **lamina** designates a millimetric to centimetric unit defined as **bedding** that can be parallel to the wall or roof or to **oblique stratal joints**, which is termed oblique stratification. Different parts of these oblique S-shaped folia are sometimes termed **topset**, **frontset**, or **bottomset**, which is evidence of the progradation of the hydraulic dune. These stratifications may differ considerably in different deposits (Fig. 4.13). This concept means that a laterally prograding deposit of variable age in space all belongs to the same bed (Fig. 4.13).

Lamina or **lamination** is a term used for elementary folia of millimetric thickness.

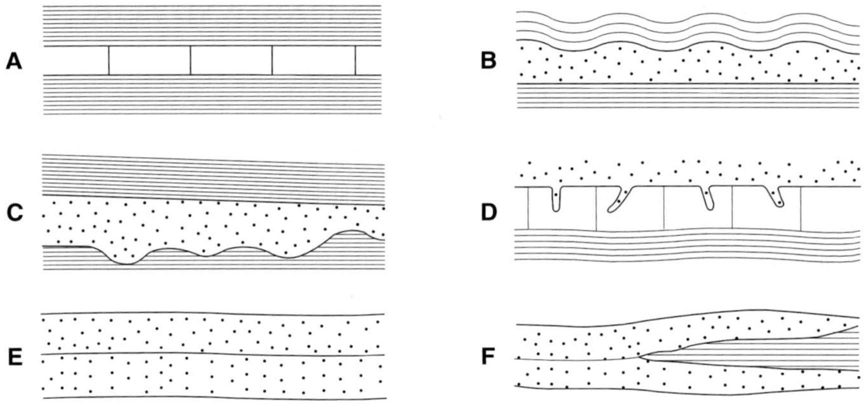

Fig. 4.12 Different stratal surfaces.
A. Regular stratal surface between different lithologies. **B.** Current ripples on the bed roof. **C.** Erosion surface on the bed base. **D.** Stuffed burrow holes in the roof of the bed. **E.** Dry joint between two beds of very similar lithology. **F.** Amalgamated banks.

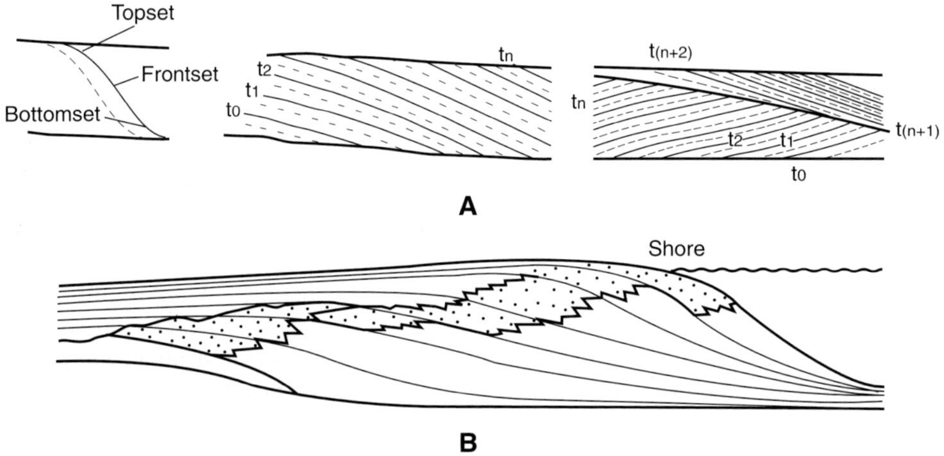

Fig. 4.13 Oblique stratification: lateral progradation in time.
A. Banks of successive regular or nested arenaceous avalanches with surface discontinuity. The lateral migration of the folia defines a prograding deposit of decreasing age going downstream. These elementary examples represent very brief durations (flood, tide, storm). **B.** On the scale of a geological formation, a prograding system (the example here is a delta) can represent evolution over considerable geological time scales of up to several tens of millions of years.

The original arrangement of flat or oblique lamina and folia can be disturbed after deposit, and sometimes completely so, by bioturbation, microslumping or water escape (Fig. 4.14). The bedding can be entirely obliterated and results as a chaotic set instead of a sequence of banks, as was described in the examples of resedimentation. In other cases, the horizontal bedding pattern is erased by all of the later processes of **diagenesis** (see Chapter 5). This is common in biochemical deposits (limestone, salts) which are often described as **massifs** with no mention of beds. This can also be true of built-up carbonate formations.

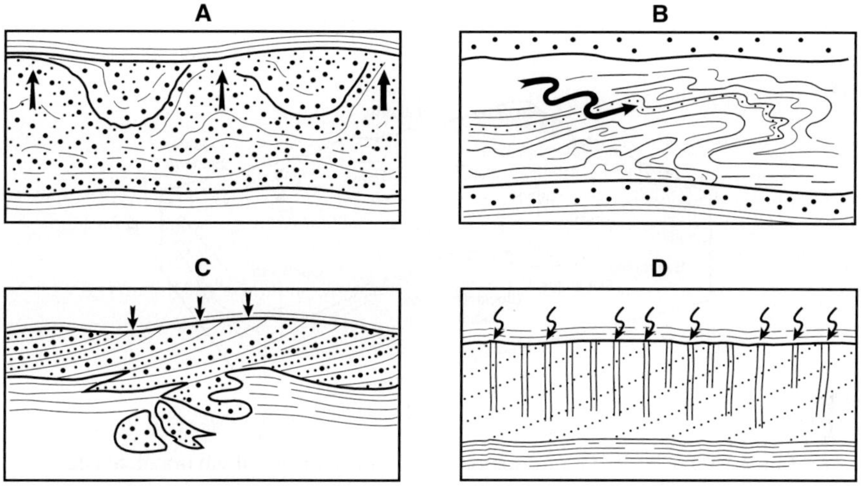

Fig. 4.14 Precocious disturbances of the structure of a sediment.
A. Water escape. **B.** Slumping. **C.** Overloading. **D.** Bioturbation.

In conclusion, a stratum, bed, or bank generally designates a precise **lithology** different from the beds above and below it.

The geometric sequence of several beds of different lithologies then constitutes the **stratification** or **bedding** which, as will be seen later on, is the subject of one of the components of stratigraphy called **lithostratigraphy**.

The same bed can often exhibit **vertical variations**, such as the vertical grading of turbidites or folia of different fluvial sand grain sizes, and **lateral variations** (Fig. 4.15). Any lateral variation is referred to as a **facies change**. If the bed is very continuous laterally on the scale of the basin, then the principle of **bed continuity** can be used to establish **correlations** from one outcropping or drillhole to another.

Lastly, the initial arrangement of beds can also be greatly disturbed by later tectonics, and way-up criteria are then used to recognize their walls and roof (Fig. 4.16).

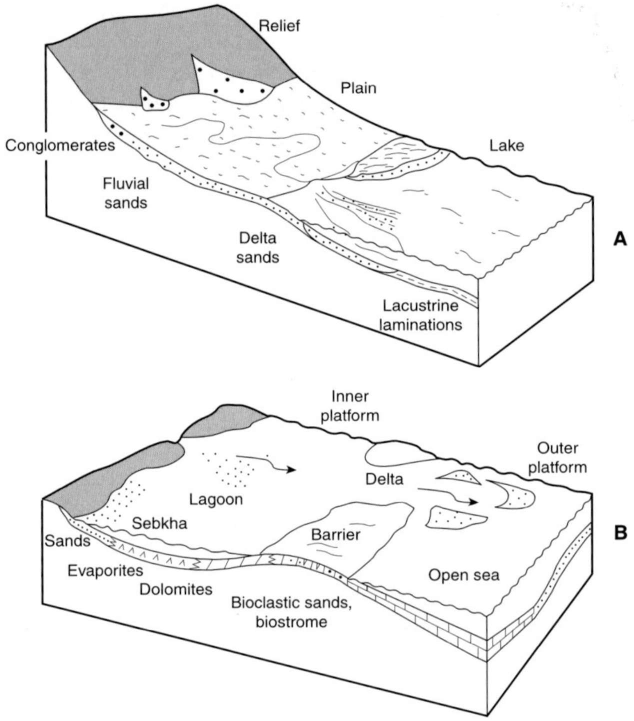

Fig. 4.15 Facies variations: expression of the diversity of environments in a given period.
A. Continental environment with transitions between different types of fluvial and lacustrine deposits. **B.** Coastal domain with more or less restricted tidal domain, shallow water or barriers, and open sea.

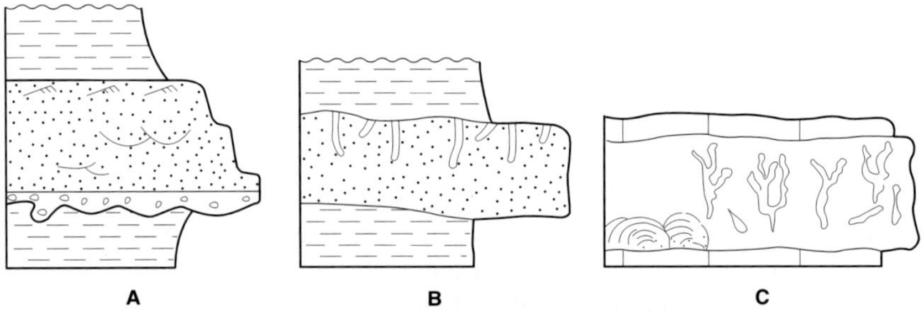

Fig. 4.16 Examples of way-up criteria at the wall or roof of a bed.
A. Basal erosion and associated figures. **B.** Bioturbation affecting the upper part of the bank. **C.** Builder organisms in life position.

4.2.2 Thicknesses and Rates of Deposit, Idea of Time and Sedimentary Cycle

Stratum thicknesses vary from a few millimeters to several tens of meters, but a large thickness is probably deceptive because of our inadequate resolving power, which depends on the observation tool. An internal bedding of lamina and dry joints may be detected under a magnifying glass or a microscope, but not by field observation. Logging may also bring out a special physical or chemical feature that will be designated as a level or particular reference.

Instantaneous well logs recorded during the drilling, or offline logs developed afterward, are physical and chemical measures of the petrographic and petrophysical characteristics of the various beds and the fluids they contain. Ordinary measurements like the natural gamma ray log, resistivity log, porosity and density logs are now complemented by high-resolution (of the order of a centimeter) logs that define the textural or geochemical parameters better, and obtain precise images.[1]

The example of the horizons identified in seismics gives a better idea of this resolving power, but it should be said that seismic horizons are not individual beds but sets of beds, or a major discontinuity (Fig. 4.17).

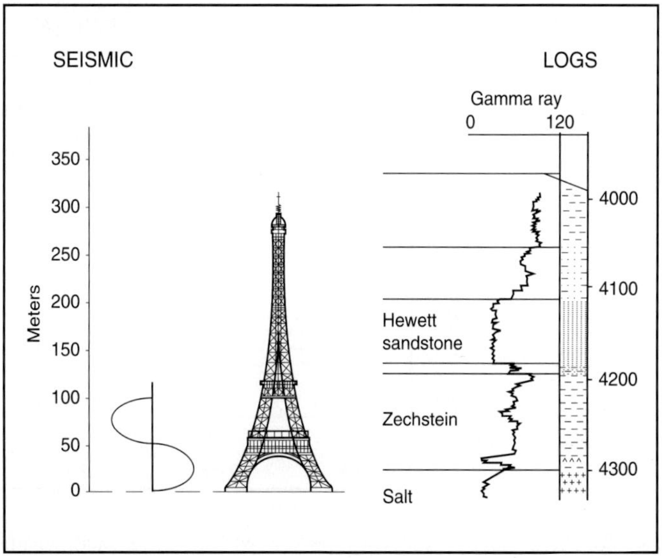

Fig. 4.17A Stratigraphic resolving power with logging and seismics. General scheme: the seismic signal has much less resolution than any log (from Flèche, 1990).

1. S. Boyer and J. L. Mari, Sismique et diagraphies, Éditions Technip, 1994.

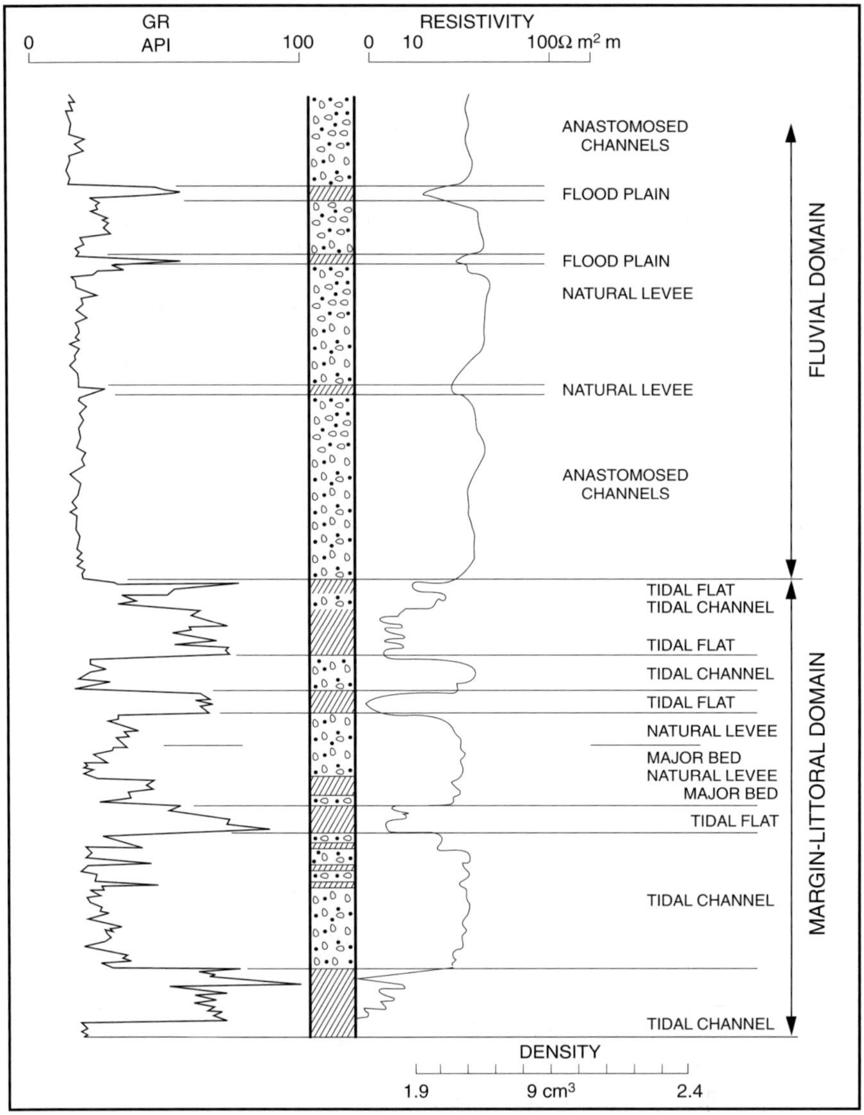

Fig. 4.17B Stratigraphic resolving power with logging and seismics. Traditional curves used in logging have enough resolution to make out facies sequences on the metric scale in the marine and continental domains. The curves here are gamma ray, resistivity, and density.

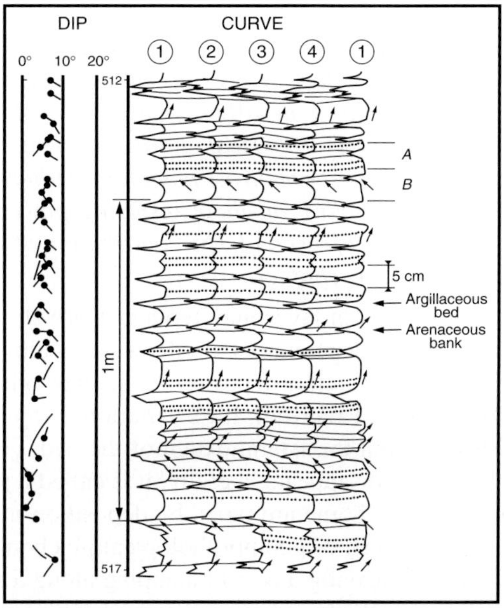

Fig. 4.17C Stratigraphic resolving power with logging and seismics. Decimetric or centimetric sequences can also be defined by special tools–here, the HDT dipmeter log.

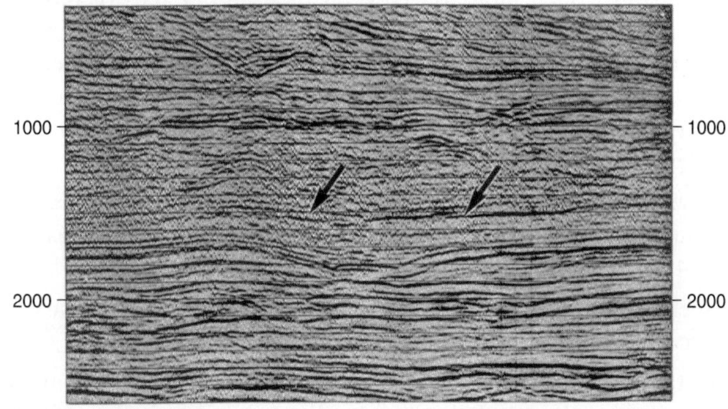

Fig. 4.17D Stratigraphic resolving power with logging and seismics. A conventional seismic section (vertical scale in meters) has less resolution but is capable of picking out events (here, a channel with clay fill) (from Chambre Syndicale de la Recherche et de la Production du Pétrole et du Gaz Naturel, 1986).

*A **seismic marker** or horizon is a reflector covering a certain area, laterally. Reflectors are usually interpreted as being bed surfaces or discontinuities. However, there are also various kinds of "diachronous" surfaces consisting of the same rock but having different ages in different areas. A reflector generally indicates a lithological contrast between two different beds of contrasting petrophysical characters such as those that might indicate contact between fluids, or the permafrost boundary. Different seismic devices will detect different characters depending on their resolving power, which varies greatly from a few meters to several tens of meters.[1] The contribution of seismics to stratigraphy will be taken up again later in this chapter.*

The thickness and facies of a given layer may remain relatively constant over long distances (several kilometers) if the depositional mode was a calm environment (such as pelagic drapings or lacustrine decantation deposits), or it may otherwise vary quickly in space (Fig. 4.18), exhibiting **lateral variation**. If its variation is rapid and the floor and roof are no longer parallel, this will limit the lateral extent of the bed, giving it a **lens** shape, or **lenticular structure**. When a bed disappears laterally, it is called a **pinchout** (Fig. 4.19), of which there are a number of types. Some are typed by deposition mode (e.g., coastal sandbars laid in stratigraphic, faciological, or geomorphological pinchouts) while others are due to later erosion or tectonic events leaving a bevel onlapping along a salt dome, for example. This idea will be taken up again in the last chapter on petroleum systems.

Referring to Chapter 3 on sedimentary mechanisms and depositional environments, it can easily be understood why there is no simple relation between the thickness of a deposit and the time it takes the deposit to develop. On the geological time scale, a turbidite flow or tsunami are practically instantaneous events lasting a few hours, while constructing and maintaining a littoral bank requires much more time (a thousand years or more), and a draping of pelagic limestone much more yet (several thousand or tens of thousands of years, especially if the surface productivity is low). In special cases, the deposition rates of the strata can be quantified (see the previous chapter where the idea of **sedimentation rate** was defined). This is true of millimetric varves, for example, which are deposited in seasonal alternations. Semi-annual and annual variations can also be determined by isotope measurements using modern laboratory methods.

The Milankovitch type cyclicity due to periodic variations in the Earth's astronomic parameters also suggests that certain bank alternations may occur in cycles of 20 ky, 40 ky, and 100 ky (Fig. 4.9).

As will be seen a little further on, there are a number of complementary time measurement methods used for **dating** a set of beds.

The bed does not always contain a record of the entire deposition time period. There can be **sedimentation gaps**, with or without erosion (the term sedimentation **hiatus** or **discontinuity** is also used), and many stratification joints or **diastems** also appear in which the deposit is briefly interrupted. That is, the physical sedimentation process is not actually

1. Lavergne, 1986; Henry, 1994.

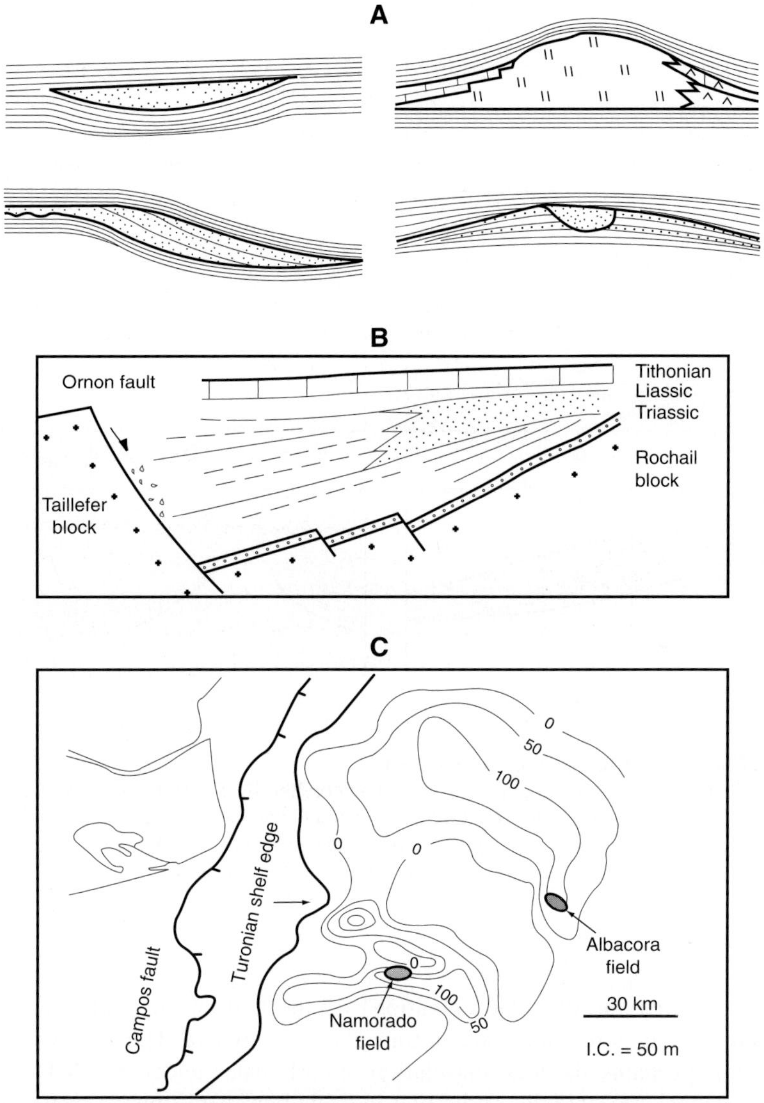

Fig. 4.18 Bed thickness variations.
A. Schematic representations of a lens, reef, wedge, and channel. **B.** Reconstruction of bed variations in the Liassic of the Alps (Bourg d'Oisans region). The beds here thicken rapidly along an active synsedimentary fault. **C.** As seen on an isopach chart, showing the Upper Cretaceous sandstones of the Campos basin offshore Brazil (from O. Gomez de Souza, 1997).

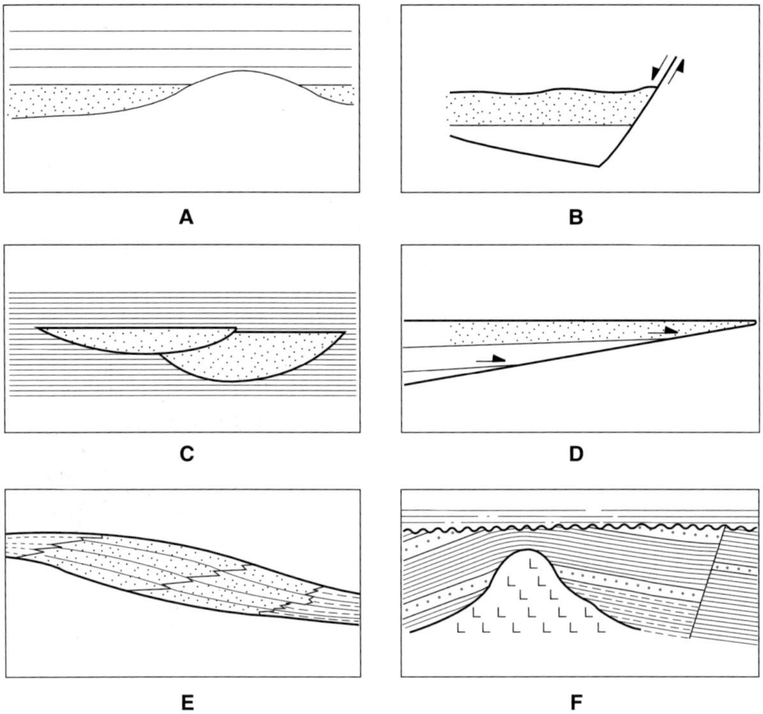

Fig. 4.19 Different types of pinchouts.
A. Geomorphological pinchout: buried paleorelief. **B.** Syntectonic pinchout: effect of an active fault. **C.** Genetic pinchout: two channels in an alluvial plain. **D.** Stratigraphic pinchout: deposit transgression. **E.** Faciological and diagenetic pinchout. **F.** Structural pinchouts: under discordance, on diapir or fault.

interrupted, but rather the dynamic conditions on the bottom are such that the material deposited is immediately remobilized and transported elsewhere. The sole case of real non-sedimentation is probably the total dissolution of carbonates under the CCD. Over longer time scales, the many repeated discontinuities in a thin bank set is more a matter of sedimentation **condensations**.

For example, the Lias and Dogger set (50 My) on the Ardeche basin border can be represented by several meters of condensed deposits.

When it can be shown that a bank was everywhere constructed during the same period of time, it is said to be **isochronous**, while it is **diachronous** if time lines are found to cut through its floor and roof (Fig. 4.20). The facies will generally be diachronous on the scale of several beds, and especially in very dynamic prograding or retrograding systems. Differ-

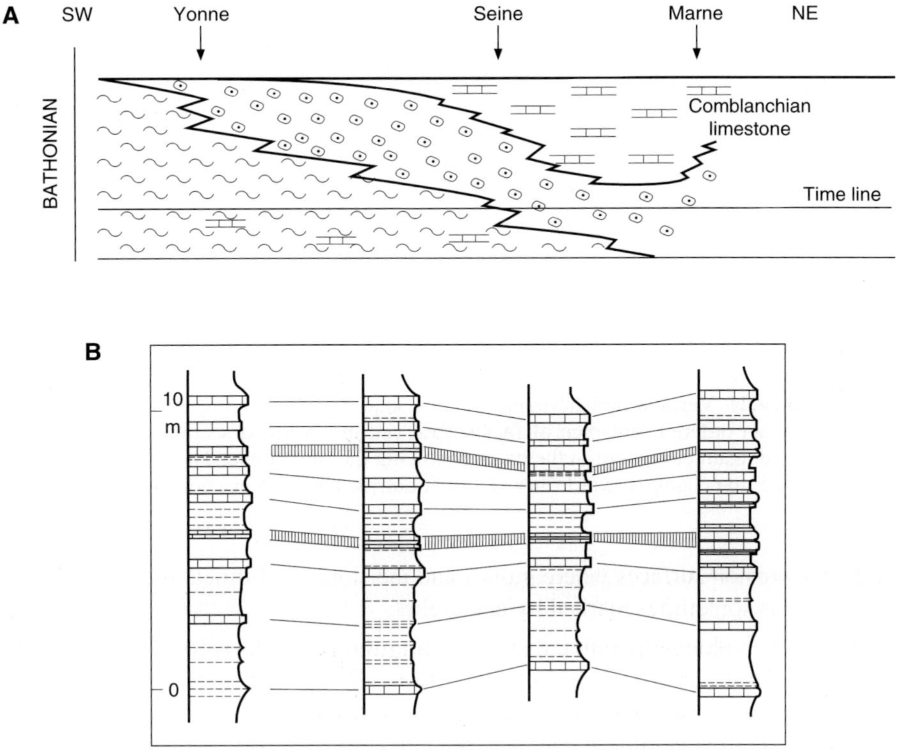

Fig. 4.20 Diachronous and isochronous beds.
A. A diachronous layer is cut by a time line. It corresponds to a facies that migrates laterally in time. The example here is the Bathonian of the Parisian basin (from Rey et al., 1997). **B.** An isochronous layer corresponds to a deposit of uniform facies over a great distance on the scale of a basin. Here, the example is that of bank bundles in the Valangian of the Vocontian basin (from Cotillon, 1988).

ent facies, if superposed in a geological series, do succeed each other in time but may also occur side by side in space at the time of deposit during a given period. This is **Walther's law** of facies sequences with no major breaks (Fig. 4.21).

A sedimentation gap, regardless of duration, may be relatively discrete, exhibiting an **apparent concordance**. The actual **unconformity** or paraconformity will be evidenced elsewhere than in the geometric relation of the beds—by a major environmental change, for example, with hardground, traces of emersion, desiccation, local erosion, or paleosoil formation. If there is much erosion and the unconformity surface is secant to the face (and even if the paleosurfaces are still parallel), there is an **erosional unconformity** (such as in the Saharan glacial erosion surfaces) (Fig. 4.22).

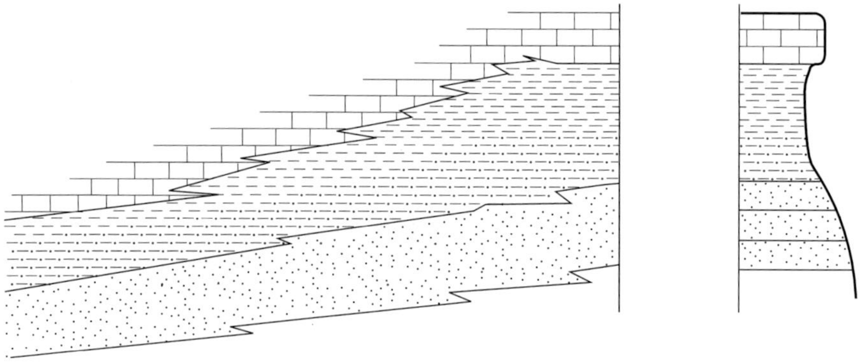

Fig. 4.21 Vertical and horizontal facies sequences illustrating Walther's law: an ordered sequence of facies observed in a vertical section can be found again elsewhere in the basin following the same logic. The example here is the change toward finer and finer detritic and carbonaceous series.

Unlike the erosion surfaces where gullies and incisions are formed, the erosional unconformity surface is sometimes marked by paleorelief.

If the beds are no longer parallel, they are in **angular unconformity**. The angle indicates a tilting of the oldest series (Fig. 4.22), with erosion and a sedimentation gap. The youngest beds then transgress or onlap the older series. **Chart discordance** is an angular discordance where the angle is so small that it can only be seen at the outcrop, or by drilling, or even in the seismic (depending on the profile orientation). It is the plot on the geological chart that identifies this discordance (Fig. 4.22).

Sedimentation unconformities do exist by themselves in autocyclic variations, but they are mainly amplified by external allocyclic factors. Tectonics creates deformations, followed by erosion. Eustatic variations cause environmental changes due to the variation in depth and the distance from the coast. These different changes have been named transgression and regression.

Transgression is an invasion of the continental environment by the marine, with a gradual progression of the sea beyond the previous coastline. A transgressive **onlap** bed is one that extends over a substratum from which it is separated by an angular or other discordance, or over an older sedimentary bed separated by an unconformity. The typical example is the marine bed deposited on an eroded continental basement with or without base conglomerate, and with a sequence of **onlapping pinchouts** (Fig. 4.23). In the marine environment, the transgressive stage will be evidenced by **positive grading** with paleobathymetric growth (a decrease in deposition energy).

The opposite phenomenon is **regression**, in which the sea retreats from the previously invaded domain with a concomitant advance of emerged lands and coastline seaward (Fig. 4.23B). Regressive **onlaps** exhibit **negative grading**, i.e., the facies become less and

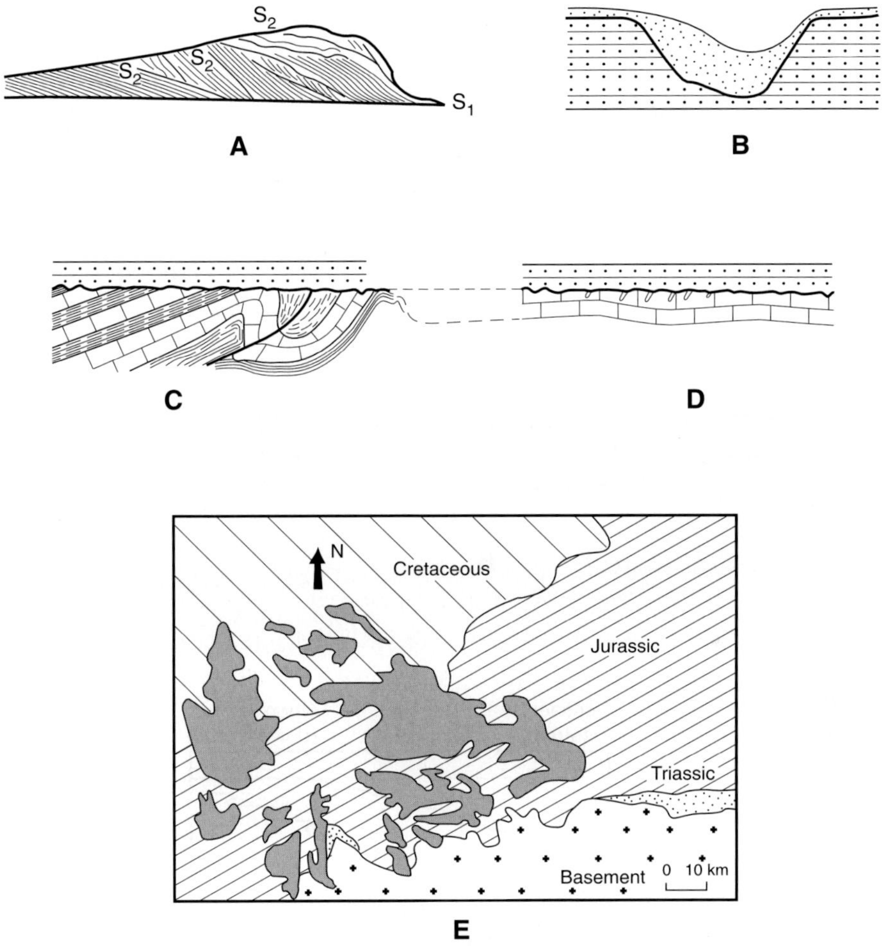

Fig. 4.22 Deposition unconformities and discordances.
A. Break in sedimentation: A tidal dune migrates over an unconformity S_1 with internal erosion or reactive surfaces S_2 (genetic discontinuity). **B.** Discontinuity due to brief erosion, either as part of the sedimentation process (subglacial channel, for example) or due to an environmental variation (tectonic, climatic) (stratigraphic discontinuity). **C.** Angular unconformity: Beds are tilted or folded before the erosion surface is cut and new "discordant" beds are laid (unconformity). **D.** Disconformity: The beds lie in sequence with no visible angular unconformity, but a large time hiatus results in a major disconformity. **E.** Map discordance: No discordance is evidenced on the scale of the outcrop, but a major one is found in the geological chart. Here, the Tertiary (in gray) overlies different terms of the Mesozoic and then the basement, in the Berry region (from the geological chart of France).

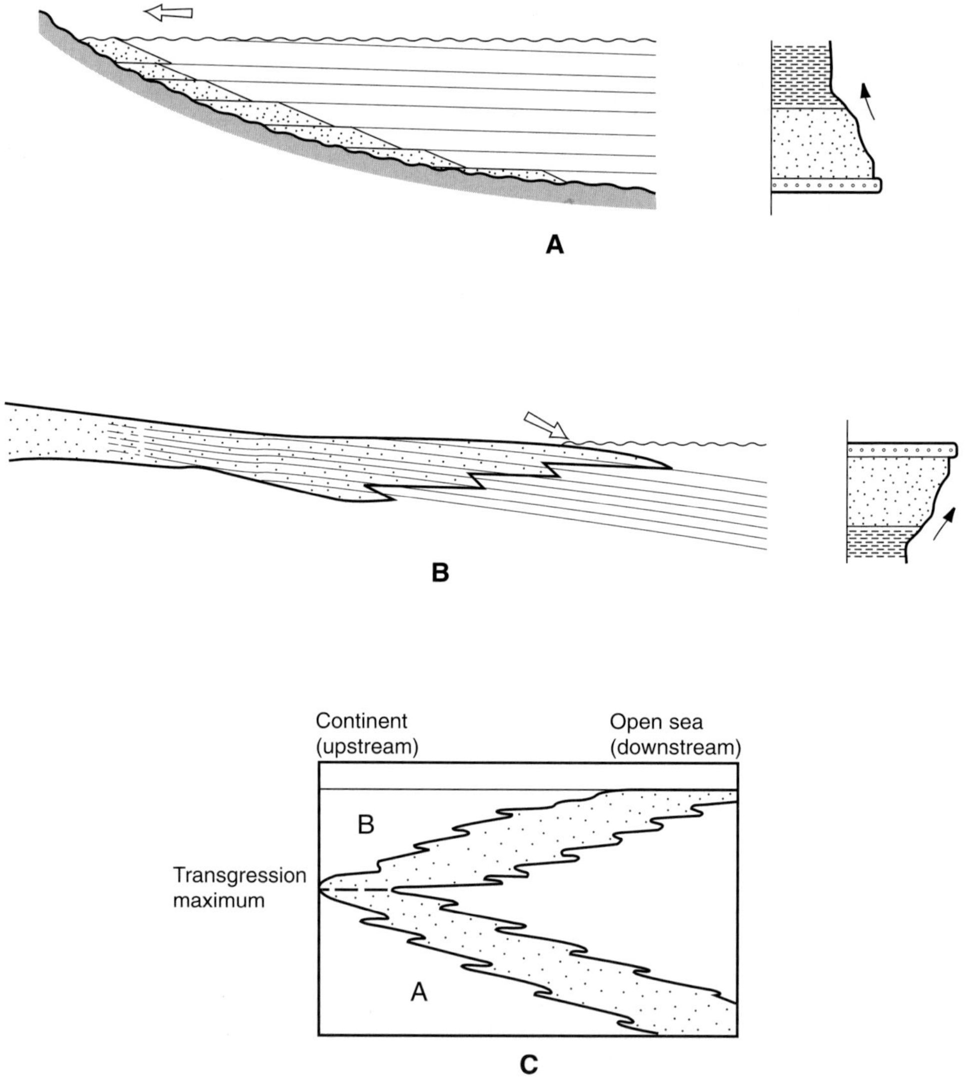

Fig. 4.23 Sea transgression and regression.
Transgression **A** is indicated by a landward progression of marine beds up to the point of maximum transgression or flooding. Regression **B** is the corresponding retreat of the sea and seaward extension of continental deposits. A series of fluctuations is often observed on different observation scales. The transgression, or marine incursion, is evidenced by a vertical sequence of positively graded deposits, and regression by a negative grading (the diagrams are theoretical). Transgressive and regressive systems found together (**C**) are evidence of a sedimentary cycle.

less marine or more and more proximal or energetic (Fig. 4.23B) in shallower water where wave and tidal actions are clearly pronounced.

A major **sedimentary cycle** is a period that begins with a transgression on the scale of an entire basin, and ends with a regression. If tectonic control is predominant, a major sedimentary cycle will be observed between two discordances; but in the stable cratonic domain, the cycles will be defined by eustasy. We will return to this aspect later on in the chapter when defining sequential stratigraphy.

4.2.3 Facies, Depositional Sequences, Lithostratigraphic Units

Stratigraphy, insofar as it can be defined as the study of stacked strata, considers patterns encompassing more than the elementary stratum: a set of beds making up a vaster lithostratigraphic set. Because of the interest of ordered depositional sequences in petroleum geology, the idea of **genetic depositional units** will be taken up later. The older term **cyclothem** used to designate a logical sequence of different facies in oil basins (Fig. 4.24) but is now deprecated as the word conjured up the idea of an **autocyclic control** due to the system itself, whereas the idea of external **allocyclic** forcing is predominant today.

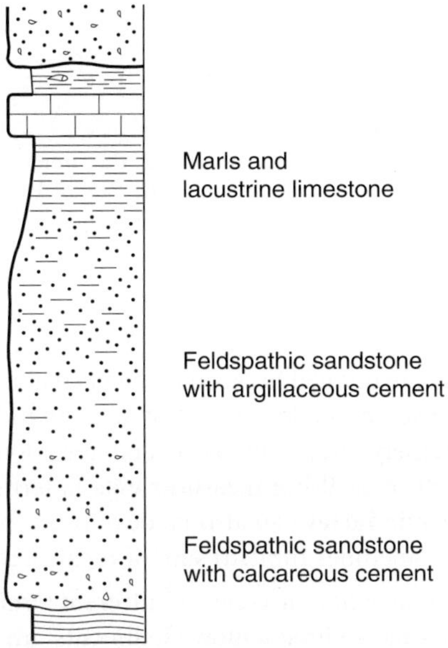

Fig. 4.24 Cyclothem: a logical sequence of facies.
The example of this depositional sequence, in the Aquitanian of Swiss molasse, illustrates an ordered facies sequence.

Without going into detail here, the **depositional sequence** idea is an old one based on the idea of **facies**. A facies is the sum of characters of a rock or set of strata that characterize its appearance and specify its origin. It is the set of lithological characters (making a "lithofacies"), biological characters (biofacies), or sedimentological characters (marine/continental, deep/littoral, confined or restricted facies). Paleogeographic reconstructions can be developed by studying the facies distributions (see the end of this chapter) in **isopic zones** (or sometimes isopic lines) to define zones of comparable facies and develop **facies maps** (Fig. 4.25).

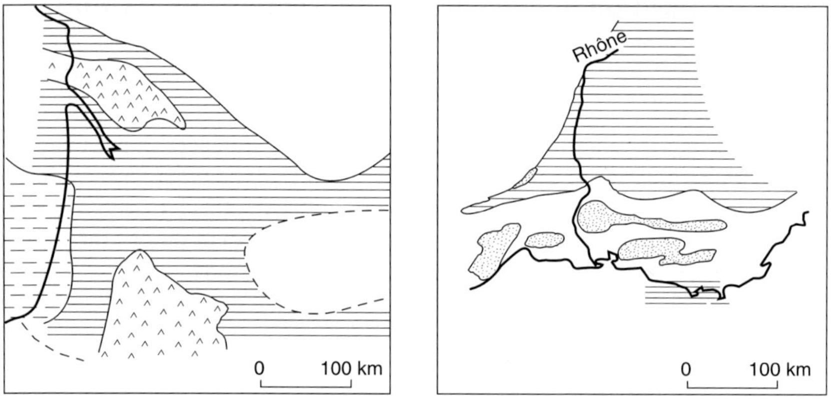

A B

Fig. 4.25 Facies distribution maps.
A. The Aquitaine Kimmeridgian, showing the inner shelf (horizontal lines) with evaporitic basins opening on to the Bay of Biscay with its outer shelf facies (dashed lines). **B.** Upper Cretaceous in the basin of southeast France. The continental facies of the Durancian isthmus here separate the marine facies of the Vocontian basin from those of the Ligurian domain.

Lithofacies and **biofacies** are defined by visible objects that are further detailed in **microfacies** or even **nanofacies** observed under optical or scanning electron microscope. Microfacies descriptions are extremely useful in basin and reservoir geology (Fig. 4.26). The indirect geophysical approach of well-log measurements is termed **electrofacies** (or sometimes electrosequences). **Seismic facies** can also be developed from reflection shooting. But in all cases, it is always the (sometimes limited) sum of identification characters that is used.

A **sequence** is a vertical association of facies of different terms following a definite order with no major interruption in the sedimentation. Sediments are said to be transgressive or regressive, for example, if the facies become more marine or less so going up. If the logical order repeats itself cyclically, the deposits are in continuous or discontinuous cycles (Fig. 4.27). It will be seen later how these ideas came to the fore as the concepts of **sequence**

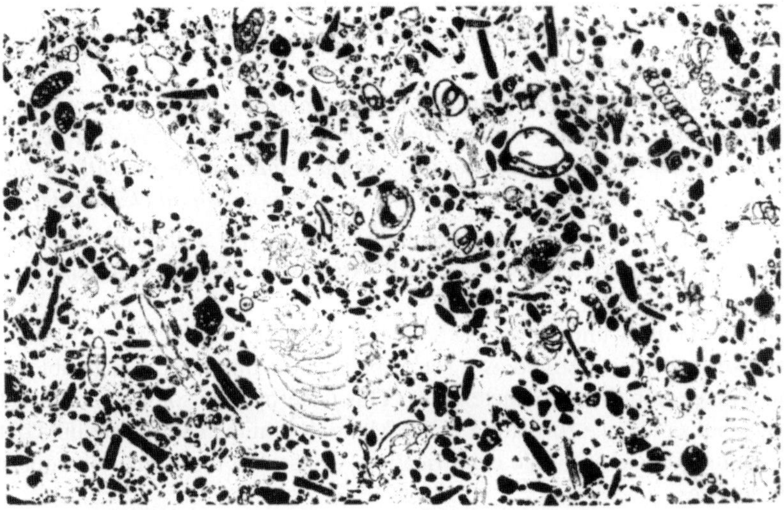

Fig. 4.26 Lagoonal microfacies of milliolids and peneroplids (foraminifers typical of this depositional environment). Enlargement H × 15 (from Purser, 1980).

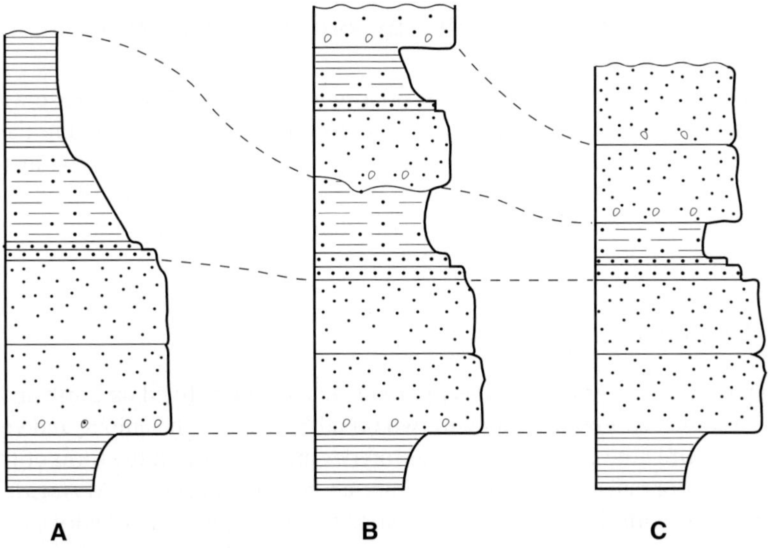

Fig. 4.27 Continuous and discontinuous facies sequences.
A. Complete facies sequence. **B.** The upper part of the lower sequence is truncated. **C.** Several of the upper sequences are amalgamated.

stratigraphy were introduced on the scale of the basin, offering a more dynamic approach with tectonic, eustatic, and sedimentary controls than with a simple analytical description of the deposit sequence or sedimentary cycle.

The sequence designation may thus be lithological, geochemical, mineralogical, or diagenetic sequences, depending on the approach. The sequence idea can also be used to define a theoretical vertical evolution in a ***virtual sequence****. As we will see later, different sequences can be defined at different scales, starting with an* ***elementary sequence****.*

Although modern perspectives have breathed new life into stratigraphic interpretation, it is still useful to recall a few terms that are still extensively used in American literature and by petroleum geologists:

- **Bed**, the elementary unit.
- **Formation**, which is defined as a set of beds extending over a mappable geographic region with certain lithic characteristics. This is defined by some typical location, such as the Annot sandstone formation or the Mano dolomite. The formation's definition is based on its lithostratigraphic classification. Generally, the formation consists of several sequences.
- **Members** are the next lower-ranking lithostratigraphic subdivision of a formation and often correspond either to real deposition sequences or to portions of them (Fig. 4.28).
- **Groups** are the next higher-ranking lithostratigraphic unit above formations, containing a number of them, as the name implies.

4.2.4 Relative Dating in Paleontology and Biostratigraphy

The basic lithostratigraphic principle of considering the age of beds to increase with depth is insufficient. Local lithostratigraphic scales have proliferated, as have the unit names, numbering in the thousands. "It should be possible to reconstruct a continuous history of the Earth"[1], and a global one too, we might add. When correlations are made and events interrelated, geology enters the field of dating. The need for common references appeared very early in geology.

4.2.4.1 Paleontology

Paleontology is the study of the remains of organisms that once lived on Earth and were fossilized in the sediments. **Fossils** belong to two broad kingdoms: the animal and vegetal. The original forms of life are very old. Proto-karyotypes are common in the form of the remains of anaerobic microorganisms in early Precambrian beds (Achaean, 35 My). Life evolution went hand in hand with the gradual evolution of the atmosphere and hydrosphere and the

1. Cotillon, 1988.

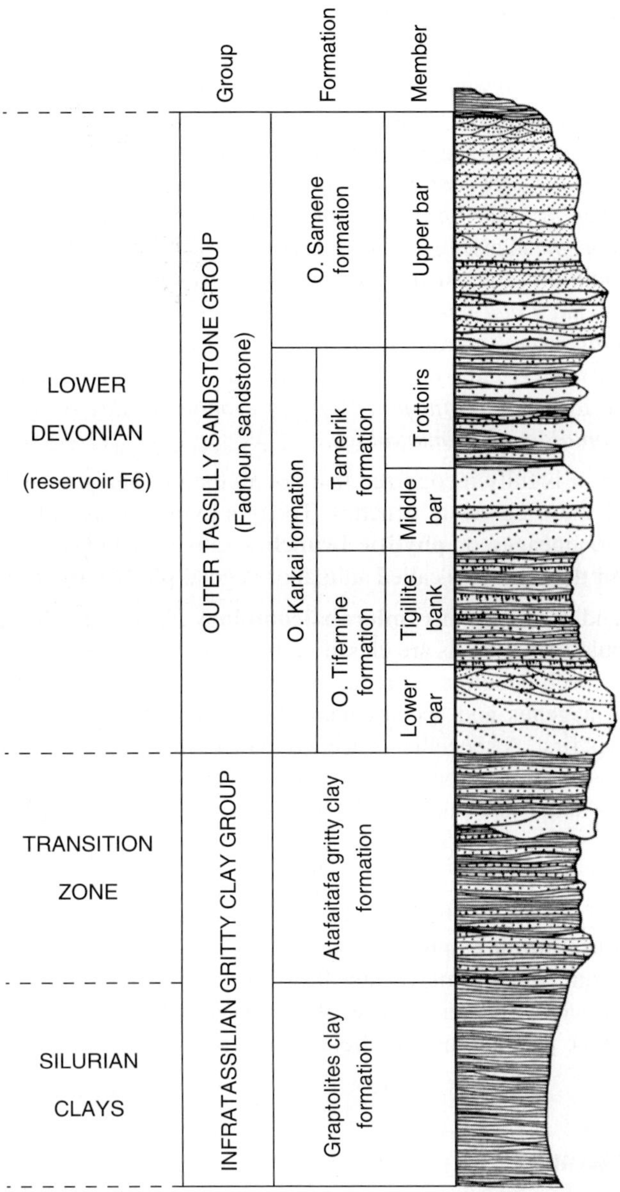

Fig. 4.28 Lithostratigraphic terminology (group, formation, member) applied to the example of the Saharan Upper Devonian (from Dubois et al., 1964).

gradual increase in the ocean's oxygen content and salinity, and wide temperature variations.

Though no scientific consequences were drawn from it, marine fossils were found on continental relief in Antiquity. The idea that these remains bore witness to the occupation of different sites by the sea in more ancient times was first issued in the Renaissance. But it was only in the 18th century that Buffon laid the foundations of paleontology by bringing out different fossil forms that were related to organisms of today. To Cuvier's mind, a little later, the evolution of living organisms was impossible: the observed variations and discontinuities were explained by successive creations, since each species was considered to be fixed and immutable. This was the fixism doctrine. The opposing concept of ***evolutionism*** *had already been thought of by the Ancients and was toyed with during the Renaissance, but was really developed only by Lamarck and Darwin and is usually referred to as "Darwinism". Proof of evolution today is based on the life sciences, and paleontology and embryology.*

Today, however, it is fully recognized that not all the mechanisms of evolution are well understood, despite the progress in genetics. Today's living forms in the animal and plant kingdoms belong to lineages or **phylitic branches** (phylum) linked to fossil forms. The gradual evolution of these links is called animal or vegetal **phylogeny** (Fig. 4.29).

Paleontology and, more recently, **micropaleontology** (for the smallest fossils) study and describe fossil populations. Fossils are classified by major faunal and floral branches of protozoans, poriferans, echinoderms, coralla, brachiopods, and are designated by Linne's nomenclature with a genus and species name followed by the name of the author who described the species first. A specialist is needed to determine a fossil exactly. The great lineages and major fossil groups characteristic of different periods of Earth history are thus established, and this data is used to date different strata in **biostratigraphy**.

Paleontology is closely related to zoology and the biological sciences. Today, this field goes beyond the global phylogenetic reconstructions to describe the history of **biodiversity**, and is addressing the question of **biological crises** with its phenomena of species **extinction** or adaptation by **mutation**. It emphasizes the developments of new lineages with the questions of **speciation** and **species migration**. It offers keys for interpreting environments, as was seen in the previous chapter: paleoecology, paleobathymetry, general paleoenvironment, and stratigraphic paleontology are fields where paleontogical knowledge of the population is essential.

4.2.4.2 Biostratigraphy

Paleontology and micropaleontology determine fossil, animal or plant **species** or **taxons** (applied to any rank of organisms such as species, family, or class, whence the term taxonomy). Each species has a given life duration, the limits of which are not always precisely known. Generally, biostratigraphy will **associate** and **correlate** different species for dating purposes. A bed is defined stratigraphically by its faunal and floristic content. Two beds

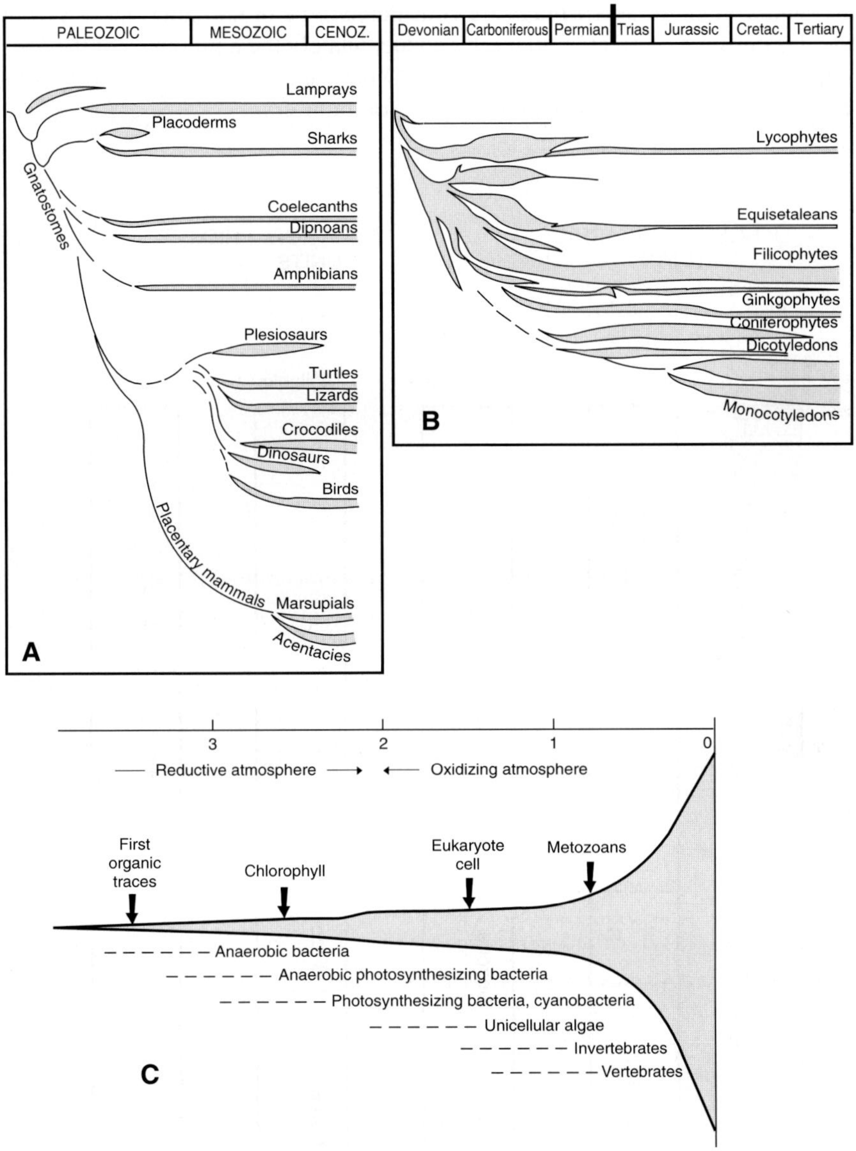

Fig. 4.29 Examples of phylogenetic relations and the ways they are represented.
A. Main vertebrate groups in chronological framework (from Chaline, 1987). **B.** Main vascular plant groups since the Devonian (from Caron et al., 1992). **C.** General diversification scheme of the living world: very slow in the Precambrian, with an explosion at the beginning of fossiliferous times (from Devillers and Chaline, 1989).

having the same content or **assemblage** are considered to be of the same age (Fig. 4.30). So certain fossil groups are of great stratigraphic value, namely the ammonites of the Mesozoic and the planktonic foraminifers in the Tertiary. Others are much less so, but will then be used as environmental indicators, such as benthic foraminifers.

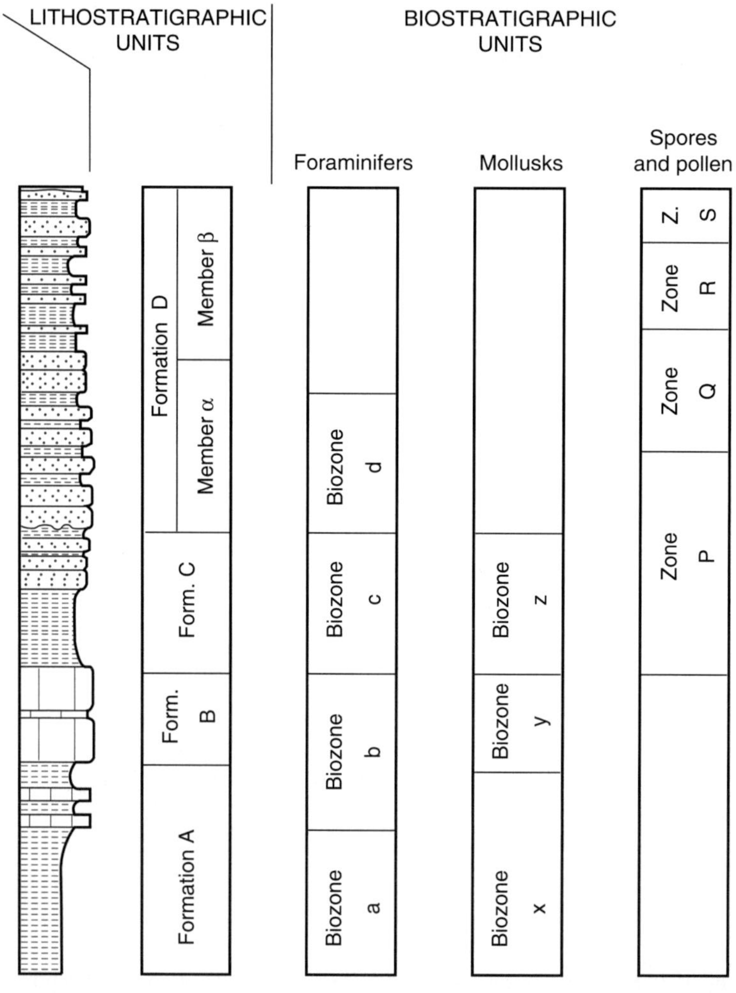

Fig. 4.30A Biostratigraphic assemblages and correlations based on them. The faunal content or assemblage is analyzed according to the various groups of fossils present. The theoretical example here considers different biozones defined by foraminifers, mollusks, spores, and pollen (from Rey, 1997).

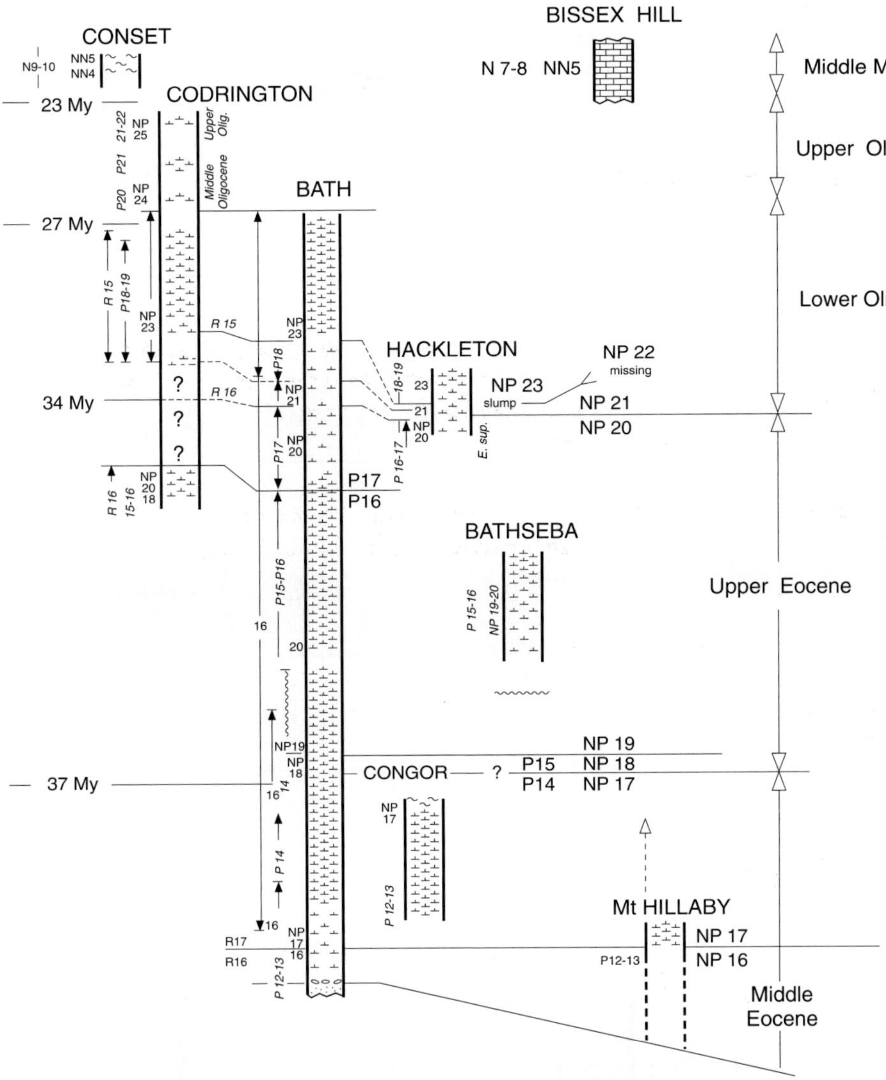

Fig. 4.30B Biostratigraphic assemblages and correlations based on them. Example of correlations based on different assemblages. Here, planktonic foraminifers (zones marked P), calcareous nanoplankton (NP zones) and radiolarians (R zones) are used to adjust sections of more or less siliceous pelagic marly limestones (Bath Cliff formation, from Biju-Duval et al., 1985).

The usefulness of fossils increases with their rate of dispersion over a very broad geographic area. This is why marine planktonic species are so useful, like the ammonites that can be found in the Jurassic of Germany and ranging over neighboring regions. Oppel conceived this as a **biozone** in which one or more species or taxons is associated with a time interval, and takes the name of the most typical fossiliferous species of the zone (such as the *orbulina suturalis* zone), even if this fossil is not found everywhere locally (Fig. 4.31). The

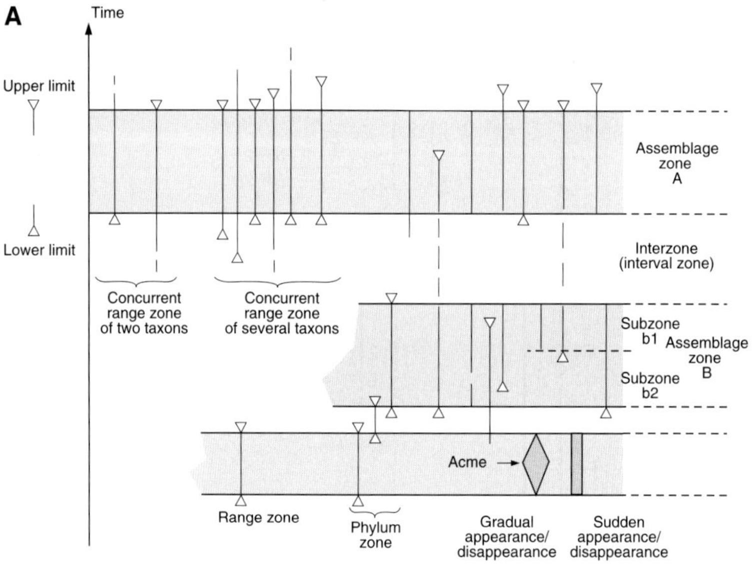

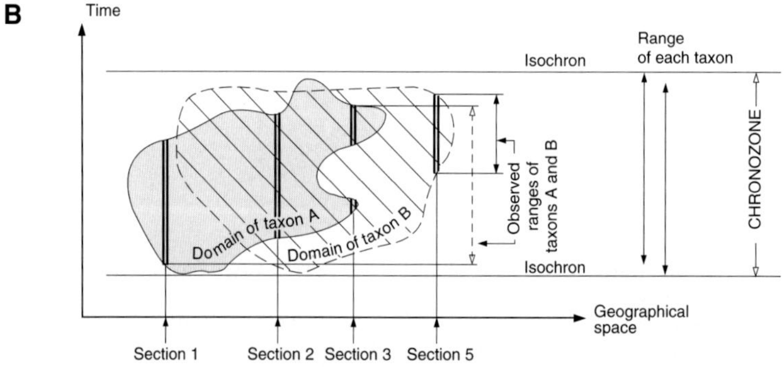

Fig. 4.31 Extent, taxon assemblage, assemblage zone, biozone, chronozone. **A.** Biozones defined by assemblages of different taxons with their lower and upper limits. **B.** Analysis of geographic distribution of several taxons (only two here) to narrow down the chronozone (from Hedberg, 1979).

acme is the point of greatest abundance of the given species. The biozone is the concurrent range of several species, of which the lower and upper appearance limits should be noted (Fig. 4.31). Intervals or interzones without fossils may occur between different association zones. A fossil of special importance will be called a **marker**. It will be seen later that other types of markers exist in addition to the biological, namely mineralogical, geochemical, isotopic, and paleothermal markers. All of the fossil markers taken together constitute the **population** which, beyond the biozone, can be used to define the ecological life conditions or **biotope**.

While certain associations of rapidly evolving taxons have a worldwide geographic range, the horizontal range of biozones is sometimes limited in space to a **faunal province**. Examples of this are the Boreal and Tethyan provinces (Fig. 4.32). These provinces have also been observed to vary notably in the course of geological time.

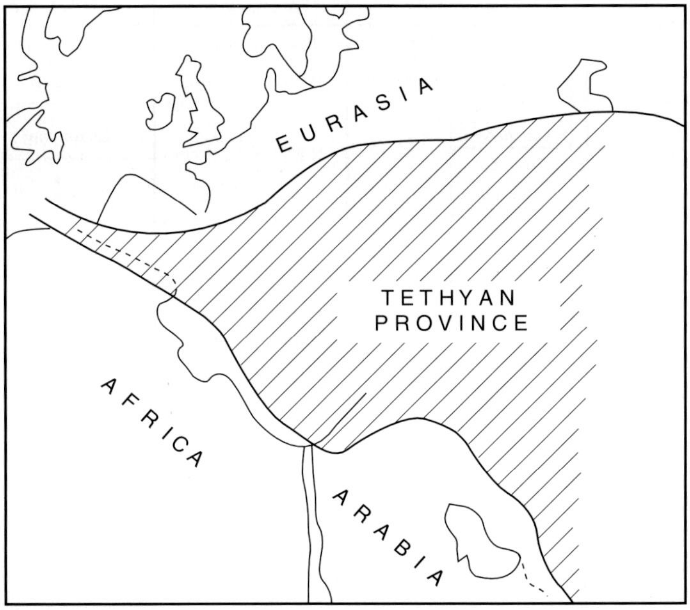

Fig. 4.32 Example of faunal provinces.
The idea is valid for a given period (the Middle Jurassic, here). The province defines a zone where certain assemblages dominate with respect to neighboring zones, with connections that are more or less uncertain and variable in time. The Tethyan province separates the (boreal) European cratonic provinces from the (southern) Ethiopian province.

With the different biozones taken in sequence (Fig. 4.33), biostratigraphy generates a veritable **biochronological scale** with a series of events characterized by fossil remains. Actually, while the time reference seems obvious, an absolute measurement of the time is not at all so. Like lithostratigraphy, biostratigraphy only describes an historical sequence with no absolute time reference defined in millions of years. It is only **relative dating**, even though time-rock stratigraphy (addressed immediately hereafter) appeared at the same time. It should be emphasized that biostratigraphy can be used only for those periods of Earth history for which fossil remains exist. These are called **Phanerozoic** eons starting at the Paleozoic, as opposed to the older **Cryptozoic** periods of the Precambrian where fossil remains are quite rare. Biostratigraphy is not applicable either if the rock, whatever its age, is metamorphosed.

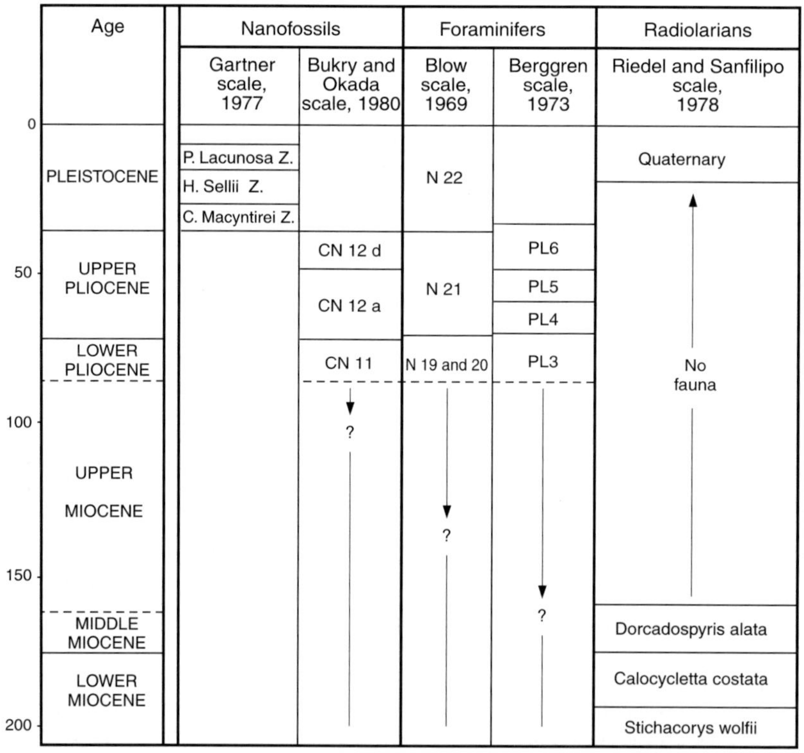

Fig. 4.33 Sequence of biozones of planktonic organisms defining a biostratigraphic scale. Example of the IPOD 543 drillhole offshore Barbados (from Biju-Duval and Moore, 1984).

While biostratigraphy is sometimes wrongly considered as a traditional, naturalistic discipline reserved for specialists in paleontology, it is actually extremely useful in applied geology. In petroleum geology, biostratigraphy is a major technique for characterizing biozones and establishing correlations on the scale of the basin and also in the fine analysis of reservoirs (high-resolution zoning), reconstruction of paleoenvironments (species abundance, diversity, and range) and the organic content of the facies. Microfauna are studied quantitatively and statistically by the study of spores and pollen in micropaleontology and palynology. Microfauna used to be the only fossils used, but they are generally inadequate because they are too rare when working with drill-core samples.

4.2.5. Chronostratigraphy, Geological Time Scale

Chronostratigraphy is the part of stratigraphy in which bed ages are established and ordered in time. Geological beds of various ages are organized in distinct units in a chronostratigraphic **dating system** based on their age. The different terms of **chronostratigraphic** units are bounded by time contour surfaces (isochrons) in a hierarchy where their relative importance depends on the time interval considered. A **chronozone** is the most elementary zonal subdivision in this hierarchy. A **chronostratigraphic horizon** is an **isochronous** surface or a stratigraphic interface. Several processes are said to be **synchronous** if they are of the same age everywhere. This term is generally (though improperly) used to designate a very thin interval. A reference time surface is also referred to as a **datum horizon**.

A **stage**, corresponding to a **geological age**, is the next higher unit of time above the chronozone. It is a fundamental working tool that serves as a kind of typical **reference** for a given time interval, and is sometimes broken down into substages. A stage is named after some locality where the type was defined, such as the Burdigalian (Latin for Bordeaux), Senonian (Sens), or Kimmeridgian (Kimmeridge). The original type found in these locations is called the **stratotype** serving to define it (and this can also be applied to any type of stratigraphic unit) because of the fossiliferous wealth at that point (Fig. 4.34). Generally, a stage covers a time interval of two to ten million years.

Stages are then grouped into series corresponding to longer durations called **epochs**. The Lias (or Liassic), the Dogger and the Malm are defined as such. Any longer **periods** are grouped into **systems** such as the Triassic, Jurassic, or Cretaceous (Fig. 4.35, Tables 4.1, 4.2, and 4.3).

Lastly, systems are themselves grouped into **erathems** corresponding to geological **eras** which, in the Phanerozoic, include the Primary or Paleozoic, Secondary or Mesozoic, Tertiary and Quaternary or Cenozoic.

Epochs, periods, and eras correspond to major Earth-scale faunal renewals (Fig. 4.35), such as the general disappearance of species, groups, and entire faunal branches at the end of the Devonian and Permian. The most popularly known of these great extinctions is still that of the dinosaurs (with ammonites and globotruncanas, among many others) at the end of the

Table 4.1 Geological distribution of main organic groups.

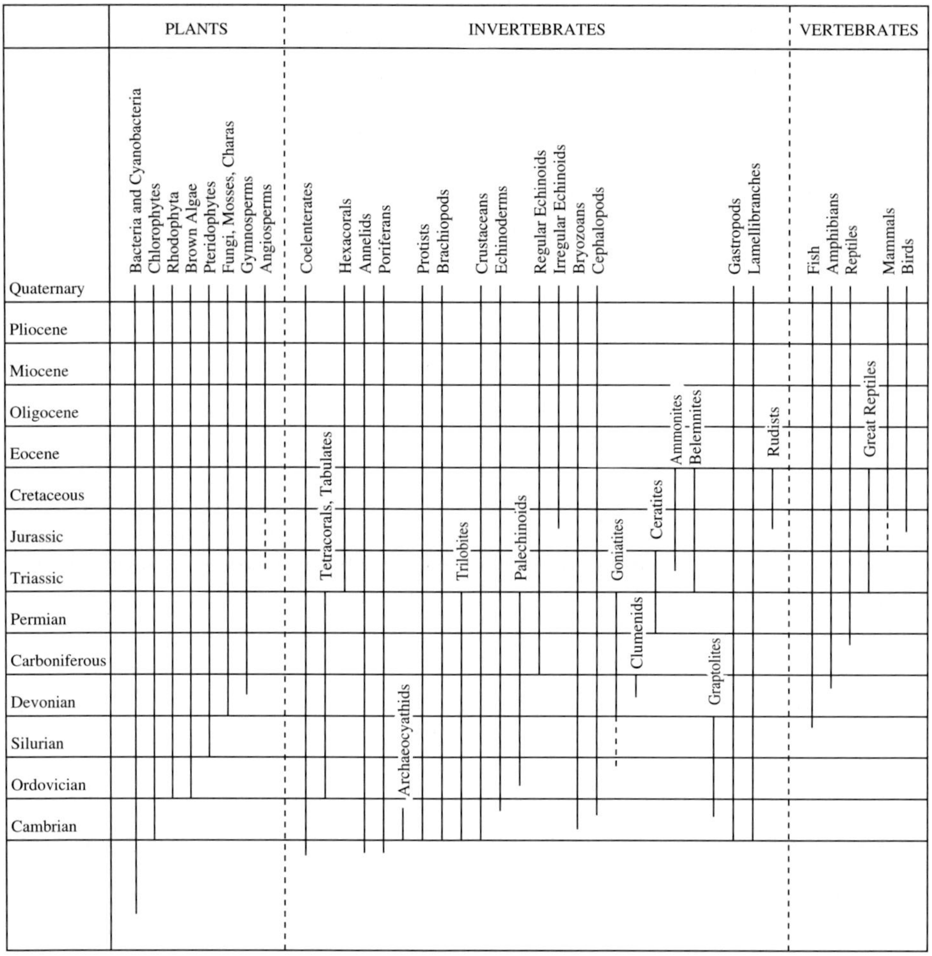

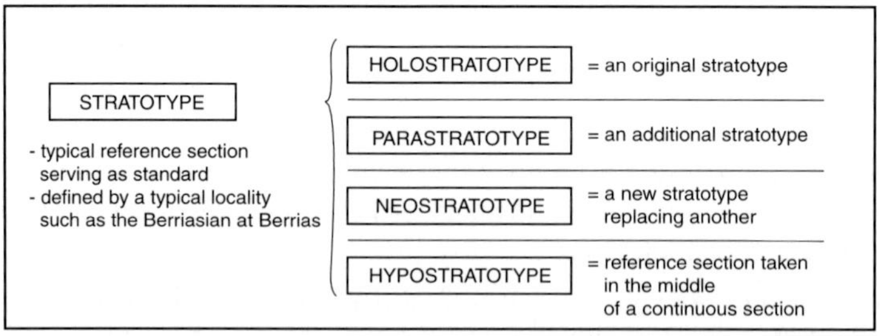

Fig. 4.34 Stratotypes and complementary reference sections.
The stratotype can serve as reference for a sequence, formation, or stage.

LITHOSTRATIGRAPHY	Group Formation Member Bed/bank	
BIOSTRATIGRAPHY	Biozones: Assemblage zone Range zone (several categories) Acme zone Interval zone Other biozone categories	
CHRONOSTRATIGRAPHY	Eonothem Erathem System Series Stage Chronozone	Eon Era Period Epoch Age Chron

Fig. 4.35 Different stratigraphic definitions (from Hedberg, 1979).

Cretaceous. The rapidity of these great extinctions and the subsequent renewals are still being discussed today.

The breakdown into stages is often not quite as sharp. Stages are more regional in reference, so a number of stratigraphic scales often exist over different geographical ranges (Fig. 4.36). The European scale, for example, is not entirely recognized in the United States (a North American stratigraphic nomenclature commission decides on this).

The global stratigraphic chart serves as a reference standard scale and is periodically revised, so old terms die out and are replaced by others.

In **geochronology**, then, time is measured by intervals defined by special dating methods that are not fixed once and for all. As research advances, the geochronological basis evolves, so the scale of reference has to be mentioned. For example, Odin and Odin (1990) will refer to the general dating table as revised by these authors.

The major breaks between geological eras thus correspond to great upheavals in the living world, but also to **events**. Suess (19th century) and more recently Stille, have emphasized the great phases of lithospheric deformation on the global scale, and stratigraphic charts commonly refer to **orogenic phases or cycles** (see the following chapter on tectonics). In this context, the beginning of the Paleozoic corresponds to the end of Precambrian times, marked by a phase of major deformation followed by major erosion. In the same way, the Triassic generally follows the baseleveling (peneplanation) erosion that follows the Hercynian phase. This kind of "event" stratigraphy marked by great transgressive and regressive cycles cannot, however, be generalized. The Tethyan ocean was not necessarily affected by the Hercynian phases, as there is a spatial variation in the intensity of tectonic manifestations. The events may indeed be isochronous on a certain scale, but are no longer so when

Table 4.2 Reference scale fossils.

Age (My)	Era		System		Reference Scales
	CENOZOIC	QUATERNARY		HOLOCENE	
3				PLEISTOCENE	
7		TERTIARY	Neogene	PLIOCENE	
25				MIOCENE	
37			Paleogene	OLIGOCENE	
65				OLIGOCENE	
141	MESOZOIC	SECONDARY		CRETACEOUS	
195				JURASSIC	
225				TRIASSIC	
20	PALEOZOIC	PRIMARY		PERMIAN	
345				CARBONIFEROUS	
395				DEVONIAN	
440				SILURIAN	
500				ORDOVICIAN	
570				CAMBRIAN	
600	PRECAMBRIAN	IV		EOCAMBRIAN	
1000					
2000		III			
3000		II			
		I			

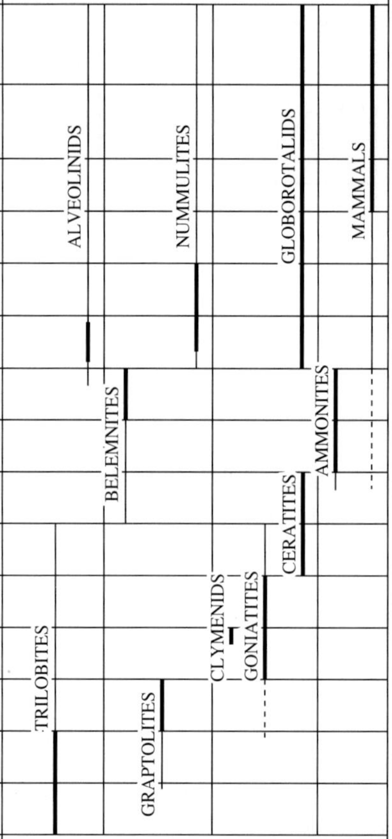

Table 4.3 Numerical scale of geological time (from G.S. Odin and C. Odin, 1990).

Era	System	Series	Stage	My
CENOZOIC	IV	HOLOCENE		0.01
		PLEISTOCENE	CALABRIAN	1.65
	NEOGENE	PLIOCENE	PLAISANCIAN	3.4
			ZANCLZEAN	5.3
		MIOCENE	MESSINIAN	6.5
			TORTONIAN	11
			SERRAVALLIAN	14.5
			LANGHIAN	16
			BURDIGALIAN	20
			AQUITANIAN	23.5
	PALEOGENE	OLIGOCENE	CHATTIAN	28
			RUPELIAN	34
		EOCENE	PRIABONIAN	37
			BARTONIAN	40
			LUTETIAN	46
			YPRESIAN	53
		PALEOCENE	THANETIAN	59
			DANIAN	65

Era	System	Series	Stage	My
				65
MESOZOIC	CRETACEOUS	UPPER	MAASTRICHTIAN	72
			CAMPANIAN	83
			SANTONIAN	87
			CONIACIAN	88
			TURONIAN	91
			CENOMANIAN	96
		LOWER	ALBIAN	108
			APTIAN	114
			BARREMIAN	116
			HAUTERIVIAN	122
			VALANGINIAN	130
			BERRIASIAN	135
	JURASSIC	UPPER (MALM)	TITHONIAN	141
			KIMMERIDGIAN	146
			OXFORDIAN	154
		MIDDLE (DOGGER)	CALLOVIAN	160
			BATHONIAN	167
			BAJOCIAN	176
			AALENIAN	180
		LOWER (LIASSIC)	TOARCIAN	187
			PLIENSBACHIAN	194
			SINEMURIAN	201
			HETTANGIAN	205
	TRIASSIC	UPPER	(RHETIAN)	
			NORIAN	220
			CARNIAN	230
		MIDDLE	LADINIAN	235
			ANISIAN	240
		LOWER	SCYTHIAN	245

Table 4.3 (cont'd) Numerical scale of geological time.

ERA	SYSTEME	SERIE		ETAGE	Ma
PALEOZOIC	PERMIAN				245
		UPPER	THURINGIEN	TATARIAN	250
				KAZANIAN	258
		LOWER	SAXONIEN	KUNGURIAN	265
				ARTINSKIAN	275
			AUTUNIEN	SAKMARIAN	285
				ASSELIAN	295
	CARBONIFEROUS	SILESIAN	STEPHANIEN	GZHELIAN	
				KASIMOVIAN	305
			WESTPH.	MOSCOVIAN	
			WESTPH. / NAMUR.	BASHKIRIAN	315
			NAMUR.	SERPUKHOVIAN	325
		DINANTIAN	VISEEN	BRIGANTIAN	
				ASBIAN	
				HOLKERIAN	
				ARUNDIAN	350
				CHADIAN	
			TOURNAISIEN	IVORIAN	
				HASTARIAN	360
	DEVONIAN	UPPER		FAMENNIAN	365
				FRASNIAN	375
		MIDDLE		GIVETIAN	380
				EIFELIAN	385
		LOWER		EMSIAN	390
				PRAGUIAN	
				LOCHKOVIAN	410
	SILURIAN	PRIDOLI		PRIDOLIAN	415
		LUDLOW		LUDFORDIAN	
				GORSTIAN	425
		WENLOCK		HOMERIAN	
				SHEINWOODIAN	430
		LLANDOVERY		TELYCHIAN	
				AERONIAN	
				RHUDDANIAN	435

ERA	SYSTEME	SERIE	ETAGE	Ma
PALEOZOIC	ORDOVICIAN			435
		ASHGILL		445
		CARADOC		455
		LLANDEILO		
		LLANVIRN		470
		ARENIG		485
		TREMADOC		500
	CAMBRIAN	UPPER	TREMPEALEAUIAN	
			FRANCONIAN	
			DRESBACHIAN	
		MIDDLE	MAYAIAN	
			AMGAIAN	
		LOWER	LENIAN	
			ATDABANIAN	530
			TOMMOTIAN	540

ERA	SYSTEME	SERIE	ETAGE	Ma
PROTEROZOIC				540
		NEO-PROTEROZOIC	NEOPROT.III	650
			CRYOGENIAN	850
			TONIAN	1000
		MESO-PROTEROZOIC	STENIAN	1200
			ECTASIAN	1400
			CALYMMIAN	1600
		PALEO-PROTEROZOIC	STATHERIAN	1800
			OROSIRIAN	2050
			RHYACIAN	2300
			SIDERIAN	2500
ARCHEAN				

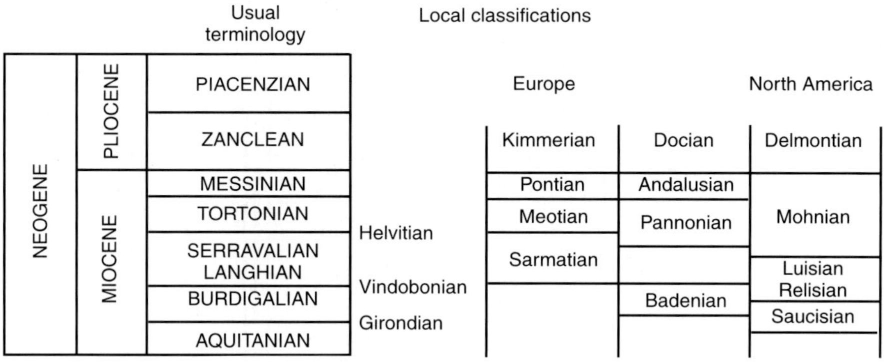

Fig. 4.36 Stratigraphic classification and local terminologies.
The example of the Neogene illustrates the variety of regional nomenclatures. A few of the main terms in use are listed here.

viewed in high resolution. Deformations are more generally **diachronous** (of different ages at different points), often gradual (such as the formation of today's modern accretion wedges, in spurts). Even if global tectonics is drawing attention to the world-scale driving forces (with the fragmentation of Pangea in the early Mesozoic, periods of accelerated accretion of the ocean ridges, for example), the Wilson cycle has no real value to stratigraphy.

It will be seen later on in this chapter how special event-type elements may at times be of general value, when speaking of magnetic field reversals and geochemical and volcanic markers.

A **chronostratigraphic** chart diagrams the ordered stratigraphic variations on a time scale where deposits, non-deposits, and erosion periods are all represented (Fig. 4.37). This is widely used to recapitulate the history of the sedimentary basin (see Chapter 6, for example).

4.2.6 Absolute Age Measurements: Geo- and Radiochronology, Isotopic Stratigraphy

It has previously been said that time is a tricky dimension to quantify in geology, firstly because the geologist deals with events of widely varying durations (ranging from the "instantaneous" earthquake to the formation of a mountain chain, from a tidal cycle to a glacio-eustatic phenomenon), but also because the different methods that have been refined for measuring absolute age have not been successfully applied to all types of rocks and sediments.

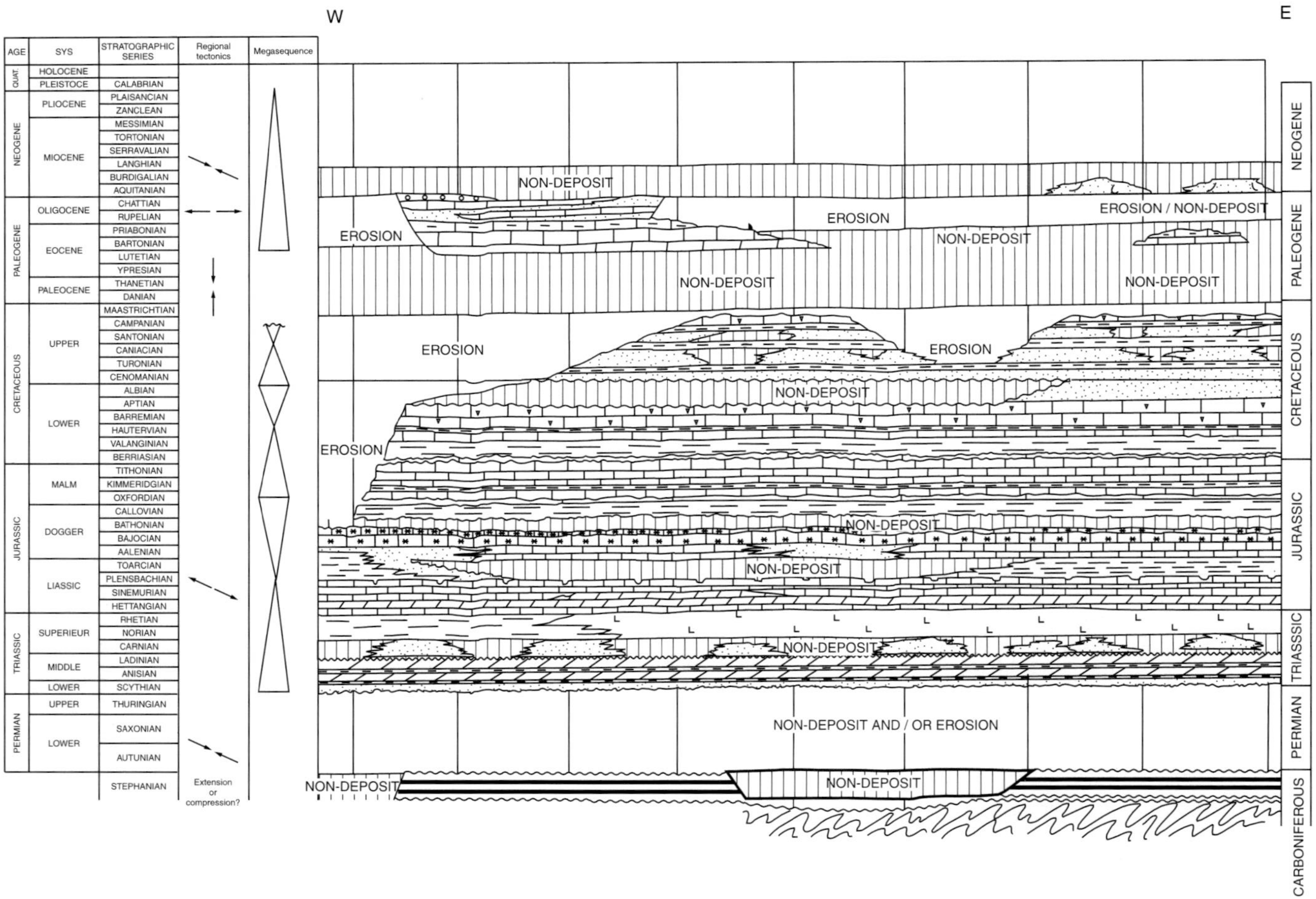

Fig. 4.37 Chronostratigraphic diagram.
This example illustrates the stratigraphic sequence of the Cevenole edge of the Ales basin (from works of ENSPM students).

Geochronology is the science of dating events in Earth history, while **geochronometry** is the branch of geochronology that measures geological time in thousands or millions of years (Ky or My, as the case may be). Geochronology is a form of chronostratigraphy; but instead of expressing time in chronozones, stages, and series, this history is broken down into chrons, ages, and epochs (Fig. 4.35). Geochronology has also been called **radiochronology** because it is based on analysis of the **radioactive elements** present in the minerals that gradually disintegrate into new radiogenic elements. This is the specialized field of isotopic geochemistry, which has lent its name to **isotopic stratigraphy** since the chronology is based on the evolving abundance or isotopic ratio of certain elements.

The principle of isotopic geochemistry is a simple one, in which the ratio between the concentration of an unstable radioactive element and its stable radiogenic product corresponds directly to the total disintegration time, as the disintegration half-life of the element present in the mineral is known. The principle is more difficult to apply in practice, because a number of conditions must be met. Namely, the decay constants must be well known, the system containing the analyzed element must be closed, and there must be no major intervening diagenetic transformations. There is no major problem dating crystalline rocks. Ages are given with a margin of error. A granite can be dated, or rock metamorphism, as long as there are no later re-crystallizations or special thermal phenomena that will "reset the clock" and thereby erase or reduce the true age. It is estimated that the precision is good and the analytical errors are of the order of 0.1 to 2%.

The (usually silicate) minerals of use in sedimentary rocks are detritic minerals inherited from a basement that is older than the deposit where they are sedimented. So these minerals do not give the age of the sedimentation in basin, but that of the parent rock. With authigenic minerals, though, like glaucony formed in situ with no other reworking, dating is possible. The elements most often used are carbon, potassium, and rubidium.

Carbon is of use for recent deposits (Quaternary) because its decay period from ^{14}C to ^{14}N is short (about 5730 years). When the carbon from CO_2 is incorporated in living matter or used for fabricating test, it has a known $^{14}C/^{12}C$ ratio. After the organism's death (closure of the system), there is a gradual decay of ^{14}C with time that can be used to date deposits from 100 to 100 000 years.

The potassium-argon and the rubidium-strontium methods are also commonly used for dating sedimentary rock. The periods are longer, ranging up to several hundred million years, and the dates are usable if certain precautions are taken (Fig. 4.38).

Generally, all of today's radiochronology techniques (including U/Pb for granites and neodimium for magmas) are combined with other methods.

For example, the **fission track** technique measuring the radiation-path damage caused by spontaneous fission of ^{238}U is a method of dating the crystallization of minerals that can provide a complementary measurement, even though it is not always very precise and reliable for absolute dating.

A

^{87}Rb →	^{87}Sr	$4.7–5 \cdot 10^{10}$
^{232}Th →	^{208}Pb	$13.9 \cdot 10^{9}$
^{40}K →	^{40}Ar	$11.9 \cdot 10^{9}$
^{238}U →	^{206}Pb	$4.6 \cdot 10^{9}$
^{235}U →	^{207}Pb	$7 \cdot 10^{8}$
^{234}U →	^{230}Th	250 000
^{230}Th →	^{226}Ra	75 200
^{14}C →	^{14}N	5 568

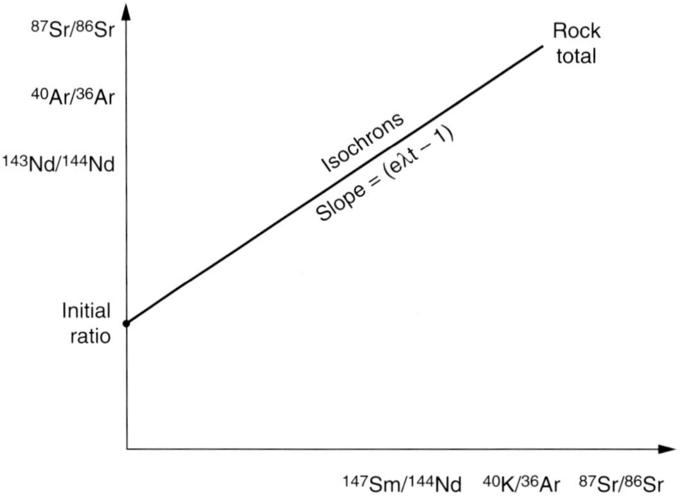

B

Fig. 4.38 Radioactive elements and isochron diagrams.
A. Elements and their decay processes and periods: the principle transformations used in radiochronology. **B.** Different ratios can be used. If all of the samples analyzed are of the same age (with the same initial isotope ratio), the analytical points will define a line whose slope is a function of the age (from N. Clauer, in J. Rey, 1997). The total rock is considered to be a closed system. The line obtained with the isotope ratios determines the rock age starting from the time of magma crystallization. The time is computed from the line slope.

New methods now exist using the isotope ratios of He and Ne to estimate the time rocks have been exposed to radiation at the surface. This can be of use in approximating questions relative to erosion surfaces. New **thermoluminescence** and **electron spin resonance** techniques are now being applied to date series ranging from a thousand to a million years, both in archaeology and in geological studies of the Quaternary, and to record the effects of the diagenesis.

This leads to the frontier of **chemostratigraphic** techniques (Fig. 4.39) and the objects studied in diagenesis (variations in stable isotopes as time signals of temperature, salinity, and other parameters), which will be addressed in the following chapter.

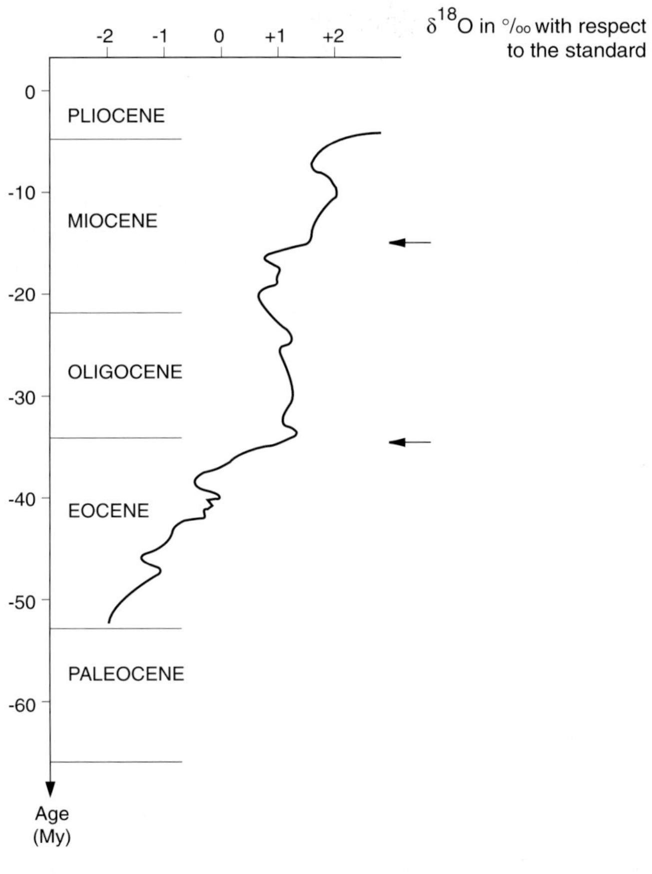

Fig. 4.39 Chemostratigraphy: example of isotopic stratigraphy.
The time variation in the isotopic composition of oxygen, carbonates, and planktonic foraminifer shells, is used to define an evolution in the course of the Cenozoic with events (rapid changes) indicated here by arrows (from Vergnaud Grazzini, 1979, and Oberhaensli, 1986).

4.2.7 Mineralogical and Geochemical Markers, Chemostratigraphy

4.2.7.1 Mineralogical Markers

Mineralogical **markers** have long been used in stratigraphy. The minerals used are mainly **heavy minerals** and sometimes clays. Though mineralogical approaches may be of great importance for local or regional **correlations**, they are of dubious value in the stratigraphy of areas beyond the regional scale. The input is indirect, so the observed assemblage can provide information concerning climatic or dynamic conditions over a given period. It can also be said that the mineralogical marker idea is used extensively in **well logging** (indispensable to petroleum geology), where highly radioactive clays serve as markers on gamma ray curves (Fig. 4.40).

4.2.7.2 Tephrachronology

Tephrachronology is a branch of stratigraphy that uses levels of volcanic **ash** or **tephra** as event markers. The example already mentioned of the catastrophic eruptions of Pinatubo and Mount Saint Helena or the older Santorin illustrate how volcanic eruptions will eject pyroclastic materials which briefly transit through the upper atmosphere, are disseminated by the wind, and are sedimented on a world scale. The ocean then gives us an archive of a unique, spectacular event: a level of **cinerite** (ash tuff or tephra) corresponds to a sudden eruption. Such levels of cinerite, preserved or transformed by diagenesis into tuffite, serve as valuable **chronohorizons** of the geological past, making it possible to establish correlations between continental (lacustrine or palustrine) and oceanic environments. They are very important for lateral correlations, especially at the regional level (Fig. 4.41).

This discussion cannot be closed without reference to the catastrophic volcanic eruptions of the Deccan, which are proposed as an explanation for the great faunal extinctions at the end of the Cretaceous. Without taking sides on this issue, let us simply mention that the Earth's great magmatic cycles—like the tectonic cycles mentioned previously—can be used in a very broad stratigraphic framework (Fig. 4.8).

4.2.7.3 Chemostratigraphy

While fossils are valuable indicators in the relative dating of geological beds, in that they bear witness to the evolution of living beings, geochemical traces are also indicators of time variations in ocean composition (Fig. 4.39). Chemostratigraphy is a new science (the term did not exist in the international stratigraphic guide of 1979) and a new correlation tool. The basic idea is that, while the ocean acts as an enormous regulator, it still fluctuates, and greatly so. Long periods of Earth history are characterized by certain types of deposits.

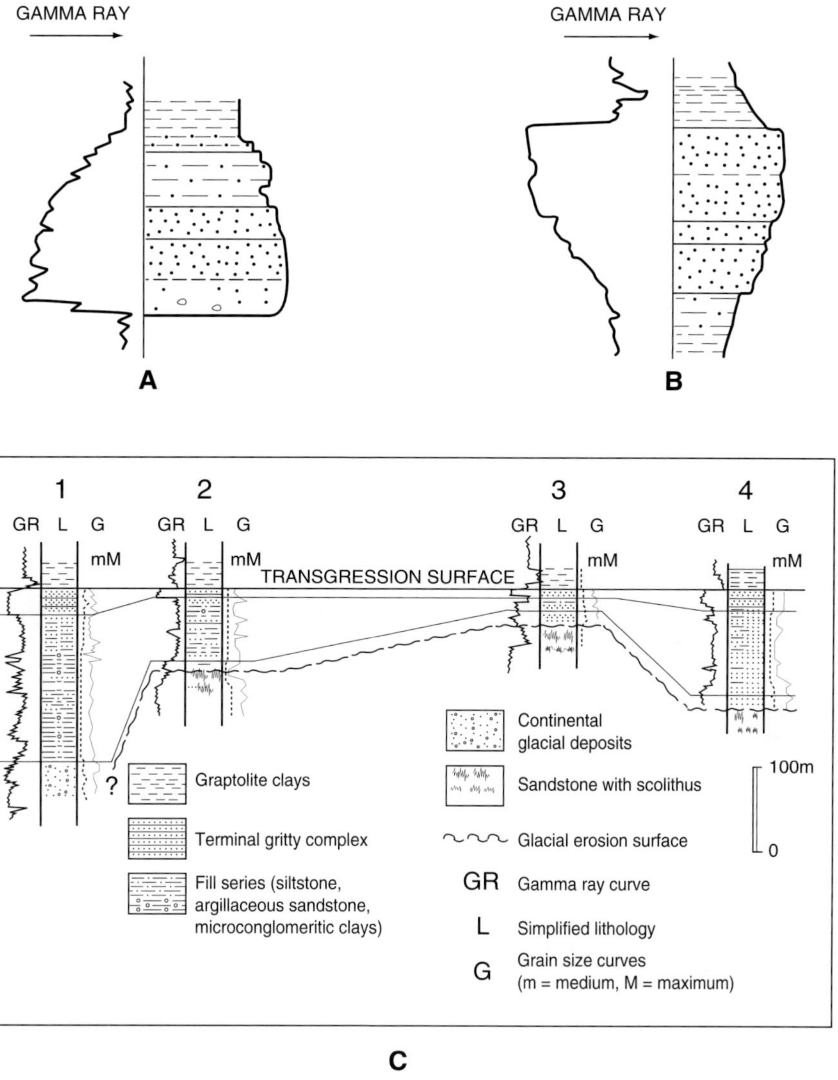

Fig. 4.40 Gamma ray curve indication of shaliness.
In example **A**, the sandstones specific to the base gradually turn to argillaceous sandstone, silts, and then clays. In example **B**, the sequence is negative: a sudden break is observed at the roof of the sandstone, with an especially sharp radioactive peak in an argillaceous level rich in organic matter. Example **C** represents the Silurian transgression with highly radioactive clays at the roof of the glacial formations of the Saharan Upper Ordovician (from Beuf et al., 1971).

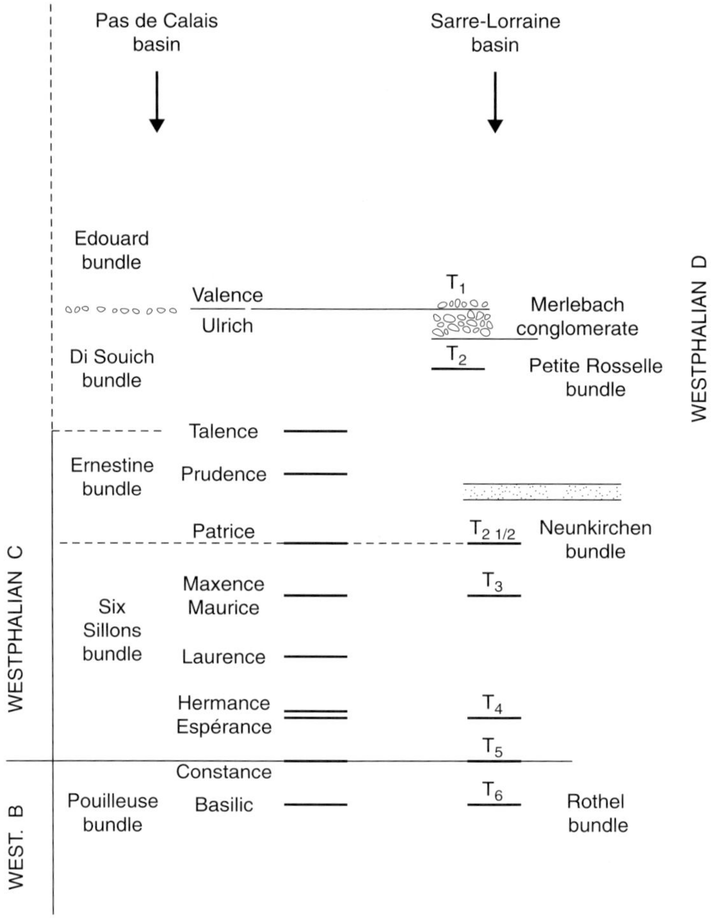

Fig. 4.41 Volcanic horizon correlations.
Here, the tonsteins (ancient ash levels) of the Paleozoic in the coal basins of northern France are used to compare different coal-producing bundles (from Pomerol, 1982).

4.2.7.4 Variations in the Major Elements

Variations in the chemical composition of the atmosphere and hydrosphere over the course of geological time, and climatic variations too, all leave certain periods imprinted with a particular lithological signature. Moreover, the theoretical constituents participating in the ocean's biological cycles (oxygen, carbonates, nitrates, phosphates) have fluctuated with the evolution of the major phyla.

The sediments first deposited during the Achaean were basic volcanoclastic debris with an abundance of heavy metals such as copper and iron. It is not known if the warm seas (60° to 90°C) were acid or, on the contrary, sodic or alkaline with very low concentrations of calcium and magnesium. The atmosphere was doubtless reductive.

Then the oxygen content increased, and the first deposits of evaporite appeared toward the end of the Protozoic, with carbonate formations, and then a high Mg/Ca ratio (relative importance of dolomites) (Fig. 4.42). But it should be noted that this is more of an imprecise definition of **paleoenvironments** than it is a stratigraphic analysis. Starting with the Cambrian, global variations are finer and require a more detailed approach with the definition of geochemical markers having rather good resolution (including extraterrestrial events such as meteor impacts).

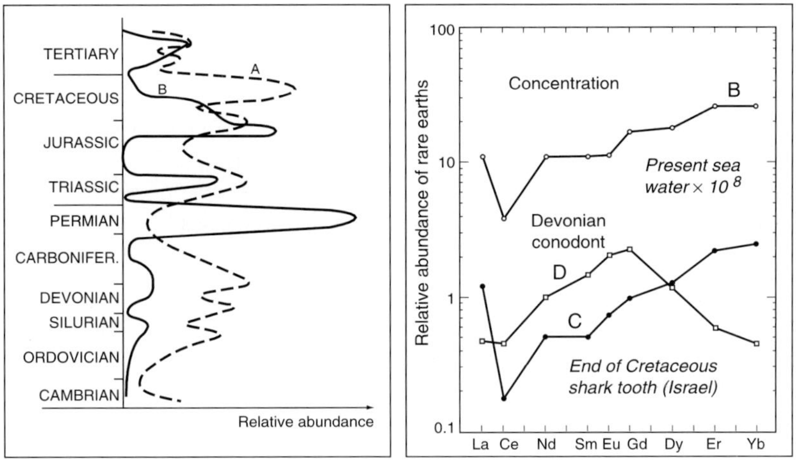

A **B**

Fig. 4.42 Stratigraphic variations in the chemical composition of deposits. **A.** The abundance of carbonates (A) and evaporites (B) is variable in sedimentary deposits, which are archives of ancient chemical composition. The very general curves express tendencies (from many authors). **B.** The normalized distribution of rare earths in present sea water compared with ancient fossil contents: shark tooth from the end of the Cretaceous (C) and from the Devonian (D) (from Alabrede, 1990).

Some euxinic (closed marine environment) events are also recognized as geochemical markers. Beds rich in organic matter from the Silurian, Devonian, Toarcian, Kimmeridgian, and Cenomano-Turonian boundary, often called **black shales**, are spectacular events that depend on global climatic conditions (Fig. 4.43). This will be seen again in the petroleum systems of Chapter 6.

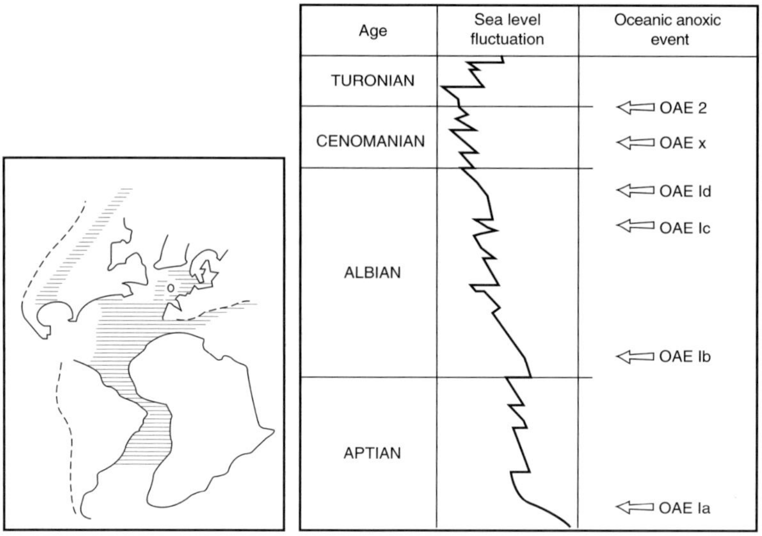

A B

Fig. 4.43 Black shales of the Tethyan and Atlantic domains.
A. Global anoxic environments at the Cenomano-Turonian limit in the Atlantic Ocean (from P. C. de Graciansky et al., 1986). **B.** Stratigraphic position of a certain number of oceanic anoxic events between the Aptian and Turonian in the Tethyan and North Atlantic domains. Sealevel fluctuations are diagrammed roughly from Vail and Haq's curve.

The present distribution of **rare earths** in today's sea water and in recent phosphated debris differ from what it was in the past, according to analyses performed on ancient phosphated debris as with the conodonts, both in absolute quantity and in relative proportions (this may be due to major changes in the distribution of nutrients in the water column, which confirms that the food chain was not the same in the Paleozoic as it is today).

Mention should be made of abnormal contents in a given element over short periods of time. Namely, iridium concentrations at the Cretaceous-Tertiary and Eocene-Oligocene boundaries are notable **geochemical events** of world-scale stratigraphic value, even though the exact significance of these anomalies is still controversial (Fig. 4.44).

4.2.7.5 Isotopic Variations

Another stratigraphic technique is based on the analysis of stable isotopes of certain elements.[1]

1. From the works of Emiliani, 1955.

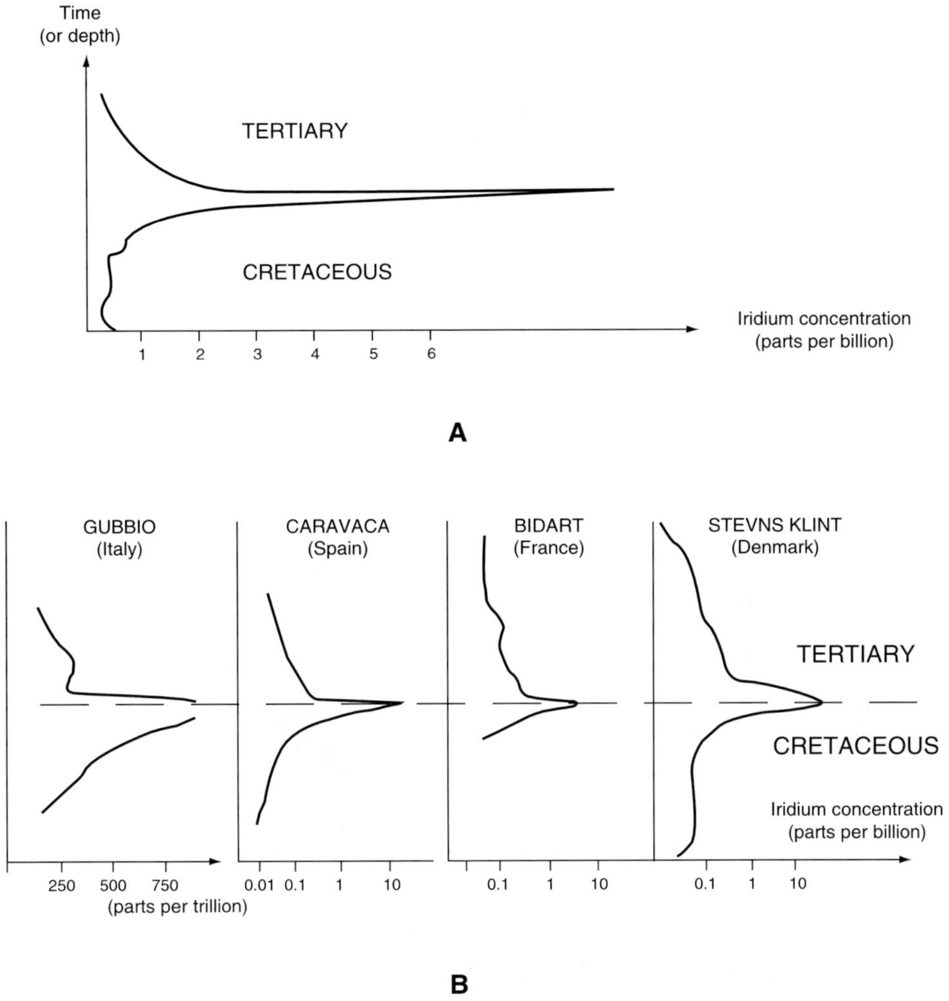

Fig. 4.44 Geochemical event: iridium anomaly at the Cretaceous-Tertiary boundary.
A. Detailed analysis of a sample in a geological section reveals a remarkable anomaly of high concentration, compared with average values. **B.** This event is observed in many sections, proving that the event was global (from V. Courtillot, 1993). Its cause, however, is still being debated (meteorite impact? volcanism?).

This is **isotopic stratigraphy**. The ^{13}C and ^{12}C contents, for example, have varied synchronously (in a ratio $^{13}C/^{12}C$ expressed in $\delta^{13}C$) in the ocean over time, even though the genetic reason for this is once again controversial (there is a tie-in with global eustasy and the extent of the euphotic zone in times of transgression, for example). This enters upon the field of paleo-oceanography and paleoclimatology (Fig. 4.45).

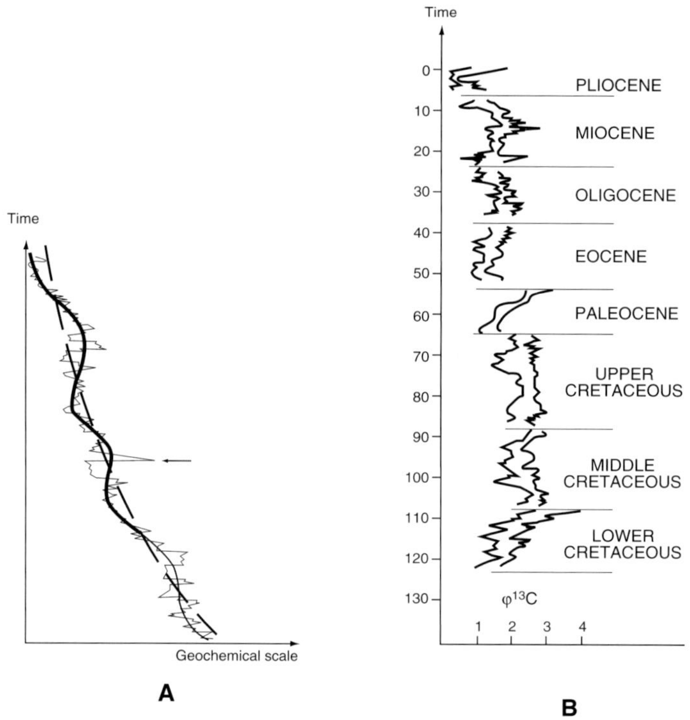

Fig. 4.45 Isotopic stratigraphy: time fluctuation.
A. Types of variation depending on the scale of analysis. The general long-term trend is the dashed line, the short-term trend defining cycles is the heavy solid line, and the high-frequency variation is the thin line. The geochemical event is marked by an arrow. **B.** Time variations in the isotopic ratio of carbon in pelagic carbonates (from M. Renard, 1985).

The ratios most commonly used (including in petroleum geology) to calibrate correlations between genetic units are the oxygen isotopes. The $^{18}O/^{16}O$ ratio of a carbonate precipitated in sea water increases inversely with temperature, so this gives us a precise **paleoclimatic indicator** of great stratigraphic quality today for synchronizing glacio-eustatic variations: this type of data is extensively used for calibrations in the study of the recent Quaternary. Results are still good when it is used for the Cenozoic, but they become more difficult to interpret for older geological series because of the re-crystallization phenomena that occur in diagenesis, when the rock is buried.

Other elements like strontium ($^{87}Sr/^{86}Sr$) and sulfur ($^{34}S/^{32}S$) are also used.

The following chapter will show us that isotopic values can also be used to characterize and sometimes date post-sedimentary events occurring either in the burial history or when the rock rises closer to the surface or is bared to the open air.

4.2.8 Paleomagnetism and Magnetostratigraphy

The first chapter on the Earth's structure made mention of the magnetic reversals recorded in basic magmas produced in the ocean ridges, from which the evolution of the oceanic domains has been retraced (Fig. 1.33).

The characteristics of the Earth's magnetic field at a given time can be read in the thermoremanent magnetization of certain ferromagnetic minerals, giving us the intensity, inclination, and declination of the field. A number of **reversals** in the course of time have thus been recorded on the world scale. These are considered to be near-instantaneous on the geological time scale. This yields a stratigraphic scale of magnetic polarities with periods of **normal polarity** and others of **reverse polarity** (Fig. 4.46). These reversals are frequent in certain periods (Lower Cretaceous, Miocene, Quaternary), while other long periods (the Callovian and Albian-Cenomanian) are characterized by a high level of stability. This method of dating and correlating events is called **magnetostratigraphy**, which can at times be a high-resolution tool, depending on the frequency of the reversals. A period of time corresponding to a given polarity interval is called a **magnetopolarity zone**.

Paleomagnetism can be measured in sedimentary rock, but care should be taken to distinguish between what is due to the orientation of the magnetic particles at the time of sedimentation, called **detritic remanent magnetization**, and what is due to secondary **diagenetic effects** (chemical remanent magnetization).

In conclusion, **magnetopolarity zones** offer a good relative scale which, in combination with biostratigraphic and radiochronological data, is coming into greater and greater use in sedimentary geology.

4.2.9 Other Methods

It is clear that no method is sufficient all by itself but that relative dating is possible and local, regional, and global correlations can be made if many different methods are used.

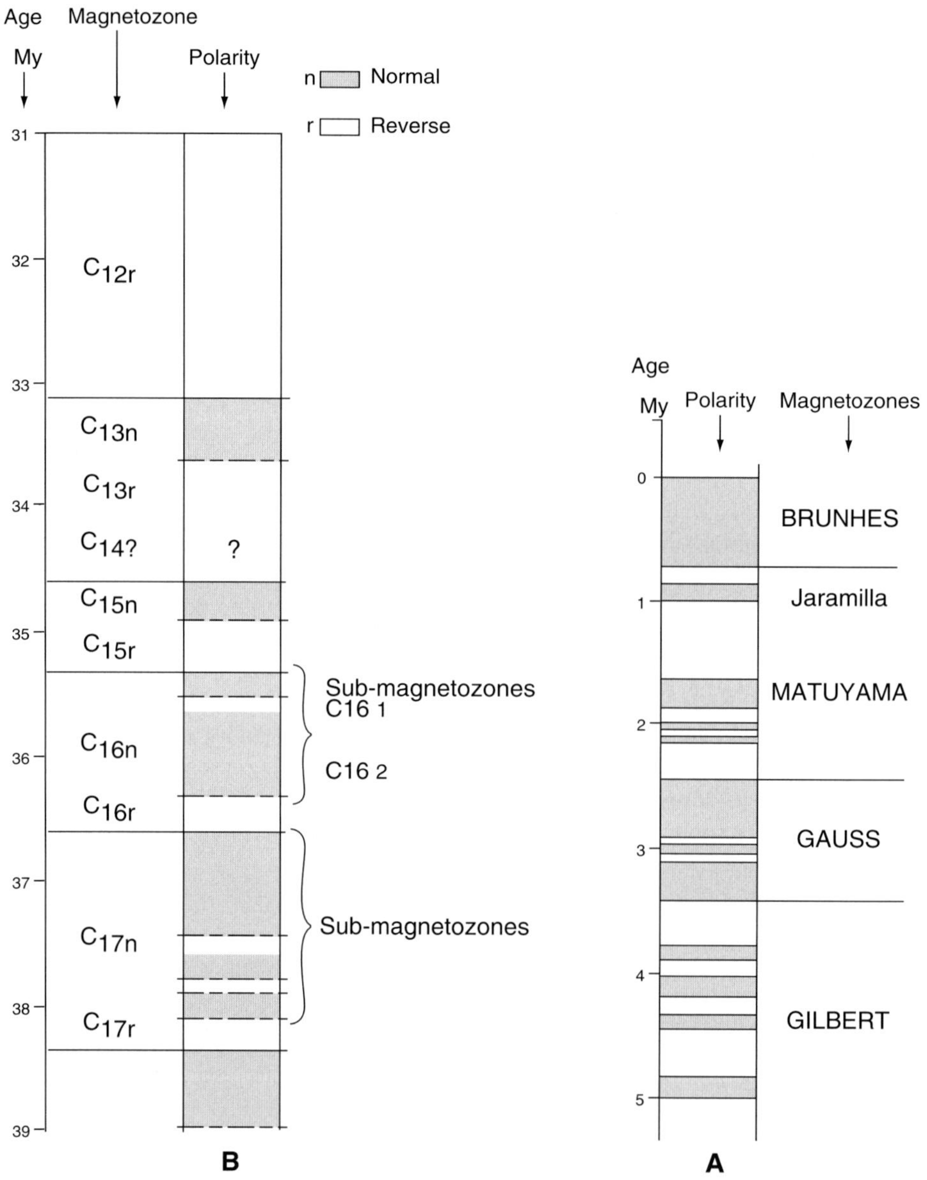

Fig. 4.46 Principle of magnetostratigraphy: a sequence of normal (n) and reversed (r) polarities in geological time.
A. Recent scale. **B.** Portion of the Tertiary. Note that a magnetopolarity zone can itself be subdivided.

To get good absolute dating, it is often necessary to use a number of techniques in combination. The geologist at times uses complementary techniques, based on working hypotheses or concepts, or special tools.

An example of a working concept is the indicator of tectonic deformations (age of folding manifested by **discordances**, age of metamorphosed minerals, age of erosion periods), but we have emphasized the rarely isochronous and normally regional or even local character of this indicator. Climatic changes can now be identified if the changes are found to be synchronous on the global scale. This holds for the glacial events of the Quaternary, and also for those of the Permian, the Upper Ordovician, and the Precambrian. Here again, though, direct indicators (glacial or periglacial facies, paleocirculation, reconstructions) do not by themselves yield fine resolution.

Especially in petroleum geology, stratigraphic data are very often complemented by seismic data in a technique called **seismic stratigraphy**, with a further complement from well logging. These are remarkable correlation tools when used together. Much of the rest of this chapter covers this.

4.3. SEISMIC, SEQUENTIAL, GENETIC STRATIGRAPHY

4.3.1 Seismic Stratigraphy

Seismic stratigraphy can be defined as the use of reflection shooting to recognize and analyze space-time relations in a sequence of geological beds.

Seismic stratigraphy is based on two simple postulates:

- Primary seismic reflections are produced when a seismic wave passes through a **discontinuity** producing a **contrast in acoustic impedance**. The impedance is the product of the density of the beds the wave passes through, and the wave propagation velocity.
- The discontinuity surfaces thus act as mirrors reflecting different densities and elastic properties, and therefore constitute locuses of discontinuity or discordance at stratum boundaries (Fig. 4.47).

Without going into details on the geophysical technique,[1] advances in seismic acquisition and processing have now made it possible to use these sections not only for the structural aspect, but also for recognizing sedimentary units in increasing detail, dating sequences, and identifying their general paleoenvironment (paleobathymetry, paleogeography). It should be emphasized that one of the strong points of this seismic technique is its capacity to obtain **continuous data** over tens or even hundreds of kilometers, and sometimes in three dimensions. So this is a rather global stratigraphic approach that is rarely pos-

1. Lavergne, 1986.

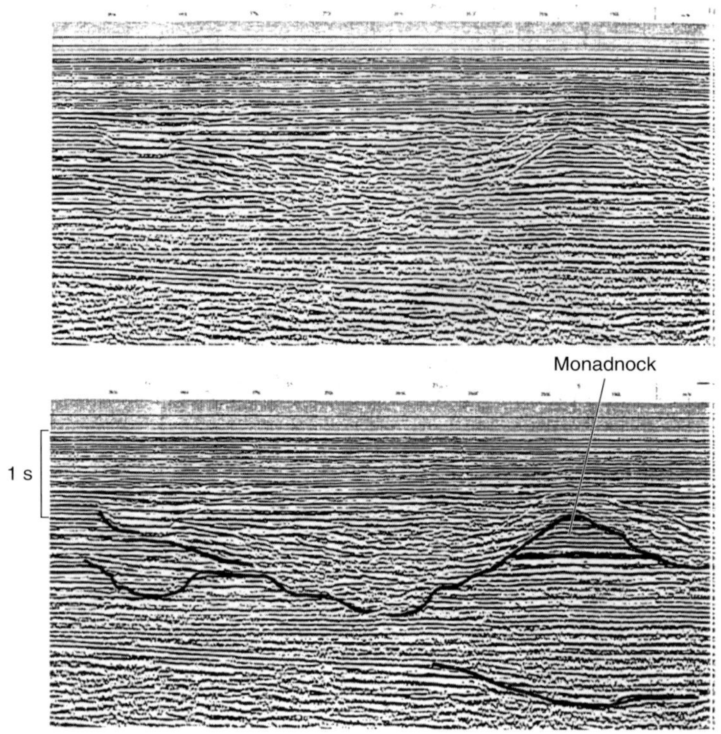

Fig. 4.47 Seismic reflectors: stratigraphic horizons and discontinuity surfaces.
The graph shows continuous reflectors corresponding to a regular stratigraphic sequence, along with horizons cut by an irregular or erosion surface (example offshore the Indus, from Ravenne, ENSPM document).

sible when working only on outcrops. This is why advances in seismic stratigraphy (strictly speaking, it should be called stratigraphic interpretation of the seismic) has been used to recognize **different types of deposition sequences** and **correlate** these sequences, and has prompted progress in **stratigraphic concepts** themselves.

4.3.1.1 Seismic Facies Analysis

A **seismic facies** analysis is a description and geological interpretation of reflection shooting parameters, which are the reflection pattern, amplitude (high or low), frequency (high or low), continuity (strong or weak), and interval velocity (characteristic of the interval between two reflectors, expressed in meters per second).

Seismic facies units are groups of reflections with parameters that are similar to each other but different from those of adjacent units. The configurations are usually apparent because they are viewed in two dimensions, but 3D units also exist.

The **external shape** of the seismic facies units and their **reflection configuration** are used to define **seismic sequences**. The main types of patterns observed (Fig. 4.48) are:

- Parallel or near-parallel
- Divergent or wedge-shaped

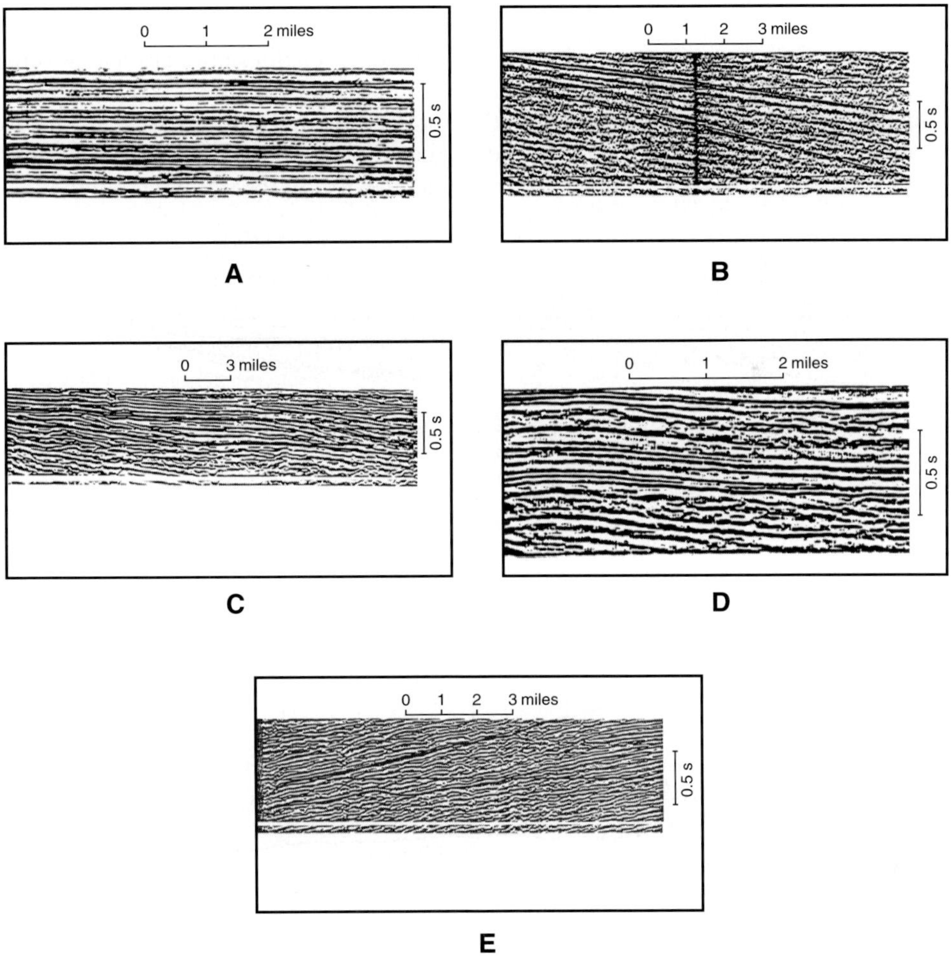

Fig. 4.48 A few typical seismic horizon configurations.
A. Parallel with good continuity. **B.** Divergent. **C.** Oblique or S-shaped.
D. Hummocky. **E.** Chaotic (from Ravenne, ENSPM document).

- Prograding oblique S-shapes or more complex shapes due to gradual lateral development of deposition surfaces, called clinoforms
- Chaotic
- "Transparent" configurations with no reflections.

The advantage of being able to recognize the external shapes of the units is that the sedimentation conditions can often be determined quite finely, in draping, subaquatic dunes, or pinnacle reefs, for example (Fig. 4.48).

4.3.1.2 Seismic Deposition Sequence

A deposition sequence can be recognized by the geometric relationships between the different strata. The sequence has a chronostratigraphic significance because it is defined by discontinuities (Fig. 4.49) of differing extent at its roof and floor. These surfaces and hiatuses indicate erosion or non-deposit, and often a discordance (corresponding to a geological time period which will be easier to evaluate from the lateral extent of the seismic profile).

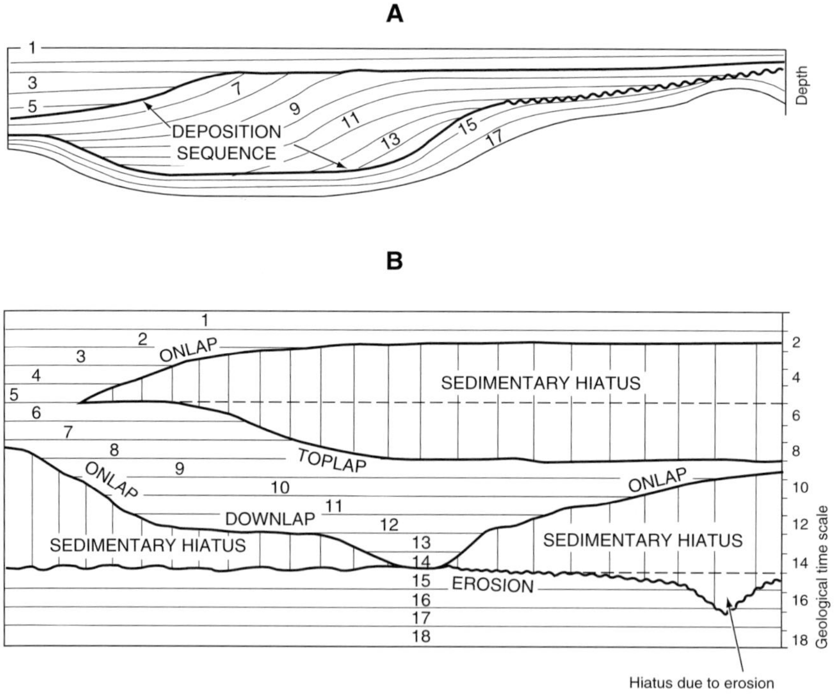

Fig. 4.49 Seismic deposition sequence.
A. Simplified geological section. **B.** Chronostratigraphic section (from ENSPM document, 1986).

The thicknesses and the correlative sequence durations may vary greatly, from a few meters to several hundred. It must be kept in mind that the definition of the seismic facies, the reflection profile, the external shape of the units, and the sequence definition, all depend entirely on the seismic resolving power. When this is high, the stratigraphic variations can be detailed very finely. The profile of the Gulf of Lions (Fig. 4.50) is a remarkable example of this.

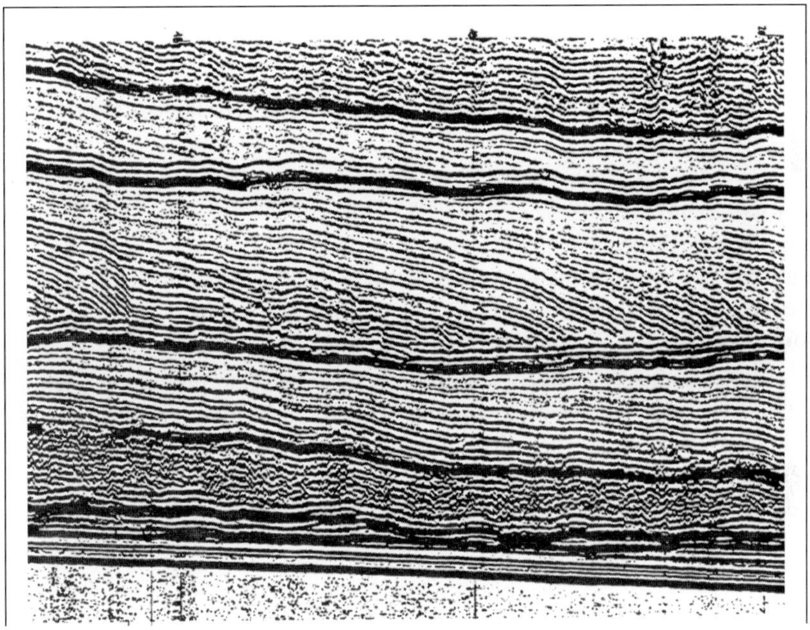

Fig. 4.50 Deposition architecture illustrated by high resolution seismic in the Gulf of Lions (from Chihi, 1997), showing progradation systems.

Seismic acquisition techniques do not permit high resolution simultaneously with great depth penetration. In ordinary petroleum seismics, the required depth of several thousand meters will allow a resolution only of the order of some ten meters.

Important **geometric relations** (Fig. 4.51) can be found among the seismic reflectors (which will be interpreted here as seismic markers of strata) indicating:

- **Concordance** of continuous strata
- **Onlap**, which is the horizontal termination of a stratum at an older topography coming up from underneath
- **Downlap**, which is the downward termination of a stratum on an older surface that is horizontal or less inclined.

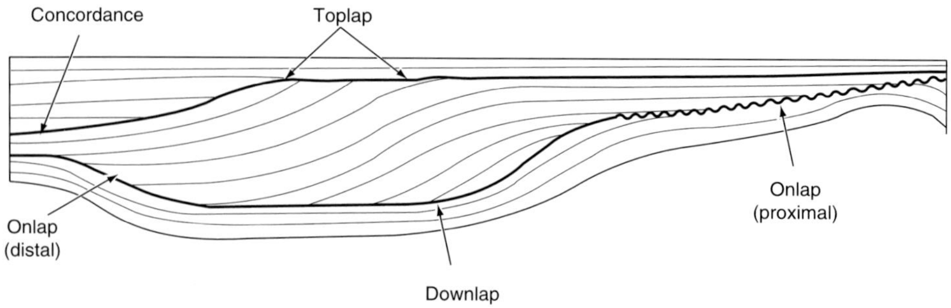

Fig. 4.51 Geometric relations of seismic strata and discontinuity terms.

The roof will exhibit:

- Stratum **concordance**
- **Toplap**, an upward termination of a stratum under the horizontal surface of a younger sequence
- **Truncation** under an irregular surface.

This is purely geometric analysis with no preconceived interpretation in sedimentological terms. However, sure figures of downlap are clearly indicative of prograding bodies, and truncation is obviously a sign of erosion (Fig. 4.52). Insofar as the observed geometries can be interpreted thusly as deposition sequences, they feed the field of sequential stratigraphy with a model for interpreting the organization of sedimentary series.

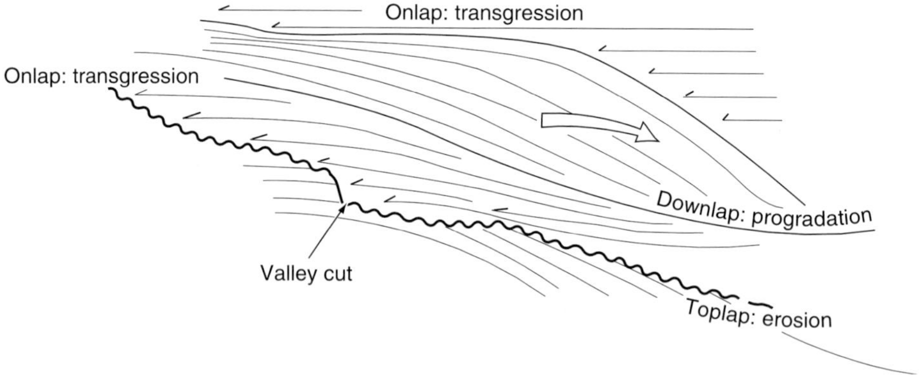

Fig. 4.52 A few geometric figures interpretable in seismics.

4.3.2 Sequential Stratigraphy

Sequential stratigraphy grew out of the methods of seismic stratigraphy and the concepts of the general Exxon model[1]. The basic principle is that sedimentary series are organized in a logical series of sequences that were deposited essentially under the control of **relative ocean level fluctuations**. These relative ocean level variations are the combined result of absolute variations in the average sea level and crustal deformations (tectonics) (subsidence and uplift, see Chapter 5). Eustasy has a precise stratigraphic value because it affects the entire oceanic bowl in all marine basins, controlling their lateral extent in eustatic transgressions and regressions. It also affects the continental equilibrium profiles, and thereby continental sedimentation. Tectonics, as has already been said, is often more regional in character than is eustasy. But on the scale of a basin or several basins resulting from the same geodynamic process, the relative sealevel fluctuations it causes can be considered as synchronous events. The environmental factors of climate, ocean dynamic, and sediment flow, were originally considered to be secondary factors. Actually, these various parameters are highly interdependent. The major role of concentrated or diluted sedimentary inputs is being stressed today, even if its effects are difficult to measure.

In early perspective views, it was also postulated that the response to relative sealevel variations was very similar for all types of environments, whether detritic, siliciclastic, carbonaceous, or combined.

The major point that can be made on the basis of seismic data is that the cycles of relative sealevel variations can be reconstructed by analyzing the geometries of seismic facies sequences. Seismic reflectors correspond to deposition surfaces, i.e., to elementary beds, so they are not only significant in sedimentology but also represent time lines, or **isochrons**. Each corresponds to a particular situation in the basin. The hiatuses, which reflect the fact that the sedimentation in a basin is a discontinuous phenomenon, are not represented by a reflector, but form composite surfaces, i.e., are the sum of several reflectors. So certain surfaces represent different lapses of time (Fig. 4.53). The chronology of eustatic variations can then be represented in the same way as the time spaces represented on chronostratigraphic charts, with the sequence of deposits and sedimentation hiatuses. This is the Wheeler diagram of Fig. 4.54.

4.3.2.1 Allocyclic Variations, Sea Level, Accommodation

Eustasy, tectonics, and sedimentary inputs are by definition **allocyclic processes** in that they are independent of and external to (allochthonous) the deposition system itself, as opposed to **autocyclic** processes that are defined by deposition system mechanisms such as periodic tidal current oscillations or lateral migrations of fluvial channels.

1. Vail et al., 1977.

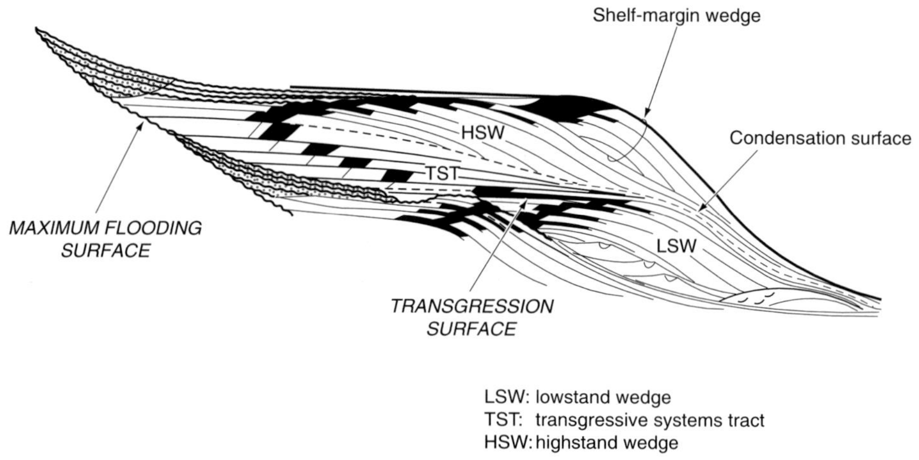

Fig. 4.53 Stratigraphic profile (from Posamentier et al., 1988).

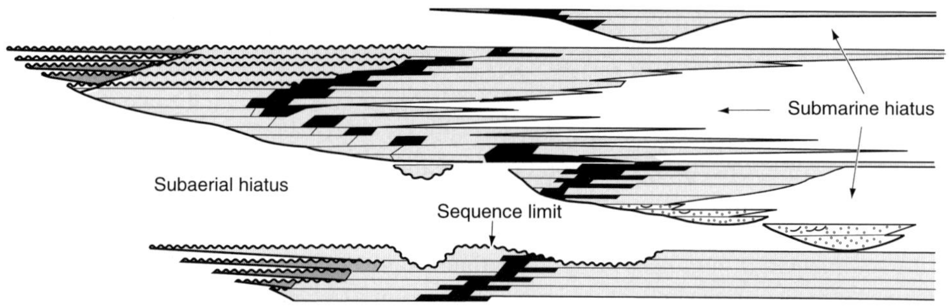

Fig. 4.54 Chronostratigraphic diagram of a deposition sequence, with the various deposits represented in a space-time diagram (from Posamentier et al., 1988).

Sea level seems to be a simple idea. But it should be remembered that the general reference is an **average sea level**, which is suggestive of sealevel variations in space. For example, the discussion of the geoid showed that the ocean's surface is not regular. Anomalies can be found in the deep ocean structure and in the ocean's own dynamic. "Average sea level" also says that the ocean's level varies in time, including the very small time steps corresponding to tidal oscillations. Oceanographers use the concept of **bathymetry**, which means the water depth between the free surface and the ocean bottom (Fig. 4.55). In this book, the **eustatic level** or absolute sea level is the height of the sea's mean level with respect to a fixed datum. The **relative level** is not the bathymetry but the distance between the free surface of the sea and the substratum of the sedimentary basin.

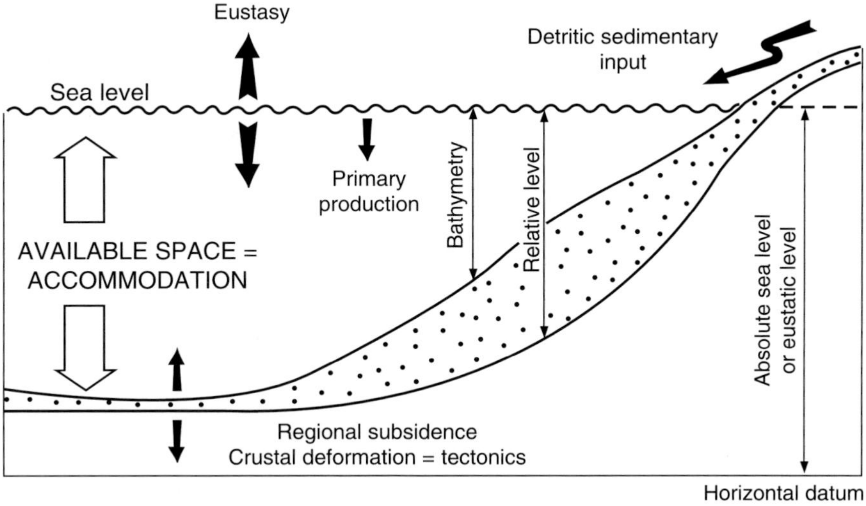

Fig. 4.55 Available space or accommodation.
Accommodation in a basin can be defined as the resultant of eustasy and tectonics (subsident or uplifting) in light of the sedimentary inputs (primary production and detritic flows). Fluctuations in the relative level determines how much space will become available.

Eustatic and tectonic variations can increase or decrease the **space actually available** for sedimentation, i.e., the theoretically open space (Fig. 4.55), which is called **accommodation**[1] space or power. Accommodation is defined as the space between the substratum and the new sea level after the change.

Strictly speaking, then, this space is not fully available. For simple physical reasons, the space (or volume) of the water column cannot be filled entirely with sediments, so only a fraction of this space is available for sedimentation at any time.

It is more important to remember the **rate** of creation or decrease in the available space, which is also called accommodation potential (Fig. 4.56). (A distinction can be made between accommodation that is "filled" at each time step, which corresponds to the thickness of sediments accumulated during this time step, and "unfilled" accommodation which would be the space actually available in the strict sense of the term: the bathymetry).

4.3.2.2 Deposition Sequence: Definitions

A **deposition sequence** is a stratigraphic unit comprising a regular sequence of relatively concordant, genetically related beds lying between two unconformities. Strictly speaking, a

1. See Jersey, 1988.

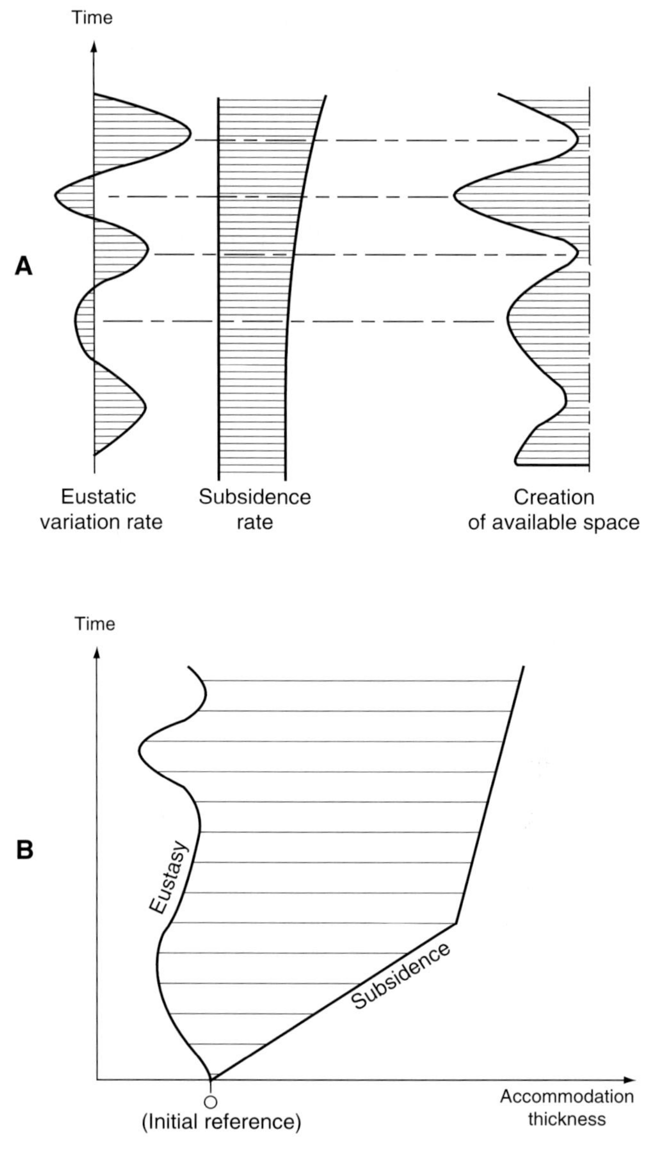

Fig. 4.56 Accommodation potential: rate of creation of available space. If the subsidence curve varies and eustatic fluctuations too, then the both variables will determine accommodation potential. **A** and **B**. Two ways of representing this.

deposition sequence should be a set formed in the course of a full relative sea-level variation cycle (Fig. 4.57); but the choice of boundaries is arbitrary because this variation is actually a continuous phenomenon. For the Exxon school, the easiest boundary to recognize in the seismic is the unconformity. The deposition sequence then, by this definition, comprises a chronological suite of **systems tracts** exhibiting the following, from the base (beginning) to the top (end):

- A **basal erosion surface**, discordance, or major unconformity formed when the sea level falls, witnessing to a major modification in the **equilibrium profile** of the streams as they readjust to a new **baselevel**. The baselevel is generally considered to be the zero level of the seas (Fig. 4.57). Concretely, canyons and **cut valleys** will be dug out of this level (Fig. 4.58).
- A **lowstand system tract** sedimented during the drop in level and the low level period. Different types of deposits can be found (deep-sea cones, hemipelagites) depending on the sedimentation rates. The resulting deposit set is also called a **lowstand wedge** (LSW).
- The **transgressive system tract** (TST) that develops during the sea-level rise phase. This transgression phase may be evidenced in a **flooding surface**, but also by a diachronous erosion surface, as will be seen later on. The sediments then exhibit aggradation and gradual retrogradation. At the maximum sealevel rise, this surface becomes the **maximum flooding surface**.
- The **highstand system tract** or **highstand wedge**, which corresponds to all of the deposits of the high level phase until the fall begins. Here again, different types of aggradational and prograding deposits occur offshore (shelf edge wedge).

Initially defined for terrigenous environments, these elementary deposition tracts are more or less identifiable on the seismic by their position in the bathymetric profile. It is clear that lowstand wedges preferably develop downstream (Fig. 4.59) while only the upper parts of the transgressive tract and the lower part of the highstand tract will be developed upstream. In other words, the stratigraphic record of the elementary sequence varies greatly in space. The Vercors outcrops illustrate this type of lateral variation, which can also be observed in the field in carbonaceous environments (Fig. 4.60).

Accommodation will be controlled by eustasy or by tectonics, depending on which varies faster (Fig. 4.61). Rigorous glacio-eustasy was the main variation parameter in the recent variations of the Quaternary (Fig. 4.62A), while tectonic control seems to have been preponderant in the example of the Mesa Verde Cretaceous (Fig. 4.62B). **Forced regression** is used, for example, to define a situation that is guided largely by tectonic effects (Fig. 4.63).

4.3.2.3 Different Types of Sequences

The elementary pattern that has just been described corresponds to cyclic variations in the relative sea level. It is known that several tens of millions of years are being observed on the geological time scale (Fig. 4.64). Also, cyclic fluctuations are being detected today at very much shorter and very regular periods, such as the cyclic Milankovitch periods.

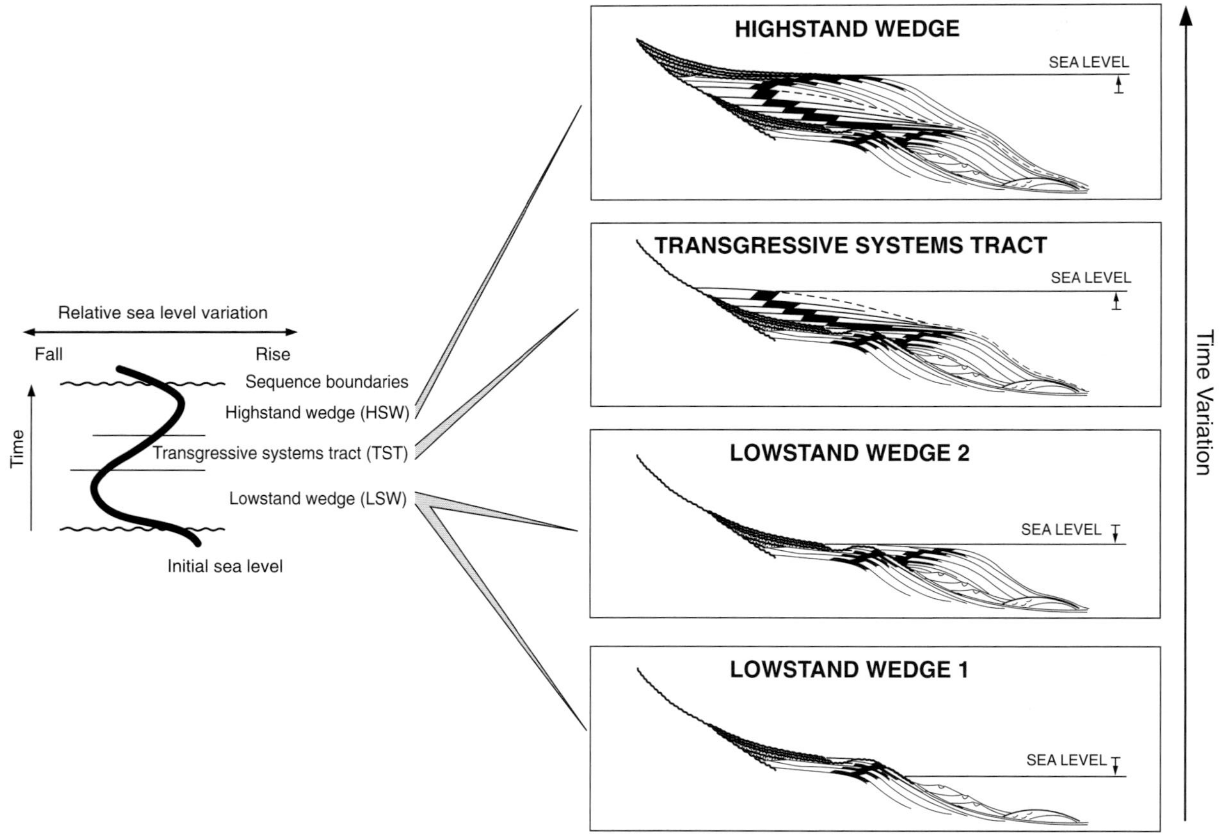

Fig. 4.57 Vail deposition sequence.
The fluctuation in the relative sea level between two unconformities (constituting the sequence boundaries) determines the distribution of the sediments on the margin in lowstand wedges, transgressive systems tracts, and highstand wedges.

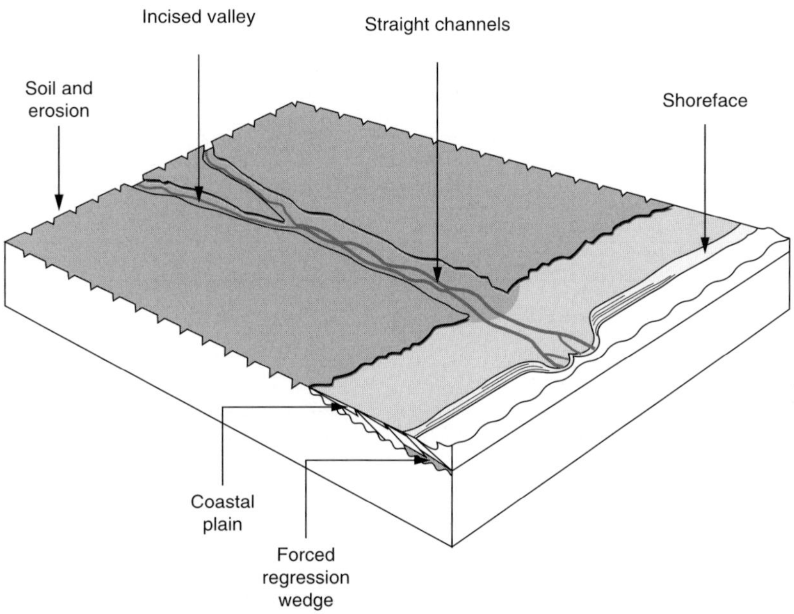

Fig. 4.58 Incised valley.
With the fall in sea level, the equilibrium profile is changed, the valley is cut out and its erosion products form downstream in a "forced regression" wedge (from an IFP document).

Following the Exxon conventions, deposition sequences are broken down by duration into six orders:

Order	My	Order	My
1	> 50	4	0.08 to 0.5
2	3 to 50	5	0.03 to 0.08
3	0.5 to 3	6	0.01 to 0.03

The elementary pattern that is generally clearly observed in conventional seismics corresponds to third-order sequences. But it is clear that sequences of higher order will be defined as the resolution of the seismic increases (and the increased detail of the terrain or subsurface studied). This point will be discussed in the following section on **genetic units**, which are considered to be the elementary deposition sequences.

It will mainly be remembered that oscillations, whether periodic or not, can combine with each other. The sedimentary record then becomes a complex one to decode in reality.

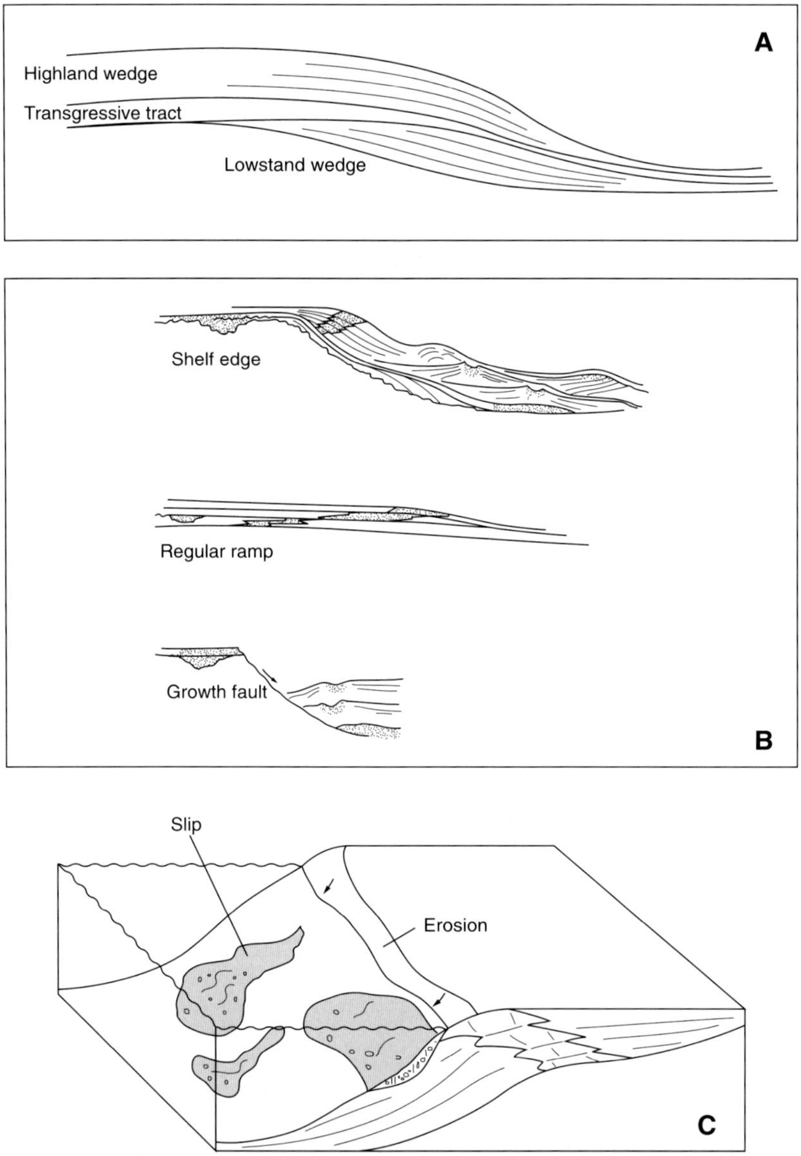

Fig. 4.59 Lowstand wedge.
A. Diagram of landward lowstand and highstand wedges. **B.** Different lowstand wedges in the detritic environment (from Vail). **C.** Example of a rapid fall in sea level in a carbonate environment (from Sarg, 1988).

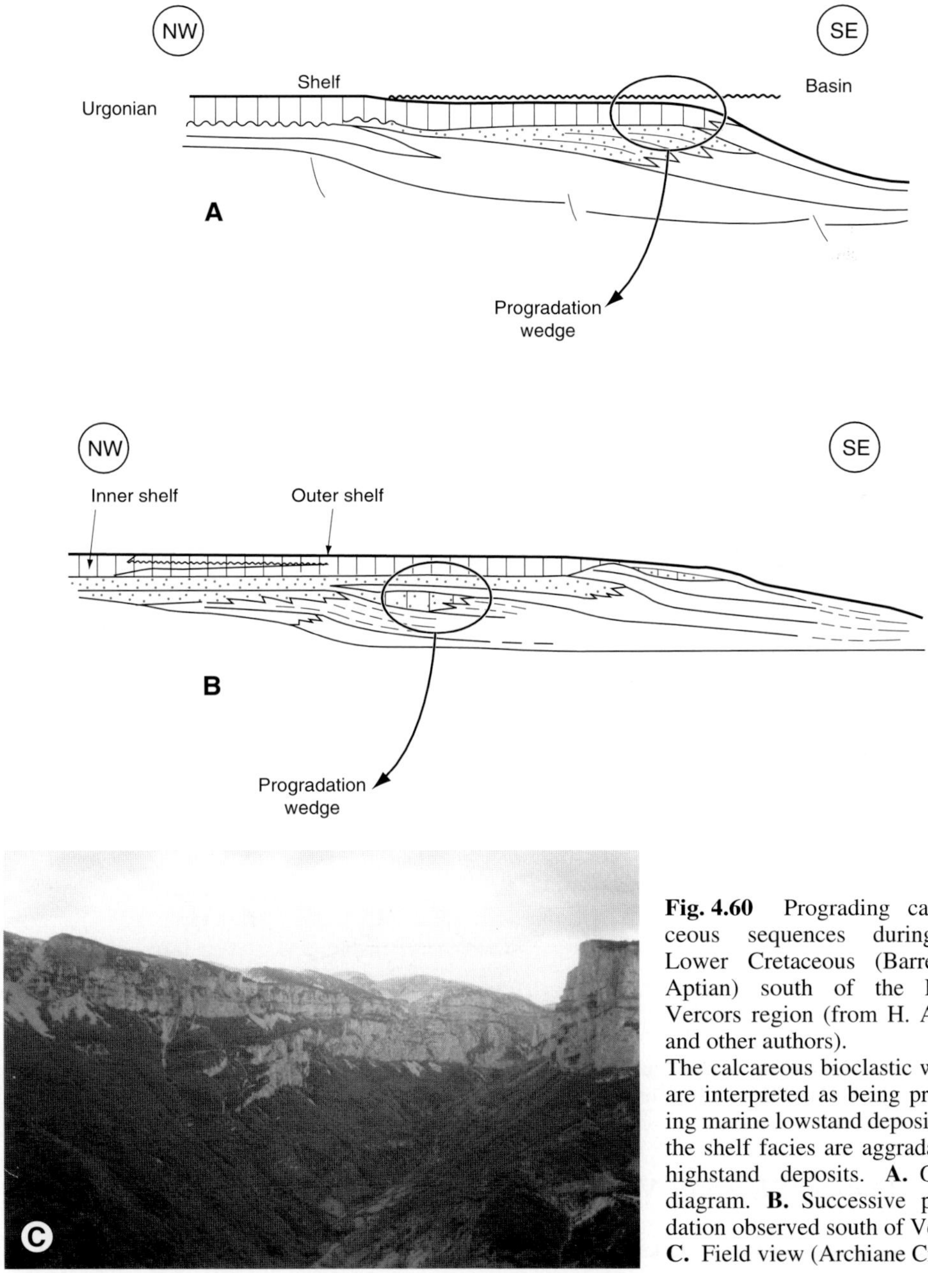

Fig. 4.60 Prograding carbonaceous sequences during the Lower Cretaceous (Barremian-Aptian) south of the French Vercors region (from H. Arnaud and other authors).
The calcareous bioclastic wedges are interpreted as being prograding marine lowstand deposits, and the shelf facies are aggradational highstand deposits. **A.** Overall diagram. **B.** Successive progradation observed south of Vercors. **C.** Field view (Archiane Circus).

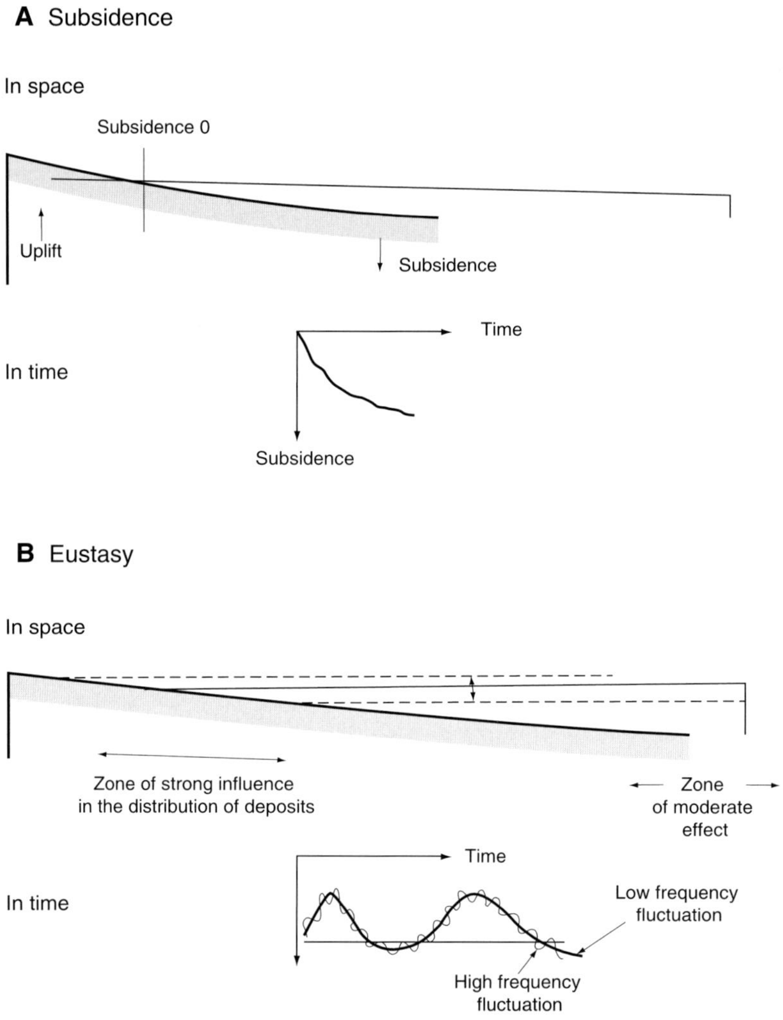

Fig. 4.61 Effects of subsidence and eustasy in time and space.
A. Subsidence. **B.** Eustasy.

4.3.2.4 Causes of Eustatic Variations, Glacio-Eustasy

General sealevel variations depend essentially on two parameters: the available volume of the ocean bowls and the volume of water they contain. The gravity field variation due to mass anomalies at depth (geoid eustasy) can be neglected, as it is more local in character. The main mechanisms are tectono-eustasy and glacio-eustasy. Tectonic plate motions can

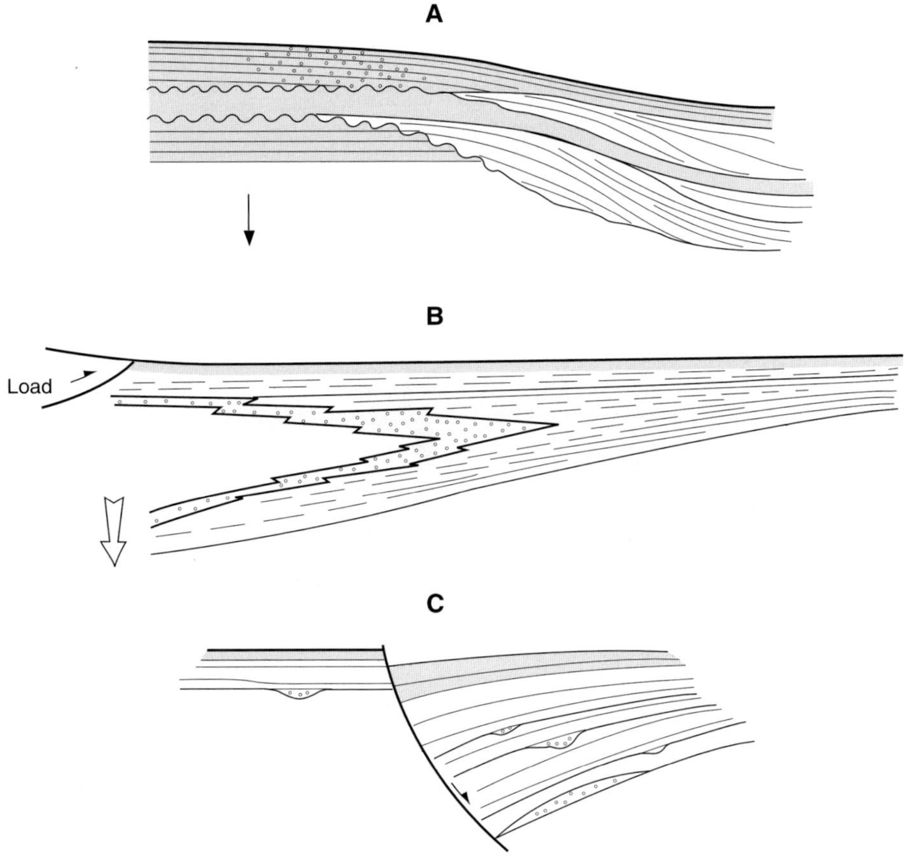

Fig. 4.62 Examples of records of eustatic and tectonic factors.
A. Shelf with moderate subsidence. Highstand wedges will be protected preferentially and eustasy will leave a record of many unconformities. **B.** Flexural foreland basin with strong subsidence. Eustatic fluctuations are diluted in the tectonic signal due, here, to a more or less regular active subsidence. **C.** Half-graben bounded by a listric fault.

generate considerable eustatic variations with rapid expansion of ocean ridges, subduction, or collision. The effects are generally very slow, of the order of a few centimeters per millennium, while glacio-eustatic variations are much faster as they are subject to the amount of ice stored at the poles. This is well known in the Quaternary and certain other geological periods. These variations may follow a Milankovitch type cycle, but their cause and form are still controversial today.

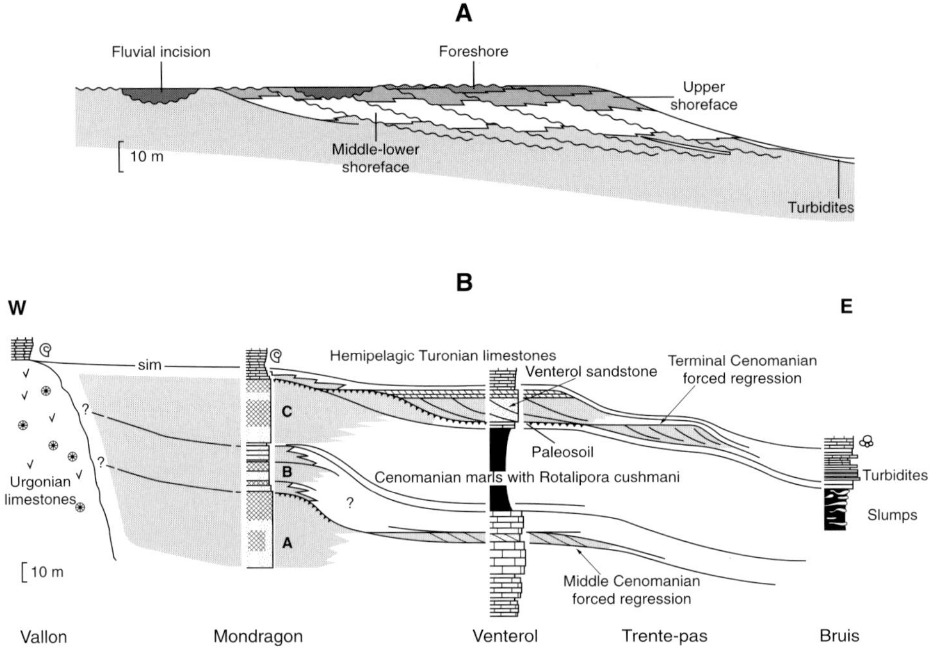

Fig. 4.63 Examples of forced regression.
A. General scheme. Fluvial deposits dominate, the coastal plain is not preserved, the shoreface units are thin and extensive, with steep clinoforms. **B.** The case of Cenomano-Turonian in the Vocontian basin (from Malartre, 1993).

4.3.2.5 Record for Carbonate Environments

Considering the major role of productive organisms in carbonate environments, the sedimentary response to sealevel variations is surely not as simple as the passive response in siliciclastic environments. Here, depending on the climate, but also on dynamic shelf conditions and the rapidity of the eustatic change, the environment may adapt differently to the fluctuations. In a period of low level, erosion may generate detritic calcareous sediments on the slope, but there may also be a change toward evaporitic or even siliciclastic formations on the shelf, where the inversion can lead to intense diagenesis and karstification. In a transgressive or high-level period, the carbonate production potential may perfectly compensate the fastest glacio-eustasy. Two types of response of have been defined according to this production rate (which in turn is affected by the slope and micritic content): a "keep-up" system of rapid deposition, and the slower "catch-up" system[1].

1. See Sarg, 1988, and Schlager, 1981.

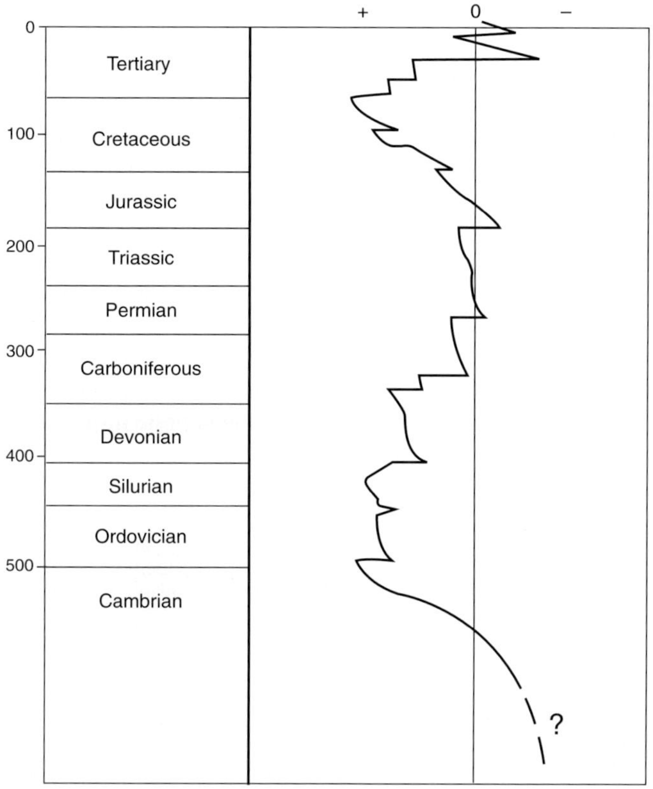

Fig. 4.64 General simplified curve of eustatic variations through geological time (from Haq et al., 1987).

In the continental environment[1], the eustatic variation brings with it a baselevel variation and the idea of available space becomes more difficult to use, depending on whether the environment considered is fluvial, evaporitic and eolian, or lacustrine, and becomes even trickier with the often difficult stratigraphic calibration that is necessary and complementary to any correlation. Allocyclic controls are often interdependent variables that are not easy to characterize well in the emerged domain.

1. See discussion in Shanley and McCabe, 1994.

4.3.3 High Resolution Genetic Stratigraphy

The idea of deposition tracts referenced to a global eustatic chart and to the shelf edge geometry, as applied in sequential stratigraphy on the seismic scale, is a remarkable tool of stratigraphic analysis on the sedimentary basin scale and is routinely used in petroleum geology. Today, however, it is relayed by **genetic stratigraphy**. Genetic stratigraphy identifies small stratigraphic units bounded by isochrons. These units are rarely visible in conventional seismics. They are defined by analyzing the sedimentary facies in the field or from well logging, allowing very high resolution for both. These genetic units are considered to be the elementary building blocks of stratigraphy.[1] While sequential stratigraphy applies more to sedimentary sets covering a few million years (the scale of geological stages) or a few hundred thousand years, the sequences encountered on the scale of a whole basin are generally shorter, ranging from tens to hundreds of thousands of years. These are the sequences most especially studied when describing oil fields. The elementary unit is the **genetic unit** (Fig. 4.65).

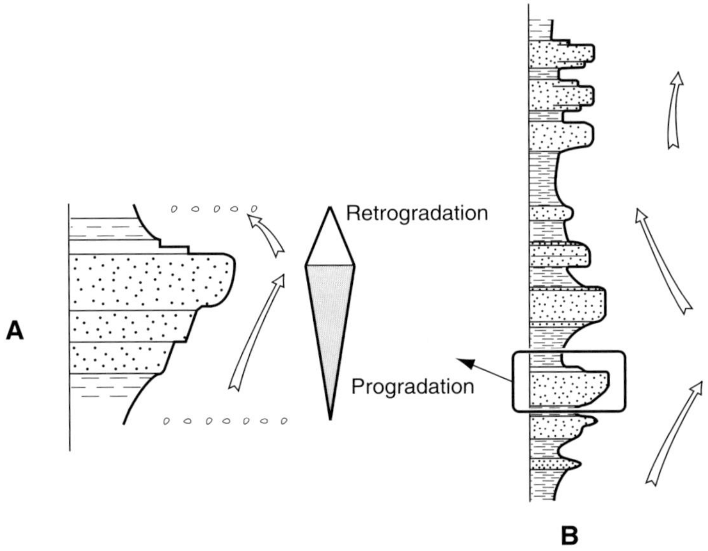

Fig. 4.65 Genetic deposition unit: the smallest elementary pattern in a logical sequence of deposits in a variable environment.
A. Elementary genetic unit. **B.** Several genetic units in different stacking patterns: prograding, retrograding, aggradational.

1. Cross, 1988; Homewood et al., 1992.

A genetic unit, which can be called a high-frequency sequence or **parasequence**, is no more or less than the old **cyclothem**. Among the elementary sedimentation mechanisms used to explain these cycles or genetic units, the high-frequency forced-oscillation allocyclic controls have primacy over autocyclic processes. Most models refer to the variations in the Earth's orbital parameters (**Milankovitch cycles**) at periods ranging from 20 Ky to 400 Ky, which correspond well to the duration of the genetic units. As the cycles continue, they construct a **genetic stacking** or **stratal pattern**. The physical processes that individuate the units may be high-frequency eustatic fluctuations, direct climatic effects, or glacio-eustasy; but whatever these processes are, the elementary sequences record global, isochronous, stratigraphic variations. **Accommodation** is still important because the quality of the record will increase with greater subsidence and greater sedimentary input. If the subsidence rate is low (such as in an intracratonic basin), no very high-frequency record would be possible, or at least it would be very much disturbed.

The (arbitrary) boundaries between genetic units differ from those used in sequential stratigraphy. Here, the stratigraphic reference is a remarkable time line, the **maximum flooding surface**, between two of which the genetic unit is defined (Fig. 4.66). Contrary to sequential stratigraphy, the ideas of marine highstand and lowstand tracts are not defined with respect to the continental shelf edge here. It is not the bathymetric profile geometry that is important (although the deposition tract datum may be the shelf profile itself, with energy gaps) but rather sedimentological criteria. The essential field of application of the genetic unit, though, is still the continental shelf and the transition to the coastal plain. The very great interest of this physiological approach, moreover, resides in the study of oil fields (extent and lateral variability of the reservoirs). Genetic stratigraphy therefore requires a strict definition of the facies and the depositional environments.

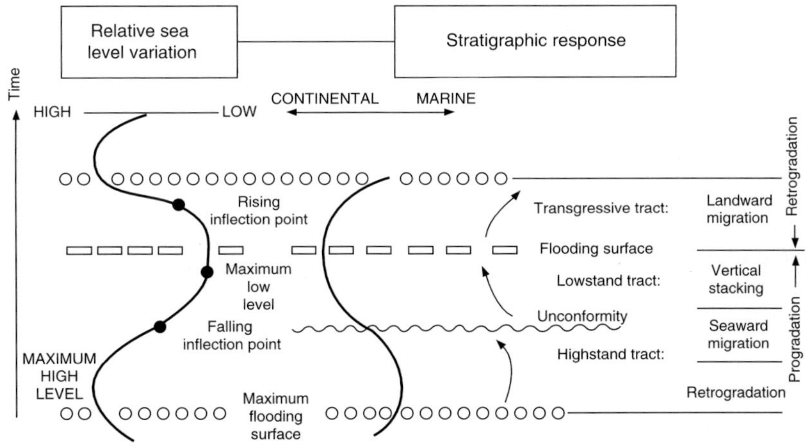

Fig. 4.66 Genetic unit boundary: stratigraphic response to sealevel variations (from Homewood et al., 1992).

By definition, the high level follows immediately upon the maximum flooding (Fig. 4.66). What is usually seen is a progradation phase comprising a downward shift of the littoral facies seaward. If the input is large, the accommodation power decreases rapidly (Fig. 4.67). Then, as the level falls, an unconformity can be created with the deposits accumulating seaward in prograding steps. As the level begins to rise, there is a slight vertical aggradation and then a transgression or **flooding surface** is created. The coast is then subject to erosion by gullying that extends over time with the rise in relative level. A diachronous surface is created (at high resolution) (Fig. 4.68), and then the transgressive tract is constructed in gradual landward steps, i.e., retrograding up to the period of maximum flooding.

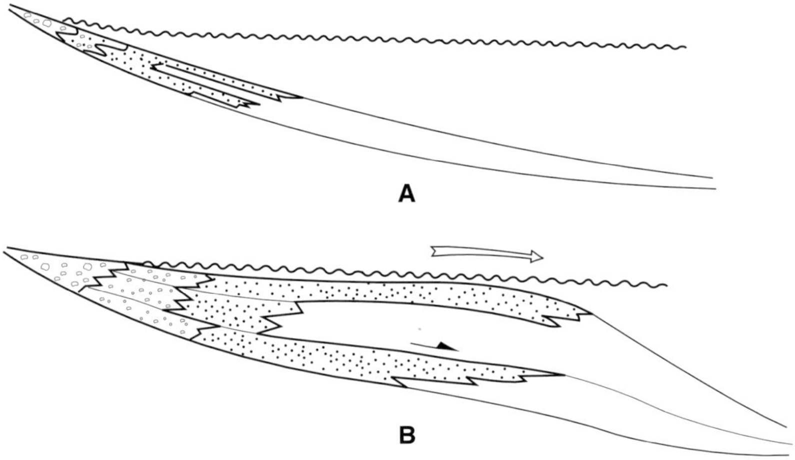

Fig. 4.67 Seaward migration of littoral facies.
A. Low sedimentary flow configuration. **B.** Large detritic flow with rapid seaward migration of the facies.

The main consequence of this time evolution is that the same depositional environment will be preserved differentially, partitioning the sediments into volumes (Fig. 4.69). This **volumetric partitioning** is important because it is accompanied by grain-size partitioning from upstream to down. The varying sandiness of the littoral domain can be explained and even predicted by model.

Depending on the combination of cycles at different frequencies, elementary units can be stacked into higher-order patterns as in sequential stratigraphy (Fig. 4.70). High-frequency progradation phases are amplified during low-frequency progradation phases. The high-frequency retrogradation phases, on the other hand, will be amplified during low-frequency retrogradation phases. This ***distortion of the genetic units*** *intensifies the volumetric partitioning, now on the scale of the stack.*

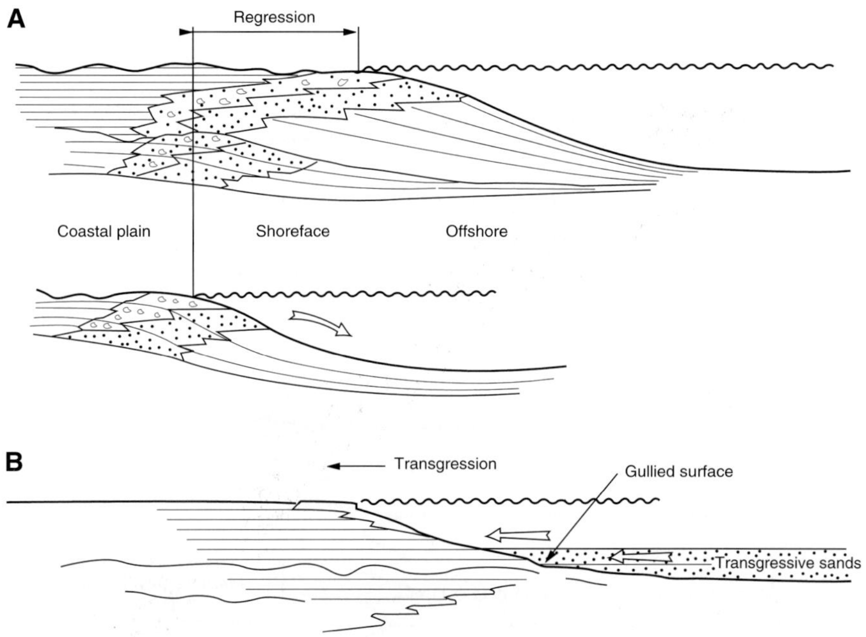

Fig. 4.68 Time evolution of genetic unit stacks along a profile (from Homewood et al., 1992).
A. Seaward progradation of a littoral wedge, i.e., regression of the sea (two stages are shown). **B.** Retrogradation of the littoral wedge, diachronous transgression with stepped onlaps over a gullied surface.

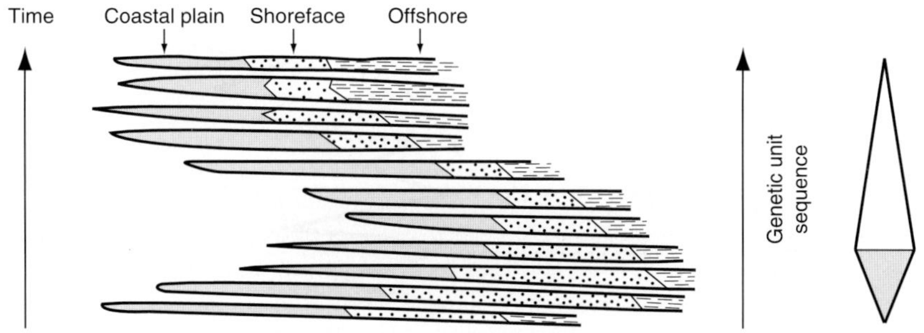

Fig. 4.69 Facies partitioning and time evolution.

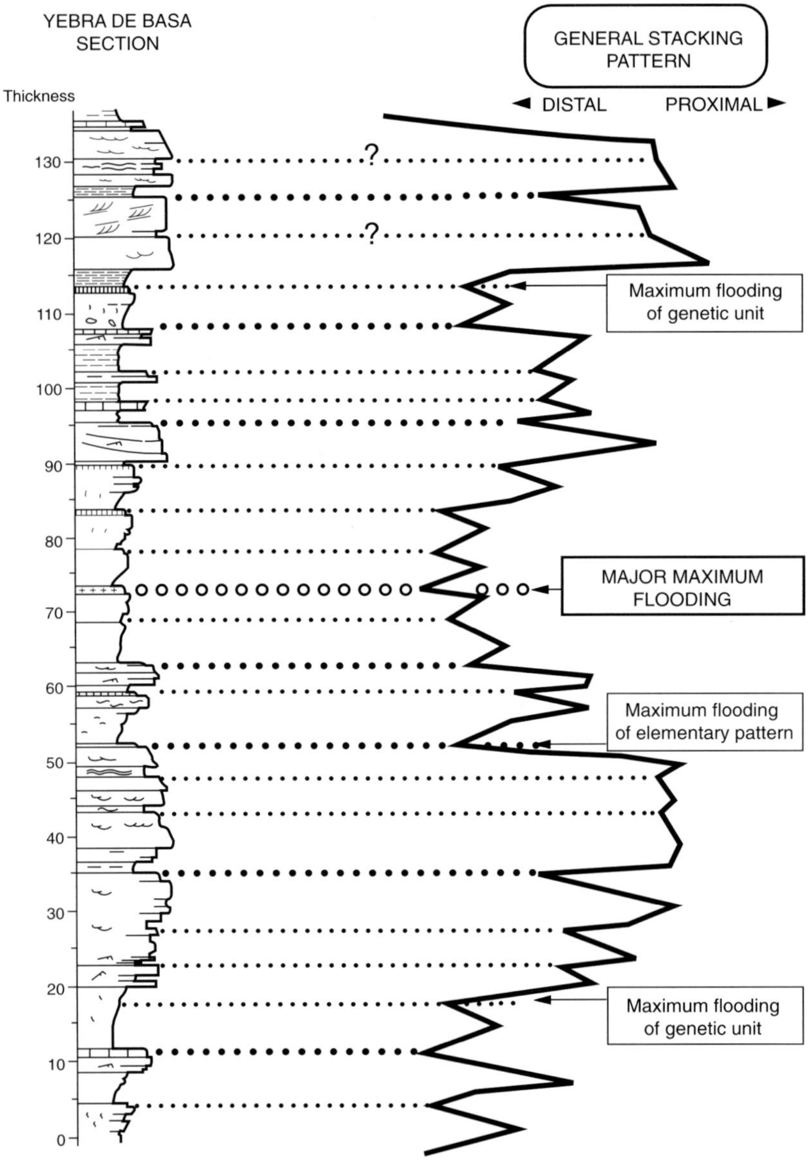

Fig. 4.70 Genetic unit stacking patterns.
Several genetic units in succession define elementary patterns that may themselves be grouped into higher-order patterns. Section of Yebra de Basa in Spain's Jaca basin (from F. Lafont, 1994).

The advantage of this approach is that it provides a consistent analysis of the continental domain from the coastal plain to the littoral, and then to the deeper offshore zones, strongly integrating sedimentology with stratigraphy. It can be seen that the approach is based largely on a thorough analysis of the facies represented, for example, in a facies substitution diagram (Fig. 4.71).

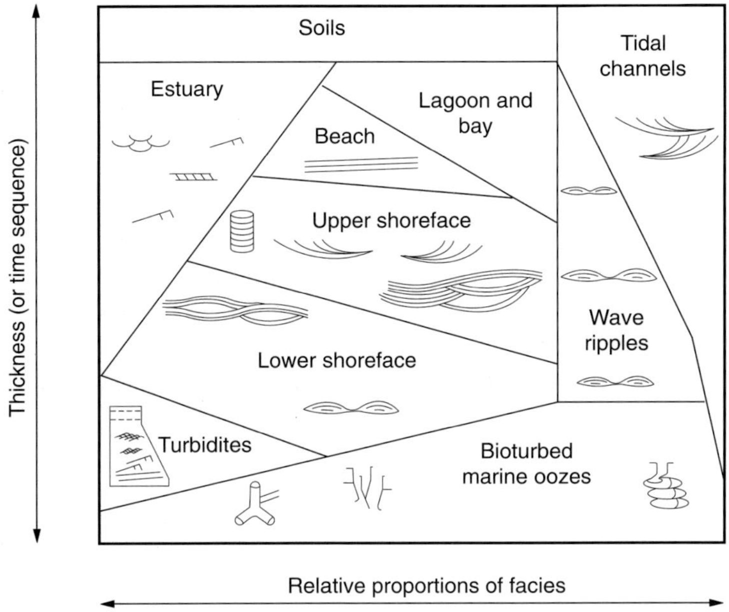

Fig. 4.71 Facies substitution diagram of a littoral domain dominated by terrigenous sedimentation.

4.4. STRATIGRAPHIC CORRELATIONS, PALEOGEOGRAPHIC RECONSTRUCTIONS

4.4.1 Stratigraphic Correlations and Facies, Cartographic Expression

It was said at the outset of this chapter that the stratigraphic principle of younger beds overlying older in the edifice was subject to exceptions. And all through the chapter it has been shown on the basis of sedimentary mechanisms that the second great principle of stratigraphy, which is the lateral continuity of the beds, was even more limited and even inherently faulted. In particular, the ideas of **transgression/regression** and then **sequential stratigra-**

phy have shown that the types of deposits (and therefore the sedimentary bodies that define the strata) may exhibit good lateral continuity without having any stratigraphic significance, by lateral prograding or retrograding migration of the same facies in time, or **lateral variation of facies** within a given period (Fig. 4.72). It has been said that **isopic lines** mark equivalent facies and that formations are not parallel to isochron surfaces.

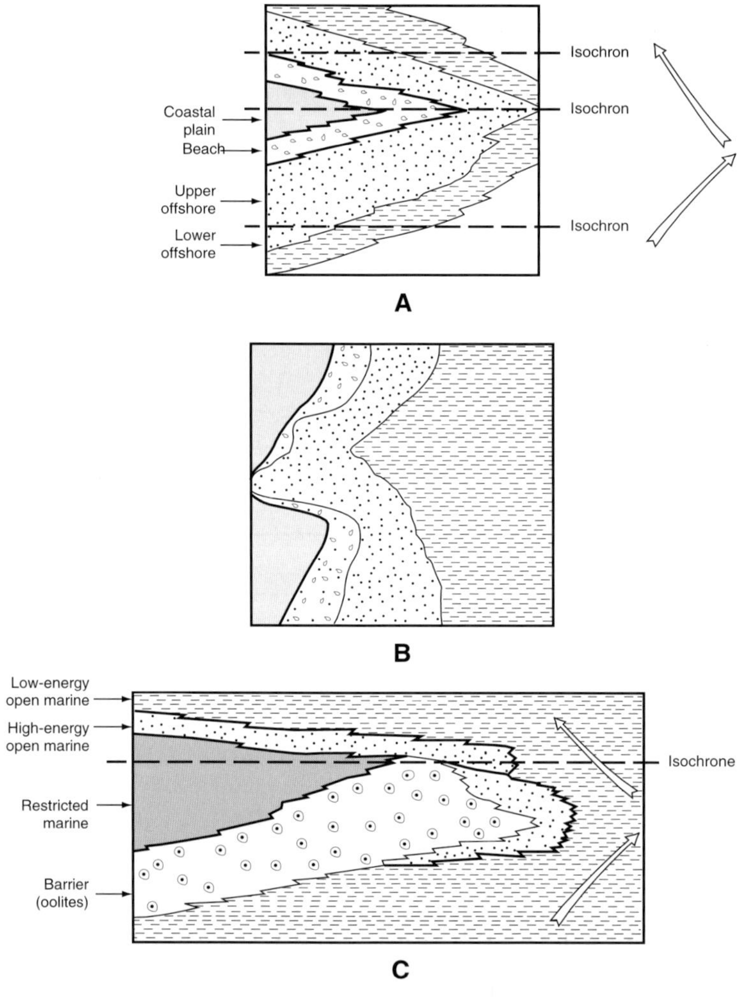

Fig. 4.72 Lateral facies migration.
A. Theoretical vertical time evolution profile with isochrons and facies variations arranged along heterochronous lines. **B.** Two-dimensional map corresponding to one of the isochrons. **C.** Vertical profile of Bathonian facies variations in the Paris basin (from J. Rey, 1997).

Remember that **stratigraphic correlations** are based on different **markers** that sometimes reflect major events. There are lithological and biostratigraphic markers, geochronological calibration markers, seismic and well logging markers. Horizons, or **datum surfaces**, serve as references. So, precise correlations can be established on small or large scale, at the level of a basin or reservoir.

Using methods of fine sedimentological analysis of the facies, reconstructing the deposition environments, and then associating the facies in sequence, the high-resolution correlations of genetic stratigraphy are a great improvement over those based solely on lithological analysis. This is because more and more reference levels are chosen (where biostratigraphy comes in) when constructing the stack of genetic unit sequences (Fig. 4.72).

Quantitative analysis of biostratigraphic data are what make these high resolution correlations possible today.

On this basis, **facies maps** (Fig. 4.25) can be established on different scales for a given time interval, geological stage, or megasequence. These maps are classical stratigraphic products: maps of paleoenvironments or **faunal provinces** (Fig. 4.73) on the global scale or that of several basins; maps of sand shale ratios for a given formation; maps of sandiness or petrophysical characters on the scale of a given field (Fig. 4.74).

A stratigraphic interval can also be represented by its thickness variations (thickness between two datum planes), reflecting the **power** of the interval in an **isopach chart** of lines or curves of equivalent thickness (Fig. 4.75A). The depth at which a datum plane (roof of the sequence or a stage, for example) is observed can be represented in an **isobath map** of curves of the same depth (Fig. 4.75). This idea will be taken up again at the end of the following chapter because this type of information gives some indication of the deformation a horizon has undergone (it is generally assumed that a horizon is deposited on a low slope). The structural map (Fig. 4.76) is another mode of representation that is commonly used.

4.4.2 Paleogeographic and Palinspastic Reconstructions

The maps drawn up for a given stratigraphic interval are often presented as **paleogeographic maps** expressing the distribution of emerged lands and the marine domain, the main sources of input, the major types of sediments deposited in different environments, and basin geometry (Fig. 4.77). Generally, the precision of these maps decreases the further back in time we go, partly because of the quality of the stratigraphic control, but also because of the global plate movements, the image of which is relatively well in focus only for recent periods. A paleogeographic map generally shows the relative position of a given geographic space. At times, it expresses a global view on the planetary scale (Fig. 4.78).

These maps can be refined into **palinspastic maps** reconstructing plate motion, i.e., global geodynamic evolution. One major element in the reconstruction of absolute position is then the **paleomagnetic data**. The movements of the lithospheric plates can be described by analysis of **magnetic anomalies** existing symmetrically to either side of the ocean floor

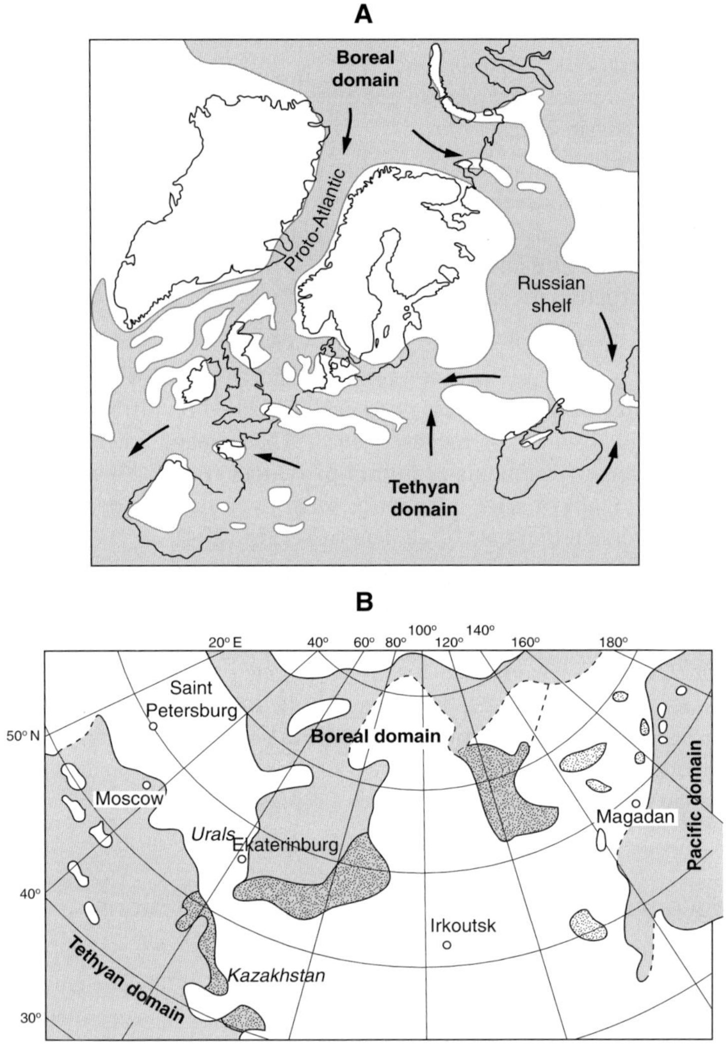

Fig. 4.73 Faunal provinces.
A. In the Upper Jurassic, the Tethyan domain communicated with the Boreal, and paths of migration can be imagined (from Ziegler, 1988, and Rey, 1997). **B.** In the Upper Cretaceous, the Siberian basins communicated neither with the Tethyan nor the Pacific (from Naidin, 1986).

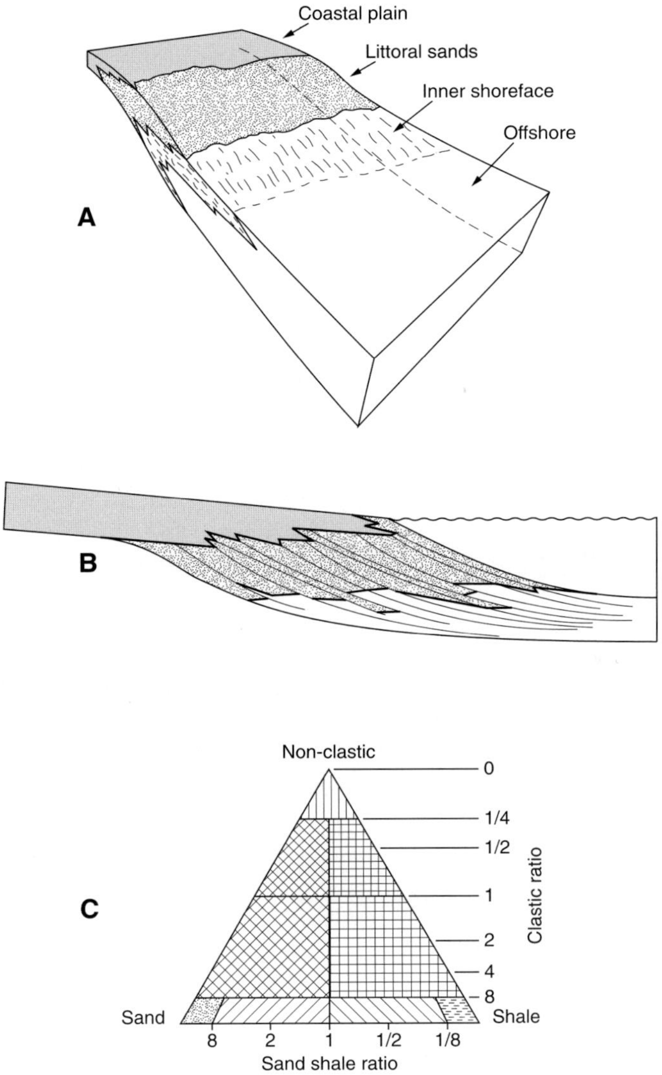

Fig. 4.74 Deposition environment and shaliness.
The sand-shale ratio in the continental and inner marine domain can be understood by reconstructing the deposition environment. **A** is a 3D modeling of a system ranging offshore from the alluvial flood plain through the beach (foreshore) and littoral domain (shoreface) (IFP document). The 2D modeling in **B** gives the sand-shale ratio as it appears in the deposit profile (IFP document). **C** is the usual sand-shale ratio diagram.

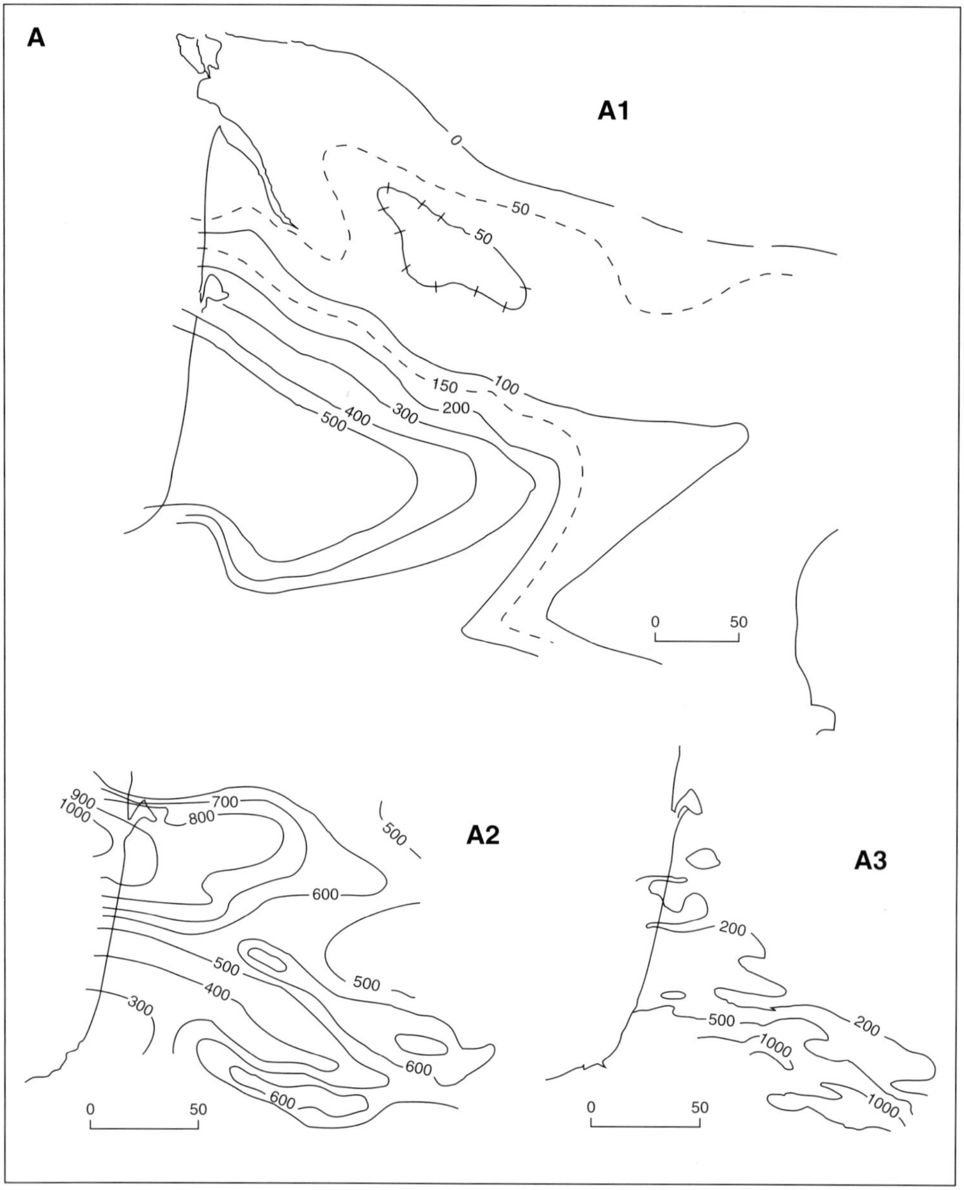

Fig. 4.75 Isopach and isobath maps.
A. Isopach maps of different stratigraphic levels in the Aquitaine basin.
A1 Lodesse shales of the Upper Triassic (from Guillemot, 1986);
A2 Lower Kimmeridgian; **A3** Lower Eocene (from Boillot, 1984).

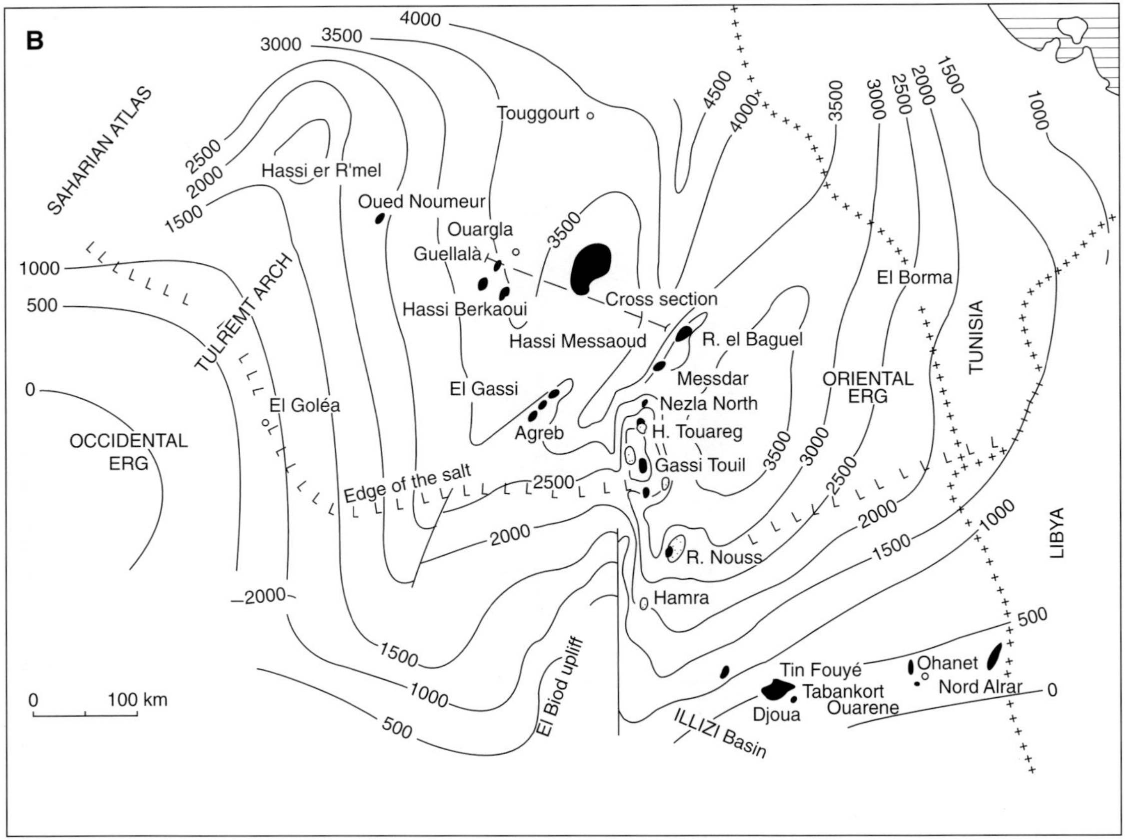

Fig. 4.75 (cont'd) Isopach and isobath maps.
B. Isobath map of the Hercynian discordance in the Sahara (from Perrodon, 1983).

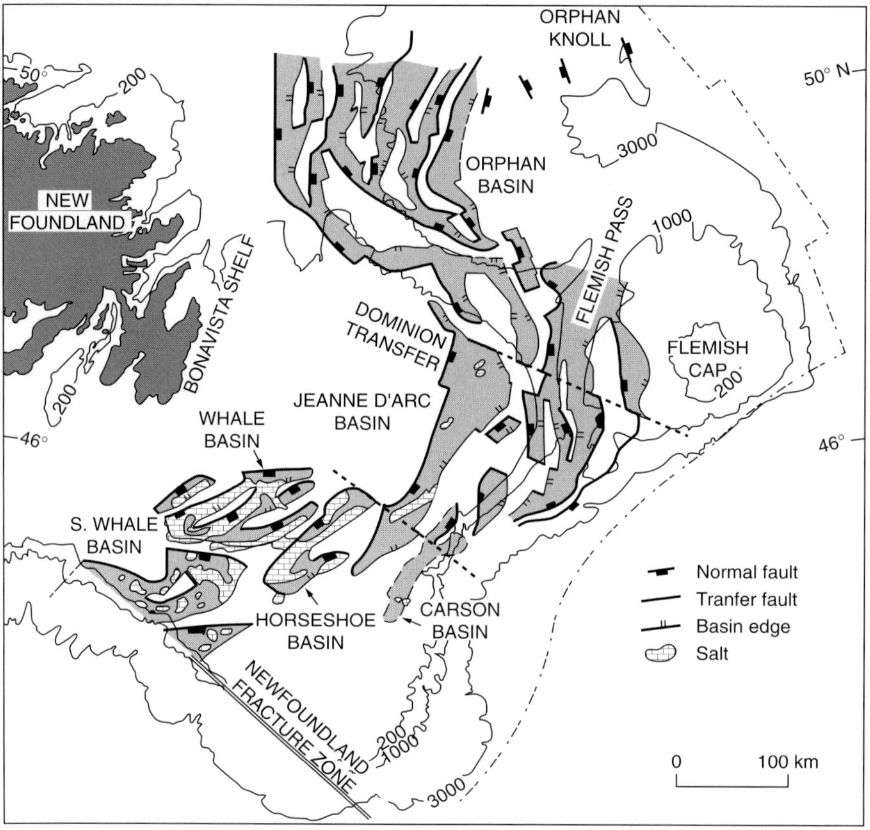

Fig. 4.76 Structural map.
Here, the architecture of offshore Newfoundland (from Rankard et al., 1989).

ridges (see Chapter 1, Fig. 1.33). In plate kinematics, a fixed reference framework is generally taken and then the relative motions of the neighboring plates is analyzed. The method is then refined by paleomagnetic data from rocks found on the continents, giving evidence of their paleoaltitude at the moment of deposit. This data has clearly shown the apparent drift of the pole in the course of geological time. Today, the talk is about the absolute motion of the continents, and paleomagnetic data is being used to reconstruct paleoequators and paleoaltitudes. A series of maps for different stratigraphic horizons show the time evolution of large portions or the whole Earth in atlas form (Fig. 4.79). Overall reconstructions are now accepted for the Mesozoic and the Cenozoic, but they are still subject to discussion for older periods.

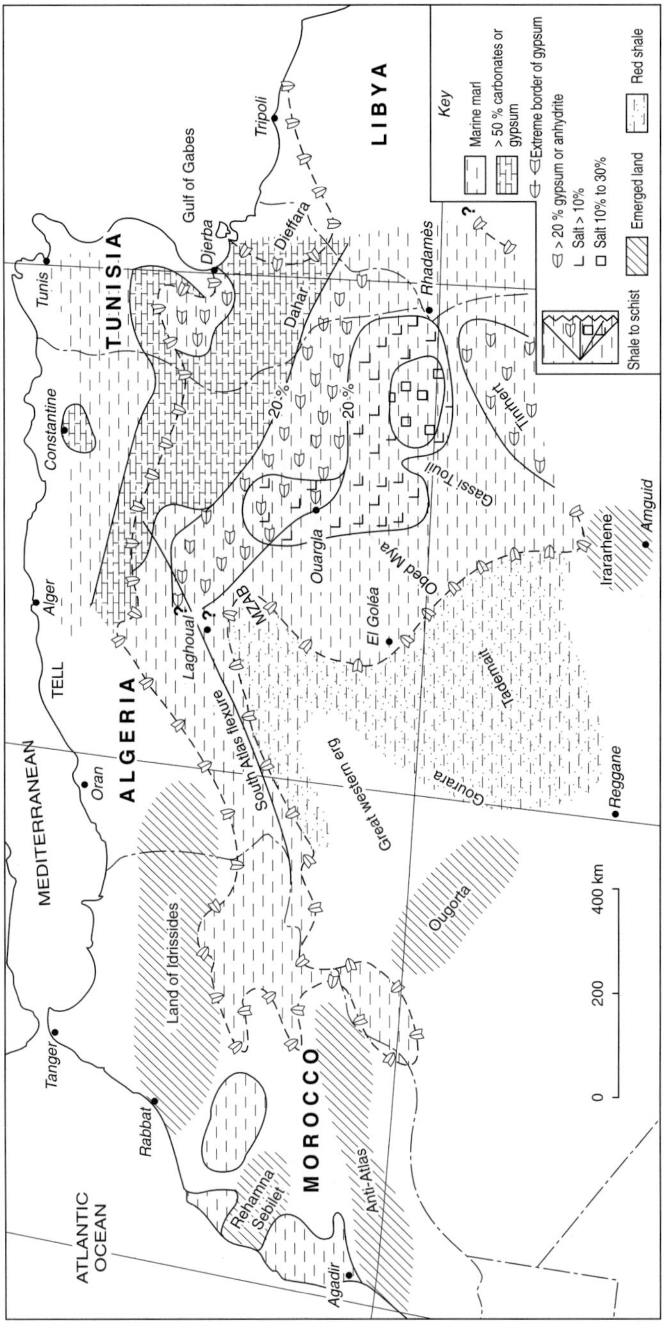

Fig. 4.77 Paleogeographic map showing the distribution of different facies and deposition environments. The example here is of the Saharan shelf in the Cenomanian (from Busson, 1972).

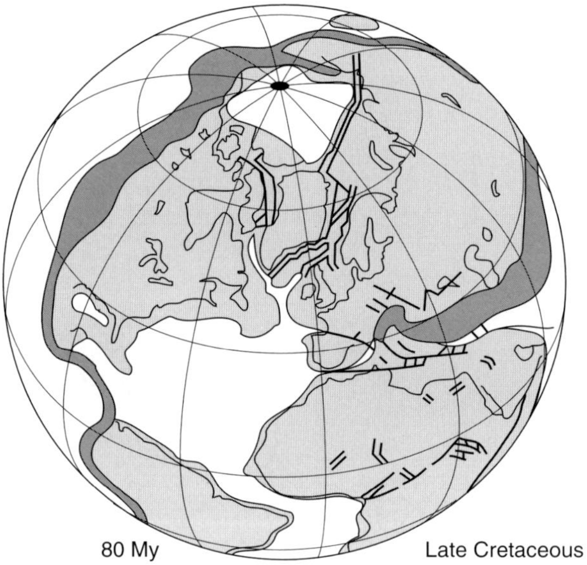

Fig. 4.78 Global palinspastic map (from Ziegler, 1993).

Palinspastic reconstructions have two limitations that should be noted. Firstly, it happens that small continental systems (actually microplates) detach from the main plates and rapidly move to stick (called ***collage****) up against another continental system. The kinematics of these small systems is generally poorly constrained and the intermediate steps in their migration are difficult to reconstruct precisely. Also, maps can do no more than express a general tendency for a given stratigraphic stage. The Kimmeridgian map, for example, covers a lapse of time within which notable time variations cannot be represented. To obtain a good reconstruction, as many "time slices" would be needed as there are events.*

Map graphics makes it possible to indicate structural data such as crust, continental or oceanic thickness, the existence of hot spots, or subduction and uplift zones.

With today's **modeling** possibilities, palinspastic maps are becoming better yet. A basin's depth is estimated on the basis of its age. Then, knowing the geometric pattern of the continents and their absolute position, the great atmospheric and oceanic gyres can be modeled for different periods, along with the corresponding surface streams and deep currents (Fig. 4.80). An attempt is then made to narrow down special events such as the formation of upwelling currents that might have been of considerable importance in the formation and accumulation of organic matter on the edge of a basin.

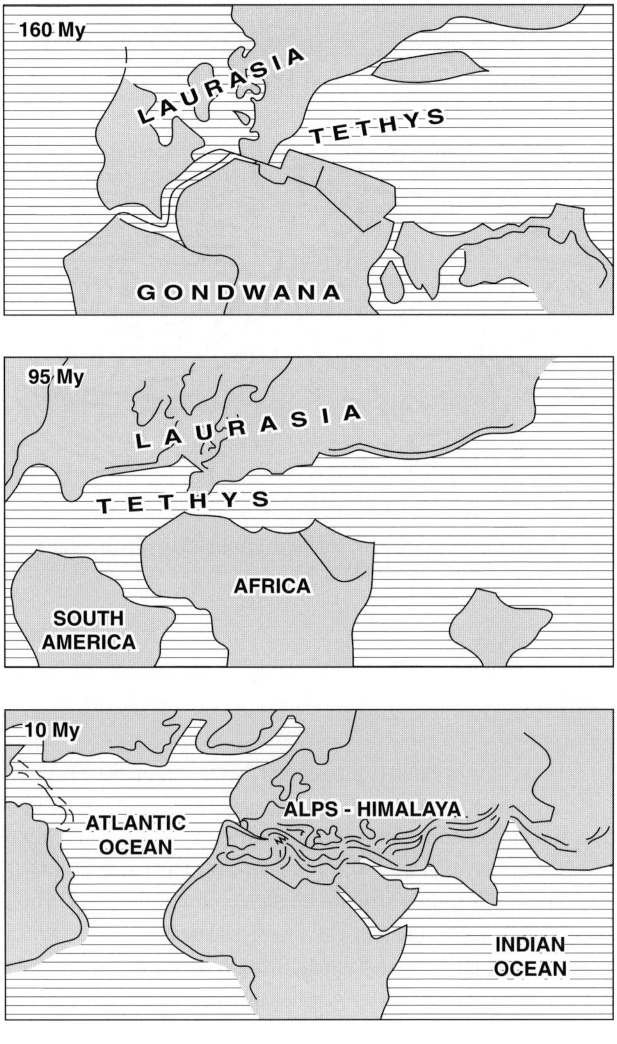

Fig. 4.79 Palinspastic reconstructions of the Tethyan domain up to the formation of the Alps.

Last mention should be made of cartographic representations in stratigraphy: horizon slice maps that make use of modern 3D-seismic techniques currently used in petroleum geology, for structural reconstructions.

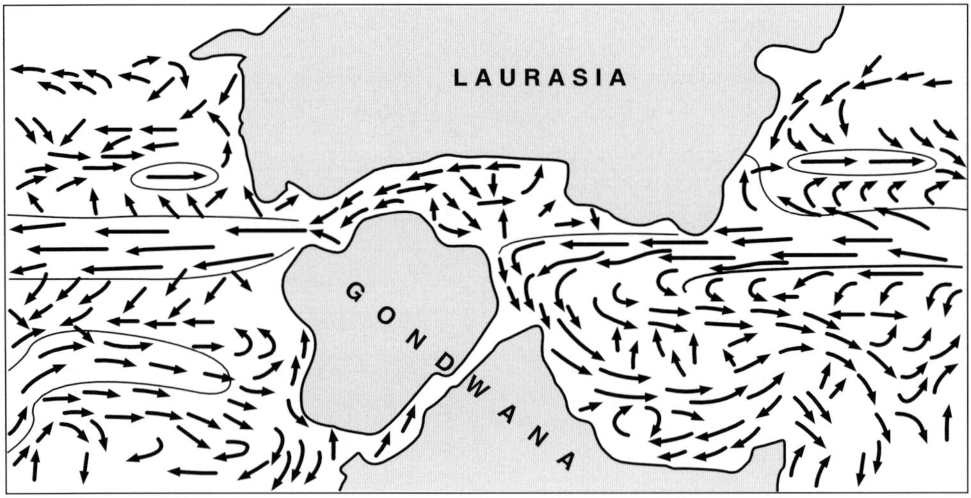

Fig. 4.80 Map modeling ocean surface circulation in the Late Jurassic (simplified from Cottereau, 1992).

BIBLIOGRAPHY

✧ Albarede F (1990) Les anciens océans. Courrier du CNRS, 76, pp 50-51.

✧ Arnaud H, Arnaud-Vanneau A Plateformes du Crétacé inférieur et stratigraphie séquentielle dans le massifs subalpins septentrionaux. Université J. Fourier. Unpublished, Grenoble.

✦ Beuf S, Biju-Duval B, Charpal O et al. (1971) Les grès du Paléozoïque inférieur au Sahara : sédimentation et discontinuités. Évolution structurale d'un craton. Éditions Technip, Paris.

✧ Biju-Duval B (1994) Océanologie. Dunod, Géosciences, Paris.

✧ Biju-Duval B (1985) The terrigenous and pelagic series of Barbadas island: Paleocene to middle Miocene slope deposits accreted to the lesser Antilles margin. Mascle A. Géodynamique des Caraïbes. Éditions Technip, Paris, pp 187-198.

✧ Biju-Duval B, Moore C (1984) Initial reports DSBP. US government printing office, Washington.

✦ Boillot G, Montadert L, Lemoine M et al. (1984) Les marges continentales actuelles et fossiles autour de la France. Masson, Paris.

✦ Boyer S, Mari JL (1994) Sismique et diagraphies. Éditions Technip, Paris.

✦ Busson G (1972) Principes, méthodes et résultats d'une étude stratigraphique du Mésozoïque saharien. Éditions du muséum, Muséum d'histoire naturelle, mémoires, série C, sciences de la terre 26, Paris.

✦ Caron JM, Gauthier A, Schaaf A et al. (1989) Comprendre et enseigner la planète Terre. Ophrys, Paris.

✦ Chaline J (1987) Paléontologie des vertébrés. Dunod, Paris.

✦ Chambre Syndicale de la Recherche et de la Production du Pétrole et du Gaz Naturel (1986) Corps sédimentaires. Exemples sismiques et diagraphiques. Éditions Technip, Paris.

✧ Chihi H (1997) Modélisation 3D des limites stratigraphiques et simulation des faciès sismiques dans la marge du golfe du Lion. Thèse de doctorat, université Pierre et Marie Curie, Paris.

✦ Cotillon P (1988) Stratigraphie. Dunod, Géosciences, Paris.

✧ Cottereau N (1992) Reconstitutions paleobathymétriques de la Tethys au Jurassique terminal : méthodes et conséquences. Thèse de doctorat, université Pierre et Marie Curie, Paris.

✧ Courtillot V (1993) Une éruption volcanique ? Buffetaut, Dossier, Pour la science, Paris.

✦ Cramez C (1990) Glossaire de stratigraphie séquentielle, anglais/français. Revue de l'Institut français du pétrole 45, 3, pp 435-453.

✦ Cross TA (1988) Controls on coal distribution in transgressive-regressive cycles, upper Cretaceous, western interior, USA. Lidz BH. Sea level changes: an integrated apporach. Society of economic paleontologists and mineralogists. SEPM, special plublication 42, Tulsa, pp 371-380.

✧ Davies PJ, Symonds PA, Feary DA et al. (1989) Controls on carbonate platform basin development. Crevello PD, Wilson JL, Sard JF et al. Controls on carbonate platform and bassin development. Society of economic paleontologists and mineralogists. SEPM, special publication 44, Tulsa, pp 233-258.

✦ Devillers C, Chaline J (1989) L'histoire de la vie. Dunod, Paris.

✧ Dubois P, Beuf S, Biju-Duval B (1967) Lithostratigraphie du Dévonien inférieur gréseux du Tassili N'Ajer. Colloque sur le Dévonien et ses limites, 16-24 septembre 1964, Rennes. Bureau des recherches géologiques et minières, Orléans, BRGM, mémoire 33, pp 29-31.

✦ Emiliani C (1956) Oligocene and Miocene temperatures of the equatorial and subtropical Atlantic ocean. Journal of geology 64, pp 281-288.

✧ Flèche JC (1992) Traitement sismique. Cours de l'ENSPM. Unpublished.

✧ Gomez de Souza O (1997) Stratigraphie séquentielle et modélisation probabiliste des réservoirs d'un cône sous-marin profond (champ de Namorado, Brésil). Thèse de doctorat, université Pierre et Marie Curie, Paris.

✧ Graciansky (de) PC, Poag CW, Montadert L et al. (1986) Evidence for changes in Mesozoïc and Cenozoïc oceanic circulation on the southwestern continental margin of Ireland: DSPD/IPOD leg 80. Summerhayes CP, Schackleton NJ. North Atlantic paleoceanography. Geological society, special publication 21, London, pp 17-33.

✦ Haq BU, Hardenbol J, Vail P (1987) Chronology of fluctuating sea levels since the Triassic. Science 235, 4793, pp 1156-1167.

✧ Hardie L (1996) Secular variation in seawater chemistry: an exploration for the coupled secular variation in the mineralogies of marine limestones and potosh evaporites over the past 600 My. Geology 24, 3, pp 279-283.

✧ Hay WW, Brock JC (1992) Temporal variation in intensity of upwelling off southwest Africa. Summerhayes CP, Prell WL, Emeis KC. Upwelling systems: evolution since the early Miocene. Geological society. Geological society, special publication 64, London, pp 463-498.

✦ Hedberg HD (1979) Guide stratigraphique international: classification, terminologie et règles de procédures. Doin, Paris.

✦ Henry G (1994) Géophysique des bassins sédimentaires. Éditions Technip, Paris.

✦ Homewood P, Guillocheau F, Eschard R et al. (1992) Corrélation haute résolution et stratigraphie génétique : une démarche intégrée. Bulletin des centres de recherches exploration production, Elf Aquitaine production 16, 2, pp 357-381.

✦ Jervey MT (1988) Quantitative geological modeling of siliciclastic rock sequences and their seismic expression. Lidz BH. Sea level changes: an integrated apporach. Society of economic paleontologists and mineralogists. SEPM, special publication 42, Tulsa, pp 47-70.

✧ Lafont F (1994) Influences relatives de la subsidence et de l'eustatisme sur la localisation et la géométrie des réservoirs d'un système deltaïque : exemple de l'Eocène du bassin de Jaca (Pyrénées espagnoles). Mémoire de géosciences 54. Géosciences, Rennes.

✧ Larson RL (1991) Geological consequences of super plumes. Geology 19, pp 963-966.

✦ Lavergne M (1986) Méthodes sismiques. Éditions Technip, Paris.

✦ Miall AD (1992) Exxon global cycle chart: an event for every occasion? Geology 20, pp 787-790.

✦ Odin G, Odin C (1990) Échelle numérique des temps géologiques. Géochroniques, 35, pp 12-21.

✦ Perrodon A (1983) Dynamics of oil and gas accumulations. Elf Aquitaine, bulletin des centres de recherches exploration-production, Elf Aquitaine, mémoire 5, Pau.

✦ Pomerol C, Babin C, Lancelot Y et al. (1980) Stratigraphie et paléogéographie : principes et méthodes. Doin, Paris.

✦ Posamentier HW, Vail PR (1988) Eustatic controls on clastic desposition II - sequence and systems tracts models. Wilgus CK, Posamentier H, Hastings BS et al. Sea level changes: an integrated approach. Society of economic paleontologists and mineralogists. SEPM, special publication 42, Tulsa, pp 125-154.

✦ Purser BH (1980) Sédimentation et diagenèse des carbonates néritiques récents. Tome 1 : Les éléments de la sédimentation et de la diagenèse. Éditions Technip, Paris.

✦ Purser BH (1983) Sédimentation et diagenèse des carbonates néritiques récents. Tome 2 : Les domaines de sédimentation carbonatée néritiques récents : application à l'interprétation des calcaires anciens. Éditions Technip, Paris.

✧ Ravenne C (1990) Cours ENSPM. Unpublished.

✧ Renard M (1984) Géochimie des carbonates pélagiques : mise en évidence des fluctuations de la composition des eaux océaniques depuis 140 Ma : essai chimiostratigraphique. Thèse de doctorat, université Pierre et Marie Curie, Paris.

✦ *Books or articles of general interest.*

✧ *Source of one of the figures used, cited in the figure caption.*

Chapter 5

FROM SEDIMENTS TO SEDIMENTARY BASIN ROCKS AND MOUNTAIN CHAINS

Chapter 3, on sedimentary mechanisms and environments, explained the variety of deposition modes and sedimentary bodies, and the following chapter (4) on sedimentary sequences showed how gradual time variations and series of events provide a record of global and local fluctuations in the stratigraphic filling of sedimentary basins.

The purpose of the present chapter is to explain post-depositional history in a review of the mechanisms by which sediments are gradually transformed into sedimentary **rock** in the stratigraphic pile of basins and mountain chains. A **sedimentary rock** is therefore an old sediment that has been subjected to mechanical and chemical modifications during its burial under younger sediments. All of these mechanisms taken together are called **diagenesis**, which can be defined more formally as the set of physical and chemical processes that affect the sediment until it **lithifies** (from the Greek "lithos" for stone).

If the conditions in a given sedimentary basin go beyond normal, with burial of 5 to 12 km, or if the tectonic stresses are large, then the rocks will begin a metamorphosis that belongs to the field of **structural** or **tectonic geology**. Nevertheless, we will discuss the most essential principles of deformation because of their importance for oil geology. These high-pressure, high-temperature transformations were briefly presented in Chapter 3 but will not be covered again here.

5.1 BURIAL AND DIAGENESIS

5.1.1 Burial and Subsidence

The general scenario is as follows. A given type of deposit such as sand, deposited in fluvial regime at a precise epoch (such as the Triassic, 230 My) is found in the field or by drilling today in **consolidated sandstone** observed at the base of a stratigraphic set several hundred meters thick. As sediment, it was originally deposited at zero level (sea level), but today it is found at depth or perhaps it has returned to the surface or can even be found at an altitude of several hundred meters in a mountain chain (Fig. 5.1). It has thus experienced a history of **burial** in geological time, and in certain cases has returned to the surface with perhaps an accompanying horizontal translation prompted by tectonic movements.

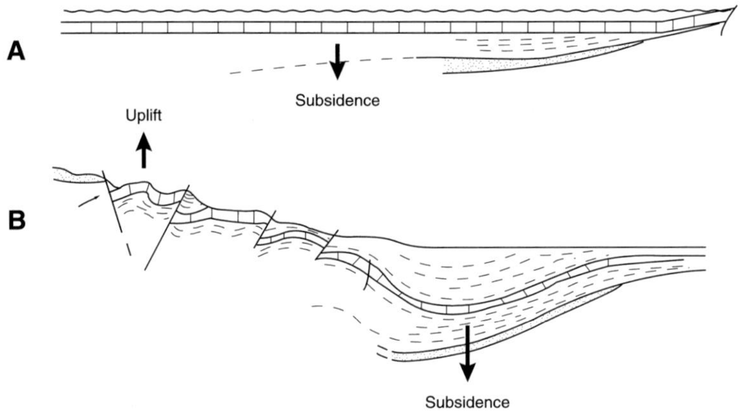

Fig. 5.1 Deposit, burial, subsidence, and uplift.
A. In the life of the basin, successive deposits of sands, clays, and carbonates accumulate and settle, and are gradually buried as subsidence continues. **B.** Subsidence may continue in one part of the basin while uplift occurs in another, bringing previously buried rock to the surface and baring it by erosion.

The burial phase, as was said in the chapter on basins, is essentially due to **subsidence** and to the gradual **accumulation** of sediments (Fig. 5.2). If subsidence is active but there is little sedimentary input, the sediment would not be buried deep. This is what happens in the broad oceanic plains where terrigenous input from the continent is extremely small and pelagic organic productivity is low too. If the sedimentation rate is high with a fast input of bulk sediment, on the other hand, the burial phenomenon will be effective and the sedimentary overload will abet the subsidence. In some cases, the space initially available will be filled up and the sediments will transit to another deposition domain.

Burial is at its maximum in rift systems and on the continental margins, especially in the vicinity of a major river mouth allowing high sedimentary flow. It will be seen later on that techniques exist for evaluating this burial, and the subject will be taken up again in the following chapter on petroleum systems, when discussing the burial and transformation of organic matter.

Accumulation rates are highly variable, as was seen in Chapter 3 (see Fig. 3.28). Generally, the sedimentary basins of interest for petroleum are filled with several kilometers of

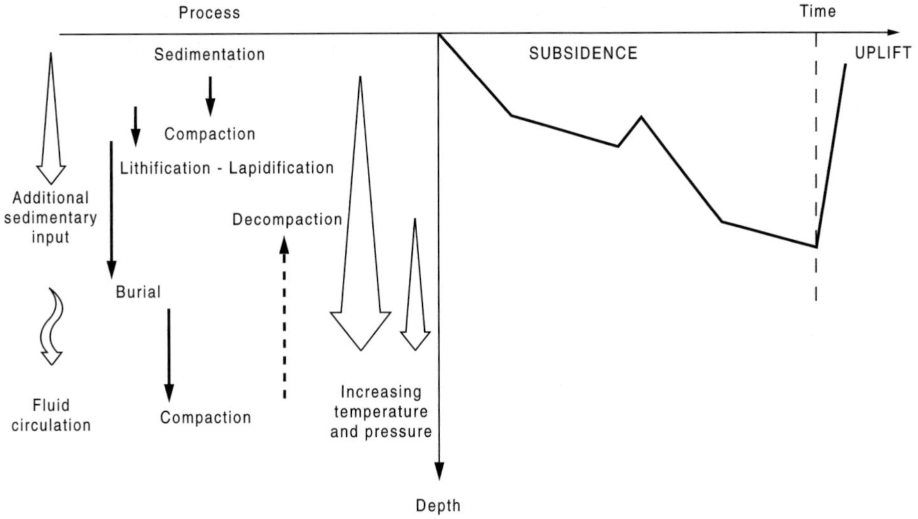

Fig. 5.2 Subsidence and associated processes.
In the course of time, the sediments accumulating and settling (by subsidence) in a basin will bury any previously deposited sediments, which thereby undergo transformations under changing temperature, pressure, and fluid conditions. If the subsidence stops and uplift begins, these processes are blocked.

sediments (Fig.2.24). Let it simply be said here that these thicknesses do not correspond to the original deposition thickness, nor to any fixed period of time. Some basins have a history of several hundred million years, and others have a much shorter life. Also, a stratigraphic horizon does not necessarily have the same burial history within the same basin, considering the possibility of differential subsidence and later uplift phenomena.

The A.A.P.G. documents (Memoir No. 37, 1984, and Studies No. 36, 1993), along with the book by Purser (1980) are useful complements to this very general presentation.

5.1.2 Diagenesis

Diagenesis, then, is the sum of biological, chemical, and physical processes affecting sediments after their deposition in the course of lithification during burial, under time-varying ambient temperature and pressure conditions.

The petroleum geologist will distinguish between **mineral diagenesis**, which by definition concerns the mineral constituents of the porous network, from the **organic diagenesis** concerning the transformations of organic sediments and its hydrocarbonated derivatives.

This latter theme will be addressed more specifically in the following chapter on petroleum systems, but the general agents of diagenesis discussed here are valid for all transformations.

Also, in the lithification process, mineral diagenesis gradually turns into **metamorphism**. The boundary between the two during burial is normally situated at depths of less than 8 km, which corresponds to 250°C under average geothermal conditions. Below this depth is the domain where the compaction is such that the interstitial water in the pores gradually loses the major effect it has on material transfers. The different types of metamorphism will not be discussed here.

The type and intensity of diagenetic processes will vary considerably from one medium to another, but they are active in all of today's sedimentary environments on the continent and from the littoral domain to the great ocean depths, beginning as soon as the initial deposition environment is modified, and especially their initial fluids (which is generally water).

5.1.2.1 Agents of Diagenesis

Diagenesis is the combined result of a number of strongly coupled processes.

First are the **biological** and microbiological factors. The action of living organisms, and especially bacteria and algae, is certainly effective in the first stages following the deposit of a sediment at shallow burial depth. The most striking example is that of **bioturbation**, in which the sedimentary layer is often subject to mechanical disturbances stemming from the life forms within it and from the movements of benthic organisms (Fig. 5.3), to the point where the texture of the initial layer is at times completely transformed. Less spectacular but no less certain are the **biochemical** factors such as the change in pH of the interstitial waters, causing a reduction in sulfates or fermentation of the organic matter (see following chapter). On the whole, organisms act both as catalysts favoring diagenesis and as inhibitors.

Then the combined **physical** factors of **pressure** and **temperature** come into play as the sediments gradually accumulate. The **lithostatic pressure** due to the weight of the deposits will **mechanically compact** the particles. The lithostatic pressure (2 or 3 bars for 10 m of burial) is generally distinguished from the **hydrostatic pressure** in the voids between the grains where only the interstitial water pressure (1 bar at a depth of 10 m) is at play. The pressure and temperature increase in parallel (Fig. 5.4). When the grains are packed, they are rearranged and the texture is changed, at times notably reducing its volume (compaction). Aside from its mechanical effects, this packing also expels the interstitial fluids, and chemical compaction occurs at the points of contact between the grains. The combined mechanical and chemical effect is **pressure-dissolution**.

It will be seen later in this chapter that such effects are not restricted to the sedimentary pile, nor are they due solely to the main vertical lithostatic stress. Similar effects can be found in stress regimes due to tectonic deformations, where the main stress is sometimes horizontal.

Fig. 5.3 Organisms, the first agents of diagenesis.
In practically all environments, the organisms living in or colonizing the sedimentation environment are effective transformation agents.
A. Submarine view: burrowing in recent oozes, observed at a depth of more than 2600 meters in the Mediterranean (IFP-Ifremer photo).
B. Worm burrows (scolites or tigilites) have obliterated the oblique stratified structure of delta sands (Saharan Paleozoic) (IFP photo).

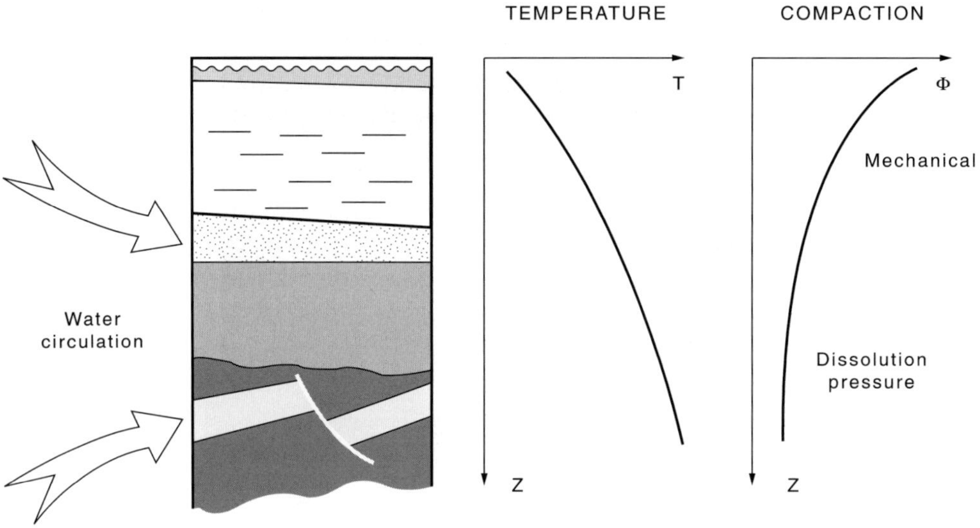

Fig. 5.4 Temperature and pressure variation in the course of burial.
Note that the thermal and mechanical effects come in addition to those of surface and deepwater circulation (with their leaching, cementing, and draining effects).

The temperature, which increases according to a **geothermal gradient** of about 30°C/km of burial depth, modifies the solubility of certain ions and generally affects the chemical balances or, more precisely, the **thermodynamic balances** among the minerals in the particles and interstitial solutions.

Chemical factors. Depending on the original composition of the sediment (siliceous, argillaceous, calcareous, evaporitic) and the type of initial water (pH, dissolved oxygen content and CO_2 pressure in lacustrine, marine, or deep ocean deposits), it can be seen that the chemical reactions that take place will differ with temperature variations and fluid circulations. These reactions may be rapid, if the original environment is porous and permeable, and reach an equilibrium point in accordance with thermodynamic laws. They will be slower in a closed environment with a sequence of transitory states according to chemical kinetics, with a number of diagenetic stages in succession having different histories, and depend on the porous medium and the fluids circulating through it.

Fluid circulation and **pore pressure** are especially important. Beyond the purely mechanical compaction process, it is the fluids that will play the major role. This enters upon the broad field of **hydrology**, and particularly that part of hydrology called **hydrodynamism**, in which fluids (especially water) are seen to play a number of roles, namely, that of mechanical stress transmission (in compaction, assisted deformation, hydro-fracturing), heat transfer (convection), as dissolved element transport vector, and the chemical role in the special geochemical environment they create and where the bulk of diagenetic reactions between fixed porous medium and mobile solutions operate.

Different types of solutions can be considered: fresh water, meteoritic water, brines, hydrothermal, magmatic, and saline waters. Fluid circulation operates at all scales in a sedimentary basin's diagenetic evolution. At the time of deposit, the sediment is generally waterlogged (up to 80% in pelagic oozes). **Initial expulsion** is a major phenomenon. Very early reactions can occur here. The initial water loss is sometimes recorded in the sediment's internal structure by flames, bird eyes and other figures. Then different types of **circulation** occur in the course of burial (Fig. 5.5): gradual **burial expulsion** (movement of generally salty **connate waters** already present at the time of deposit), invasion by recent or fossil **meteoritic waters**, igneous or hydrothermal **juvenile waters**. The fluids follow **migration paths** through permeable porous layers by horizontal diffusion, often over long distances, or by vertical diffusion along active faults.

There are different types of water bodies (Fig. 5.6) in valley or alluvial **sheets** also called the free-water table near the surface) and confined groundwater aquifers. A sedimentary basin generally includes one or more levels of these confined water sheets (Fig. 5.7). If the pressure in the sheet is high, it is an artesian aquifer, the **aquifer** being the bed through which the sheets circulates. Another important idea that helps to explain diagenetic phenomena is **saturation**. At the surface, a saturation zone is where the pores are filled and an unsaturated aerated zone is one where **vadose water** will infiltrate and percolate and some-

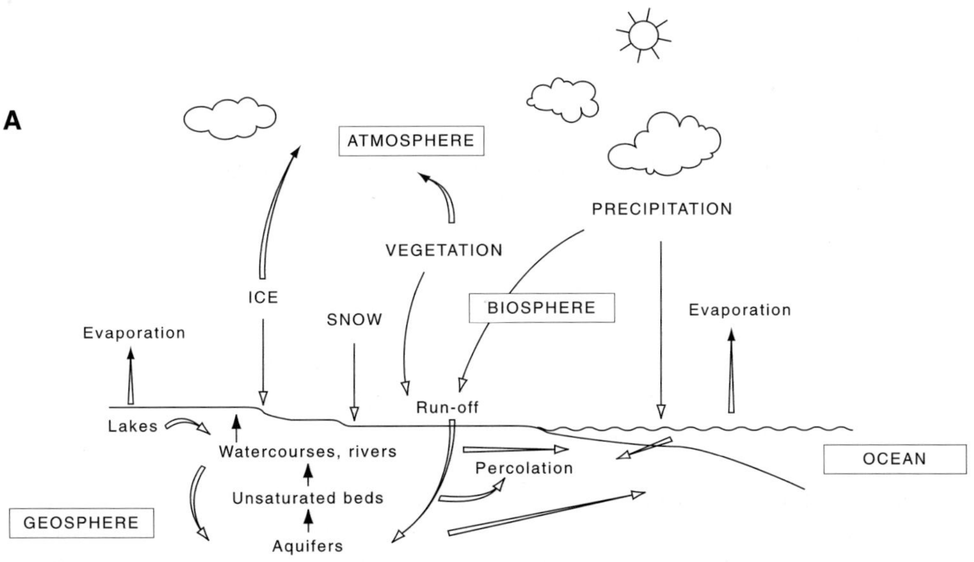

Fig. 5.5 Water circulation and evolution in a sedimentary basin.
A. General water cycle scheme.

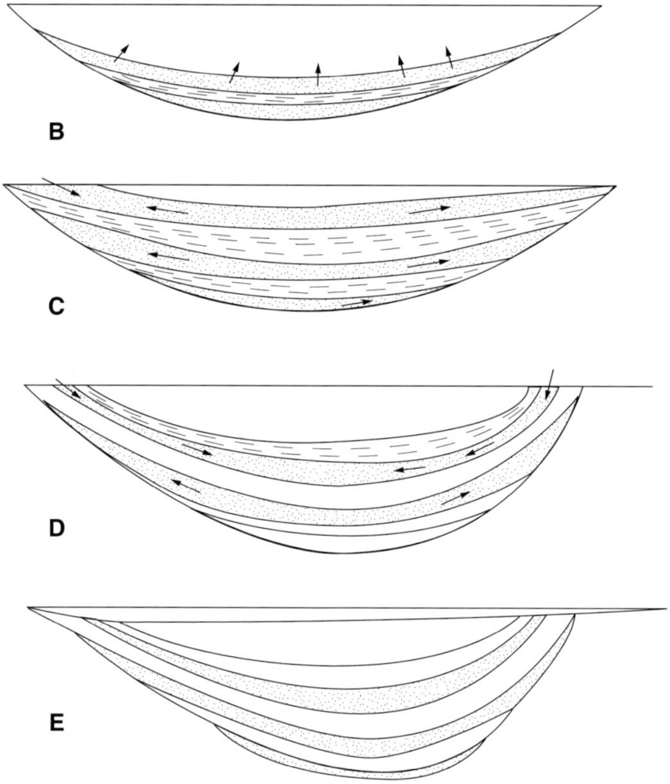

Fig. 5.5 (cont'd) Water circulation and evolution in a sedimentary basin. Time evolution in a basin. **B.** Initial expulsion: from ooze to sediment. **C.** Juvenile basin, burial expulsion, compaction expulsion. **D.** Invasion by meteoritic waters, replenishment and flushing. **E.** Aging, stabilized basin.

times **evaporation** can occur (Fig. 5.8). These **vadose** zones may be continental or marine. **Leaching** is an imprecise term that is sometimes used to designate different types of water circulation. The major consequence water circulation has for petroleum reservoirs will be seen later on in this chapter and in the next one.

5.1.2.2 Length of Diagenesis

It has been said, mainly as concerns biological factors, that diagenetic transformations can occur very quickly, and that the transition to the metamorphic domain was then gradual. It should be stressed that diagenesis is rather long and continuous.

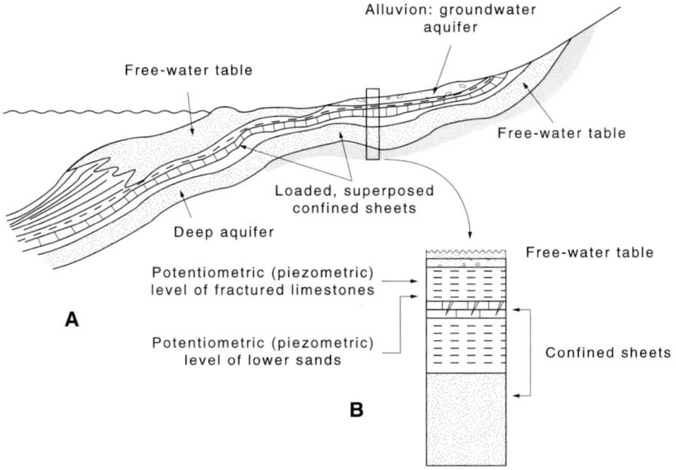

Fig. 5.6 Different types of piled water sheets.
A. General scheme: free and confined sheets. **B.** Different potentiometric (piezometric) levels. If the pressure is high in an aquifer, the water will come naturally to the surface, making an artesian well. Otherwise, the water will rise only to some potentiometric (piezometric) level.

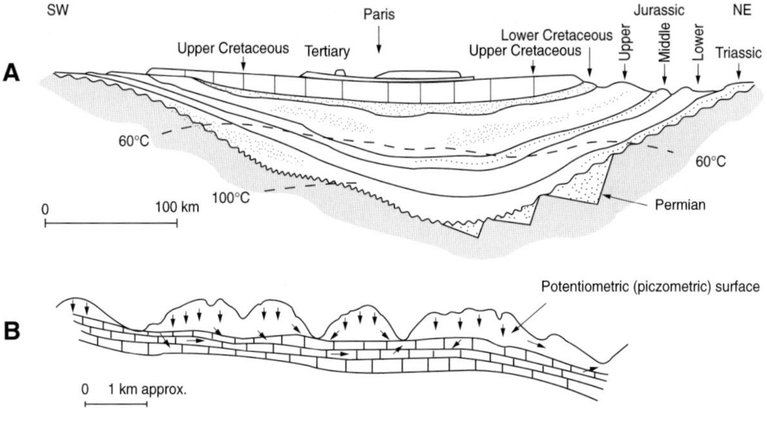

Fig. 5.7 Aquifers in a sedimentary basin.
A. Example of the Paris basin where several sheets lie atop each other (Triassic, Jurassic, Cretaceous). The simplified cross-section with heights greatly exaggerated (modified from Dercourt and Paquet, 1990) illustrates the fact that the water in the deep sheets is hot and the pressure is applied upstream (green sands of Champagne and Lorraine). **B.** The example of a karstic sheet in China's Yalongjing river basin near Sechwan (modified from Gang, 1996). The water is supplied here from the karstic surface network.

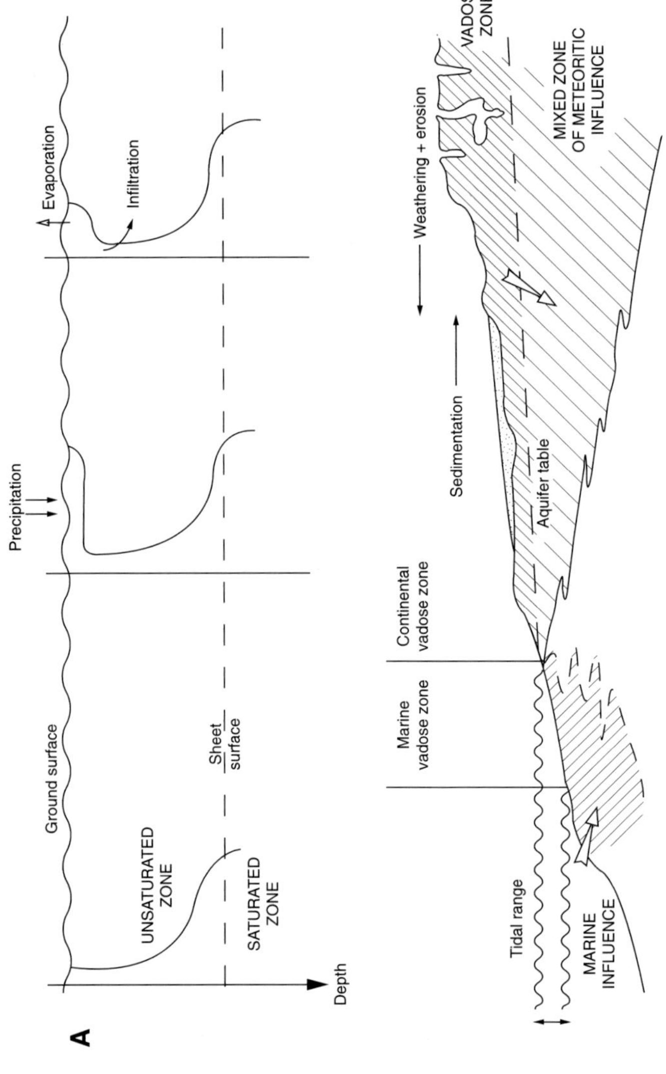

Fig. 5.8 Vadose zones and domains of combined continental or marine influence.
A. General moisture profile scheme between the ground surface and sheet surface at a depth that is, of course, variable: equilibrium profile; saturation surface after heavy precipitation; gradual return to equilibrium. These profiles will differ greatly in different types of soil and rock. **B.** Longitudinal profile scheme between the continental and marine domains. The vadose zone is the unsaturated space between the surface and the top of the aquifer.

The initial transformation processes are called **early diagenesis**. This begins with **synsedimentary diagenesis** for immediate phenomena or those occurring on the scale of a few years or tens of years. In a way, the example of the dissolution of carbonates in the deep oceanic environment can be taken as an ultra-precocious process. More generally, as was seen in Chapter 3 on sedimentary deposition environments, littoral calcareous environments and certain "evaporitic" environments are extremely favorable to diagenesis. Even the deposits themselves are sometimes marked by synsedimentary diagenesis, as illustrated by dolomites and sulfates from lagoonal environments (Fig. 5.9).

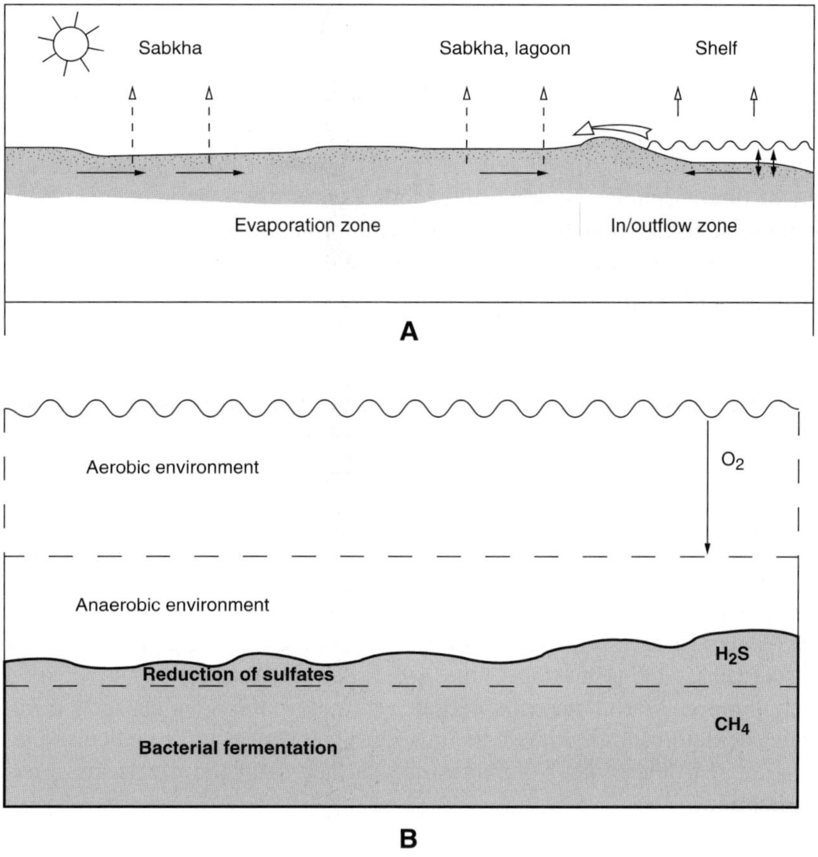

Fig. 5.9 Synsedimentary diagenesis.
A. Example of evaporitic deposits in sabkhas and lagoons in arid regions.
B. Example of anoxic environments where organic matter is conserved and sulfates and bacterial activity reduced.

The presence of debris from human activity in certain beach rocks and hardgrounds certifies to the rapidity of a geological process on the human scale. Early cementation of this kind in the marine environment can be found in the continental environment too, with dissolutions and recrystallization. Another illustration of the instability of certain sedimentary constituents is the oxidation and the rapid transformation of organic matter, which will be discussed in Chapter 6.

The idea of relative sea-level variation by eustasy and tectonics, as addressed in the previous chapter, shows how sediments deposited in shallow water can emerge quickly and be subjected to the action of meteoritic waters. Precipitated oxides, sulfides, or silicates can then act on the still-soft sediment and form opal, zeoliths, or clays in the process of **authigenesis**.

Early diagenesis is still active in the first stages of burial, and is especially rapid with the circulation of the initial fluids.

Late diagenesis, or **epigenesis**, refers to all the exchanges occurring in the life of the sedimentary basin, decreasing by an exponential law as is known statistically, with temperature and pressure variations. It was seen in the discussion of fluids, though, that these processes may be discontinuous, marking **diagenetic events**. Tectonic movement, for example, on a regional scale for a whole basin may subject a formation to meteoritic water circulation after a certain amount of burial has already occurred (Fig. 5.10).

Stages of disequilibrium and return to equilibrium may succeed each other in time, as evidenced in the example of sparry calcitic cement zoning. Early creation of hydrocarbons may inhibit a later diagenetic phenomena, as may also occur in early cementation.

The next chapter will explain how the formation of hydrocarbons and their migration to an accumulation zone is result of this late diagenesis in burial.

The ultimate stage of diagenesis is **metamorphism**, as has been said.

But another form of late diagenesis is **telogenesis**, where rocks that have already undergone burial diagenesis are returned to the surface by tectonic effect. The processes then occurring can resemble those of early diagenesis. The example of **karsts** illustrates this type of weathering of ancient series by meteoritic agents (Fig. 5.11). Everything that occurs at the surface has to do with the formation of **soils**, which is the field of **pedology**. The **discontinuities**, discordances and **paleosoils** witnessing to these returns to the surface are therefore of prime importance. Marine deposits may be returned to the open air early in diagenesis, with eustatic oscillations, or late, with the orogenic cycles. It may be difficult to distinguish between one or the other type. Sometimes the surface diagenic effects are referred to as **continentalization**.

This sequence of continuous processes punctuated with events affecting sediments and then rocks is the often complex **diagenetic history** which, when well analyzed, makes it possible to reconstruct the history of the sedimentary basin more finely. This is why the definition of this diagenetic history is of such concern today, especially in a framework of sequential stratigraphy. In petroleum geology, it is very important to reconstruct all of the diagenetic modifications in time. Not only does it affect the genesis of hydrocarbons, but it has a fundamental influence on the petrophysical parameters of the reservoirs.

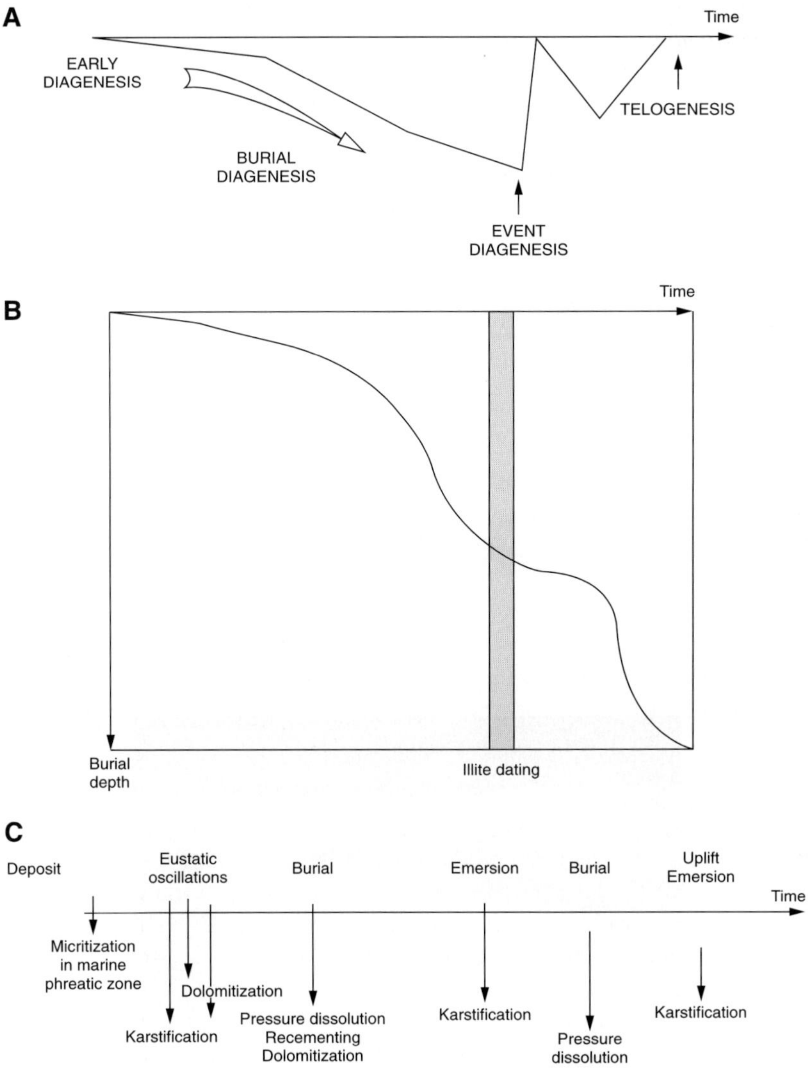

Fig. 5.10 Burial diagenesis and diagenetic events.
A. Time evolution of diagenesis. **B.** Scheme of a short-lived diagenetic episode. Here, the precise age of a diagenetic event is defined by precise illite dating by the potassium-argon method. **C.** Example of Urgonian limestones of the Vercors shelf (from Arnaud, 1993, and Mass and Tucker, 1995): polyphase diagenetic history.

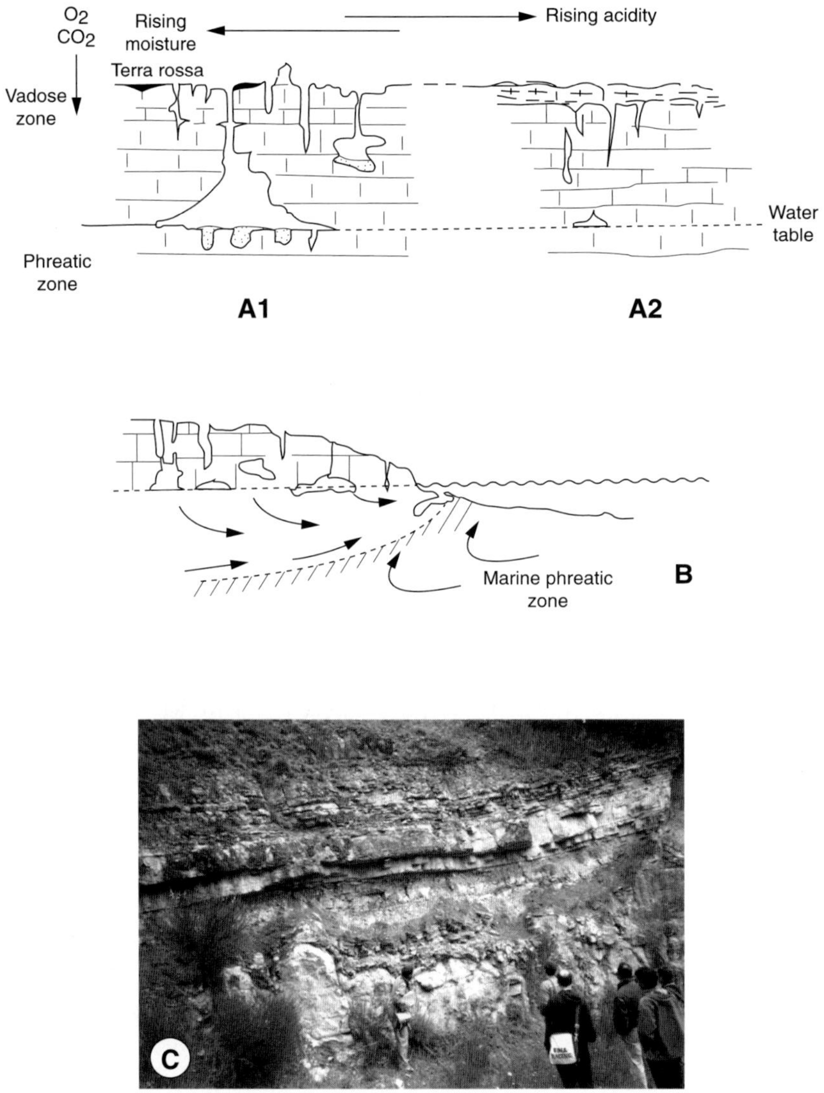

Fig. 5.11 Karstification and karst.
A. Karstification: vadose actions in limestones in a wet climate (A1) or a more arid one (A2). **B.** Interaction of vadose and marine water tables. **C.** Natural example of paleokarst with filling (southeast basin of France) (IFP photo).

5.1.2.3 Effects of Diagenesis

For the sake of simplicity, the various effects of diagenesis will be treated separately here, though it should be remembered that they can occur concomitantly.

A. Mechanical Changes

Expulsion of water is the most rapid initial phenomenon. The resulting reduction in volume is already extremely large.

The effect of weight and pressure results in **settling** or **mechanical compaction**, followed by **mechanical and chemical compaction**. "Normal" compaction is a simple reduction in the total volume, beginning with the reduction of the liquid volume by expulsion of the water, while the solid volume does not vary at the beginning of burial. The grains are then mechanically compacted with no chemical or physical variation, reducing the porosity as a monotonic, continuous function of depth. Normal compaction occurs in sandy and argillaceous sediments, and especially in well-drained environments at hydrostatic pressure, but not in calcareous environments where dissolution and secondary precipitations occur, nor in very poorly drained environments where overpressures exist but water expulsion is blocked. Here, the compaction is said to be abnormal. Mechano-chemical compaction occurs when the solid volume is subjected to the reactions of a return to equilibrium in the remaining fluid. While the compaction generally increases with burial (Fig. 5.12A), it does depend on the type of sediment and initial texture. Compaction reduces the pore space, i.e., increases the density and decreases the **porosity** and **permeability**, which then limits the circulation of fluids. Thick under-compacted series have been found in special cases with very high sedimentation rates. This **undercompaction** is frequent in delta formations (Fig. 5.12B). Differential compaction (Fig. 5.13) is an irregular loading in space, applied to different lithological series.

When starting with the present state of a basin and reconstructing its past situation, its different stages of evolution have to be **decompacted** (Fig. 5.14). An amount of **natural decompaction** may occur by elasticity in an ancient series if its initial load is released.

One intense expression of the mechanical-chemical effect is compaction **stylolites** (Fig. 5.15). These **stratiform** formations are recognizable by the insoluble residues that line the **stylolite joints**. These may occur during burial but also early on, as it is thought that a sedimentary load of a few dozen meters would be enough. Tectonic stylolites are different, and will be discussed in the second part of this chapter. If the transfers are large, dissolution schistosity (cleavage) and sometimes even microfolds can develop at burial depths of more than a kilometer.

Fracturing is another mechanical effect of burial. This may occur under the simple effect of the load without requiring any notable tectonic stresses. If the confinement or hydrostatic pressure is low, microfissuring or microfracturing will tend to propagate. If the pressure exceeds a certain threshold, the rocks remain ductile and the fracture does not occur. When a bed is sandwiched in a permeable layers, the release of the fluids can create

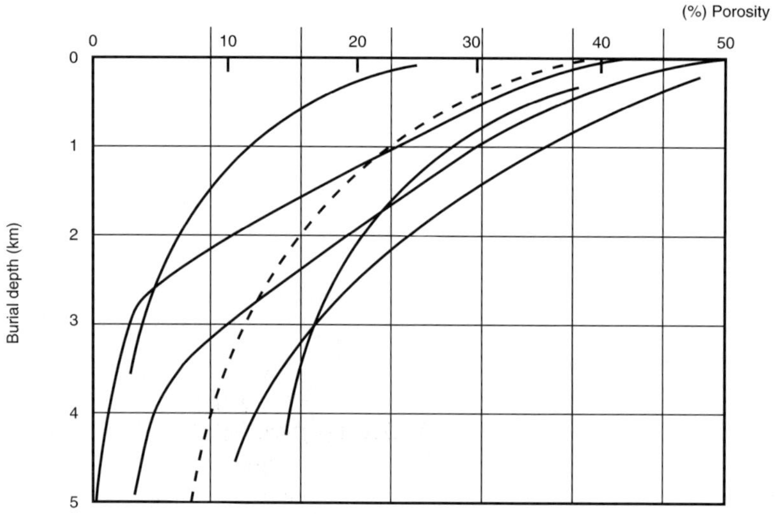

A

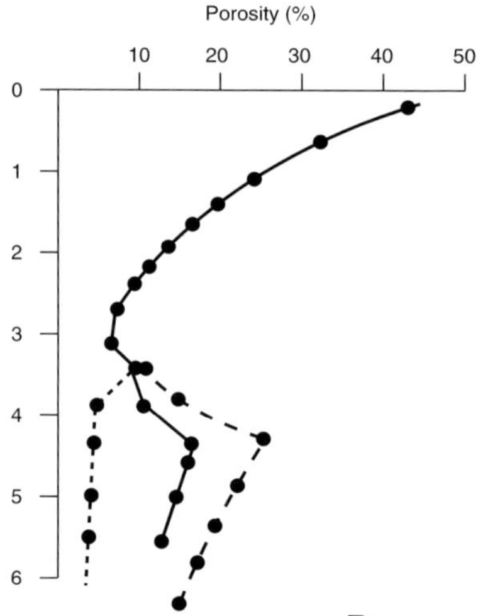

B

Fig. 5.12 Compaction and undercompaction.
A. Porosity curves of various argillaceous sediments as a function of depth reflect rapid compaction. The dashed curve represents a mean (from Guillemot, 1984). **B.** The porosity curves here show anomalous undercompacted zones at depth.

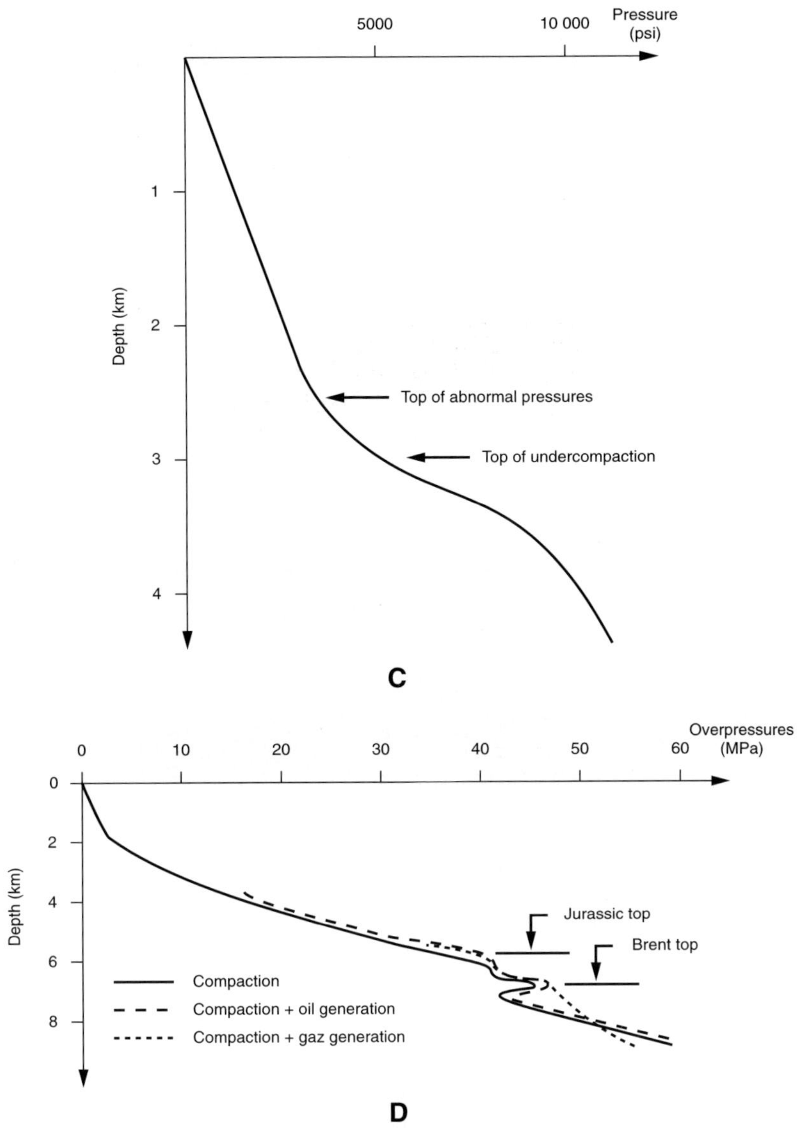

Fig. 5.12 (cont'd) Compaction and undercompaction.
C. Here, the compaction anomalies show up in pressure-overpressure curves. **D.** In this example, the anomalies are caused by the generation of hydrocarbons (from Burrus, 1996).

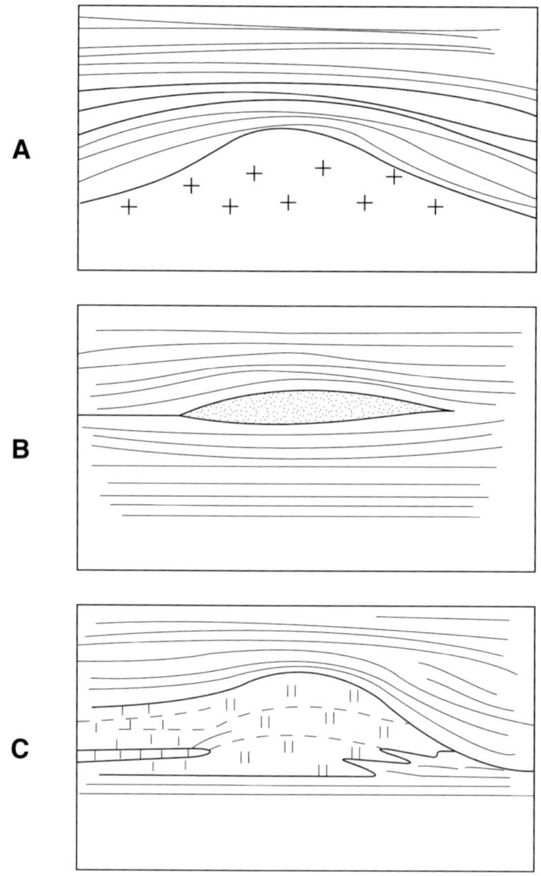

Fig. 5.13 Examples of differential compaction.
A. On a paleorelief. **B.** To either side of an arenaceous lens. **C.** On a reef edifice.

an interstitial pressure that is greater than the hydrostatic, and **hydraulic fracturing** will occur. This is part of structural or tectonic geology.

The effect the load has on a sediment depends on the type of material. The plastic behavior of salt and clay is one special case. Pillows, domes, and **diapirs** may form by **creep**, solely under the pressure of the weight of sediments (Fig. 5.16). This is again the field of deformation and tectonics.

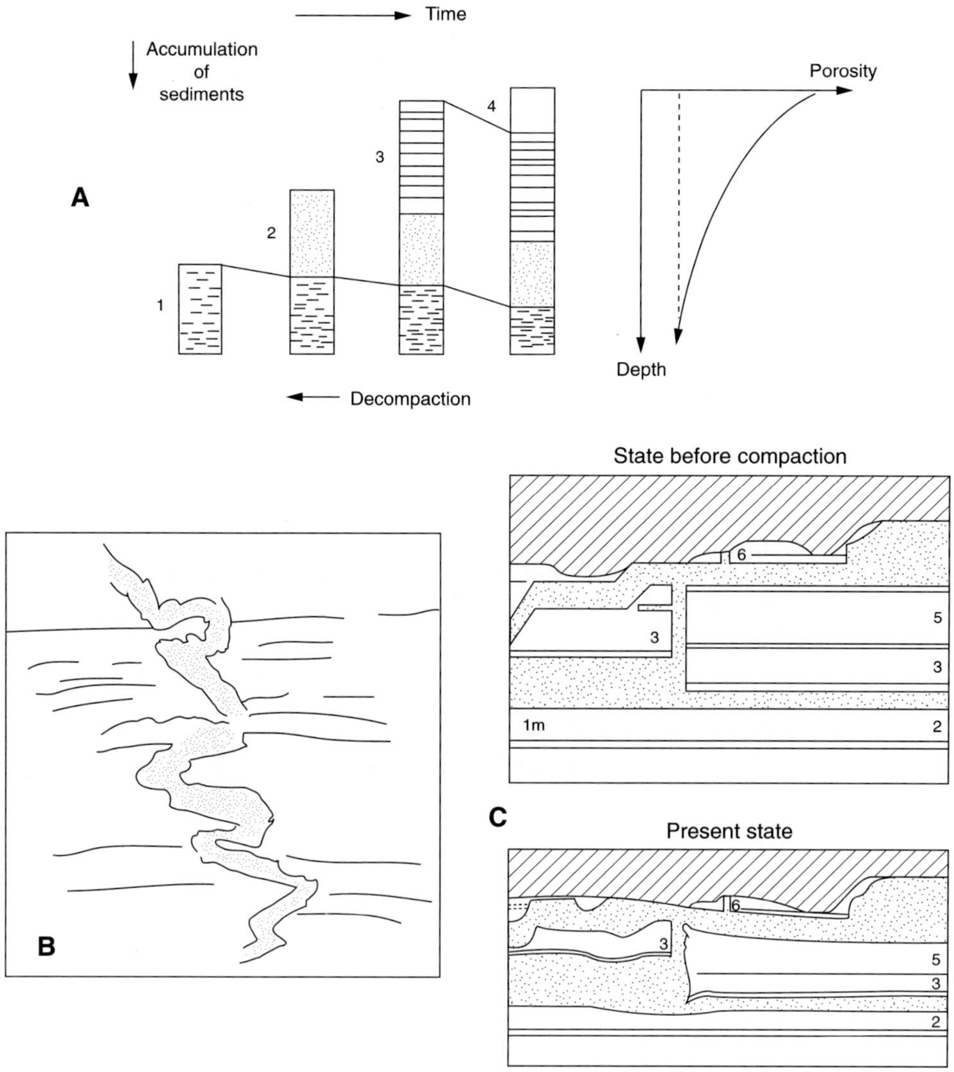

Fig. 5.14 Decompaction.
A. The time accumulation of sediments compacts the various formations. The initial thicknesses can be reconstructed by back-stripping the series using porosity curves. **B.** Example of gritty dikes deformed in a clay matrix as the clay compacts (about 50%) (from Beaudoin, 1985). **C.** Gritty silts and dikes of the Cretaceous in the Rasous region in the southeast basin of France (from Beaudoin, 1985).

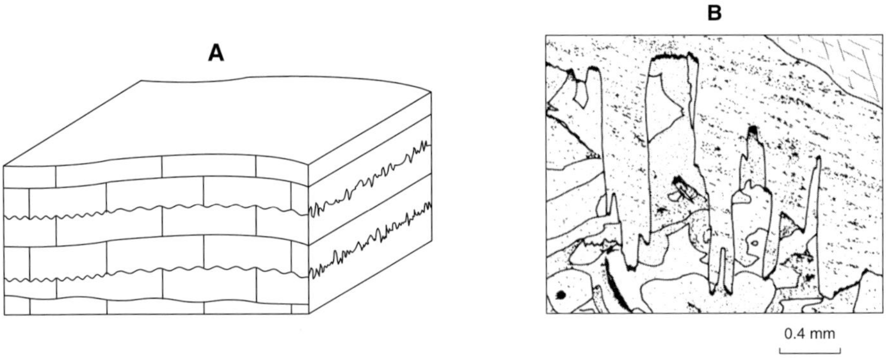

Fig. 5.15 Stratiform stylolite joints.
A. General scheme with no scale: a sequence of calcareous beds is sometimes characterized by stylolite joints that are said to be "stratiform" because they are found in the stratification planes. **B.** Enlarged view (from Guillemot, 1986) diagramming the interpenetration of two limestones of well-differentiated texture, through several centimeters. This interpenetration is due to dissolution, and a black film of insoluble residues emphasizes the stylolite points.

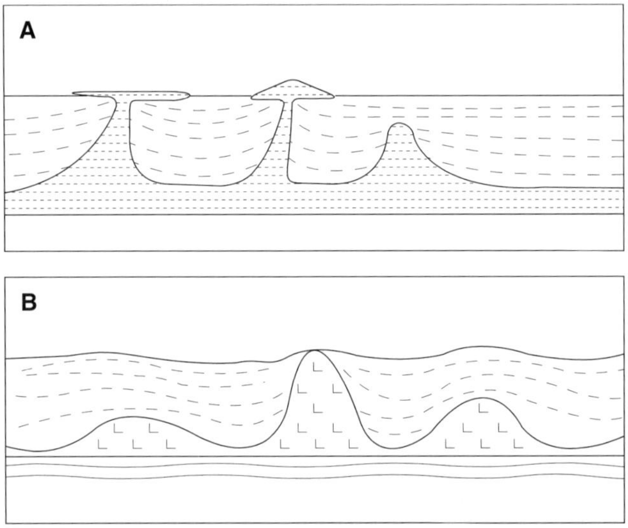

Fig. 5.16 Mud volcanoes and saliferous diapirs.
The load of the sedimentary column and tectonic forces can cause major deformations in geological series: water-swollen clay producing fluid expulsions and mud volcanoes, salt and evaporites. These are the phenomena of argillokinesis (**A**) and halokinesis (**B**) (also see Fig. 5.34).

B. Chemical Changes

Very many chemical diagenetic processes of different types can occur. The time schedule of this type of transformation depends very much on the lithology of the initial sediments. The transformation is very early for the most unstable of them (sulfates and carbonates), but requires burial depths of 1000 to 2000 m for clays. The type of fluid, its pH, and the CO_2 pressure are decisive. Certain elements like calcium and magnesium are highly mobile, while others (iron, potassium, manganese) are more passively concentrated. Chemical phenomena occurring at the surface come under **weathering** (acidolysis, halmyrolysis, oxidation, karstification, or pedogenesis). **Mineral transformations** are designated by other terms like authigenesis, segregation, or neoformation.

The previous chapter showed the importance of physical and chemical signals in sedimentary series for magnetostratigraphy and chemostratigraphy. It should be remembered that diagenesis can disturb the signals greatly with the mineral transformations occurring during burial, and care should be taken to avoid mistaken interpretations.

Cementation. A sediment generally includes particles of different sizes (grains and microgranular matrix) with a pore space that may or may not be filled with fluid. Depending on the temperature, pressure, and ionic balance, the interstitial solutions may precipitate a mineral **cement** that will bind the particles together. Cementation will thus reduce the pore space (Fig. 5.17). As the conditions evolve, this "secondary" cement (secondary as opposed to the primary binder of the matrix) may become stable, but analysis of intergranular cementing sometimes shows several stages in the process, and sometimes dissolution phases too. Cements will vary widely depending on the rock and fluids: silicification of sands and then of sandstones; quartzification by development of a siliceous cement; sandstone with calcareous cement; different types of carbonates. One special case of crystallization of a new mineral in the pore space is **authigenesis**, a term generally used for the neoformation of argillaceous minerals.

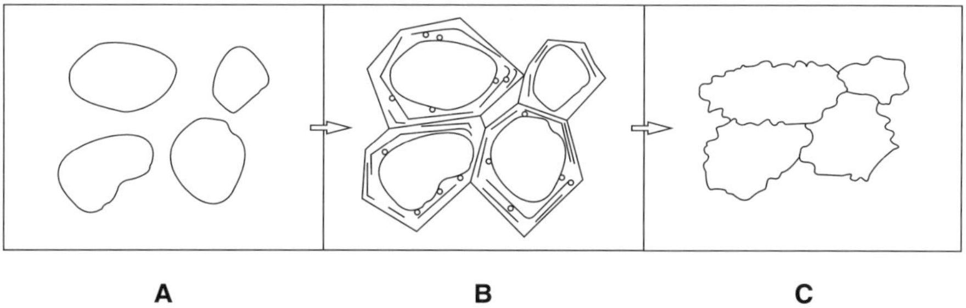

Fig. 5.17 Secondary cementation with disappearance of the pore space. **A.** Initial state: separate grains of sand (a 2D view, as grains would be seen jointed in 3D). **B.** Sandstone: successive rings of secondary quartz with fluid inclusions present around siliceous sands. **C.** Final state after pressure dissolution phenomena and total disappearance of the pore space.

Recrystallization is not the same as cementation, but it is sometimes difficult to distinguish between the two. Before recrystallizing, the initial grains are transformed by ion exchange or dissolution (Fig. 5.18). Argillaceous minerals are very sensitive to this phenomenon, as are previously crystallized cements.

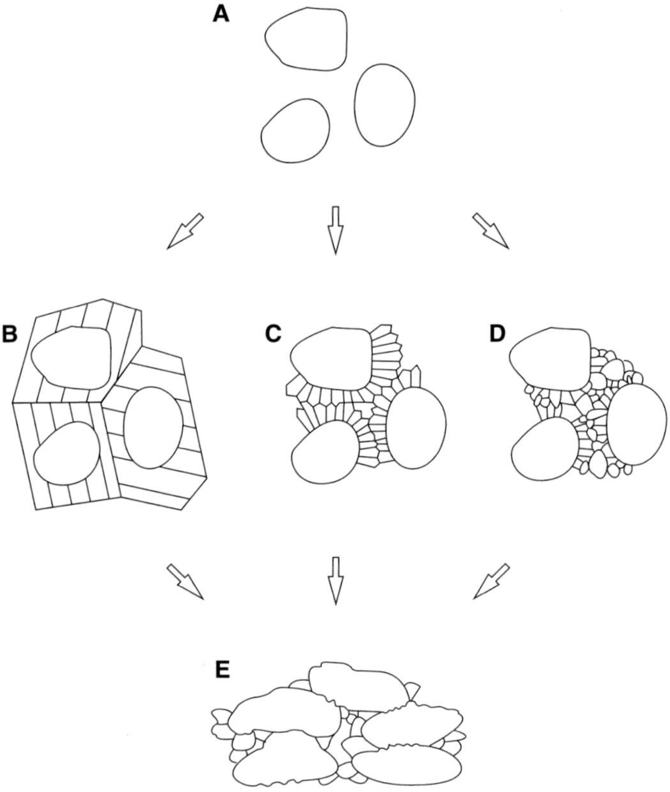

Fig. 5.18 Development of sparry cement around carbonate grains. **A.** Initial state. **B, C,** and **D.** Different crystallization patterns. **E.** Compaction will vary depending on the type of limestone and its diagenetic history, and the process can change radically with any major pressure dissolution and recrystallization phenomena.

Metasomatism. This is a phenomenon of ionic transfer between the solution and the mineral particles, resulting in major mineral replacements. The best known example of it is in the **dolomitization** of limestone (Fig. 5.19). The dolomite is said to be **secondary** or **penecontemporaneous** if the diagenesis is early, or **late** if the processes occur in the later history of the basin, such as in tectonic fracturing.

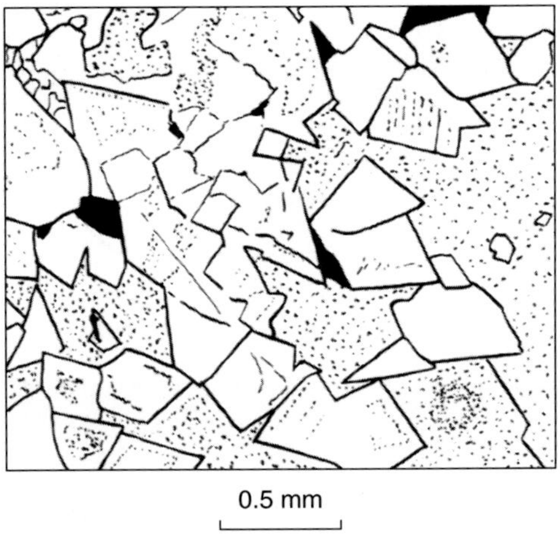

Fig. 5.19 Example of metasomatism: dolomitization of limestone. Diagram of rhombohedrans of dolomite in a cryptocrystalline limestone (from Guillemot, 1986).

Dolomitization is common but still poorly explained. There is a question about its kinetics, because it does not precipitate naturally even when in oversaturated sea water. There is a thermodynamic problem of formation at low temperature, and the volumetric question of crystallization in thick, extensive massifs. Although more stable than limestone, dolomite is sometimes subject to de-dolomitization. Three major variables are at play: Ca^{2+}/Mg^{2+} *activity,* CO^{2+}/CO_3^{2+} *activity, and temperature.*

If there is a great deal of metasomatism, the resulting texture in the rock will differ greatly from the initial structure. In the case of dolomitization, there will be greater porosity, while the rock will take on the appearance of breccia in other cases like cellular dolomite or **cargneule**.

Certain minerals **segregate** in a sediment that might be initially homogeneous, and concentrate into concretions in the bedding, or nodules of different types, such as flint, chert nodules, carbonate nodules, or barytine (Fig. 5.20).

Porosity is also improved by **dissolution**. This may be discrete or massive, the typical case being **karstification** (Fig. 5.11). This affects limestone beds leached by aggressive acid waters that are generally meteoritic, though some cases of submarine dissolution also exist.

Micritization. This is a special phenomenon affecting the fine carbonate matrix containing fine calcareous particles of all origins. It is probably biochemical for the most part, in both early and late diagenesis.

Hardgrounds are the most spectacular products of submarine diagenesis. They are one of the forms of early diagenesis and can occur at any water depth, but preferentially in deep water (Fig. 5.21). This form of lithification often corresponds to a more or less prolonged

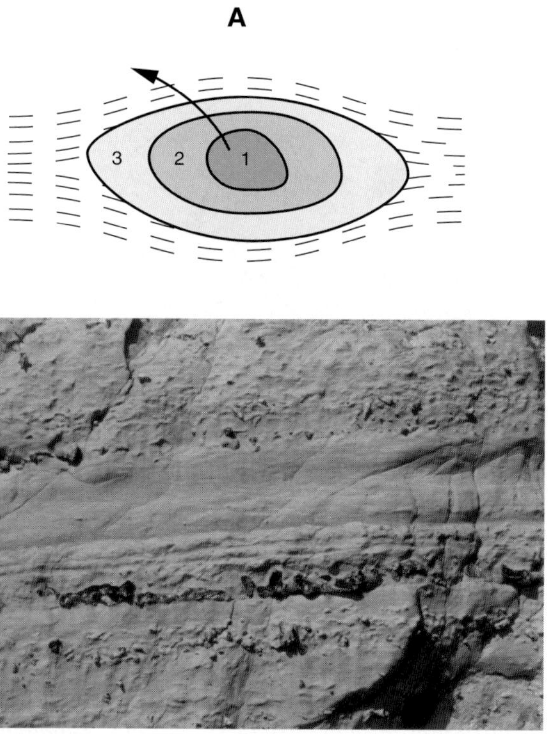

Fig. 5.20 Concretion.
A. Theoretical diagram showing how precipitation begins at the center while compaction is moderate, then continues with time as the compaction increases. The youngest part of the concretion is around the outside.
B. Cherty chalk: chert nodules or concretions in the carbonate domain. Here, centimetric flint nodules in Dorset chalk (Upper Cretaceous, England) (IFP photo).

sedimentation hiatus. **Coating** crusts can appear in both marine and continental environments.

The **desiccation cracks** of continental and intertidal environments are formed by the retreat of water (with rapid evaporation) and the swelling of an initially water-rich sediment (clays).

Cementation in the vadose zone. Circulation of meteoritic water and evaporation in the continental environment above the water table causes a **crust** or hard surface to form, under which dissolution, precipitation, and recrystallization can enter into action (Fig. 5.22). The pedological crust in a carbonate environment is called **caliche** or **calcrete**. It depends on the climate, of course, and can include concretions, cracks, and stromatolithic-type laminations,

Fig. 5.21 Hardgrounds and sedimentation hiatuses.
A. Perforated hardground. This view of a bank surface shows a fine hardened coat overlying a more crumbly limestone. The soil also shows signs of bioerosion by pholad-type organisms. **B.** General view. **C.** Close-up of sedimentation hiatus between limestones of the Urgonian and arenaceous clay of the Aptian (Ales region). The unconformity surface is an ancient hardground with coating and perforations.

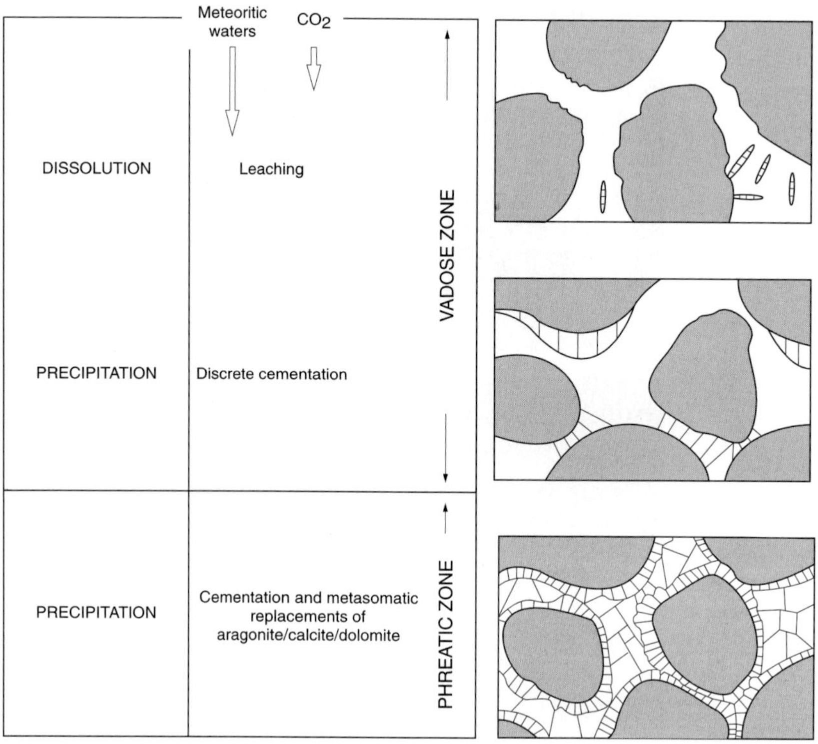

Fig. 5.22 Dissolution and precipitation in vadose and phreatic zones: differences in cementation.

and is often marked by traces of roots. The siliceous crusts in a siliceous environment are called **silcrete**. In a tropical environment with strong evaporation, a ferralitic (aluminum or iron hydrate) ground called **laterite** can form. **Bauxite** is one example of such a **residual rock** with pisolithic structure, formed on site by pedological weathering in tropical countries.

Generally, continental **weathering** depends largely on the climate. Exchanges are blocked in the higher latitudes and during glacial periods. Reaction rates will vary by the hydrological balance, between humid and arid climates. With weathering and paleoweathering providing valuable information concerning paleoclimates, the sedimentary cycle is closed and surface geology comes back into play (where erosion brings us back to the origin of sedimentary materials, discussed in Chapter 3).

5.1.3 Petrophysical Characters of Sedimentary Rocks

All of the rock formation phenomena analyzed so far have gradually contributed to changing the initial petrophysical characters, which is especially important in petroleum geology for the following two reasons.

With the experimental **reflection shooting** method used in geophysics and largely in exploration, a wave train[1] propagates through the beds in the basin's stratigraphic series in a way that depends on the lithology of the individual beds. The propagation velocity and mass density (the product of which defines the acoustic impedance) are characteristics specific to each bed or set of beds, so that the seismic response will vary according to the cementing, burial depth in the basin, and the presence of any residual porous space that may or may not be filled with fluid. So the seismic imaging of a basin will depend very much on the entire diagenetic history of burial. The petrophysical characters determined by the seismic, such as the Young's modulus, Poisson coefficient, and other special phenomena such as absorption, are due as much to the diagenesis as they are to the original composition.

The idea of pore space is essential to **reservoir geology**. Whatever the dominant mineralogical composition, the distinction is still made between **grain**, sedimentary **matrix**, and **cement** (Fig. 5.23). The residual **pore space** of more or less connected **pores** may itself be occupied by **fluids** filling these voids.

Grains are characterized by their grain size distribution, grading, and shape (or morphology). **Grain size distribution** is expressed in classes: **lutite** for grains of less than 62.5 µm, **arenites** up to 2 mm, and **rudites** beyond this (Fig. 5.24). (It is not always possible to recognize the original grains after the effects of diagenesis). Sometimes, "grain growth" or secondary growth can occur (Fig. 5.17).

Grading, which is guided by the depositional transport mechanisms as was said in the chapter on sedimentation, may be good or bad. The better it is, the better the porosity (Fig. 5.25). Grading quality is preferably expressed in cumulative grain size distribution curves rather than the usual Gauss curves. The effects of diagenesis, insofar as it affects porosity, will vary greatly depending on the initial grading.

*Knowledge of the porous environment is enhanced by the study of the grain shapes in **grain morphology**, and by scanning electron microscopy or **exoscopy**, to analyze their surface state.*

Grains are bound together by the **sedimentary matrix and cement**. The matrix is contemporary with the deposit, and in fact consists of the finer fractions. The more poorly graded the sediment, the harder it is to distinguish grains from matrix. If there is little or no matrix, the original sediment is **soft**. The **cement** resulting from diagenesis is often called secondary or neogenic and is at times difficult or impossible to distinguish from the original matrix when the matrix itself has undergone strong diagenetic effects. It is by analysis of the

1. Lavergne, 1986.

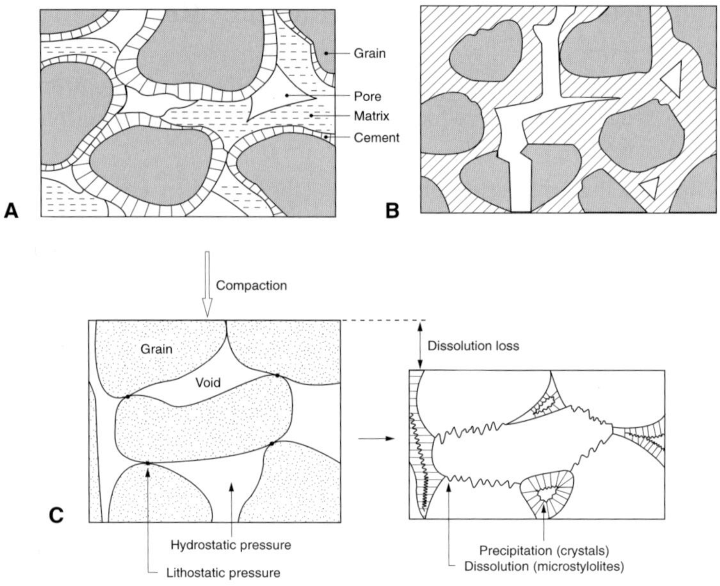

Fig. 5.23 Grains, matrix, cement, pore space.
Grains are distinguished from matrix (in the sedimentary constituents), cement (secondary) and the pore space, which may stem either from the matrix (**A**) or from deformation of cracks (**B**). **C.** Diagrams showing the loss of porosity by compaction and pressure dissolution in diagenesis.

Phi scale	10	9	8	7	6	5	4	3	2	1	0	-1	-2	-3	-4
			1/256	1/128	1/64	1/32	1/16	1/8	1/4	1/2					
	1 µm	2	4	8	16	31	62,5	125	250	500	1 mm	2	4	8	16

	Lutites		Arenites					Rudites		
			Sands							
Soft	Clays	Silts (aleurites)	Very fine	Fine	Average	Coarse	Very coarse	Granules	Gravels	Pebbles
Consolidated	Argilites	Silts, siltstones	Sandstones (same subdivisions as sands)					Conglomerates		

Fig. 5.24 Table of grain size classes (from Guillemot, 1986).
Diameters are expressed in millimeters or micrometers, in fractions of millimeters, and in Phi scale units (Phi is the negative base 2 log of the diameter, in millimeters).

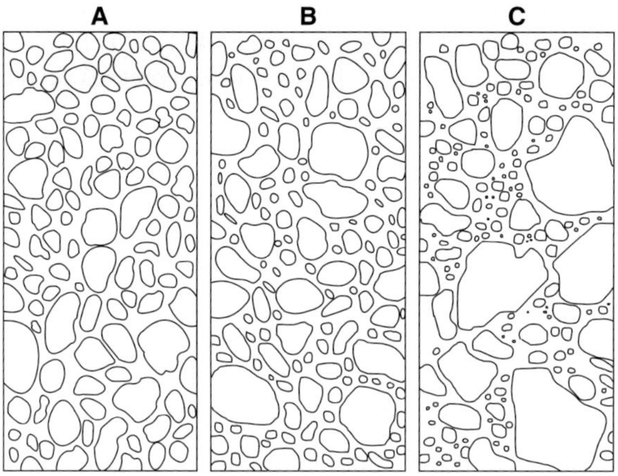

Fig. 5.25 Different types of grain size grading (from Guillemot, 1986). **A.** Good grading. **B.** Poor grading. **C.** Bad grading. Porosity diminishes as the pore space is invaded by smaller particles.

cements, of course, that the diagenetic history is described. If the cement is highly developed, the rock is said to be **indurated**.

The **pore space** is the part that interests the petroleum geologist. **Porosity** is defined as the ratio between the volume of voids (whether occupied or not) and the total volume of the rock, or:

$$\omega \text{ or } \varphi = (V_{\text{void}})/(V_{\text{total}})$$

expressed as a percent. This will vary greatly depending on the type of rock, and can be as high as 30%. The total porosity corresponds to the residual voids for granular rocks, but the cracks and diaclase joints also count. For example, the matrix porosity is very low in certain compact limestones (of the order of 1%, which is then called microporosity), but the total porosity can increase somewhat if there is any cracking. Clays are a very special case in light of their foliated mineralogical texture, which endows them with very high porosity (up to 80%!) but almost no permeability to speak of.

This hints that there are two kinds of porosity: one where the pores are not connected (intragranular porosity) and **effective porosity** where the voids are interconnected so fluids can circulate, though intergranular or intercrystalline pores. It should also be remembered that part of the fluids may be retained, so the porous medium has a retention capacity, due mainly to capillary forces.

If the effective porosity is also dynamic, with pores large enough to allow fluids to circulate, this circulation capacity is expressed as **permeability**. This is the aptitude of a medium to allow the passage of fluid, liquid or gas. Darcy's law expresses permeability in the form:

$$Q = \frac{S\,dp}{\mu\,dx}K$$

in which:

- K is the permeability
- Q the flow rate
- μ the viscosity
- S the area
- dp the pressure differential
- dx the length differential.

The permeability unit is the **darcy**, but the millidarcy (mD) is the unit commonly used because of the low values in geology. Interstitial and crack permeability are two complementary forms of permeability. For most rocks, the permeability depends on the grain size. It is very low for clays, which are often considered to be impermeable and thereby constitute barriers (or "seals", in petroleum geology terminology) despite their good porosity. The greater the cementation, the less permeability there will be. Salt beds are very effective caps (for their ductile properties).

On the scale of a set of heterogeneous beds, the idea of average permeability (see Chapter 6) has to be used because only an approximation will be possible, considering the many determining factors including the diagenetic history. On the scale of a basin, permeability may be seen as horizontal or vertical.

5.1.4 Laboratory Techniques

The study of diagenesis requires a rigorous approach and very advanced laboratory techniques (Fig. 5.26).

The types of minerals, their relative time relations, the state of the pore network, and the processes involved, are all determined by the usual techniques of **petrography** and **mineralogy**. For example, the dissolution of potassic feldspaths or the increase in the illite content of feldspathic sandstones (Hassi Messaoud, Alwin), dolomitization and dedolomitization, are determined by **X-ray diffraction microscopy** (optical or scanning), while **chemical** and **isotopic analyses** will determine the type of water and precipitation temperatures ($*^{13}C$, $*^{18}O$, Sr^{87}/Sr^{96}) and the characteristics of the secondary cements. Radiometric **dating** (K/Ar) on the illites will determine the age of formation of the diagenetic minerals. By analyzing the **fluid inclusions** in one or more phases, the minimum formation (trapping) temperature of these inclusions can be estimated, along with the salinity of the fluid. For carbonates, **cathodoluminescence** (chromatic separation of the mineral phases) will characterize the growth of the mineral species in one or more phases.

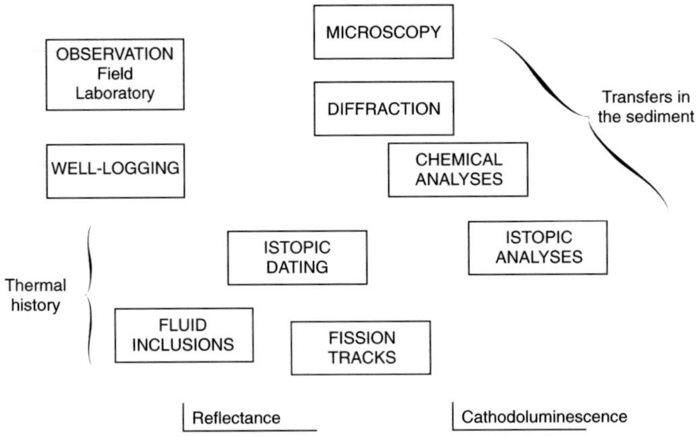

Fig. 5.26 Different techniques used in the study of diagenesis.

The fluid inclusion data will be complemented by analyzing the apatite (or zircon) **fission tracks** to find the maximum temperature reached during burial diagenesis, and the possible ages of thermal events like cooling that may occur during uplift, for example.

5.1.5 Ultimate Term of Diagenesis: Sedimentary Rocks (Table 5.1)

Chapter 3 on sedimentary environments spoke of the broad types of deposits. The following will briefly classify the resulting sedimentary rock after its diagenetic history.

5.1.5.1 Clastic or Terrigenous Detritic Rocks

Debris accumulations, which are essentially siliciclastic sediments but also calcareous sediments, as has been said, are broken down by grain size class into: rudites, arenites, and lutites (Fig. 5.27). Among the rudites, **conglomerates** of angular elements are called **breccia**, and the term **puddingstone** will be reserved for conglomerates of rounded elements. Depending on the matrix and amount of cementation, the conglomerates will be argillaceous or calcareous.

Among the arenites, **sands** and **sandstones** (which account for nearly 60% of the world's hydrocarbon-producing reservoirs) are generally characterized by their matrix and the amount of cementation. Siliciclastic sands are said to be **argillaceous** or **calcareous sandstone** depending on the secondary mineralization. They may even become quartzite if the diagenetic silication has been very intense, in which case the initial permeable porous level has lost all of its reservoir qualities. Other terms are used depending on the initial mineralogical content: **arkose** for a very feldspathic sandstone deriving from a granitic grit that

Table 5.1 Different types of sedimentary rocks

Type of sedimentation ↓ / Mineralogical content →	Grain size class	SILICEOUS ROCKS		CARBONATE ROCKS		SALIFEROUS ROCKS	OTHER
DETRITIC or terrigenous or clastic ROCKS	*Rudites*	Siliciclastic rocks					Ferriferous
		Breccia					Aluminiferous
		CONGLOMERATES	*Tillites*	Calcareous conglomerates			Phosphatic
		Puddingstone					
		Microconglomerates, microbreccia					
	Arenites			Skeletal limestone	*- Mudstones - Wackestones - Packstones - Grainstones -*		Carbonaceous
		Sands	*Arkoses*	DETRITIC LIMESTONES			
		SANDSTONES	Graywacke	CALCERINITES		*Gypsarenites*	
		Quartzites	Calcareous sandstones	Bioclastic limestones			
			Psammites	Gritty limestones *Lumachelles*			
			Argillaceous sandstones				
	Lutites	SILTS					
		Pelites		Argillaceous limestones			
		CLAY *Argilites*	MARL (Texture, composition)				
				Marly limestones			
CHEMICAL ROCKS and BIOCHEMICAL ROCKS			Siliceous limestones				
			Millstone	LACUSTRINE LIMESTONES			
				PELAGIC LIMESTONES			
			Silexites	CALCILUTITES			
			Cherts	*Micrites*			Ooliticores
				Biomicrites			Laterites
				Chalks			Crust
				OOLITIC LIMESTONES			Bauxites
				DOLOMITIC LIMESTONES			
		Radiolarites	*Jaspers*	DOLOMITES			PHOSHATES
			Lydite				PHOSPHORITES
			Phtanites		*Bound stones*	ANHYDRITE	
				Algal limestones		GYPSUM	
		Diatomites		BUILT-UP LIMESTONES			Peat
				Reefs		SALT (halite)	COALS
		Spongolites	*Gaizes*	Spicular limestones Stromatolites		POTASSIUM	BITUMENS
							Kerogens

Fig. 5.27 Different types of detritic rocks.
A. Conglomerate: fluvial deposits of the Saharan Cambro-Ordovician. **B.** Sandstones and breccia: fluvial deposits of the Upper Permian in Devonshire. **C.** Coarse sandstone: delta deposits in Tunisia's Lower Cretaceous (Boudinar formation). **D.** Fine sandstones, siltstones, shales: delta deposits of the Lower Cretaceous in Tunisia (Meloussi formation) (IFP photos).

has been transported little; **graywacke** for a sandstone that is very rich in different kinds of often ferromagnesian and volcanic debris; **psammite** for a sandstone very rich in mica lamellas; and glauconitic sandstone. Depending on the initial content, the diagenetic formations will differ, resulting in kaolinization, illitization, chertification.

Among the finest lutites (or pelites) are the **siltstones** and **shales** (generally less than 4 μm), which are characterized by their precise mineralogical content. If the proportion of $CaCO_3$ increases (35% to 65%), the rock is termed **marl**.

For sands of carbonate grains or calcerinite, the Dunham classification is used. This states the amount of fine carbonated and/or argillaceous matrix (see Chapter 3).

5.1.5.2 Carbonate Rocks

As has already been emphasized a number of times, there are two broad mineralogical classes: **limestones** and **dolomites**.

Limestones sometimes consist of detritic terrigenous material, but they are more generally made up of settled, transported, or built-up biogenic deposits (see Chapter 3, Fig. 3.32). The varied "grains" of bioclasts, oolites, pellets, or oncolites lie in a matrix that is often called **micrite**, though this is a term also used for the cement and is a contraction of "microcrystalline calcite" (**microcrystalline** is also a term used, or **cryptocrystalline** if the crystals are smaller than 16 μm, i.e., too small to be distinguished under a microscope) (Fig. 5.28). Effective diagenesis may change the original texture entirely. The **sparry** cement of coarse **sparite** crystals is often difficult to distinguish from the matrix. Dissolution and recrystallization occurring in diagenesis may completely erase not only the original structures but also the faunal content. The cements take widely varied forms (Fig. 5.29).

Dolomites, a double carbonate of calcium and magnesium (Ca or $Mg(CO_3)_2$) are largely the product of intense diagenesis (a field still being researched). They are of great interest for petroleum insofar as they generally make good oil reservoirs. These rocks are found more in ancient series than in recent ones. Several formation modes are possible, depending on various degrees of diagenesis (Fig. 5.30):

- **Evaporation** dolomites produced by very early synsedimentary diagenesis in the supratidal domain
- Dolomites formed by **infiltration** and **return flow** of hypersaline brine at a shoal edge, or by **mixing** of fresh and saline waters at the littoral fringe
- **Burial** dolomites where shelf carbonates are enriched in Mg^{2+} by fluids expelled during compaction.

Seawater is always the source of the Mg^{2+} by fluid circulation, but during burial the magnesium can also be released by smectic schists in metamorphosis. The case of non-stratiform dolomites should also be mentioned. These are developed diagenetically in blocks and channeled into fractured or faulted zones making "dolomitic mushrooms". These are tectonic **dolomites**.

Fig. 5.28 Types of calcareous rocks.
A. General view of pelagic limestone beds (Upper Jurassic in the Vocontian basin) (IFP photo). **B.** General view of prograding beds of shelf limestone (Reef edge of the Miocene, Mut basin in Turkey) (IFP photo). **C.** Calcarenites of the Miocene in the Touraine region of France (IGAL photo). **D.** Reef limestone of the Tyrrhenian at Jebel Zeit, Red Sea (IGAL photo). **E.** Calcirudite of the Pleistocene in Sicily (IGAL photo).

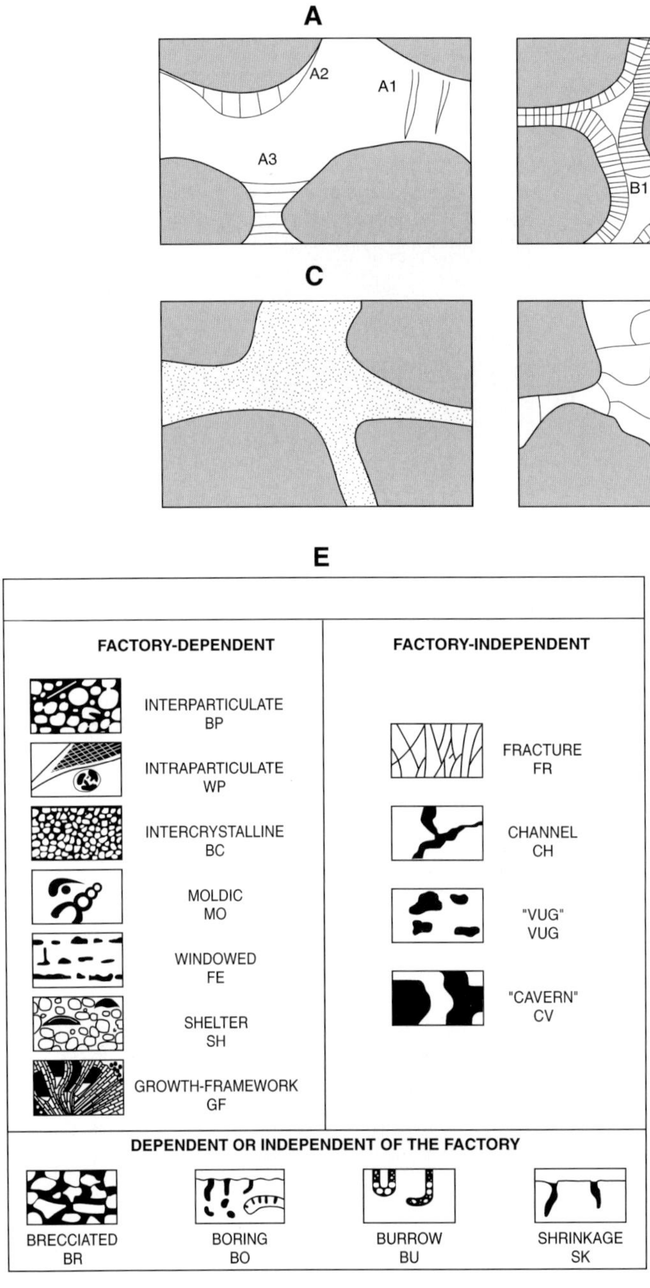

Fig. 5.29 Limestone cementing and types of porosity.
A, **B**, **C**, **D**. Types of cementation. **A.** Small, needle-like cementation (A1), hanging (A2) or miniscus (A3). **B.** Circumgranular (B1) or columnar (B2). **C.** Diffuse, micritic, or microcrystalline. **D.** Sparry, in coarse irregular crystals (diagram from various authors). **E.** Types of porosity (from Choquette and Pray, 1970), sometimes depending on the initial factory.

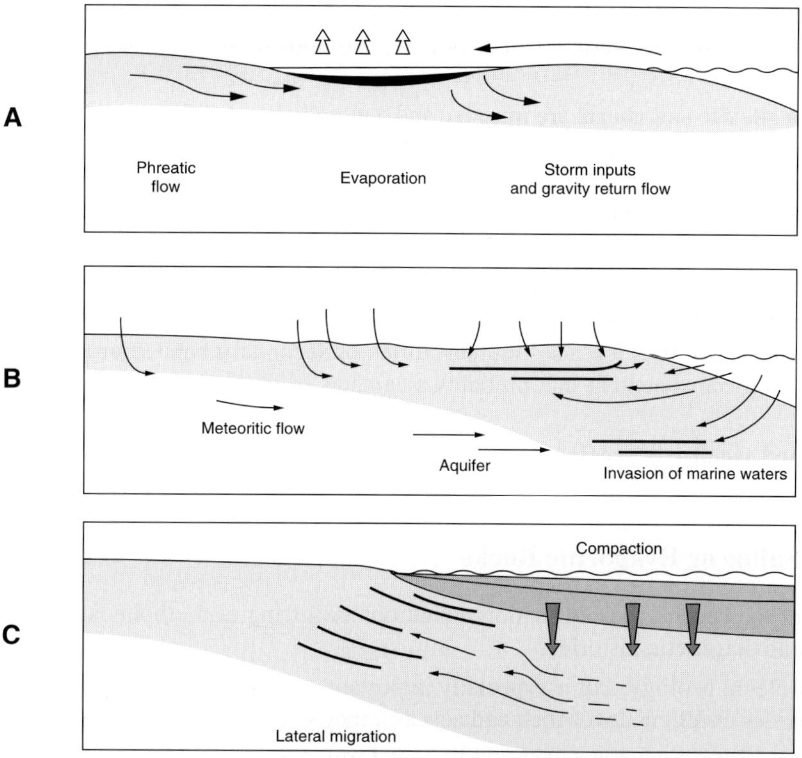

Fig. 5.30 Dolomitization: different forms of metasomatism.
A. Lagoons and sabkhas: evaporation leading to alternating calcareous and dolomitic deposits at the surface. **B.** Littoral fringe: mixtures at shallow burial depths of a few meters to several tens of meters. **C.** Compaction and deep burial: residual fluids are expelled and transfer zones are established.

It will be noted that different types of magnesian (low content) limestones also exist, and that other forms like **siderite** (Ca or Fe $(CO_3)_2$) and **ankerite** (Ca, Mg or Fe $(CO_3)_2$) also result from diagenetic effects.

5.1.5.3 Non-Detritic Siliceous Rocks

These rocks are of rare abundance in geological series and of little importance in petroleum geology. They vary by the mineralogy of the silica (opal, chalcedony, cryptocrystalline quartz), by biological origin (organism tests) and by diagenetic effects.

Diatomite and **radiolarite** are rocks made mostly of diatom and radiolarean skeletons. **Lydite** and **phtanite** are varieties of radiolarite. **Spongilite** of siliceous sponge spicula are rare.

Flint or **silexite** and **cherts** are **massive** and generally banded fine-grain rocks, usually of obscure biological or chemical origin. **Flintstones**, nodular cherts of silica with an outer cortex separated from the surrounding carbonate structure, are due to the concentration of silica in diagenesis, and are often felt to be of biological origin, as are the more diffuse clusters of **chert**.

Millstone or burrstones are typically diagenetic rocks with a complex silicified lattice in certain lacustrine limestones.

Siliceous limestones are also frequent forms of secondary banded or diffuse silicification, with a siliceous content that probably originates from an in situ biogenic pelagic fraction.

Petrified wood is a special case of fossilization by epigenesis, i.e., changes in mineral content.

5.1.5.4 Saline or Evaporitic Rocks

Remember the special rock formation conditions occurring at the boundary between sedimentary and diagenetic histories.

In petroleum geology, salt is especially important. As will be seen in the following chapter, it provides excellent cover rock and acts as a motor for structural and stratigraphic traps.

Gypsum is very unstable and is usually transformed into anhydrite at depth. Anhydrite is rare under surface conditions, and many neoformations of gypsum crystals are known.

5.1.5.5 Organic Rocks

Organic matter produces a wide variety of products as it decomposes. These carbanaceous products are generally grouped under the term of organic rocks. The study of their origin, composition, and degree of thermal evolution come within the field of organic geochemistry. The main types of organic rock are the following.

Coals (Fig. 5.31). The various types grouped under this term are sedimentary rocks formed from vegetal lignocellulose debris transformed by **carbonization** (which is a loss of oxygen and then of hydrogen, by thermal effect) into a combustible product enriched in carbon and depleted of volatile products.

The main categories of coal are:

- **Peat**, which is the sediment before it undergoes much transformation (55% carbon).
- **Lignite**, where the ligneous vegetal debris is recognizable (70% to 75% carbon). One variety is called jet.
- **Coal** (85% carbon) with a variable content of volatile material making bituminous coal, flame coal or pitch coal.

Fig. 5.31 Coal bed. Deposit in a delta environment at Mahakam (IFP photo).

- **Anthracite** is even more concentrated with 90% to 95% carbon, before becoming **graphite** of pure carbon, found in the metamorphic domain.

Different components can be found: fusain, durain, clarain, vitrain. **Vitrinite** is a gel of organic matter used (by measure of the reflectance) to determine the degree of evolution and burial (see following chapter).

Aside from the coals that are derived from ligneous plant debris, there are also the algal coals (debris of lacustrine algae) or **boghead**, and cannel coal, which differ by their composition. The various types and stages of diagenetic evolution are differentiated by the type of microscopic **maceral** organic elements found.

Hydrocarbonated products will be detailed in the following chapter: **kerogens** of complex organic molecules of large molecular size and varied origin, bitumens (asphalts, resins), crude **oils** and combustible **gases**, i.e., the hydrocarbon chain.

5.1.5.6 Other Types of Rocks

Ferriferous rocks. Most rocks contain some iron in various forms. The ferriferous minerals may be carbonates (called siderite), oxides or hydroxides (hematite, limonite), sulfides (pyrite, marcanite), or sulfates (chamosite), sometimes in association and sometimes concentrated.

The main concentrations containing more than 10% of metallic iron corresponding to **sedimentary iron ore** are mainly a result of diagenesis, in lateritic and ferralitic crusts with the reduction or development of sulfides that concentrate in the genesis and early diagenesis of oolites. Ferriferous oolites sometimes come from phosphatic deposits. In the sedimentation process, iron is transported in ionic or colloidal form (hydroxides).

Polymetallic crusts and **nodules** are very special forms of concentration in early diagenesis on the ocean bottom (Fig. 5.32).

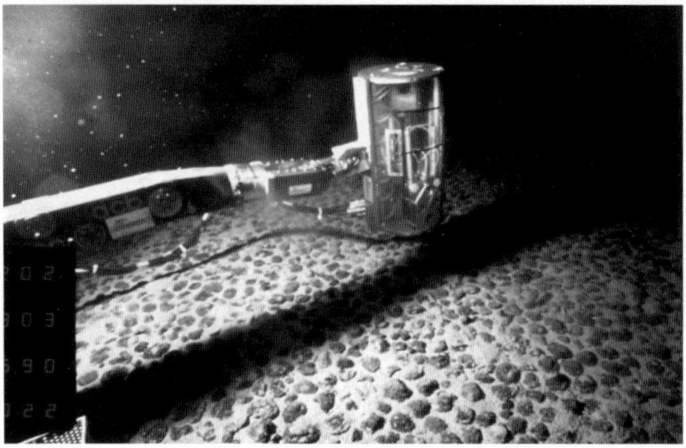

Fig. 5.32 Polymetallic nodules on the broad ocean undulations. This Ifremer photo illustrates the density of these nodules in certain deep ocean zones (the Pacific, here) (see Fig. 3.152).

Phosphatic rocks. Phosphatic beds and fields are the result of a sedimentary phenomenon in which biological productivity plays an important role. **Phosphates** in grains, nodules, crusts, or epigenetic debris (often associated with glaucony) may be primary or may be due to synsedimentary diagenesis in an oxygenated marine environment (they become soluble in a reductive environment). They may be secondary products in the continental domain: **phosphorites** can be found as leaching products in crusts in a karstic environment.

The phosphatic fields of North Africa (Morocco, Tunisia) are special situations in which enormous concentrations of phosphates.

5.1.6 Deformations Stemming from Diagenesis

Chapter 3 mentioned the different types of deformations affecting a bank or set of banks in their initial synsedimentary stages, due to deposition factors. It should be remembered from this discussion that the basal erosion figures produced on the top of a sedimentary bed by streams and the objects they carry are the first mechanical effects after deposition. These figures are of much greater interest in sedimentology (Fig. 5.33) than in diagenesis.

Slumps are a very common figure that are not necessarily indicative of their environment, as they may be the result of gravitational sedimentation or unstable tectonics. The

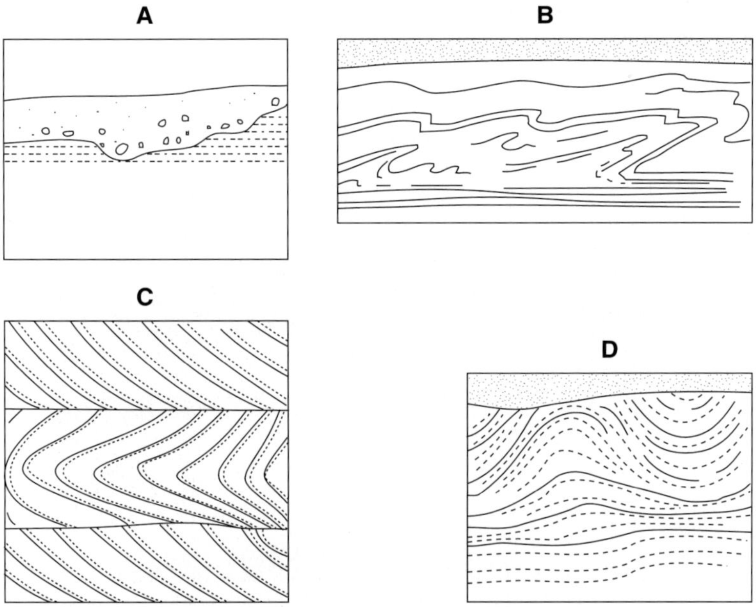

Fig. 5.33 Different types of early deformations of sediments.
A. Erosive action of a tractive current. **B.** Slumping between two undisturbed beds. **C.** Overturned oblique stratification. **D.** Fluid expulsion figure.

same slump often affects a number of banks through several meters of thickness, while the banks generally retain a good level of cohesion. This in turn implies a certain amount of lithification during early diagensis. In any case, slumping is clearly a part of post-depositional history (Fig. 5.33B).

Pressure expulsion figures. Special structures can be found on different scales indicating rebalances, with fluid migration:

- **Overturned oblique stratifications** are typical of fluvial systems in which deformation by lateral expulsion of water occurs immediately after the oblique avalanche sets form (Fig. 5.33C).
- **Mud and sand extrusions**, **convolutes** and **buckling** flames on the millimetric to metric scale are very common in rapidly sedimented gritty argillaceous series. The deformed texture of the sediment indicates liquefaction and/or **water expulsion** entraining the beds (Fig. 5.33D). The expulsion is sometimes seen in a discrete dish structure.
- **Mud volcanoes** on the decametric to kilometric scale generally involve argillaceous series that are undercompacted and therefore deposited at a high sedimentation rate. In a mud volcano, a mixture of sediments and fluids (water, hydrocarbons) is ejected sud-

denly and abundantly. There are beautiful examples of this in tectonic zones such as the Barbados accretion wedge (Fig. 5.34A).

- **Pingos** are also mound-shaped structures created by the growth of an ice lens in permafrost (Fig. 5.34B). Very young series can also be marked by other types of deformations in glacial environments like these, such as polygonal soils and ice wedges.

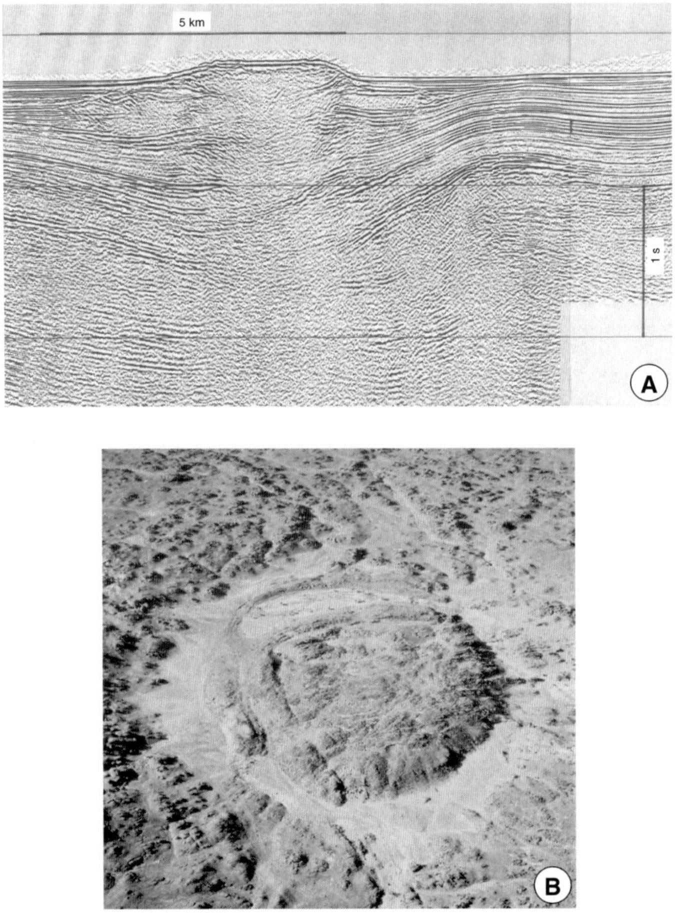

Fig. 5.34 Different forms of large scale early deformations.
A. Submarine mud volcano. The seismic profile expresses a topographic anomaly of uplifted deep clays disturbing the otherwise regular bedding. The example is the Barbados wedge. **B.** Pingo. This oblique aerial photo of Ordovician outcroppings in the Sahara shows the circular deformation of periglacial beds, which has been interpreted as the result of growth and melting of an ice lens (IFP photos).

- **Saliferous diapirs** and smaller domes are remarkable deformations amplified by tectonics. It should once again be said, though, that they are initiated in an ordinary burial history by the weight of the sediments alone.
- Sedimentary **clastic dikes** are a very special form of sedimentary material injected into a stratified series. These may result from the filling of open cracks or from an actually pressurized injection into consolidated, partly compacted sediments (Fig. 5.35A). The sediments can be injected tens of meters into vertical oblique dikes, and stratiform dikes can extend for up to a hundred meters.
- **Load casts** are downward projections of balls and pillows of decimetric scale. They usually occur early on between two banks (Fig. 3.35B). These figures may gradually

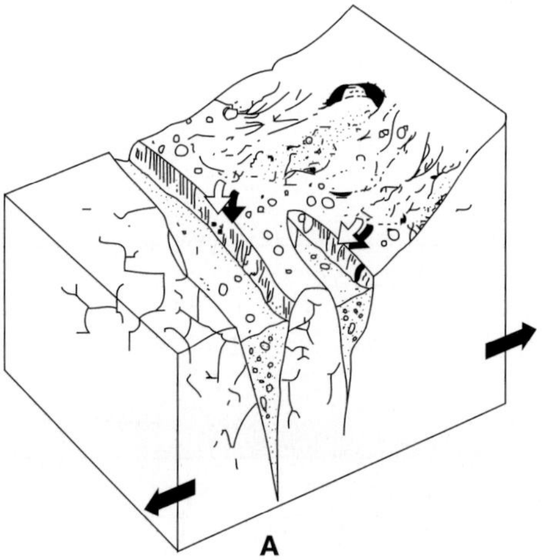

A

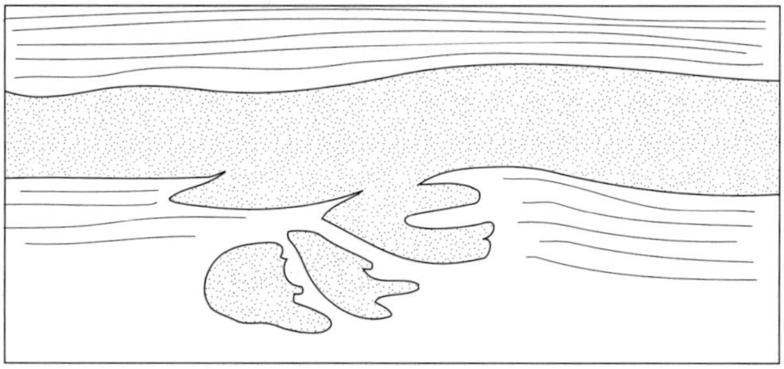

B

Fig. 5.35 Clastic dikes and load casts.
A. Clastic dike (drawing from P. Barrier, 1987). Cracks are opened to a breadth of several meters and then filled to make dikes. **B.** Load cast under an arenaceous bank.

change with the increasing load, and the ensuing **differential compaction** can be observed on the decametric to kilometric scale.

- **Mechanical and geochemical deformations**, such as mud cracks, calcareous crust, or teepees with associated breccia, from the centimetric to decametric scale (Fig. 5.36):

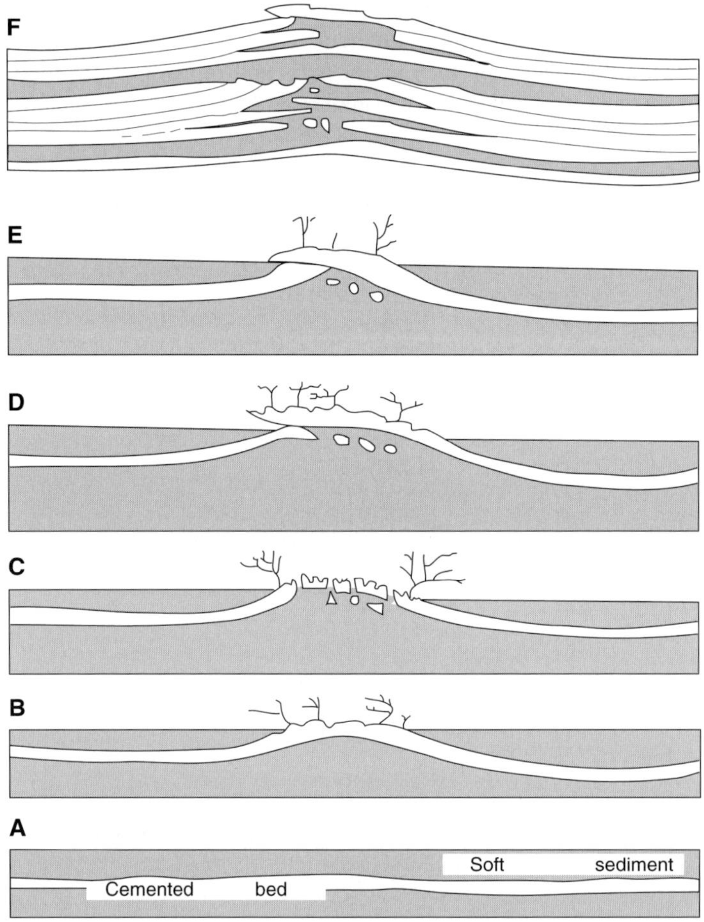

Fig. 5.36 Development of pseudoanticlines by diagenetic expansion of a submarine crust.
A. Cemented crust embedded in soft sediments. **B.** Expansion of the crust, probably due to cement crystallization forces, and formation of a pseudoanticline. **C.** Brecciation and collapse of the structure. **D.** Renewed expansion leading to micro-overthrust. **E.** Biological erosion of the "sheet". **F.** The final diagenetic structure (from E.A. Shinn, 1969, in Purser, 1980).

- **Mud cracks** appearing early in argillaceous sediments saturated in water with a retreat to open air.
- **Infratidal pseudoanticlines**. The recent strata in the core are due to a deformation between two beds by lateral expansion of a hardened crust during crystallization, with the development of breccia, polygonal lattices of tension cracks, and folds.
- **Righted slabs** or **teepees** and rims are found in beach rocks of the intertidal domain with breccia, furrowed surfaces, and mud cracks.

5.1.7 Importance of Diagenesis for Petroleum Geology

The following chapter will detail the essential question of the diagenetic evolution of sedimentary organic material that transforms into **kerogen** and generates **hydrocarbons** from the parent rock, with emphasis on the role of burial in subsidence for the maturation of oil and gas. In anticipation of this, the following is a brief summary of the role and place of diagenesis in forming and trapping hydrocarbons.

It is clear that the **petrophysical characteristics** of candidate beds for hydrocarbons (sandstones, limestones) depend on burial, compaction, and secondary cementing conditions, as these will deteriorate or improve the initial porosity and permeability.

Also, beyond the initial heterogeneity due to the sedimentary processes, the unequal development of diagenesis on the scale of a basin or a part of it will generate **vertical and lateral heterogeneity in the reservoirs**. Data collected at the surface will not always be applicable to the deeper subsurface beds.

5.2 STRUCTURAL EVOLUTION FROM BASINS TO MOUNTAIN CHAINS

Even though diagenesis is extremely important in transforming sediments into sedimentary rocks, it is less spectacular than **tectonic deformation**, or **tectogenesis** of the rocks. The simplest manifestations of this – folds and faults (Fig. 5.37) – are outstanding even to the untrained eye. The very fact that rocks can be analyzed at the surface means that there has been a deformation because, without deformation, the rock would normally be buried. Today, deformation mechanisms are observed directly on the edges of basins and in adjacent mountain chains. **Tectonics**, or **structural geology**, is a qualitative and quantitative analysis of geological objects (structures) deformed during tectogenesis. Sometimes the deformation mechanisms are referred to as **tectonophysics**, which is based on **rock mechanics** or physics. The combination of seismology (study of quakes) and tectonics (the study of tectonic structures) is **seismotectonic** analysis. Fine observation of joints, beds, and failure surfaces are termed **microtectonics**.

Fig. 5.37 A classic example of tectonic deformation: the Sassenage stretch thrust. This landscape of the Isère water gap near Grenoble shows beds that were initially horizontal and were then lifted up and folded, developing faults that were then exposed by recent erosion.

More information can be gathered on this subject from the books by Mattauer (1973), Debelmas and Mascle (1991), and Angelier et al. (1992).

5.2.1 Deformation Mechanisms

5.2.1.1 Stresses and Strains (from Rock Mechanics)

It was said at the beginning of this chapter that when sediments and strata are buried in basins, they are first subject to the law of gravity. There are volume forces to explain the isostatic balances, for example, and surface forces to express the idea of lithostatic pressure. **Stress** has already been mentioned, for example, in the formation of stratiform stylolites along stratification planes (Fig. 5.15). All of the forces acting on the surface of a rock produce internal forces, i.e., a **stress state**. What is developed hereafter is not the theory of stresses and laboratory experiments in rock mechanics, but rather the relations and resulting deformations in **rheology**, which can be defined as the response of the materials to the effects of evolving stresses.

First, a few basic terms. Tectonic forces will create a **stress state** which, at a given point, is expressed in a matrix by the **stress tensor**, or more simply by an ellipsoid called the **stress ellipsoid** whose axes correspond to the main stresses. When there are large differences between the main stress σ_1, the middle stress σ_2, and minor stress σ_3, then the triaxial stress ellipsoid (Fig. 5.38A) is said to be polyaxial. When $\sigma_2 = \sigma_3$, the ellipsoid is one of revolution and, when the stress state is homogeneous ($\sigma_1 = \sigma_2 = \sigma_3$), the stress is said to be hydrostatic. If the ellipsoids differ in shape and orientation at different points in a given volume, the stress field is heterogeneous. Otherwise, it is homogeneous and uniform. The **mean stress** $\bar{\sigma}$ designates the hydrostatic part of the tensor. The difference $\sigma - \bar{\sigma}$ is called the **stress deviator**. It is understandable that, if the stress state is hydrostatic, the volume of the rock will change but not its shape, whereas if the stress deviator increases, there would be a further deformation with change of shape. Mechanics speaks in terms of **compression** and **tension**. Geology generally refers to compression (horizontal σ_1) as opposed to **extension** (horizontal σ_3).

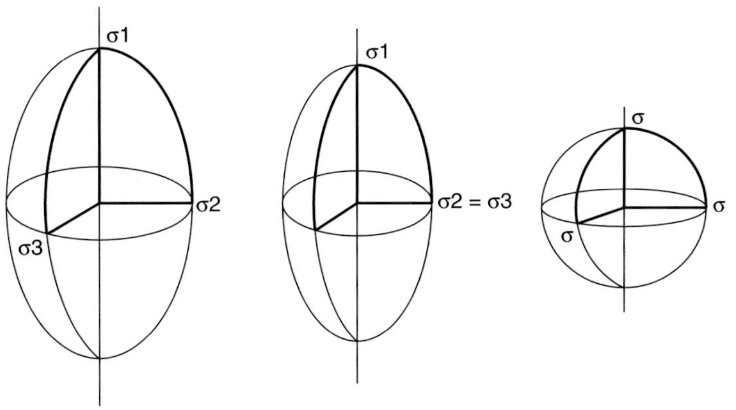

Fig. 5.38 Triaxial stress ellipsoids.
A. Polyaxial, $\sigma_1 > \sigma_2 > \sigma_3$. **B.** Revolution $\sigma_1 > \sigma_2 = \sigma_3$. **C.** Hydrostatic $\sigma_1 = \sigma_2 = \sigma_3$.

A rock set will be deformed by the **stress state** (or the size of the stress deviator). What is observed is the **resultant strain**, and the stress state causing it can then be reconstructed. Geological deformation is defined as the variation in the localization, shape, or dimension of the initial object, so there may be **translation**, **rotation** or **distortion** of the objects (Fig. 5.39). It should also be remembered that the strain can be **homogeneous** or **heterogeneous**, continuous or discontinuous (Fig. 5.40).

When the object is distorted, there is an **internal strain**. At times, tectonic geologists also use the terms pure or simple shear to describe different strain modes. The **deformation ellipsoid** is also used to represent homogeneous strain with the surface conserved, reduced, or increased. This internal strain is not always observable in the field; but if a bank contains an object whose initial shape was known, it then becomes detectable.

Experiments on rocks have also shown that different strain domains can be defined depending on rock lithology, as shown below.

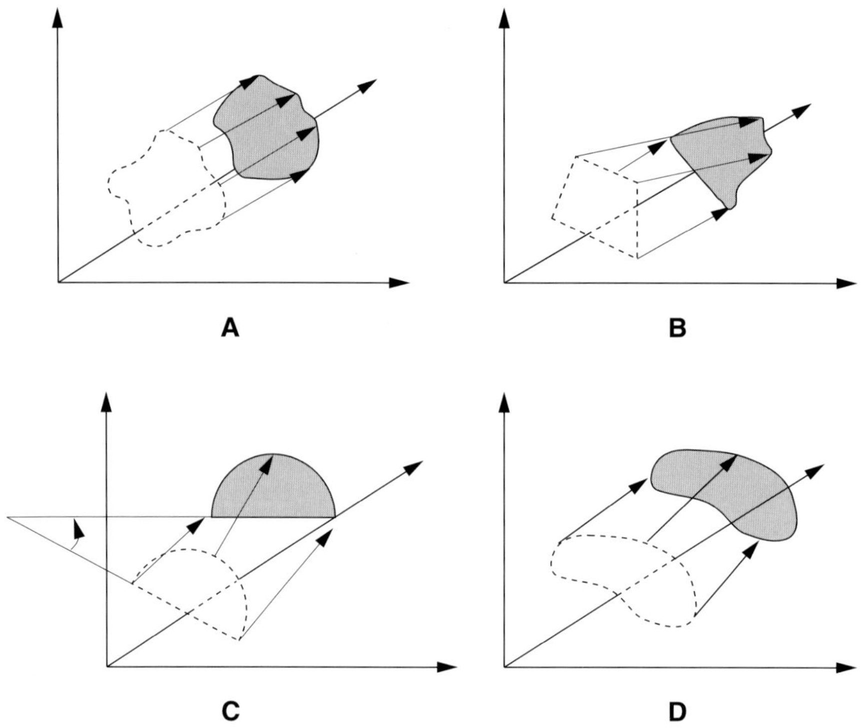

Fig. 5.39 Object deformations: resultant of several processes.
A. Simple rigid translation in space with a homogeneous displacement field. Only position changes, not the shape. **B.** Distortion in a heterogeneous displacement field. The shape changes (internal deformation). **C.** Rigid rotation with no shape change. **D.** Translation with a change of volume. (From Angelier et al., 1992.)

- The **elastic domain** is one in which the deformations are small and reversible (Fig. 5.41).
- The **plastic domain** lies beyond the elastic limit or yield point. Here, the strain will increase more quickly for a small increase in stress. If the stress is released, the strain is only partly returned, so a permanent deformation will subsist. If the stress is increased further, the material enters the **creep** domain of plastic flow.
- At a point well into the plastic domain, the rock may begin to exhibit the liquid property of **viscosity**. The plastic-to-viscous transition depends, of course, on the lithology and fluid content of the rock, the environing temperature and pressure conditions, the applied stress and the strain rate. In viscosity, the rock will deform permanently, and the strain rate becomes a direct function of the stress.

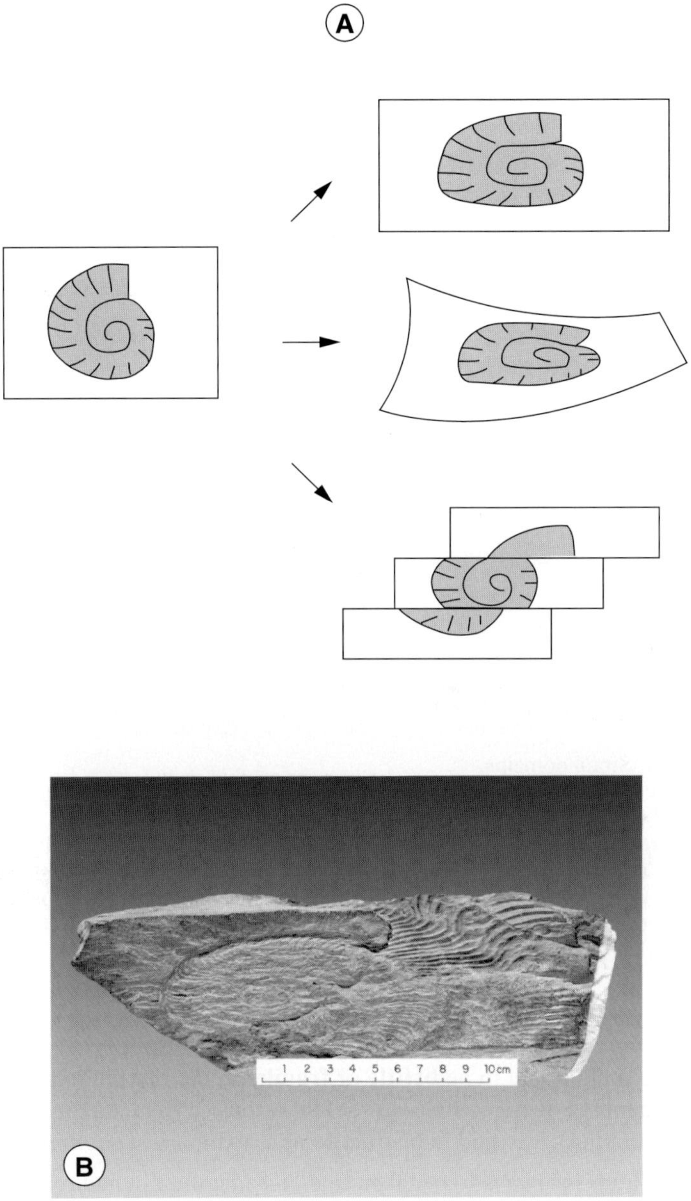

Fig. 5.40 Continuous homogeneous and heterogeneous strain and discontinuous heterogeneous strain (modified from Angelier et al., 1992). **A.** Theoretical diagram. **B.** Deformed sample.

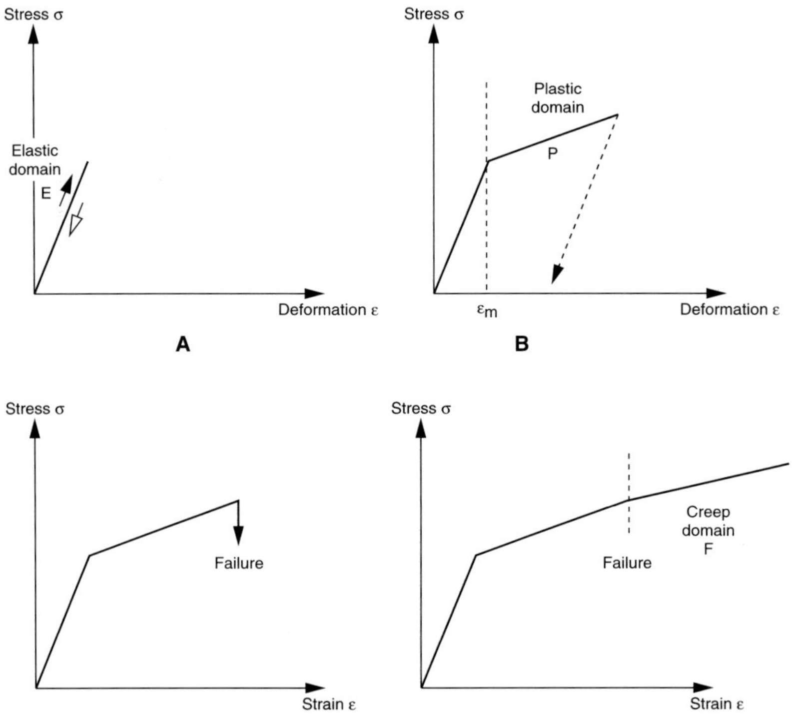

Fig. 5.41 Strain domains.
In elastic domain (E), objects will return to their initial shape if the stress is released. The strain is generally low. In the plastic domain (P) beyond the yield strength e_M, the strain quickly increases (low-sloped curve). If the stress is released, the sample will remain deformed (following the dashed line, not returning to the initial state). Under certain conditions, failure may occur. Otherwise, the material will continue into the creep or plastic flow domain (F).

- Under certain conditions, the strain will gradually lead to **failure**. If the strain is discontinuous, the rock is said to be **brittle** or **fragile**. In the plastic domain, failure may occur simultaneously with continuous deformations. When the rock continues to deform without failure, it is said to be **ductile** or to exhibit viscous creep.

The presence of fluids is always to be taken into consideration in nature. Triaxial laboratory tests of rock properties make a distinction between the **confining pressure** or stress and the **pore pressure** due to the fluid under pressure (the fluid also has other effects, such as capillary pressure and the pressure dissolution phenomena mentioned above).

Each rock is characterized by its Young's modulus or ***elastic modulus*** *ε and its* ***Poisson coefficient*** *υ defining its properties. In rock mechanics, the elasticity is linear (for com-*

pact limestones, low-porosity sandstones) or nonlinear (in grainstone or ill-consolidated sandstone, sands).

Brittle **failure** is evidence of a discontinuity in the strain beyond a certain **failure strength**. Laboratory tests have shown that rocks are fragile under tension (the failure plane is perpendicular to the main stress) but are less so under compression. Any cracks thus created play an important role in the natural environment, as will be seen later on with the idea of heritage.

The idea of brittle failure prompted by ***damage*** *due to previous elastic reaction is of use in certain applications, but not so much in structural geology.*

For natural environments, remember that the breaking strength depends on the confinement pressure, pore pressure, temperature, and the mutual ratios of the stresses σ_1, σ_2, and σ_3. If σ_2 and σ_3 are small with a large vertical σ_1, the strength is low, and microfractures, fractures, and faults will develop in **extension**. (Note that this is not tension in the sense of laboratory experiments.) The break becomes a dislocation along which there is an infinitesimal or considerable **displacement**.

Generally, the **strain rate** and time factor are difficult to define precisely and are hardly reproducible in experiments. In geology, it is known that the response to a stress state may be instantaneous, such as in a **seism** or earthquake, or may be very slow, such as in gravitational creep.

5.2.1.2 Geodynamic Aspects

Chapters 1 and 2 on geodynamics and the formation of sedimentary basins explained that the time-variable **stress regime** is controlled by **plate tectonics**. The following refers to these chapters and reviews what is essential to this approach to rock deformations.

Tectonic plates were originally defined as rigid, undeformable systems. The stresses due to the lithospheric movements were assumed to occur only along narrow boundaries. It has since been realized that the heritage and fragility of the basements (with stresses transmitted far from the boundaries) and the time evolution, unstable borders and diffuse seismicity, are evidence of **intraplate tectonics**.

Highly contrasted situations occur depending on the **plate boundaries**:

- In continental rifts, on the passive continental margins, along functional oceanic ridges, i.e., in those places where divergence is active, the general stress regime is mainly **extension** (Fig. 5.42). The prime consequence of this is crustal thinning, a set of normal faults, the creation of a sedimentary basin, and subsidence.
- In subduction or along collision zones where plate convergence is effective, the dominant stress regime is **compression** (Fig. 5.42), resulting in major crustal thickening, uplift and erosion, and arrays of reverse faults, **folding** and formation of mountain chains. Actually, the effects vary widely depending on the type of subduction.
- In strike slip zones between plates where the displacement rates vary, the stresses vary too. While these boundaries are characterized by **shear**, the oblique border geometry

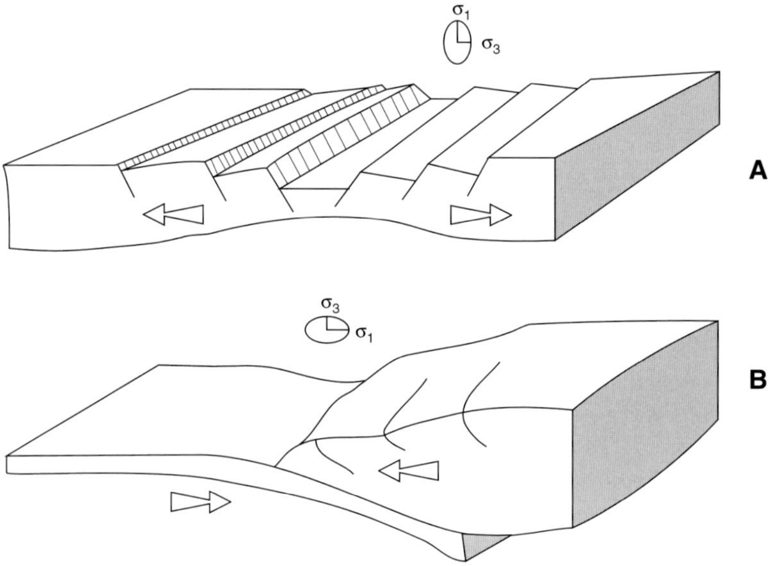

Fig. 5.42 Extension and compression.
A. In continental rifts and oceanic divergence zones (accretion on the ridges), σ_1 is vertical, faults are normal, and the whole zone is stretched and thinned. **B.** In subduction or collision zones where the plates converge, σ_1 is generally horizontal (with σ_3 or σ_2 vertical) and reverse faults and strike slip are observed, though extension faults (σ_1 vertical) are not unheard of.

will be such that the strike-slip motion will combine either with extension to make **transtension** or with compression, which is called **transpression**. Strike slip is not limited to transformation zones. Any such horizontal strike slip may occur whenever the direction of displacement is not orthogonal to the boundary. (A **pull-apart basin** is the result of such strike-slip effects).

Remember that the instantaneous stress regime of an earthquake is determined in ***seismology*** *by the* ***focal-mechanism solution*** *(Fig. 5.43). Also remember that stress regimes will vary on the same boundary on the geodynamic scale, depending on the thickness of the lithosphere. The expression of a subduction plane at the surface, for example, will differ considerably from its expression at depth. Also, very different brittle, ductile, plastic and viscous domains can be found at increasing depths, i.e., with increasing temperature and hydrostatic pressure (Fig. 5.44).*
Of course, the type of rock also has to be considered. Rock salt has plastic creep characteristics at the surface, while other rocks remain in the fragile domain down to burial depths of several thousand meters. It was said in Chapter 2 (on basin formation) that the basic rifting mechanisms affected both the rheological structure of the crust in the

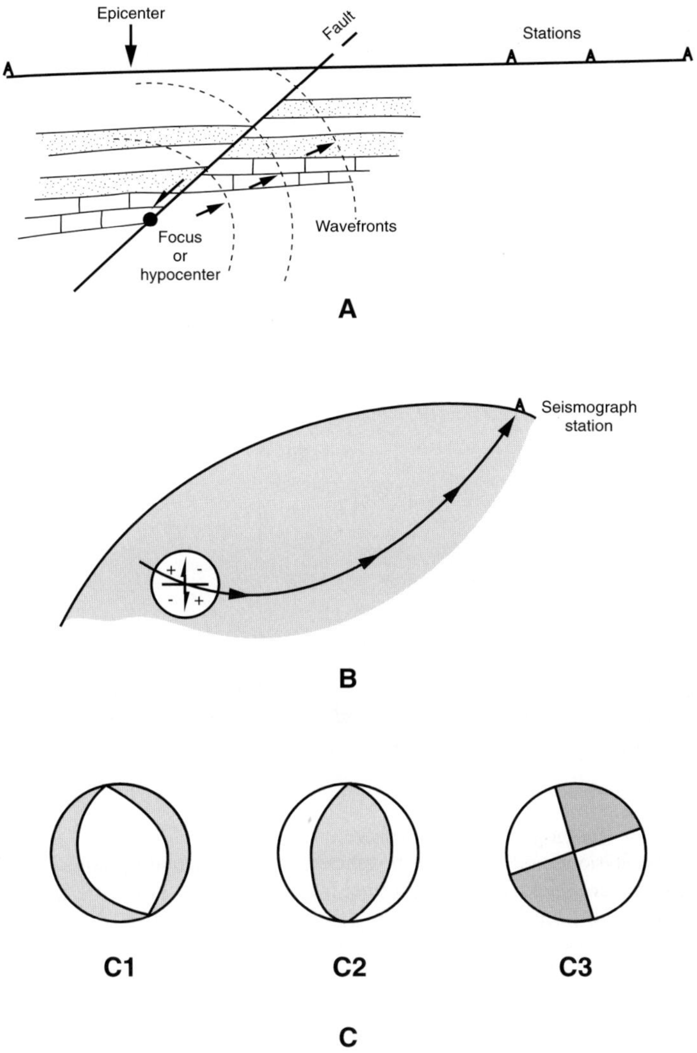

Fig. 5.43 Focal mechanism solution.
A. General diagram. Seismographs are arrayed around the focus of an earthquake to determine the type of stresses and motions associated with the seisms even if the fault does not reach the surface. **B.** Focal sphere and polarity of the P waves. **C.** Three types of focal mechanism solutions (polarity of the waves represented on the lower focal sphere). **C1**: Normal fault; **C2**: Reverse fault; **C3**: Strike-slip (horizontal) fault.

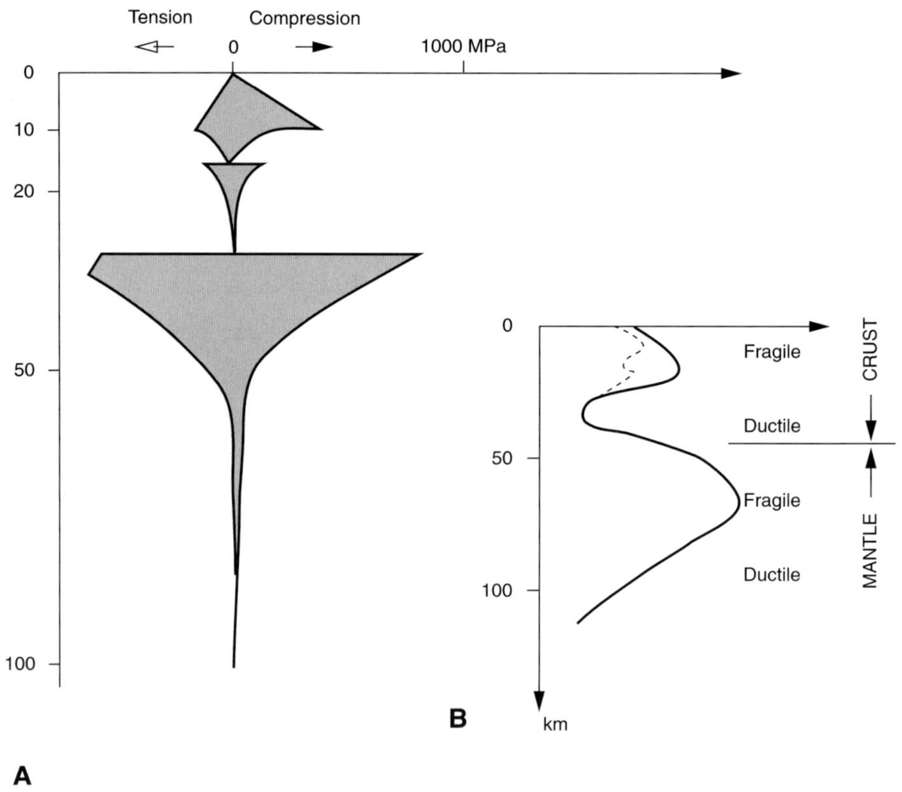

Fig. 5.44 Rheology of the lithosphere.
A. Rheological model of the lithosphere for a continental (cratonic) crust, 30 kilometers thick (Western Europe, from P. Ziegler, 1996). **B.** Another example of the mechanical properties of the lithosphere (from Caron et al., 1986). The dashed curve indicates the possibility of coming across an intermediate ductile layer.

fragile layer at the surface and a ductile, stretched layer decoupled from it, at depth. The lithosphere in the oceanic domain is generally considered to be double-decked, while it is thought to be much more heterogeneous and multi-layered in the continental domain, with two or even three fragile domains separated by zones with ductile properties. It will be seen later in this chapter that the two behaviors—discontinuous breaking (heterogeneous) and continuous ductile (homogeneous) characteristics—are often observed together, in light of the kind of rock and its fluid content. Actually, the space scale has to be considered along with the time scale. The alternation of continuous deformation and strike slip in time has been proposed for the Indian-Eurasian collision, for example.

Structural levels are defined on a rheological basis on the lithospheric scale. The idea of rigid or brittle rock is often opposed to that of brittle-ductile cover rock in structural geology, but this distinction should be taken with care because deep-buried base rock can also have a ductile behavior, as in the metamorphic domain (Fig. 5.45).

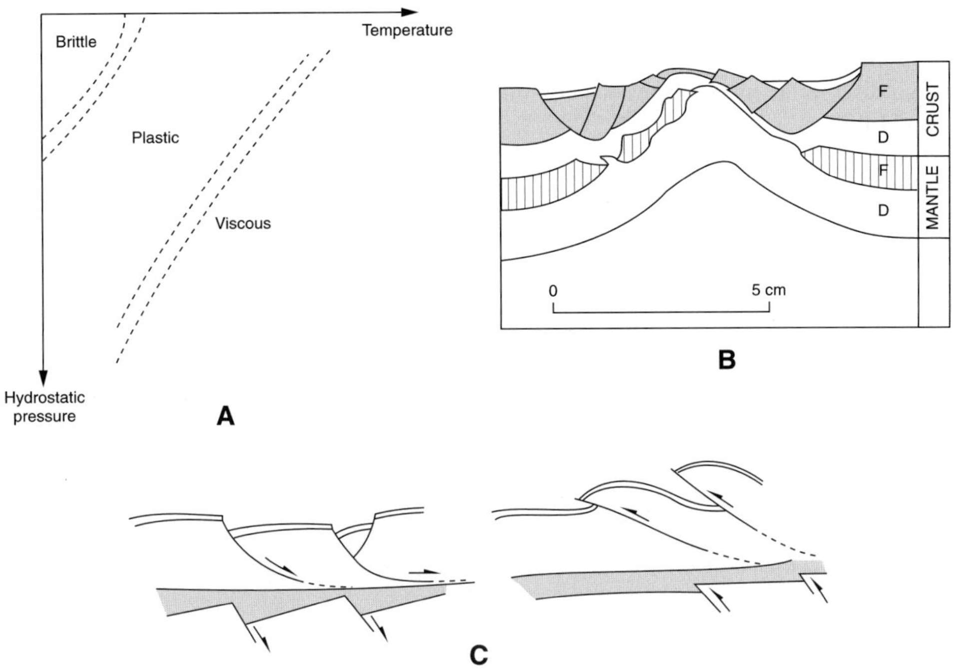

Fig. 5.45 Deformation domains at different structural levels.
Rocks do not all have the same mechanical properties. Some are brittle and others ductile, and they will react differently depending on their fluid content. An increase in water content will promote ductile behavior. **A.** Variation of properties with increasing depth (or hydrostatic pressure) and temperature. Rock near the surface is fragile with breakup deformations, becomes ductile with plastic deformations at greater depths, and enters into partial fusion at great depth, where it exhibits viscous deformations. **B.** The rheology of the beds will change their large scale behavior. This is a simulation experiment with a downscale model of a four-bed lithosphere. Two of the beds are fragile (F), with high strength (sands) while the other two are low-strength ductile beds (silicone) (from Brun, 1990). **C.** Examples of uncoupling between surface faults affecting the cover and basement faults, with an especially ductile intermediate décollement level.

The purely elastic domain is at times considered to be limited. It produces small deformations (around 1%) that are difficult to bring out in geological observations. On the lithospheric scale (100 km thickness), however, an elastic bulge may cause considerable repercussions. Remember that the raised shoulders at the boundaries of extension basins, and isostatic rebound, have been explained as viscoelastic effects.

These discrete actions in the elastic domain could be termed **epeirogeny**, which originally meant slow, large movements without folding, as opposed to orogeny, which means the formation of the large folded structures, or mountain chains (Fig. 5.46) where continuous and discontinuous deformations combine to produce the **orogenic motions** of **shortening** and **folding**.

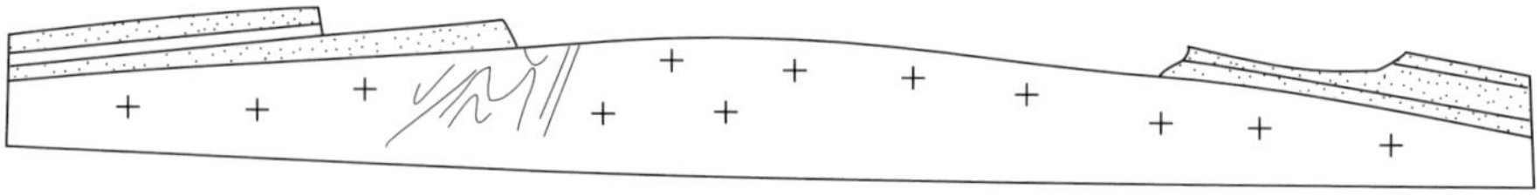

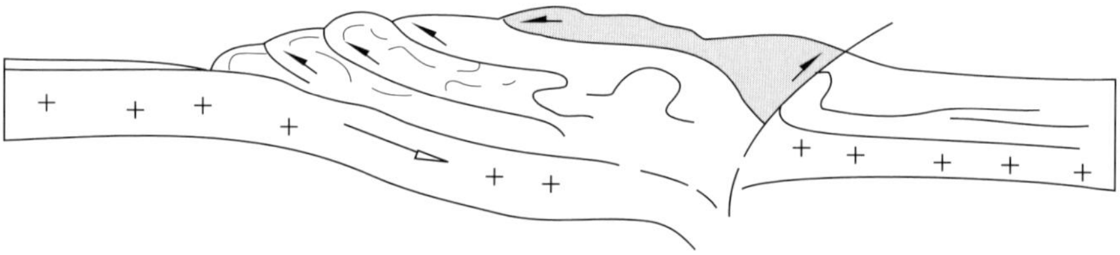

Fig. 5.46 Epeirogeny and orogeny.
A. Epeirogeny: continental deformation with faults and with folds of large radius of curvature. Here, the example is of the Saharan slab with the Hoggar shield. **B.** Orogeny: large folded, faulted mountain chain structures. Diagram of the Alps.

The time dimension is extremely important in geodynamics. On the scale of the sedimentary basins of interest here, the **Wilson cycle** is the logical sequence of events starting with the extension, which created the basin, and terminating with orogeny by subduction (and/or collision). That is, the same geographic area has evolved in a time-variable stress regime. This again shows the **multiphase** idea that will enter the discussion again later.

It should be stressed that, while the general regime in a basin or basin edge may be extension or compression through a given period of history, this should in no way be taken to mean that the region will show only one type of deformation expressing this regime. Firstly, transtension or transpression may be visible in the edge geometry or that of inherited faults. Secondly, extension is very often found in a convergence domain (such as forearc or backarc zones) in line with subduction, while compression figures can be found on the edges of a rift or continental margin (in the slumps and overlapping folds of delta wedges).

In the context of this book on applications of geology to petroleum, the very important aspects of tectonics, in which magmatic and volcanic material intrudes at depth or is placed at the surface, are left aside, as are all the accompanying aspects of metamorphism.

5.2.2 Deformation Types (Geometric Expression on the Local and Regional Scales)

The main object of tectonics is to describe the bed deformation geometry and kinematics, for the purpose reconstructing the stresses that produced the observed tectonic structures.

Although theory will have it that beds begin with an internal **homogeneous deformation** by compaction or pressure dissolution, for example, this is rarely the case in geology. Generally, the deformation is more or less continuous and **heterogeneous**.

This section of the book will begin with an examination of discontinuous deformations in the brittle domain, with the accompanying types of breakage: cracks, diaclases or joints, faults. The discussion will then go on to continuous, flexible deformations in the ductile domain which, for petroleum geologists, means essentially the folds.

Whether or not a deformation is continuous is a matter of geometry, whereas the difference between brittleness and ductility is in the mechanical properties of the rock. This is why the observation scale—in the field or under the microscope—should always be kept in mind, because a deformation represented as continuous may be no more than the sum of many discontinuous deformations when viewed at higher resolution. It will be shown that the folding process is in fact often associated with major faults.

5.2.2.1 Brittle Deformations

Brittle deformations are observable on all scales at outcrops and in the subsurface, from core samples or reflection shooting, where discontinuity surfaces can be seen passing through the banks more or less obliquely to the stratification joints. There are two broad categories of discontinuous deformations: faults, which characterize a displacement along the discontinuity, and joints or diaclases where the break generates no movement along the failure plane.

A. Diaclases and Joints

A diaclase (from the Greek *dia* for "across" and *klasis* for "break") is a break with no apparent displacement in the beds (Fig. 5.47). The joint, sometimes referred to as "tectonic" to distinguish it from stratification joints, is a term that is sometimes used in a very general sense but is misleading if the type of joint is not stated. For some authors, a "diaclase" is a split through the bed at a high angle, whereas a "joint" is a fracture parallel to the beds.

Fig. 5.47 Diaclases and tension cracks.
A. Diagram of diaclases and tension cracks. **B.** Calcitic filling of tension cracks (shifted by inclined fractures). **C.** Example of diaclase network observed on a structural surface (IFP photos).

Closed fractures observed in nature are at times termed "diaclase" in the strict sense, while "open" fractures, which are usually filled afterwards, go by their genetic interpretation as tension cracks or fractures (Fig. 5.47). Here, there is a slight (millimetric to centimetric) displacement of the blocks away from the fracture line, i.e., perpendicular to the fracture lips.

These discontinuous breaks are sometimes termed fracture, crack or microcrack. The edges to either side of the break are called **lips** or **walls**. Even if the displacement is minimal, the type and texture of the seam filling the fracture are extremely important because they give some indication of the mode of deformation and the kinematics.

Tectonic **stylolites** are discontinuities of a different type. They are formed by **pressure dissolution** as was seen in the discussion of compaction during burial. Stylolites thus indicate compressive effects. Their peaks indicate the direction of maximum local stress σ_1 (Fig. 5.48), and the height and thickness of the peaks of insoluble material give an idea of the shortening, i.e., the displacement. (It should be noted that this can be combined with filling of tension cracks and compensation of the volume dissolved by the volume precipitated.) These figures are often observed in limestones and are good markers of deformation. The bearing surfaces are generally termed stylolite joints (or planes).

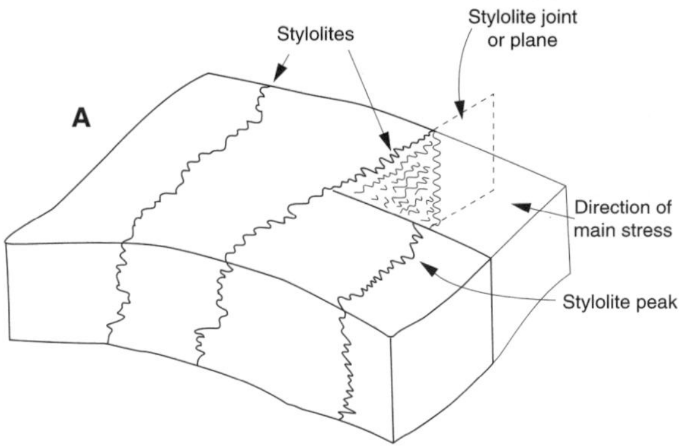

Fig. 5.48 Stylolites, stylolite joints.
A. Diagram. **B.** Outcrop with several stylolite planes.

Any joint that exhibits even the slightest displacement parallel to the crack is called a shear joint, or could even be called a microfault.

In fact, all of the transition terms exist with faults. It will be seen later that these shear joints and stylolites are of use in reconstructing the deformation, and that they are of great interest for petroleum as micromarkers. For example, an impermeable compact limestone will in fact make a very good reservoir, depending on how many diaclases are opened in it. Inversely, intense stylolite formation may mean a deterioration of the initial conditions.

B. Faults

A **fault** is a fracture with a macroscopic (ranging from a centimeter to several kilometers) displacement of the geological beds. The **blocks** to either side of the fault are clearly separated, and the lip displacement is called the **throw** (vertical) or heave (horizontal), the amount of which expresses the amount of discontinuous deformation. The **fault surface** or **plane** (Fig. 5.49) is a shear plane with an irregular, curved, or undulated surface, at times exhibiting **striations** or **flutes** witnessing to the slippage and recrystallization of the material along the shear plane and indicating the line of motion. The sense of this motion is found by analyzing the microstructures, such as the oblique stylolites observed on the fault plane, and the recrystallized calcite or silica, of which the arrangement and direction of growth on the irregularities are telltale indicators of plane motion (Fig. 5.50). If the striated surface is polished enough, it is sometimes called **fault polish**.

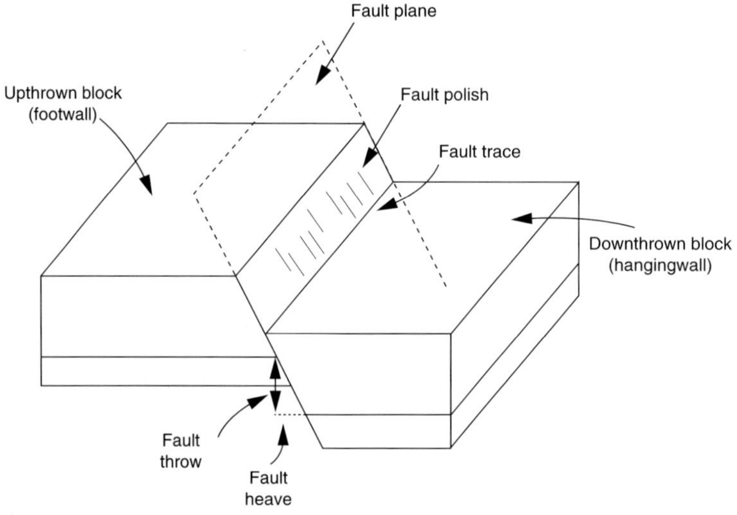

Fig. 5.49 Fault characteristics.

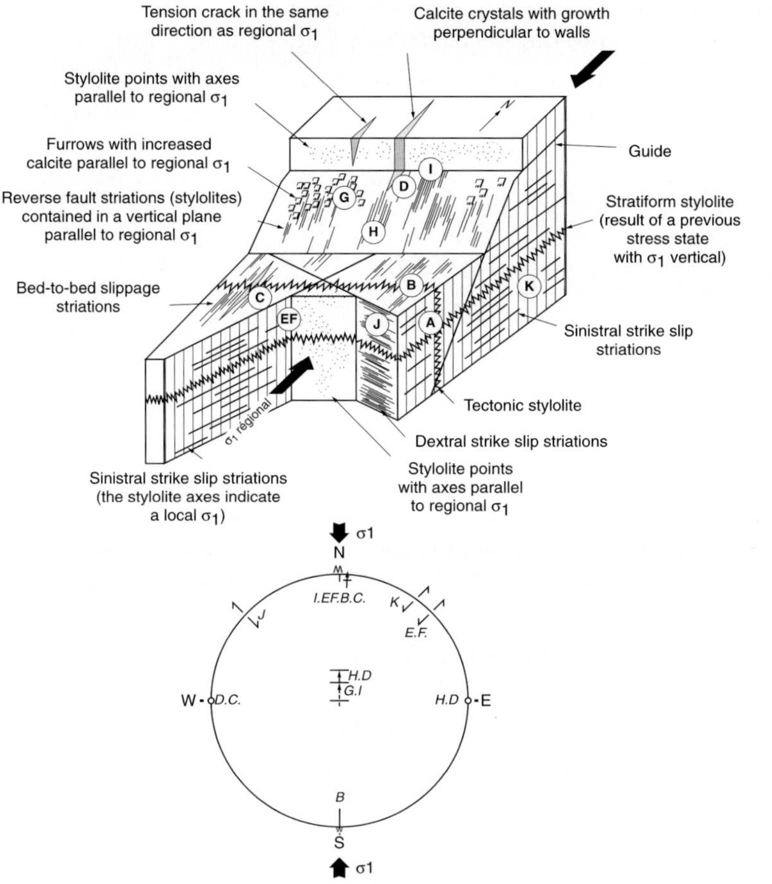

Fig. 5.50 Main objects measured in microtectonics and how they are plotted on a stereogram providing indicators of paleostresses and motion (from Trémolières, 1981).

Once the displacement has caused major internal fracturing, tectonic or **fault breccia** will develop (Fig. 5.51). In extreme cases, this tectonic brecciation process will produce **mylonization**, where the rock is completely ground up. This type of rock, affected by strong tectonics, called **dynamometamorphism**, is designated **mylonite** and the limestone or sandstone subjected to it is said to be mylonized or mylonitic if the shear in the rocks is not localized only along a discontinuity surface. Very often, these fault breccia themselves contain secondary fault polish.

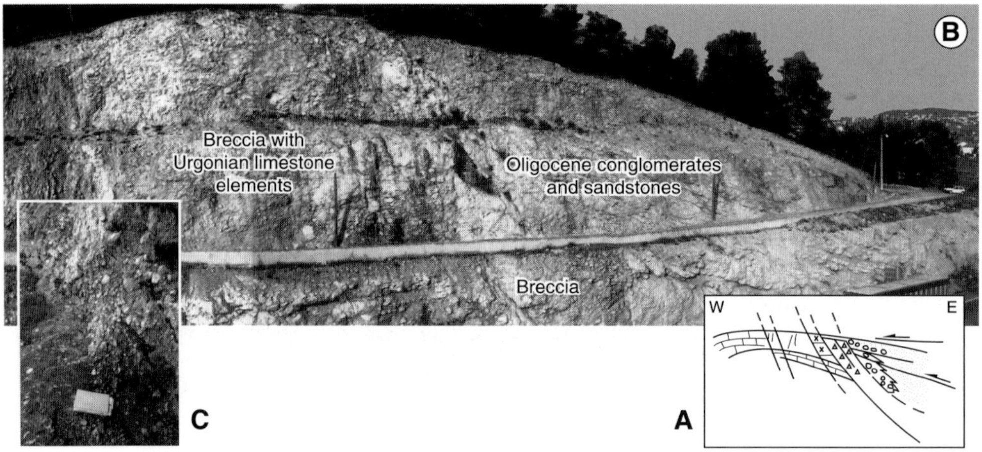

Fig. 5.51 Fault breccia.
A. General scheme. This fault on the edge of the Ales basin cuts through the limestones of the Lower Cretaceous. The Urgonian is coarsely brecciated through several meters of thickness, and the syntectonic deposits of the Oligocene onlap the fault breccia with sedimentary breccia that are nearly indistinguishable from the tectonic breccia. **B.** Field view. **C.** Close-up of Urgonian limestones on the left and breccia with rubified (reddened) cement on the right.

Faults vary widely in **dimensions**, both in terms of the length of the trace (from a hundred kilometers to the scale of centimetric samples) and their measurable heave and throw (Fig. 5.52). Major faults are generally part of a system of several faults on the lithospheric scale, and the largest are not necessarily the easiest to interpret: the faulting process is often identifiable only by criteria found along some secondary fault. This is because there is a general **reorientation of the stresses** at the approach and along the fault when major breaks initiate.

Faults will differ in **type** depending on the stress process that created them. There are three types, depending on the type of heave and throw (Fig. 5.53).

A **normal fault** is one with a steep fault plane so that the throw is mainly vertical and any slope is in the direction of the downthrown block, so the materials are separated. A normal fault is an expression of horizontal stretching or extension with the major stress σ_1 being vertical. The inclination is generally high (about 60°), which may vary with the lithology but the slope can decrease rapidly, making a listric fault.

Note that, when a drill passes through a normal fault, it automatically skips a part of the stratigraphic series (Fig. 5.54).

When the fault surface exhibits a pronounced concavity and gradually flattens out at depth toward some stratification plane where it starts, the fault is termed **listric**. This type of

Ⓐ

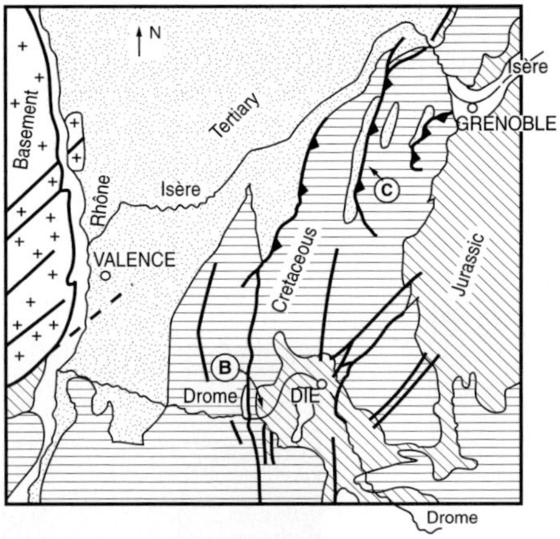

Fig. 5.52 Fault dimensions and offset. **A.** Diagram of the edge of the Alps (from the 1:1 000 000 scale map of France) indicating major faults in the basement and sedimentary cover. On this scale, the throws are large, as is the geographic extent. **B.** One of these faults, the Rencurel overthrust in the Vercors area, where the Urgonian limestones (light shade on the left) overlie silty and marly clays from the Miocene (dark on the right). **C.** Beds of the Upper Jurassic in Diois near the major Saillans fault. The small faults here cannot be represented on the map for reasons of scale: and the fault throw is only in the 1- to 10-meter range.

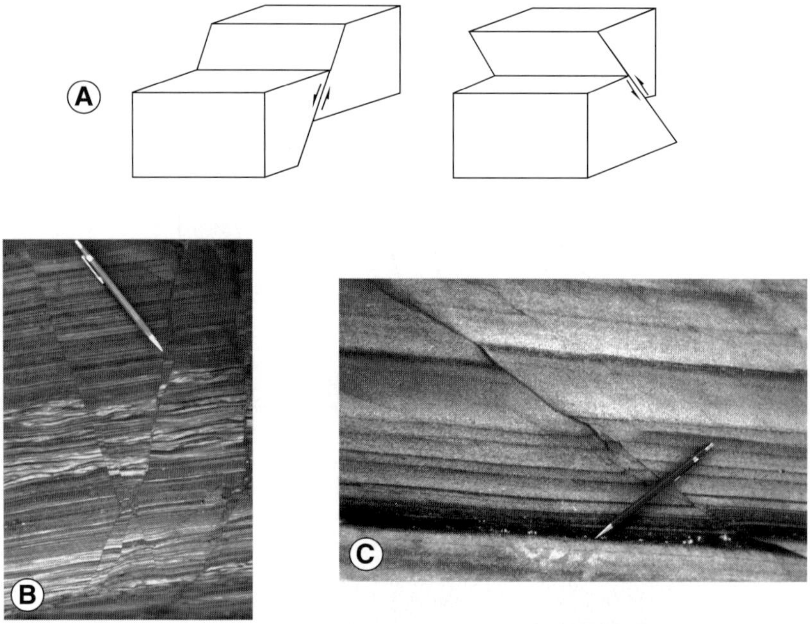

Fig. 5.53 Normal and reverse faults.
A. Diagram. **B.** Normal fault. **C.** Reverse fault.

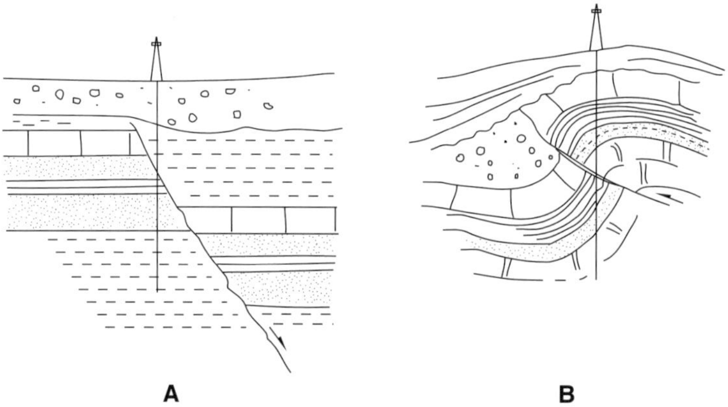

Fig. 5.54 Stratigraphic anomalies on faults.
A. Normal fault. The stratigraphic gap is an apparent one, as the drill does not pass through the upper limestone layer and hardly touches the underlying sands. **B.** Reverse fault. The well passes twice through the series of clays, sands, and lower limestones.

fault can be found in various geological environments, such as on the scale of the tilted rift blocks or delta series with high sedimentation rates (Fig. 5.55), in which case they are referred to as **growth faults**. It is clear that listric and growth faults are not expressions of the same process, as there has been crustal extension along a stretched level in the former, whereas the latter is a gravity-driven extension along a settling slope; but in both cases, the fault is damped at depth on a décollement level where fluids play an important role.

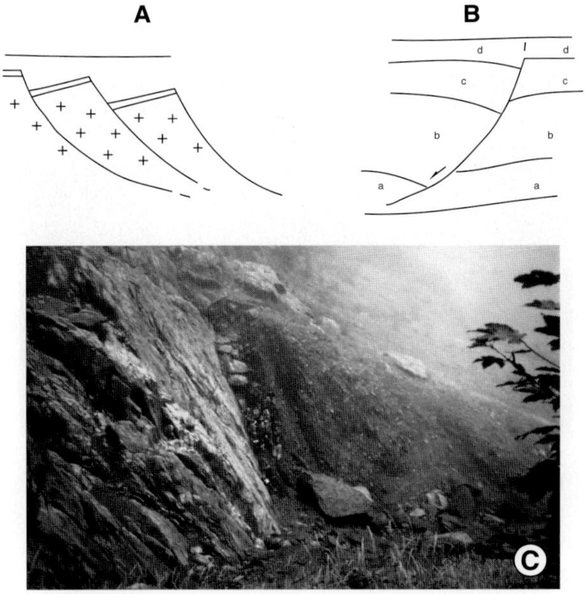

Fig. 5.55 Examples of listric faults.
A. Listric fault in the basement. **B.** Synsedimentary growth fault **C.** Example of the Ornon fault. The basement is on the left and the Liassic marls on the right.

It should be noted that the faulting process induces or is concomitant with a ductile-type deformation. This leads to **drag faults** as in Fig. 5.56, or to the large-scale **rollover anticline** formed by the slippage of beds along a listric fault (see further on).

A **reverse fault** is one in which the thrust is generally less than the heave, and the resultant movement of which is a shortening. One block moves up over the other (Fig. 5.53) as the result of compression, with the principle stress σ_1 being horizontal. The fault plane generally lies at a lower angle (20° to 35°) than that of normal faults, depending on the materials involved and the confinement pressure. If a drillhole passes through a reverse fault, it will pass through a part of the stratigraphic series twice (as in Fig. 5.54), contrary to what happens with the normal fault.

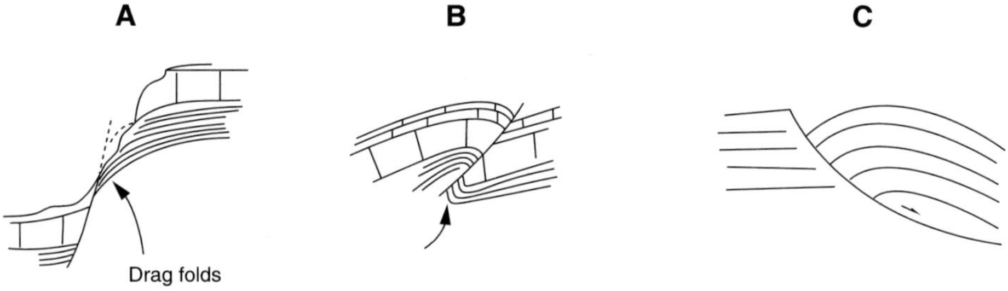

Fig. 5.56 Deformations near faults.
A. and **B.** Drag folds. **C.** Rollover anticline. **D.** Field example of the Kimmeridgian of Diois.

There are various types of reverse accidents in tectonics. When the fault angle approaches the horizontal and the displacement is considerable, it is called a **thrust** or **overthrust** fault. This transports sedimentary stacks in folded **thrust sheet** systems, called **nappes**, over considerable distances (Fig. 5.57) atop a thrust plane.

A reverse fault can also develop from a **décollement level** into a **ramp**, such as can be found today in folded foreland regions. This is always related with ductile deformation and folding (Fig. 5.58). The importance of this décollement level is that the discontinuities take root in a level where the mechanical properties are different, as is true of certain normal faults.

There may be one or more décollement levels depending on the composition and thickness of the sedimentary series.

In the same way as for normal faults, one of the blocks is defined as the **upthrown block** and the other as the **downthrown block**, or **hanging wall** and **footwall**, respectively, for reverse faults (Fig. 5.59).

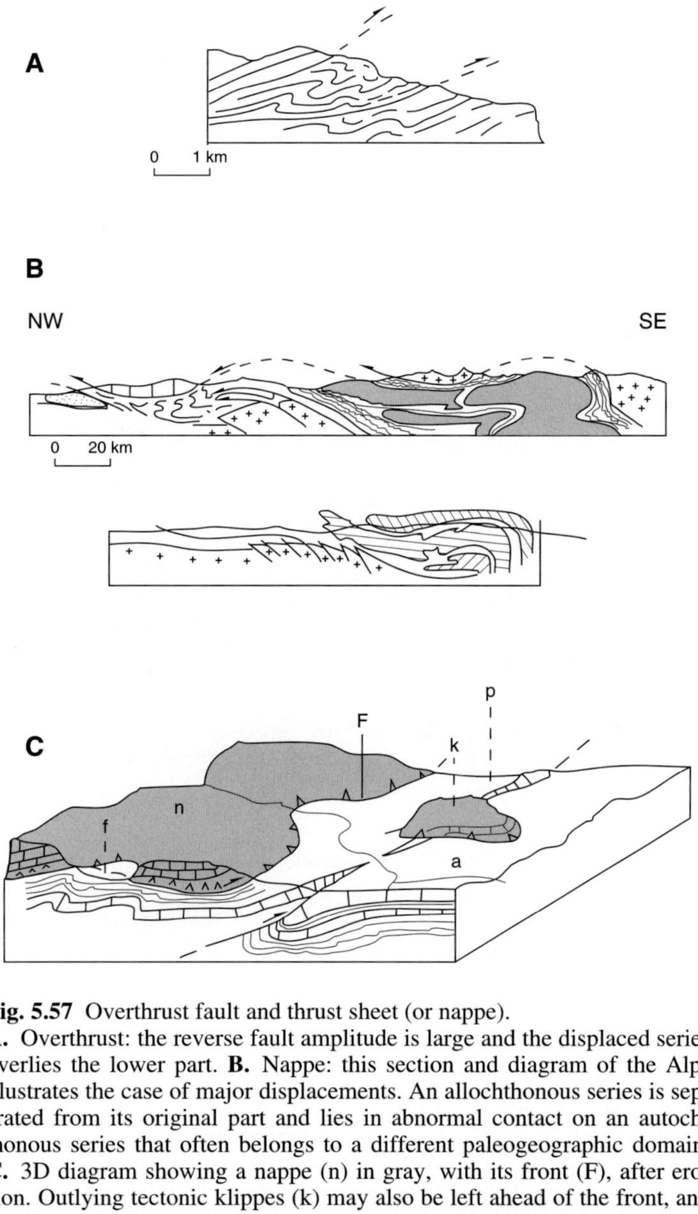

Fig. 5.57 Overthrust fault and thrust sheet (or nappe).
A. Overthrust: the reverse fault amplitude is large and the displaced series overlies the lower part. **B.** Nappe: this section and diagram of the Alps illustrates the case of major displacements. An allochthonous series is separated from its original part and lies in abnormal contact on an autochthonous series that often belongs to a different paleogeographic domain. **C.** 3D diagram showing a nappe (n) in gray, with its front (F), after erosion. Outlying tectonic klippes (k) may also be left ahead of the front, and erosion can also open up windows (f) where intermediate parautochthonous rock (p) can be seen overlying the autochthonous domain (a).

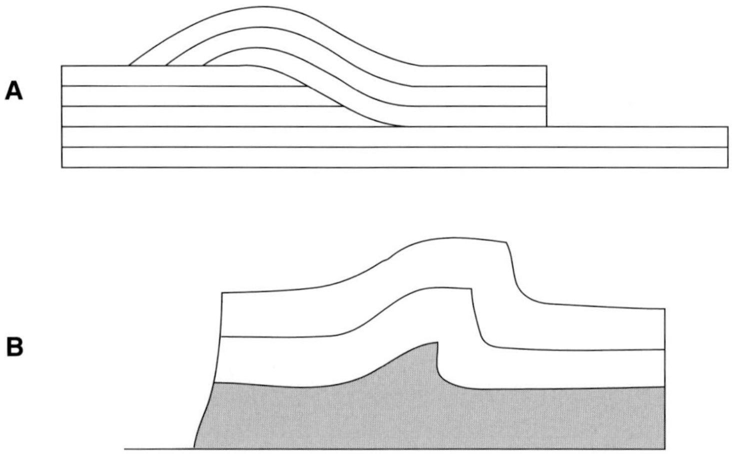

Fig. 5.58 Ramp and décollement level.
A. Ramp anticline. **B.** Décollement level.

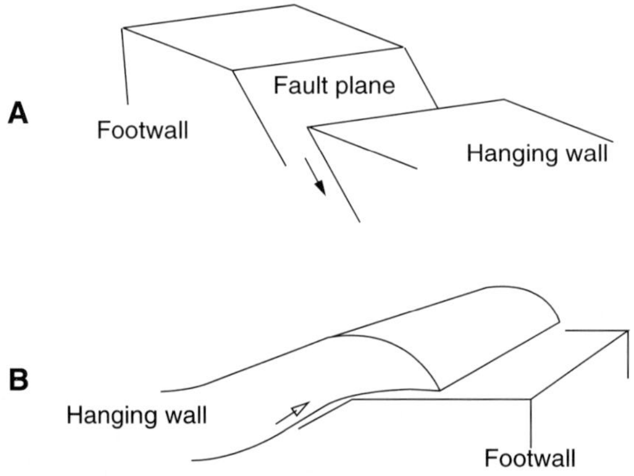

Fig. 5.59 Fault blocks.
A. Normal fault. The wall resting on the fault is the hanging wall, which is the downthrust side when the fault is normal, while the footwall block on which it rests is thrust upward. **B.** Reverse fault in which the hanging wall is upthrust over the footwall.

The fault displacement is sometimes called **transfer**, and the resulting **duplication** of a stratigraphic series by two or more reverse faults, creating imbricated **slices**, is more and more often being called **duplex** (Fig. 5.60). As a two-dimensional representation is generally inadequate, the idea of a **lateral ramp transfer** is often used.

A

B

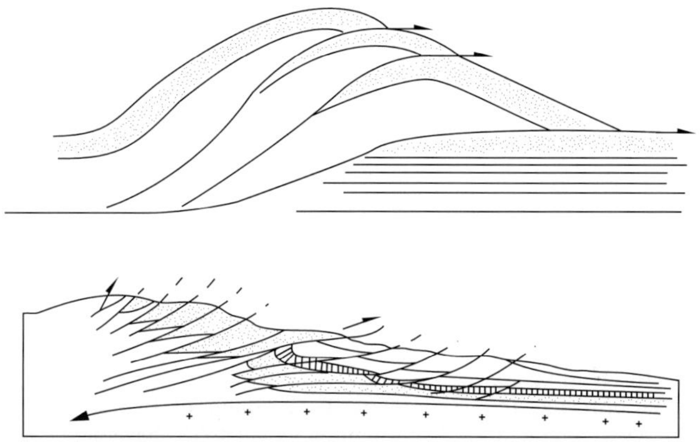

Fig. 5.60 Sequence of peel-thrusted imbricated slices.
A. Diagram of a duplex system. **B.** Diagram of Canadian Rockies (from Debelmas and Mascle, 1993).

When a reverse fault in a folded chain induces a local displacement in the direction of general motion (i.e., toward the foreland), the fault or fold inclination plane **vergence** is normal (Fig. 5.61). **Backthrusting** can also occur in the reverse direction. Both types of vergence can be seen on all scales. When two reverse faults of different directions are directly associated, they cause a **pop-up** (or sometimes a **pop-down**) structure (Fig. 5.62).

In certain cases of complex tectonics, reverse fault displacements may differ between the surface and depth, but generally the vergence is consistent throughout any zone deformed in compression.

A **strike-slip fault** is one in which lateral heave dominates (Fig. 5.63), evidencing transpression or transtension. The strike slip is further defined as **right-slip** (dextral) or **left-slip** (senestral) depending on whether the block opposite the observer has moved to the right or left, respectively.

In the same way as for normal and reverse faults, the dominant thrust is the one considered. If there is real strike slip as in true normal and reverse faults, the general motion can sometimes be broken down into a vertical and horizontal motions. Often the faults include a large strike slip, which can be detected by the attitude of the striations in the fault plane. The **pitch**, which is the inclination of the striation with respect to the fault plane, indicates the strike slip component (Fig. 5.64). If the pitch is large, the fault will be termed "right-normal" or "left-reverse", for example.

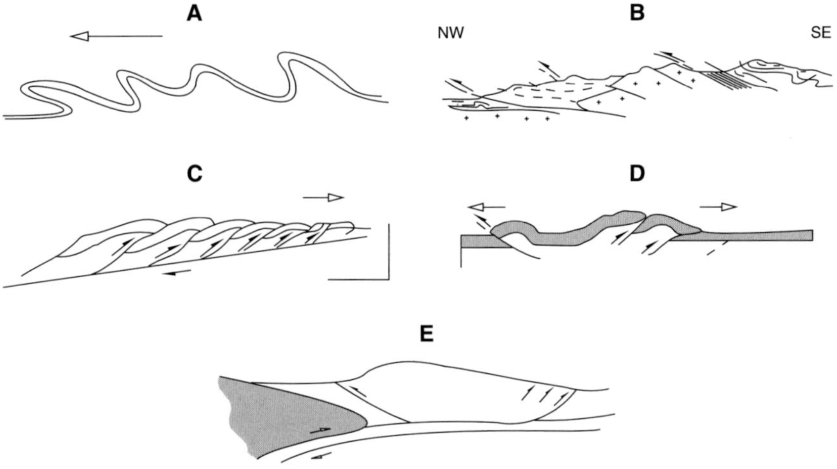

Fig. 5.61 Fold vergence.
A. Fold unfurling in one direction. **B.** Example of a chain (western front of the Alps) where all of the structures (cover folds, basement, sheets), are folding in the same direction. **C.** Regular vergence of an accretion wedge. **D.** Opposing vergence, with backthrust. **E.** Opposing vergence: diagram of the Mediterranean wedge with a backstop generating backthrust.

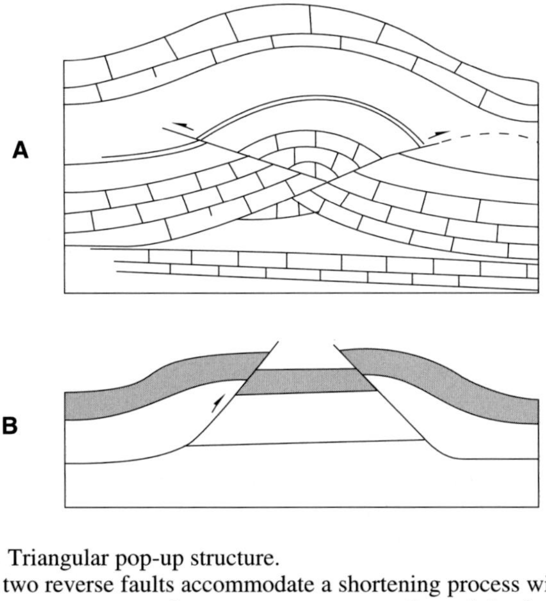

Fig. 5.62 Triangular pop-up structure.
A. Here, two reverse faults accommodate a shortening process with opposite displacements, prompting a pop-up structure. One of the faults takes root on a décollement level. **B.** Triangular zone.

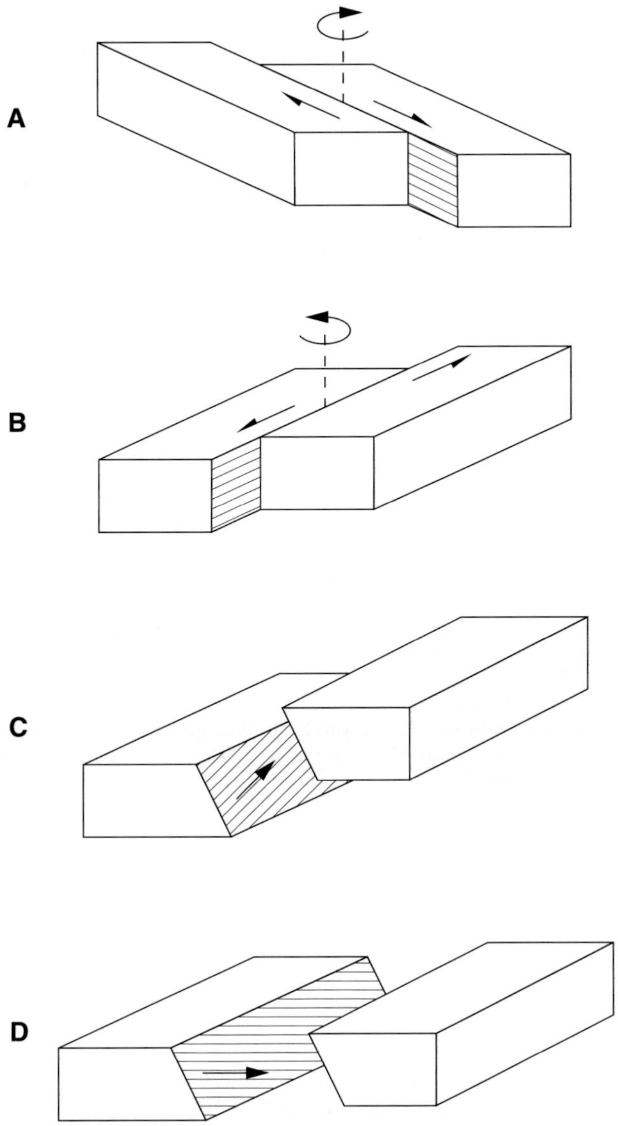

Fig. 5.63 Strike slip fault.
When the dominant thrust is strike slip, the striations are horizontal.
A. Right-lateral (or dextral) fault. **B.** Left-lateral (or sinistral) fault. **C.** and **D.** Left-lateral reverse (**C**) and normal (**D**) faults in which the striations are inclined, indicating the direction of motion by their pitch.

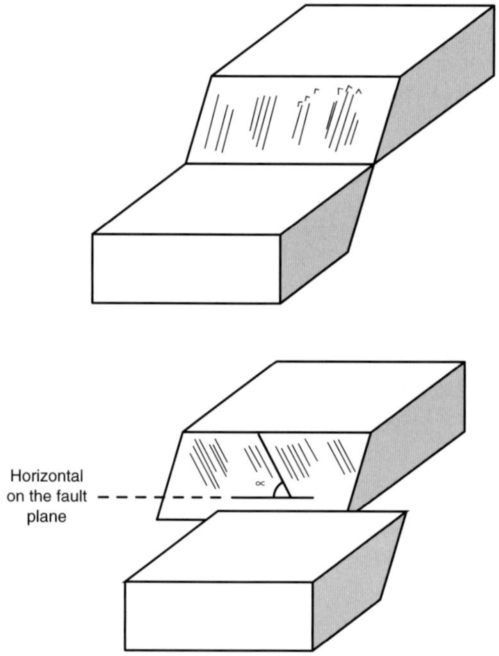

Fig. 5.64 Striation pitch.
A. Pure normal faulting with 90° pitch. **B.** A slight strike slip generates an oblique motion and the striations are at an angle of less than 90° with the horizontal, measured on the fault plane.

The idea of ***apparent displacement*** *should be emphasized in strike slip faults. Cartographically, a fault may seem to shift a stratigraphic sequence of beds greatly, but post-faulting erosion can accentuate the actual shift or even mask the real (normal or reverse) motion.*

One special case of horizontal faulting is the transform fault defined in Chapter 1. This is the result of slippage due to a differential oceanic accretion rate in the blocks to either side of the fault (Fig. 1.24). On the scale of a basin, the major strike slip systems are linked either to these transform faults (Spitzberg), or to continental punching (southeast Asia).

C. Fault Chronology

For faults of any type, the history of the motion along the fault plane may very well be a complex one if the stress regime varies as it normally will in the course of geological history. This history leaves a heritage of small discontinuities that may serve as a path for a new fault, or a neogenic fault may occur which avoids these pre-existing paths. Once a discontinuity exists in an otherwise homogeneous environment, the stress regime may change and the fault may be reactivated in another direction, and propagate (Fig. 5.65). **Recurrent faulting** like this creates a **tectonic heritage**. The position, direction, and recurrence of a fault are conditioned by the original structural configuration, and especially that of the basement. If its history is long enough, the same fault may then alternately act as normal, reverse, and strike-slip fault (Fig. 5.65). Its history can be reconstructed by closely analyzing its microstructure.

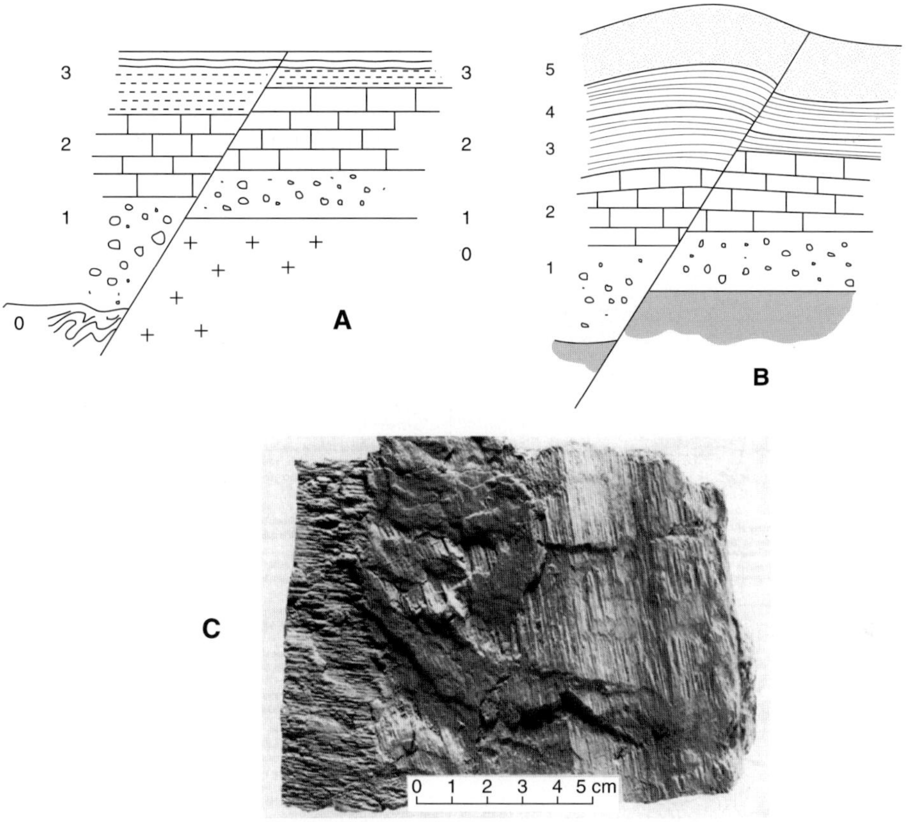

Fig. 5.65 Multiphase fault heritage.
A. An existing discontinuity in the basement (0) acts as a normal fault while the first bed (1) is being deposited. The fault then ceases activity while the limestone (2) is deposited, then recurs as a normal fault as the clays (3) are deposited. **B.** In the course of geological time, a normal fault (periods 1 and 3) may act as a reverse fault (period 5), with intervening periods of inactivity (2 and 4). This is an example of tectonic inversion (from P. Trémolières, 1981). **C.** A striated plane may be evidence of different faulting sequences. The example here shows sideslip and normal faulting (from P. Trémolières, 1981).

When a given fault exhibits reverse, then normal, then reverse motion (or vice-versa), it has gone through **tectonic inversion**. This reversal phenomenon is very important in petroleum geology. It is essential to establish a good chronology, which can be done by analyzing striations and mineral growth.

Faulting can last hundreds of millions of years or more, and is fortunately detectable in any sediments that are associated with the faults by **synsedimentary tectonics**, as stratigraphic arguments then exist to determine the chronology (Fig. 5.66). A **synsedimentary fault** is one in which the trace is filled by a correlative deposit, accompanied by "syntectonic" sedimentation in the downthrust block. A post-sedimentary fault will cut across entire series without leaving a marker to define the time at which it operated.

Various types of faults (and diaclases too) can be considered as ante-, syn-, or post-folding faults (an ante-folding fault would be tilted with the fold). The passage from brittle to ductile domains thus depends on many factors beyond the theoretical aspects.

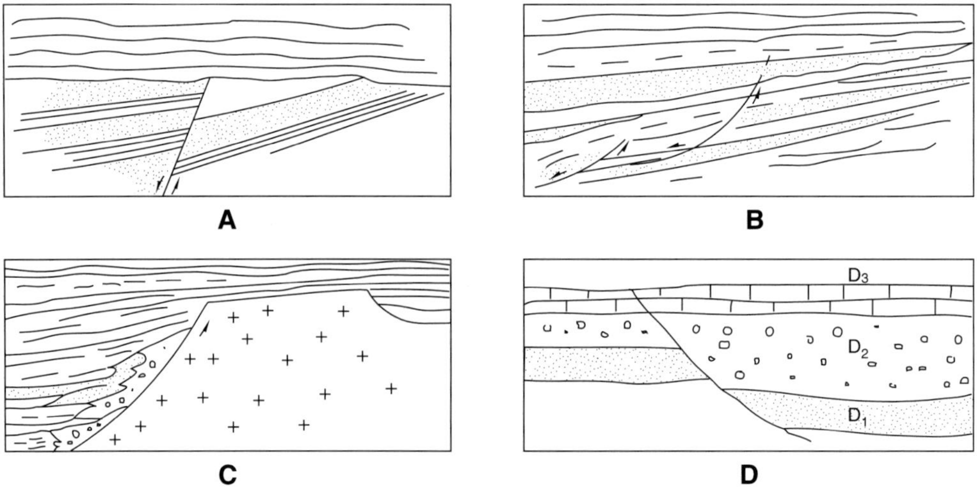

Fig. 5.66 Fault ages.
A. A discontinuity cuts across the fault, and post-tectonic beds postdate the faulting. **B/C.** The fault here is synsedimentary. Growth faults develop in an argillo-arenaceous delta in **B**, while a listric basement fault in **C** influences the sedimentation in the lower block where more or less coarse detritic deposits can be found. **D.** The fault occurs after the sand bed (D_1) is laid. It is active while the conglomerates (D_2) are being deposited, and inoperative during the calcareous deposition (D_3).

D. Regional Organization of Faults

A fault at a given location is defined by a shear plane. In some cases, it comprises a set of discontinuities in association over a breadth of one or more meters. Sometimes **fault breccia** develops from an in situ mechanical fragmentation of the rock in which the fault system develops. Depending on the observation scale, one or more elementary faults may be found.

Faults are never isolated but are associated in **conjugate fault systems** on all scales (Fig. 5.67). The system can be analyzed on a vertical plane, along a geological section, or on a horizontal geological map, but they should be interpreted in space. Normal, reverse, and sideslip faults are associated into conjugate systems, from which the paleo-stress orientation and direction of shortening or elongation can be determined. A **microtectonic structural analysis** can be developed by analyzing measurements made on structural elements—usually microstructural tectonic objects—and then plotting the different attitudes and motions of the planes bearing the tectonic figures on a stereogram (Fig. 5.67). The time history of paleo-stresses can then be retraced in a well-analyzed stratigraphic set.

Synthetic and antithetic faults will occur in an extensive regime where high-dip normal faults predominate. Synthetic faults dip in the same direction as the beds, while antithetic faults dip counter to them (Fig. 5.68), creating graben structures that are rarely symmetrical. The structure is generally guided by a bundle of major faults often referred to as a half-graben or tilted block. The opposite case of an upthrown block is a horst. As has already been said, combined faulting generally allows flexible accommodation.

Faults that dip in the same direction as the beds in normal and reverse accidents are said to be **conformable**, whereas faults dipping contrary to the beds are **unconformable** faults (Fig. 5.69).

The faults that appear in a compressive regime in association with folds indicative of a shortening process are arranged in a non-random field. They occur in imbricate slices, duplexes, and ramps, with level flats between the ramps. A major reverse fault may be accompanied by a number of secondary reverse faults. Arches also develop with normal faults (Fig. 5.70A) with tension cracks on the convex extrados side of the arch (by expansion-traction). These can often be found on the roofs of banks in compressive structures (Fig. 5.70B).

Just as the extensive structures of grabens are rarely linear but are organized in transfer relays (Fig. 5.71A), reverse fault systems cut through the folding in compressive systems and jut like bayonets along **oblique ramps** (Fig. 5.71B). The oblique ramp pattern is often due directly to the of lateral variation and heritage the underlying beds.

Reverse faults can develop in regular order in space and time in a **progradational piggy-back sequence** from the inner tectonized domain upstream to the outer foreland domain downstream (Fig. 5.72), as in accretion wedges. If they develop by cutting across previously active faults, on the other hand, the system is “retrograde” and **out of sequence**.

Some faults develop concomitantly with folds, according to a number of different models. If the mechanical discontinuity exists already, the anticline is an arch overhanging a

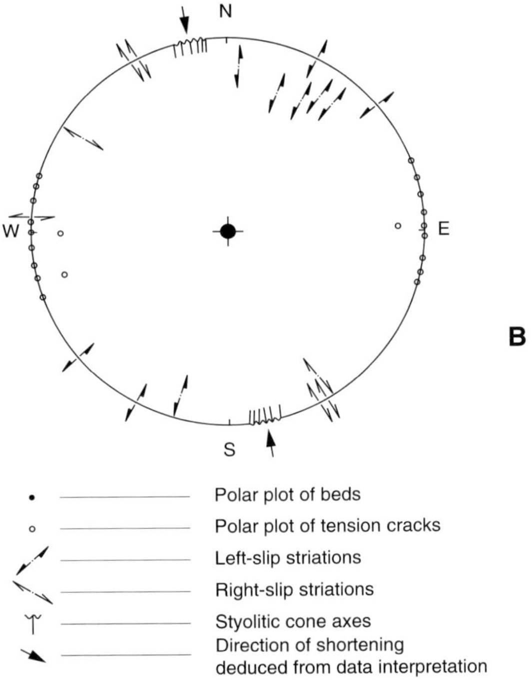

Fig. 5.67 Association of faults.
A. Natural example: group of normal and slip faults (playa de Los Muertas, Andalusia) (IGAL photo). **B.** Use of various microtectonic markers observed in a field station. The markers are plotted on a stereogram (upper reference hemisphere). This example of the Paris basin (from Trémolières, 1981) indicates an N-S oriented compression phase among the various displacements.

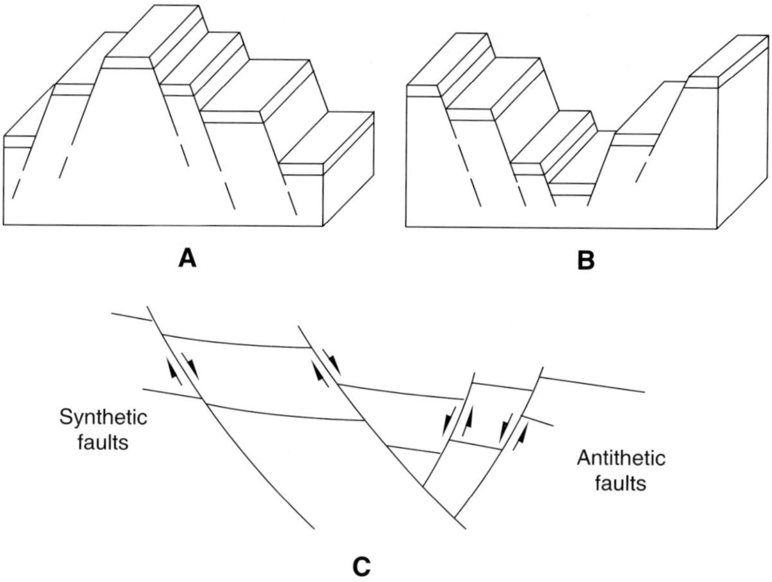

Fig. 5.68 Horsts and grabens.
A. Horst. **B.** Graben. **C.** Synthetic and antithetic faults (from Guillemot, 1986).

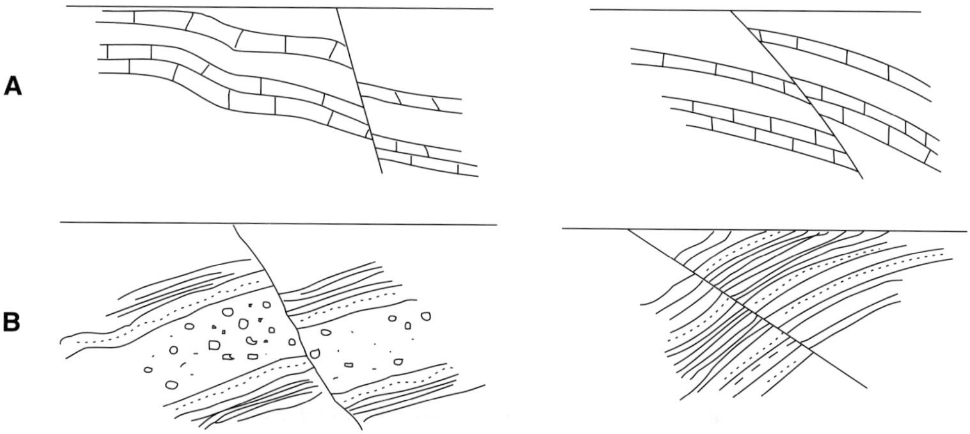

Fig. 5.69 Conformable and unconformable faults.
A. In (normal and reverse) conformable faults, the fault dip is oriented conformably with that of the displaced beds. **B.** Unconformable (normal and reverse) faults: the fault dip here opposes that of the shifted beds.

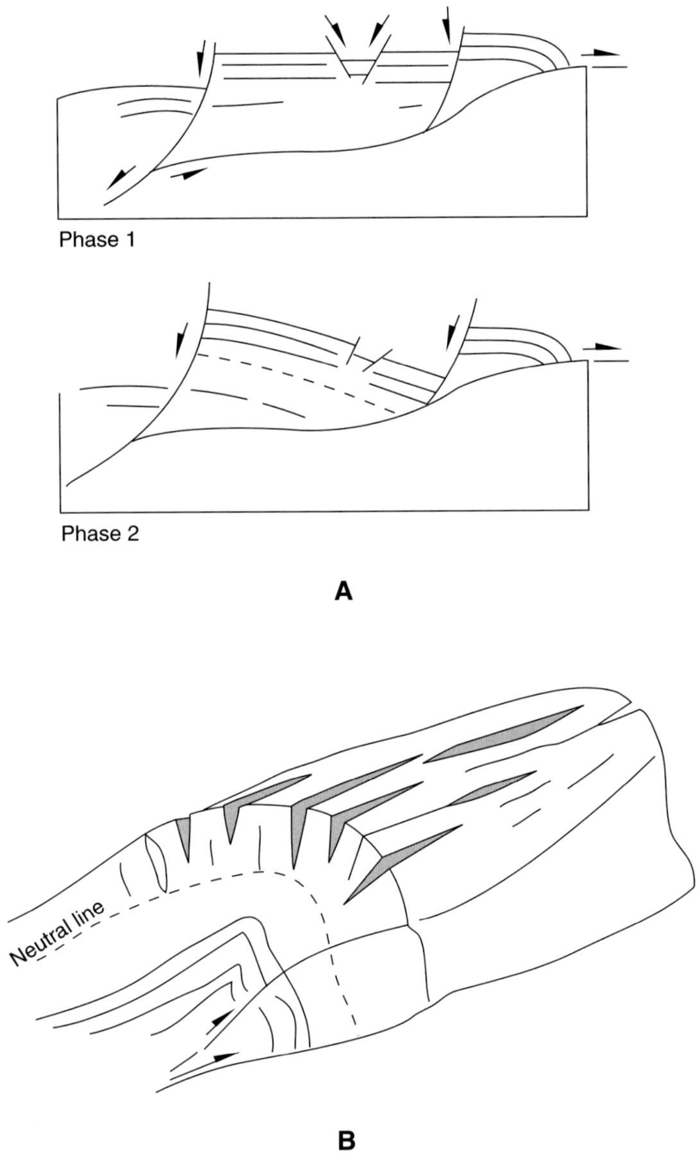

Fig. 5.70 Sequence of ramps, reverse faults and minor extensive structures.
A. Section of a portion of the southern Apennines (from Roure et al., 1981). **B.** Extrados cracks in the proximity of a reverse fault (no scale).

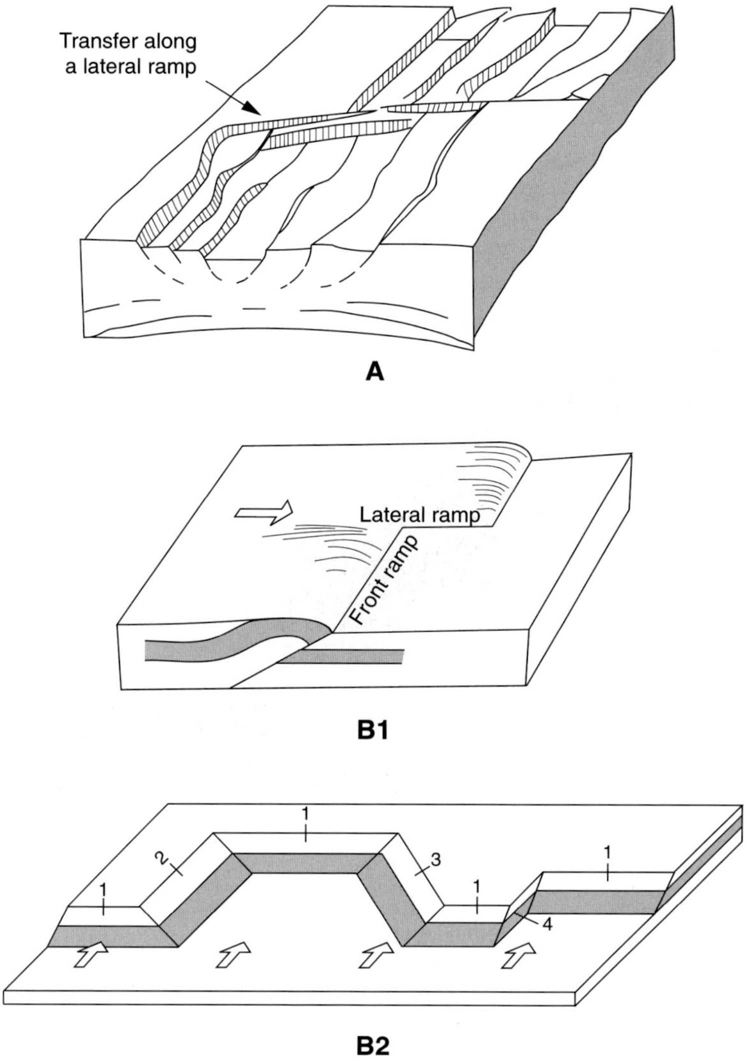

Fig. 5.71 Lateral ramps.
A. The tilted blocks of a rift are not always continuous over long distances, but shift along transfer faults. **B.** The overthrust front ramp is shifted by an oblique "lateral" ramp (B1). If this relay structure becomes vertical, the fault is referred to as a strike-slip fault. The lateral ramp thus lies parallel to the direction of overthrust transport (B2, 1), otherwise the ramp is oblique or transverse (B2, 3, 4) (diagrams from Mc Klay, 1992).

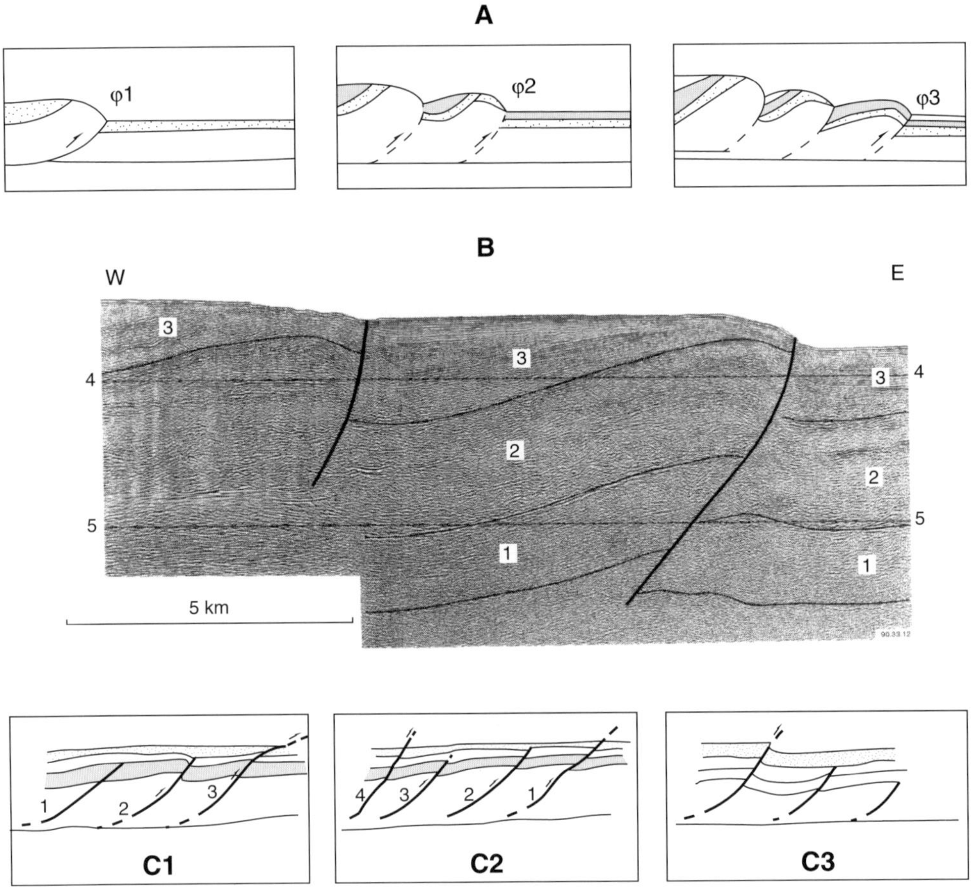

Fig. 5.72 Repeated overthrusts and piggyback basins.
A. Theoretical diagram of repeated events with gradual tilting of the small basins carried atop the overthrust material. **B.** Seismic profile (front of Barbados wedge). The basins are very small here (IFP photo). **C.** Overthrusts: Piggyback (**C1**), out of sequence (**C2**) and reverse sequence (**C3**).

ramp and is called a **fault-bend fold**. If the fault develops and propagates with the fold, it is a **fault propagation fold** or **ramp** with no fault associated with it at the surface. It is then often called a **blind overthrust** (Fig. 5.73). Other models have also been proposed, but it should be remembered that nature's arrangements are generally more complex.

Beyond the scale of the outcrop, a distinction often has to be made on the regional scale between two broad categories of tectonics: accidents concerning the basement, which come

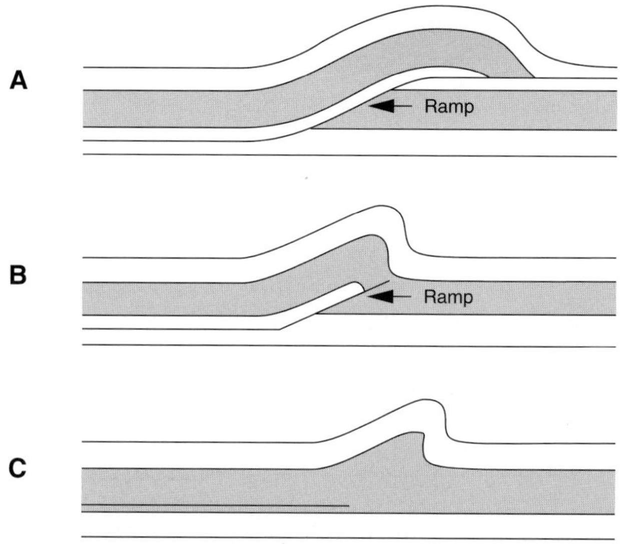

Fig. 5.73 Different types of folds.
A. Fault-bend fold. **B.** Fault-propagation fold. **C.** Ductile décollement (from Jamison, 1987).

under **thick skin** tectonics of deep-seated structures, and deformations of the sedimentary cover, which are called **thin skin tectonics** (Fig. 5.74). Both these styles of deformation may be coeval in time and/or space.

E. Shearing Zones

Brittle shearing zones on the scale of the outcrop or region are accompanied underneath by a continuous deformation with fractures, tension cracks, folded ductile and cleaved beds, and at depth this gradually becomes ductile shear (Fig. 5.75) affecting a large part of the crust.

F. Schistosity

When rocks are subjected to high pressure at depth, parallel micro-cracks develop and their mineralogical constituents are rearranged. The discontinuity planes produced are generally orthogonal to the main stress, and are termed **cleavage**. **Schistosity** is the sum of elementary cleavages. There is fracture cleavage, flow cleavage, and **foliation** (schistosity by metamorphism). Schistosity is generally not essential in the study of petroleum basins, even though is does develop spectacularly along certain reverse fault planes. The subject is discussed no further here.

Certain linear structures like corrugation, stretching, or boudinage, are special forms witnessing to the heterogeneity of the deformation.

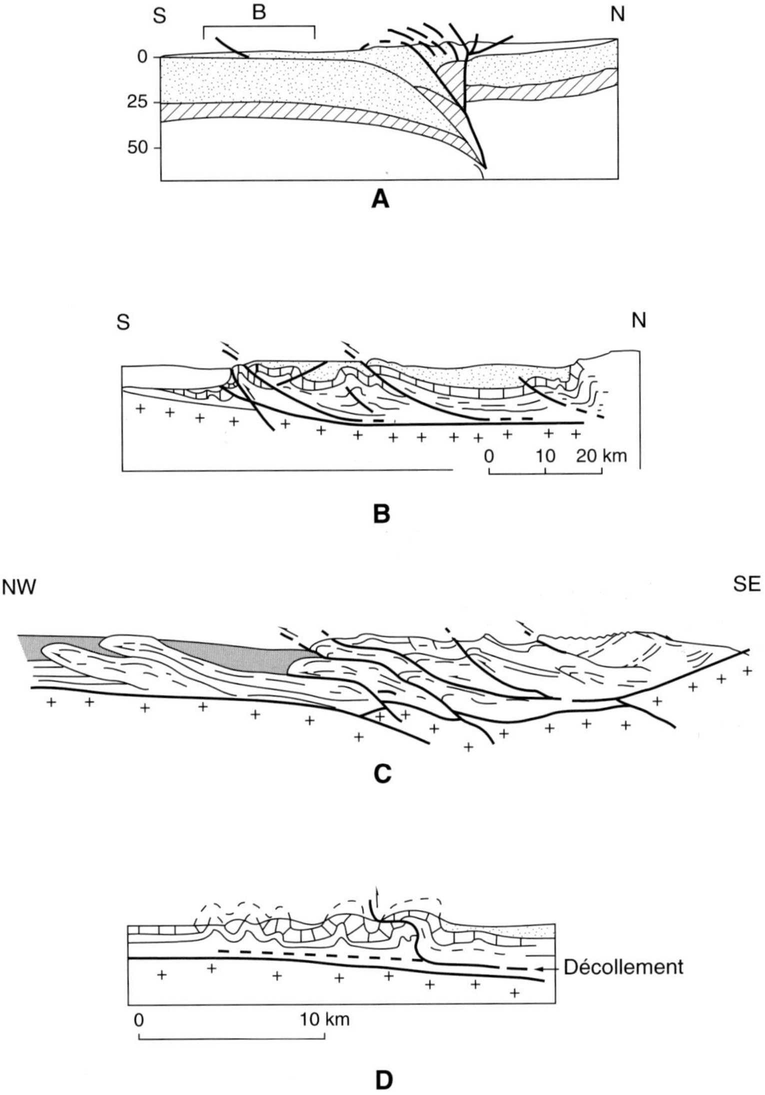

Fig. 5.74 Basement and cover tectonics.
A. Section of Pyrenees crust (from ECORS – Pyrenees profile, from Mattauer, 1990). The whole crust is involved. **B.** Cover décollement on the south side of the Pyrenees. **C.** The Alps: the basement and cover are both involved here too, but are mostly de-coupled. **D.** Décollement in the Jura area (from various authors).

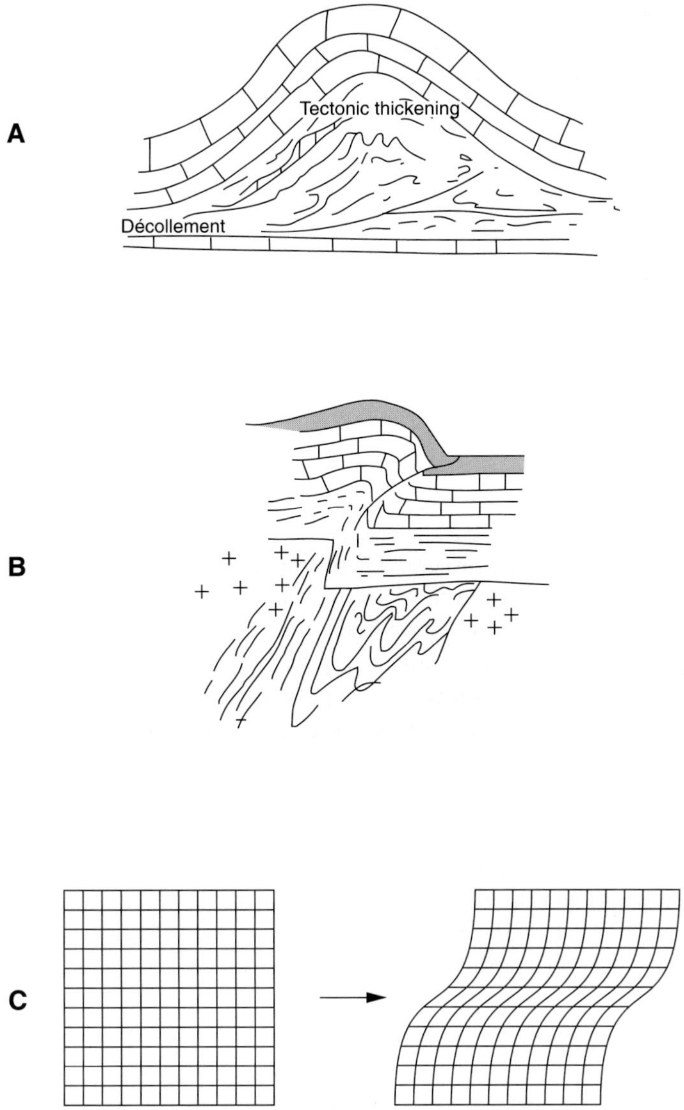

Fig. 5.75 Shear and ductility.
A. "Incompetent" ductile rock (adjacent to more brittle rock) is often observed in surface levels, forming tectonic thickening over detached décollement zones. **B.** Fragile surface beds becoming more and more ductile as they are driven to depth. **C.** Diagram of ductile shear (from Ramsay, 1967).

5.2.2.2 Flexible Deformations

A. Deformations within the Bed

The importance of early deformation was mentioned in the part of the present chapter on burial diagenesis. The mechanical settling and the pressure dissolution effects that lead to **compaction** are the simplest and earliest forms of deformation. While this compaction and the accompanying reduction in volume is generally developed along the vertical axis in the sedimentary pile, another form of compaction can affect the strata before flexible deformation, when the main stress tensor lies in a horizontal plane. This results in objects (generally fossils) being deformed in the beds, expressing a shortening of as much as 30%, which is considerable. Tectonic **pressure dissolution** generates spectacular **tectonic stylolites** (Fig. 5.48), reflecting a shortening due to a horizontal main stress, σ_1.

B. Bed Folding

Shortening, which usually occurs in compressive regime, results in bed undulations of a form, amplitude, and frequency that depend on the bed thickness and lithology, the stress intensity, and the confinement pressure. These undulations are called **folds**. **Folding** can designate either the fold generating mechanism, the fold or set of folds themselves, or the deformation period—Hercynian or Alpine folds, for example.

Folds are observable by following the visible boundaries of strata. A fold will include an upward convex curve called an **antiform** and an upward concave part called the **synform**, which describe the geometric attitude. Another general definition is that the **anticlinal** portion of the curve is the one in which the older strata are inside the curvature, and the **synclinal** portion is the one enveloping the younger strata (Fig. 5.76).

A fold has an **axis**, one or more hinges where the curvature is maximum, and **flanks** to either side of the hinge. The **fold direction** is that of the fold axis. The highest point is called the **culmination** and the lowest the **saddle**. If the fold is cylindrical, its axis is the same as its hinge. The fold hinge can be measured on a series of beds, and the plane passing through these various hinges defines the **axial plane** (or surface), which can be vertical or inclined.

An anticlinal that is severely eroded to the point where the deep beds are exposed is called an **inlier** (Fig. 5.77).

A fold does not extend infinitely. It terminates laterally in a **pericline** where the anticline vanishes (often only the surface expression!) and the syncline rises up (watch for erosion).

Considering a single bed as it folds and bends, the maximum deformation at the hinge may reflect an **extrados** in extension and an **intrados** under compression (Fig. 5.78) with a neutral surface or fiber running between them, or the deformation may be heterogeneous with internal shear and creep.

A creased fold in which there is a sudden change in the angular dip of the beds, often with conjugate diaclases, is often termed a "kink" (Fig. 5.79).

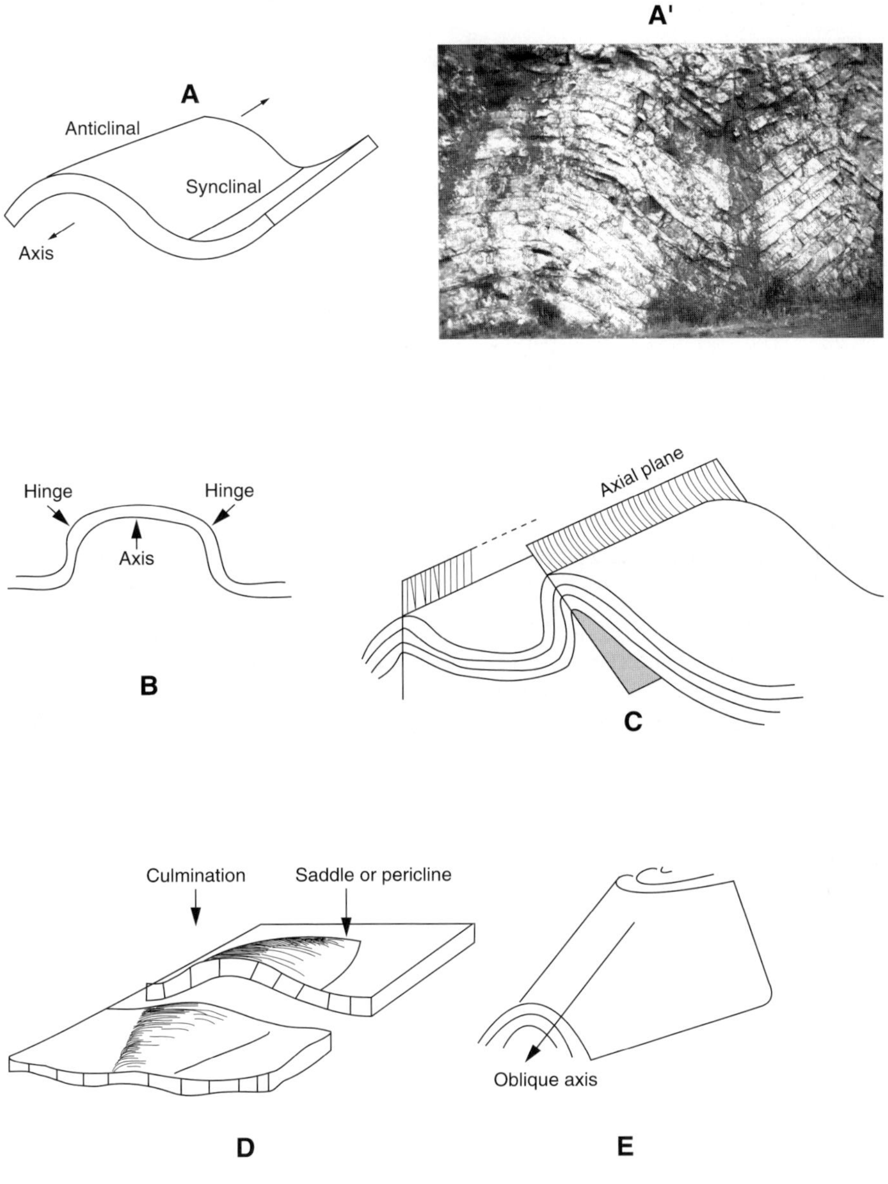

Fig. 5.76 Definitions of anticlinal and synclinal.
A. Cylindrical anticlinal and synclinal structure. **B.** Fold axis and hinge (sometimes the same, as in A). **C.** Axial plane joining all of the fold axes at different levels. **D.** Culmination and saddle of an anticlinal fold. **E.** Diagram of an oblique plunge in the fold axis.

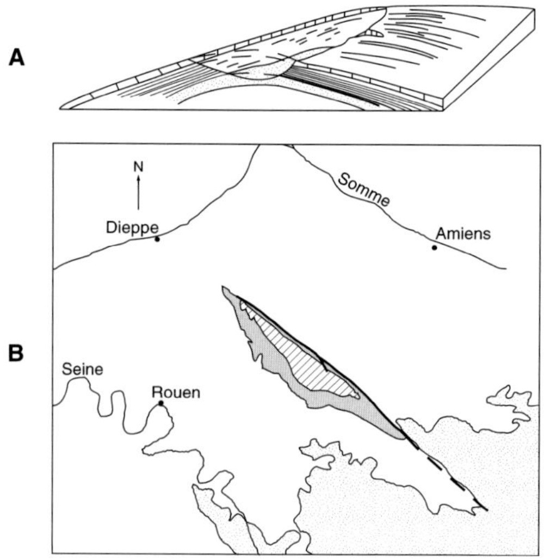

Fig. 5.77 Inlier at the top of an anticlinal.
A. General diagram. **B.** Example of an inlier in the Bray area between the Somme and Seine rivers. The Lower Cretaceous (in gray) and the Jurassic (hatched) outcrop from under the subtabular beds of the chalky Upper Cretaceous, along a fault. The boundary of the Tertiary outcrop is parallel to the fault.

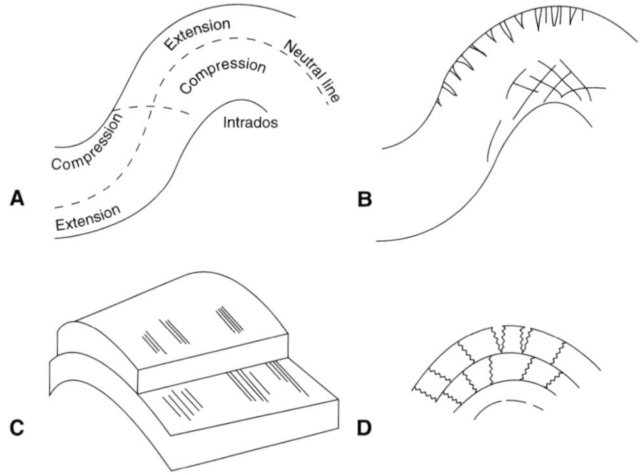

Fig. 5.78 Mechanical properties and cracking associated with a fold.
A. Theoretical diagram (from Ramsay, 1967). **B.** Extension cracks and friction joints. **C.** Multi-bed shear with telltale striations in the bed planes. **D.** Tension cracks on the extrados are often preceded by the formation of stylolites deformed with the fold and reused.

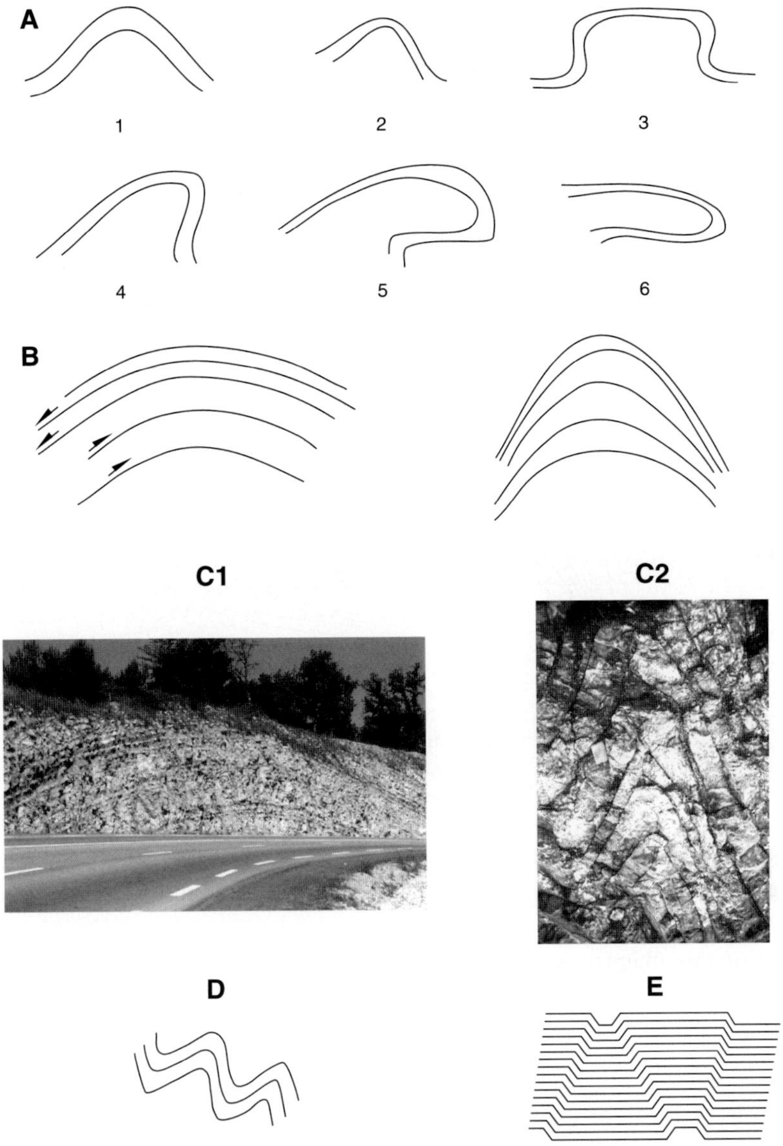

Fig. 5.79 Different types of folds.
A. Upright (1), asymmetric (2), box (3), overturned (4), inverted (5) and recumbent (6) folds. **B.** Concentric or parallel fold and a similar or non-parallel fold. **C.** Real examples: Jurassic (Ardèche), Cretaceous (Alès). **D.** Herringbone. **E.** Kinks.

Folds vary in **form** from upright to asymmetric box, inverted, or recumbent (Fig. 5.79), when the flanks are referred to as **reverse** and **normal**. If the bed thicknesses are constant, the fold is **concentric** or **parallel** (in deeper structural stages where the beds are of the same thickness when measured parallel to the axial plane). When flattening or stretching is observed in the flanks, the folds are said to be **similar** or non-parallel.

At depth, the folding becomes **disharmonic** as the plastic properties of the extrados and intrados vary with bed lithology and produce simultaneous folds of differing wavelengths (Fig. 5.80). This is sometimes called **thickening**. With enough disharmonic folding, some of the beds may be subject to **décollement**. All folds associated with a thick, ductile **décollement level** are disharmonic folds. Those occurring along strike-slip faults sometimes curl into **drag** or **en-echelon folds** (Fig. 5.81). The rotation that can occur more specifically along major listric faults and growth faults in extensive regime can produce **roll-over faults** (Fig. 5.56).

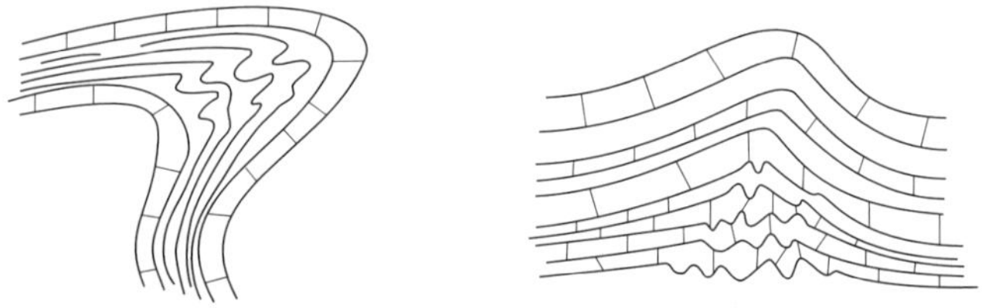

Fig. 5.80 Examples of disharmonic folding.
The geometry on the scale of a structure can be highly disharmonic if beds of different lithologies and competence entrain each other.

Different faulted fold models have been found in compressive regime: **fault-bend folds** and **fault-propagation folds**, or a damping anticline on a blind thrust, which are simplified forms of folds (Fig. 5.73).

The end of this chapter will again stress the importance of décollement levels on a regional scale.

C. Diapirs

When thick plastic levels are subjected to loading (σ_1 vertical) or compressive (σ_1 horizontal) forces, a particular kind of décollement can occur in which these plastic beds penetrate the overlying structure, making a **diapir** (from the Greek *diapeiren*, meaning "to pierce"). Two phenomena of this type occur in sedimentary series.

Shale diapirism is the piercing of under-compacted shale series that have not yet expelled their water. This is commonly found in prodelta series on the continental margins where the sedimentation rates are high. It can be triggered by the sedimentary load alone. It develops in the spectacular form of temporary **mud volcanoes** in accretion wedges like

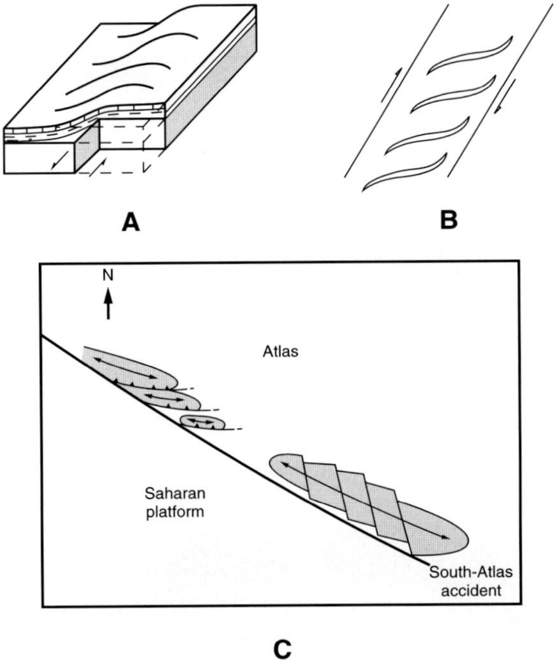

Fig. 5.81 En-echelon folding.
A. Strike-slip in the basement and en-echelon folding in the cover. **B.** Riedel shearing. **C.** En-echelon folding and faults along the south-Atlas accident (from Letouzey et al., 1995).

those of Barbados or the Niger delta (Fig. 5.82). Some of these are rich in hydrocarbons. The extreme case is the tar pool.

Clay can entrain portions of banks or blocks in its motion, forming chaotic series and mixtures.

The other type of diapirism is **halokinesis**, which is saliferous deformation and penetration. The very great plasticity of evaporitic formations, halite, and to lesser extent sulfates, make them disharmonic levels of décollement. While salt beds form the floor of overthrusts and sheets, they also form **diapirs**. **Diapirism** can take varied forms (Fig. 5.83).

For reasons of relative densities, the salt may begin to move very early, under the effect of the sedimentary load. The motion may be continuous and rapid, temporary, or persistent. When the piercing occurs, surface **dissolution** may occur or **cap rock** may form. In some cases, the slippage may continue laterally and **salt rafts** or **salt glaciers** may develop and migrate over large distances.

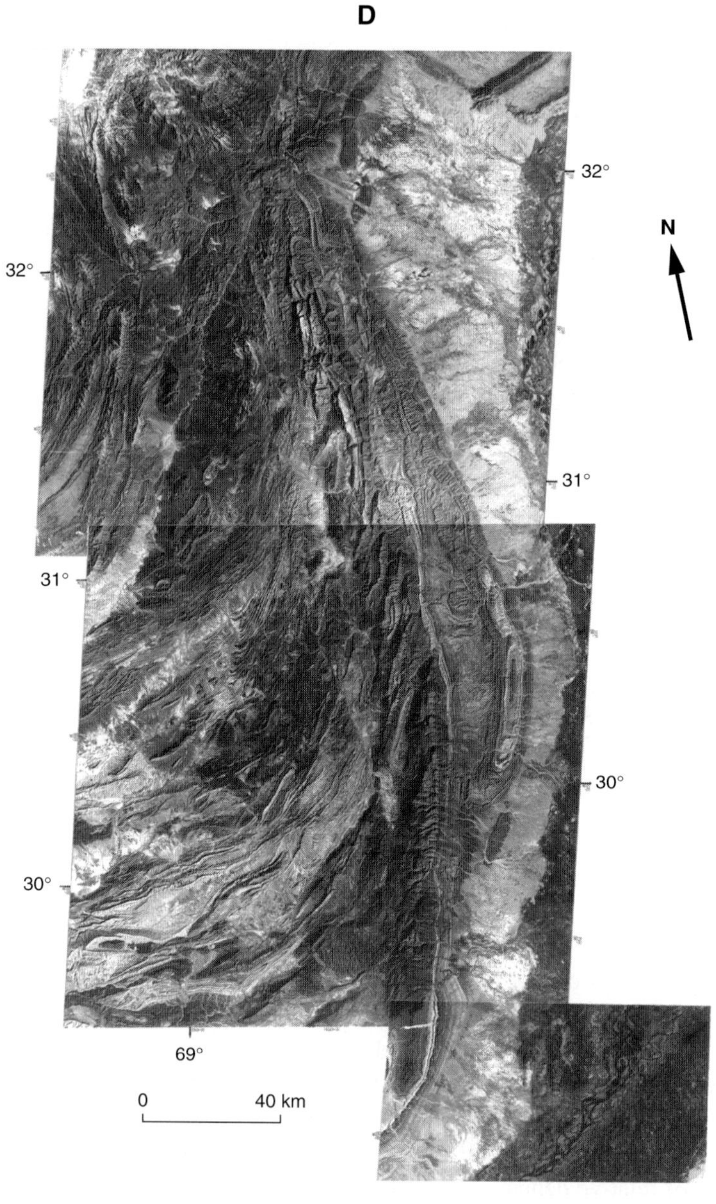

Fig. 5.81 (cont'd) En-echelon folding.
D. Example of chains in east Sulaiman, Pakistan (satellite photo).

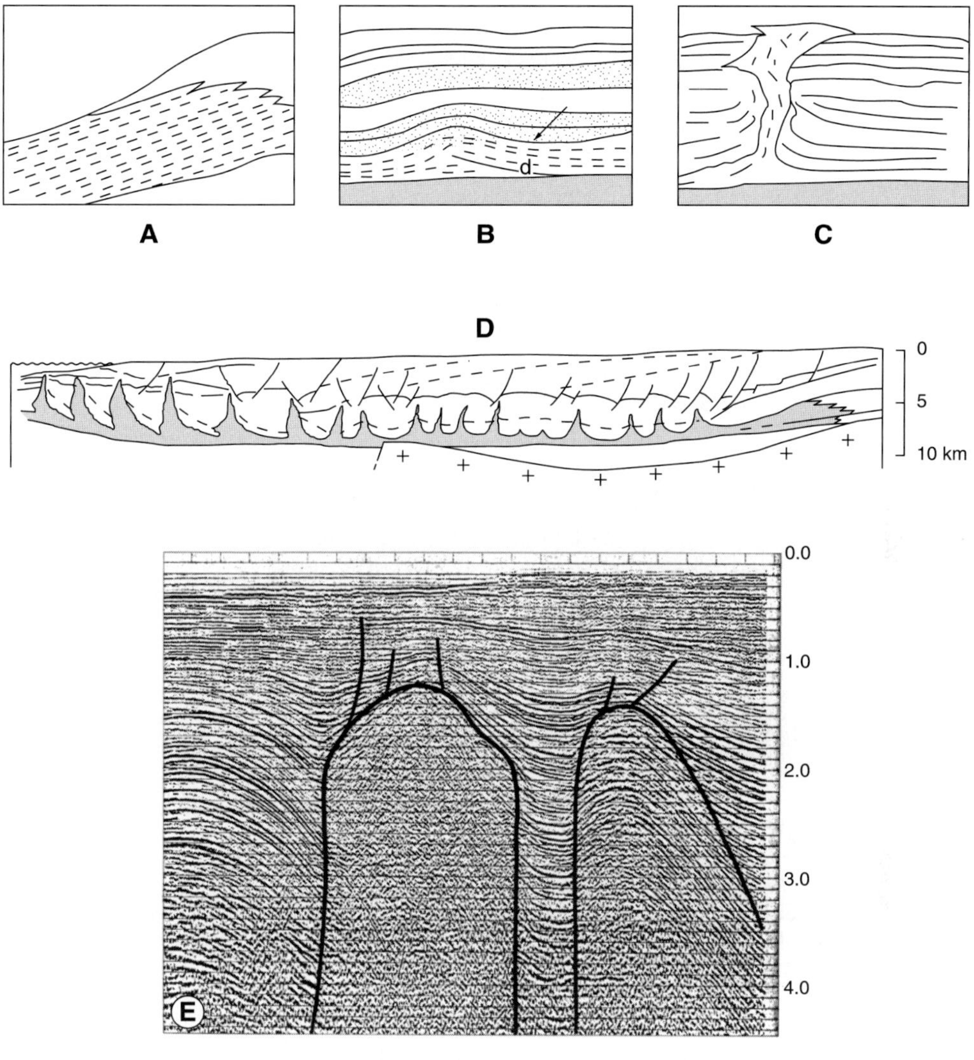

Fig. 5.82 Shale diapirism.
A. Favorable depositional environment: prodelta clay where accumulation is rapid and the sediment is waterlogged and undercompacted. **B.** A décollement level forms (d) by sedimentary load (σ_1 vertical) or lateral compression (σ_1 horizontal or oblique). **C.** Vertical expulsion. **D.** The classical example of the Niger delta. **E.** Example of seismic expression.

A

B

C

D

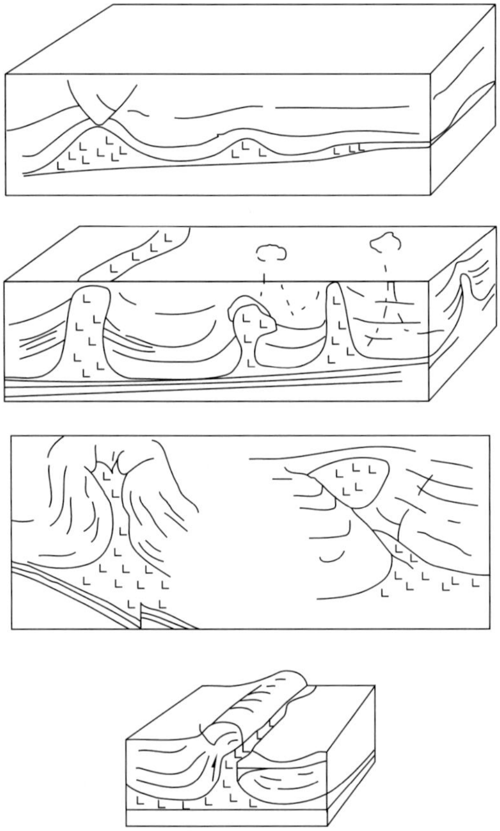

Fig. 5.83 Varied forms of halokinesis.
A. Saliferous domes and anticlines (pillows). **B.** Piercing diapirs and salt walls. **C.** Piercing salt anticline, rafts, and glaciers. **D.** Sole of a stretch thrust (décollement level).

Diapiric deformations are very important in petroleum geology because the quick movement of the salt can generate early structures; the formation of impervious hardpan and domes can produce pinchout on the flanks; the cap rock itself can make a good reservoir (see following chapter); and synclines can develop on the rim of the diapirs overhead the zone where the salt has withdrawn.

D. Thrust Nappes and Folded Nappes

The décollement levels of evaporites and clays at the surface, and the ductile shear zones at depth, allow horizontal displacements of large amplitude producing the **thrust sheets** that form mountain chains (Fig. 5.57). The displacement at the front of the folded zone is generally superficial, while the basement may be involved in the internal zones, which generally exhibit peel thrusting between superposed shearing effects. After one or more **nappes** are placed, the edifice may undergo further folding. The nappe front is generally an **erosion front**, which should not be confused with the **deformation front** that is generally clearly observable in present accretion wedges. If much of the sheet is eroded, part of it may remain ahead of the erosion front in a tectonic **klippe** (to be distinguished from a sedimentary klippe, which is an allochthonous element resedimented in a younger matrix). If the series underlying the sheet is bared by this erosion, the outcropping is called an **inlier**, fenster or window.

Generally, any series displaced in a shortening process are said to be **allochthonous** as opposed to the beds that have stayed in place, which are **autochthonous** (or parautochthone if they have moved very little, as is often the case between the two series).

The sheets may consist of relatively consolidated material, i.e., a series of beds of varied lithologies having undergone extensive diagenesis. But sometimes the material is softer or even highly plastic, and gravity effects can then generate **slump sheets** (olistostrome). If the deformation mechanism is gravitational slippage on a plastic floor, then these can be called **gravity nappes**.

Some large mountain chain sheets like the flysch nappes in the Alps are cover delaminations uprooted from their original site. If the allochthone involves the deep basement, two granitic beds may lie atop each other as in the Himalayas. There is then a **collision** and the continental crust thickens.

Such thickening occurs in all chains, and is sometimes followed by relaxation and extension (as in the Basin and Range) and stripping of the uppermost levels by erosion, revealing the deeper metamorphic core complex. The mechanisms that produce these situations are poorly known and are currently being studied.

5.2.3 Successive Paleostresses and Deformation Dating

Any early part of rock deformation belongs to diagenesis, and the later parts to **tectogenesis**. Dating these deformations and the processes that generate them is tricky at times, but very important. The following discusses a few basic ideas on this subject.

5.2.3.1 Synsedimentary Tectonics

A. Basins in Extension and Compression

Chapter 2 tried to show the various geodynamic situations that originate sedimentary basins. The **paleostresses** at various scales will differ completely depending on whether extension predominates, with rifting, continental margin, and pull-apart, or compression due to plate convergence. Normal synsedimentary faults, for example, will develop in the case of extension, while compression will construct a piggyback basin with fault-related folds and reverse faults and the concomitant sedimentary variations. Generally, basin **subsidence** is the most observable form of the initial tectonics, while **uplift** is generally more difficult to characterize because it is usually accompanied and followed by erosion, which disrupts the original process. These slow continental movements used to be called **epeirogeny** (from *epeiros* for "continent"), as opposed to the **orogeny** (from *oros* for "mountain"), which is still used to define the formation of vigorous "orogenic" mountain chain relief. Deformation thus occurred at different stages in the history of the basins in the course of the Hercynian and Alpine orogeny.

B. Signs and Effects of Synsedimentary Tectonics

Faults are the most spectacular example of a delimitation between two blocks of differing sedimentary fill (Fig. 5.84). The end of the faulting process can be dated by the age of the oldest bed sealing the fault (see also Fig. 5.66).

Salt domes are another example of deformations inducing sedimentary variations on different scales: stratigraphic pinchout on the flanks and structural cups on the scale of a portion of the basin (Fig. 5.85).

Other very spectacular examples on the seismic scale are the **accretion wedge fronts**, piggyback basins, and the shale diapirism mentioned before.

The signs are sometimes much more discreet. When tectonics is active during sedimentation, it varies the deposit profile, i.e., the slope and, more generally, the sedimentary environment. The entire volume available for sedimentation is in fact modified. Chapter 4 on sequential stratigraphy spoke of **accommodation**, saying that the eustatic effect is important but that the tectonic complement is sometimes a major one, more than just a disturbance in the eustatic effect. Subsidence accelerations (or decelerations) in the great foreland basins have been known to control major sedimentary processes. Tectonic figures such as faults, folds, and diapirs localize the sedimentary systems as they form, and on another scale the tectonics in the broad sense arrange the facies in order (Fig. 5.86) by controlling the **relative sea level**.

The role of tectonics in **erosion** on the scale of a basin deserves greater emphasis. The volume of sediments is mainly the result of tectonics, which directly controls the **sedimentary input**. Even if there is no direct indicator of tectonics, for example, a generalized resumption of a siliceous detritic sedimentation means that there is a new stock of such clastic particles upstream, and therefore a **relief rejuvenation** (Fig. 5.87). Tectonics creates relief and slopes and thereby generates special features in the sedimentary flow. The development of sedimentary **breccia** is often a discreet mark of resumed erosion following a deformation.

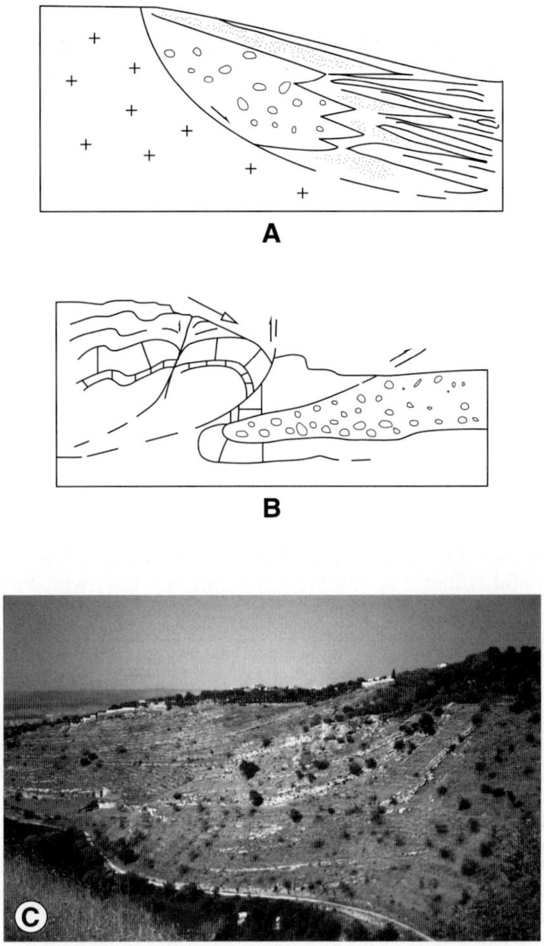

Fig. 5.84 Synsedimentary faults.
A. Development of a listric fault with breccia and correlative detritic deposits. **B.** Erosion of relief as the fault propagates, with the development of conglomerates picked up in the overthrust. **C.** Example of synsedimentary tectonics: fan of Plio-Quaternary deposits along an active fault in Sicily (IGAL photo).

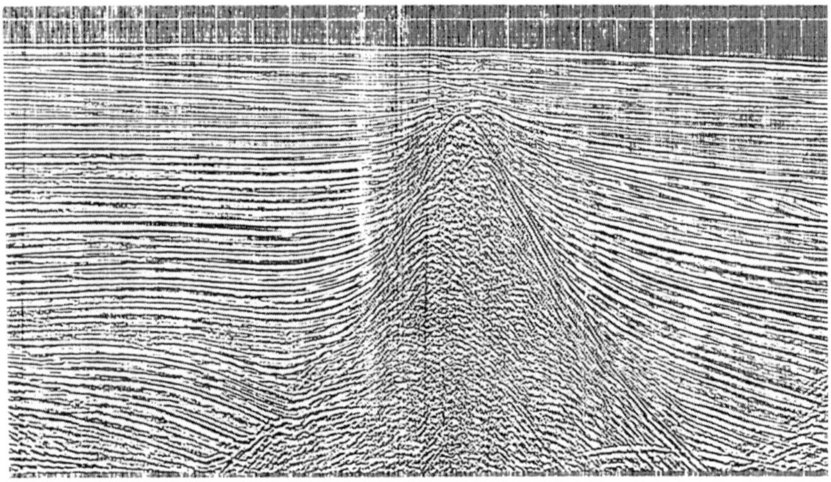

Fig. 5.85 Example of the synsedimentary effects induced by the rise of a salt diapir.
The peripheral collapse is observed in a rim syncline, with thinning and pinchout, and creation of slopes generating progradations. The most recent beds remain undisturbed (IFP photo).

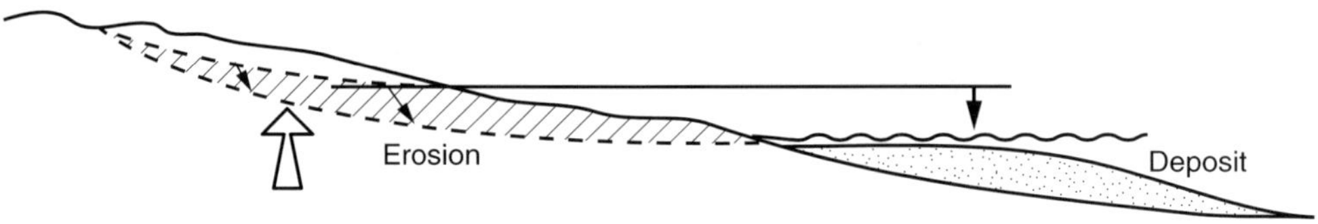

Fig. 5.86 Tectono-eustasy: evident impact on sedimentation.
The uplift of relief will result in a relative fall in sea level and change the fluvial equilibrium profile. A large part of the sediments will be carried toward the basin where the detritic deposits will leave evidence of this tectonic effect.

Faciological or **stratigraphic pinchout** is another form in which the deformation leaves a record on the small scale in a stable region (Fig. 5.88) or an unstable one, or on a larger scale such as along a fault or diapir (Fig. 5.85).

C. Examples of Synsedimentary Deformations

Slumps, water flow figures in extrusions or dragged stratifications, hydrolaccolith, and load figures have been mentioned at the beginning of this chapter on diagenesis.

A

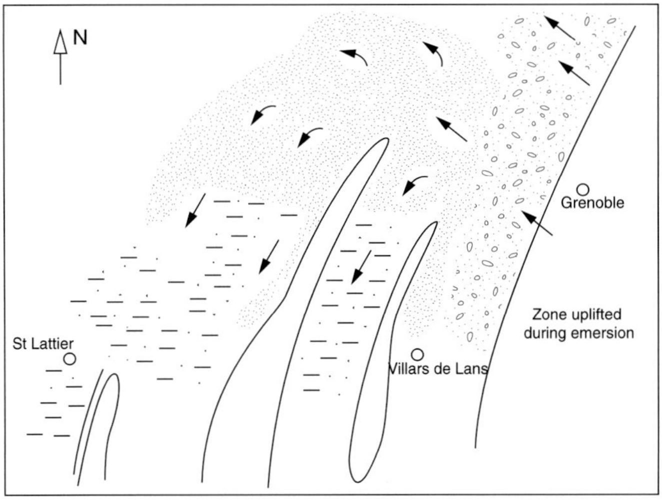

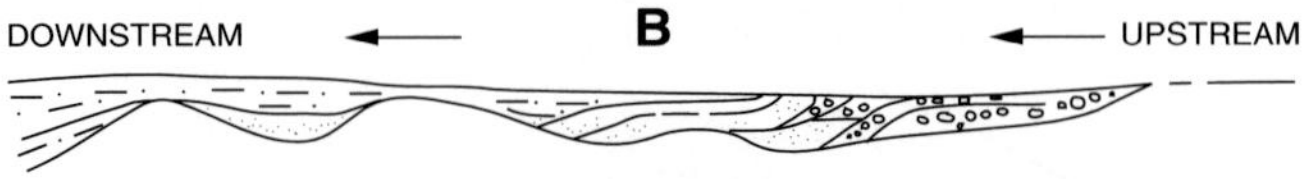

Fig. 5.87 Tectono-genesis and syntectonic detritic deposits.
A. Deposit distribution map (conglomerates, sandstones, siltstones) in the Dauphiné region during the Alpine uplift in the Middle Miocene. **B.** General schematic section. **C.** Distal delta gritty and silty facies (St Lattier).

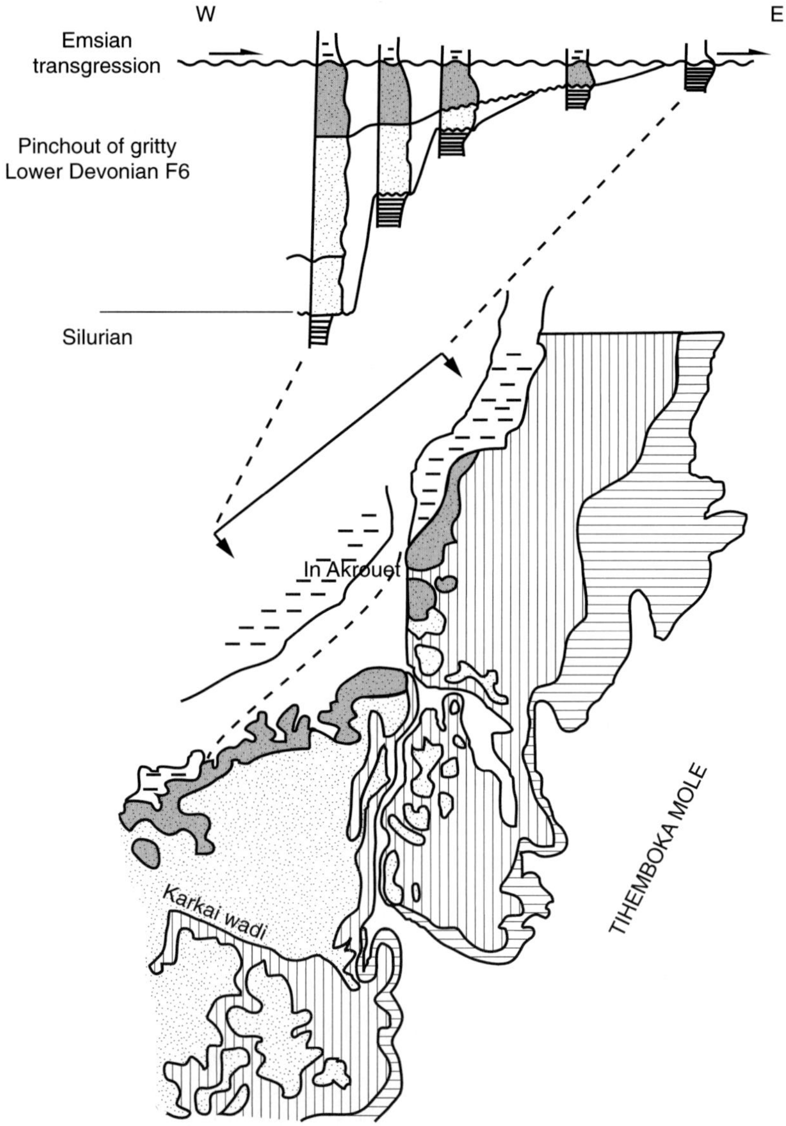

Fig. 5.88 Stratigraphic pinchout.
The thinning of the Early Devonian sandstones in the Tassily of the Ajjers (Sahara) is expressed in a sequence of sections where the Emsian clays can be seen onlapping the Silurian gritty shale with a gradual disappearance of sandstone (F6). This can also be seen on the map between Karkai wadi and In Akeouet (from Beuf et al., 1971).

Growth faults are listric faults that develop on all scales at the subsident basin rim, especially in major delta formations (Fig. 5.89). They form very attractive oil structures (Niger, Alaska, Gulf Coast). They are characterized by a special stress regime that does not involve the crust, and can operate over variable time periods.

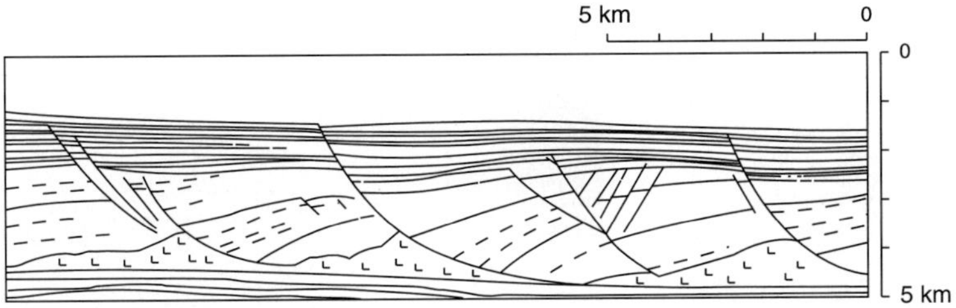

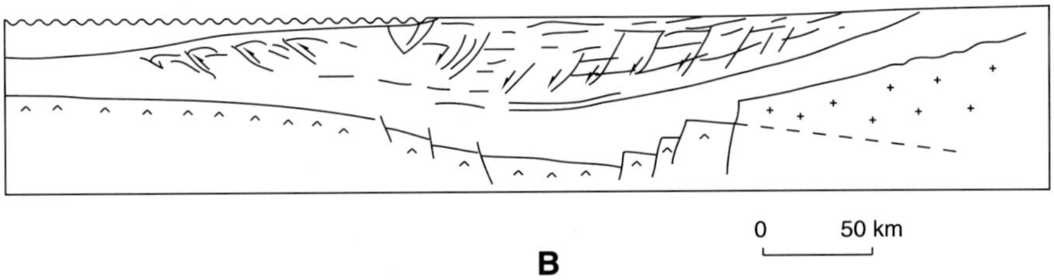

Fig. 5.89 Listric growth faults.
A. Example of the Plio-Quaternary series of the Gulf of Lions, deep Rhone delta. The faults here take root in the Messinian salt level. The infrahaline series are not involved in deformation. **B.** Example of the deep Niger delta.

D. Resedimentation by Gravity

Olistolites are objects ranging from a centimeter to a kilometer in size. They are resedimented in an unstable environment. The largest bodies are termed **sedimentary klippes** by analogy with the tectonic klippe, to indicate a large allochthonous transported mass. These can be found isolated at the foot of a normal fault escarpment or dispersed along the front of an overthrust (Fig. 5.90). Generally, they are linked with the faulting process.

Olistostromes, as was seen in Chapter 3's discussion of gravity-driven deposits, are chaotic accumulations of many olistolites in a poorly stratified argillaceous matrix. Abundant examples exist at the foot of folded chains like the Apenine, Sicily, the Betic Cordilleras, and the Caribbean (Fig. 5.91). Special **block series** and **melanges** have been found of such tectonic sedimentary records. It is not always easy to distinguish them from certain cohesive debris flows, because the gravity mechanism is the same. The matrix in the olistostrome

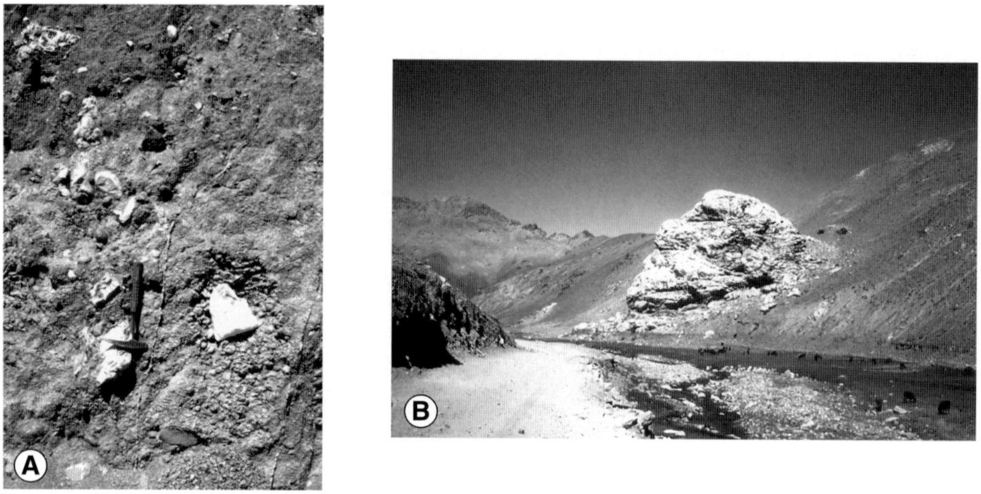

Fig. 5.90 Examples of olistolites.
A. Small calcareous fragments (angular, breccia) resedimented in an argillaceous-arenaceous matrix. Eocene of the Sierra El Numero, Dominican Republic (IFP photo). **B.** Enormous block enclosed in the Waras flyschoids in Afghanistan (IGAL photo).

itself exhibits slump deformations, but often the olistostrome's synsedimentary deformation is itself taken over by the continued compression with shearing, or by continued slippage of the gravity sheet. The tectonics can be dated if the fauna in the matrix itself is determined.

Synsedimentary tectonics is generally very discrete in large cratonic basins where it is localized along particular structural lines and is usually recorded in the sedimentation with no spectacular deformation.

5.2.3.2 Later Tectonics, in the Strict Sense

Once a functional sedimentary basin has undergone a geodynamic evolution marked by lithospheric or surface stresses, it enters a quiescent phase during which it becomes a residual basin (with little subsidence), and then a structural (deformed) basin. That is, the tectonics become such that their original operation is entirely eradicated. Some basins retain their cup shape, which is often quite different from their initial morphology (such as in the Paris basin), while others are largely tectonized (Alaska), and still others are transformed into folded chains (Alps), entering the domain of **orogeny** (Fig. 5.92).

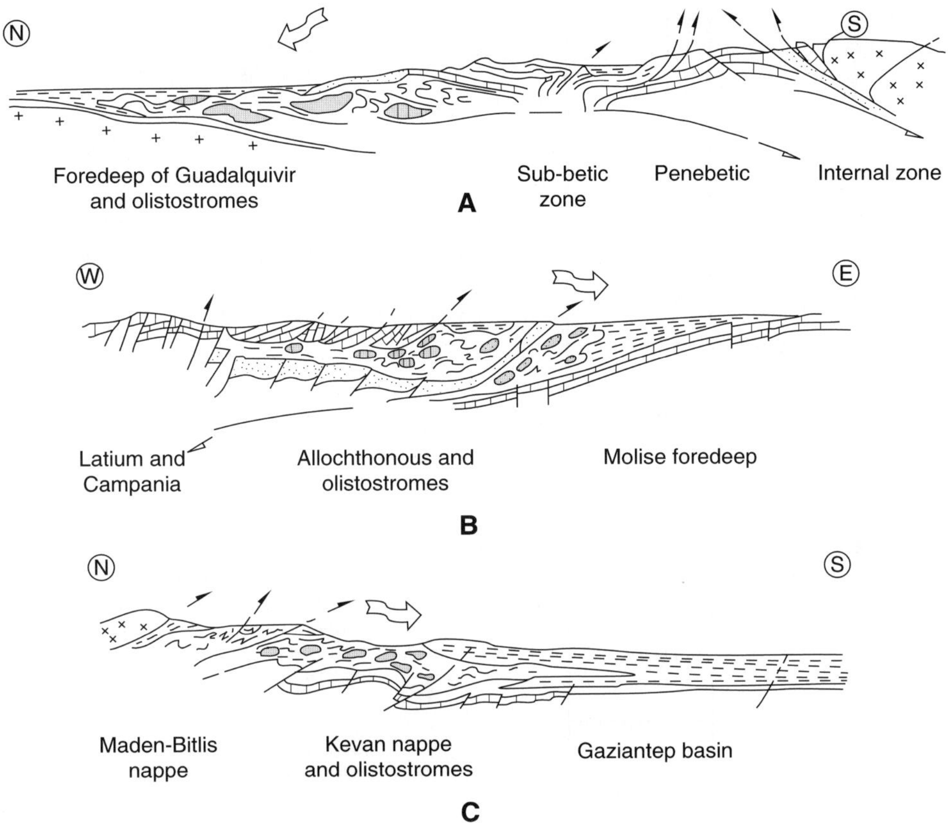

Fig. 5.91 Foreland of folded chains where the olistostromes are installed by gravity slippage in the subsident basin.
A. Betic Cordillera front. **B.** South Apenine front. **C.** Southern front of the Taurus (from Biju-Duval, 1974).

A. Tectonic Phase

The **tectonic phase** idea is an important one. It encompasses all of the deformations that will, after a long and relatively calm period of a few tens of millions of years, affect a number of basins in large segments of the Earth's crust over a relatively brief lapse of time (a few million years). The age of the tectonic phase is defined by the first beds that are not affected by the deformation, and are in discordance and thus posterior to it.

The Caledonian phase (end of the Silurian), the Hercynian (end of the Carboniferous), the Pyrenean-Provence phase (Middle-Upper Eocene), and the Alpine (Miocene) thus define different folding phases that have affected Western Europe (Fig. 5.93). A phase is generally

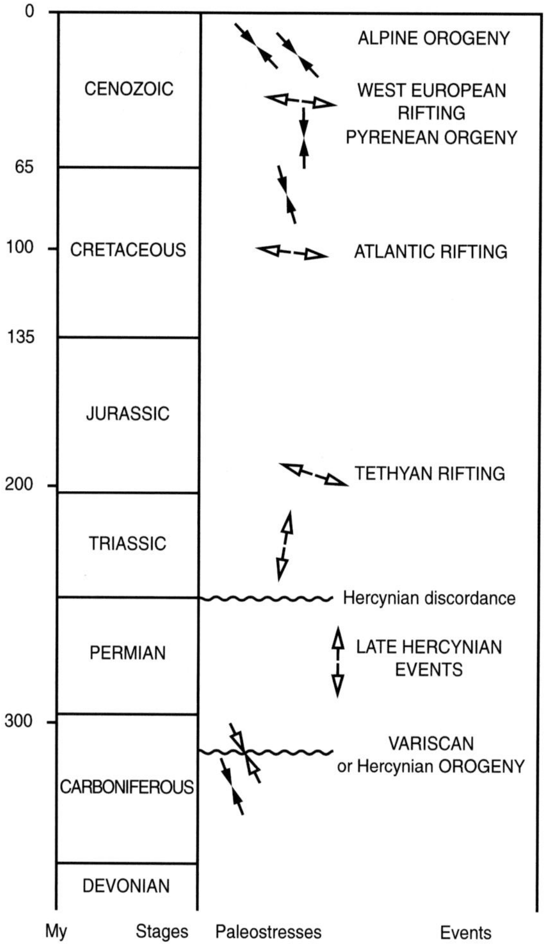

Fig. 5.92 Sequence of orogenic phases and tectonic events in the Alpine domain. The periods of compression (converging solid arrows) are separated by periods of extension (diverging open arrows).

named after the region (such as the Alpine, Andine) where it was defined, but there are also a number of local names to watch out for.

On such a time scale, this may be referred to as an **orogenic cycle**. After the sedimentation period in a subsidence area, the deformation paroxysm closes the basin history. This idea is used in the **Wilson cycle** concept in plate tectonics.

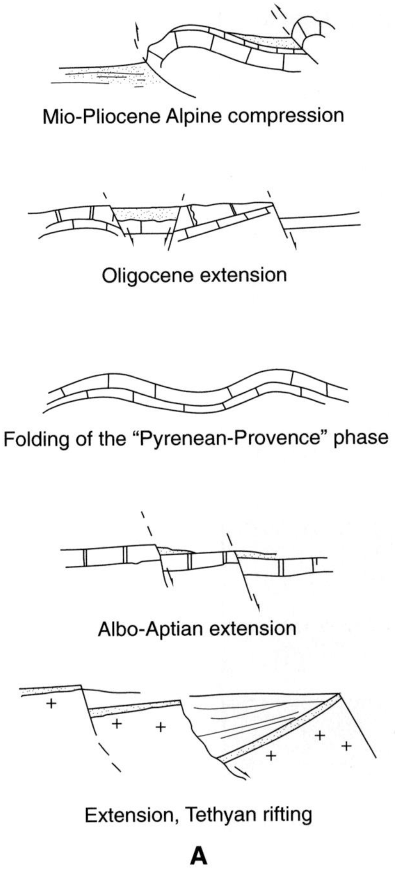

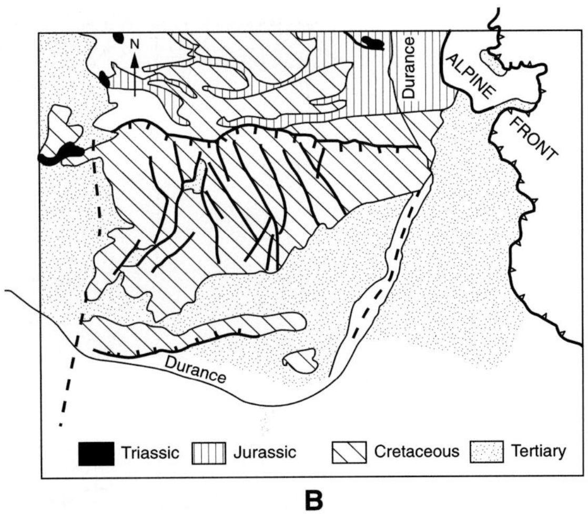

Fig. 5.93 Sketch of paleostress sequences recorded in structures (deformations).
A. Scheme of phases recognized in the southeast basin of France. **B.** Cartographic expression. The Tethyan rifting is not apparent, but the north-south compression has generated large east-west anticlines (Luberon, Ventoux-Lure, Baronnies). The Oligocene extension is characterized by the faults of the Durance and Valreas. The Alpine front is evidence of the Late Alpine deformation. The Triassic diapirs in black reflect the mobility of the salt.

B. Major Discordance

The idea of **major discordance** goes along with that of the tectonic phase. After folding, very extensive territories are eroded and the relief is gradually flattened (leading to the general term of peneplanation or baseleveling). Large volumes of material are oblated and resedimented in another basin. New deposits can then be made on this continually evolving new topographical surface modeled under the prevailing climatic conditions, and may be conserved as long as another period of subsidence occurs (new tectonic cycle). These new sediments will be discordant on the erosion surface. This is true of the Upper Triassic beds on the Hercynian basement, and the Cambro-Ordovician beds on the African shield.

The elementary expression is the **cartographic discordance**. The angularity of the contacts betrays the **angular unconformity** that can at times be seen on the scale of the outcrop or seismic line (Fig. 5.94).

Fig. 5.94 Angular unconformities.
A. The example of a major discordance after orogeny and a great deal of erosion. These are Lower Liassic beds overlying mica schist (IGAL photo). **B.** Angular unconformity in an intermontane basin of the Alpine chain. Here, the Miocene overlies tilted beds of the Mesozoic (Guercif basin) (IGAL photo). **C.** Sedimentation discontinuities. There is an angularity in the formation again, but due to the cutting (and filling) of a channel, with no tectonic deformation. Lower Cretaceous on the edge of the Vercors zone.

The discordance surface may correspond to a hiatus of several million years, regardless of the intensity of the discordance, because this is independent of the hiatus duration. Some very small discordances may correspond to very long periods of erosion and denudation, while other shorter discordances correspond to a very short period of time. Everything depends on the amplitude of the deformation.

C. Gradual Deformation

This simple perspective on tectonic phases should be explained and qualified somewhat. That is, in a modern geodynamic perspective, the evolution of the stresses and resulting deformations have to be determined. Are the tectonics discontinuous and of the event type, which would hint at the idea of phases, or are they rather continuous? Field data pleads in favor of the following distinctions.

It has been shown from the study of **accretion wedges** in **plate convergence** domains that the deformation resulting from the stress of horizontal shortening is a gradual one. This **gradual deformation** migrates normally and gradually from the deformed zone toward the undeformed foreland (Fig. 5.95). The sum of the discontinuous deformations constitutes a sequence with time variations. While the deformation is active at the front, it has already been sealed in the rear with piggyback basins in discordance, becoming older and thicker going upstream. The time differences between these deformations is further evidenced by the existence of accidents out of sequence. Can this still be called a phase? If the stratigraphic resolution power is not high enough, it will probably be difficult to distinguish these regularly ordered sequences of events.

This idea of a sequence of events relaying each other in time and space can again be found on the scale of a basin or the rim of a mountain chain. **Unconformities** and **discontinuities** are superimposed or imbricated on different scales, such as the Cretaceous and the Eocene of the Alps. Not all of them are necessarily of great geodynamic value, and some are known to be local in character, such as the Senonian of Devoluy (Fig. 5.96).

Here, another look needs to be taken at the deformation drive mechanism itself: the lithospheric plate motion that determine the **stress field orientations**. While it is true that the plate motion is continuous and generates a deformation continuum at the boundaries (such as accretion wedges), the blockages that occur during collision also have to be considered, as do the changes in the relative motions of two plates or rapid variations in the subduction rates. Each of these phenomena may cause what geologists call an orogenic phase, which is an acceleration of the process in a relatively short period. For many geologists, this phase would in fact be evidence of continental collisions. In a way, the tectonic phase is the punctuation ending a sentence, and the stress distribution will change radically in the next.

D. Paleostress Records

Continuing with the above approach, in which the major tectonic phase is seen in the continuum of the plate motions and deformations it generates, the next step is the physical record

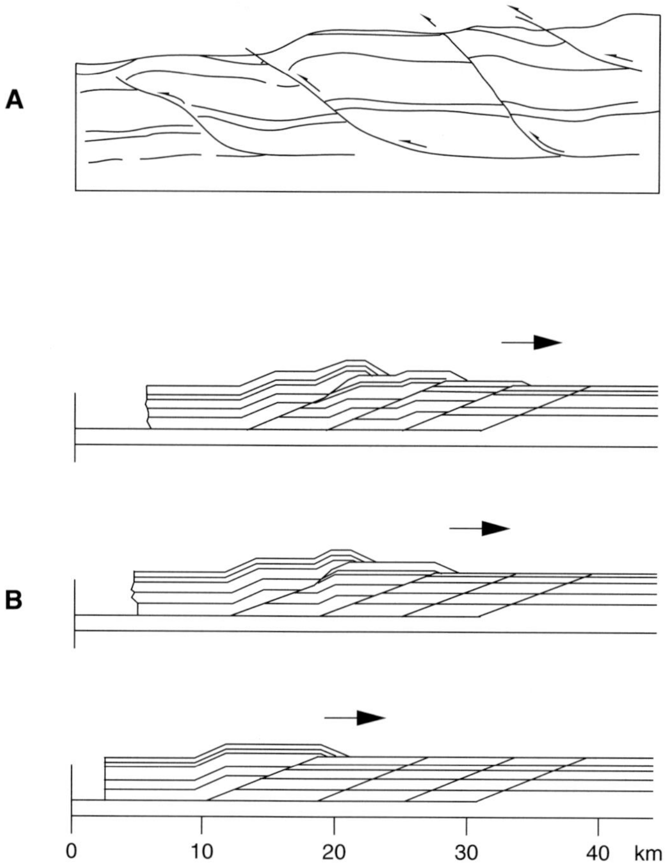

Fig. 5.95 Gradual deformation of sediments.
A. Successive slices (or duplex) advance in time, starting at the décollement level, at the front of the accretion wedges, in sequence, such as along the Nankai trough or in the Barbados. **B.** Modeling of the frontal accretion (from Mascle et al., 1990).

of the stresses in the course of time. More precisely, these major tectonic phases are recorded by microtectonic indicators showing the orientation of these paleostresses. The great tectonic events affecting the Tethyan domain from the arc of Gibraltar to Turkey during the Eocene were perfectly recorded this way. The general north-south compression corresponding to the relative movement between the African and the Eurasian plates can be detected not only in the chains and its immediate boundaries, but also in the adjacent sedimentary basins (Fig. 5.97).

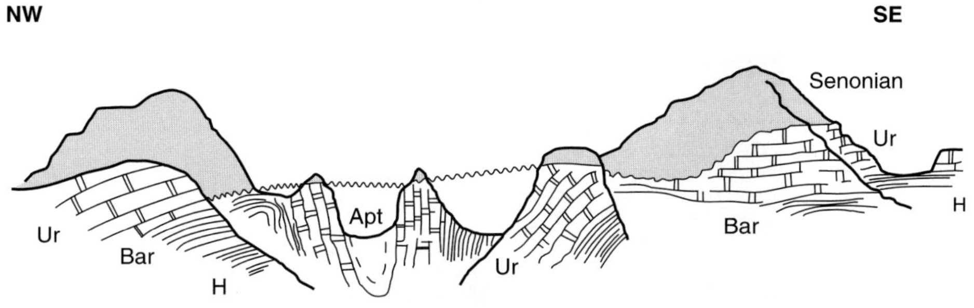

Fig. 5.96 Discordance of the Senonian on the folded Lower Cretaceous in the Lus needles of Croix Haute.
Apt: Aptian; Ur: Urgonian; Bar: Barrenian; H: Hauterivian (from Debelmas et al., 1970).

The **microtectonic indicators** of the compressive or extensive process are the orientation and direction of striated fracture planes, stylolites, mineral growths like calcite, and tension cracks, which can reveal either a compressive or an extensive process.

E. Multiphase History and Tectonic Inversion

The geological process of building a mountain chain out of a basin can hardly be viewed as **monophase history**.

Different terms are commonly used to indicate the diversity of sequential events in a given chain. Early or late events are designated **paleotectonic** and **late tectonic**, respectively, often with a fuzzy distinction between the two. Another term that requires a more precise definition is **neotectonic**, which originally referred to renewed deformations affecting an already-formed orogeny (for the Alpine chain). This term has often been taken to mean extensive fault tectonics, but it is clear that the most recent deformations of a given chain may include different normal, reverse, or strike-slip motions. It is often used today to mean something like **seismotectonics**, where breakup, extension-compression, or strike-slip are determined by the type of **focal-mechanism solution**.

Other terms have also been used for defining a **multiphase history**. In the Alps, the **multiphasing** evidenced in the superimposition of several major deformation events within a continuum of shortening is broken down into **eoalpine** (Late Cretaceous/Paleocene), **mesoalpine** (Late Eocene) and **neoalpine** phases. This can of course lead to some confusion (the Eoalpine in Yugoslavia's Dinarides is from the Jurassic!). However, local habits will have it that terms like Pyrenean-Provencian, Alpine, or Laramide, are still used. The chronological reference should be specified each time.

Paleostresses are generally analyzed in a chronological sequence of many extensive and compressive events, some of which are relatively minor and others of major significance both in the chains and in their adjacent basins (Fig. 5.97).

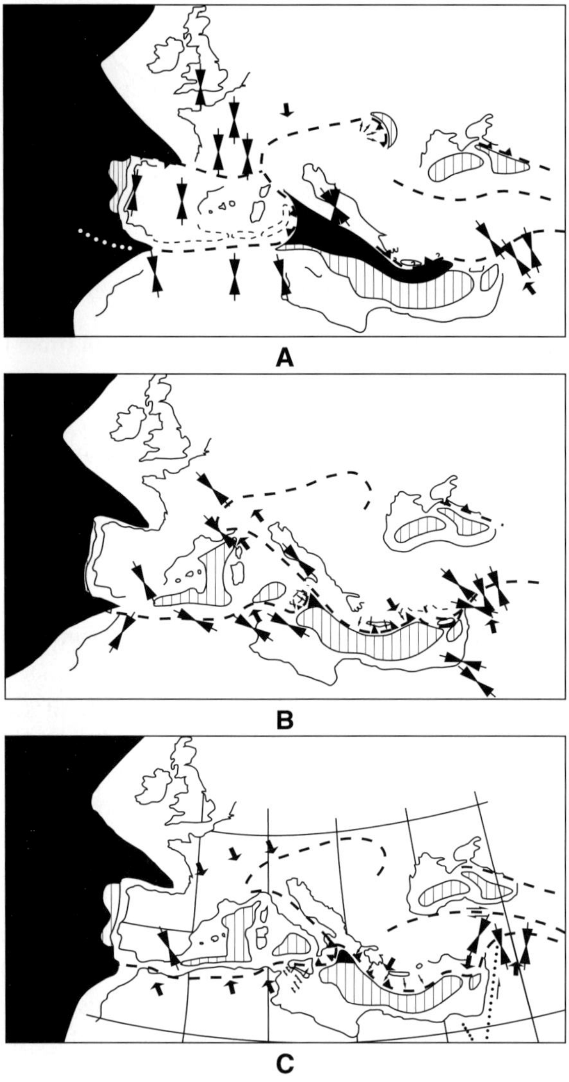

Fig. 5.97 Evolving orientation of paleostresses in the Perimediterranean domain.
Simplified general diagram (from Letouzey and Trémolières, 1980).

It is extremely important to reconstruct this process of **successive paleostresses**, especially for adjacent sedimentary basins, because it can explain the phenomena of **tectonic inversion**. **Inversion** means a reversal in the faulting motion, and more generally in the motion of the geological structures on the scale of the basin. A normal fault subjected to a compressive stress will begin to operate as a reverse fault, or a subsident basin will be uplifted and form a positive structure (Fig. 5.98). Reversal is thus one of the phenomena due to multiphase history. The displacement of strike-slip faults, often evidenced by **flower structures** (Fig. 5.99), is another expression of this tectonic rearrangement.

When the period considered is long enough, multiphasing will lead to **superimposed basins**. The southeast basin of France is a good illustration of this, with the Vocontian Mesozoic basin deformed by the early Alpine compressions and with the Valence basin established after the Oligo-Miocene extension, which was itself affected by the more recent compressive Mio-Pliocene deformations.

5.2.4 Mountain Chains and Adjacent Basins, Orogeny

Until recently, the veteran concept of the **geosyncline** was used to define the area of sedimentation, burial, and subsidence that was later to undergo reversal and deformation, folding, and uplift to form an **orogeny**. It was thought that **metamorphism** developed in the deepest parts underneath this, because of the high temperature and pressure conditions. Different authors thus distinguished between different types of geosynclines and chains, and chains were generally considered as participating in the **growth of continents** in a series of orogenic cycles since the Late Precambrian.

With today's global perspective of plate tectonics as presented in Chapters 1 and 2, mountain chains and the associated fold basins can be grouped into a few broad categories defined by the geodynamic framework. The oldest chains of the Achaean, which are still the subject of various theoretical explanations, will not be discussed here because they are of no importance to petroleum geology. We will simply summarize the major types of orogeny and extend them to the basins associated with them (refer also to Chapter 2). It should be remembered that, in one way or another, the chain structures always involve the base, the continental or ocean crust, and the sedimentary cover.

5.2.4.1 Intracontinental Chains

The thick continental crust can be affected by deformation. This generally occurs along major crustal lineaments inherited from a previous history and then reactivated, for example, by rifting or in aulacogens, or strike-slip on the continental scale after a collision. An **intracontinental chain** is an edifice that is folded during episodes of compression that affect this continental crust. There is, of course, a direct relation with the convergence of the plates, and often with the other types of chains too.

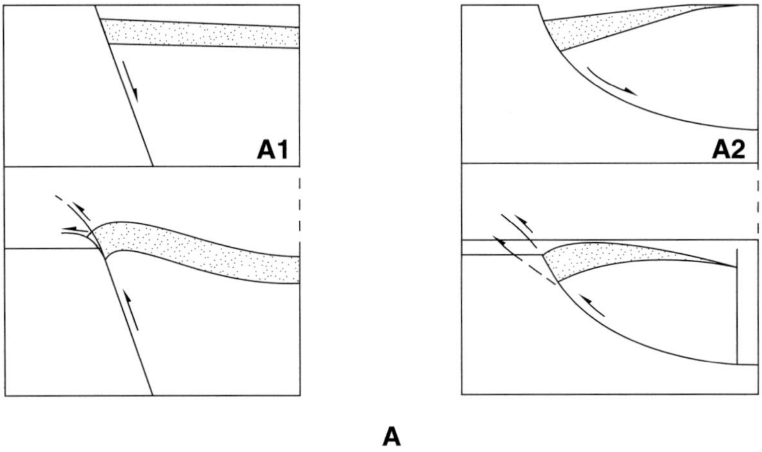

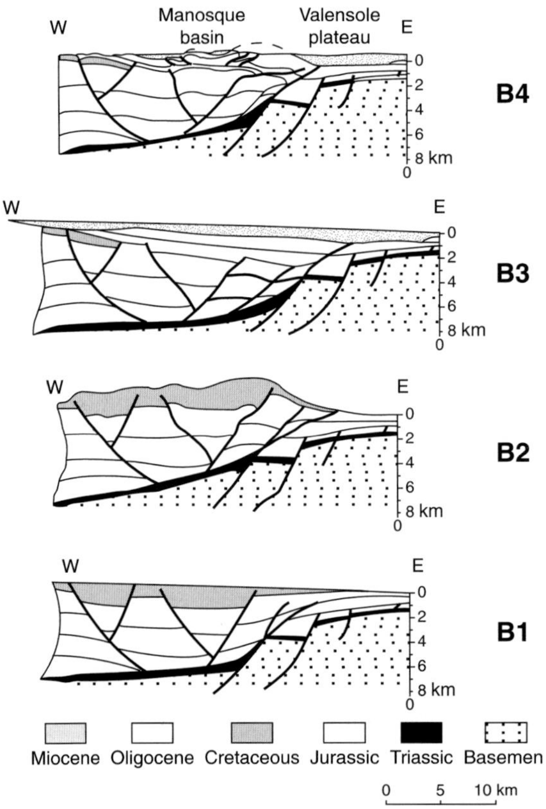

Fig. 5.98 Tectonic inversion.
A. Normal plane fault (A1) or listric fault (A2) inverted by compression. The initial plane is used again, but neoformed short-cuts can appear. **B.** Reverse and multiphase tectonic structure: the Durance fault. (B1) The Triassic-Liassic rim is rifted with normal faults at the Albian Cenomanian. (B2) The whole system is folded in the Eocene. (B3) The system is extended again in the Oligocene. (B4) The Alpine orogeny reactivates the whole in compression.

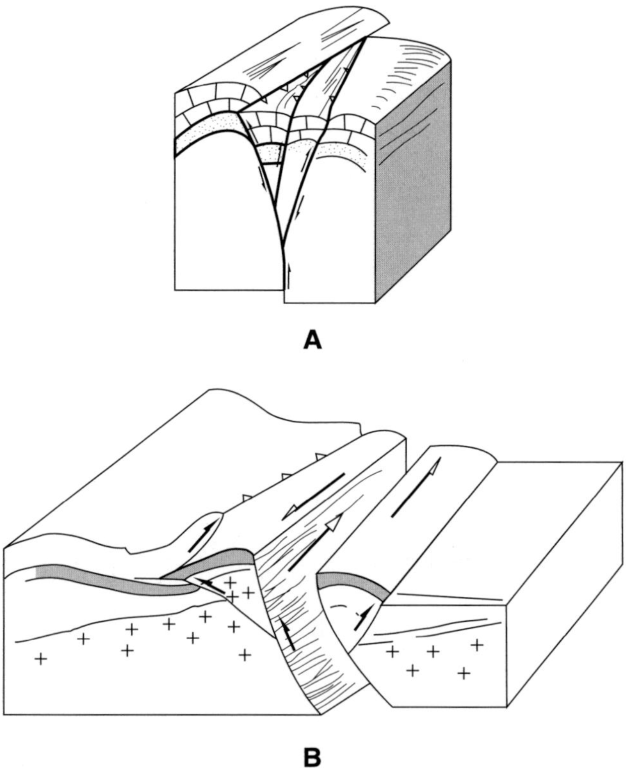

Fig. 5.99 Flower structures.
The shear faulting in transpression mode can create a fan of faults or positive flowers as seen in the cover (**A**), but often rooted together in the basement (**B**).

Intermontane chains are of various types:

- Relatively **linear bulges** largely involve the basement. There are sometimes overlapping double-discharging structures of opposite vergence, which may be the result of crustal lineaments reactivated in reverse. This is how north Africa's Atlas chains are built, for example (Fig. 5.100), with major variations and décollements of the cover (Atlas front). The folding involves thick sedimentary series, and multiphase strike-slip faulting can complicate the edifice, as is true in Venezuela's Andes de Merida. These chains are often defined as reversal chains.
- **Folded cover** is separated from the base, as in the Palmyrides and Negev chains. These are shifted by the intracontinental Levant transform fault, which itself generates the Lebanese and the Ante-Lebanese chains (Fig. 5.101).

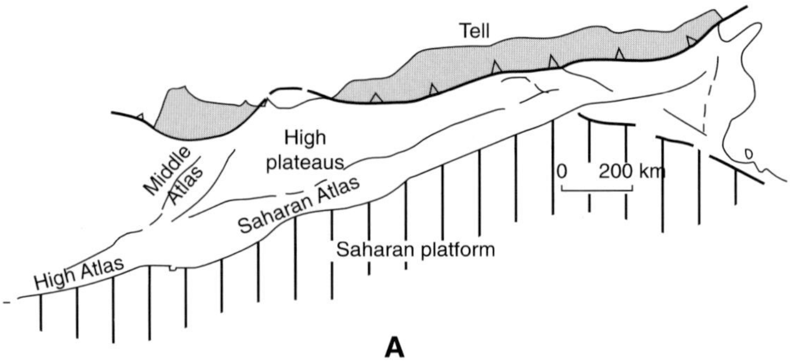

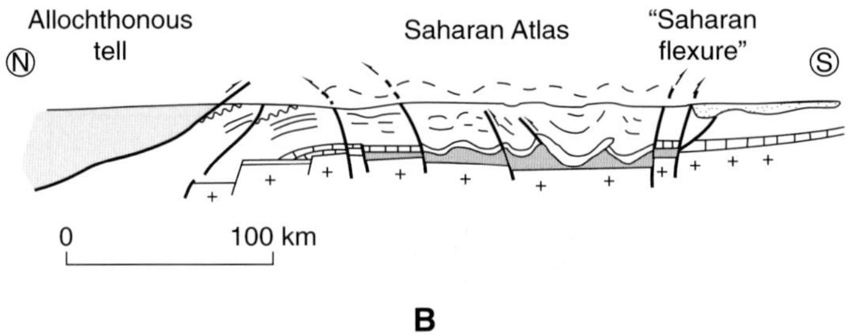

Fig. 5.100 The Saharan Atlas chain: example of an intracontinental chain resulting from a reversal.
A. General map. **B.** Meridian section in the central portion (from Letouzey et al., 1995).

- **Vast intracratonic folds** involve extensive continental domains. One example is the Saharan Paleozoic with its folded anticlises and syneclises, segmented by major basement accidents (Fig. 5.102). Many ancient petroleum basins fall within this category, with more recent tectonic faulting reactivating ancient structures. (Vast intracontinental arches can be formed by thermal effect without major deformation, as in France's Massif Central, which has been uplifted since the Miocene over a lithospheric plume).

This is fully within the continental domain. The boundary is fuzzy between vast intracratonic folds and **broad folded structures** that cannot be called chains insofar as they do not or have not generated major relief. These are still remarkable tectonic systems on the cratons, stemming directly from an effective subduction process at the craton edge. This is true of all of the **foothill folds** in flexural basins. The example of the Middle East with its spectacular and **immense arches** is simpler than the Western rim of the American Cordilleras (Fig. 5.103) or Alaska, where many slices are stacked up by décollement into **foothills** at the front.

All of these foothill folds are the external element of collision chains.

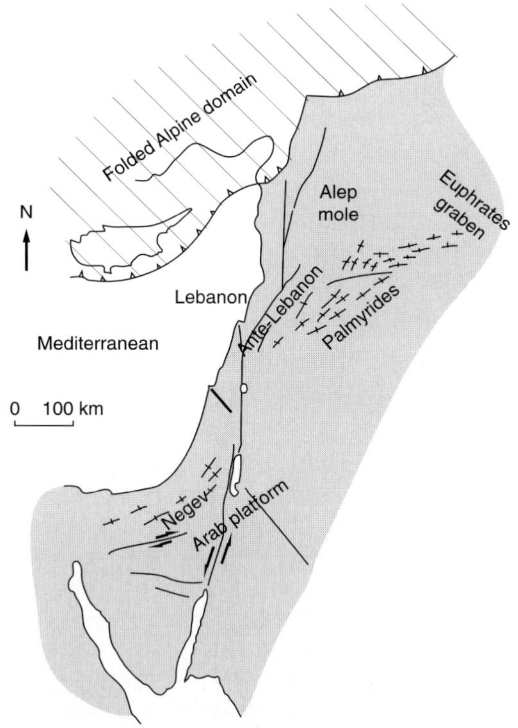

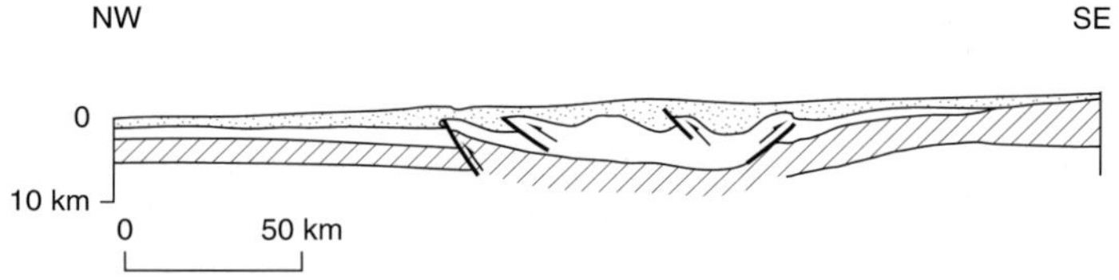

Fig. 5.101 Lebanon, Ante-Lebanon, Palmyrides, Negev, and Levant fault. **A.** Map: mount Lebanon and Ante-Lebanon are independent structures to either side of the Great Levant strike-slip fault, making independent anticlinal structures of the folded Negev and Palmyrides systems. **B.** Schematic section of the Palmyrides (from McBride et al., 1990). Note the heavy thickening of the Mesozoic and the Paleogene (white and dotted portions). The multiphasing of the folds (end of Mesozoic and Mio-Pliocene) is not shown.

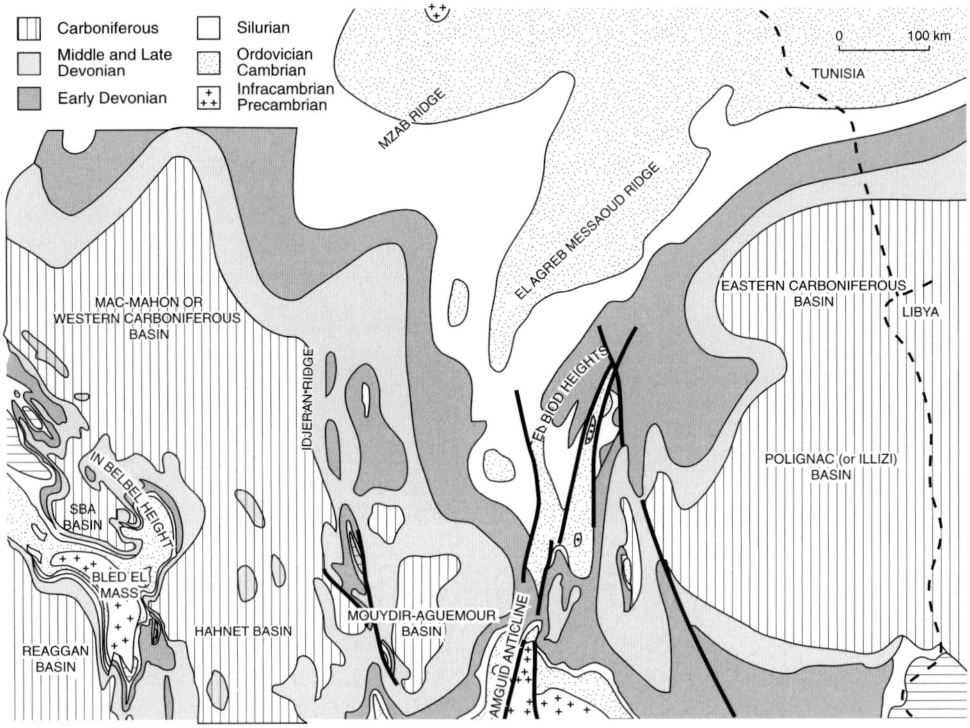

Fig. 5.102 Peel map of the Ante-Mesozoic in the Algerian Sahara.
A. Great basins (syneclises) are separated by basement accidents and ridges, oriented roughly north-south (from Beuf et al., 1971).

The case of the Pyrenees is a special one, with a history linked to the oceanic expansion of the Gulf of Biscay. A large part of the edifice constitutes a true intracontinental chain (between the Iberian and Eurasian plates) with lithospheric collision at its oceanic end (the structure is still poorly known) (Fig. 5.104). The system once operated in a typical basin **inversion** of previously thinned continental crust, with extensive shortening and double discharge driven by a lithospheric subduction mechanism. The highly tectonized surface basins may have been carried piggyback.

5.2.4.2 Chains Resulting from Subduction Processes

The **orogenic belts** formed at the Earth's surface during the major orogenies were each constructed on a continental margin in plate convergence zones where active subduction developed. These chains are **megastructures** of varying ages: Caledonian, Hercynian, or Alpine depending on the major orogenic cycles. The **chains** themselves, on which **intermontane** or

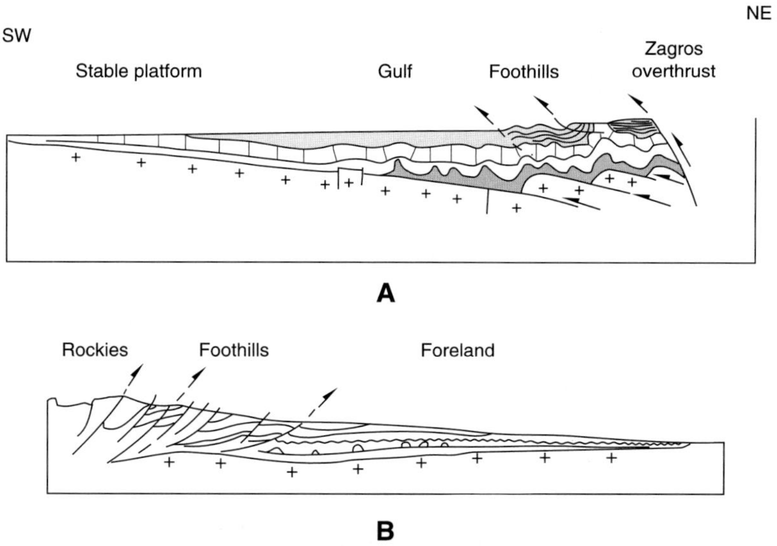

Fig. 5.103 General sections of the Middle East (**A**) and the Canadian Rockies (**B**).
A. The stable platform is deformed little, but foothill folds develop in the flexural foreland at the Zagros overthrust front. **B.** The stable platform includes Devonian reef structures. The foothill area consists of a series of slices. (from Perrodon, 1985).

episutural basins have developed, are different from mobile foothill belts resulting from the deformation of **flexural** or **perisutural basins** (Fig. 5.105).

These structural edifices can be classified in a number of categories depending on the type of crust involved in the subduction, but it should first be pointed out that arc **volcanism** is of major importance in these chains (Fig. 1.22).

A. Chains Stemming from Oceanic Subduction (Subduction B)

Oceanic subduction of the "active margins" can affect the oceanic domain (such as the Mariannas-Bonnin, Tonga, Kermadec, Lesser Antilles arc) or the continental borderland (Japan, Indonesia, Andes) (Fig. 5.106). The former case is of little interest to petroleum geology, but the latter is of greater importance because of the amount of sedimentary filling in the basins that develop, either in front or behind the arc. Depending on the geographic position, different types of structures—overthrust folds and extensive structures—may be attractive for exploration. Chapter 1 mentioned that the Pacific borderland exhibited either the development of a great marginal basin in extension (Japan), or continental underthrust (Andes) with a flexural type foreland.

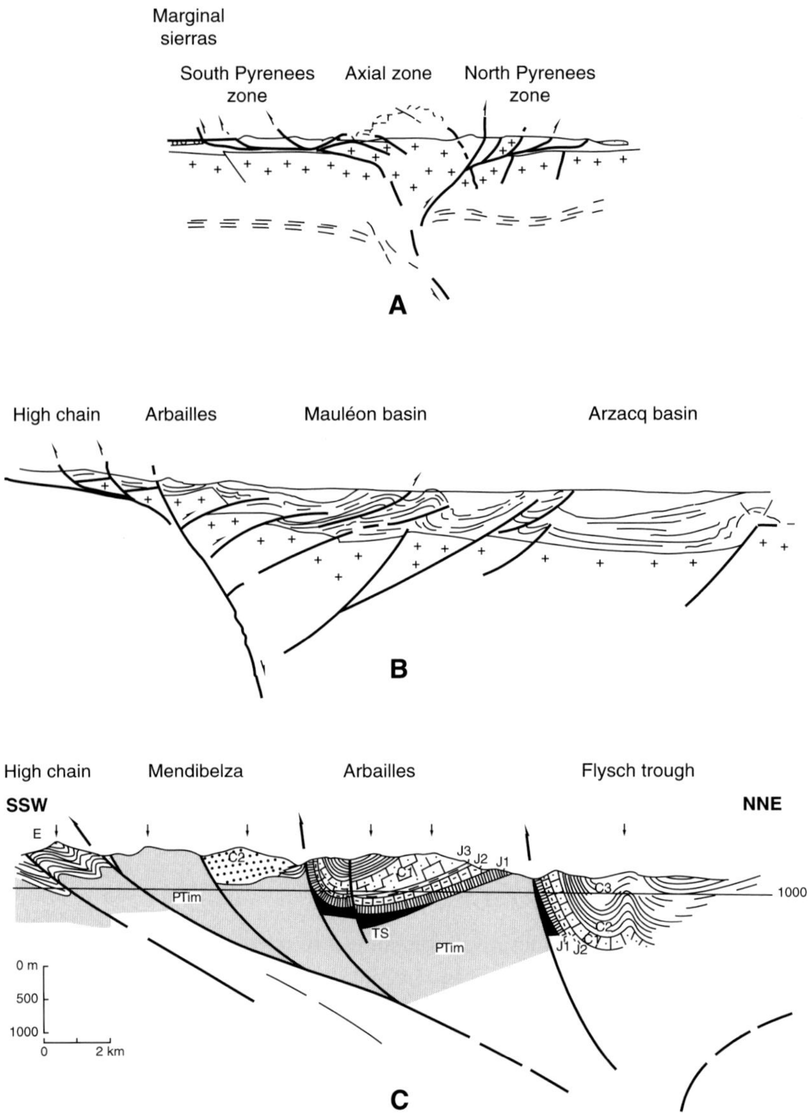

Fig. 5.104 Double tectonic front of the Pyrenees between Iberia and Europe.
A. General lithospheric section (from ECORS profile). **B.** Section to the west of the chain where the tilted blocks of the Mesozoic rifting can be recognized. **C.** Details of the Arbailla zone (from Canérot, ENSPM document).

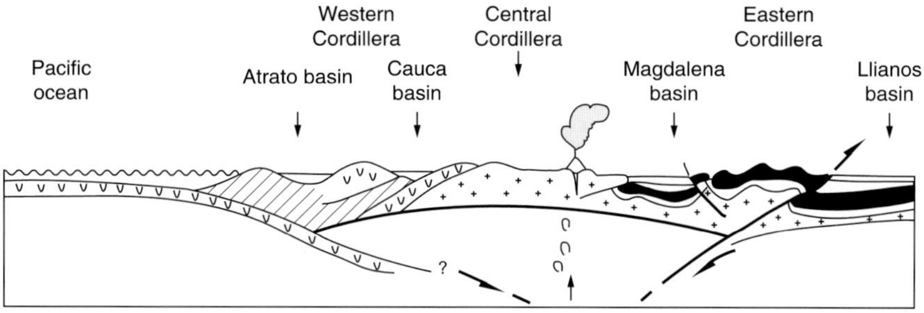

Fig. 5.105 Episutural or flexural basins associated with mountain chains. The example of the Columbian Andes basins, which are recent basins deformed by the Plio-Quaternary uplift of the Andes (from Colletta et al., 1990).

B. Chains Characterized by Obduction

Peel thrusting of the oceanic crust may not always be evidence of subduction, but rather of an overthrust foreland zone. This is an obduction process where basins can be found covered with ophiolitic nappes, as in Oman (Fig. 5.107), Papouasie and New Caledonia. Highly tectonized sedimentary series in slices or nappes can be found under the oceanic crust overthrust.

C. Collision Chains (Subduction B and Subduction A)

If the subduction process continues long enough, a crustal **collision** may occur either between two continental crusts (Alpine and Himalayan arcs) or between an arc (or portion of it) and the continent (Andes, Taiwan, American Pacific chains).

The crustal collision is a **continental collision**. Once the oceanic crust is absorbed (subduction B), a continental subduction A occurs with the greater and greater lithospheric plunge of the continental crust. The oceanic series either disappear completely by subduction or are ejected in superimposed nappes, leaving the ancient continental rim highly faulted. Only the foreland will conserve allochthonous or a parautochthon of folded sedimentary series, while the deformation propagates very far onto the adjacent autochthonous platform (Fig. 5.108). This type of chain used to be called **biliminar** to indicate overthrusts of opposite vergence to either side of the axial zone. But this arrangement varies greatly depending on the initial geometry of the plate boundaries. Small intermontane basins may develop, or a generalized back-arc type extension with a development of large episutural basins (Pannonian basin, Aegean basin of the Alpine arc) or, as in the case of the Himalayan hypercollision, a lateral expulsion by **punching** of large continental blocks.

These examples of the relatively young Alpine arc are easier to identify than those of older orogenies. The Hercynian (or Variscan) chain so well studied in Western Europe is also a chain resulting from complex collisions between Laurasia and Gondwana

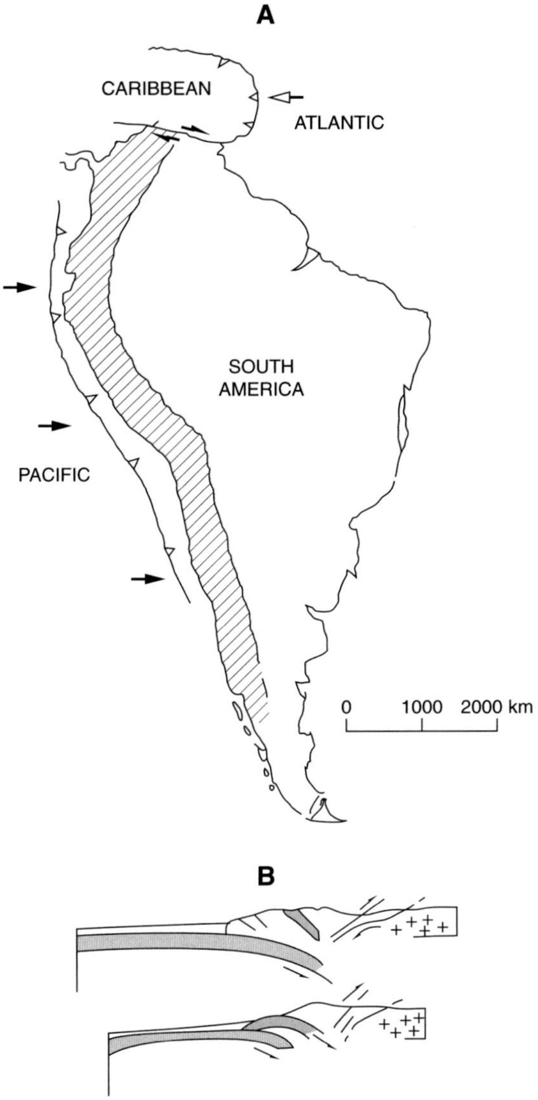

Fig. 5.106 Andes and Caribbean subduction chains.
The Pacific ocean subducts under the South American continent, creating the Andes Cordillera by a process that differs from north to south. To the north of the strike-slip accidents affecting the Venezuelan Cordilleras, the Atlantic ocean subducts in the opposite direction in the Antilles arc, giving rise to a string of volcanic islands. **A.** General map. **B.** Diagrams representing the subduction of the Andes and the underthrusting of the Brazilian shield.

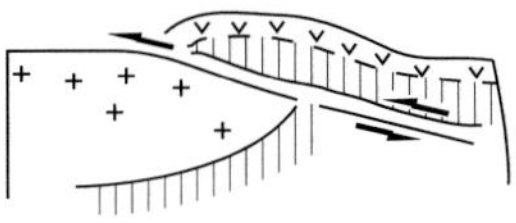

A

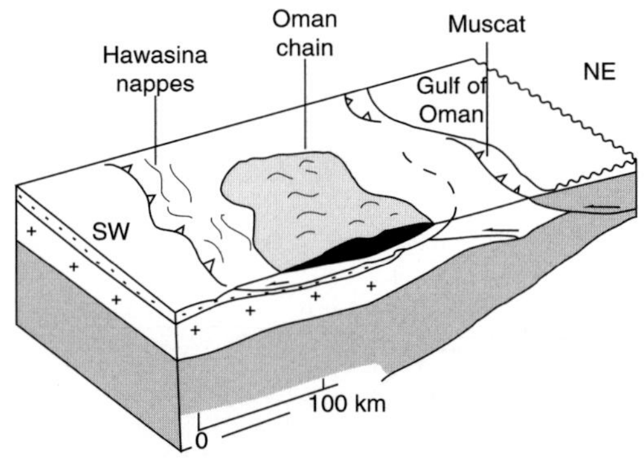

B

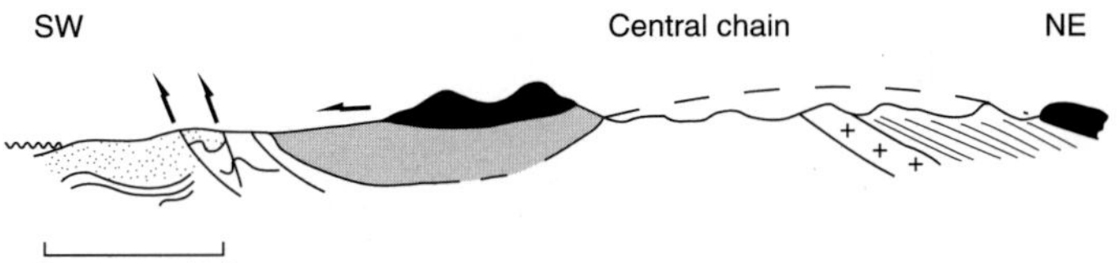

C

Fig. 5.107 Chains with oceanic crust obduction.
A. Theory. With the oceanic crust, the ophiolitic series, peridotites, and ultrabasic rock of the upper mantle are overthrust on the neighboring continental crust. **B.** Oman chains: The oceanic crust and upper mantle (gray) are overthrust on the Arabian shield with a sole of folded sedimentary series (oceanic basalts and radiolarites) (from Mattauer, 1995). **C.** New Caledonia's peridotite nappe: Peridotites and gabbros (black) and basalts (gray) overlap the Mesozoic and Primary series (from Debelmas and Mascle, 1993).

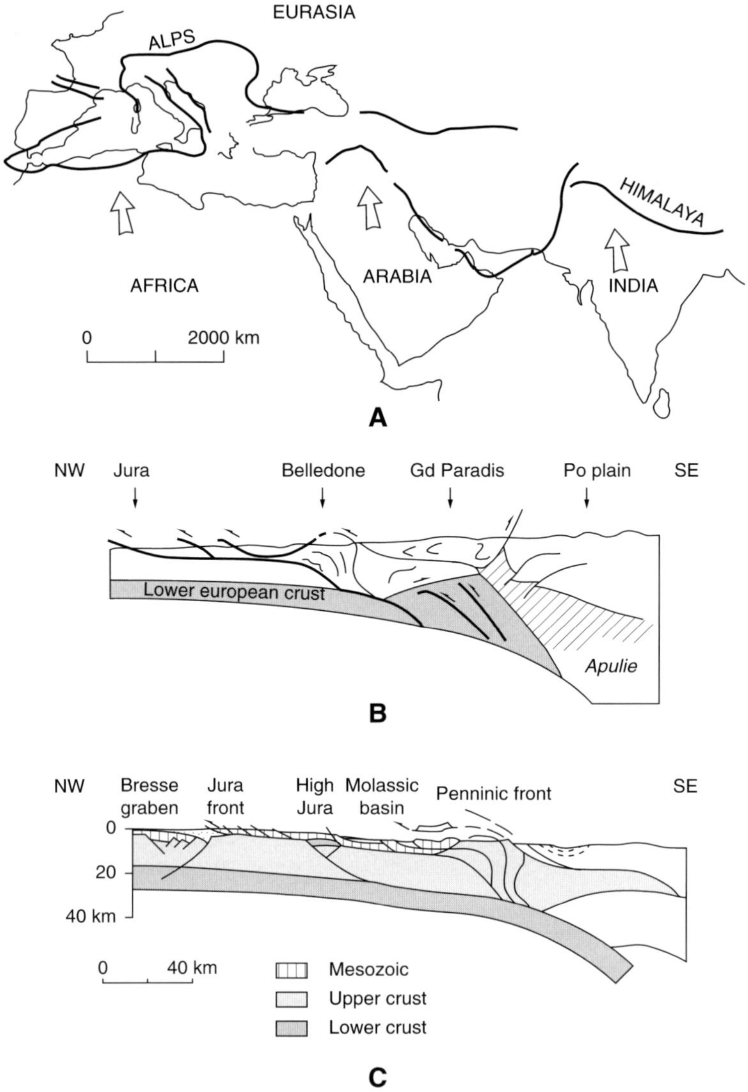

Fig. 5.108 The Alps: a collision range between Europe and the Apulian in the North of Africa.
A. General map of the Alpine-Himalayan system: collision between Eurasia and Africa, Arabia, and the Andes. **B.** Section of the western Alps from the Jura to the Penninic front (from Roure and Colletta, 1996). **C.** General section (from Ziegler and Roure, 1996).

(Fig. 5.109), and it has already been said how large an area was affected by the shortening process. The Permo-Carboniferous basins on this tectonic edifice illustrate the chain's multiphase history ("Late Hercynian" movements) with a post-collisional collapse of the axial part of the chain.

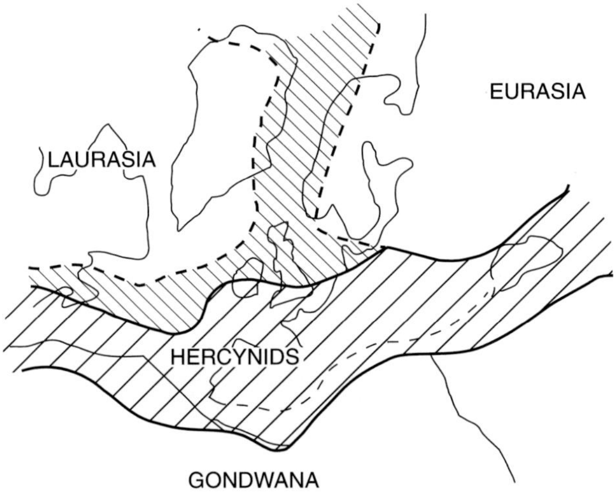

Fig. 5.109 Hercynian chain: an orogenic belt uniting Gondwana (to the south) with all of Laurasia-Eurasia to form the Pangea Mega-continent. Note that the folded Hercynian domain overlies the Caledonian range (tight hatchmarks).

Liminar ranges are different in that the collision with the continent affects narrower strips of various types (oceanic crust, volcanic arc, continental micro-blocks). The range is largely asymmetrical and various types of basins can be left over or develop. Depending on the initial geometry of the intraplate boundary and the direction of motion, strike-slip may be observed (San Andreas fault) along with fairly narrow basins, sometimes of the pull-apart type. Some crustal systems (terranes) drift over considerable distances (according to paleomagnetic data) and are integrated (stock) by **collage** in the range, as in the south of Alaska.

5.2.5 Role of Tectonics in Reservoir Geology

Leading into the next chapter on petroleum systems and the question of reservoirs and traps, the following briefly summarizes a few essential points concerning the major role of tectonic deformations in the quality of reservoirs and traps.

It goes without saying that the **general architecture and distribution** of reservoirs depends greatly on tectonic processes. Early synsedimentary tectonics controls the arrangement of sedimentary bodies directly by creating the available space for the accumulation of deposits (tilted blocks, structures and anticlines, growth faults, piggyback basins). The general architecture determines what traps there can be. It will be seen that most oil fields have structural control.

Faults and fractures on any scale then play a major role for certain lithologies such as compact limestones. The **fracturing density** and **hydraulic quality** (permeable or impermeable) of the fractures will affect:

- The development of early or late diagenesis, by controlling fluid circulation
- The positioning of the hydrocarbons, as will be seen in the following chapter
- Production, in industrial operations.

Lastly, bed displacements of different proportions along faults will **partition** the oil reservoirs, i.e., divide them up with permeability barriers.

BIBLIOGRAPHY

✦ Angelier J, Bartintzeef JM, Chauve P et al. (1992) Enseigner la géologie au collège et au lycée. Nathan, Paris.

✦ Angelier J, Colletta B (1983) Tensional fractures and extensional tectonics. Nature, 301, p. 49.

✧ Arnaud-Vanneau A, Arnaud H, Cario-Schaffhauser E et al. (1993) Anomalies diagénétiques et séries réduites dans l'Urgonien des chaînes subalpines : implications tectoniques. Congrès français de sédimentologie, 4, 17-19 novembre, 1993, Lille. S.l. : Association des sédimentologistes français, ASF, publication 19, pp 17-20.

✦ Bally AW, Bernouilly D, Davis FA et al. (1981) Listric normal fault. Oceanologica acta, special issue, pp 87-101.

✧ Barrier P, Montenat C, Ott d'Estevou P (1987) Exemples de relation entre tectonique et sédimentation dans le Plio-Pleistocene du détroit de Messine, troncatures sous-marines et dépôt-centres. Documents et travaux de l'Institut géologique Albert de Lapparent, 11, pp 143-151.

✦ Beaudoin B, Cojan I, Fries G et al. (1995) Loi de décompaction et approche d'évolution du taux de sédimentation dans les forages pétroliers du sud est de la France : programme de géologie profonde de la France. Orléans, Bureau de recherches géologiques et minières, BRGM, document 95/11, thème II, pp 133-148.

✧ Beuf S, Biju-Duval B, de Charpal O et al. (1971) Les grès du Paléozoïque inférieur au Sahara : Sédimentation et discontinuités. Évolution structurale d'un craton. Éditions Technip, Paris.

✧ Bizon G, Biju-Duval B, Letouzey J et al. (1974) Nouvelles précisions stratigraphiques concernant les bassins tertiaires du sud de la Turquie (Antalaya, Mut, Adana). Revue de l'Institut français du pétrole 29, 3, pp 305-325.

✧ Brosse E, Deroo G, Herbin JP et al. (1986) Caractérisation des couches riches en matière organique dans l'océan Atlantique au Mésozoïque. Les couches riches en matières organiques et leurs conditions de dépôts. Tours, 14-15 novembre, 1985. Bureau de recherches géologiques et minières, Orléans, BRGM, mémoire 110, pp 281-282.

✦ Brun JP (1990) Comment la lithosphère continentale s'amincit dans la terre. La terre de l'observation à la modélisation. Courrier du CNRS, dossiers scientifiques 76, juillet 1990.

✦ Caron JM, Gauthier A, Schaaf A 1986 (1989) Comprendre et enseigner la planète Terre. Ophrys, Paris.

✦ Choquette PW, Pray LC (1970) Geologic nomenclature and classification of porosity in sedimentary carbonates. AAPG bulletin 54, pp 207-250.

✧ Colletta B, Hebrard F, Letouzey J et al. (1990) Tectonic style and crustal structure of the eastern cordillera (Colombia) from a balanced cross section. In: Petroleum and tectonics in mobile belts (Letouzey J, Ed). Collection colloques et séminaires, Éditions Technip, Paris, 47, pp 81-100.

✧ Debelmas J (1983) Alpes du Dauphiné. Masson, Guides rouges géologiques, Paris.

✦ Debelmas J, Mascle A (1993) Les grandes structures géologiques. Masson, Enseignement des sciences de la terre, Paris.

✦ Dercourt J, Paquet J (1990) Géologie : objets et méthodes. Dunod, Paris.

✧ Gang W (1996) Study on hydrogeological conditions of karst strata in Jinping II hydropower project of Yalongjing river. 30th international geological congress.

✦ Goguel J (1965) Traité de tectonique. Masson, Paris.

✦ Guillemot J (1986) Éléments de géologie. Éditions Technip, Paris.

✦ Horbury AD, Robinson G (1993) Diagenesis and basin development. American association of petroleum geologists, AAPG studies in geology 36, Tulsa.

✦ Jamison WR (1987) Geometric analysis of fold development in overthrust terranes. Journal of structural geology 9, 2, pp 207-219.

✦ Lavergne M (1986) Méthodes sismiques. Éditions Technip, Paris.

✧ Letouzey J, Colleta B, Vially R et al. (1995) Evolution of salt related structures in compressional settings. Jackson MPA, Roberts DG, Snelson S. Salt tectonics: a global perspective. American association of petroleum geologists, AAPG memoir 65, Tulsa.

✧ Letouzey J, Trémolières P (1980) Paleo-stress fields around the Mediterranean since the Mesozoic derived from microtectonics: comparisons with plate tectonic data. Géologie des chaînes alpines issues de la Téthys. 26e congrès géologique international. Éditions Technip, Paris.

✦ Marsily (de) G (1981) Hydrogéologie quantitative. Masson, Sciences de la terre, Paris.

✧ Mascle A, Edingnoux L, Chennouff T (1990) Frontal accretion and piggyback basin development at the southern edge of Barbadas ridge accretionary complex. Ocean drilling programme 110, pp 17-26.

✦ Mattauer M (1973) Les déformations des matériaux de l'écorce terrestre. Hermann, Méthodes, Paris.

✧ McBride J, Barazangi M, Best J (1990) Seismic reflexion structure of intra cratonic palmyride fold-thrust belt and surrounding arabian platform, Syria. AAPG bulletin 74, 3, pp 238-259.

✧ McDonald DA, Surdam RC (1984) Clastic diagenesis. American association of petroleum geologists, AAPG memoir 37, Tulsa.

✦ McKlay KR (1992) Thrust tectonics. Chapman and Hall, London.

✦ Mercier J, Vergely P (1992) Tectonique. Dunod, Géosciences, Paris.

✦ Nicolas A (1988) Principes de tectonique. Masson, Paris.

✦ Perrodon A (1985) Géodynamique pétrolière : genèse et répartition des gisements d'hydrocarbures. Elf aquitaine, Bulletin des centres de recherches exploration production, Elf Aquitaine, mémoire 2, Pau.

✦ Purser BH (1980) Sédimentation et diagenèse des carbonates néritiques récents. Tome 1 : Les éléments de la sédimentation et de la diagenèse. Éditions Technip, Paris.

✦ Purser BH (1983) Sédimentation et diagenèse des carbonates néritiques récents. Tome 2 : Les domaines de sédimentation carbonatée néritiques récents : application à l'interprétation des calcaires anciens. Éditions Technip, Paris.

✦ Ramsay JG (1967) Folding and fracturing of rocks. McGraw-Hill, International series in the earth and planetary sciences, New York.

Roure F, Casero P, Vially R (1991) Growth processes and melange formation in the southern accretionnary wedge. Earth and planetary science letter 102, pp 395-412.

Roure F, Colletta B (1996) Cenozoic inversion structures in the forland of the Pyrenees and Alps. Ziegler PA, Horvath F. Peri-Tethys memoir 2: structure and prospects of alpine basins and forelands. Muséum national d'histoire naturelle, Paris, pp 173-210.

✦ Trémolières P (1981) Mécanismes de la déformation en zones de plate-forme : méthodes et applications du bassin de Paris. Revue de l'Institut français du pétrole 36, 4, pp 579-593.

✦ Tucker ME, Wright VP (1990) Carbonate sedimentology. Blackwell Scientific, Oxford.

✧ Wilson JT (1973) Mantle plumes and plate notions. Tectonophysics 19, pp 149-164.

✧ Ziegler PA, Roure F (1996) Architecture and petroleum systems of the Alpine orogen and associated basins. Ziegler PA, Horvath F. Peri-Tethys memoir 2: structure and prospects of Alpine basins and forelands. Museum national d'histoire naturelle, Paris, pp 15-46.

✦ *Books or articles of general interest.*

✧ *Source of one of the figures used, cited in the figure caption.*

Chapter 6

PETROLEUM SYSTEMS

The preceding chapters have traced the history of sedimentary basins in formation, from a global geodynamic perspective to basin genesis, sedimentary filling, stratigraphic sequence, burial, and deformation. Throughout this history, the organic products participating in the sedimentation process follow an evolution of their own which, in favorable cases, will lead to the formation of an oil or gas field.

For this last chapter, the reader will find it useful to consult the general works by Tissot and Welte (1984), Durand et al. (1980), Bordenave et al. (1993), and the references given in the figures and footnotes.

6.1 PETROLEUM COMPOUNDS. DEFINITIONS

The word **petroleum** (from medieval Latin, meaning "rock oil") designates a set of products and common natural derivatives that have been used since the remotest of ancient times.

The Bible's story of the Flood speaks of Noah's Arc coated inside and out with bitumen. This was a common way of caulking vessels among the ancient peoples of the Mediterranean. The Chinese used bamboo tubes and bronze pipes to drill for petroleum long before the Christian era. Various applications like that of bitumen in the pharmacies of the Middle Ages, for example, preceded the 19th century's great ***black gold*** *rush at the dawn of the industrial era.*

Liquid petroleum is usually called oil, crude oil, or just "crude". It includes a variety of products. **Light oil** refers to those oils consisting largely of **hydrocarbons** (generally up to 99%), while **heavy oil** designates those of high viscosity, which is due to a high proportion of resins and **asphaltenes**.

Hydrocarbons are molecules that contain only carbon and hydrogen.

Natural gas is the gaseous fraction of hydrocarbons (C_1 to C_5) under normal temperature and pressure conditions. This petroleum gas generally contains low proportions of carbon dioxide (CO_2), nitrogen (N_2), and hydrogen sulfide (H_2S).

There are different types of liquid hydrocarbons (Figs. 6.1 and 6.2):

- **Saturated hydrocarbons** from C_5 to C_{40} account for 15 to 35% of crude oils, alkanes, and iso-alkanes, including isoprenoids (C_9 to C_{25}). Acyclic chains and the naphthene-type cyclic products are among these.
- **Aromatic hydrocarbons**, which are cyclic and unsaturated, and which account for 30 to 35% of crude oils.
- **Naphthene-aromatic hydrocarbons**, which combine complex saturated cycles (cycloparaffins) and unsaturated cycles, and which account for more than 30% of crude oil products. The term "naphtha" is now archaic.

Asphalts and **bitumens** are highly viscous or solid petroleum products with a very high proportion of resins and asphaltenes.

Resins and **asphaltenes**, sometimes called NSOs (standing for compounds of nitrogen, sulfur, and oxygen), are hetero-atomic compounds of high molecular weight (higher in resins than in asphaltenes), often characterizing degraded oils. Resins are soluble in n-heptane, but asphaltenes are not.

The **sulfated compounds** already mentioned are most abundant in the form of H_2S (as much as 14% in the Lacq gas field). Traces of oxygenated or organometallic compounds, which are also considered to be "geochemical fossils", can be found here.

Biomarkers of various sorts derive from cellular membranes (especially from unicellular organisms) providing proof of the organic origin of petroleum. These are heptanes, steranes, carotenoids, isoprenoids, and prophyrins (Fig. 6.3).

The term **kerogen** designates the insoluble organic material in a rock, i.e., the fraction of the initial organic material in a sediment that has not (or not yet) been transformed into hydrocarbons (Fig. 6.4).

Oil shales are sedimentary rocks containing a large proportion of kerogen that can produce oil by thermal effect. **Tars** are generally sedimentary rocks having a high proportion of kerogen and/or heavy products.

Oils can also be classified by various criteria such as specific gravity, molecular weight, or refractive index. Producers use specific gravity, which generally varies from 0.75 to 0.80. **API gravity** or degree is another way of defining this specific gravity:

$$API^\circ = \frac{141.5}{d\ 60^\circ F} - 131.5$$

The **fluorescence** of oils examined under ultraviolet light can also be a criterion of recognition and classification. Oils can be detected in infinitesimal quantities this way. The tint depends on the proportion of aromatic nuclei. As the nuclei increase, the fluorescence turns from green to yellow and then to red. Paraphinic crude is not fluorescent.

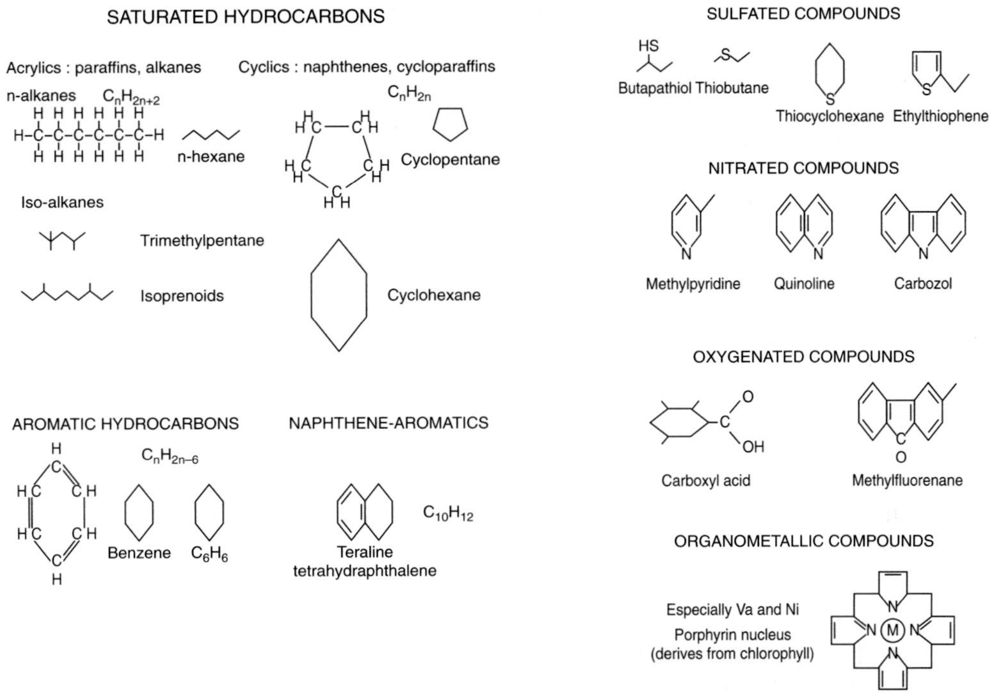

Fig. 6.1 Different types of hydrocarbons by chemical composition.

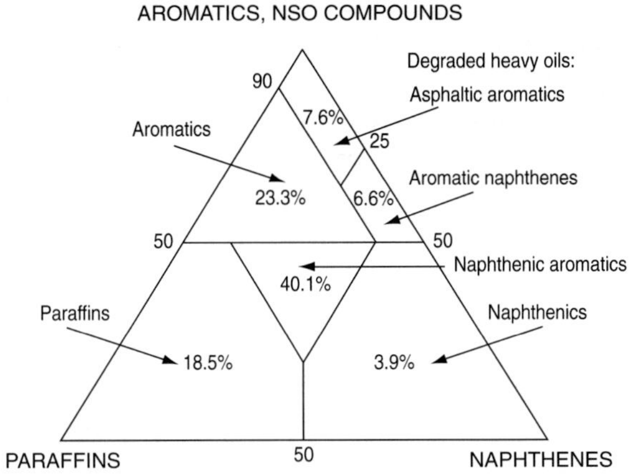

Fig. 6.2 Classification of crude oils (according to Tissot and Welte, 1984).

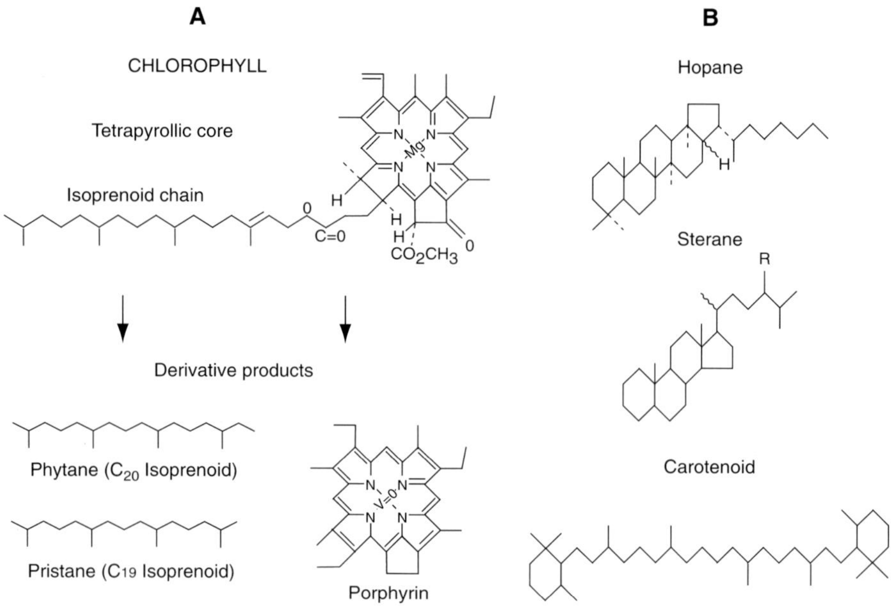

Fig. 6.3 Geochemical fossils and biomarkers (from Bordenave, 1993). **A.** Porphyrins and isoprenoids are considered to be direct derivatives of chlorophyll. **B.** Hopanes, with pentacyclic skeleton, derive from cellular products of bacteria and blue-green algae. Steranes come from the cellular membranes of planktonic organisms. Carotenoids come from carotens.

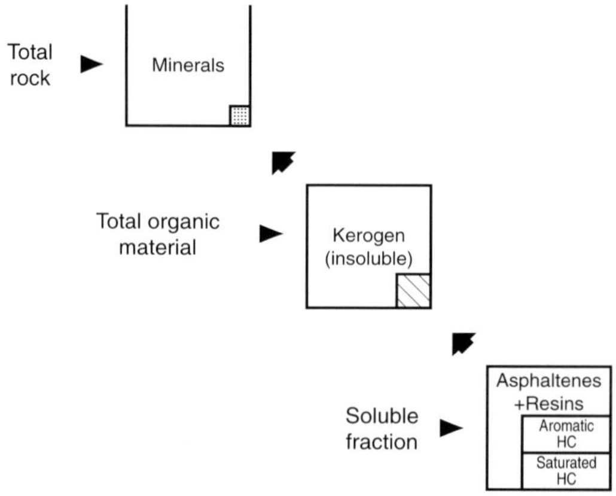

Fig. 6.4 Different fractions of organic products in rocks (from Tissot and Welte, 1984).

6.2 ORIGIN AND GENERATION OF OILS AND NATURAL GAS

It is widely agreed among the community of geochemists today that liquid and gaseous petroleum products, like coals are of **organic origin**.

For a long time after Mendeleyev, a Russian school defended the idea that hydrocarbons, resins, and asphaltenes were of inorganic origin, stemming from simple reactions involving CO, CO_2, H_2N, *and* H_2O. *This viewpoint is nearly forgotten now. More recently, the idea that methane (and even oil) was generated inside the Earth from remnants of the primitive atmosphere, as might be thought on the basis of the carbonated products found in meteorites, has also been defended.*

Natural gas can be produced by the bacterial metabolism by the action of bacteria on organic substances present in sediments. This bacterial fermentation could doubtless explain 20% of natural gas reserves in **bacterial gas**, **dry gas**, or **biogenic gas**. It has never been proven that bacteria play any role in the formation of oil.

But most natural gas comes from the transformation of a sedimentary organic material under the effect of temperature and pressure during burial. This is **thermogenic gas**.

Chemical transformations occur slowly during burial, with a gradual enrichment in carbon and hydrogen atoms and depletion (by leakage) of other elements such as oxygen and nitrogen.

Oils mainly consist of **hydrocarbons** (i.e., molecules containing only carbonate hydrogen) but also of **resins** and **asphaltenes**. Oils all result from the transformation of organic products that are sedimented and then buried at depth. As has been said, they generally include other components like carbon dioxide, hydrogen sulfide, nitrogen, and a small amount of helium. They include **biomarkers** that are very close to molecules existing in living organisms.

6.2.1 Sedimentation of Organic Matter

Hydrocarbons, which cover most petroleum products, thus derive from the transformation of carbonaceous products, i.e., essentially from the lipids in living matter. In the carbon cycle, there is a mineral carbon cycle and a shorter organic carbon cycle (Fig. 6.5). In this organic cycle, all the carbon of sedimentary rocks derives from atmospheric dioxide (originally coming from the mineral carbon cycle) and the hydrogen comes from water. These two elements are incorporated and combined in vegetal and animal organic material with the necessary energy coming from the sun's radiation. It is by **chlorophyll assimilation** that atmospheric CO_2 is extracted by photosynthetic marine and land organisms. Much of it is returned by the breathing of macro- and microorganisms, but part of the carbon does not return to the atmosphere in the form of dioxide. It is trapped and, after diagenesis, is transformed into carbonaceous and hydrocarbonaceous products. It is by this "leak" in the organic carbon cycle that fossil fuels, coals, petroleum, and gas are constituted in the course of geological time.

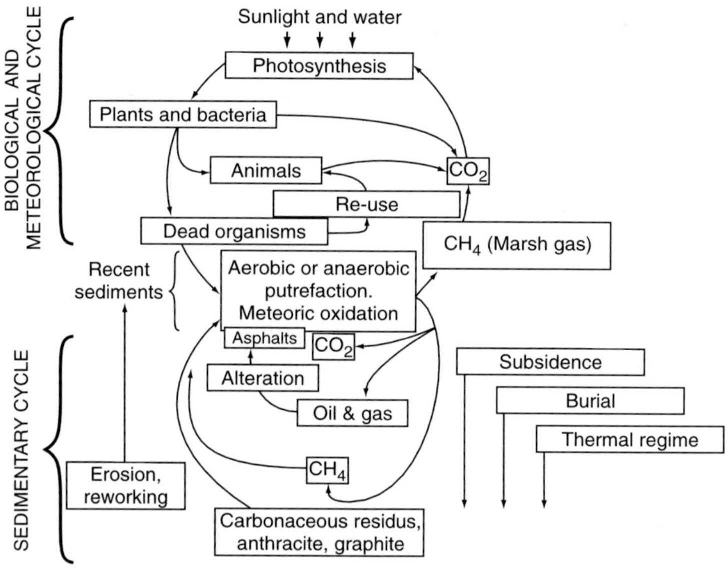

Fig. 6.5 Organic carbon cycle (from Durand, 1987).

6.2.1.1 Origin of Initial Organic Matter, Biomass Production, Role of Photosynthesis[1]

Coal and petroleum should not be separated too categorically from each other in light of the similar origins of their constituents (the petroleum potential of coal is recognized today, especially for certain gas fields), but what is of special interest here is the chain that leads from living matter to hydrocarbons.

The vegetal **biomass** (higher plants, algae, phytoplankton, photosynthetic bacteria) accounts for about 90% of the total biomass at the Earth's surface today (about 10^{12} tons of carbon), while the animal kingdom accounts only for 10%. It is by their **photosynthetic** activity that plants assimilate carbon from dioxide in the atmosphere (also about 10^{12} tons), which thus constitutes a major initial source.

The time evolution of the major phyla is such that today's situation cannot reflect exactly what was happening 500 My ago, and even less so a billion years ago. The balances between marine and continental biomasses, in particular, were quite different.

Today, the continental domain where the **primary productivity** of higher plants depends on the climate (Fig. 6.6) is seen to differ from the marine or oceanic domain where this biomass productivity is influenced not only by climate, but also by continental inputs, continental shelf structure (Fig. 6.7), and ocean currents such as upwelling.

1. See Pelet, 1985; Huc, 1988.

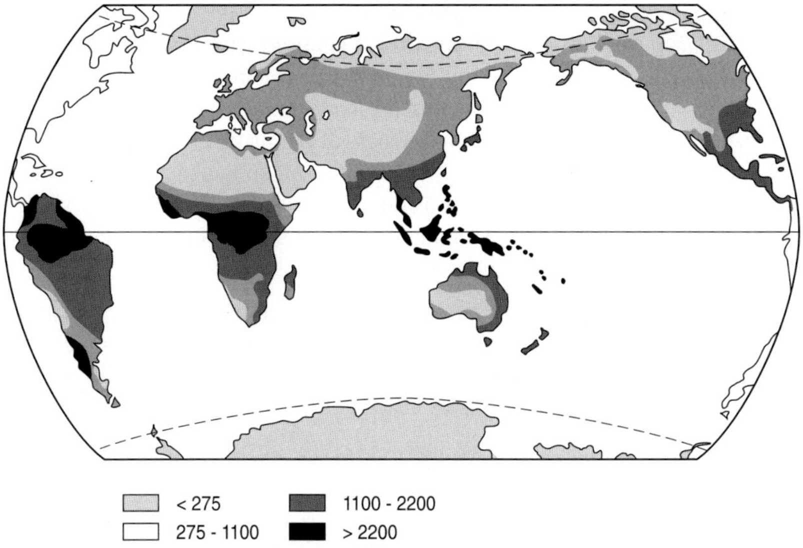

Fig. 6.6 Distribution of primary productivity on the continents (from Lieth and Whittaker, 1975).

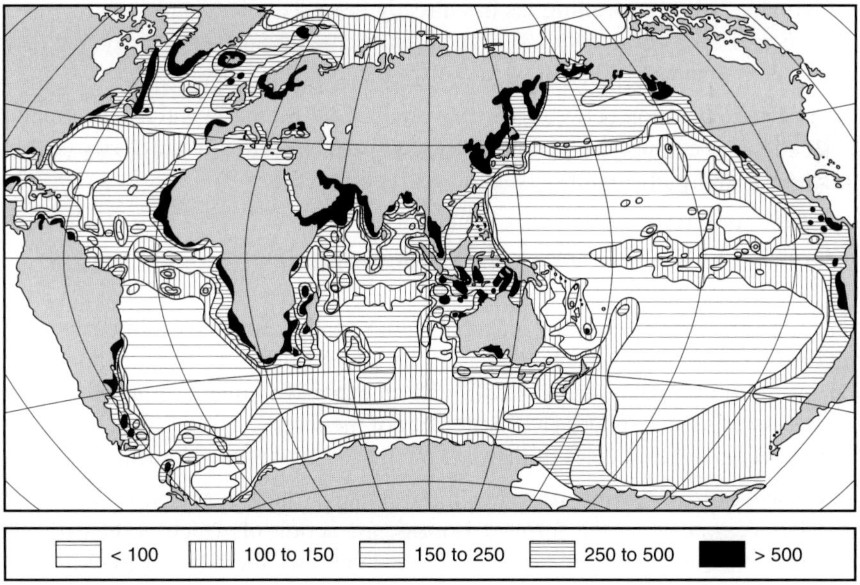

Fig. 6.7 Phytoplankton productivity (mg C(m²H day) in the oceans (from Pelet, 1985).

The production of organic matter by **photosynthesis** is active on land and in the euphotic zone of the oceans (up to depths of 100 meters). This production is estimated at about 10^{11} tons of carbon per year, which is distributed about equivalently between the continents and the oceans. In the ocean, carbon is concentrated mostly in the phytoplankton. The zooplankton that use this nutrient thus already recycle a part of the production at the beginning of the complex food chain. The important role of unicellular algae is again seen in lakes. Chapter 4, on the time variations that have to be considered in geology, showed how seasonal variations, climatic events, and longer-term changes in the transfers could affect primary production on all scales.

From the chemical viewpoint, organisms contain different types of molecules: glucides or carbohydrates (sugars are the first products of photosynthesis), protides (amino-acid chains) and lipids. The higher plants of the continent manufacture cellulose and lignin, which are tissues richer in oxygen than those of marine plants.

Some of the initial terrestrial organic matter is either destroyed or recycled on site, with no chance of being incorporated in sediments, while a small part will be carried to a lacustrine or marine sedimentary basin and participate in the sedimentation. At sea, however, all of the biomass produced and more or less recycled in the food chain constitutes a non-negligible portion of the sedimentation. But this productivity is not equally distributed. It is largely controlled by the availability of sufficiently renewed nutrients (like phosphates and nitrates), by illumination, and by temperature. The coastal environments (estuaries, lagoons, continental shelves in general) are thus the preferential zones. Very turbid environments are not very favorable, while **upwelling**, as has already been seen, is one of the very favorable factors (Fig. 6.8). Lakes cover only 1% of the surface that oceans cover today, but their primary activity is as much as 100 times greater than that of the oceans.

From the general relation between primary production (biomass) and the geographic and climatic situation, an attempt can be made to reconstruct the **paleoproductivity** of different epochs on the basis of global palinspastic reconstructions (Fig. 6.9).

6.2.1.2 Sedimentation, Recycling, and Conservation of Organic Matter

The remains of animal and vegetable organisms after death constitute **sedimentary organic matter**. This is of various types depending on the origin (phyto- or zooplankton, algae tissues, ligneous detritus, spores, and pollen), with an extensive range of organic products on the chemical level. Organic tissues are fragile. Part of them serve as nutrients for other organisms and part become assemblages of thermodynamically unstable biological molecules.

Various **mechanical actions** also alter these remains during transport. On continents, they will first accumulate in the form of **humus**, a fraction of which is then transported down the rivers. Organic matter is generally fragmented and altered very quickly, and then transits in the form of debris or colloids, or is even dissolved.

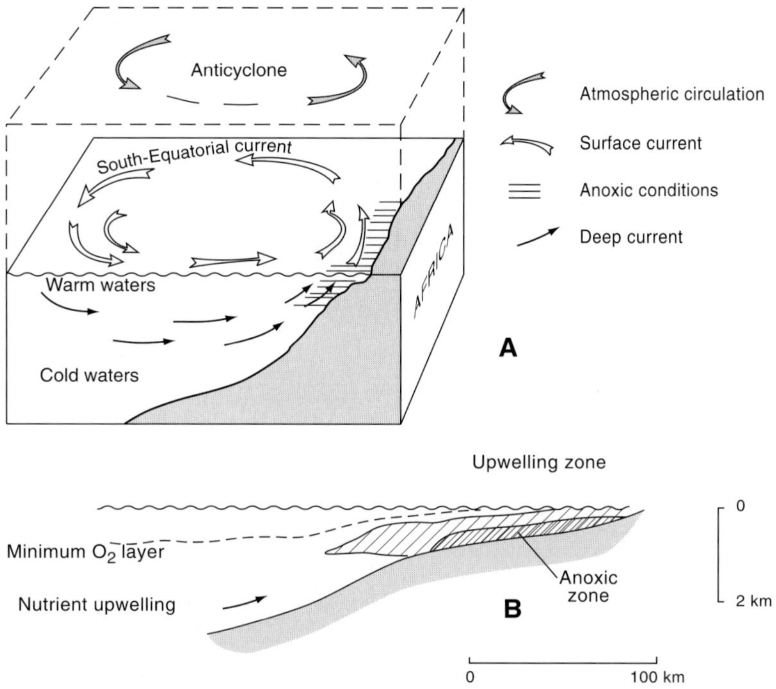

Fig. 6.8 Upwelling model.
A. General scheme. Ocean currents, determined by atmospheric gyres, allow a proliferation of organisms that produce a large amount of sedimentary organic material. Intense bacterial activity then occurs. The zone of anoxic conditions can range beyond a hundred kilometers. **B.** Schematic section recalling the chemical stratification of waters on the continental rim (also see Fig. 3.96).

Peat bogs and marshes are exceptional areas where the organic matter can accumulate with very little or no transport at all, allowing high concentrations and a geochemical confinement that promotes good preservation of the organic matter.

This is the typical environment for the formation of **coals** (Fig. 6.10).

In the aquatic environment, transport may be vigorous (in rivers, deltas, tides, as explained in Chapter 3), in which case the mechanical effects of the currents in fragmenting and also concentrating the material will be effective. If transport is limited (as it is in lakes, confined lagoons, or the open ocean), the organic detritus will decant. The very low density of organic matter limits its sedimentation rate greatly and thereby favors its use by other organisms, which recycle it in the food chain or degrade it biochemically. If the physical and chemical processes create flocs, though, or if fecal pellets are produced, these phenomena will favor faster sedimentation of organic matter.

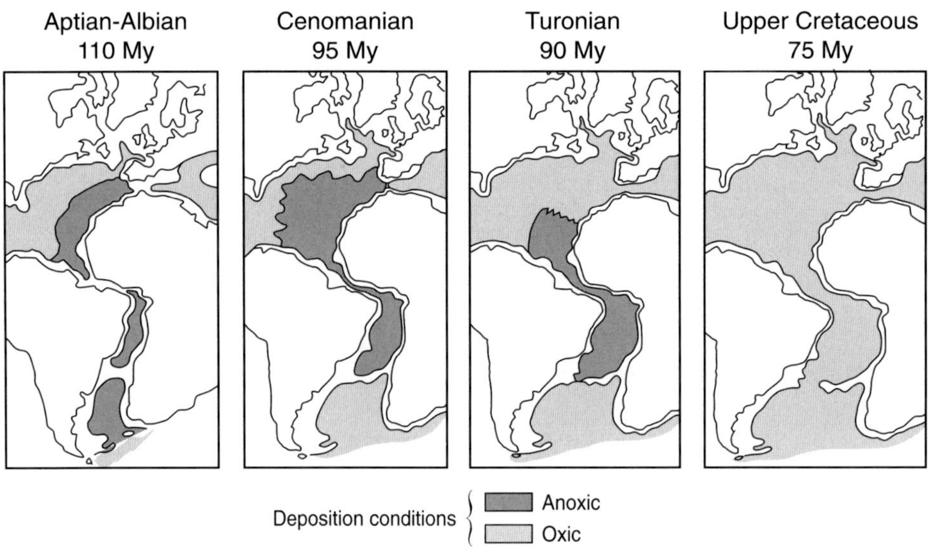

Fig. 6.9 Attempt to reconstruct anoxic deposition environments in the course of the Atlantic opening (from Tissot and Welte, 1984).

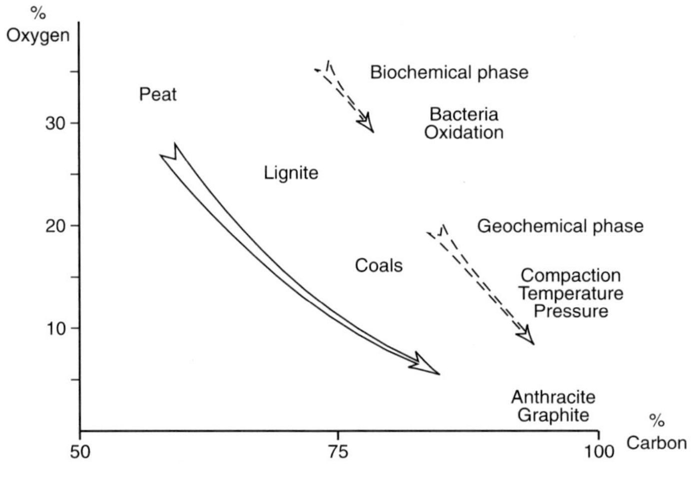

Fig. 6.10 Evolution of chemical compositions of the various products in the coal lineage.

One important factor reducing the quality of organic material that reaches the bottom is its **recycling** in a complex food chain. The tissues and their products derived from biochemical actions become a source of food for plankton, lekton and, lastly, the benthos at the bottom.

Chemical alterations during sedimentation are another factor to be considered because, as has already been said, biomolecules are unstable and decompose very quickly into several products (H_2O, CO_2, CH_4, NH_4, among others). This degradation is largely activated by bacterial action. In the continental environment, degradation is very active in humus and soils and then in streams, because of the oxidizing environment. Organic matter is poor in hydrogen. In a lacustrine or marine environment, the organic material is highly reactive and degrades almost exponentially with increasing depth and residence time (Fig. 6.11). However, the chances of conserving the initial products are better than in the continental domain, especially if the waters are calm and not stirred very much, and thus are not highly oxygenated.

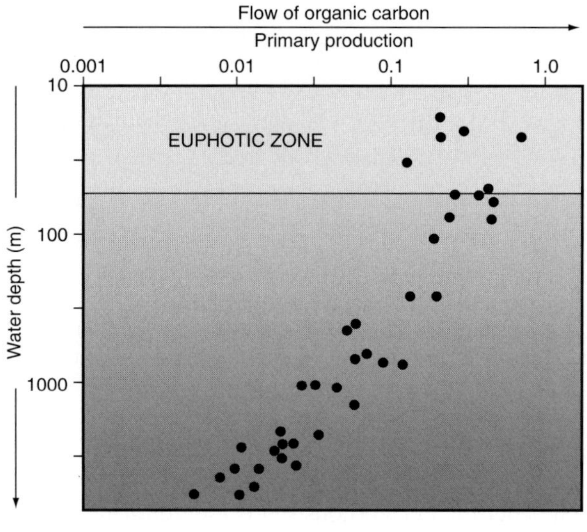

Fig. 6.11 Rapid degradation of organic matter in the ocean's water column (from Suess, 1980, in Huc, 1995).

It is clear that surface waters, in that they are the richest in oxygen because of transfers with the atmosphere and the photosynthesis that occurs there, are the ideal locus of degradation. With hydrodynamic motions, though, the oxygen-rich surface waters can be entrained toward the bottom of abnormally impoverished basins. Chapter 3 showed how waters of different densities, with the further effect of atmospheric circulation, can form stream currents. This is true of the higher latitudes, which are doubtless the most striking example with cold, high-density surface waters plunging to depths and initiating the bottom circulation, thereby providing good ventilation. The development of anoxic conditions, on the other hand, is favored by an (unstable) stratification of the water column and an irregular bottom topography with thresholds and depressions (Fig. 6.12). These isolation or confinement conditions go hand in hand with a reduction of mechanical degradations.

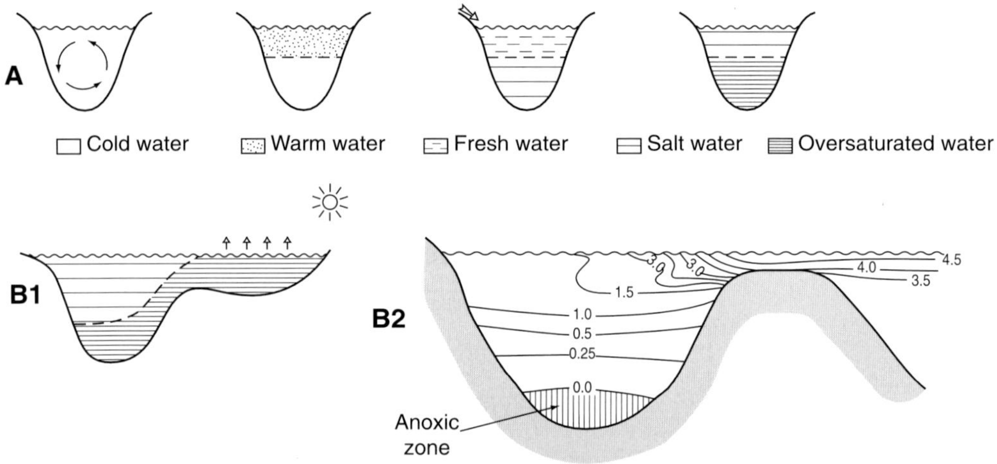

Fig. 6.12 Basin water stratification models.
A. General schemes from an open, ventilated system to a concentration of brines in the bottom of the basin (from Huc, 1995). **B.** Threshold effects. (**B1**) Relation between two bays with different evaporation level. (**B2**) Distribution of dissolved oxygen (cm^3/m) in Indonesia's Kaoe bay.

It can be seen that the average particle size of the sediments making up the deposition environment (99%) is decisive. In an impermeable argillaceous environment, the chances of preservation are much better than in a porous sandy environment.

With the transport, mechanical fragmentation action, colloidal formation, biological recycling, dissolution, and chemical alterations, the amount of organic matter actually input into the sediment is very small. It is estimated that an average of less than 1% of primary production is incorporated in sediments, though this may exceptionally reach a larger fraction (8% at the bottom of lake Tanganyika). Generally, though, it is found that the distribution of organic carbon content in sediments corresponds well to that of primary productivity on the global scale (Figs. 6.7 and 6.13).

Looking more closely, now, two major factors will condition the possibility of **conserving** and **concentrating** the organic matter in the sediment: the **anoxic environment** and the relation between the **sedimentation rate** and subsidence.

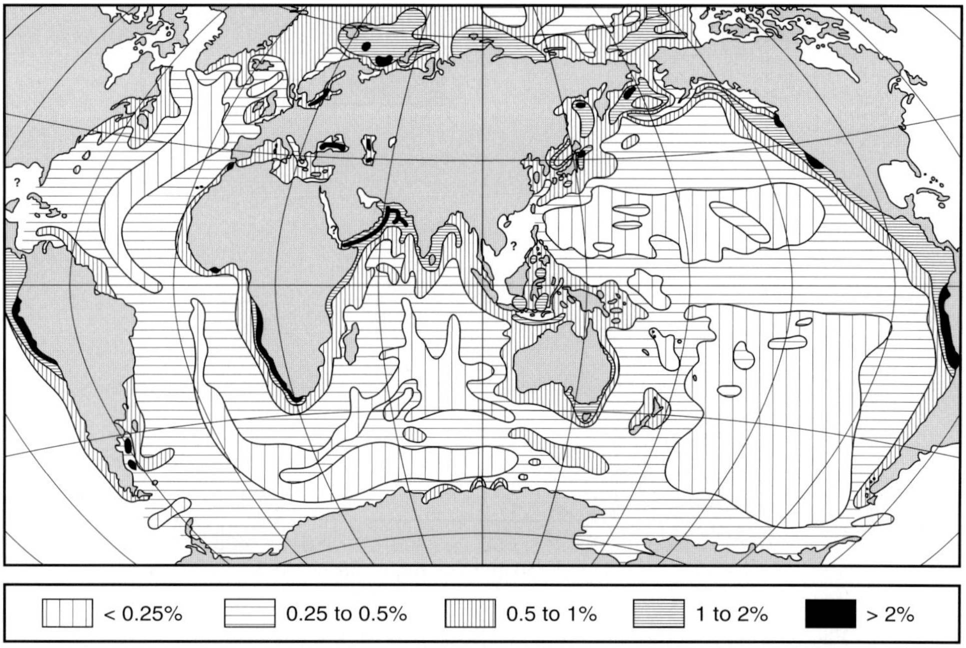

Fig. 6.13 Organic carbon content (by weight) in today's marine sediments (from Pelet, 1985).

In the oxygenated aerobic environment at the bottom of the basin, which is generally the case of marine environments, the chances of preservation are low because the action of burrowing organisms and bacteria of the benthos is effective with time (Fig. 6.14). The organic matter is then recycled until the sediment, with burial, reaches the reductive level (Fig. 6.15). In an **anaerobic** environment depleted of oxygen at the water's bottom, the chances of conservation are better because the benthos is poor and the organic matter then sediments into a reductive environment. But an **anoxic** (or anaerobic) environment is really needed in order to allow preservation of animal **tissues** and their organic derivatives during sedimentation. The benthos is then lacking. This is a **euxinic** environment (from *Pontus Euxinus*, the ancient name of the Black Sea where such conditions were described).

The type of environment is obviously governed largely by the circulation and renewal of water masses. When there is no mixing, the environment is said to be **confined**. This is a situation that can occur in different bathymetric situations.

The other important factor is the **sedimentation rate**. If the rate is high, with adequate accommodation such as occurs with active subsidence, the (particulate) organic matter can be partly preserved even under oxidizing conditions (Fig. 6.16). On the other hand, high sedimentation rates can be an inhibiting factor by the effect of dilution, unless co-variant factors come into play, such as water depth and distance to the coast.

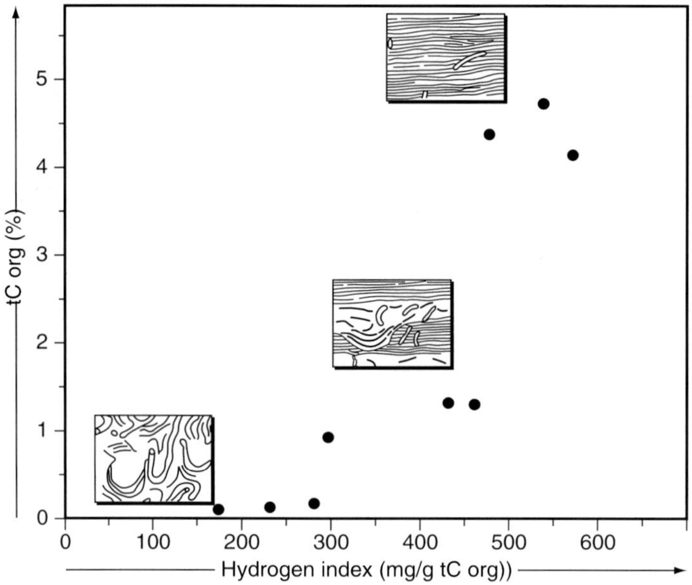

Fig. 6.14 Role of the benthos in the degradation of organic matter. When the benthos can develop under normally oxygenated conditions, bioturbation is abundant and the preserved organic carbon ratio decreases, as does the hydrogen index (from Pratt, 1984, in Huc, 1995).

Generally, the sedimentation and concentration conditions affect the sediment quantitatively and qualitatively, i.e., in its **petroleum potential** (high hydrogen indices, discussed later).

The importance of deepwater **upwelling** has long been recognized (Fig. 6.8). In these zones of high productivity, the high oxygen demand results in oxygen depletion while the circulation of water masses is large.

6.2.2 Geological Perspectives: Rock Source and Kerogens

After the sedimentation process, the portion of the initial organic matter (OM) that has escaped the biological, enzymatic, and other actions is preserved in a deposit. This material is of widely varying sorts: ligneous particles from continental plants, spores and pollen, various cellular tissues, algal debris, amorphous material (amorphous under an optical microscope, but in which membrane structures can sometimes be found) (Fig. 6.17). The biological diversity of this organic matter is very extensive; but blue algae, diatoms, and higher plants can sometimes be recognized as the origin.

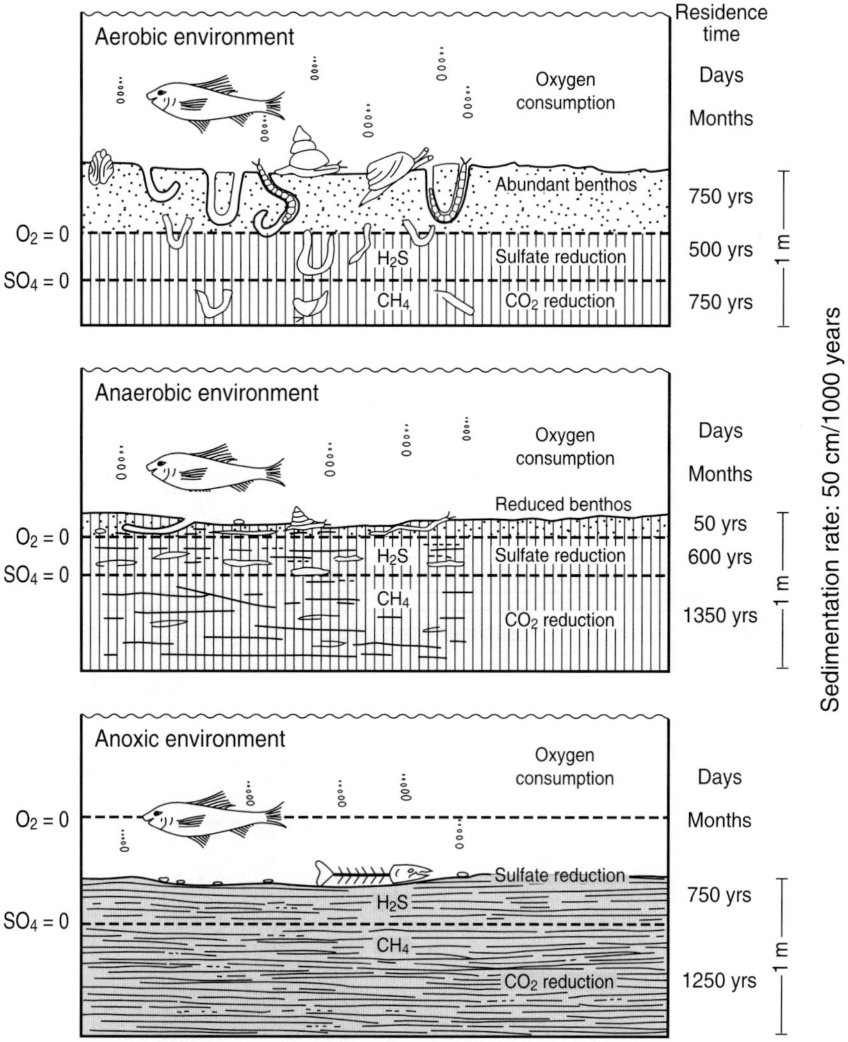

Fig. 6.15 Different types of environments: aerobic, anaerobic, anoxic (from Huc, 1995).

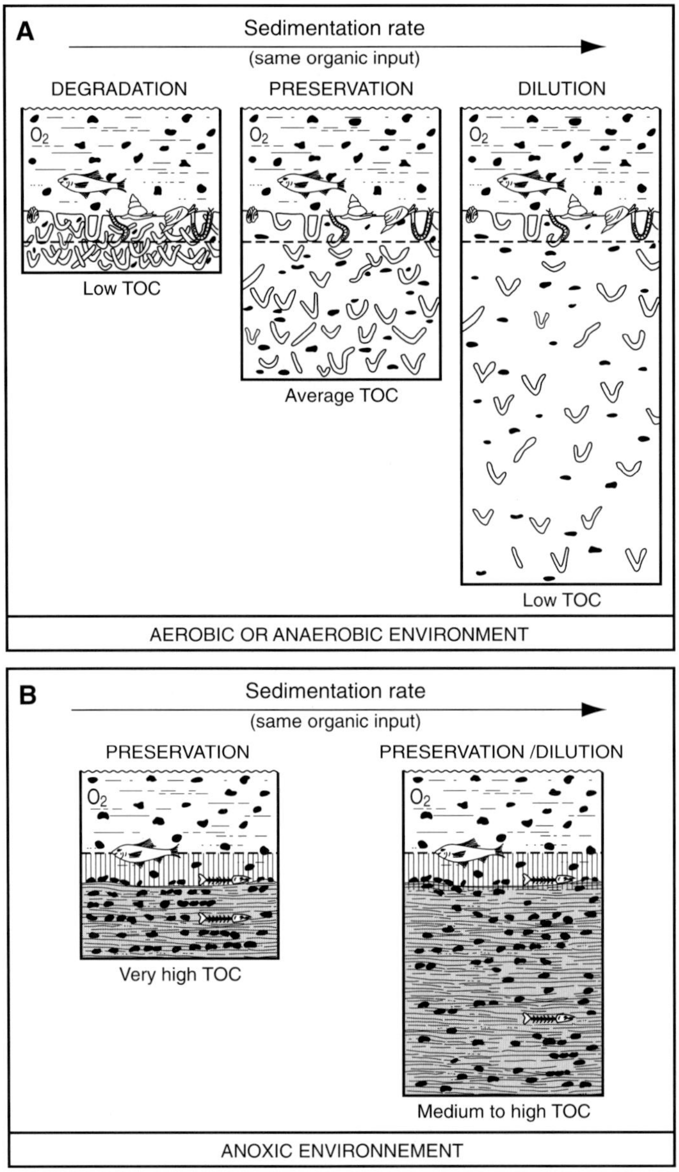

Fig. 6.16 Preservation or dilution of organic matter in relation with sedimentation rates. (from A. Huc, 1995)
A. Aerobic or weakly anaerobic environment. **B.** Anoxic environment.

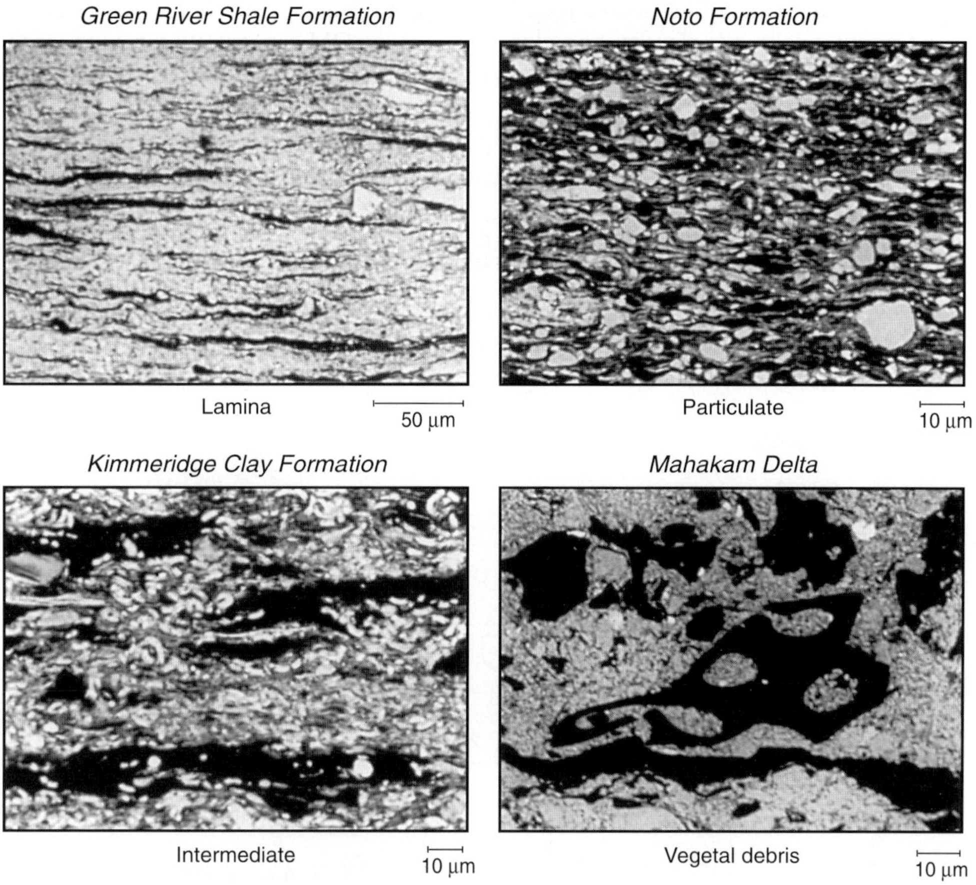

Fig. 6.17 Different types of organic matter in sediments (from Huc, IFP photos).

A few molecules such as n-alkanes are incorporated with little change, while others (steroids, pentacyclic triterpenoids) are already the result of minor early transformations. These molecules are considered to be veritable ***geochemical fossils*** *left over from some subtle early diagenesis (Fig. 6.3). Others are much more altered and transformed.*

The average organic content of sedimentary rocks is 0.5%. When the organic matter concentration becomes large (4% to 12%), this element is then considered to be potential **source rock**, i.e., a rock likely to generate hydrocarbons.

Source rock is the first essential link in the **petroleum system**. It should immediately be said that, in order to speak of **petroleum potential**, the qualitative and quantitative aspects have to be considered. Qualitatively, depending on the type of organic matter, and essentially the H/C and O/C ratios, source rocks are likely to generate different types of gas-oil fields. Quantitatively, it is clear that a sediment where the OM concentration is high (of the order of 10%, for example) will have a high **genetic potential** but, if it is limited in volume (lateral extent, bed thickness), its petroleum potential will be negligible.

On the scale of a basin, source rock is subject to **lateral heterogeneity**. This is made so mainly by hydrodynamism, bathymetric barriers, and the particle size distribution of the enclosing sediments.

The content distribution in the closed Black and Caspian Sea basins of today are distributed concentrically, with a centripetal increase toward the bottom of the basin (Fig. 6.18). This is also true for certain ancient basins. But often, this simple model does not exist and the lateral distribution of the organic facies is highly heterogeneous.

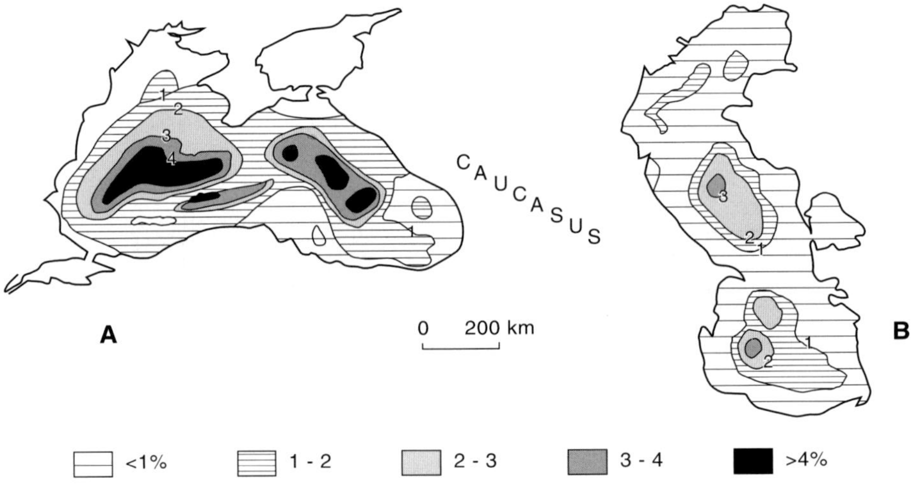

Fig. 6.18 Concentrations of organic matter in closed basins.
A. Black Sea. **B.** Caspian Sea. Concentrations are expressed in organic carbon content of the sediments (from various authors).

The lateral extent of a source rock determines the quantity of hydrocarbons formed: the greater the extent, the more effective the future "kitchen" will be.

The highly variable **stratigraphic range** of source rock horizons reflects the time variations of the sedimentary basins, which was discussed in Chapter 4. This goes beyond the domain of sedimentology and enters stratigraphy. On the world scale and throughout geological time, certain periods have been particularly favorable to the deposition of source rock. Final analysis also shows that those sedimentary intervals corresponding to source rock in fact exhibit vertical heterogeneities that result from time variations in the deposition conditions (Fig. 6.19).

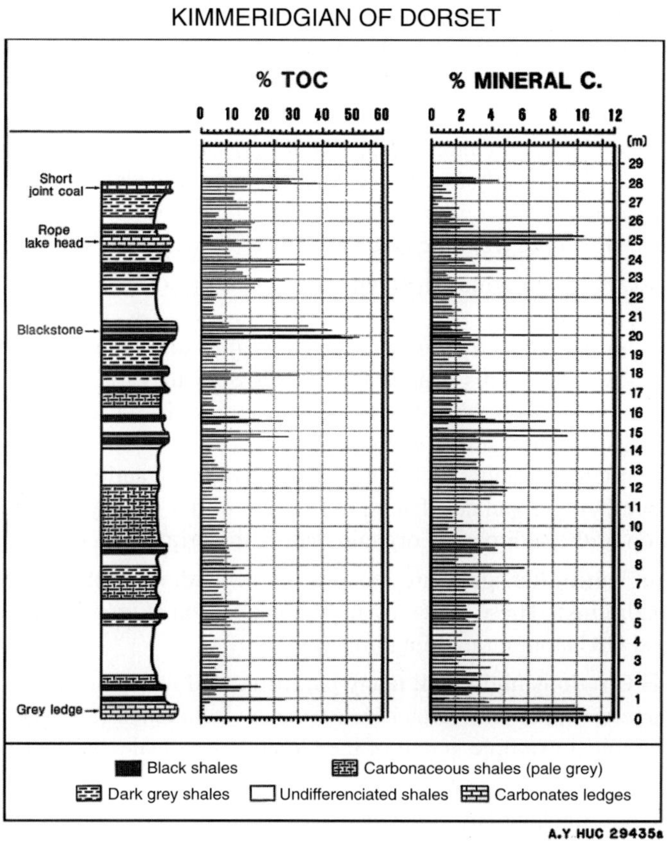

Fig. 6.19 Vertical variations in organic carbon content (TOC) in a formation considered to be source rock (from A. Huc).

It is thought that confinements, beyond the topographical control of the local geographic configuration, have also been influenced by (absolute or relative) sea-level variations. On the global scale, periods when the sea level has risen are considered to be particularly favorable intervals.

What characterizes source rock—quality (genetic potential), quantity (petroleum potential), maturation (geological history)—is defined by studies of petroleum basins. The very term "source rock" implies a certain diagenetic history, since time has turned the sediment into rock.

The beginning of the burial history is characterized by **early diagenesis** in an anoxic environment. This whole process is dominated by bacterial metabolism. The degradation of sedimented organic material leads to the formation of the **kerogen**, in which polymers form.

Polymers are heavy macromolecules (molecular weight >10 000) formed from cyclic nuclei bounded by hetero-atoms or paraffinic chains. Under certain anaerobic conditions, a large share of this **initial kerogen** undergoes hydrolysis by acids and biochemical methane and can be released early (as in marshes). This is the first stage in the genesis of hydrocarbons.

The term "kerogen" was originally used to define the organic matter in bituminous shale likely to produce oil by pyrolysis. Other meanings have since been given to the term[1]*, as in humic products with a high proportion of macerals, amorphous fractions, or insoluble organic material resulting from the condensation of lipids.*

Geochemists today have defined different products commonly contained in source rock (Fig. 6.4).

- **Kerogen** is the insoluble fraction of organic matter after the chloroform or dichloromethane is extracted. This usually constitutes 80% to 95% of the organic content. The sedimentary matrix is difficult to isolate, attacked as it is by hydrochloric and hydrofluoric acids in limestones and silicates.
- The **extract**, micropetroleum or bitumen, is the soluble fraction dispersed in the rock matrix (before being expelled, if it is, during the burial history) and produced by the first evolution of the sedimented organic matter, the primordial kerogen.

Three types of initial kerogens are recognized, depending on the initial biomass (and sedimentation conditions). These differ by their composition (Fig. 6.20) and by their position in the H/C and O/C atomic ratio diagrams (Fig. 6.21).

Type I. The H/C ratio is high, with many paraffinic and few aromatic chains. This type of kerogen is characteristic of fresh water lacustrine environments. It derives from algal and bacterial lipids (a typical example is that of the Green River Shales of the Cretaceous in the United States).

Type II. With intermediate H/C and O/C ratios, there are more aromatic and naphthenic products. This type of kerogen is characteristic of anoxic marine environments. It is mainly of planktonic origin but is sometimes mixed with products from higher plants (clays of the Toarcian in the Paris basin or Germany are often given as examples).

Type III. The O/C ratio is high and more or less oxygenated polyaromatic products are observed. This type of kerogen comes from vegetal products of continental origin (transitory form with coals) and the typical example is that of the Miocene clays of Mahakam in Indonesia.

This classification is obviously a simplified expression of an often more complex reality where mixtures due to sedimentation and to varying degrees of alteration lead to quite a broad range of products.

1. See discussion in Durand, 1980.

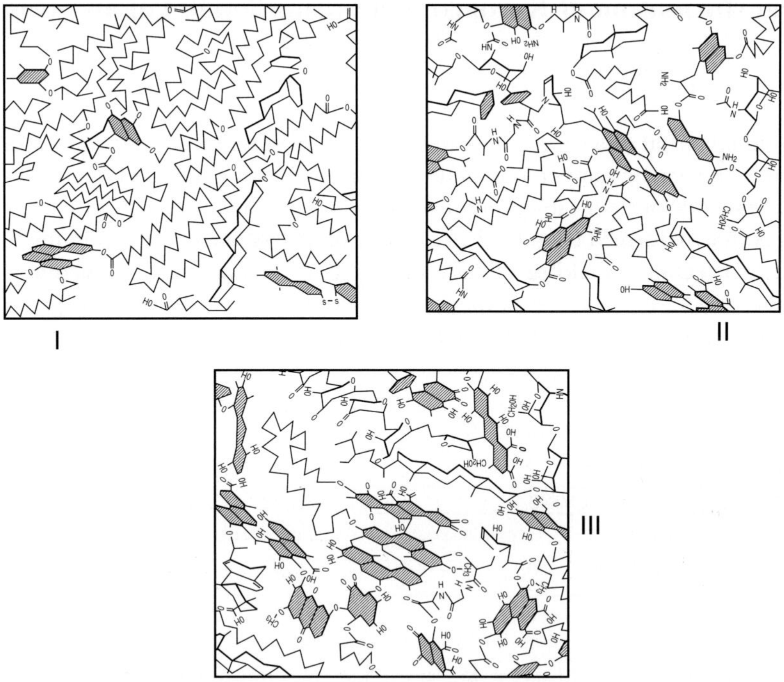

Fig. 6.20 Varied composition of three types of kerogens (I, II, III) (from Béhar and Vandenbrouke in Bordenave et al., 1993).

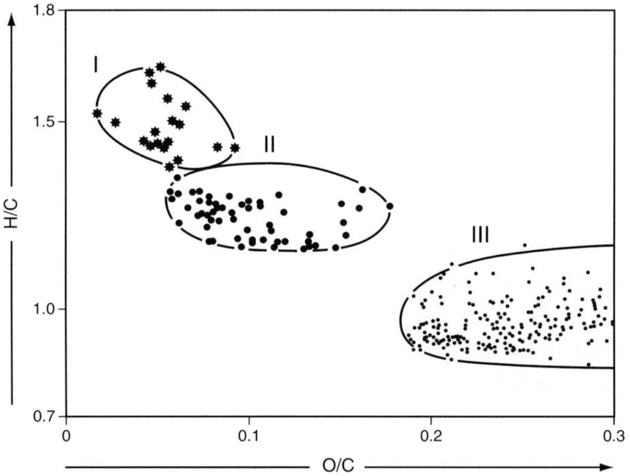

Fig. 6.21 Van Krevelen diagram: different types of kerogens (from Bordenave et al., 1993).

6.2.3 Transformation of Kerogen and Formation of Oil and Gas

6.2.3.1 Successive Stages of Kerogen Transformation to Petroleum

With the gradual burial of a source rock in the course of geological time, the **initial kerogen** is subjected to increasing temperature and, with the successive stages of degradation, will be transformed into various petroleum products: **oils** and **gases**, and more or less **residual kerogen** that is not transformed (Fig. 6.22). The **genesis of hydrocarbons** from the initial kerogen is due to **thermal cracking** of the initial products (Fig. 6.23). **Thermogenesis** thus describes the evolution from the initial **immature** product to the **mature** petroleum. The formation of biogenic gas will be discussed later.

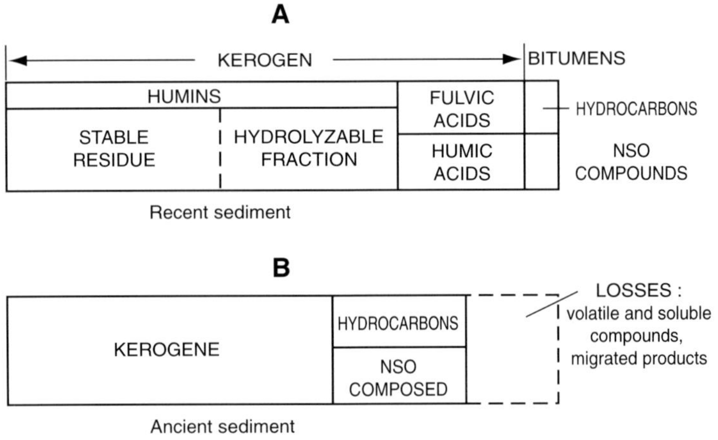

Fig. 6.22 Evolution from initial kerogen to residual kerogen (from Bordenave, 1993).

As was mentioned in the previous chapter concerning the diagenesis of sediments, the burial history of sedimentary organic matter can be broken down into several phases (Fig. 6.24):

- Very early **initial diagenesis** during which the bacterial effects can be quite effective in generating "biogenic" methane gas.
- **Burial diagenesis** during which the three types of kerogen begin to lose their oxygenated compounds in the form of water and carbon dioxide (the decrease in the O/C ratio is faster than the H/C).
- **Catagenesis**, corresponding to increasing burial depths, is a stage in which oil and light hydrocarbon form (gas to condensates). While the O/C hardly decreases at all, the

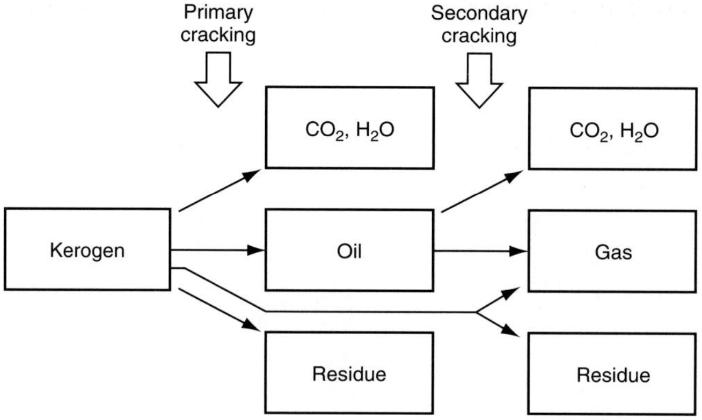

Fig. 6.23 Thermal cracking of kerogen. The initial kerogen is degraded in several cracking stages (from Tissot and Welte, 1984).

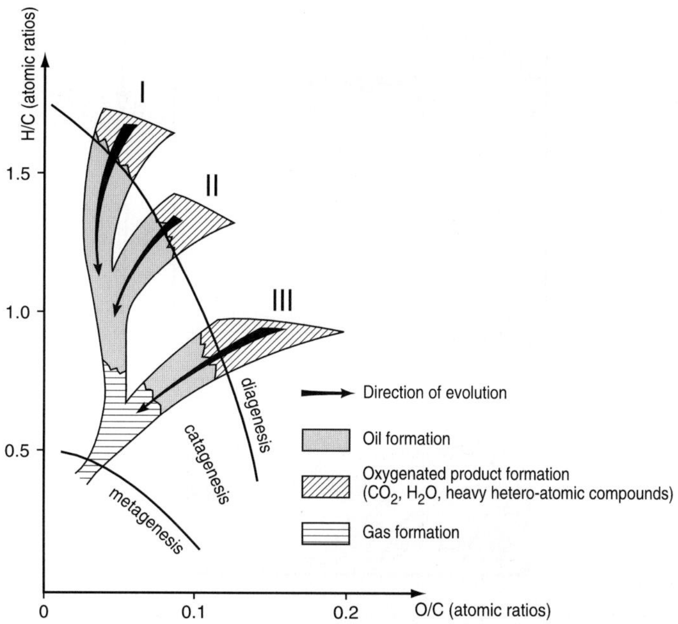

Fig. 6.24 Paths of kerogen evolution (from Tissot and Welte, 1984).
The main types of kerogens in the Van Krevelen diagram are clearly separate to begin with, but their paths of evolution converge in the course of thermal maturation.

H/C decreases quickly. **Thermal cracking** is efficient. The molecular weight of the hydrocarbons released decreases with burial (Fig. 6.25). The depth at which this genesis of hydrocarbons occurs is called the **oil window** (Fig. 6.26). The threshold at which catagenesis begins varies from 60°C to 100°C and from 500 m to 4000 m, depending on the geothermal gradients of the sedimentary basins.

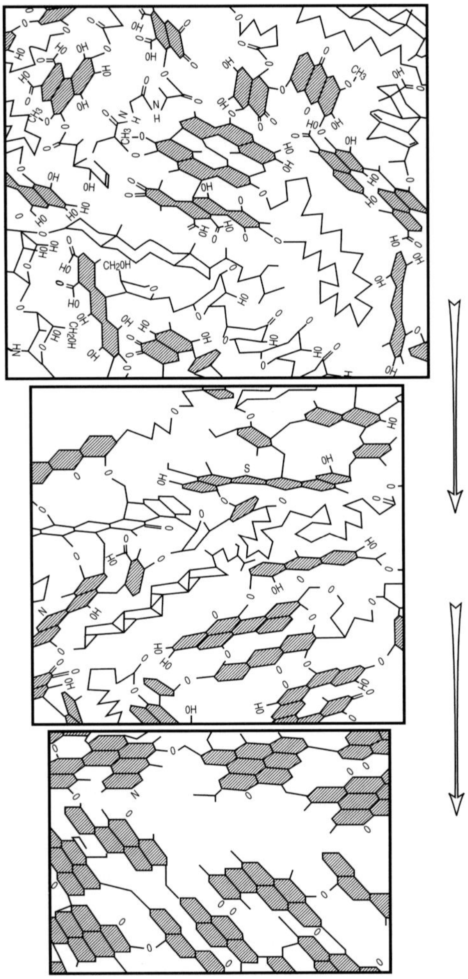

Fig. 6.25 Evolution of kerogen structures in the course of burial (from Béhar and Vandenbrouke, in Bordenave et al., 1993). This is a type III kerogen (see Fig. 6.20).

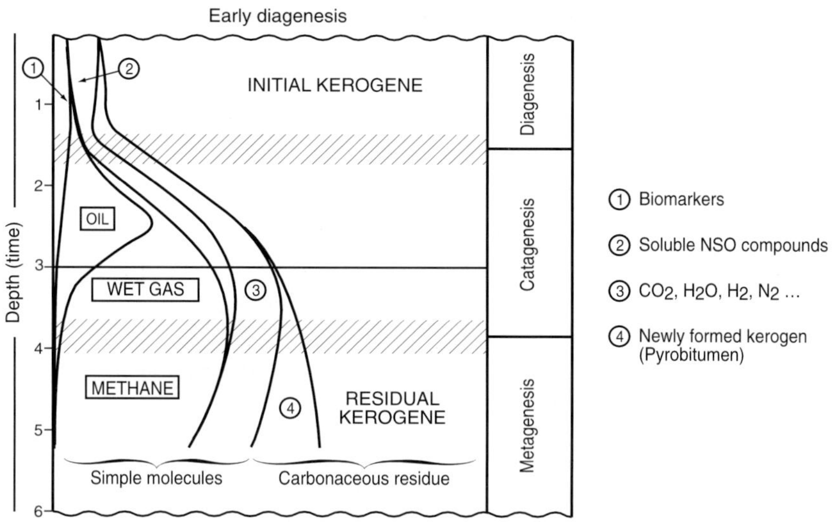

Fig. 6.26 Oil window stage in the formation of hydrocarbons (from Perrodon, 1985).

From the study of a great many sedimentary basins, it can be said without hesitation that the evolution of kerogens varies with depth and age (cracking time). As the depth increases, the thermal degradation is such that the liquid hydrocarbons are transformed more and more into light products. The first hydrocarbons formed are gradually diluted, along with biomarkers, and molecular fossils.

The residue, or pyrobitumen, is highly aromatic and analogous to refinery coke or to the initial kerogen. This is sometimes called neoformed kerogen.

- **Metagenesis** is the last phase of kerogen evolution in which dry gas (methane) or thermogenic gas forms by cracking of the previously formed hydrocarbons and the residual kerogen. This is called the **gas window**, generally starting at depths of 3000 m (Fig. 6.26).

6.2.3.2 Products Generated and Evolution Paths, Maturity of Source Rock

The three initial kerogen types undergo the same thermal evolution, but the composition of the products formed from them is conditioned by their characteristics (Fig. 6.27). The burial history will thus define **evolution paths** with increasing temperature profiles.

Actually, while the thermal effect is major, the initial composition, the pressure, and the effects of mineral catalysis are also at play, as is the time factor. Geochemists will analyze the genesis of petroleum in terms of chemical kinetics.

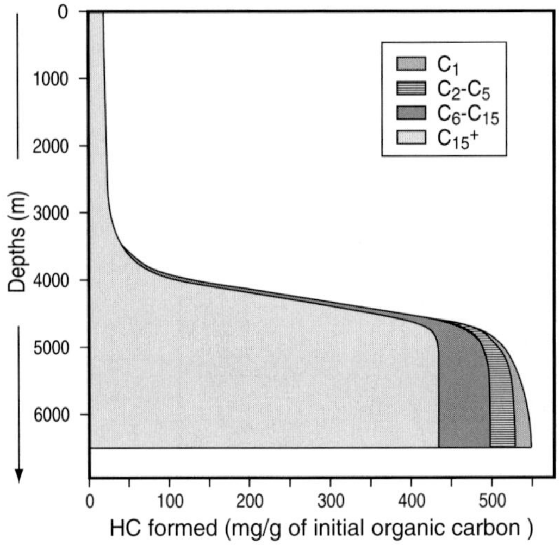

A

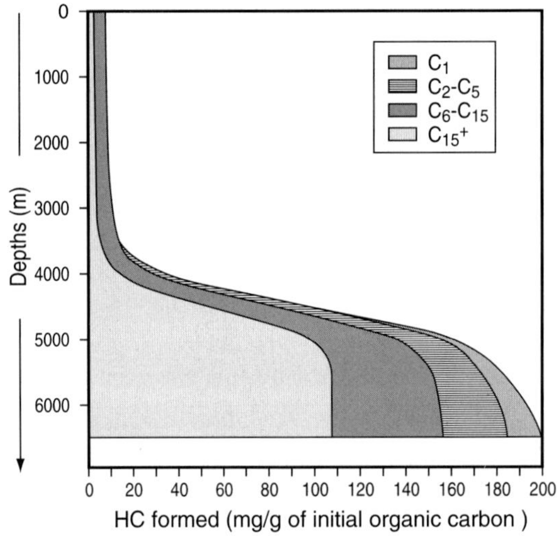

B

Fig. 6.27 Composition of products formed, depending on the initial type of kerogen.
A. Path of a type II kerogen from Viking Graben (North Sea), with normal geothermal gradient and burial at a rate of about 50 m/My. **B.** Case of Brent, under analogous conditions (from Huc, 1995).

The amount of CO_2 formed depends on the initial O/C ratio, and therefore decreases from type III to type I, while the quantity of hydrocarbons depends on the initial H/C ratio and therefore increases from type III to type I.

The situation is often complicated by the fact that the products formed can be degraded by the action of surface agents, especially by bacterial action. Secondary **degradation** then leads to heavy, viscous compounds, and to pasty to solid asphalts.

It is extremely important to estimate the degree of evolution, or **maturity** of a source rock. A source rock may be called **potential source rock** at the beginning of its burial, but may or may not become an **effective source rock** depending on the degree of burial. Different maturity parameters can be defined by various techniques using examples from the field or well core samples:

- **Vitrinite reflectance**. The light reflectance of a carbonaceous material indicates its molecular structure. Reflectance is measured by optical methods. Vitrinite is a maceral that is abundant in organic matter of continental origin (this analytical technique is derived from the one used in the coal industry). So this method is well suited for defining the evolution of type III kerogens (Fig. 6.28). The vitrinite reflectance (Ro) increases with the temperature the rock has reached. The oil window occurs at Ro values of between 0.5% and 1.35%.

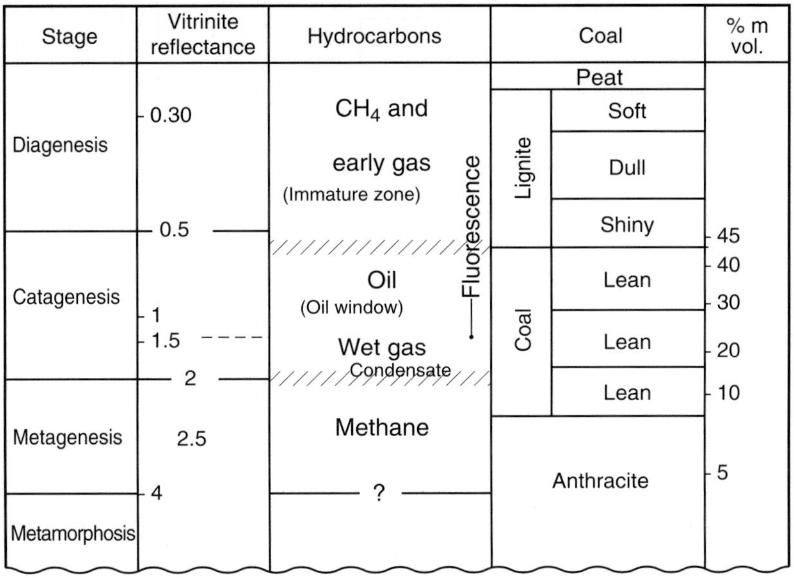

Fig. 6.28 Vitrinite reflectance, a parameter indicating the diagenetic transformation of hydrocarbons and coal (from Perrodon, 1985).

However, estimates are often subjective. Also, there is no vitrinite in type I kerogens.

- The most precise and complete method is **pyrolysis**. This is a cracking technique (not of the kerogen, considering the complexity and cost of preparing it, but of the rock) for characterizing the free hydrocarbons (gas and oil) contained in the rock sample, along with the oxygenated (CO_2) and hydrocarbonated compounds and the total organic carbon (TOC), found by oxidizing the residual organic matter in air.[1]

  The IFP developed the Rock-Eval instrument (linked to a micro-computer and specific software) as a "petroleum evaluation station" to make all these measurements. This instrument will not be described here, but let it simply be said that this pyrolysis technique can locate any sample on its evolution path and generate precise petroleum potential data.

The basic parameters expressed in milligrams per gram of rock (or kilos per ton) are the following (Fig. 6.29):

- The quantity of free hydrocarbons volatilized at 300°C for 2 min: this is peak $\mathbf{S_1}$.
- The quantity of hydrocarbonated products from cracking the kerogen between 300°C and 600°C: this is peak $\mathbf{S_2}$.
- The quantity of CO_2: peak $\mathbf{S_3}$.

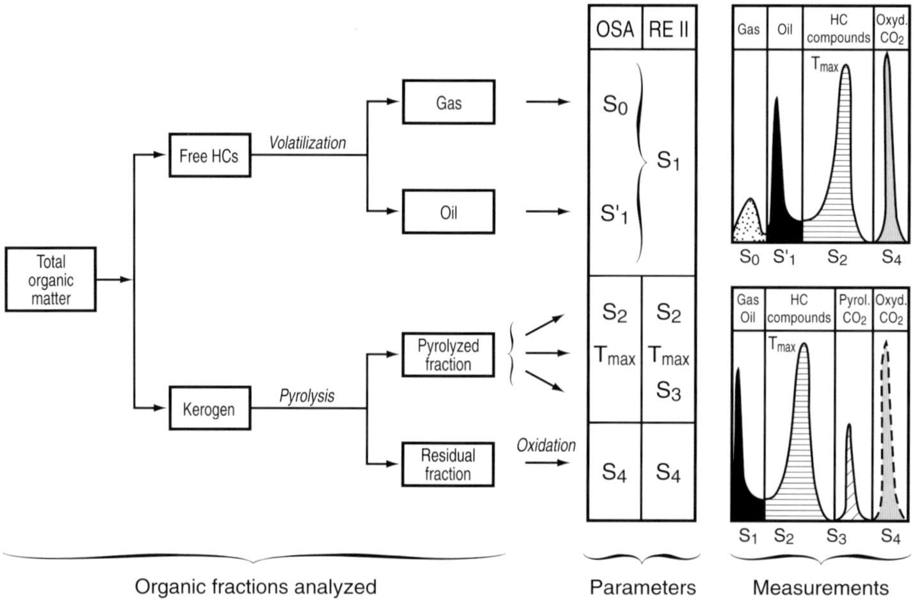

Fig. 6.29 The pyrolysis technique and the parameters used (from Espitalié et al., 1995).

1. See Espitalié, 1985-1986.

- The temperature reached at the top of peak S_1, called **Tmax**, is the main indicator of maturation provided by pyrolysis, but it depends on the original organic matter and it will be of use only for types II and III.
- The residual organic carbon content (percent by weight) obtained by combustion at 600°C (the resulting CO_2 corresponds to peak S_4).

From this data, the following are defined:

- The **total organic carbon content** (TOC), expressed as a percent by weight (residual organic carbon + pyrolyzed organic carbon deduced from peaks S_1 and S_2).
- The **hydrogen index** (HI) and oxygen index (OI), which are calculated from S_2, S_3, and the TOC.
- The **production index**, which is the ratio $S_1/(S_1 + S_2)$, is difficult to use because of the migration phenomena (see further on).

  Pyrolysis techniques can also be used to compare the evolution of kerogens with natural extracts.
- **Other methods**. Palynofacies analysis by optical method, infrared spectroscopy, and especially nuclear magnetic resonance (NMR) are used for more in-depth research. Well logging is commonly used in the industry to attempt to evaluate the quality of source rocks.

6.2.3.3 Generator System Dimensions

From a quantitative viewpoint, the dimensions of the system that has been plunged to sufficient depth are extremely important. The **oil window** has to concern a sufficiently large area and volume in order for the system to be efficient. The oil or gas **kitchen** is the volume of source rock taken to an effective burial depth, and helps determine the **petroleum potential** of a basin (Fig. 6.30).

The kitchen dimensions are often smaller than the extent of the source rock. In the course of geological history, the situation has evolved and the size of the kitchen changes too. Several generation stages may succeed each other and the kitchen may stop being functional (e.g. by uplift).

6.2.4 Biogenic, Bacterial Gas

At least 30% of all natural gas has been generated by the decomposition of organic matter at low temperature and low depth, by archaebacteria-type anaerobic bacteria.[1] The favorable formation conditions can be summarized as follows: strictly anoxic environment, low sulfate content, optimum temperature between 35°C and 45°C, abundance of organic matter, state of the organic matter (most ancient fields derive from type III, although type II is easier to

1. See Rice, in Vially, 1992.

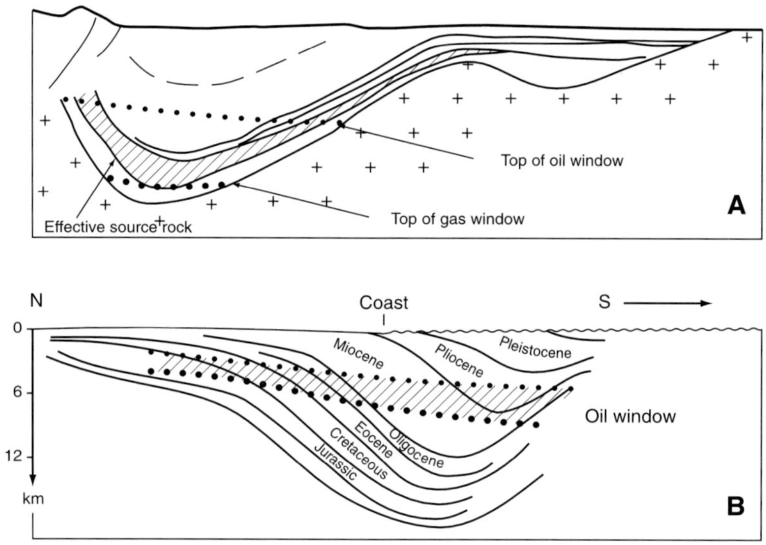

Fig. 6.30 Oil window and oil kitchen; gas window.
A. Diagram of a source rock in a sedimentary basin. The part of the strata under the oil window is the kitchen, the size of which is capital for generating economic quantities of hydrocarbons (from Magoon and Dow, 1994).
B. Example of the Louisiana margin. The various stratigraphic levels get through the oil and gas windows in a highly subsident basin with a high sedimentation rate (from Dow, 1978).

degrade by bacterial activity), adequate open pore space for the development of the bacteria, average sedimentation rate. The reduction and decarboxylation of CO_2, which can be called **fermentation**, are recognized paths of evolution, though their importance doubtless differs depending on the case. More than 99% of the hydrocarbons generated are methane, whence the term "dry gas". Accumulations are often found in young series (since the Cretaceous), at shallow depths in thermal immature sediments of various types. Those habitats that are especially favorable are deltas, turbiditic lobes, and clastic or calcareous platforms, though many fields also exist in the continental environment (coastal plain, marshes). Many gas reserves in Italy are of this type. One way of recognizing them is by comparing the C and H isotopes of methane, and the carbon dioxide.

6.2.5 Gas Hydrates

Gas hydrates are a very special case. Large quantities of methane are sometimes trapped in a crystalline ice matrix in the form of **methane hydrates**, which are sometimes called **clath-**

rates or **cryohydrates**.[1] The methane may be of biogenic origin (forming between 200 and 1000 m) or thermogenic (deeper, from 2000 to 9000 m). The concentration in a permeable porous environment is due to migration of the gas (see following pages) in a drain (fracturing can be very important here). Then, under low-temperature, high-pressure conditions, if the gas concentrations are high, the methane and other light products can be trapped in a dodecahedral or tetradecahedral lattice of frozen water forming "cage" molecules. The stability domain of gas hydrates is very limited, but it is effective enough under certain conditions to constitute enormous gas resources. There are two very different domains:

- Under continental permafrost (permanently frozen arctic zones) such as in Siberia where the gas is of thermogenic origin (Messoiakh field, for example).
- At shallow depths under the sea bottom, along passive or active continental margins, where clathrates can be identified on seismic profiles (Fig. 6.31), and where it seems that the gas is generated essentially biogenically.

The lateral extent of gas hydrates is such that these beds can serve as cap rock for oil and gas as they dysmigrate (see further on).

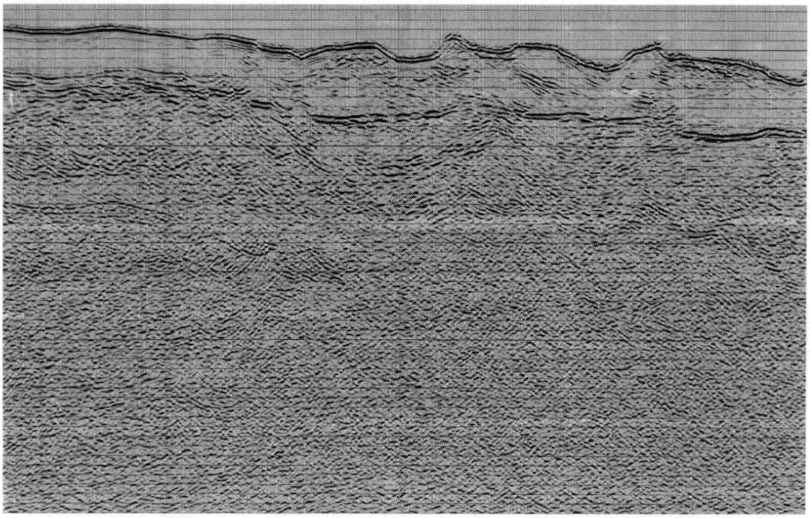

Fig. 6.31 Presence of gas hydrate in continental margin sediments.
A powerful bottom simulator reflector (BSR) can be found at a certain depth under the seabed. With the special temperature/pressure conditions at this burial level, this is known to be due to the presence of gas hydrate as it corresponds to the sharp seismic contrast existing at the base of the gas hydrate level.

1. See Yseux, in Vially, 1992, and Krason, 1994.

6.3 HYDROCARBON MIGRATION

The hydrocarbons and other petroleum products formed during the gradual maturation of a source rock sometimes stay in place. That is, if the initial permeability is and remains very small in the course of geological evolution, as is true of many **bituminous shales** (the Toarcian of the Paris basin, for example), very large quantities of oil remain trapped and dispersed in the rock matrix, without really becoming concentrated. From a commercial viewpoint, the genesis of hydrocarbons is of no use for this reason. The micropetroleum cannot be produced.

But fortunately, there are many cases where the fluids formed (gas and oil) can be drained efficiently and the petroleum can then concentrate into a commercial **accumulation**. Generally, this motion is through a porous, permeable medium (and because of these very characteristics) that is not initially a source rock, as the sedimentary organic matter in it was destroyed (oxidized) early on. This movement of the hydrocarbons is called **migration**. Migration is conventionally broken down into three phases (Fig. 6.32):

- **Primary migration** is when the oil and gas are **expelled** from the source rock to a porous, permeable drain.
- **Secondary migration** is when the hydrocarbons move through one or more drains toward an accumulation zone (reservoir) where they will concentrate if there is a trap (see further on).
- **Tertiary migration** designates the phenomenon by which initially trapped hydrocarbons sometimes move toward the surface or toward another trap, in a movement called **dysmigration**.

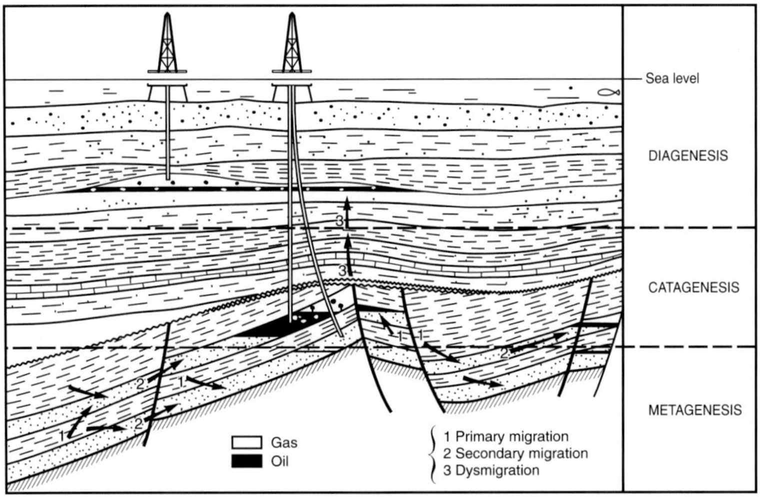

Fig. 6.32 Scheme of main phases of migration (from Pelet, 1985).

6.3.1 Primary Migration or Expulsion

Primary migration was the subject of much discussion for years.[1]

While solubility in the water expelled during rock compaction may facilitate the primary migration of the lightest products, this is not sufficient. Nearly all of the initial water in the sediments has already been expelled anyway, before entering the "oil window".

Molecular diffusion is of no consequence either.

The major cause of the oil being expelled in an argillaceous environment is **compaction**. With the gradual loss of porosity, once the bulk of the water has been expelled, the hydrocarbons can increasingly saturate the porous space and form a veritable network. Primary migration is a **two-phase flow** of water and hydrocarbons, separately.

This expulsion depends on the initial organic richness, its spatial distribution in the rock, the pore saturation, and the capillary forces. If the TOC is low, the saturation is insufficient to allow primary migration (Fig. 6.33). Expulsion will always depend on the pore pressure, which can be much higher in an argillaceous source rock (fine grain, low permeability). The expulsion rate depends on the permeability and pressure gradient in the source matrix and drain zone. It is felt that expulsion is facilitated by the generation of the gas, and various secondary factors such as mineralogical transformation and thermal effect can enhance excessive pore pressure.

Expulsion models now exist, but authors generally recognize that the exact physics of expulsion requires further research.[2]

Natural microfacturing also plays a role in facilitating the expulsion of hydrocarbons. This is doubtless an essential condition in calcareous environments where "explosive" migration sometimes occurs.

6.3.2 Secondary Migration

Secondary migration is the movement of hydrocarbons from their place of expulsion from the source rock to their locus of accumulation in a reservoir rock which, under certain conditions, will provide a trap (a cap rock or seal is needed).

The expulsion from fine-grain rock (all source rocks) to a reservoir assumes there are **migration paths** along **drains**.

Secondary migration mechanisms are relatively well understood. The various fluids (water and relatively insoluble petroleum constituents) expelled from the rock are lighter than the rock and will circulate by buoyancy in accordance with their density, following a

1. See discussion in Vandenbrouke, 1993, and Durand, 1988.
2. See Ungerer, for example, in Bordenave, 1993.

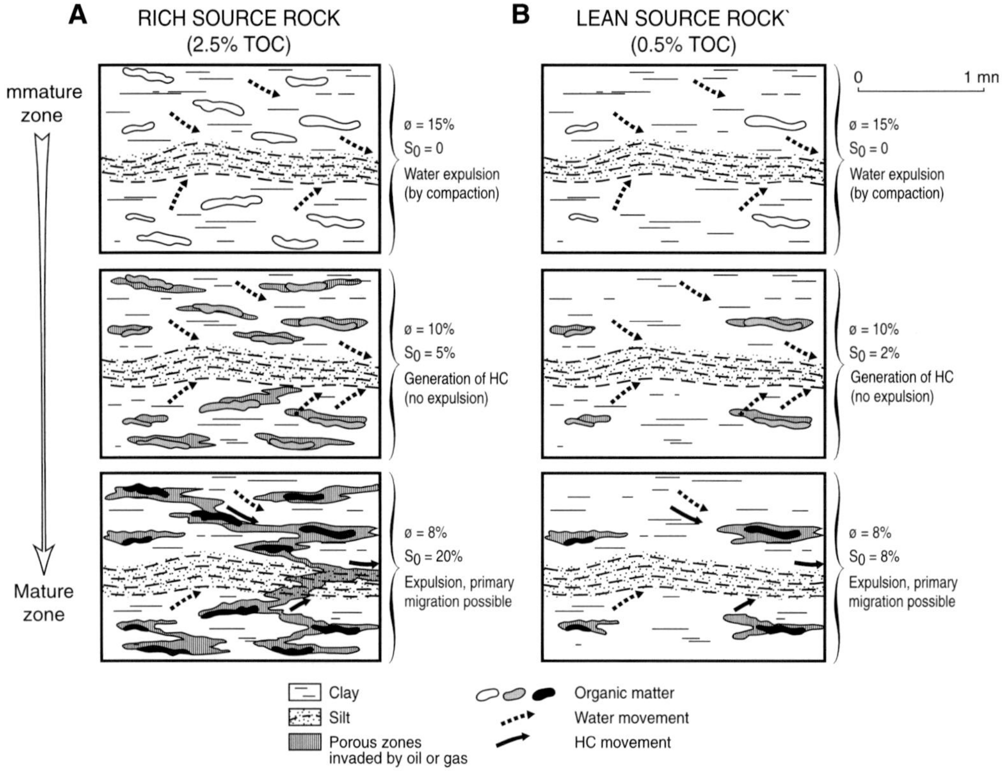

Fig. 6.33 Expulsion of hydrocarbons: conceptual scheme of two-phase primary migration (from Durand, 1996).
A. Rich source rock (~2.5% TOC). **B.** Lean source rock (0.5% TOC) (from A. Huc).

drain toward the surface in distinct phases. This circulation depends on the capillary forces, and the displacement is formalized in the multiphase Darcy law.

Each of the phases has its infiltration velocity, relative permeability to water, oil, or gas, and relative pore pressure.

Migration paths are determined by:

- The **type of drains**, which are either of porous permeable rock (sand, sandstone, limestone, such as grainstone) with the same characters as the reservoir rock (see further on), or a system of open cracks (faults and microcracks) (Fig. 6.34A), or bounding surfaces, (unconformities).
- The migration **distances** will depend on the proximity of a trap. In a few rare cases (such as bituminous shales), the distance may be infinitesimal or very moderate (Fig. 6.34B). However, very long migration paths of up to several hundred kilometers

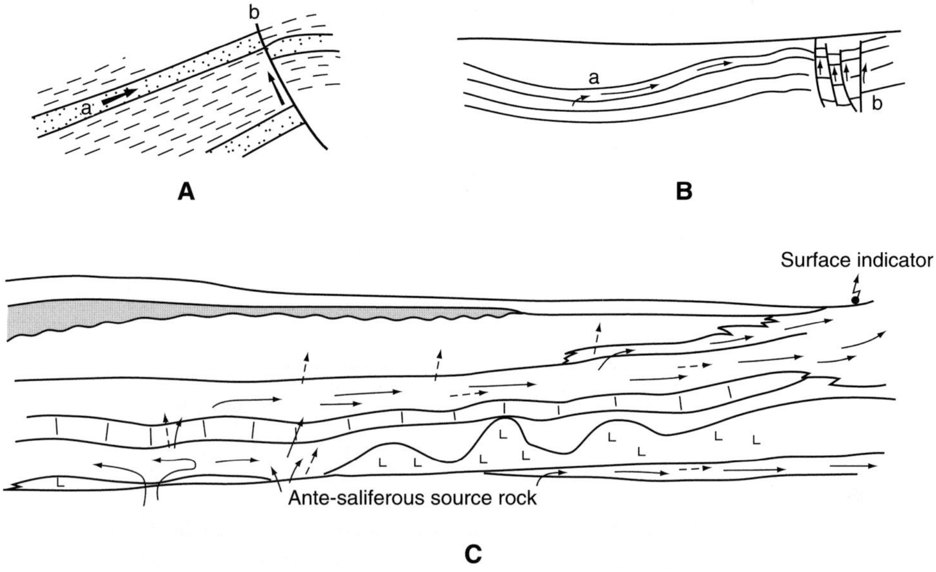

Fig. 6.34 Migration paths.
A. Types of drains: migration within a bed (a) and along a fault (b). **B.** Migration distances. (a) long, horizontal; (b) short, vertical. **C.** Coastal basin of the Congo with long migration paths of oil (solid arrows) and gas (dashed arrows) (from EAP document, Augier, in Chiarelli and Du Rouchet, 1977).

have been recognized today, and their history can sometimes be complex, depending as it does on the geological history (Fig. 6.34C).

- **Migration styles**. Vertical or horizontal displacement will dominate, and the style of drainage path depends on the structural context.
  Two-dimensional models are being developed today to determine migration paths and timing. The circulation model is two-phase, with the oil and gas separate from the water.
- The basin's **general hydrodynamism** may be important. It will be more or less active and vary through the basin's geological history.
- **Migration timing** is a very important consideration when reconstructing the history of a sedimentary basin and evaluating its petroleum potential (this will be discussed again later in the chapter). If the migration occurs before any traps are formed, it is clear that no hydrocarbon concentration will be possible. Expulsion phenomena continue to operate over several million years during secondary migration (Fig. 6.35).

The **drainage area** is another dimension to be considered in the petroleum evaluation. This means the geometry and size of the system that has reached maturity for generation and expulsion.

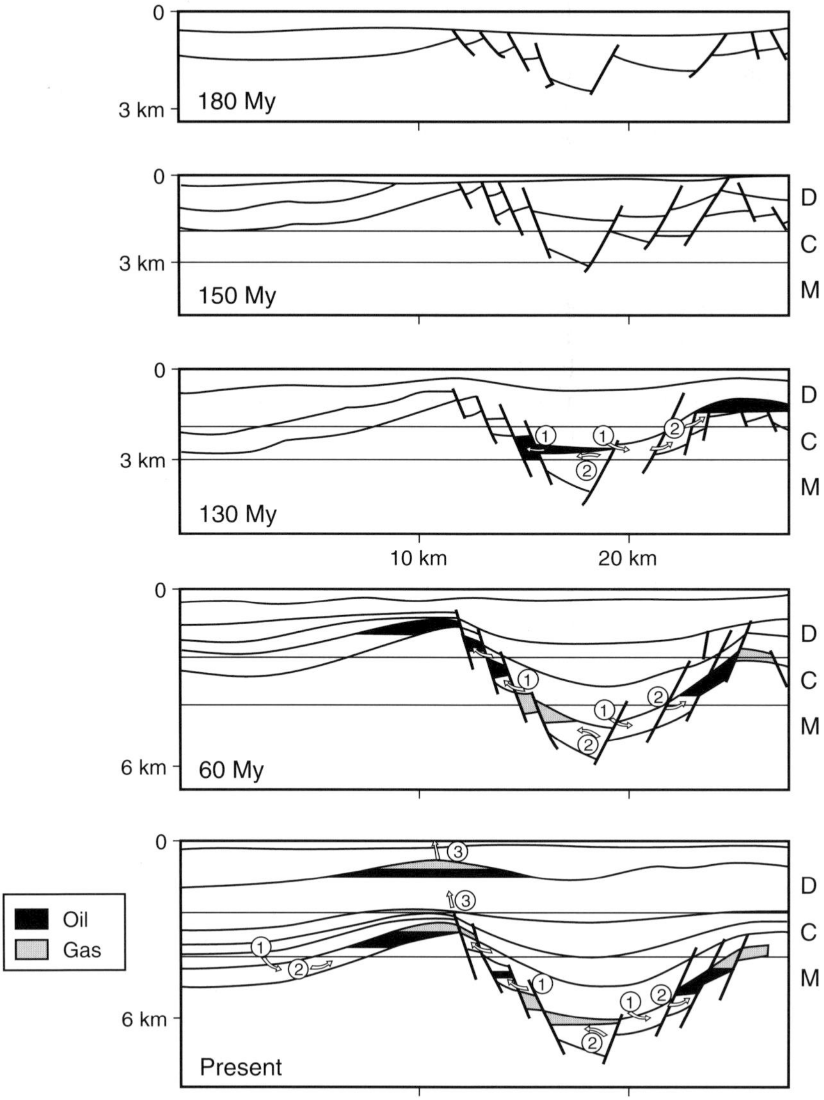

Fig. 6.35 Duration of expulsion and migration.
In this example of the North Sea, expulsion (1) and migration (2) began 130 My ago and continued for millions of years. Today, dysmigration (3) is sometimes active (from Durand, 1996). D, C, M are domains of diagenesis, catagenesis, and metagenesis.

The locus of accumulation is called the **trap**. There are different types of traps, which will be described later. The number of them, their dimensions, and their formation timing will determine the migration distances and the size of the drainage areas. In all configurations, a trap consists of a **reservoir** and a **cap rock** to seal it. If these conditions are not (or are poorly) met, the migration will continue into tertiary migration, or dysmigration.

6.3.3 Dysmigration

Tertiary **dysmigration** is the "leakage" of hydrocarbons which, starting from a trap where they have accumulated and concentrated, then move toward the surface or toward another trap, often called a secondary field (Fig. 6.36). The sealing effect of a trap or field depends on the impermeability of its cover. In favorable cases, this prevents the petroleum from following its normal upward motion. If the seal is poor, a dysmigration will occur. This will happen, for example, along an open (permeable) fault or through heterogeneous cover in which the capillary pressure barrier becomes weaker than the buoyancy forces.

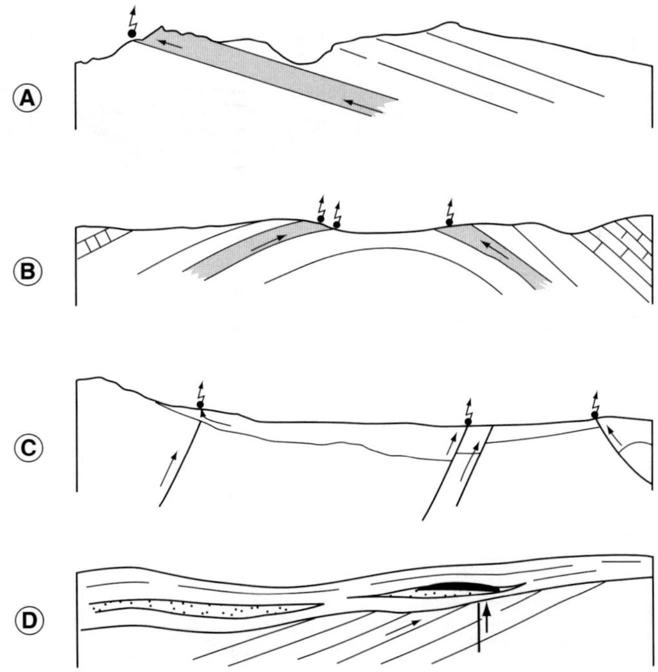

Fig. 6.36 Dysmigration: oil shows and secondary fields.
A. Surface show on a homocline ramp. **B.** Surface show on an anticline. **C.** Surface show along faults (normal and reverse) and recovery trap overhead a fault. **D.** Secondary field over an unconformity.

When hydrocarbons reach the soil surface or sea bottom, there is **seepage**.

The oil show that drillers speak of when they find oil or gas seepage does not necessarily have the same meaning, but often refers to dispersed micropetroleum or a small concentration of hydrocarbons.

When many points of seepage are found in the exploration process, this information is important in the analysis of the petroleum basin.

Lastly, as was seen before, dry gas can be trapped as it rises to the surface. Gas hydrate accumulations then form under the permafrost or at shallow depths on the ocean margins.

6.3.4 Alteration, Degradation

The crude oil that accumulates by migration into a reservoir is generally associated with gas (Fig. 6.37). After this migration, it is subject to secondary alteration that can generate vola-

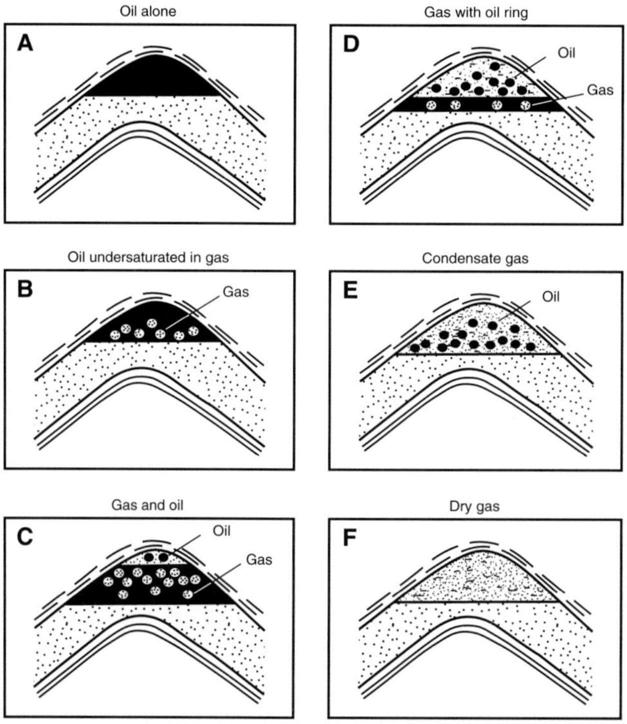

Fig. 6.37 Different types of oil and gas associations in a reservoir (from Durand).

tile products. Light petroleum gases under pressure and temperature conditions at depth are called condensates, as distinguished from light oil.

The oils found in a reservoir are generally of a different chemical composition than the initial oil because of the fractionation occurring during the expulsion and secondary migration phases. Only the biomarkers remain as good tracer of the source rock (Fig. 6.38). Moreover, the oils may **mature** greatly after accumulating in the reservoir (Fig. 6.39). Thermal effects can increase both the gaseous products and pyrobitumens in a complementary cracking process within the reservoir. **Deasphalting** occurs when gaseous hydrocarbons are dissolved and asphaltene is precipitated, making the oil lighter.

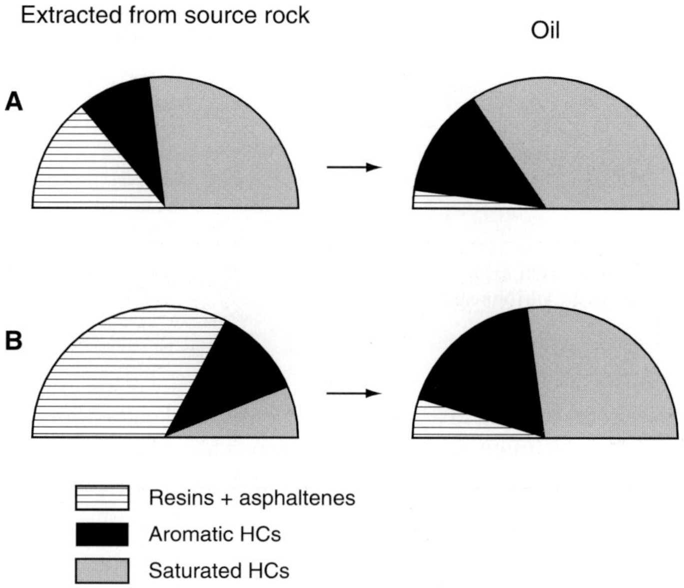

Fig. 6.38 Comparison between the composition of oils and that of the source rocks they come from.
A. Sandy clay series (Algerian Devonian). **B.** Calcareous series (French Jurassic). The resins and asphaltenes are in very low proportion (from Sallé, Debyser, 1976).

But the petroleum products can also undergo a secondary physical and chemical alteration by washing with the water in the bed, segregation by gravity, selective dysmigration, and biodegradation. The effects of **biodegradation** are doubtless the most important. The action of bacteria is sufficiently effective to gradually oxidize the hydrocarbons. The various processes depend on the conditions within the reservoir, the amount of nutrients, and the aerobic or anoxic situation.

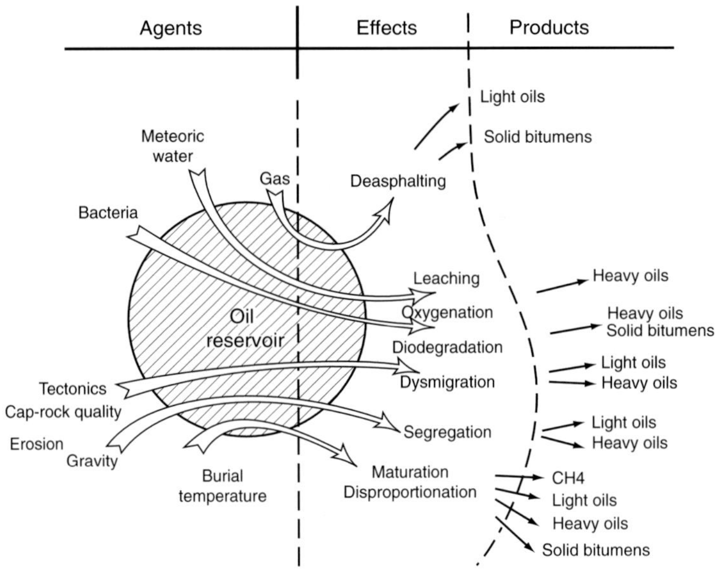

Fig. 6.39 Different effects of alteration agents after the oil is trapped in a reservoir (from various authors).

Most alteration processes lead to an increase in density and end up producing the derivative products **asphalt** or **bitumen**, which are viscous residues that obstruct the reservoir.

Under special conditions, the degradation may be accentuated and **erosion** may occur. This will gradually destroy the initial field.

6.4 RESERVOIRS, TRAPS, AND OIL FIELDS

6.4.1 Reservoir Rock

6.4.1.1 General Characters and Physical Properties

The previous chapter showed how a sediment lithifies into rock gradually under the effect of diagenetic actions during burial. All rocks include a major mineral fraction of grains, matrix, and cement, with voids or **pore space** which may be filled with fluids (Fig. 6.40). The way these voids are interconnected defines the intergranular space, and certain physical properties of the rock determine its **porosity** and **permeability**.

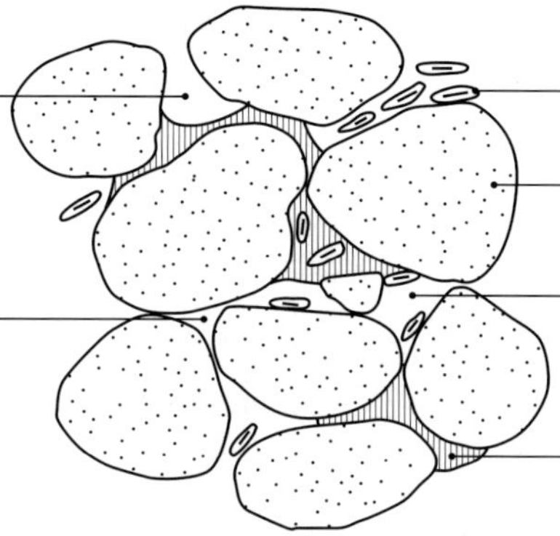

Fig. 6.40 Pore space in a rock.
This is a cross-section of a granular rock where the voids, or pores, are easily identifiable. Of course, the pore space has to be imagined in three dimensions since the porosity is not always only intergranular.

In oil geology, reservoir rock is a rock whose pore space is such that a large quantity of petroleum can be accumulated in it. So it is defined primarily by its **effective porosity**, ω or φ (designating the connection between the voids), and its **permeability** ***k***.

Generally, the relation between porosity and permeability is neither simple nor direct. These two petrophysical characteristics depend on many parameters relating to the primary characters of sedimentation and the secondary characters of diagenesis.

Remember that porosity is defined as the ratio between the volume of pores (occupied by a fluid) and the total volume of the rock:

$$\frac{V_{\text{pores}}}{V_{\text{total}}} = \frac{V_{\text{total}} - V_{\text{solid}}}{V_{\text{total}}} = \varphi$$

Any porosity less than 5% is small, while any beyond 20% is good. This is measured either on a sample in the laboratory, or from logging records at the well. The **matrix porosity** is defined by the voids in a matrix and is thus different from the **fracture porosity** measured with respect to the microfacturing in the reservoir.

Remember that matrix porosity is independent of the grain size but does depend on the grain shape, grading, arrangement. It mainly depends on burial depth and the diagenetic effects (Fig. 6.41).

Permeability characterizes the aptitude of a rock to allow fluids to circulate. This is expressed in darcys by Darcy's law, but rock permeability is so low that the millidarcy (mD) is commonly used. Generally, on the scale of the reservoir, the environment is not isotropic: the physical properties vary greatly, both vertically and laterally. So the usual reference is to an average or equivalent permeability (Fig. 6.42).

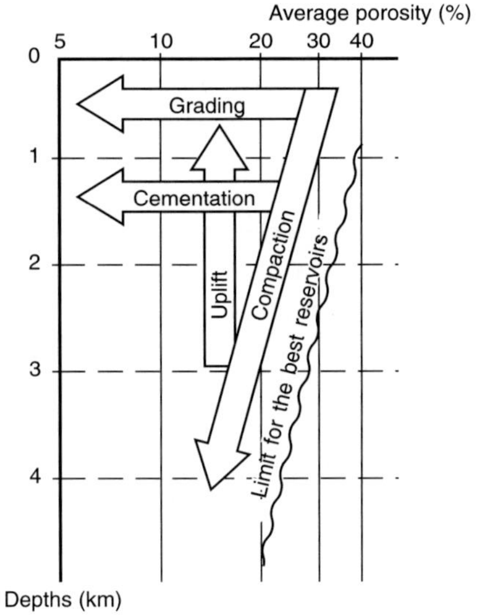

Fig. 6.41 Effects of burial on porosity.
Mechanical compaction and mineral diagenesis are the main factors that will reduce porosity in the course of burial. The diagram here is for sandstone reservoirs (from Perrodon, 1980).

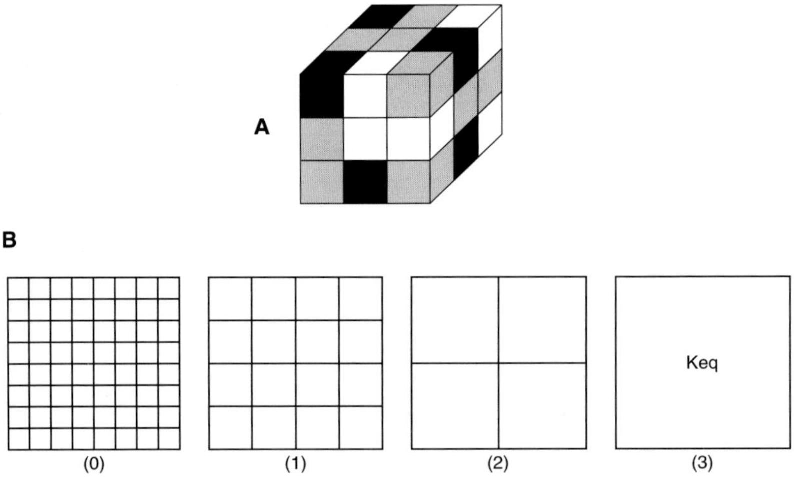

Fig. 6.42 Permeability: heterogeneous on the scale of the reservoir.
A. Common representation of a heterogeneous environment. **B.** Square grid scheme used for calculating equivalent or average permeability (from Kruel Romeu, 1994).

Also, permeability is a relative idea. The intrinsic permeability measured with a single fluid does not correspond to the situation in the field, where water, oil, and gas exist together. The relative permeability for one of these fluids will depend on its saturation and, depending on this, the product may be anhydrous (water-free) oil, or a mixture of oil and water, or just water.

Wettability, capillary pressure, interface tension, and compressibility, are also important parameters that vary greatly depending on the different types of rock.

The idea of drainage is also important in a multiphase environment. A non-penetrant fluid can displace a penetrant one, such as gas displacing oil. This property is used in secondary recovery processes.[1]

A rock with good reservoir characteristics (high φ and k) may not contain any hydrocarbons at all (as is often the case!); but as the petroleum is generated and migrates, hydrocarbons can accumulate under certain conditions, which will be discussed later in this chapter. The volume of pores is then more or less filled with fluids, which is called **saturation** because the voids are more or less saturated.

There are different expressions for saturation: saturation in oil (*S*o with the "o" standing for "oil"), in gas (*S*g), or in water (*S*w), defined by:

$$So = \frac{Vo}{V_{\text{pores}}} \qquad Sg = \frac{Vg}{V_{\text{p}}} \qquad Sw = \frac{Vw}{V_{\text{p}}}$$

If the oil and gas saturations are known, the volumes accumulated in the reservoir can be calculated.

A reservoir rock is thus characterized by a set of **petrophysical properties** and by the fluids it contains. It is always analyzed from two viewpoints: static and dynamic.

6.4.1.2 Different Types of Reservoirs

Reservoirs are usually classified in one of two broad categories: terrigenous detritic and calcareous.

Among **detritic reservoirs** of sands and sandstone, the reservoir quality will depend on the particle size grading and **shaliness**, which affect the permeability considerably. The sand-shale ratio is often used to define reservoir quality or a set that is likely to behave as reservoirs (Fig. 6.43). Geological history gives many examples from the Cambrian to the Neogene, and in different types of environments (continental, littoral, deep ocean) to illustrate this type of reservoir (Fig. 6.44). Sands and sandstones account for about 60% of reservoir rock discoveries.

The internal architecture of siliciclastic detritic reservoirs is determined by the deposition systems. Eolian and fluvial deposits can make very good reservoirs (such as the Sarir field of the Cretaceous in Libya), as can lacustrine sands (China). The most attractive reser-

1. See Cossé, 1988.

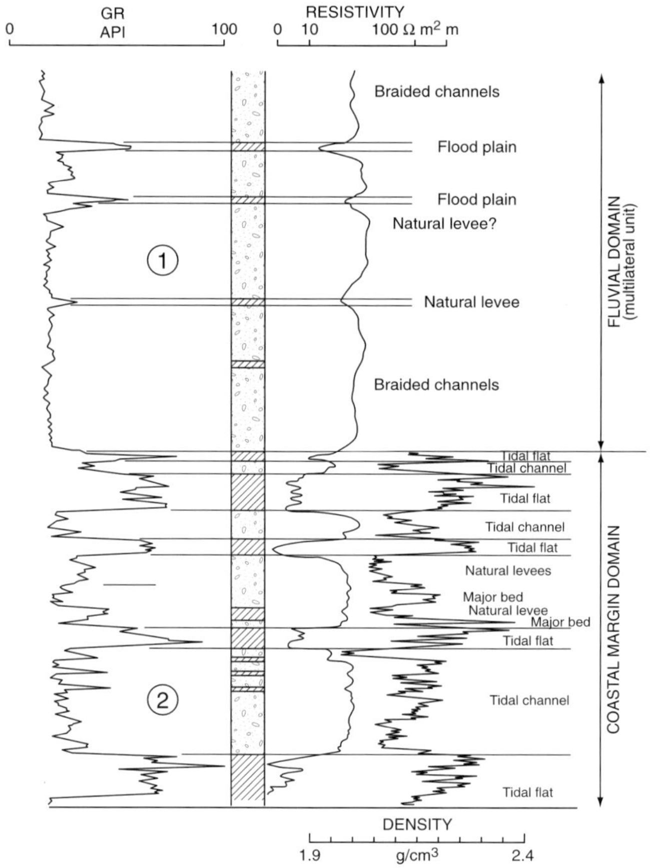

Fig. 6.43 Gritty clay series.
The shaliness is well defined by the gamma ray (GR) logging measurement (from the Chambre Syndicale de la Recherche et de la Production du Pétrole et du Gaz Naturel, 1986).

voirs are surely the delta and coastal sands, of which there are various examples (Indonesia, North Sea, Wyoming), but deep gravity deposits (Brazil offshore, North Sea) should also be considered.

Calcareous reservoirs are also extremely varied, with lacustrine limestone, platform deposits, chalk, dolomite, and reef constructions (Fig. 6.45). These account for more than 30% of known reservoirs in oil fields. Their primary characteristics are very important (reefs, for example), but the still pronounced diagenesis plays a considerable role in determining both the porosity and permeability. Many calcareous reservoirs have good petrophysical characteristics only if they have sufficient **fracturing**.

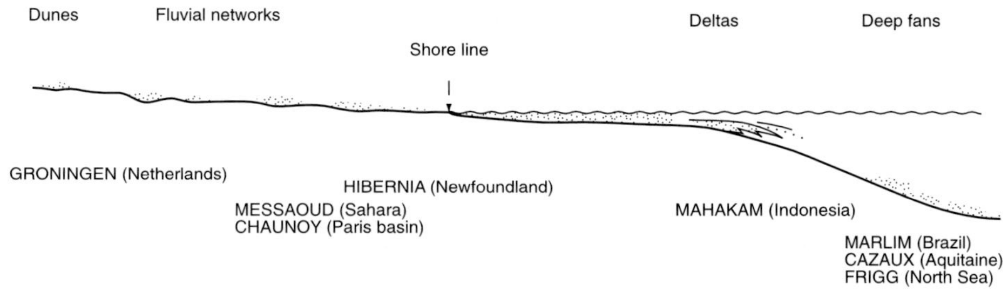

Fig. 6.44 Clastic environments and oil reservoirs.

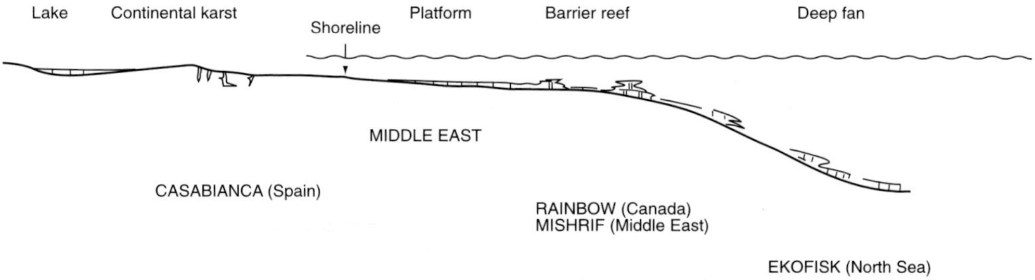

Fig. 6.45 Calcareous environments and oil reservoirs.

Sedimentologists classify the lithofacies in different ways, and generally use a classification by porosity type in carbonates (Fig. 6.46). One or another type will predominate depending on the type of carbonate.

Statistically, the petrophysical characters (φ and k) vary depending on the initial situation of the deposition environment.

6.4.1.3 Architecture, Heterogeneity

On the scale of an oil field, reservoirs generally exhibit heterogeneities. These variations in their petrophysical characteristics depend closely on the initial deposition conditions, and the reader is referred to Chapter 3 to understand the architectures of each of the oolithic reservoirs, deep submarine fans, or reef constructions, each of which have their own characteristics.

The extent of lateral continuity is doubtless one of the simplest parameters to define, but it is also the one that varies the most. Certain gritty reservoirs of the large Paleozoic shelves (Saharan Ordovician in the Libyan and Algerian fields, for example) have remarkable continuity over tens of kilometers (Fig. 6.47). The sandy lenses of the Mahakam delta and the

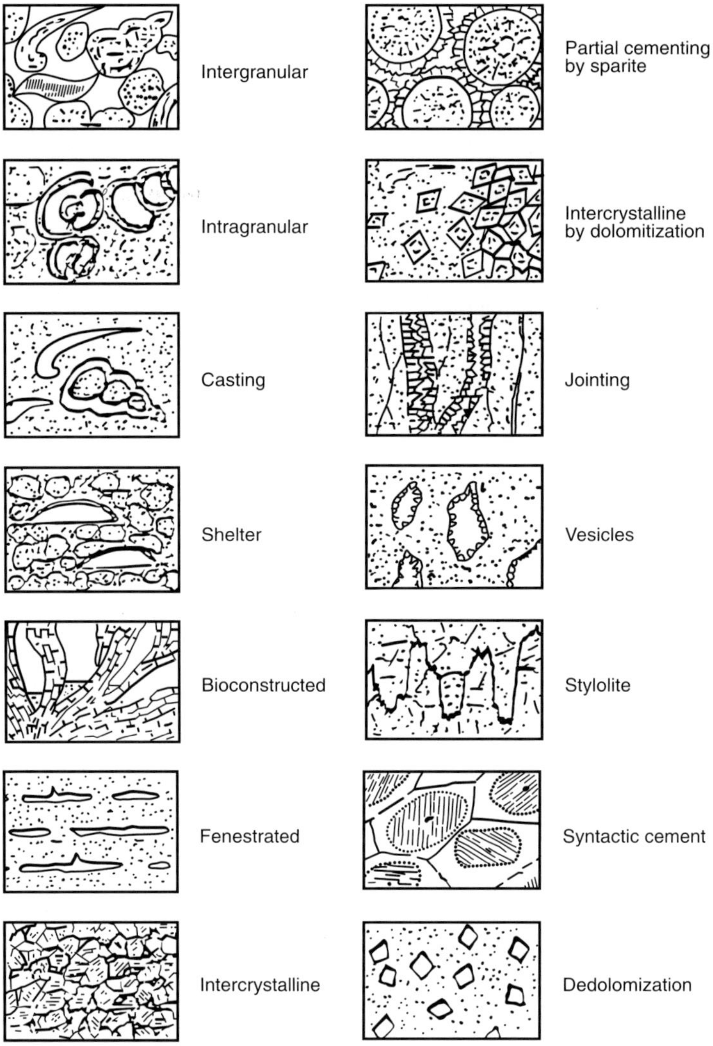

Fig. 6.46 Porosity classification of carbonates (modified from Choquette and Pray, 1970).

reefs of Alberta, on the other hand, are small. Under very favorable conditions, the total **power** (thickness) is considerable, as in certain cases of the Middle East (Fig. 6.48), but it also may be very small.

Generally, reservoir heterogeneity should be viewed as both lateral and vertical. A reservoir will be termed **monolayer** if it is a single reservoir corresponding to a homogeneous

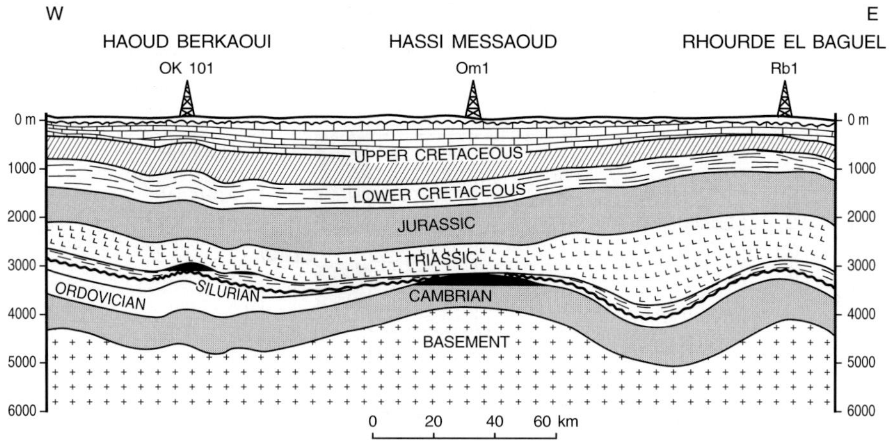

Fig. 6.47 Geological section of the Hassi Messaoud field: example of a gritty reservoir of large lateral extent.

sedimentological and stratigraphic episode. Sometimes banks are joined and constitute a single reservoir. The banks are then amalgamated, but the reservoir is monolayer. In a **multilayer** reservoir, the banks are close to each other and may have similar or differing characteristics, but are separated by an impermeable layer. In all cases, the original facies variations and (especially for calcaraceous reservoirs) the diagenetic effects will lead to major heterogeneities that need to be evaluated for bringing in an oil field. The sedimentological and structural analysis needed to evaluate these heterogeneities today need to be as fine as possible in order to define the most porous drains, estimate the permeability barriers, and the production rate (Fig. 6.49).

Reservoir production models are being established today with grids of different fineness using real well data, when available, and analogical models.

6.4.2 Traps and Sealing Rock

Hydrocarbons migrate through drains before accumulating in the reservoir, where they will **accumulate** and **concentrate** only if there is a **trap**. If a trap is to form, there must first of all be a **sealing** rock (Fig. 6.50), which is an impermeable envelope on the top of the reservoir preventing further migration toward the surface. In this rock, the pore entrance pressure has to be greater than the buoyancy pressure on the hydrocarbons circulating in the reservoir.

The most common seals are fine-grain rock, clay and silty clay, evaporites, halite and anhydrite and, more rarely, carbonates (mudstone type limestones, for example). The effectiveness of the seal varies with the type of rock. It is very good for evaporites as their plastic-

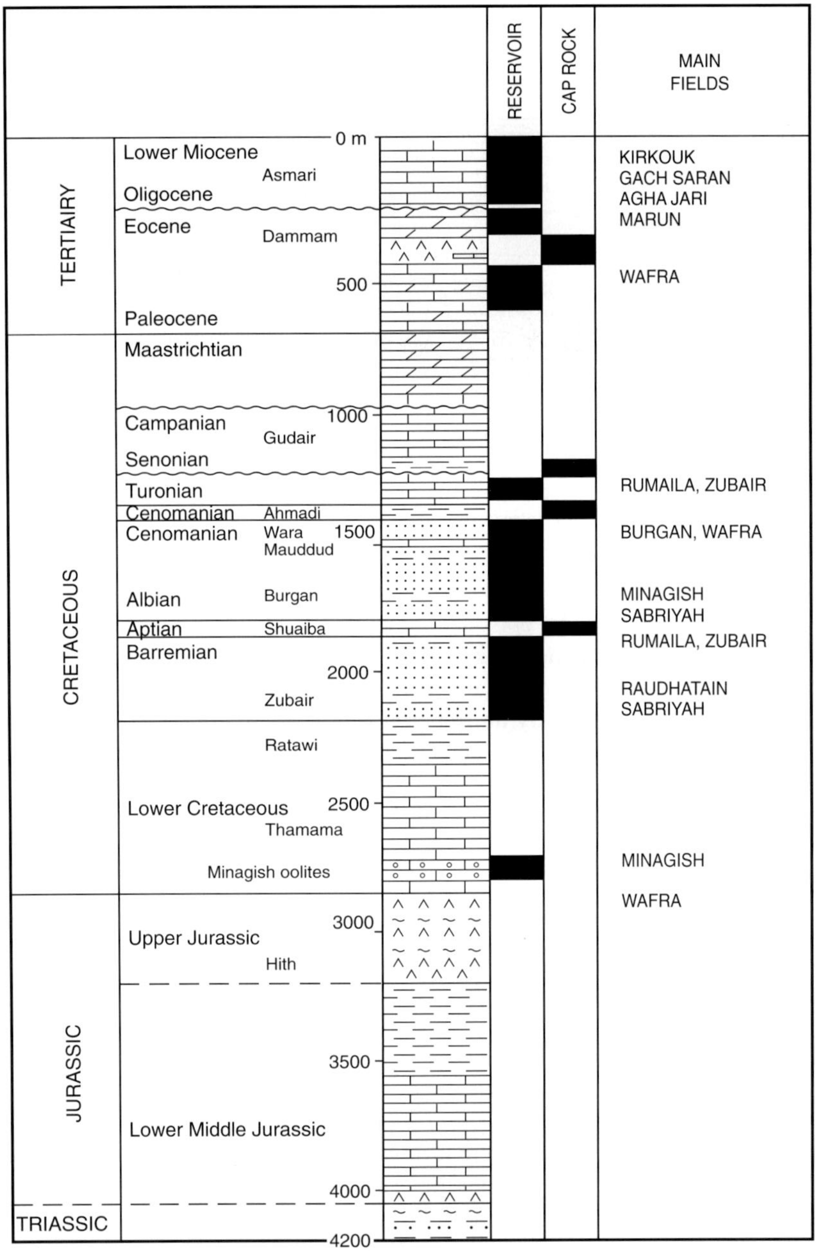

Fig. 6.48 Typical series of the foreland zone in the Middle East (Iraq–Kuwait): illustration of the superimposition of many high-power reservoirs (from Perrodon, 1980).

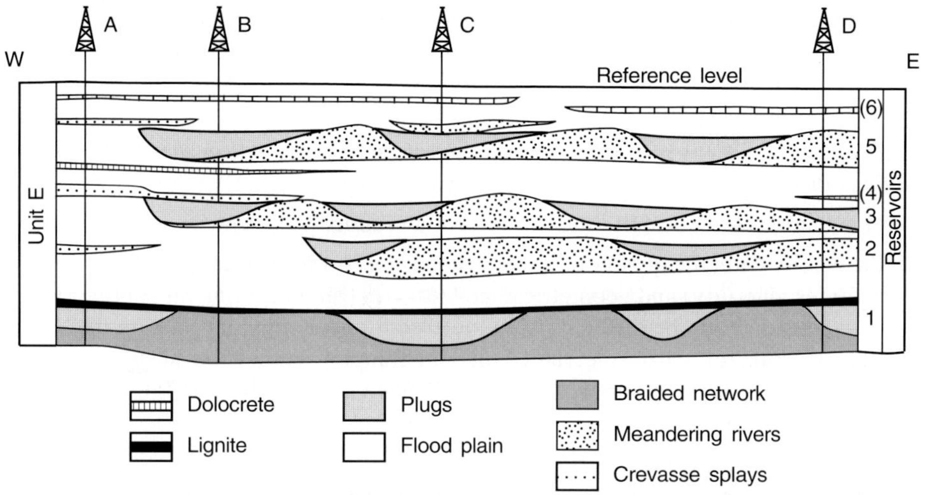

Fig. 6.49 Lateral heterogeneity of arenaceous reservoirs. Example of the Keuper in the Paris basin (from Mathieu et al., 1993).

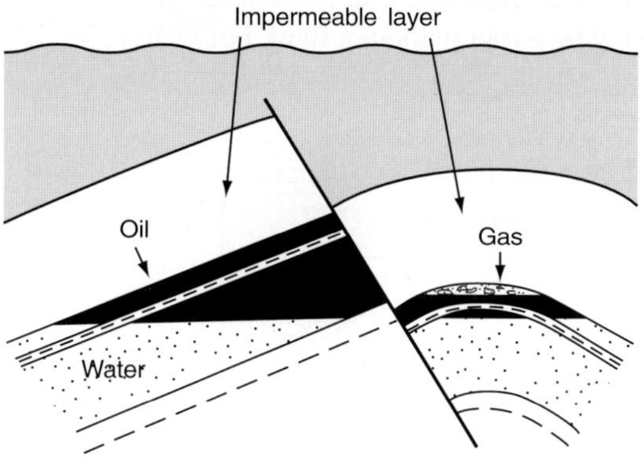

Fig. 6.50 Trap and seal.
Petroleum can accumulate only if there is a trap (a faulted anticline here). In order for the trap to be effective, the reservoir has to be sealed by an impermeable rock.

ity limits the fracturing effects. If the seal rock power is small, its effect as a permeability barrier would be small and the lateral continuity will be extremely important. But if the power is high, the rock will serve as good sealing even if it is not totally impermeable.

The idea of seal depends on the fluids considered. Almost no seal rock is totally impermeable to gas. This is why it is the measure of the gas content in the drilling mud and cuttings that indicate that a field is approaching.

Certain clay seals are also the source rock for an immediately underlying reservoir.

On the whole, then, the quality of a seal depends on its physical properties (permeability, capillary forces, ductility) and geometric characteristics (thickness, lateral continuity). Local seal rock is considered to be different from more continuous regional rock on the scale of a basin. The question of what role faults play in trapping cannot be answered generally because, under different conditions, faults may behave as cover by trapping the petroleum, or may allow it to leak out.

A seal rock atop a reservoir is not a sufficient condition to allow hydrocarbons to accumulate into a **field**. There must also be a special geometric **trap** arrangement ensuring enclosure that will stop the migration. The trap may generally be defined as an anomaly in the geometry of the seal-store cover, making it concave downward (Fig. 6.51). But many types of traps exist in nature. Their size depends on their **closure**. The vertical amplitude of the anomaly and its lateral extent in three dimensions conditions the volume of hydrocarbons that potentially can be held in place. Detecting, localizing, and evaluating closure is one of the prime objectives of oil exploration. The 3D seismic techniques of today are extremely useful for this. There may be a trap without a field, but there is no field without a trap.

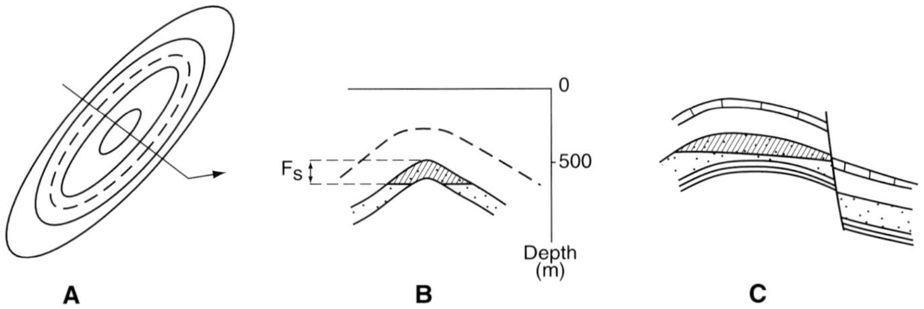

Fig. 6.51 Trap closure.
A. Cartographic expression: the curves represent the depths of the reservoir roof. The dashed line is the limit of the impregnated zone. All of the curves close regularly. The closure is visible only in 3D. **B.** 2D cross-section clearly expressing the value of the structural closure, F_s. **C.** Example of closure associated with a sealed fault.

In a normal field, the fluid densities are such that they are stratified, with the lighter-than-water gases on top, oil underneath (forming a ring in the case of anticlines), and water at the bottom (Fig. 6.37). This arrangement is clearly defined in large traps, but sometimes varies when the hydrodynamic factor, for example, plays a role in the trapping process.

There are different types of traps (Fig. 6.52):

- **Structural traps** (Fig. 6.53), defined by the deformation of the beds, depend on the basin's tectonic history. There are of two kinds of these:
  - **Anticlinal traps** (more than 30% of all fields) are the commonest type. They are the most efficient (nearly 80% of the World's recognized reserves).
  - **Fault traps**, though common (7%), are generally not major traps (1% to 2% of reserves).

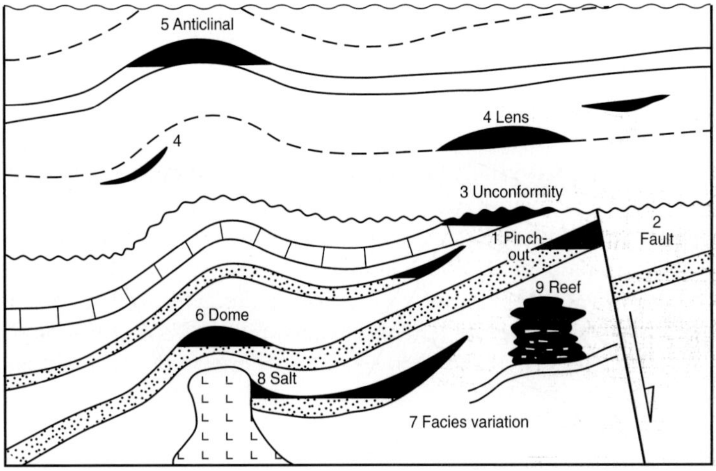

Fig. 6.52 Different types of traps.

This distinction is somewhat simplified because traps are usually a **combination of folds and faults** (Fig. 6.54). The rollover anticline is a classical example of this, but others illustrate the complexity of the situation. They may be found in intense compressive systems like the foothill folds of the Middle East where they often produce large asymmetrical anticlines, or in complex mountain chain structures (Columbian Andes), but they can also be found in an extension systems along listric faults (African margin, North Sea) where tectonic reversal is often of importance.

- **Salt domes** deserve special mention. Their structuring may be circular or elongated, buried or penetrating (diapir), with several types of traps associated (Fig. 6.55):
  - In rocks in anticlinal deformation over the salt (Parentis, Ekofisk).
  - In cap rock directly overhead the salt.
  - In reservoir beds gradually deformed as the salt column rises.

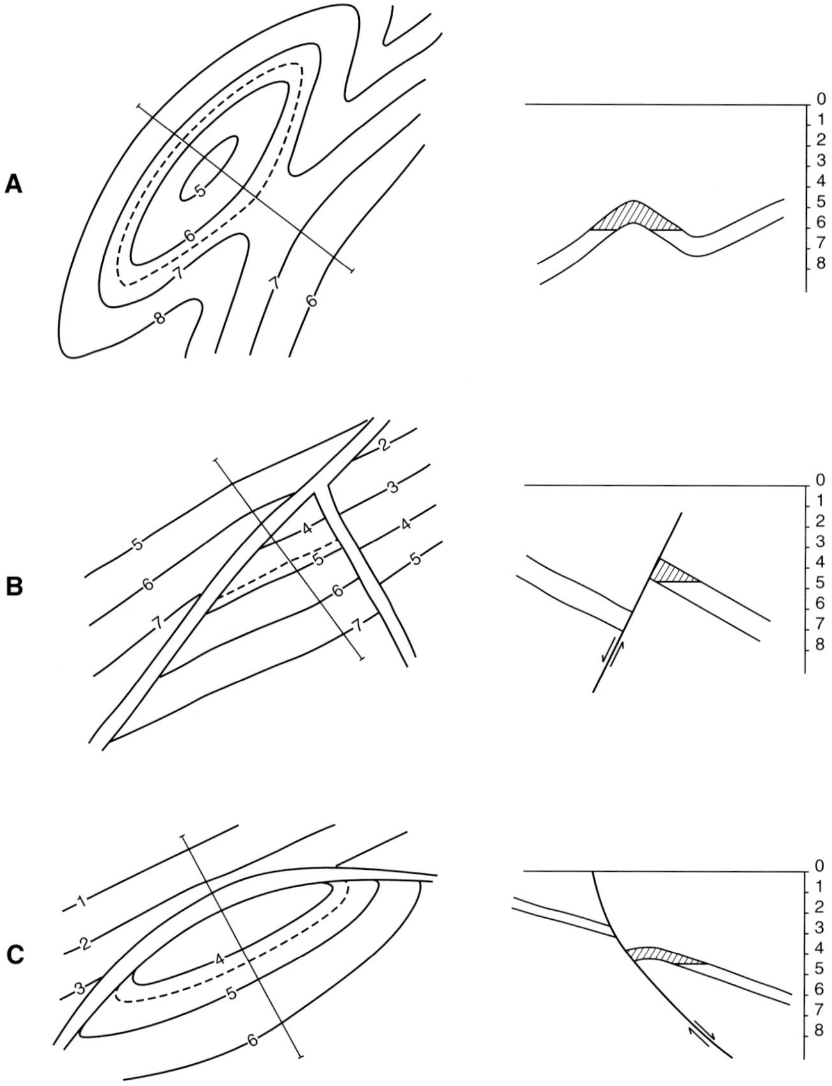

Fig. 6.53 Structural traps (from Guillemot, 1986).
A. Anticlinal. **B.** Fault. **C.** Fault and rollover anticline.

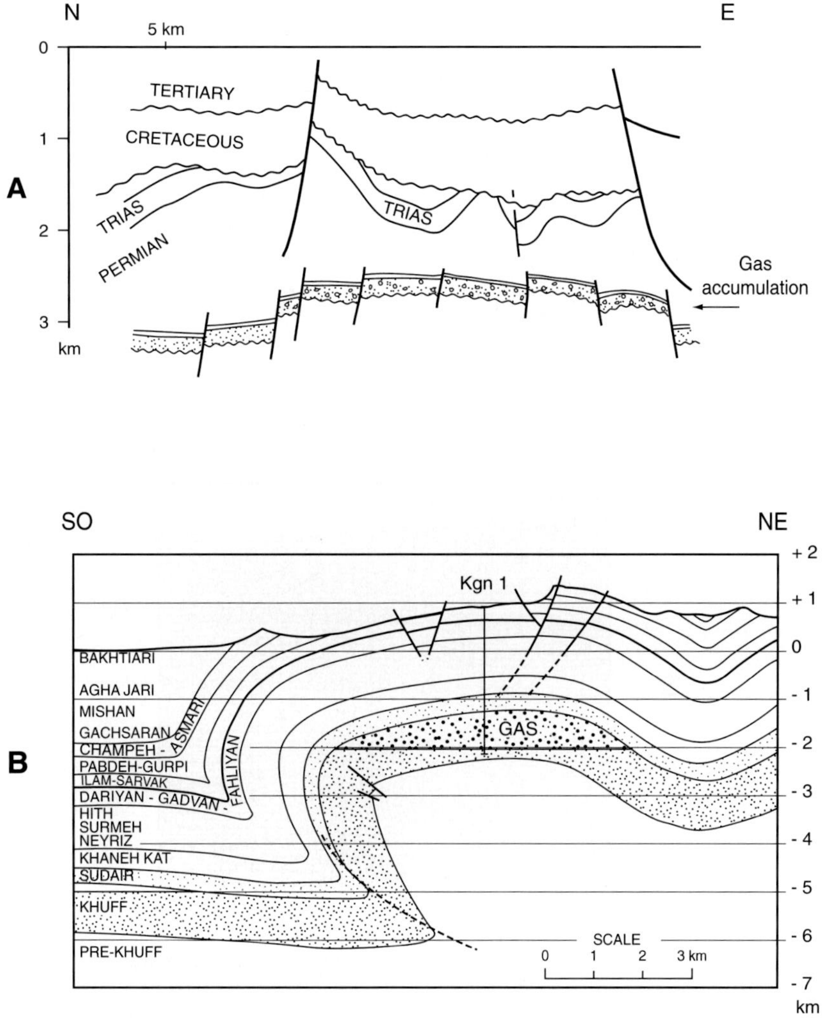

Fig. 6.54 Examples of structural traps combining folds and faults.
A. Section of the Groningen field (Netherlands) (from Stouble and Milios, 1970, in Perrodon, 1985). **B.** Section of Iran's Kangan field (from Reyre, 1975, in Perrodon, 1985).

A

B

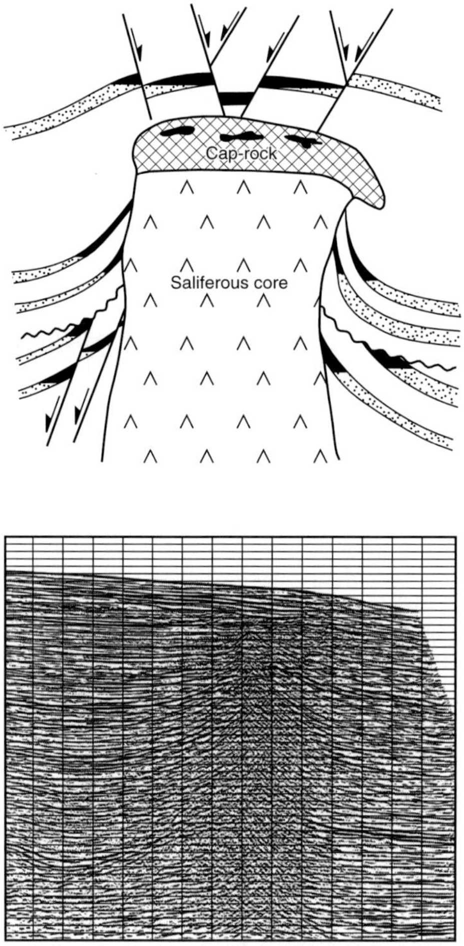

Fig. 6.55 Traps resulting from salt domes.
A. Schematic (from Levorsen, 1956). **B.** Seismic view.

- **Stratigraphic traps** are mainly geometric anomalies is stemming from the basin's stratigraphic history. These traps are very often mixed and the structural shape is necessary, especially for lateral closure. There are different types (Fig. 6.56):
  - **Unconformity traps** have originated some of the largest oil fields (Fig. 6.57). The reservoir here is truncated by erosion and ends in an erosion pinchout. (This type of trap could be likened to structural traps, since it is a matter of deformation occurring before the erosion surface is created.)
  - **Stratigraphic pinchout and facies variations**. The deposition conditions, with their transgressions and regressions, generate discontinuous bed formations. These

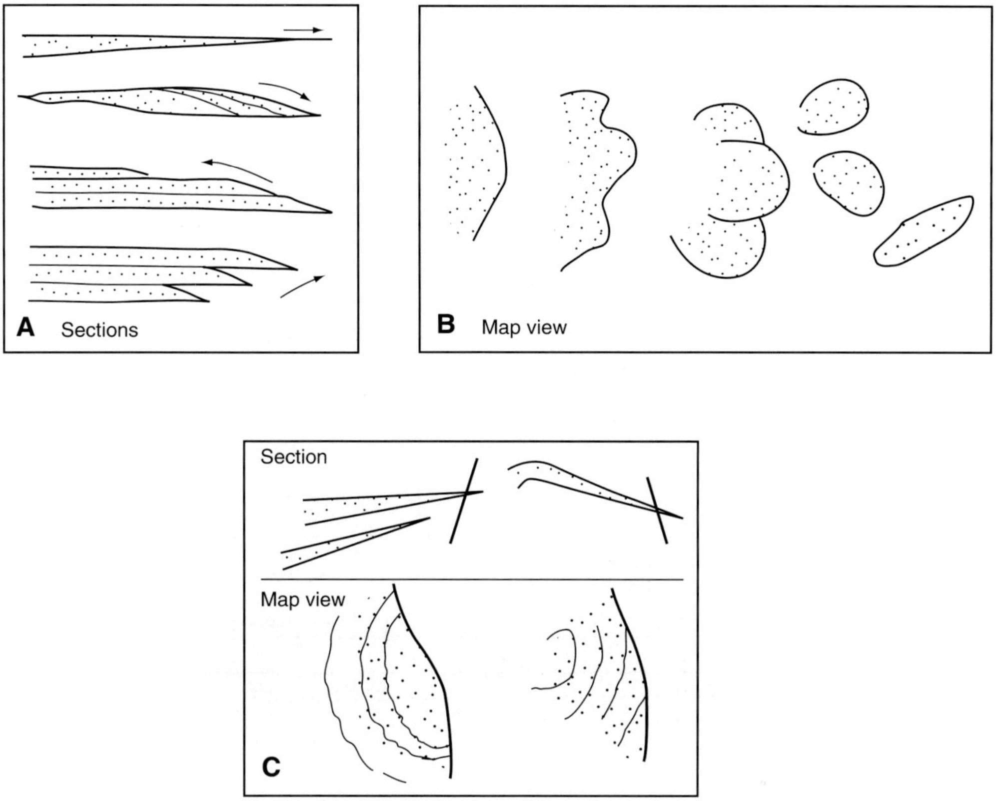

Fig. 6.56 Stratigraphic traps.
A. Vertical sections of different configurations. **B.** Corresponding configuration in a map view. **C.** Combination with a structure.

are stratigraphic or faciological pinchouts. Many variants produce traps: fluvial and tidal channels, delta bars and fans, deep lobes (Fig. 6.58). While arenaceous facies are the most typical, carbonates also produce such traps. Faciological trapping is usually accentuated by diagenetic effects.

- **Reef traps**. This is a very special category of fields (Fig. 6.59) resulting from the construction of vegetal and (usually) animal constructions that are often dolomitized. Reefs are organized in constructions of highly varied sizes, from the simple reef to the pinnacle, atolls, fringing reef, and continuous great barriers.
- **Geomorphological traps** are quite rare but remarkable. This can be found in buried paleorelief of a geometry recalling that of reef bodies (Fig. 6.60).

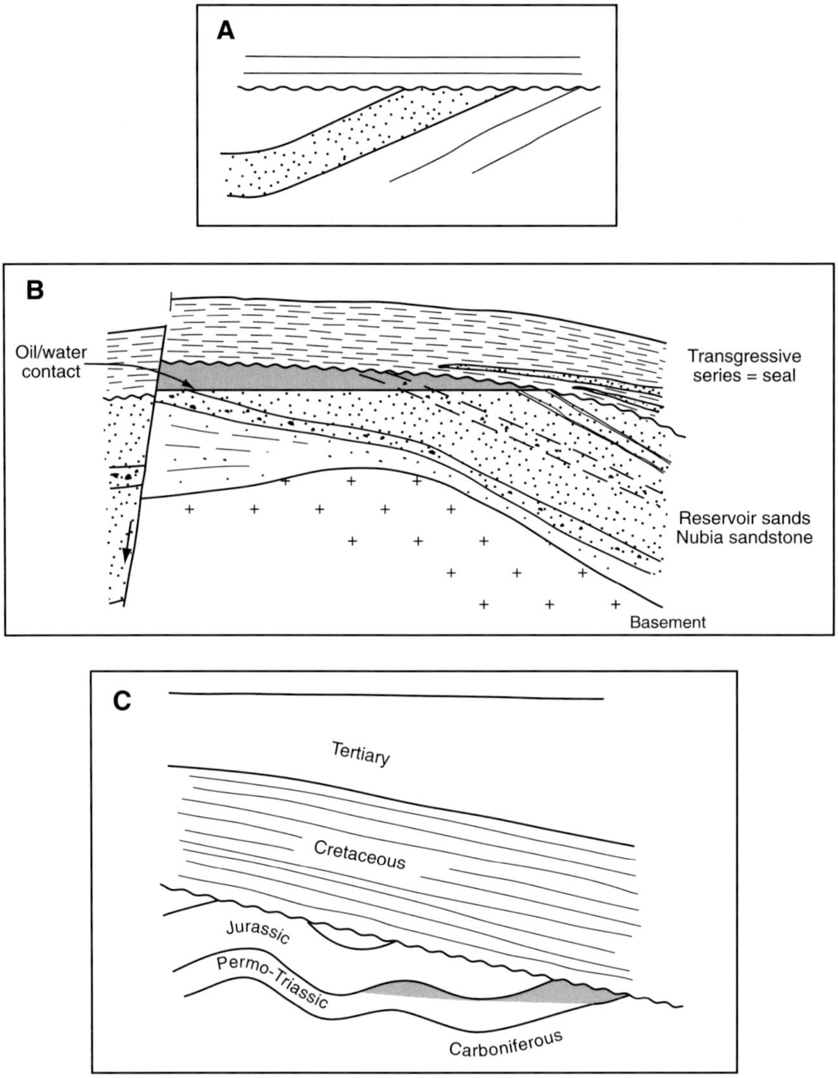

Fig. 6.57 Unconformity traps.
A. General diagram. **B.** Example of Libya's Sarir field. A potential stratigraphic trap can also be seen in the transgressive series overlying the unconformity. **C.** Example of the Prudhoe field in the Antarctic. This is a dual-trap scheme with an anticline and homocline under the unconformity (from Perrodon, 1985).

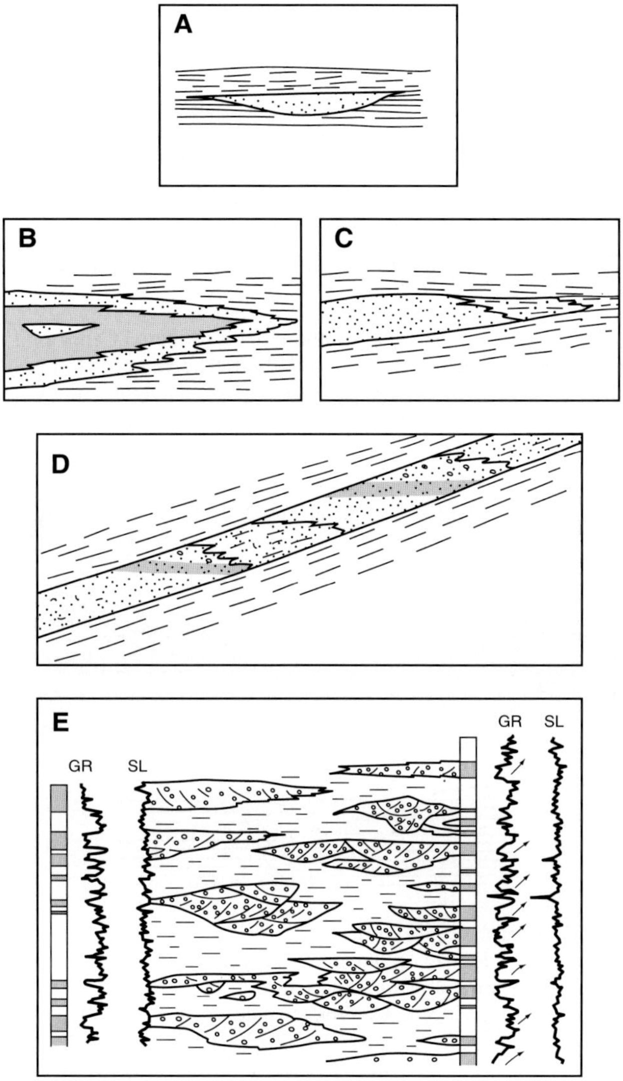

Fig. 6.58 Stratigraphic and faciological pinchouts.
A. Arenaceous lens. **B.** Prograding, retrograding littoral wedge. **C.** Delta front. **D.** Differential trapping scheme along a monocline with facies variations. **E.** Example of delta channels reconstructed from well data (Niger delta, ENSPM document, 1986).

Fig. 6.59 Example of reef trap: Rainbow fields (Alberta, Canada).
Several pinnacles, isolated from the reef barrier, contain oil (in gray) and gas (small circles) above a water table that varies in height with the various edifices.

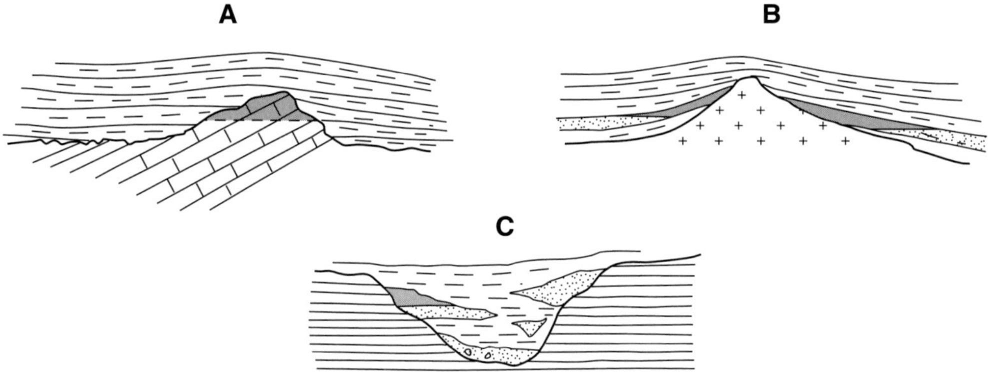

Fig. 6.60 Geomorphological traps.
A. Paleorelief buried in a positive structure under discordant seal. **B.** Paleorelief creating a positive structure in the sedimentary cover (role of differential compaction). **C.** Glacial paleo-valley with heterogeneous filling.

- **Other traps** are found in special fields where the trapping factors depend on the diagenetic and structural history:
  - **Diagenetic traps**, where the reservoir quality is modified by cementation and dissolution (Fig. 6.61), include the special cases of karsts (karstic field) and entrapment by asphalt plugs.
  - **Fracturation traps**. Beyond the structural traps described above, where fracturing necessarily occurs, fracturing sometimes provides the necessary trapping if major

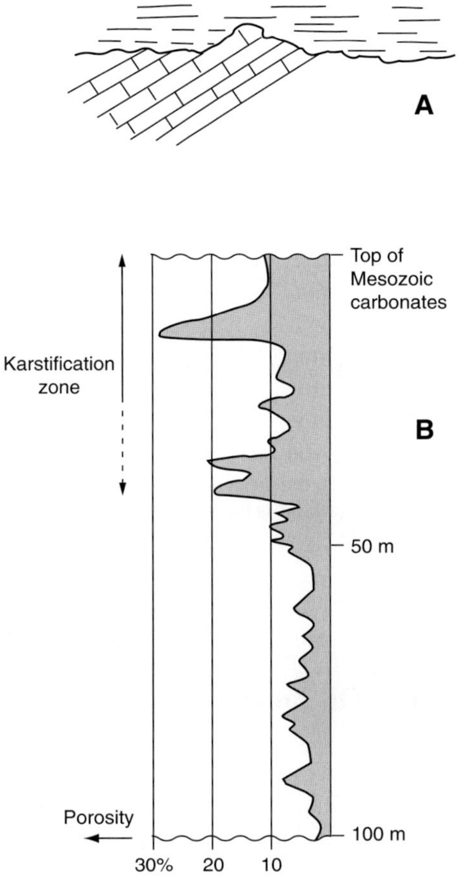

Fig. 6.61 Geomorphological karstic reservoir: the Casabianca field (Valence basin, Spain).
A. General scheme. The limestones of the Mesozoic were deformed during the Tertiary and etched by erosion, leaving a geological ridge buried by Miocene clastics. **B.** Continental alteration prior to Miocene transgression considerably improved the initial porosity by developing a major karst in line with the relief (from Watson, 1992).

structural deformation is lacking. Special cases of fields with fractured basements or volcanic intrusions fall within this category.

- **Hydrodynamic traps**. Under certain conditions, trapping is determined by the **hydrodynamic** process (Fig. 6.62), which, even if the closure is not good, can retain the oil and gas in an abnormal geometric situation.

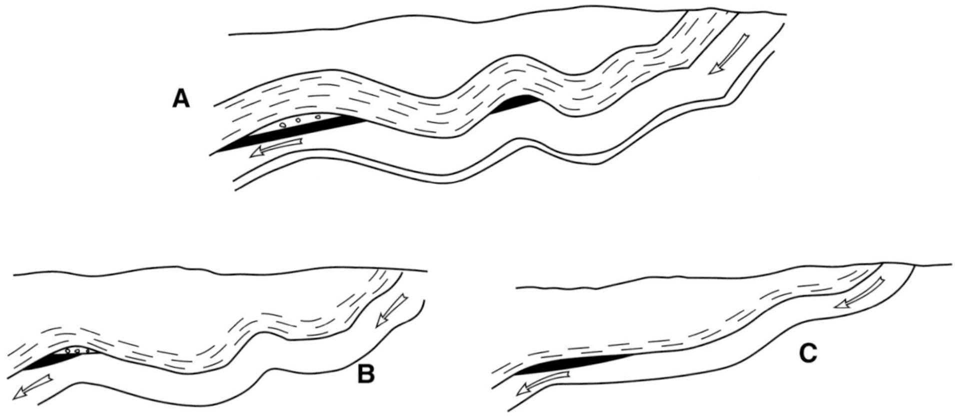

Fig. 6.62 Hydrodynamic traps.
The structural trap generally exists, but active hydrodynamics can entrain part of the oil (**A**). If the gradient is high, the oil can be flushed out entirely (**B**). Sometimes, a simple bench with no structural closure can form an effective trap.

6.4.3 Oil Pools and Fields, Oil Zones

When the trap is filled with hydrocarbons, it is an oil pool. After a **pool** is discovered, it is defined by a development program, and it becomes a **field** for **development**. The idea of a field is generally larger, because it could include the production of several localized pools in different reservoir horizons (Fig. 6.63).

A **pool** is defined in terms of a commercial accumulation. In other words, beyond the concentration of hydrocarbons in a trap, there is the idea of **mining** it. A petroleum accumulation is the initial quantity of hydrocarbons (oil and/or gas) in the site. The **reserves** are the recoverable quantity of hydrocarbons, or the final cumulative production or **recoverable reserve**, and the recoverable coefficient is the reserve/accumulation ratio.

Once the height of the deposit is known, with the impregnated zone and closure, a first estimate of the reserves can be made by **delineation**.

Optimizing reserves calls for a sequence of operations in developing the field. Before production starts, a development scheme is established corresponding to production forecasts (with production tests and production model with well locations). Then the field's characteristics (pressures, flow rates of water, oil, and gas) are analyzed systematically through the life of the field and actions are programmed on the producing wells or complementary well program.

An estimated average of 25% of the oils is recovered, and at least 75% of the gases.

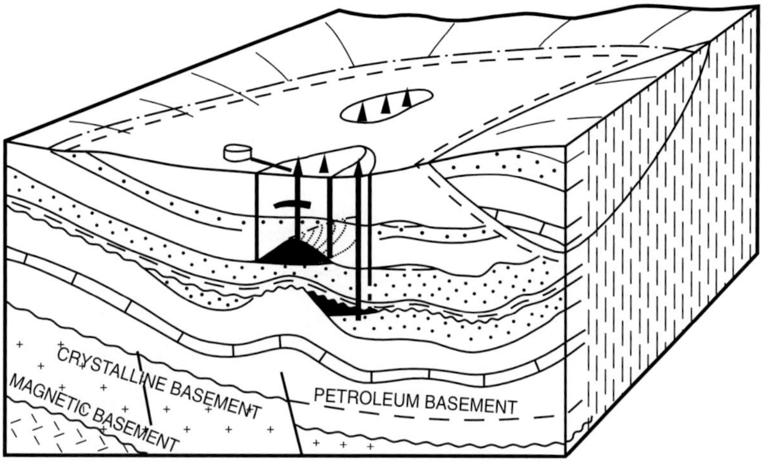

Fig. 6.63 Main definitions of oil system elements.

These average recovery values are constantly being improved, on the basis of considerable research over the past few years both in reservoir geology, with a major contribution from high-resolution seismics (Fig. 6.64), and in field engineering, with more and more sophisticated production models.

Oil and gas fields come in considerable variety, as do traps, and of course the size of the accumulations. Basins of all types have stratigraphic and structural traps: broad continental shelves, rifts, oceanic margins, flexural basins and basins associated with mountain chains (Fig. 6.65). They concern the entire stratigraphic series from the Cambrian to the Neogene.

Petroleum zones and systems can be defined on the basis of geographic and stratigraphic distribution of the basins. A **petroleum zone** is defined as a vast region where many oil and gas fields have been found. Such petroleum zones are the Middle East, to either side of the Persian Gulf, the North Sea, the North Saharan platform, and the rim of the Andes, among others.

In many cases, oil fields are located in different reservoirs at distinct stratigraphic levels, depending on the geological history of the region.

Using the guidelines of an older study[1], let us emphasize the following:

- Relations with the global geodynamic history (shelves and mobile belts), especially the concentration (69%) of reserves along the line of Mesogea (or Tethys).
- Relations with paleo-oceanographic and climatic variations and events that regulate the formation of source rock and reservoirs (and especially the most effective cap rock of evaporites).

1. See Bois et al., 1979.

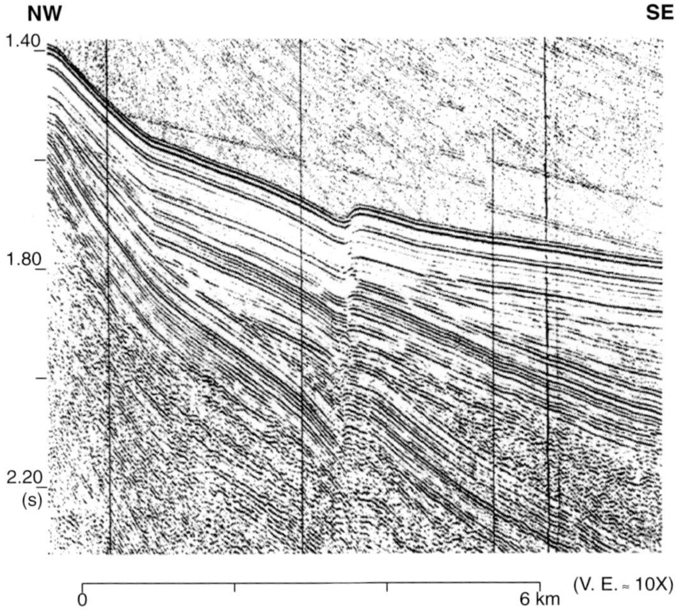

Fig. 6.64 Example of high-resolution seismic profile (from Bally, 1989). Various sequences and pinchouts can be picked out from various reflectors, and an active fault can be seen.

This global approach has also led to thinking on how to define a **petroleum system**, from the initial organic source to the production of a series of fields.

6.5 PETROLEUM SYSTEMS

6.5.1 Definitions and Review

The maturation of hydrocarbons in the gradual course of burial in a sedimentary basin leads to the genesis of petroleum. This petroleum is expelled from its source rock and migrates toward a trap in a reservoir where a field is formed under special geological conditions. The existence of petroleum provinces with special characteristics raises the question of why and how such provinces were generated, and what ties them in with the history of sedimentary basins. The **petroleum system** concept is based on this perspective of a sequence of events. The idea of a special relationship among the parameters that condition the existence of a

Geodynamic provinces			Some examples
Shelf	Stable or unstable	Continental or marine: Calcareous; Reef; Sand-sandstone; Mixed	Alberta Middle East Michigan Sahara American continent
Rift	Continental Oceanic	Pre- and synrift deposits	Brazil Gulf of Suez
Divergent continental margin		Pre- or synrift deposits Deep post-rift deposits	Gulf of Mexico African margin Brazilian margin
Convergent margins, orogenies	Foreland flexural basins (perisutural)	Foreland Foredeep	Middle East Venezuela Andes Carpathians
	Marginal basins	Back-arc	Pannonian basin
	Intermontane episutural basins		California Andes

Fig. 6.65 Petroleum zones and geodynamic units.

petroleum deposit was first introduced for oil deposits and then generalized to all petroleum fields.[1]

Generally, the petroleum system is analyzed in a global perspective on the scale of the sedimentary basin. In its initial definition, the petroleum system is all of the geological criteria, (sedimentary and geodynamic processes) governing the distribution of the fields, and especially the combined presence of source rock, reservoir, and seal rock, with a certain geographical extent in which a family of oil fields is observed to form. This is a way of saying that several deposits or fields will belong to the same oil system in a sedimentary basin, i.e., they depend on a very comparable sequence of geological events. An oil system then defines a certain quantity of hydrocarbons produced from a source rock, and also a set of reservoirs, traps, and sealing rock capable of containing accumulations.

1. See Dow, 1972; Perrodon, 1980, 1992; Demaison, 1984; Magoon, 1988; Magoon and Dow, 1994; Magoon and Sanchez, 1995.

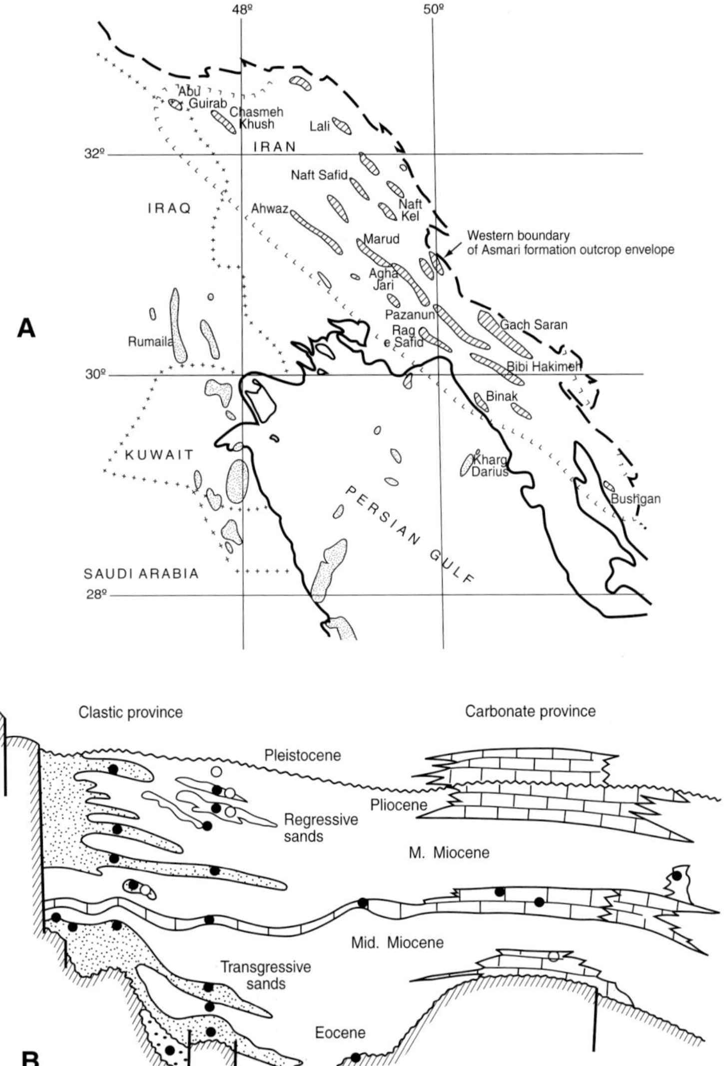

Fig. 6.66 Petroleum province and petroleum systems.
A. General map of the northern end of the Persian Gulf showing the producing fields in the Asmari and Ghar limestones (within the Fars salt extent boundaries) in the Middle East province, and fields producing from older formations (from Perrodon, 1985). **B.** Example of the complex petroleum system in Indonesia.

The term "petroleum system" is more restrictive than "petroleum zone" or "**petroleum province**" used above, which are geographical spaces corresponding to a group of sub-basins. Each province may, in fact, include one or more petroleum systems depending on the basin's richness in source rock and reservoir horizons (such as the Middle East) (Fig. 6.66). Within each basin, considering a particular petroleum system, the exploration geologist then defines two other ideas, which are **play** and **prospect**. Respectively, these designate a set of objectives of similar characters and a particular objective for exploration in the basin. The petroleum system is analyzed at the level of the basin and is generally identified by the age of the source rock and productive reservoirs, while the objective of play and prospect is primarily defined by the reservoir and trap. In analyzing the oil system, great importance is attached to the **critical moment** (the capital geological period for resource production), while in looking for plays, it is rather the present conditions that are essential for defining the economic aspect of the objective.

To summarize, there are four levels of petroleum investigation (Fig. 6.67):
- Sedimentary basin, petroleum zone
- Petroleum system
- Play
- Prospect.

6.5.2 Calendar, Critical Moment

The petroleum system is first defined by the existence of oil or gas in a basin. Whether in abundance or in infinitesimal traces, this presence is evidence of an ordered sequence of geological processes that have led to an accumulation. The petroleum system idea is therefore to define the calendar of events precisely in their time and stratigraphic aspects, along with the geographical area concerned.

A petroleum system is thus characterized by:
- Source rock (leading to the generation of thermal or biogenic gas, oil or condensate, asphalts), which is the *sine qua non* of any system
- Reservoir rock
- Sealing rock
- Burial, which defines the possibility and rate of maturation
- Trap formation age, which depends on the structural evolution
- The presumed age of expulsion-migration-accumulation
- The preservation or retention time (during which alteration, degradation, or even erosion may occur).

All of these aspects are recorded in a stratigraphic table to define the **critical moment** (Fig. 6.68), which corresponds to the migration-accumulation period. This moment is critical because, at this time, traps must exist in the basin to retain an accumulation. If there are

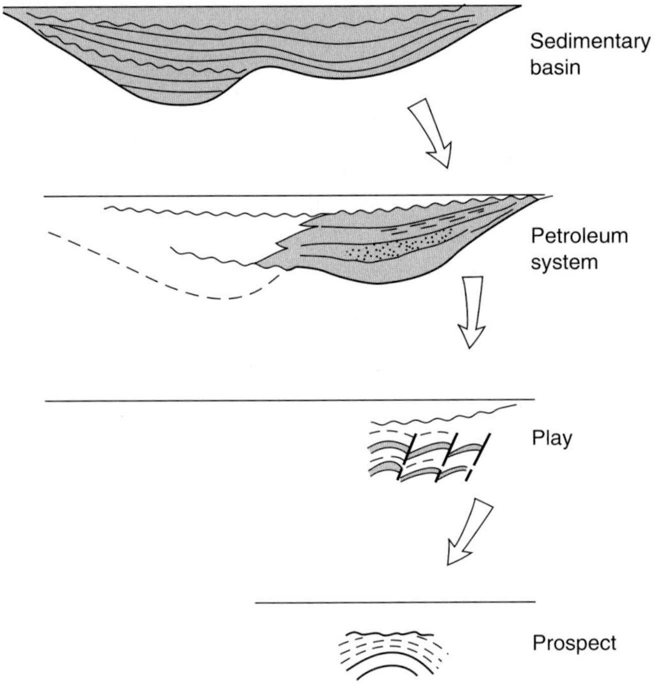

Fig. 6.67 Levels of petroleum investigation: sedimentary basin, petroleum system, play, prospect.
While the geological aspects are important for exploration in the first two levels, economics is the overriding concern for the latter two.

no traps, no pool will form. The calendar of different events must thus be considered carefully. The subsidence curves and burial history developed will be of capital importance for defining this critical moment (Fig. 6.69). The geological sections reconstructed for this period will visualize the relationships among the various petroleum system elements. This calendar is generally considered within the geodynamic history of the basin (an accurate sequence of events). The system geography (geometry) also has to be considered, whence the need to reconstruct maps for the different periods. The calendar may not be the same for different sub-basins if their burial histories are different, for example.

6.5.3 Different Petroleum Systems, Efficiency

As has been said since the preliminary discussion of deposits in speaking of their **habitat**, petroleum systems have been classified variously by a number of authors, each with their own approach.

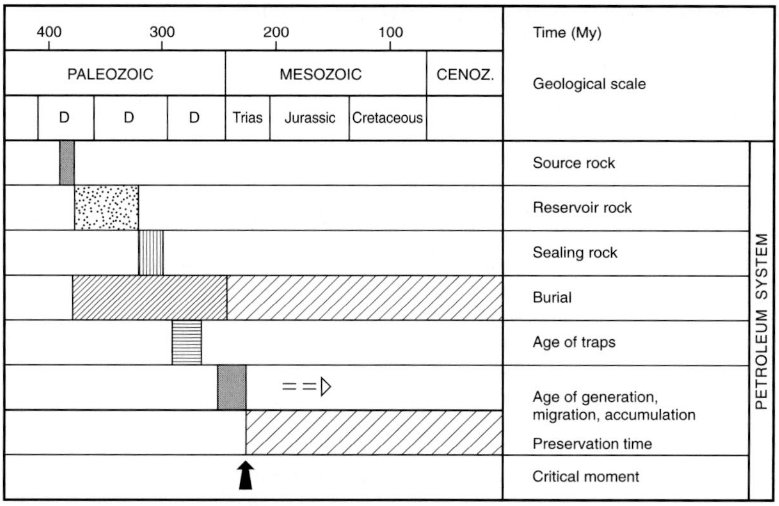

Fig. 6.68 Chart of essential elements in the petroleum system and critical moment (from Magoon and Dow, 1994). It must be remembered that oil (and gas) maturation and migration can be extended in time if the source rock has a cracking potential.

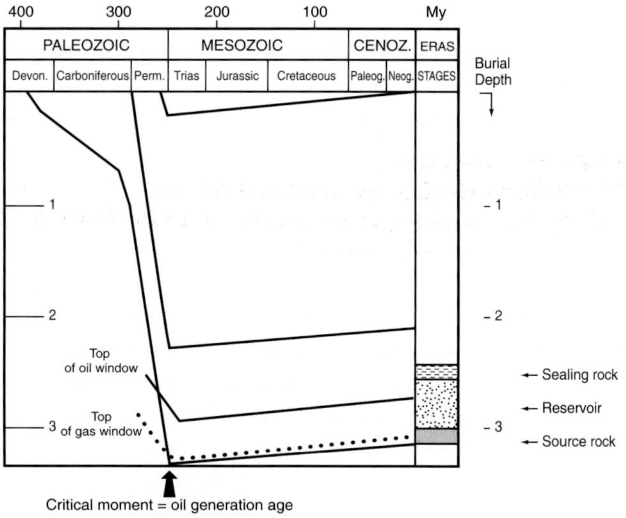

Fig. 6.69 Burial history and critical moment.
The epoch of oil maturation and expulsion is located in a burial diagram. The theoretical case here (from Magoon and Dow, 1994) shows rapid subsidence at the end of the Paleozoic. The existence of an early trap is indispensable in order for the accumulation to occur.

From the perspective of basin genesis on the scale of global tectonics, three broad categories are proposed.[1]

Extension basin systems of the continental rift type, with many peri-tethyan examples stemming from the fracturing of Pangea. Pull-apart and back-arc basins can be grouped into this category, which has the following characteristics:

- High subsidence and sedimentation rates, good chances of confined environments, high heat flow, e.g., favorable conditions for the formation and quick maturation of source rock
- Varied, heterogeneous, and often dispersed reservoirs
- Mainly vertical migration along active faults
- Traps on faults, horsts, or draping anticlines and en-echelon faults along strike-slip faults
- Problems due to easy dysmigration under frequent overpressures, and to the sealing rock quality.

Cratonic shelf systems (and those of certain divergent margins), with the following main characteristics:

- Relatively low subsidence rate, many unconformities in the series, moderate heat flow distributed over a long period, e.g., slow maturation compensated by large extent (and sometimes large volumes) of source rock
- Many quality reservoirs, sometimes of large geographic extent
- Long lateral migration toward relatively simple stratigraphic and structural traps (including large domes)
- Moderate pool pressures.

It should also be noted that several systems may overlie or dovetail with each other, depending on the stratigraphic position of the source rock.

Compressive orogenic domain systems of foreland regions up to the inner part of the chains. This category exhibits the following characters:

- Subsidence activity by overload and tectonics, but a low thermal gradient with rare source rock and irregular sedimentation
- Varied reservoirs and dispersed habitats, with complex migration paths
- Abundant structural traps of simple (foreland) and complex (chain) kinds
- Risks of dysmigration and destruction by erosion, and sometimes very high pressures.

It should also be noted, as was seen in the chapter on basins, that the time evolution of a pericratonic basin (or even a rift) may bring it to a foreland situation or in a position to be incorporated into a chain, in which case the system would come within the previous category.

With this overall perspective, analysis can be used directly in the exploration process for estimating the chances of success and defining new objectives for a given petroleum system.

1. Perrodon, 1992.

This approach addresses the qualitative ideas of the original organic facies, and also the quantities of hydrocarbons formed and felt to have been trapped. The useful thickness of the source rock likely to expel hydrocarbons, the size of the oil kitchen, the potential losses during migration, are also important quantitative elements. This idea of a petroleum system's **efficiency** is presented differently in a genetic classification[1], which considers three factors that will be summarized briefly hereafter: load, migration, and trapping.

The **load** is the amount of petroleum available to be trapped, e.g., the petroleum expelled minus the losses during migration. This represents the richness of the source rock and the volume having reached maturity in burial, or the petroleum potential.

The source potential index (SPI) is the quantity of hydrocarbons generated in a column of source rock of 1 m^2 sectional area. This is the indicator of the potential volume of reserves.

The difference between **proven reserves** and **potential reserves** (Fig. 6.70) should be remembered here. Proven reserves are calculated from exploration data while potential reserves are estimated from petroleum system analysis and the estimated **regional load**.

From the viewpoint of the load, then, there are:

- Super-loaded systems
- Normally loaded systems
- Underloaded systems.

The genetic potential is expressed by the ratio of HC to rock (in kilos).

Migration styles and drainage paths are the second qualification factor. This depends on the basin's stratigraphic and structural arrangement:

- Predominant **vertical migration** characterizes systems containing many faults, such as rift basins, saliferous provinces, and highly tectonized zones.
- **Lateral migration**, on the other hand, dominates in stable provinces such as cratonic and foreland basins.

The **trapping style** is defined by the **impedance**, which defines the efficiency of the sealing rock in terms of its degree of resistance to migration toward the surface:

- High **impedance** designates a trap of sealing rock with good continuity and moderate deformation.
- **Low impedance** indicates either an inadequate seal on the regional scale coupled with a high or low structural deformation, or good regional continuity of the sealing rock but with too low a degree of deformation.

Taking these three factors in combination, we get a "genetic" classification as diagrammed in Fig. 6.71.

1. Demaison and Huizinga, 1994.

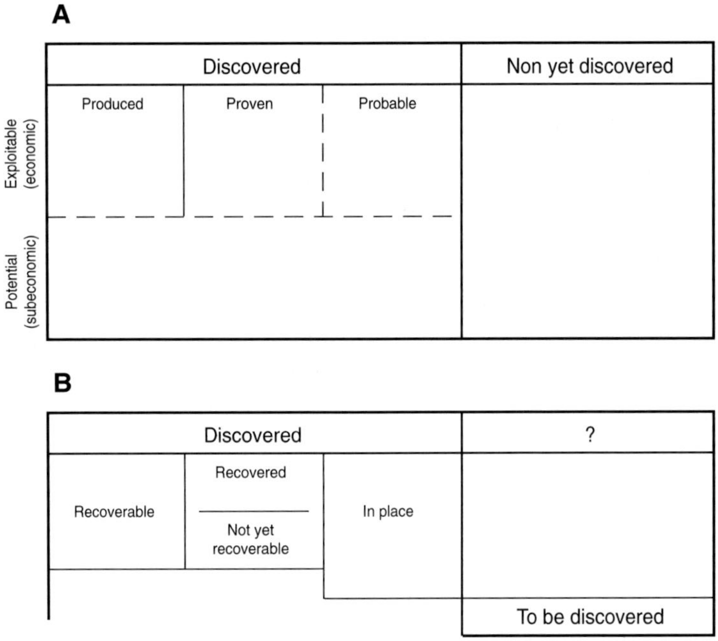

Fig. 6.70 Different types of reserves (from Perrodon, 1985).
A. Produced reserves are known. Proven reserves are measured (toward 20%). Probable reserves are estimated (by probability). **B.** Recoverable reserves means the quantities that can be produced according to the state of the art. This is only a part of the reserves in place.

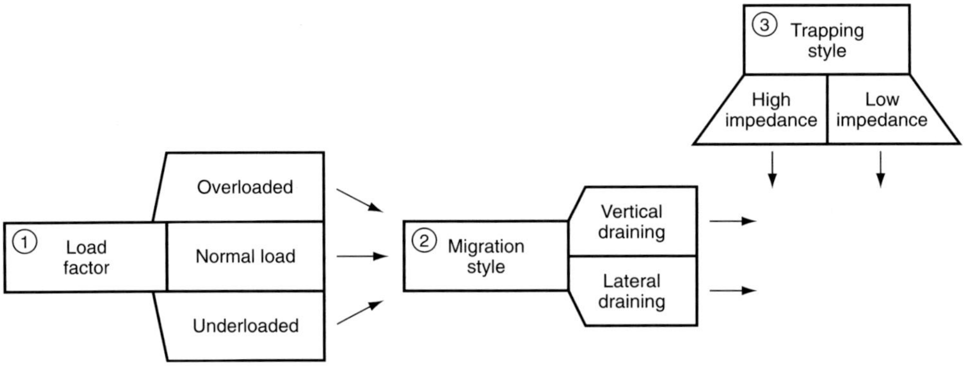

Fig. 6.71 Genetic classification of oil systems (from Demaison and Huizinga, 1994).

BIBLIOGRAPHY

✧ Al-Husseini MI (1994) Dhahran geological society; American association of petroleum geologists. The Middle East petroleum geosciences. GEO'94. April 25-27, 1994. Bahrain. Volumes 1 and 2. Gulf Petrolink, Mananma.

✦ Alpern B (1988) The application of the "Alpern scientific classification of solid fossil fuels" to quality of Gondwana coals from different basins. Publication of Museum, laboratorio de mineralogia and geologia. Faculta de ciencas, Porto.

✧ Bally AW(1989) Atlas of seismic stratigraphy. American association of petroleum geologists, AAPG, studies in geology 27/3, Tulsa.

✧ Belin S (1992) Distribution microscopique de la matière organique disséminée dans les roches mères. Techniques d'étude : interprétation des conditions de dépôts et de diagenèse. Thèse de doctorat. Éditions Technip, Paris.

✦ Bois C (1975) Petroleum - Zone concept and the similarity analysis. Contribution to resource appraisal. Methods of estimating the volume of undiscovered oil and gas resources. American association of petroleum geologists, Tulsa, AAPG, studies in geology 1, pp 87-89.

✦ Bois C, Bouche P, Pelet R (1982) Global geologic history and distribution of hydrocarbon reserves. AAPG bulletin 66, 9, pp 1248-1270.

✦ Bordenave ML (1993) Applied petroleum geochemistry. Éditions Technip, Paris.

✦ Burrus J (1986) Thermal modeling in sedimentary basins. Institut français du pétrole exploration research conference. June 3-7, 1985. Carcans, France. Éditions Technip, Paris.

✧ Chiarelli A, Rouchet (du) J (1977) Importance des phénomènes de migration verticale des hydrocarbures. Revue de l'Institut français du pétrole 30, 2, pp 189-203.

✧ Choquette PW, Pray LC (1970) Geologic nomenclature and classification of porosity in sedimentary carbonates. AAPG bulletin 54, pp 207-250.

✦ Connan J (1987) Regard sur la géochimie organique pétrolière d'aujourd'hui. Bulletin des centres de recherches exploration production, Elf Aquitaine 11, 2, pp 181-219.

✦ Cossé R (1988) Techniques d'exploitation pétrolière : le gisement. Éditions Technip, Paris.

✧ CSRPPGN - Comité des techniciens (1986) Corps sédimentaires : exemples sismiques et diagraphiques. Éditions Technip, Paris.

✦ Demaison G (1984) The generative basin concept. Demaison G, Murris RJ. Petroleum geochemistry and bassin evaluation. American Association of Petroleum Geologists, Tulsa, AAPG memoir 35, pp 1-14.

✦ Demaison G, Huizinga BJ (1994) Genetic classification of petroleum systems using three factors: charge, migration and entrapment. Magoon LB, Dow WG. The petroleum system from source to trap. American Association of Petroleum Geologists, Tulsa, AAPG memoir 60, pp 73-89.

✧ Dow WG (1972) Application of oil correlation and source rock data to exploration in Williston bassin. AAPG bulletin 56, 615.

✧ Dow WG (1974) Application of oil correlation and source rock data to exploration in Williston bassin. AAPG bulletin 58, 7, pp 1253-1262.

✦ Dow WG (1978) Petroleum source bed on continental slopes and rises. AAPG bulletin 62, 9, pp 1584-1602.

✧ Durand B (1985) Diagenetic modification of kerogens. Pholisophical transactions of the Royal society of London, A, 315, 1531, pp 77-90.

✦ Durand B (1980) Kerogen: insoluble organic matter from sedimentary rocks. Éditions Technip, Paris.

✧ Durand B (1996) Origin, generation and accumulation of oil and natural gas. Cours ENSPM. Unpublished.

✦ Durand B (1988) Understanding of HC migration in sedimentary bassins. Mattavelli, L and Novelli, L Advances in organic geochemistry 1987. International meeting on organic geochemistry 13. September 21-25, 1987. Venice, Italy. Pergamon press, Oxford.

✦ Durand B, Durand-Souron C, Huc AY et al. (1980) Kerogen, insoluble organic matter from sedimentary rocks. Éditions Technip, Paris.

✦ Durand B, Espitalié J (1973) Évolution de la matière organique au cours de l'enfouissement des sédiments. Comptes rendus de l'Académie des sciences, D, 276, pp 2253-2256.

✧ Espitalié J, Deroo G, Marquis F (1985) La pyrolyse Rock-Eval et ses applications (1re partie). Revue de l'Institut français du pétrole 40, 5, pp 563-579.

✧ Espitalié J, Deroo G, Marquis F (1985) La pyrolyse Rock-Eval et ses applications (2e partie). Revue de l'Institut français du pétrole 40, 6, pp 755-784.

✦ Espitalié J, Deroo G, Marquis F (1986) La pyrolyse Rock-Eval et ses applications (3e partie). Revue de l'Institut français du pétrole 41, 1, pp 73-89.

✦ Huc AY (1995) Cours ENSPM. Unpublished.

✦ Huc AY (1988) Sedimentology of organic matter. Frimmel FH, Christman RF. Humic substances and their role in the environment. Wiley and sons, Chichester, pp 215-243.

✧ Huc AY, Durand B (1977) Occurence and signification of humic acids in ancient sediments. Fuel 56, 1, pp 73-80.

✦ Iseux JC (1992) Gas hydrates: a new source of natural gas. Vially R. Bacterial gas. Éditions Technip, Paris, pp 205-222.

✦ Krason J (1994) Study of 21 marine basins indicates wide prevalence for hydrates. Offshore magazine 54, 8, pp 34-35.

✧ Kruel Romeu R (1994) Écoulements en milieux hétérogènes : prise de moyenne de perméabilité en régimes permanent et transitoire.Thèse de doctorat Paris 6 / IFP. Éditions Technip, Paris.

✦ Levorsen AI (1995) Geology of petroleum. Freeman, Geology series, New York.

✧ Lieth H, Whittaker RH (1975) Primary productivity of the biosphere. Springer, Ecological studies. Berlin, p 14.

✦ Magoon LB (1988) The petroleum systems a classification scheme for research resource assessment and exploration. US geological survey, pp 2-15.

✦ Magoon LB, Dow WG (1994) The petroleum system. Magoon LB, Dow WG. The petroleum system from source to trap. American association of petroleum geologists, Tulsa, AAPG memoir 60, pp 3-24.

✦ Magoon LB, Sanchez RM (1995) Beyond the petroleum system. AAPG bulletin 79, 12, pp 1731-1736.

✧ Mathieu Y, Verdier F, Houel P et al. (1993) Reservoir heterogeneity in fluviatile Keuper facies: a subsurface and outcrop study. Eschard R, Doligez B. Subsurface reservoir characterization from outcrop observations. Éditions Technip, Paris, pp 145-160.

- ✧ Malartre F (1993) Stratigraphie séquentielle des systèmes sédimentaires mixtes : le Crétacé supérieur du bassin vocrutien et de sa bordure occidentale. Thèse de doctorat. Université de Lyon I, Lyon.
- ✦ Mattavelli L, Novelli L (1988) Advances in organic geochemistry 1987. International meeting on organic geochemistry 13. September 21-25, 1987. Venice, Italy. Pergamon press, Oxford.
- ✦ Ourisson G, Albrecht P, Rohmer M (1984) The microbial origin of fossil fuels. Scientific American 251, 2, pp 44-51.
- ✧ Pelet R (1985) Évaluation quantitative des produits formés lors de l'évolution géochimique de la matière organique. Revue de l'Institut français du pétrole 40, 5, pp 551-562.
- ✦ Perrodon A (1980) Géodynamique pétrolière : genèse et répartition des gisements d'hydrocarbures. Paris, Elf Aquitaine, Bulletin des centres de recherches exploration-production Elf Aquitaine, mémoire 2.
- ✦ Perrodon A (1985) Géodynamique pétrolière : genèse et répartition des gisements d'hydrocarbures. Pau, Elf Aquitaine, Bulletin des centres de recherches explortation-production, Elf Aquitaine, mémoire 2.
- ✦ Perrodon A (1992) Petoleum systems models and applications. Journal of petroleum geology 15, 3, pp 319-326.
- ✦ Pratt LM (1984) Influence of paleo environments factors of the preservation of organic matter in middle cretaceous Greenhary formation, Pueblo, Colorado. AAPG bulletin 68, pp 1146-1159.
- ✧ Pratt LM, Kauffman EG, Zelt FB (1985) Fine-grained deposits and biofacies of the cretaceous western interior seaway: evidence of cyclic sedimentary pricesses. Society of Economic Paleontologists and Mineralogists, SEPM, field trip guidbook 4, Tulsa.
- ✦ Rice DD (1992) Controls, habitat, and resource potential of ancient bacterial gas. Vially R. Bacterial gas. Éditions Technip, Paris, pp 91-118.
- ✦ Rice DD, Claypool GE (1981) Generation accumulation and resource potential of biogenic gas. AAPG bulletin 65, pp 5-25.
- ✦ Rojey A (1994) Le gaz naturel. Éditions Technip, Paris.
- ✦ Sallé C, Debyser J (1976) Formation des gisements de pétrole : étude des phénomènes géologiques fondamentaux. Éditions Technip, Paris.
- ✦ Suess E, Bolin B, Degens ET et al. (1980) The global carbon cycle. Marine chemistry 9, 2, pp 153-155.
- ✧ Tissot B, Espitalié J (1975) L'évolution thermique de la matière organique des sédiments : application d'une simulation mathématique. Revue de l'Institut français du pétrole 30, 5, pp 743-777.
- ✦ Tissot BP, Welte DH (1984) Petroleum formation and occurrence. Springer, Berlin.
- ✧ Ulmishek GF, Klemme HD (1990) Depositional controls, distribution, and effectiveness of world's petroleum source rocks. US government printing office, Geological survey, bulletin 1931, Denver.
- ✦ Ungerer P, Bessis F, Chenet PY et al. (1984) Geological and geochemical models in oil exploration, priciples ans practical exemples. Demaison G, Murris RJ. Petroleum geochemistry and basin evaluation. American Association of Petroleum Geologists, Tulsa, AAPG memoir 35, pp 53-77.
- ✦ Ungerer P (1990) State of the art of research in kinetic modelling of oil formation. Advances in organic geochemistry 16, pp 1-25.
- ✦ Van Krevelen DW (1984) Coal Amsterdam, Elsevier.

✦ Vandenbroucke M (1993) Migration of hydrocarbons. Bordenave ML. Applied petroleum geochemistry. Éditions Technip, Paris, pp 123-148.

✧ Watson HJ (1982) Casablanca field offshore Spain, a paleogeomorphic trap. Halbouty MT. The deliberate search for the subtle trap. American Association of Petroleum Geologists, Tulsa, AAPG, memoir 32, pp 237-250.

✦ *Books or articles of general interest.*

✧ *Source of one of the figures used, cited in the figure caption.*

INDEX

D

E

F

G

H

I

S

U

V

W

X

Y

Z

Achevé d'imprimer en juin 2002 par l'Imprimerie EMD à Lassay-les-Châteaux.
N° d'impression : 9501 - N° d'éditeur 1065 - Dépôt légal : juin 2002